Linear Programming
with Game Theory

Useful for

- Undergraduate and Post Graduate students of Engineering
- Students of MCA, B.Sc. and M.Sc.
- All Entrance Examinations for Admission in M.Sc., M.Phil. and Ph.D. courses
- NET / JRF aspirants

Dr. SUDHIR KUMAR PUNDIR

M.Sc., M.Phil, NET (JRF), Ph.D.
HEAD,
Department of Mathematics
S.D. (P.G.) College,
Muzaffarnagar (U.P.)

CBSPD

CBS Publishers & Distributors Pvt Ltd

New Delhi • Bengaluru • Chennai • Kochi • Kolkata • Lucknow • Mumbai
Hyderabad • Jharkhand • Nagpur • Patna • Pune • Uttarakhand

LINEAR PROGRAMMING
with Game Theory

ISBN: 978-93-89396-33-1

Copyright © Author and Publisher

First Edition: 2020

Reprint: 2023

Published by **Satish Kumar Jain** and produced by **Varun Jain** for

CBS Publishers & Distributors Pvt Ltd

4819/XI Prahlad Street, 24 Ansari Road, Daryaganj, New Delhi 110 002, India
Ph: 011-23289259, 23266861   Website: www.cbspd.com
e-mail: delhi@cbspd.com

Corporate Office: 204 FIE, Industrial Area, Patparganj, Delhi 110 092, India
Ph: 011-4934 4934   Fax: 011-4934 4935   e-mail: publishing@cbspd.com;
publicity@cbspd.com

Branches

- **Bengaluru:** Seema House 2975, 17th Cross, KR Road, Banasankari 2nd Stage, Bengaluru 560 070, Karnataka, India
  Ph: +91-80-26771678/79   Fax: +91-80-26771680   e-mail: bangalore@cbspd.com
- **Chennai:** 7, Subbaraya Street, Shenoy Nagar, Chennai 600 030, Tamil Nadu, India
  Ph: +91-44-26680620, 26681266   Fax: +91-44-42032115   e-mail: chennai@cbspd.com
- **Kochi:** 42/1325, 1326, Power House Road, Opp KSEB, Power House, Ernakulum Kochi 682 018, Kerala, India
  Ph: +91-484-4059061-65,67   Fax: +91-484-4059065   e-mail: kochi@cbspd.com
- **Kolkata:** 147, Hind Ceramics Compound, 1st Floor, Nilgunj Road, Belghoria, Kolkata-700056, West Bengal, India
  Ph: +033-25633055, 033-25633056   e-mail: kolkata@cbspd.com
- **Lucknow:** Basement, Khushnuma Complex, 7 Meerabai Marg (Behind Jawahar Bhawan),Lucknow-226001, UP, India
  Ph: +91-522-4000032   e-mail: tiwari.lucknow@cbspd.com
- **Mumbai:** PWD Shed, Gala no 25/26, Ramchandra Bhatt Marg, Next to JJ Hospital Gate no. 2, Opp. Union Bank of India Noorbaug, Mumbai-400009, Maharashtra, India
  Ph: 022-66661880/89   e-mail: mumbai@cbspd.com

Representatives

- Hyderabad   0-9885175004
- Patna   0-9334159340
- Jharkhand   0-9811541605
- Pune   0-9923910676
- Nagpur   0-9421945513
- Uttarakhand   0-9716462459

Printed at: Glorious Printers, Jhilmil Industrial Area, Delhi, India

Preface

The book entitled **"LINEAR PROGRAMMING WITH GAME THEORY"** meet the needs of engineering and science students of UG and PG levels. Besides, it will also be very useful for students preparing for various competitive examinations.

The approach of the book is student friendly as it enables easy understanding of the fundamentals. The contents of the book have been chosen after careful persual of syllabi of UG and PG courses of different universities. The book explains the fundamental principles of the various concepts with illustrations. A list of objective questions are given at the end of each chapter.

I express my gratitude to the authors and publishers of various books I consulted during the preparation of the book.

I wish to sincerely thank **Sh S.K. Jain** and **Sh. Varun Jain**, Managing Director, CBS Publishers and Distributors, New Delhi for encouragement and help in bringing out this publication in a present nice form.

My special thanks to Sh. B.M. Singh, Sh. Sunil Dutt, Sh. Suresh Sharma and entire team of CBS Publishers and Distributors, New Delhi whose encouragement and unstinted support enabled me to complete my book. Mr. Peeyush Goel, M/s Dreamshapers also deserve special mention for nice type setting.

I must also record my appreciation due to my wife Dr. Rimple, daughter Rijuta and son Shrish for their understanding and love during the long period that I have taken to complete this book.

Above all I am thankful to The Almighty God, without whose grace nothing is possible for any one.

Readers are welcomed to point out errors, if any and send their valuable suggestions for improving the quality of the book.

Dr. Sudhir Kumar Pundir
email : skpundir05@yahoo.co.in

Contents

Some Mathematical Preliminaries and Convex Sets

1.1 INTRODUCTION

Linear Programming problem (LPP) is an integral part of Operation Research which is an important branch of mathematics.

So, before discussing the basic concepts of linear programming problems, let us recall some mathematical concepts which are very useful in LPP.

1.2 MATRIX

A set of mn numbers either real or complex arranged in the form of a rectangular array in which there are m rows and n columns, is called a matrix of order $m \times n$ which is denoted by $[a_{ij}]_{m \times n}$ where $i = 1, 2, 3, ..., m$ represents the number of rows and $j = 1, 2, 3, ..., n$ represents the number of columns and thus a matrix of order $m \times n$ is usually written as

$$[a_{ij}]_{m \times n} = \begin{bmatrix} a_{11} & a_{12} & \cdots & a_{1n} \\ a_{21} & a_{22} & \cdots & a_{2n} \\ \vdots & \vdots & \vdots & \vdots \\ a_{m1} & a_{m2} & \cdots & a_{mn} \end{bmatrix}_{m \times n}$$

☛ **REMARK**

- Sometimes, a matrix is a rectangular array of numbers enclosed in double straight lines shown as '|| ||' or enclosed in parenthesis '( )'.

1.3 TYPE OF MATRICES

1.3.1 NULL MATRIX (OR ZERO MATRIX)

A matrix of order $m \times n$ is called a *null matrix* if it contains all mn elements zero. It is denoted by O and is usually written as

$$O = \begin{bmatrix} 0 & 0 & \cdots & 0 \\ 0 & 0 & \cdots & 0 \\ \vdots & \vdots & \vdots & \vdots \\ 0 & 0 & \cdots & 0 \end{bmatrix}_{m \times n}$$

1.3.2 ROW MATRIX

A matrix having only one row and n columns is called a *row matrix* of order $1 \times n$.

For example : $A = \begin{bmatrix} a_{11} & a_{12} & a_{13} & \cdots & a_{1n} \end{bmatrix}_{1 \times n}$

1.3.3 COLUMN MATRIX

A matrix having m rows and only one column is called a *column matrix* of order $m \times 1$.

$$\text{For example :} \qquad A = \begin{bmatrix} a_{11} \\ a_{21} \\ a_{31} \\ \vdots \\ a_{m1} \end{bmatrix}_{m \times 1}$$

1.3.4 HORIZONTAL MATRIX

A matrix having more columns than the number of its rows, is called *Horizontal matrix*.

$$\text{For example:} \qquad A = \begin{bmatrix} a_{11} & a_{12} & a_{13} \\ a_{21} & a_{22} & a_{23} \end{bmatrix}_{2 \times 3}$$

1.3.5 VERTICAL MATRIX

A matrix having more number of rows than its columns, is called *vertical matrix*.

$$\text{For exmaple:} \qquad A = \begin{bmatrix} a_{11} & a_{12} \\ a_{21} & a_{22} \\ a_{31} & a_{32} \end{bmatrix}_{3 \times 2}$$

☞ REMARK
- Row matrix is also a horizontal matrix and column matrix is also a vertical matrix.

1.3.6 SQUARE MATRIX

A matrix having a number of rows equal to number of columns, is called *square matrix*.

$$\text{For example :} \qquad A = \begin{bmatrix} a_{11} & a_{12} & a_{13} \\ a_{21} & a_{22} & a_{23} \\ a_{31} & a_{32} & a_{33} \end{bmatrix}_{3 \times 3}$$

Here, the matrix A has 3 rows and 3 columns, so it is a square matrix. Also the elements a_{11}, a_{22}, a_{33} are placed in the diagonal, so these elements are known as *diagonal elements*.

1.3.7 DIAGONAL MATRIX

A matrix of order $n \times n$ is called a *diagonal matrix* if it contains all its off diagonal elements equal to zero.

Suppose $A = [a_{ij}]_{n \times n}$ and if $a_{ij} = 0$ for all $i \neq j$, then A is a diagonal matrix. Diagonal matrix of order $n \times n$ is usually written as Diag $[a_{11} \quad a_{22} \quad a_{33} \quad \cdots \quad a_{nn}]$

$$\text{For example:} \qquad A = \begin{bmatrix} 1 & 0 & 0 \\ 0 & 2 & 0 \\ 0 & 0 & 3 \end{bmatrix}_{3 \times 3} = \text{Diag } [1 \ 2 \ 3]$$

1.3.8 SCALAR MATRIX

A diagonal matrix whose diagonal elements are all equal but not equal to 1 is called a *scalar matrix*.

$$\text{For example:} \qquad A = \begin{bmatrix} k & 0 & 0 \\ 0 & k & 0 \\ 0 & 0 & k \end{bmatrix}, k \neq 1$$

1.3.9 UNIT MATRIX

A square matrix of order $n \times n$ having all off-diagonal elements equal to zero and each of the diagonal elements equal to 1, is called a *unit matrix*. It is usually denoted by I_n and is written as

$$I_n = \begin{bmatrix} 1 & 0 & \cdots & 0 \\ 0 & 1 & \cdots & 0 \\ 0 & 0 & \cdots & 0 \\ \vdots & \vdots & \vdots & \vdots \\ 0 & 0 & \cdots & 1 \end{bmatrix}_{n \times n}$$

☛ **REMARK**

• Unit matrix can also be denoted by I.

1.3.10 TRIANGULAR MATRIX

A matrix in which the elements lying above or below principal diagonal are all zero, is called a *triangular matrix*.

There are two kinds of triangular matrix.

(a) Upper triangular matrix : A matrix of order $n \times n$ is called an *upper triangular matrix* if it contains all its elements below the diagonal elements equal to zero.

Suppose $A = [a_{ij}]_{n \times n}$ and if $a_{ij} = 0$ for all $i > j$, then A is an upper triangular matrix.

For example : $\qquad A = \begin{bmatrix} 2 & 3 & 4 \\ 0 & 1 & 5 \\ 0 & 0 & 3 \end{bmatrix}_{3 \times 3}$

is an upper triangular matrix of order 3×3.

(b) Lower triangular matrix : A matrix of order $n \times n$ is called a *lower triangular matrix* if it contains all its elements above the diagonal elements equal to zero.

Suppose $A = [a_{ij}]_{n \times n}$ and if $a_{ij} = 0$ for all $i < j$, then A is called lower triangular matrix.

For example : $\qquad A = \begin{bmatrix} 1 & 0 & 0 \\ 3 & 4 & 0 \\ 5 & 6 & 7 \end{bmatrix}_{3 \times 3}$

is a lower triangular matrix of order 3×3.

1.4 OPERATION ON MATRICES

1.4.1 ADDITION OF MATRICES

Suppose A and B are two matrices of same order, then the addition of these two matrices is obtained by adding corresponding elements of A and B. It is denoted by $A + B$. If the order of A and B is $m \times n$, then the order of $A + B$ will be $m \times n$.

Suppose $\qquad\qquad A = [a_{ij}]_{m \times n}$ and $B = [b_{ij}]_{m \times n}$

then $\qquad\qquad A + B = [a_{ij} + b_{ij}]_{m \times n}$

For example: If $\qquad A = \begin{bmatrix} 1 & 2 & 3 \\ 5 & 1 & 4 \\ 7 & 8 & 9 \end{bmatrix}$ and $B = \begin{bmatrix} 1 & 3 & 5 \\ 5 & 0 & 1 \\ 3 & 2 & 12 \end{bmatrix}$

then $\qquad A + B = \begin{bmatrix} 1 & 2 & 3 \\ 5 & 1 & 4 \\ 7 & 8 & 9 \end{bmatrix} + \begin{bmatrix} 1 & 3 & 5 \\ 5 & 0 & 1 \\ 3 & 2 & 12 \end{bmatrix}$

$$= \begin{bmatrix} 1+1 & 2+3 & 3+5 \\ 5+5 & 1+0 & 4+1 \\ 7+3 & 8+2 & 9+12 \end{bmatrix} = \begin{bmatrix} 2 & 5 & 8 \\ 10 & 1 & 5 \\ 10 & 10 & 21 \end{bmatrix}$$

☞ **REMARK**

• If the orders of the matrices are different, then they are not conformable for addition.

1.4.2 SUBSTRACTION OF MATRICES

Suppose A and B are two matrices of same order, then the substraction of A and B, i.e., A–B is obtained by substracting each element of B from the corresponding element of A. If A and B are of order $m \times n$, then $A - B$ will be of order $m \times n$.

Let $\qquad A = [a_{ij}]_{m \times n}$ and $B = [b_{ij}]_{m \times n}$

then $\qquad A - B = [a_{ij} - b_{ij}]_{m \times n}$

For example: If $\quad A = \begin{bmatrix} 1 & 2 & 3 \\ 3 & 4 & 5 \\ 5 & 6 & 7 \end{bmatrix}$ and $B = \begin{bmatrix} 0 & 5 & 2 \\ 3 & -2 & 2 \\ 5 & 7 & 8 \end{bmatrix}$

then $\qquad A - B = \begin{bmatrix} 1 & 2 & 3 \\ 3 & 4 & 5 \\ 5 & 6 & 7 \end{bmatrix} - \begin{bmatrix} 0 & 5 & 2 \\ 3 & -2 & 2 \\ 5 & 7 & 8 \end{bmatrix}$

$$= \begin{bmatrix} 1-0 & 2-5 & 3-2 \\ 3-3 & 4-(-2) & 5-2 \\ 5-5 & 6-7 & 7-8 \end{bmatrix} = \begin{bmatrix} 1 & -3 & 1 \\ 0 & 6 & 3 \\ 0 & -1 & -1 \end{bmatrix}$$

☞ **REMARK**

• If the order of matrices are different, then they are not conformable for substraction.

1.4.3 MULTIPLICATION OF A MATRIX BY A SCALAR

Suppose A is a matrix of order $m \times n$ and k is a scalar, then the multiplication of A by k, i.e. kA is obtained by multiplying each element of A by k.

Let $\qquad A = [a_{ij}]_{m \times n} \ \forall \ 1 \le i \le m$ and $1 \le j \le n$, then $kA = [ka_{ij}]_{m \times n}$

For example : If $\quad A = \begin{bmatrix} 1 & 2 & 3 \\ 4 & 5 & 6 \\ 7 & 8 & 9 \end{bmatrix}$ and $k = 3$,

then $\qquad 3A = 3 \begin{bmatrix} 1 & 2 & 3 \\ 4 & 5 & 6 \\ 7 & 8 & 9 \end{bmatrix}$

$$= \begin{bmatrix} 3 \times 1 & 3 \times 2 & 3 \times 3 \\ 3 \times 4 & 3 \times 5 & 3 \times 6 \\ 3 \times 7 & 3 \times 8 & 3 \times 9 \end{bmatrix} = \begin{bmatrix} 3 & 6 & 9 \\ 12 & 15 & 18 \\ 21 & 24 & 27 \end{bmatrix}$$

1.4.4 EQUALITY OF MATRICES

Two matrices are said to be equal if both have same order and having same corresponding elements.

For example : The matrices $A = \begin{bmatrix} 1 & 2 \\ -3 & 4 \end{bmatrix}$ and $B = \begin{bmatrix} x & y \\ z & 4 \end{bmatrix}$ are said to be equal if $x = 1$, $y = 2$ and $z = -3$.

1.5 PROPERTIES OF MATRIX ADDITION

1.5.1 COMMUTATIVE LAW

If A and B are two matrices of same order $m \times n$, then $A + B = B + A$

Proof. Let $A = [a_{ij}]_{m \times n}$ and $B = [b_{ij}]_{m \times n}$ where $1 \le i \le m$ and $1 \le j \le n$. Then

$$A+B = [a_{ij}]_{m \times n} + [b_{ij}]_{m \times n}$$
$$= [a_{ij} + b_{ij}]_{m \times n} \quad \text{(By definition of addition)}$$
$$= [b_{ij} + a_{ij}]_{m \times n}$$
$$(\because \text{Addition of Real numbers are always commutative})$$
$$= [b_{ij}]_{m \times n} + [a_{ij}]_{m \times n}$$
$$= B + A$$

Hence, $\quad A + B = B + A$

1.5.2 ASSOCIATIVE LAW

If A ,B and C are three matrices of same order $m \times n$, then
$$(A + B) + C = A + (B + C)$$

Proof. Let $A = [a_{ij}]_{m \times n}$ and $B = [b_{ij}]_{m \times n}$ where $1 \le i \le m$ and $1 \le j \le n$. Then

$$(A+B)+C = ([a_{ij}]_{m \times n} + [b_{ij}]_{m \times n}) + [c_{ij}]_{m \times n}$$
$$= [a_{ij} + b_{ij}]_{m \times n} + [c_{ij}]_{m \times n}$$
$$= [(a_{ij} + b_{ij}) + (c_{ij})]_{m \times n}$$
$$(\because \text{Addition of numbers are always associative})$$
$$= [a_{ij}]_{m \times n} + ([b_{ij} + c_{ij}]_{m \times n})$$
$$= [a_{ij}]_{m \times n} + ([b_{ij}]_{m \times n} + [c_{ij}]_{m \times n})$$
$$= A + (B + C)$$

Hence, $\quad (A+B) +C = A + (B+C)$

1.5.3 ADDITIVE IDENTITY

If A is a matrix of order $m \times n$ and O is a null matrix of the same order $m \times n$, then
$$A + O = A = O + A$$

Proof. Let $A = [a_{ij}]_{m \times n}$ and $O = [0]_{m \times n}$, then

$$A + O = [a_{ij}]_{m \times n} + [0]_{m \times n}$$
$$= [a_{ij} + 0]_{m \times n}$$
$$= [a_{ij}]_{m \times n} = A$$

Also $\quad O + A = [0]_{m \times n} + [a_{ij}]_{m \times n}$
$$= [0 + a_{ij}]_{m \times n}$$
$$= [a_{ij}]_{m \times n} = A$$

Hence $A + O = A = O + A$

Therefore, the null matrix O is treated as an additive identity.

1.5.4 ADDITIVE INVERSE

If A is a matrix of order $m \times n$ and $-A$ is the negative of A, so its order is also $m \times n$, then

$$-A + A = O \quad \text{(null matrix)}$$

Here, $-A$ *is the* additive inverse *of A.*

1.5.5 CANCELLATION LAW

If A, B and C are three matrices of order $m \times n$ then

   (i) $A + B = A + C \Rightarrow B = C$ (Left cancellation law)

   (ii) $B + A = C + A \Rightarrow B = C$ (Right cancellation law)

Proof.

   (i) It is given that

$$A + B = A + C \qquad \qquad \text{...(1)}$$

   Adding $- A$ to the left of both sides, we get

$$- A + (A + B) = - A + (A + C)$$

$\Rightarrow \qquad\qquad (-A + A) + B = (-A + A) + C$ \qquad (By associative law)

$\Rightarrow \qquad\qquad O + B = O + C$ \qquad\qquad\qquad (By additive inverse)

$\Rightarrow \qquad\qquad B = C$ \qquad\qquad\qquad\qquad (By additive identity)

   Similarly, we can prove that if $B + A = C + A$, then $B = C$.

1.6 PROPERTIES OF MULTIPLICATION OF MATRIX BY A SCALAR

   (i) **Distribution law of scalar multiplication over matrix addition :** *If A and B are two matrices of order $m \times n$ and k is any scalar, then $k(A + B) = kA + kB$*

   Proof. Let $A = [a_{ij}]_{m \times n}$ and $B = [b_{ij}]_{m \times n}$, then

$$k(A + B) = k([a_{ij}]_{m \times n} + [b_{ij}]_{m \times n})$$
$$= k([a_{ij} + b_{ij}]_{m \times n})$$
$$= [k(a_{ij} + b_{ij})]_{m \times n}$$
$$= [ka_{ij} + kb_{ij}]_{m \times n}$$
$$= [ka_{ij}]_{m \times n} + [kb_{ij}]_{m \times n}$$
$$= k[a_{ij}]_{m \times n} + k[b_{ij}]_{m \times n}$$
$$= kA + kB$$

   Hence $k(A + B) = kA + kB.$

   (ii) *If A is a matrix of order $m \times n$ and a, b are two scalars, then $(a + b)A = aA + bA$*

   Proof. Let $A = [a_{ij}]_{m \times n}$, then

$$(a + b)A = (a + b)[a_{ij}]_{m \times n}$$
$$= [(a + b)a_{ij}]_{m \times n} \qquad \text{(By scalar multiplication)}$$
$$= [aa_{ij} + ba_{ij}]_{m \times n} \qquad (\because \text{Numbers are distributive})$$
$$= [aa_{ij}]_{m \times n} + [ba_{ij}]_{m \times n}$$

$$= a[a_{ij}]_{m \times n} + b[a_{ij}]_{m \times n}$$
$$= aA + bA$$

Hence $\qquad (a + b)A = aA + bA$

(iii) *If A is a matrix of order* $m \times n$ *and a, b are two scalars, then* $a(bA) = (ab)A$

Proof. Let $A = [a_{ij}]_{m \times n}$, then

$$a(bA) = a(b[a_{ij}]_{m \times n})$$
$$= [a(ba_{ij})]_{m \times n} \qquad \text{(By scalar multiplication)}$$
$$= [(ab)a_{ij}]_{m \times n} \qquad (\because \text{Numbers are associative)}$$
$$= (ab)[a_{ij}]_{m \times n}$$
$$= (ab) A$$

Hence $\qquad a(bA) = (ab)A.$

(iv) *If A is a matrix of order* $m \times n$ *and k is any scalar, then* $(-k)A = -(kA) = k(-A)$

Proof. Let $A = [a_{ij}]_{m \times n}$, then

$$(-k)A = (-k)[a_{ij}]_{m \times n}$$
$$= [(-k)a_{ij}]_{m \times n} \qquad \text{(By scalar multiplication)}$$
$$= [-ka_{ij}]_{m \times n}$$
$$= -[ka_{ij}]_{m \times n}$$
$$= -(kA)$$

Now $\qquad (-k)A = (-k)[a_{ij}]_{m \times n}$
$$= [(-k)a_{ij}]_{m \times n}$$
$$= [k(-a_{ij})]_{m \times n}$$
$$= k[-a_{ij}]_{m \times n}$$
$$= k(-A)$$

Hence $\qquad (-k)A = -(kA) = k(-A).$

1.7 MULTIPLICATION OF MATRICES

Let A and B be two matrices of order $m \times n$ and $n \times p$ respectively. Then a matrix C of order $m \times p$ is obtained by multiplying each row of A to each column of B.

Suppose $A = [a_{ij}]_{m \times n}$, $B = [b_{jk}]_{n \times p}$, then $C = [c_{ik}]_{m \times p}$ is known as the multiplication of A and B if

$$c_{ik} = \sum_{j=1}^{n} a_{ij} b_{jk}$$

and hence we can write $\qquad C = AB$

WORKING PROCEDURE

First we check whether the matrices are conformable for multiplication or not. For this we check that if the number of columns of first matrix is equal to the number of rows of the second matrix, then the matrices can be multiplied. Multiplication is operated by the rule (row × column). In this rule, we first put the first row of the first matrix next to the first column of the second matrix and the corresponding elements are now multiplied and then summed up which gives the first element of the first row of the product matrix. This process runs till the first row of the first matrix is operated to all columns of the second matrix. After that the first process is applied to the second, third etc. rows of the first matrix.

For example : If $A = \begin{bmatrix} 2 & 1 & 5 \\ 6 & 2 & 3 \end{bmatrix}_{2 \times 3}$ and $B = \begin{bmatrix} 3 & 4 \\ 5 & 6 \\ 7 & 8 \end{bmatrix}_{3 \times 2}$, then

$$AB = \begin{bmatrix} 2 & 1 & 5 \\ 6 & 2 & 3 \end{bmatrix} \begin{bmatrix} 3 & 4 \\ 5 & 6 \\ 7 & 8 \end{bmatrix}$$

$$= \begin{bmatrix} 2 \times 3 + 1 \times 5 + 5 \times 7 & 2 \times 4 + 1 \times 6 + 5 \times 8 \\ 6 \times 3 + 2 \times 5 + 3 \times 7 & 6 \times 4 + 2 \times 6 + 3 \times 8 \end{bmatrix}$$

$$= \begin{bmatrix} 6 + 5 + 35 & 8 + 6 + 40 \\ 18 + 10 + 21 & 24 + 12 + 24 \end{bmatrix}$$

$$= \begin{bmatrix} 46 & 54 \\ 49 & 60 \end{bmatrix}$$

☛ REMARKS

- If the number of columns of the matrix A is equal to the number of rows of matrix B, then A and B are conformable for the multiplication AB but not for BA.
- Square matrices are always conformable for multiplication in both ways.

1.8 DETERMINANT OF A SQUARE MATRIX

Let A be a square matrix. Then the determinant which is formed by the elements of matrix A is usually denoted by $|A|$.

For example : If $A = \begin{bmatrix} a_{11} & a_{12} & a_{13} \\ a_{21} & a_{22} & a_{23} \\ a_{31} & a_{32} & a_{33} \end{bmatrix}$, then its determinant is

$$A = \begin{vmatrix} a_{11} & a_{12} & a_{13} \\ a_{21} & a_{22} & a_{23} \\ a_{31} & a_{32} & a_{33} \end{vmatrix}$$

☛ REMARK

- The determinant of a matrix is reduced to a number.

1.9 PROPERTIES OF DETERMINANTS

(1) The value of a determinant is zero if all the elements of a row or column are zero.

(2) The value of a determinant remain unchanged when rows are changed into corresponding columns.

(3) If any two rows or columns of a determinant are interchanged the sign of the determinant is changed.

(4) If any two rows or columns of a determinant are identical, then the value of the determinant is zero.

(5) If every element of same columns or row is the sum of two terms then determinant is equal to the sum of two determinant are containing only the first term and other the second term only in place of each sum.

(6) If each element of a row (or column) is multiplied by a constant k, then the value of the new determinant will be k times the value of original determinant.

(7) If each element of a row (or column) of a determinant multiplied by a constant k and then added to the corresponding elements of some other row (or column) then the value of the determinant remain same.

(8) If the elements of the determinant are the polynomial in a variable x and if by putting $x = a$, the determinant vanishes then $(x - a)$ will be a factor of determinant.

1.10 EVOLUTION OF A DETERMINANT BY SARRUS DIAGRAM

$$\begin{vmatrix} a_{11} & a_{12} & a_{13} \\ a_{21} & a_{22} & a_{23} \\ a_{31} & a_{32} & a_{33} \end{vmatrix} \begin{aligned} &= a_{11}(a_{22}a_{33} - a_{32}a_{23}) - a_{12}(a_{21}a_{33} - a_{31}a_{23}) + a_{13}(a_{21}a_{32} - a_{31}a_{22}) \\ &= a_{11}a_{22}a_{33} + a_{12}a_{31}a_{23} + a_{13}a_{21}a_{32} - (a_{11}a_{32}a_{23} + a_{12}a_{21}a_{33} + a_{13}a_{31}a_{22}) \end{aligned}$$

WORKING PROCEDURE

Write the columns of the determinant and again write the first and second columns on the right side and draw the lines as shown in the following figure :

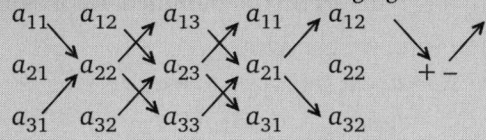

For example :

Let
$$A = \begin{vmatrix} 1 & 2 & 3 \\ 2 & 3 & 4 \\ 2 & 0 & 5 \end{vmatrix}$$

Then we have

$\Rightarrow$
$$|A| = 1 \cdot 3 \cdot 5 + 2 \cdot 4 \cdot 2 + 3 \cdot 2 \cdot 0 - (2 \cdot 3 \cdot 3 + 0 \cdot 4 \cdot 1 + 5 \cdot 2 \cdot 2)$$
$$= 15 + 16 + 0 - (18 + 0 + 20) = 31 - 38 = -7$$

1.11 MINORS AND COFACTORS

1.11.1 MINORS

In determinant,
$$\Delta = \begin{vmatrix} a_{11} & a_{12} & a_{13} \\ a_{21} & a_{22} & a_{23} \\ a_{31} & a_{32} & a_{33} \end{vmatrix} \qquad \ldots(1)$$

If we leave the row and column passing through the element a_{ij} then we obtained the second order determinant, which is called the minor of the element a_{ij}. It is denoted by M_{ij}. Therefore, in a determinant of order 3, we may get 9 minors corresponding to the 9 elements of the determinant.

For example, in determinant (1)

$$\text{Minor of } a_{21} = \begin{vmatrix} a_{12} & a_{13} \\ a_{32} & a_{33} \end{vmatrix} = M_{21}$$

and

$$\text{Minor of } a_{32} = \begin{vmatrix} a_{11} & a_{13} \\ a_{21} & a_{23} \end{vmatrix} = M_{32}$$

If we expand the determinant along the first row, then

$$\Delta = (-1)^{1+1} a_{11} M_{11} + (-1)^{1+2} a_{12} M_{12} + (-1)^{1+3} a_{13} M_{13}$$
$$= a_{11} M_{11} - a_{12} M_{12} + a_{13} M_{13}$$

Similarly, along second column, we can write

$$\Delta = -a_{12} M_{12} + a_{22} M_{22} - a_{32} M_{32}$$

1.11.2 Cofactor

If we multiply the minor M_{ij} by $(-1)^{i+j}$. Then resulting value is called cofactor of the element a_{ij}. If A_{ij} is the cofactor of a_{ij}, then we write

$$\text{Cofactor of } a_{ij} = A_{ij} = (-1)^{i+j} M_{ij}$$

$$\text{Cofactor of } a_{21} = A_{21} = (-1)^{2+1} M_{21} = -\begin{vmatrix} a_{12} & a_{13} \\ a_{32} & a_{33} \end{vmatrix}$$

$$\text{Cofactor of } a_{32} = A_{32} = (-1)^{3+2} M_{32} = -\begin{vmatrix} a_{11} & a_{13} \\ a_{21} & a_{23} \end{vmatrix}$$

Hence, cofactor of $a_{ij} = (-1)^{i+j}$ determinant obtained by leaving row and column passing through that element. Therefore, we can write

$$\Delta = a_{11} A_{11} + a_{12} A_{12} + a_{13} A_{13}$$
$$\Delta = a_{21} A_{21} + a_{22} A_{22} + a_{23} A_{23}$$
$$\Delta = a_{31} A_{31} + a_{32} A_{32} + a_{33} A_{33}$$

and

$$a_{11} A_{21} + a_{12} A_{22} + a_{13} A_{23} = 0$$
$$a_{11} A_{31} + a_{12} A_{32} + a_{13} A_{33} = 0$$

1.12 SINGULAR AND NON-SINGULAR MATRIX

Definition. *A matrix whose determinant value is zero, is said to be singular matrix. If the matrix is not singular, then it is said to be non-singular.*

For example : If $A = \begin{bmatrix} 2 & 3 \\ 6 & 9 \end{bmatrix}$, then its determinant value.

$$|A| = \begin{vmatrix} 2 & 3 \\ 6 & 9 \end{vmatrix} = 2 \times 9 - 3 \times 6 = 18 - 18 = 0$$

Thus the matrix A is singular.

1.13 TRANSPOSE OF A MATRIX

Consider a matrix $A = [a_{ij}]_{m \times n}$. Then a matrix which is obtained by interchanging the rows and columns of A is called the transpose of A. It is denoted by A' or A^T.

That is , if $A = [a_{ij}]_{m \times n}$, then $A' = [a_{ji}]_{n \times m}$.

For example : If $\quad A = \begin{bmatrix} 2 & 3 & 5 \\ 1 & 6 & 7 \end{bmatrix}_{2 \times 3}$, then its transpose is

$$A' = \begin{bmatrix} 2 & 3 & 5 \\ 1 & 6 & 7 \end{bmatrix}'$$

$$= \begin{bmatrix} 2 & 1 \\ 3 & 6 \\ 5 & 7 \end{bmatrix}_{3 \times 2}$$

☛ REMARKS
- Transpose of row matrix is a column matrix and transpose of a column matrix is a row matrix.
- If a matrix is square then its transpose will be a square matrix of same order.

1.14 PROPERTIES OF TRANSPOSE OF A MATRIX

THEOREM 1. *If A' and B' are the transpose of the matrix A and B respectively, then :*
(i) $(A')' = A$
(ii) $(A+B)' = A' + B'$, *here A and B must be of same order.*
(iii) $(kA)' = kA'$, *here k is any scalar.*
(iv) $(AB)' = B'A'$, *here AB and B'A' are conformable for multiplication.*

Proof.

(i) Let $\quad A = [a_{ij}]_{m \times n}$, then $\quad A' = [a_{ji}]_{n \times m}$

Since, $\quad (i, j)$th element in $(A')' = (j, i)$th element in A'
$$= (i, j)\text{th element in } A$$
Thus by the definition of equality of matrices, we must have $(A')' = A$.

(ii) Let $A = [a_{ij}]_{m \times n}$, $B = [b_{ij}]_{m \times n}$. So, $A' = [a_{ji}]_{n \times m}$ and $B' = [b_{ji}]_{n \times m}$, then
$$(i, j)\text{th element in } (A+B)' = (j, i)\text{th element in } (A+B)$$
$$= (j, i)\text{th element in } A + (j, i)\text{th element in } B$$
$$= (i, j)\text{th element in } A' + (i, j)\text{th element in } B'$$
$$= (i, j)\text{th element in } (A'+B')$$
Thus by the definition of equality of matrices, we get
$$(A + B)' = A' + B'$$

(iii) Let $A = [a_{ij}]_{m \times n}$ so that $\quad A' = [a_{ji}]_{n \times m}$ and k be a scalar, then
$$(i, j)\text{th element in } (kA)' = (j, i)\text{th element in } (kA)$$
$$= (i, j)\text{th element in } kA'$$
Thus by the definition of equality of matrices, we get
$$(kA)' = kA'$$

(iv) Let $A = [a_{ij}]_{m \times n}$ and $B = [b_{ij}]_{n \times p}$ then AB is conformable for multiplication and having the order $m \times p$. Therefore, the order of $(AB)'$ is $p \times m$. Since the orders of A' and B' are respectively $n \times m$ and $p \times n$ so $B'A'$ is conformable for multiplication and having the order $p \times m$.

Now (k, i)th element in $(AB)' = (i, k)$th element in AB

$$= \sum_{j=1}^{n} a_{ij} b_{jk}$$

[By the definition of multiplication of matrices]

But (k, i)th element in $B'A' = \sum_{j=1}^{n} b_{kj} a_{ji}$

$$= \sum_{j=1}^{n} a_{ji} b_{kj}$$

$$= (i, k)\text{th element in } AB$$

$\therefore$ (k, i)th element in $(AB)' = (k, i)$th element in $B'A'$

Thus by the definition of equality of matrices, we must have

$$(AB)' = B'A'$$

1.15 SYMMETRIC MATRIX

A matrix 'A' is said to be a symmetric matrix if $A' = A$, that is, the transpose of a matrix is equal to the matrix itself.

For exmaple : If $A = \begin{bmatrix} 1 & 2 & 3 \\ 2 & 4 & 5 \\ 3 & 5 & 6 \end{bmatrix}$, then $A' = \begin{bmatrix} 1 & 2 & 3 \\ 2 & 4 & 5 \\ 3 & 5 & 6 \end{bmatrix}$ so that $A' = A$

Hence, A is symmetric.

1.16 SKEW-SYMMETRIC MATRIX

A matrix 'A' is said to be a skew-symmetric matrix if $A' = -A$.

For exmaple : If $A = \begin{bmatrix} 0 & 2 & 3 \\ -2 & 0 & 4 \\ -3 & -4 & 0 \end{bmatrix}$, then

$$A' = \begin{bmatrix} 0 & -2 & -3 \\ 2 & 0 & -4 \\ 3 & 4 & 0 \end{bmatrix}$$

$$= \begin{bmatrix} 0 & 2 & 3 \\ -2 & 0 & 4 \\ -3 & -4 & 0 \end{bmatrix} = -A$$

Hence A is skew-symmetric matrix.

1.17 RANK OF A MATRIX

Let A be a matrix of order $m \times n$, then a non-negative integer r is said to be the rank of matrix A if it possesses the following two properties :

(i) There exists at least one r-minor of A which is not equal to zero.

(ii) Every s-minor of A for all $s > r$ is zero.

We denote the rank of A by $\rho(A)$.

In other words, the rank of a matrix is the order of any highest order of a non-zero minor of the matrix.

☛ REMARKS
- If the order of a matrix A is $m \times n$, then $\rho(A) \leq$ min. $\{m, n\}$
- A is a null matrix iff $\rho(A) = 0$.
- If A is any non-zero matrix, then $\rho(A) \geq 1$.
- $\rho(A) \geq r$, if there exists a non-zero r-minor of A.
- For any square matrix A of order n, $\rho(A) = n$ iff A is non-singular.
- For any square matrix A of order n, $\rho(A) < n$ iff A is singular.
- $\rho(A) \leq r$ if every s-minor of A is zero, where $s > r$.
- Every $(r+1)$- rowed minor of A can be expressed as a linear combination of its r-rowed minors, therefore if every r-minor of A is zero, then its every $(r+1)$-minor is also zero.

1.18 ECHELON FORM OF A MATRIX

A matrix A is said to be in Echelon form if :
  (*i*) every row of A has all its entries 0 which occurs below every row having a non-zero entry. and
  (*ii*) the number of zeros before the first non-zero entry in a row is less than the number of such zeros in the next row.

☛ REMARK
- The rank of a matrix is equal to the number of non-zero rows in Echelon form of that matrix.

For example: Consider a matrix $A = \begin{bmatrix} 0 & 2 & 3 & 5 \\ 0 & 0 & 3 & 2 \\ 0 & 0 & 0 & 0 \end{bmatrix}$

Clearly, A is in Echelon form which has 2 non-zero rows, hence the rank of A is 2.

THEOREM 1. *The rank of the transpose of a matrix is equal to the rank of that matrix.*

PROOF. Let A be a marix, then A' is its transpose and let $\rho(A) = r$, then there exists an r-rowed minor of A which is not equal to zero and all s-rowed minors of A are zero, where $s > r$. Let $|B|$ be a r-rowed minor of A such that $|B| \neq 0$. Since A' is the transpose of A, then $|B'|$ is the r-rowed minor of A' but $|B'| = |B| \neq 0$, therefore $\rho(A') \geq r$. Suppose there is an s-minor $|C|$ of A' such that $|C| \neq 0$, where $s > r$, then $|C'|$ will be an s-minor of A such that $|C'| = |C| \neq 0$, therefore $\rho(A) > r$ which is a contradiction, hence $\rho(A') = r$.

Solved Examples

EXAMPLE 1. *Find the rank of the following matrices :*

  (i) $\begin{bmatrix} 3 & 0 & 0 \end{bmatrix}$

  (ii) $\begin{bmatrix} 1 & 2 & 3 \\ 2 & 4 & 5 \end{bmatrix}$

  (iii) $\begin{bmatrix} 1 & 2 & 3 \\ 3 & 4 & 5 \\ 4 & 5 & 6 \end{bmatrix}$

  (iv) $\begin{bmatrix} 1 & 5 & 2 & 4 \\ 0 & 1 & 3 & 1 \\ 0 & 0 & 1 & 3 \end{bmatrix}$

SOLUTION. (i) Let $A = \begin{bmatrix} 3 & 0 & 0 \end{bmatrix}$, then A is the non-zero rowed matrix, thereoofore $\rho(A) \geq 1$. Also A is a matrix of order 1×3, then $\rho(A) \leq 1$, hence $\rho(A) = 1$.

(ii) Let $A = \begin{bmatrix} 1 & 2 & 3 \\ 2 & 4 & 5 \end{bmatrix}$

The order of A is 2×3, then $\rho(A) \leq 2$.

Also there is a 2-minor $\begin{vmatrix} 2 & 3 \\ 4 & 5 \end{vmatrix}$ of A which is not equal to zero, then $\rho(A) \geq 2$, hence $\rho(A) = 2$.

(iii) Let $A = \begin{bmatrix} 1 & 2 & 3 \\ 3 & 4 & 5 \\ 4 & 5 & 6 \end{bmatrix}$. The order of A is 3×3, then $\rho(A) \leq 3$.

Now $|A| = \begin{vmatrix} 1 & 2 & 3 \\ 3 & 4 & 5 \\ 4 & 5 & 6 \end{vmatrix} = 1(24-25) - 2(18-20) + 3(15-16) = 0$ '

$\therefore$ The only 3-minor $|A|$ of A is zero, thus $\rho(A) < 3$. Further, there is a 2-minor $\begin{vmatrix} 1 & 2 \\ 3 & 4 \end{vmatrix}$ of A which is not equal to zero, hence $\rho(A) = 2$.

(iv) Let $A = \begin{bmatrix} 1 & 5 & 2 & 4 \\ 0 & 1 & 3 & 1 \\ 0 & 0 & 1 & 3 \end{bmatrix}$

The order of A is 3×4, then $\rho(A) \leq 3$.

Now there is a 3-minor $\begin{vmatrix} 1 & 5 & 2 \\ 0 & 1 & 3 \\ 0 & 0 & 1 \end{vmatrix}$ of A which is not equal to zero, then $\rho(A) \geq 3$.

Hence $\rho(A) = 3$.

1.19 INVERSE OF A MATRIX

Let A be a non-singular matrix of order $n \times n$. Then it is said to be invertible if there exists a non-singular square matrix B of order $n \times n$ such that

$$AB = I_n = BA$$

where I_n is the unit matrix of order $n \times n$.
The matrix B is called the inverse of A, we write $B = A^{-1}$.

  Solved Examples

EXAMPLE 1. *By using elementary row-transformations find the inverse of the following matrices:*

(i) $\begin{bmatrix} 1 & 2 \\ 3 & 7 \end{bmatrix}$
(ii) $\begin{bmatrix} 1 & 2 \\ 2 & -1 \end{bmatrix}$

SOLUTION. (i) We write

$$A = I_2 A$$

or

$$\begin{bmatrix} 1 & 2 \\ 3 & 7 \end{bmatrix} = \begin{bmatrix} 1 & 0 \\ 0 & 1 \end{bmatrix} A$$

Applying $R_2 \to R_2 - 3R_1$, we get

$$\begin{bmatrix} 1 & 2 \\ 0 & 1 \end{bmatrix} = \begin{bmatrix} 1 & 0 \\ -3 & 1 \end{bmatrix} A$$

Again applying $R_1 \to R_1 - 2R_2$, we get

$$\begin{bmatrix} 1 & 0 \\ 0 & 1 \end{bmatrix} = \begin{bmatrix} 7 & -2 \\ -3 & 1 \end{bmatrix} A$$

$\Rightarrow$ $\qquad\qquad I_2 = BA$

$\Rightarrow$ $\qquad\qquad A^{-1} = B = \begin{bmatrix} 7 & -2 \\ -3 & 1 \end{bmatrix}.$

(ii) We write

$$A = I_2 A$$

or $\qquad\qquad \begin{bmatrix} 1 & 2 \\ 2 & -1 \end{bmatrix} = \begin{bmatrix} 1 & 0 \\ 0 & 1 \end{bmatrix} A$

Applying $R_2 \to R_2 - 2R_1$, we get

$$\begin{bmatrix} 1 & 2 \\ 0 & -5 \end{bmatrix} = \begin{bmatrix} 1 & 0 \\ -2 & 1 \end{bmatrix} A$$

Applying $R_2 \to -\dfrac{1}{5} R_2$, we get

$$\begin{bmatrix} 1 & 2 \\ 0 & 1 \end{bmatrix} = \begin{bmatrix} 1 & 0 \\ 2/5 & -1/5 \end{bmatrix} A$$

Applying $R_1 \to R_1 - 2R_2$, we get

$$\begin{bmatrix} 1 & 0 \\ 0 & 1 \end{bmatrix} = \begin{bmatrix} 1/5 & 2/5 \\ 2/5 & -1/5 \end{bmatrix} A$$

$\Rightarrow$ $\qquad\qquad I_2 = BA$

$\Rightarrow$ $\qquad\qquad A^{-1} = B = \begin{bmatrix} 1/5 & 2/5 \\ 2/5 & -1/5 \end{bmatrix}$

EXAMPLE 2. *Find the inverse of the matrix*

$$A = \begin{bmatrix} 1 & 2 & 1 \\ 3 & 2 & 3 \\ 1 & 1 & 2 \end{bmatrix}$$

by using elementary row-transformation.

SOLUTION. We write $\qquad\qquad A = I_3 A$

or $\qquad\qquad \begin{bmatrix} 1 & 2 & 1 \\ 3 & 2 & 3 \\ 1 & 1 & 2 \end{bmatrix} = \begin{bmatrix} 1 & 0 & 0 \\ 0 & 1 & 0 \\ 0 & 0 & 1 \end{bmatrix} A$

Applying $R_2 \to R_2 - 3R_1$, $R_3 \to R_3 - R_1$, we get

$$\begin{bmatrix} 1 & 2 & 1 \\ 0 & -4 & 0 \\ 0 & -1 & 1 \end{bmatrix} = \begin{bmatrix} 1 & 0 & 0 \\ -3 & 1 & 0 \\ -1 & 0 & 1 \end{bmatrix} A$$

Applying $R_2 \to \dfrac{-1}{4} R_2$, we get

$$\begin{bmatrix} 1 & 2 & 1 \\ 0 & 1 & 0 \\ 0 & -1 & 1 \end{bmatrix} = \begin{bmatrix} 1 & 0 & 0 \\ 3/4 & -1/4 & 0 \\ -1 & 0 & 1 \end{bmatrix} A$$

Applying $R_3 \to R_3 + R_2$, we get

$$\begin{bmatrix} 1 & 2 & 1 \\ 0 & 1 & 0 \\ 0 & 0 & 1 \end{bmatrix} = \begin{bmatrix} 1 & 0 & 0 \\ 3/4 & -1/4 & 0 \\ -1/4 & -1/4 & 1 \end{bmatrix} A$$

Applying $R_1 \to R_1 - 2R_2$, we get

$$\begin{bmatrix} 1 & 0 & 1 \\ 0 & 1 & 0 \\ 0 & 0 & 1 \end{bmatrix} = \begin{bmatrix} -1/2 & 1/2 & 0 \\ 3/4 & -1/4 & 0 \\ -1/4 & -1/4 & 1 \end{bmatrix} A$$

Applying $R_1 \to R_1 - R_3$, we get

$$\begin{bmatrix} 1 & 0 & 0 \\ 0 & 1 & 0 \\ 0 & 0 & 1 \end{bmatrix} = \begin{bmatrix} -1/4 & 3/4 & -1 \\ 3/4 & -1/4 & 0 \\ -1/4 & -1/4 & 1 \end{bmatrix} A$$

$\Rightarrow \qquad\qquad\qquad I_3 = BA$

$\Rightarrow \qquad\qquad A^{-1} = B = \begin{bmatrix} -1/4 & 3/4 & -1 \\ 3/4 & -1/4 & 0 \\ -1/4 & -1/4 & 1 \end{bmatrix}$

 Exercise- 1.1

1. Are the following pairs of matrices equivalent?

(i) $\begin{bmatrix} 4 & 0 & 2 \\ 3 & 1 & 0 \\ 5 & 2 & 0 \end{bmatrix}, \begin{bmatrix} 3 & 9 & 0 & 2 \\ 7 & -2 & 0 & 1 \\ 8 & 1 & 1 & 5 \end{bmatrix}$

(ii) $\begin{bmatrix} 2 & -1 & 3 & 4 \\ 0 & 3 & 4 & 1 \\ 2 & 3 & 7 & 5 \\ 2 & 5 & 11 & 5 \end{bmatrix}, \begin{bmatrix} 1 & 0 & -5 & 6 \\ 3 & -2 & 1 & 2 \\ 5 & -2 & -9 & 14 \\ 4 & -2 & -4 & 8 \end{bmatrix}$

Determine the rank of the following matrices:

2. $\begin{bmatrix} 1 & 1 & 1 \\ 2 & 2 & 2 \\ 3 & 3 & 3 \end{bmatrix}$

3. $\begin{bmatrix} 2 & 1 & 3 \\ 4 & 7 & 13 \\ 4 & -3 & -1 \end{bmatrix}$

4. $\begin{bmatrix} 4 & 5 & 6 \\ 5 & 6 & 7 \\ 7 & 8 & 9 \end{bmatrix}$

5. $\begin{bmatrix} 1 & 2 & 3 \\ 2 & 3 & 4 \\ 3 & 5 & 7 \end{bmatrix}$

6. $\begin{bmatrix} 2 & 3 & 7 \\ 3 & -2 & 4 \\ 1 & -3 & -1 \end{bmatrix}$

7. $\begin{bmatrix} 3 & -1 & 2 \\ -6 & 2 & -4 \\ -3 & 1 & -2 \end{bmatrix}$

8. $\begin{bmatrix} 1 & 2 & 3 & 1 \\ 2 & 4 & 6 & 2 \\ 1 & 2 & 3 & 2 \end{bmatrix}$

9. $\begin{bmatrix} 1 & 3 & 4 & 3 \\ 3 & 9 & 12 & 9 \\ 1 & 3 & 4 & 1 \end{bmatrix}$

10. $\begin{bmatrix} 1 & 2 & -1 & 4 \\ 2 & 4 & 3 & 5 \\ -1 & -2 & 6 & -7 \end{bmatrix}$

11. $\begin{bmatrix} 1 & 2 & -4 & 5 \\ 2 & -1 & 3 & 6 \\ 8 & 1 & 9 & 7 \end{bmatrix}$

12. $\begin{bmatrix} 1 & -1 & 3 & 6 \\ 1 & 3 & -3 & -4 \\ 5 & 3 & 3 & 11 \end{bmatrix}$

13. $\begin{bmatrix} 1 & 2 & 3 & 0 \\ 2 & 4 & 3 & 2 \\ 3 & 2 & 1 & 3 \\ 6 & 8 & 7 & 5 \end{bmatrix}$

17. $\begin{bmatrix} 0 & 1 & -3 & -1 \\ 1 & 0 & 1 & 1 \\ 3 & 1 & 0 & 2 \\ 1 & 1 & -2 & 0 \end{bmatrix}$

14. $\begin{bmatrix} 2 & 3 & -1 & -1 \\ 1 & -1 & -2 & -4 \\ 3 & 1 & 3 & -2 \\ 6 & 3 & 0 & -7 \end{bmatrix}$

18. $\begin{bmatrix} 1 & 2 & -1 & 3 \\ 4 & 1 & 2 & 1 \\ 3 & -1 & 1 & 2 \\ 1 & 2 & 0 & 1 \end{bmatrix}$

15. $\begin{bmatrix} 1 & 2 & 1 & 2 \\ 1 & 3 & 2 & 2 \\ 2 & 4 & 3 & 4 \\ 3 & 7 & 4 & 6 \end{bmatrix}$

19. $\begin{bmatrix} 1 & 0 & 2 & 1 \\ 0 & 1 & -2 & 1 \\ 1 & -1 & 4 & 0 \\ -2 & 2 & 8 & 0 \end{bmatrix}$

16. $\begin{bmatrix} 3 & -2 & 0 & -1 \\ 0 & 2 & 2 & 1 \\ 1 & -2 & -3 & 2 \\ 0 & 1 & 2 & 1 \end{bmatrix}$

20. $\begin{bmatrix} 8 & 0 & 0 & 1 \\ 1 & 0 & 8 & 1 \\ 0 & 0 & 1 & 8 \\ 0 & 1 & 1 & 8 \end{bmatrix}$

Answers

1. (i) Not equivalent			(ii) Not equivalent		**2.** 1	**3.** 2	**4.** 2	**5.** 2
6. 1	**7.** 2	**8.** 2	**9.** 2	**10.** 3	**11.** 3	**12.** 3	**13.** 3	**14.** 3
15. 4	**16.** 2	**17.** 3	**18.** 3	**19.** 4	**20.** 2			

1.20 CONVEX SET

A set of points is said to be convex if for any two points in the set, the line segment joining these points is also in the set, *i.e.*, the set is said to be convex if convex combinations of any two points in the set is also in the set. [MEERUT–2009, 10, 11, 14, 17; GORAKHPUR–2007, 11, 18]

Mathematically, for any two points x_1, x_2 in the set, if every point $x = \lambda x_1 + (1 - \lambda)x_2$, $0 \le \lambda \le 1$ is also in the set, then set is convex.

Following are the examples of convex sets.

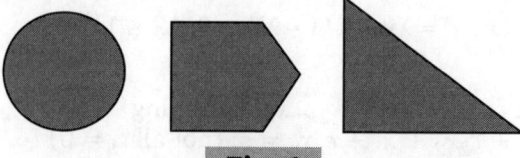

Fig. 1

Similarly some non-convex sets are given below:

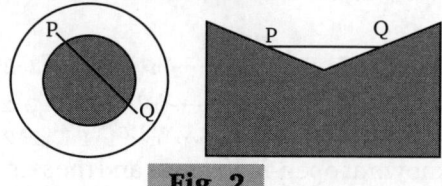

Fig. 2

☛ Remarks
- The convex combination of any number of points in the convex set also belongs to the set.
- A set of one point is always convex.

1.21 SOME RELATED DEFINITIONS

1.21.1 Point Set

A set whose elements are points or vectors in R^n is said to be point set.

For example:

1. A line $a_1x_1 + a_2x_2 = b$ represents a line in two dimensions which may be considered as a set of these points (x_1, x_2). Therefore, set of points can be written as
$$S = \{(x_1, x_2) : a_1x_1 + a_2x_2 = b\}$$

2. If we consider the set of points lying inside a circle of unit radius with centre at the origin in two dimensional space. Clearly, the points (x_1, x_2) of this set satisfy the inequality
$$x_1^2 + x_2^2 < 1$$
Therefore, set of points is given by
$$S = \{(x_1, x_2) : x_1^2 + x_2^2 < 1\}$$

1.21.2 HYPERSPHERE

In n-dimensional space, a hypersphere, with centre a and radius $r(>0)$ is the set of points
$$X = \{x : |x - a| = r\}.$$
The equation of hypersphere in E^n (or R^n) is given by $\Sigma(x_i - a_i)^2 = r^2$

☛ REMARKS
- The set of points inside the hypersphere is the set $X = \{x : |x - a| < r\}$
- The set of points lying inside the hypersphere with centre a and radius $\epsilon > 0$ is said to be ϵ-neighbourhood about the point a.

1.21.3 LINES AND LINE SEGMENTS

Let x_1, x_2 be two distinct points in n-dimensional space E^n, then the line through the points x_1 and x_2 is defined to be the set of points given by
$$X = \{x : x = \lambda x_1 + (1 - \lambda)x_2, \text{ for } \lambda \in \boldsymbol{R}\}$$
and the line segment joining two points x_1 and x_2 in E^n is defined to be the set of points given by
$$X = \{x : x = \lambda x_1 + (1 - \lambda)x_2, 0 \le \lambda \le 1\}$$

1.21.4 HYPERPLANE

It is defined as the set of points $(x_1, x_2,...x_n)$ satisfying
$$c_1x_1 + c_2x_2 + ... + c_nx_n = z, \text{ (not all } c_i = 0)$$
for prescribed values of $c_1, c_2,..., c_n$ and z

We clearly observe that a hyperplane divides the whole space into three mutually disjoint sets as given below:
$$X_1 = \{x : \boldsymbol{cx} > \boldsymbol{z}\} = \{(x_1, x_2,...,x_n) : c_1x_1 + c_2x_2 + ... + c_nx_n > z\}$$
$$X_2 = \{x : \boldsymbol{cx} = \boldsymbol{z}\} = \{(x_1, x_2,...,x_n) : c_1x_1 + c_2x_2 + ... + c_nx_n = z\}$$
$$X_3 = \{x : \boldsymbol{cx} < \boldsymbol{z}\} = \{(x_1, x_2,...,x_n) : c_1x_1 + c_2x_2 + ... + c_nx_n < z\}$$

Here the set X_1 and X_3 are known as open-half spaces and the sets $\{x : \boldsymbol{cx} \le \boldsymbol{z}\}$ and $\{x : \boldsymbol{cx} \ge \boldsymbol{z}\}$ are known as closed-half spaces.

☛ REMARKS
- For optimum value of z, the hyperplane $\boldsymbol{cx} = \boldsymbol{z}$ is called optimal hyperplane.
- The vector $\boldsymbol{c}$ is known as vector normal to the hyperplane
- The value $\pm\dfrac{c}{|\boldsymbol{c}|}$ are called unit normals.
- The hyperplanes are always closed sets.

In an LPP, the objective function represents a hyperplane and each constraints($\leq$ or $\geq$) is a closed-half space produced by the hyperplane given by the constraints by taking ($=$) sign in place of $\leq$ or $\geq$.

1.21.5 PARALLEL HYPERPLANES

Two hyperplanes $c_1 x = z_1$ and $c_2 x = z_2$ are said to be parallel if they have the same unit normals, *i.e.*, if $c_1 = \lambda c_2$ for some λ, $(\lambda \neq 0)$

1.21.6 CONVEX COMBINATIONS

A convex combinations of a finite number of points $x_1, x_2, ..., x_n$ is defined by the point
$$x = a_1 x_1 + a_2 x_2 + ... + a_n x_n, \; a_i \in R, \; a_i \geq 0 \; \forall \; i \text{ and } \Sigma a_i = 1$$

In particular, the convex combination of two points x_1, x_2 be given by $x = a_1 x_1 + a_2 x_2$ such that $a_1, a_2 \geq 0$ and $a_1 + a_2 = 1$ which can also be written as $x = a x_1 + (1 - a) x_2$, $0 \leq a \leq 1$

☛ **REMARK**
- The line segment of two points x_1 and x_2 is the set of all possible convex combinations of two points x_1 and x_2.

1.21.7 EXTREME POINT OF A CONVEX SET

A point x in a convex set C is an extreme point of C if it does not lie on the line segment of any two points, different from x in the set, *i.e.*, it can not be expressed as a convex combinations of any two distinct points x_1 and x_2 in C.

Mathematically an extreme point can be defined as follows:

A point x is said to be an extreme point of a convex set if there do not exist other points $x_1, x_2 (x_1 \neq x_2)$ in the set such that $x = \lambda x_1 + (1 - \lambda) x_2$, $0 < \lambda < 1$

☛ **REMARKS**
- A convex set may also have infinite number of extreme points.
- The polygons which are convex, have the extreme points as their vertices.
- An extreme point is a boundary point of the set.
- All boundary points of a convex set are not necessarily extreme points.
- A point of a convex set C, which is not an extreme point, is said to be internal point of C.

1.22 CONVEX HULL
[MEERUT–2006]

The set of all convex combinations of sets of points from the set X of points is called convex hull, *i.e.*, the intersection of all convex sets containing X in n-dimensional space is called the convex hull of X. Hence, the convex hull of a set $X \subseteq E^n$ is the smallest convex set containing X.

For example: If X is the boundary of a circle, then the convex hull $C(x)$ is the whole circle.

1.23 CONVEX FUNCTION AND CONVEX POLYHEDRON [MEERUT–2006; GORAKHPUR–2009, 10, 18]

Definition 1. *A function $f(x)$ is said to be strictly convex at x if for any two other distinct points x_1 and x_2*
$$f\{\lambda \, x_1 + (1 - \lambda) x_2\} < \lambda f(x_1) + (1 - \lambda) f(x_2), \; 0 < \lambda < 1$$

Definition 2. *The set of all convex combinations of finite number of points is said to be the convex polyhedron generated by these points.*

In other words, we can say that if the set X consist of a finite number of points, the convex hull of X is said to be the convex polyhedron with vertices at those points.

For example: The set of the area of a triangle is a convex polyhedron of its vertices.

☛ **REMARK**

- A function $f(x)$ is said to be strictly concave if $-f(x)$ is strictly convex.

RELATED THEOREMS

THEOREM 1.   *The hyperplane is a convex set.*   [MEERUT–2007]

PROOF. Let $X = [x : cx = z]$ be a hyperplane and $x_1, x_2 \in X$ then,

$$cx_1 = z \text{ and } cx_2 = z \qquad \text{(By definition)}$$

Now, if $x_3 = \lambda x_1 + (1 - \lambda)x_2, \ 0 \le \lambda \le 1$

Then, $cx_3 = \lambda c \cdot x_1 + (1 - \lambda)cx_2$

$$= \lambda z + (1 - \lambda)z$$

$$= z$$

$\Rightarrow$ $x_3 = \lambda x_1 + (1 - \lambda)x_2 \in X$

$\Rightarrow x_3$ is also a point in X

Hence, X is a convex set.

THEOREM 2.   *The closed half spaces* $H_1 = \{x : cx \ge z\}$ *and* $H_2 = \{x : cx \le z\}$ *are convex sets.*

PROOF. Let $x_1 \in H_1$ and $x_2 \in H_2$. Then by definition of H_1, we can write $cx_1 \ge z : cx_2 \ge z$

Now, if $0 \le \lambda \le 1$, then we have

$$c[\lambda x_1 + (1 - \lambda)x_2] = \lambda c \cdot x_1 + (1 - \lambda)cx_2$$

$$\ge \lambda z + (1 - \lambda)z = z$$

Therefore, $x_1, x_2 \in H_1$ and $0 \le \lambda \le 1$ implies $\lambda x_1 + (1 - \lambda)x_2 \in H_1$

Hence, H_1 is a convex set

Similarly, we may prove that H_2 is a convex set.

☛ **REMARK**

- In a similar way (as above) we may prove that the open half spaces $\{x : cx > z\}$ and $\{x : cx < z\}$ are convex sets.

THEOREM 3.   *Intersection of two convex sets is also a convex set.*   [MEERUT–2007, 08, 12,15]

PROOF. Let X_1 and X_2 be two convex sets. We have to prove that $X_1 \cap X_2$ is also convex.

If $x_1 \in X_1 \cap X_2 \Rightarrow x_1 \in X_1$ and $x_1 \in X_2$

$x_2 \in X_1 \cap X_2 \Rightarrow x_2 \in X_1$ and $x_2 \in X_2$

Now, by definition of convex sets

$$x_1, x_2 \in X_1 \ \Rightarrow \lambda x_1 + (1 - \lambda)x_2 \in X_1 \ ; \ 0 \le \lambda \le 1$$

$$x_1, x_2 \in X_2 \ \Rightarrow \lambda x_1 + (1 - \lambda)x_2 \in X_2 \ ; \ 0 \le \lambda \le 1$$

Therefore, $\lambda x_1 + (1 - \lambda)x_2 \in X_1$ and $\lambda x_1 + (1 - \lambda)x_2 \in X_2$

$\Rightarrow \lambda x_1 + (1 - \lambda)x_2 \in X_1 \cap X_2$

Hence, $X_1 \cap X_2$ is a convex set.

THEOREM 4.   *Finite intersection of convex sets is also a convex set.*

PROOF. Let $X_1, X_2, ..., X_n$ be n convex sets.

We have to prove that $X = X_1 \cap X_2 \cap ... \cap X_n$ is also convex.

Let $x_1 \in X_1 \cap X_2 \cap ... \cap X_n \ \Rightarrow \ x_1 \in X_i \forall i = 1, 2, ..., n$

$x_2 \in X_1 \cap X_2 \cap ... \cap X_n \ \Rightarrow \ x_2 \in X_i \forall i = 1, 2, ..., n$

Since each X_i is convex set for $i = 1, 2, ..., n$

Therefore, $\boldsymbol{x}_1, \boldsymbol{x}_2 \in X_i$

$\Rightarrow \lambda \boldsymbol{x}_1 + (1 - \lambda) \boldsymbol{x}_2 \in X_i \quad \forall i = 1, 2, ..., n, \ 0 \leq \lambda \leq 1$

$\Rightarrow \lambda \boldsymbol{x}_1 + (1 - \lambda) \boldsymbol{x}_2 \in X_1 \cap X_2 \cap ... \cap X_n$

$\Rightarrow \boldsymbol{x}_1 \in X_1 \cap X_2 \cap ... \cap X_n$ and $\boldsymbol{x}_2 \in X_1 \cap X_2 \cap ... \cap X_n$

$\Rightarrow \lambda \boldsymbol{x}_1 + (1 - \lambda) \boldsymbol{x}_2 \in X_1 \cap X_2 \cap ... \cap X_n, \ 0 \leq \lambda \leq 1.$

Hence, $X_1 \cap X_2 \cap ... \cap X_n$ is a convex set.

☞ REMARK

- In the similar way we may extend the above result as follows:
"Arbitrary intersection of convex sets is also a convex set"

THEOREM 5. *The set of all convex combinations of a finite number of points* $\boldsymbol{x}_1, \boldsymbol{x}_2, ..., \boldsymbol{x}_n$ *is a convex set.*

PROOF. Let us define the set X of all convex combinations as follows:

$$X = \left\{ x : x = \sum_{i=1}^{n} \lambda_i \boldsymbol{x}_i, \sum_{i=1}^{n} \lambda_i = 1, \lambda_i \geq 0 \right\}$$

We have to prove that X is convex.

Let $\alpha, \beta \in \boldsymbol{X}$ such that

$$\alpha = \sum_{i=1}^{n} a_i \boldsymbol{x}_i \ ; \ \sum_{i=1}^{n} a_i = 1 \ , \ a_i \geq 0 \ \text{and} \quad \beta = \sum_{i=1}^{n} b_i \boldsymbol{x}_i \ ; \ \sum_{i=1}^{n} b_i = 1, b_i \geq 0$$

Let us consider

$$\boldsymbol{w} = \lambda \alpha + (1 - \lambda) \beta \ , \ 0 \leq \lambda \leq 1$$

$$= \lambda \sum_{i=1}^{n} a_i \boldsymbol{x}_i + (1 - \lambda) \sum_{i=1}^{n} b_i \boldsymbol{x}_i = \sum_{i=1}^{n} \left\{ \lambda a_i + (1 - \lambda) b_i \right\} \boldsymbol{x}_i$$

$$= \sum_{i=1}^{n} c_i \boldsymbol{x}_i \ \text{where} \ c_i = \lambda a_i + (1 - \lambda) b_i \qquad ...(1)$$

Now, we shall prove that

$$\sum_{i=1}^{n} c_i = \sum_{i=1}^{n} \left\{ \lambda a_i + (1 - \lambda) b_i \right\}$$

$$= \lambda \sum_{i=1}^{n} a_i + (1 - \lambda) \sum_{i=1}^{n} b_i = \lambda \cdot 1 + (1 - \lambda) \cdot 1 = 1 \qquad ...(2)$$

Further, $c_i = \lambda a_i + (1 - \lambda) b_i \geq 0 \ \forall \ i$ \qquad ...(3)

Hence, from (1), (2) and (3) we conclude that $\boldsymbol{w} = \sum_{i=1}^{n} c_i \cdot \boldsymbol{x}_i$ is a convex combination

of $x_1, x_2, ..., x_n$. Hence X is convex.

THEOREM 6. *Let S_1 and S_2 be two convex sets in E^n, then for any scalar α, β the set $(\alpha S_1 + \beta S_2)$ is also convex.*

PROOF. Let S_1 and S_2 be two convex sets. We have to prove that for any two scalars α and β, $(\alpha S_1 + \beta S_2)$ is also convex.

Let $\boldsymbol{x}, \boldsymbol{y} \in \alpha S_1 + \beta S_2$. Then these are of the following forms:

$$\boldsymbol{x} = \alpha \boldsymbol{u}_1 + \beta \boldsymbol{v}_1 \ \text{and} \ \boldsymbol{y} = \alpha \boldsymbol{u}_2 + \beta \boldsymbol{v}_2 \qquad ...(1)$$

$\boldsymbol{u}_1, \boldsymbol{u}_2 \in S_1$ and $\boldsymbol{v}_1, \boldsymbol{v}_2 \in S_2$

Now for any scalar λ, $0 \le \lambda \le 1$, we have

$$\lambda \boldsymbol{x} + (1 - \lambda)\boldsymbol{y} = \lambda(\alpha \boldsymbol{u}_1 + \beta \boldsymbol{v}_1) + (1 - \lambda)(\alpha \boldsymbol{u}_2 + \beta \boldsymbol{v}_2)$$
$$= \alpha\{\lambda \boldsymbol{u}_1 + (1 - \lambda)\boldsymbol{u}_2\} + \beta\{\lambda \boldsymbol{v}_1 + (1 - \lambda)\boldsymbol{v}_2\} \qquad ...(2)$$

Since, S_1 and S_2 both are convex, so by definition, we can write

$$\boldsymbol{u}_1, \boldsymbol{u}_2 \in S_1 \quad \Rightarrow \quad \lambda \boldsymbol{u}_1 + (1 - \lambda)\boldsymbol{u}_2 \in S_1, 0 \le \lambda \le 1 \qquad ...(3)$$
$$\boldsymbol{v}_1, \boldsymbol{v}_2 \in S_2 \quad \Rightarrow \quad \lambda \boldsymbol{v}_1 + (1 - \lambda)\boldsymbol{v}_2 \in S_2, 0 \le \lambda \le 1 \qquad ...(4)$$

Using (2), (3) and (4) we can write

$$\lambda \boldsymbol{x} + (1 - \lambda)\boldsymbol{y} \in \alpha S_1 + \beta S_2, 0 \le \lambda \le 1$$

Hence, $\alpha \cdot S_1 + \beta \cdot S_2$ is a convex set.

☛ REMARKS
- From the above theorem we may easily prove the following result:
"The sum $(S_1 + S_2)$ and difference $(S_1 - S_2)$ of two convex sets S_1 and S_2 is again convex."
- The set of all convex combinations of a finite number of points is also convex. [MEERUT–2008, 09]
- Those convex sets which are the intersection of a finite number of closed half spaces are called 'polyhedral convex sets'.

THEOREM 7. *The set of all the internal points of a convex set is convex.*　　[MEERUT–2011, 12, 15]

PROOF. Let S be the convex set and S_1 be the set of all vertices of S.

Then clearly $S - S_1$ is the set of all internal points of S.

If $\boldsymbol{u}, \boldsymbol{v} \in S - S_1$. Then $\boldsymbol{u}, \boldsymbol{v} \in S$

Suppose that $\boldsymbol{z}$ is a point on the line segment joining $\boldsymbol{u}$ and $\boldsymbol{v}$ then we can write

$$\boldsymbol{z} = \lambda \boldsymbol{u} + (1 - \lambda)\boldsymbol{v}, 0 \le \lambda \le 1$$
$$\in S \qquad (\because S \text{ is convex})$$
$$\Rightarrow \qquad \boldsymbol{z} \in S - S_1$$

Since $\boldsymbol{z}$ is not a vertex of S_1, therefore $S - S_1$ (set of all internal points) is convex.

 Solved Examples

EXAMPLE 1. *Show that* $S = \{(x_1, x_2) : 2x_1 + 3x_2 = 7\} \subset \boldsymbol{R}^2$ *is a convex set.*

[MEERUT–2003, 08; GORAKHPUR–2009; GARHWAL–2014]

SOLUTION. Let $\boldsymbol{u}, \boldsymbol{v} \in S$

Then we can write

$$\boldsymbol{u} = (u_1, u_2), \boldsymbol{v} = (v_1, v_2)$$
$$\therefore \qquad 2u_1 + 3u_2 = 7 \text{ and } 2v_1 + 3v_2 = 7 \qquad ...(1)$$

Further, suppose that $\boldsymbol{w} = (w_1, w_2)$ is a point on the line segment joining the points $\boldsymbol{u}$ and $\boldsymbol{v}$. Then we can write

$$\boldsymbol{w} = \lambda \boldsymbol{u} + (1 - \lambda)\boldsymbol{v}, 0 \le \lambda \le 1$$
$$\therefore \qquad (w_1, w_2) = \lambda(u_1, u_2) + (1 - \lambda)(v_1, v_2)$$
$$= \{\lambda u_1 + (1 - \lambda)v_1, \lambda u_2 + (1 - \lambda)v_2\}$$

which implies that

$$w_1 = \lambda u_1 + (1 - \lambda)v_1 \text{ and } w_2 = \lambda u_2 + (1 - \lambda)v_2$$

Consider $2w_1 + 3w_2 = 2[\lambda u_1 + (1 - \lambda)v_1] + 3[\lambda u_2 + (1 - \lambda)v_2]$

$$= \lambda[2u_1 + 3u_2] + (1 - \lambda)[2v_1 + 3v_2]$$
$$= \lambda \cdot 7 + (1 - \lambda) \cdot 7 \qquad \text{(By (1))}$$
$$= 7$$
$$\Rightarrow \qquad \qquad w \in S$$

Hence, S is a convex set.

EXAMPLE 2. *Show that the set $S = \{(x_1, x_2, x_3) : 2x_1 - x_2 + x_3 \le 4\} \subset R^3$ is convex.*

[MEERUT–2005, 12, 15]

SOLUTION. Let $x = (x_1, x_2, x_3)$ and $y = (y_1, y_2, y_3)$ be any two points of the given set S. Then by definition of S, we can write

$$2x_1 - x_2 + x_3 \le 4 \text{ and } 2y_1 - y_2 + y_3 \le 4 \qquad \ldots(2)$$

Let $w = (w_1, w_2, w_3)$ be a point such that

$$w = \lambda x + (1 - \lambda)y, \ 0 \le \lambda \le 1$$

which implies that

$$(w_1, w_2, w_3) = \lambda(x_1, x_2, x_3) + (1 - \lambda)(y_1, y_2, y_3)$$
$$= (\lambda x_1 + (1 - \lambda)y_1, \lambda x_2 + (1 - \lambda)y_2, \lambda x_3 + (1 - \lambda)y_3)$$
$$\Rightarrow \qquad w_1 = \lambda x_1 + (1 - \lambda)y_1$$
$$w_2 = \lambda x_2 + (1 - \lambda)y_2$$
and $\qquad w_3 = \lambda x_3 + (1 - \lambda)y_3$

Consider $2w_1 - w_2 + w_3$

$$= \lambda(2x_1 - x_2 + x_3) + (1 - \lambda)(2y_1 - y_2 + y_3)$$
$$\le 4\lambda + 4(1 - \lambda) \qquad \text{(By (1))}$$
$$\le 4$$

Thus, $\qquad w = (w_1, w_2, w_3) \in S$

Hence, S is convex.

EXAMPLE 3. *Examine the convexity of the set*

$$S = \{(x_1, x_2) \in R^2 : 4x_1 + 3x_2 \le 6, x_1 + x_2 \ge 1\}$$

[MEERUT–1997, 2009; DELHI–2009; ASSAM–2011; PATNA–2013]

SOLUTION. We have $S = \{(x_1, x_2) \in R^2 : 4x_1 + 3x_2 \le 6, x_1 + x_2 \ge 1\}$

Let $u = (x_1, x_2) \in S$ and $v = (y_1, y_2) \in S$. Then,

$$4x_1 + 3x_2 \le 6, x_1 + x_2 \ge 1 \text{ and } 4y_1 + 3y_2 \le 6, y_1 + y_2 \ge 1 \qquad \ldots(1)$$

Let $w = (w_1, w_2)$ be a point on the line segment joining the points u and v, then

$$w = \lambda u + (1 - \lambda)v$$
$$\Rightarrow \qquad (w_1, w_2) = (\lambda x_1 + (1 - \lambda)y_1, \lambda x_2 + (1 - \lambda)y_2)$$
$$\Rightarrow \qquad w_1 = \lambda x_1 + (1 - \lambda)y_1 \text{ and } w_2 = \lambda x_2 + (1 - \lambda)y_2$$

Consider, $4w_1 + 3w_2 = \lambda(4x_1 + 3x_2) + (1 - \lambda)(4y_1 + 3y_2)$

$$\le \lambda \cdot 6 + (1 - \lambda) \cdot 6 \qquad \text{(By (1))}$$
$$\Rightarrow \qquad 4w_1 + 3w_2 \le 6 \qquad \ldots(2)$$

Also, $\qquad w_1 + w_2 = \lambda(x_1 + x_2) + (1 - \lambda)(y_1 + y_2)$

$$\ge \lambda \cdot 1 + (1 - \lambda) \cdot 1 \qquad \text{(Again by (1))}$$
$$\Rightarrow \qquad w_1 + w_2 \ge 1 \qquad \ldots(3)$$

Hence, from (2) and (3) we conclude that S is a convex set.

EXAMPLE 4. *Show that the set $S = \{ \boldsymbol{x} : \boldsymbol{x} = (x_1, x_2, x_3),\ x_1^2 + x_2^2 + x_3^2 \le 1 \}$ is a convex set.*

[MEERUT–2006]

SOLUTION. Let $\boldsymbol{x}, \boldsymbol{y} \in S$ be arbitrary.

Then we write

$$\boldsymbol{x} = (x_1, x_2, x_3) \text{ and } \boldsymbol{y} = (y_1, y_2, y_3)$$

Now by definition of S, we have

$$x_1^2 + x_2^2 + x_3^2 \le 1 \text{ and } y_1^2 + y_2^2 + y_3^2 \le 1 \qquad \ldots(1)$$

Further, let $\boldsymbol{z} = (z_1, z_2, z_3)$ be a point on the line segment joining the points $\boldsymbol{x}$ and $\boldsymbol{y}$ then we can write

$$\boldsymbol{z} = \lambda \boldsymbol{x} + (1 - \lambda)\boldsymbol{y},\ 0 \le \lambda \le 1$$

$$\Rightarrow \quad (z_1, z_2, z_3) = \lambda(x_1, x_2, x_3) + (1 - \lambda)(y_1, y_2, y_3)$$

$$= (\lambda x_1 + (1 - \lambda)y_1,\ \lambda x_2 + (1 - \lambda)y_2,\ \lambda x_3 + (1 - \lambda)y_3)$$

which gives $z_1 = \lambda x_1 + (1 - \lambda)y_1,\ z_2 = \lambda x_2 + (1 - \lambda)y_2,\ z_3 = \lambda x_3 + (1 - \lambda)y_3$

Now consider $z_1^2 + z_2^2 + z_3^2$

$$= [\lambda x_1 + (1 - \lambda)y_1]^2 + [\lambda x_2 + (1 - \lambda)y_2]^2 + [\lambda x_3 + (1 - \lambda)y_3]^2$$

$$= \lambda^2(x_1^2 + x_2^2 + x_3^2) + (1 - \lambda)^2(y_1^2 + y_2^2 + y_3^2)$$

$$+ 2\lambda(1 - \lambda)(x_1 y_1 + x_2 y_2 + x_3 y_3)$$

$$\le \lambda^2 \cdot 1 + (1 - \lambda)^2 + 1 + 2\lambda(1 - \lambda)(x_1 y_1 + x_2 y_2 + x_3 y_3) \quad \text{(By (1))}$$

$$\ldots(2)$$

Using Lagrange's identity

$$(x_1^2 + x_2^2 + x_3^2)(y_1^2 + y_2^2 + y_3^2) - (x_1 y_1 + x_2 y_2 + x_3 y_3)^2$$

$$\equiv \Sigma(x_1 y_2 - x_2 y_1)^2 \ge 0$$

we can write

$$x_1 y_1 + x_2 y_2 + x_3 y_3 \le \sqrt{x_1^2 + x_2^2 + x_3^2} \cdot \sqrt{y_1^2 + y_2^2 + y_3^2} \le 1 \qquad \text{(By (1))}$$

Using these values in (2) we get

$$z_1^2 + z_2^2 + z_3^2 \le \lambda^2 + (1 - \lambda)^2 + 2\lambda(1 - \lambda) = 1$$

$$\Rightarrow \qquad \boldsymbol{z} = (z_1, z_2, z_3) \in S$$

Hence, the given set S is convex.

EXAMPLE 6. *Show that the set $S = \{ (x_1, x_2) : 3x_1^2 + 2x_2^2 \le 6 \}$ is convex.*

SOLUTION. Let $\boldsymbol{u}, \boldsymbol{v} \in S$ where $\boldsymbol{u} = (u_1, u_2)$ and $\boldsymbol{v} = (v_1, v_2)$

Then by definition of S, we can write

$$3u_1^2 + 2u_2^2 \le 6 \text{ and } 3v_1^2 + 2v_2^2 \le 6 \qquad \ldots(1)$$

Let $\boldsymbol{w} = (w_1, w_2)$ be a point on the line segment joining the points $\boldsymbol{u}$ and $\boldsymbol{v}$, then

$$\boldsymbol{w} = \lambda \boldsymbol{u} + (1 - \lambda)\boldsymbol{v};\ 0 \le \lambda \le 1$$

Therefore, $\boldsymbol{w} = (w_1, w_2) = \lambda(u_1, u_2) + (1 - \lambda)(v_1, v_2)$

$$= (\lambda u_1 + (1 - \lambda)v_1,\ \lambda u_2 + (1 - \lambda)v_2)$$

$$\Rightarrow \qquad w_1 = \lambda u_1 + (1 - \lambda)v_1 \text{ and } w_2 = \lambda u_2 + (1 - \lambda)v_2$$

Now, consider

$$3w_1^2 + 2w_2^2 = 3\{\lambda u_1 + (1 - \lambda)v_1\}^2 + 2\{\lambda u_2 + (1 - \lambda)v_2\}^2$$

$$= \lambda^2(3u_1^2 + 2u_2^2) + (1 - \lambda)^2(3v_1^2 + 2v_2^2) + 2\lambda(1 - \lambda)(3u_1 v_1 + 2u_2 v_2)$$

$$\ldots(2)$$

But we have
$$(3u_1^2 + 2u_2^2)(3v_1^2 + 2v_2^2) - 3(u_1v_1 + 2u_2v_2)^2 = 6(u_1v_2 - u_2v_1)^2 \geq 0$$
$$\Rightarrow \quad (3u_1v_1 + 2u_2v_2)^2 \leq (3u_1^2 + 2u_2^2)(3v_1^2 + 2v_2^2) \leq 6 \times 6$$
$$\Rightarrow \quad 3u_1v_1 + 2u_2v_2 \leq 6 \qquad \qquad \dots(3)$$
Finally using (1) and (3) in (2) we get
$$3w_1^2 + 2w_2^2 \leq 6\lambda^2 + 6(1-\lambda)^2 + 2\lambda(1-\lambda) \times 6$$
$$\Rightarrow \quad 3w_1^2 + 2w_2^2 \leq 6$$
$$\Rightarrow \quad \boldsymbol{w} = (w_1, w_2) \in S \; \forall \; 0 \leq \lambda \leq 1$$
Hence, S is a convex set.

EXAMPLE 7. *Show that $S = \{(x_1, x_2, x_3) : 2x_1 - x_2 + x_3 \leq 4;\ x_1 + 2x_2 - x_3 \leq 1\}$ is a convex set.*

SOLUTION. Let $\boldsymbol{x}, \boldsymbol{y} \in S$ then by definition of S we can write
$$\boldsymbol{x} = (x_1, x_2, x_3) \text{ and } \boldsymbol{y} = (y_1, y_2, y_3)$$
such that
$$\left. \begin{array}{l} 2x_1 - x_2 + x_3 \leq 4;\, x_1 + 2x_2 - x_3 \leq 1 \\ \text{and} \quad 2y_1 - y_2 + y_3 \leq 4;\, y_1 + 2y_2 - y_3 \leq 1 \end{array} \right\} \qquad \dots(1)$$
Let $\boldsymbol{z} = (z_1, z_2, z_3)$ be such that
$$\boldsymbol{z} = \lambda \boldsymbol{x} + (1-\lambda)\boldsymbol{y},\ 0 \leq \lambda \leq 1$$
$$\Rightarrow \quad (z_1, z_2, z_3) = \lambda(x_1, x_2, x_3) + (1-\lambda)(y_1, y_2, y_3)$$
$$= \{\lambda x_1 + (1-\lambda)y_1, \lambda x_2 + (1-\lambda)y_2, \lambda x_3 + (1-\lambda)y_3\}$$
$$\Rightarrow \quad z_1 = \lambda x_1 + (1-\lambda)y_1,\ z_2 = \lambda x_2 + (1-\lambda)y_2,\ z_3 = \lambda x_3 + (1-\lambda)y_3$$
Now, consider $2z_1 - z_2 + z_3$
$$= \lambda(2x_1 - x_2 + x_3) + (1-\lambda)(2y_1 - y_2 + y_3)$$
$$\leq 4\lambda + 4(1-\lambda) \qquad \qquad \text{(Using (1))}$$
$$= 4$$
$$\Rightarrow 2z_1 - z_2 + z_3 \leq 4$$
Similarly, $z_1 + 2z_2 - z_3 \leq 1$
Therefore, $\boldsymbol{z} = (z_1, z_2, z_3) \in S$
Hence, S is a convex set.

EXAMPLE 8. *If S_1 and S_2 be two non-empty disjoint convex sets and S be a set such that if $\boldsymbol{x}_1 \in S_1$, $\boldsymbol{x}_2 \in S_2$ then $\boldsymbol{x}_1 - \boldsymbol{x}_2 \in S$. Show that S is also convex and does not contain the origin.*

SOLUTION. Let us write $\boldsymbol{u} = \boldsymbol{x}_1 - \boldsymbol{x}_2$ and $\boldsymbol{v} = \boldsymbol{y}_1 - \boldsymbol{y}_2 \in S$
Then we have $\boldsymbol{x}_1, \boldsymbol{y}_1 \in S_1$ and $\boldsymbol{x}_2, \boldsymbol{y}_2 \in S_2$
If $\boldsymbol{z}$ is a point on the line segment joining $\boldsymbol{u}$ and $\boldsymbol{v}$ then
$$\boldsymbol{z} = \lambda \boldsymbol{u} + (1-\lambda)\boldsymbol{v};\ 0 \leq \lambda \leq 1$$
$$= \lambda(\boldsymbol{x}_1 - \boldsymbol{x}_2) + (1-\lambda)(\boldsymbol{y}_1 - \boldsymbol{y}_2)$$
$$= \{\lambda \boldsymbol{x}_1 + (1-\lambda)\boldsymbol{y}_1\} - \{\lambda \boldsymbol{x}_2 + (1-\lambda)\boldsymbol{y}_2\} \qquad \dots(1)$$
Further, it is given that S_1 and S_2 both are convex, therefore by definition
$$\boldsymbol{x}_1, \boldsymbol{y}_1 \in S_1 \qquad \Rightarrow \qquad \lambda \boldsymbol{x}_1 + (1-\lambda)\boldsymbol{y}_1 = \boldsymbol{z}_1 \in S_1;\ (0 \leq \lambda \leq 1)$$
$$\boldsymbol{x}_2, \boldsymbol{y}_2 \in S_2 \qquad \Rightarrow \qquad \lambda \boldsymbol{x}_2 + (1-\lambda)\boldsymbol{y}_2 = \boldsymbol{z}_2 \in S_2;\ (0 \leq \lambda \leq 1)$$

Then from (1)
$$z = z_1 - z_2 \in S \ \forall \ 0 \le \lambda \le 1$$
Hence, S is convex.

Finally, let if possible $\mathbf{0} \in S$ then there exist $x_1 \in S_1$ and $x_2 \in S_2$ such that
$$\mathbf{0} = x_1 - x_2$$
$$\Rightarrow \quad x_1 = x_2, x_1 \in S_1, x_2 \in S_2$$
$$\Rightarrow \quad S_1 \text{ and } S_2 \text{ are not disjoint, which is a contradiction.}$$
Hence, $\mathbf{0} \notin S$, i.e., S does not contain the origin.

EXAMPLE 9. *Examine the convexity of the following set*
$$C = \{z \in R^n : z = x + y, x \in A, y \in B\}$$
where A and B are convex sets in R^n.

SOLUTION. Let us suppose $z_1 = x_1 + y_1$ and $z_2 = x_2 + y_2$ be any two points in the set C.
Then $x_1, x_2 \in A$ and $y_1, y_2 \in B$
If u is the point on the line segment joining the points z_1 and z_2
Then we can write
$$u = \lambda z_1 + (1 - \lambda)z_2 : 0 \le \lambda \le 1$$
$$= \lambda(x_1 + y_1) + (1 - \lambda)(x_2 + y_2)$$
$$= (\lambda x_1 + (1 - \lambda)x_2) + (\lambda y_1 + (1 - \lambda)y_2), 0 \le \lambda \le 1 \qquad \ldots(1)$$
Further since both sets A and B are convex, therefore, we can write
$$\left. \begin{array}{l} x_1, x_2 \in A \Rightarrow \lambda x_1 + (1 - \lambda)x_2 = u_1 \in A \ \forall \, 0 \le \lambda \le 1 \\ y_1, y_2 \in B \Rightarrow \lambda y_1 + (1 - \lambda)y_2 = u_2 \in B \ \forall \, 0 \le \lambda \le 1 \end{array} \right\} \qquad \ldots(2)$$
Using (2) in (1) we get
$$u = u_1 + u_2 \in C$$
$\Rightarrow$ every point of the line segment joining z_1 and z_2 of C are also in C.
Hence, C is a convex set

EXAMPLE 10. *Express any point w inside a triangle as a convex combinations of the vertices (extreme points) x_1, x_2, x_3 of the triangle.*

SOLUTION. Consider a triangle ABC with vertices x_1, x_2 and x_3. If P is any point w inside the triangle. Join A and P and extend this line to meet the base BC at $D(u)$, a point on the line BC.

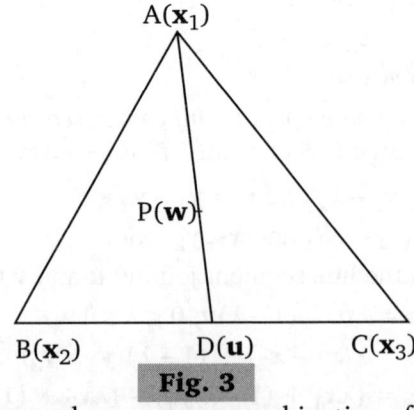

Fig. 3

$\Rightarrow$ u can be expressed as a convex combination of x_2 and x_3.
Therefore, $u = \lambda_1 x_2 + (1 - \lambda_1)x_3, 0 \le \lambda_1 \le 1$ \qquad \ldots(1)
Further, since P is a point on the line segment AD, therefore,
$$w = \lambda_2 x_1 + (1 - \lambda_2)u , 0 \le \lambda_2 \le 1$$

$$= \lambda_2 x_1 + (1 - \lambda_2)[\lambda_1 x_2 + (1 - \lambda)x_3] \qquad \text{(Using (1))}$$
$$= \lambda_2 x_1 + \lambda_1(1 - \lambda_2)x_2 + (1 - \lambda_1)(1 - \lambda_2)x_3$$
$$\Rightarrow \qquad w = \mu_1 x_1 + \mu_2 x_2 + \mu_3 x_3 \qquad \ldots(2)$$

where $\mu_1 = \lambda_2$, $\mu_2 = \lambda_1(1 - \lambda_2)$, $\mu_3 = (1 - \lambda_1)(1 - \lambda_2)$

Clearly each μ_i lies between 0 and 1, i.e., $0 \le \mu_i \le 1$, $i = 1, 2, 3$

Also, $\qquad \mu_1 + \mu_2 + \mu_3 = \lambda_2 + \lambda_1(1 - \lambda_2) + (1 - \lambda_1)(1 - \lambda_2)$
$$= \lambda_2 + \lambda_1 - \lambda_1 \lambda_2 + 1 - \lambda_1 - \lambda_2 + \lambda_1 \lambda_2 = 1$$

Thus, we conclude that $\mu_1 x_1 + \mu_2 x_2 + \mu_3 x_3$ is a convex combination of the points x_1, x_2, x_3.

Hence, the combination given by (2) is the required combination for the point w.

EXAMPLE 11. *A hyperplane is given by the equation $3x_1 + 2x_2 + 4x_3 + 7x_4 = 8$. Find in which half spaces do the following points $(-6, 1, 7, 2)$ and $(1, 2, -4, 1)$ lie.*

SOLUTION. We have

The hyperplane: $3x_1 + 2x_2 + 4x_3 + 7x_4 = 8$ $\qquad \ldots(1)$

using $(-6, 1, 7, 2)$ in (1) we get

$\qquad$ LHS $= 3(-6) + 2(1) + 7(2) = 26 > 8 =$ RHS

$\Rightarrow$ The point $(-6, 1, 7, 2)$ lies in the open half spaces $3x_1 + 2x_2 + 4x_3 + 7x_4 > 8$

Now, substituting $(1, 2, -4, 1)$ in the LHS of (1) we get

$\qquad$ LHS $= 3(1) + 2(2) + 4(-4) + 7(1) = -2 < 8 =$ RHS

Hence, the point $(1, 2, -4, 1)$ lies in the open half space $3x_1 + 2x_2 + 4x_3 + 7x_4 < 8$

EXAMPLE 12. *Sketch the convex polygon spanned by the following points in a two dimensional Euclidean space. Which of these points are vertices? Express the other as the convex linear combination of the vertices $(0,0), (0,1), (1,0), \left(\dfrac{1}{2}, \dfrac{1}{4}\right)$*

SOLUTION. Clearly, the convex combinations of the points $(0,0), (1,0)$; $(0,0), (0,1)$ and $(1,0), (0,1)$ give the line segments OA, OB and AB respectively.

Therefore, the convex combination of points $(0,0), (1,0)$ and $(0,1)$ is the interior of the $\triangle OAB$.

Thus, the points $O(0,0)$, $A(1, 0)$ and $B(0, 1)$ are the vertices and the point C is the interior point of the convex polygon spanned by the given points.

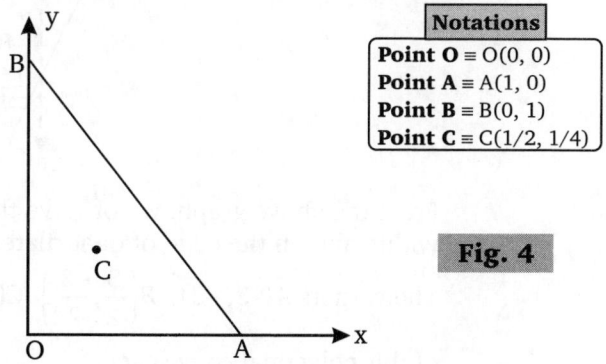

Notations
Point O $\equiv O(0, 0)$
Point A $\equiv A(1, 0)$
Point B $\equiv B(0, 1)$
Point C $\equiv C(1/2, 1/4)$

Fig. 4

Now, we have to express the point $\left(\dfrac{1}{2}, \dfrac{1}{4}\right)$ as the linear combinations of $(0,0)$,

$(0, 1)$ and $(1, 0)$

Write $\left(\dfrac{1}{2}, \dfrac{1}{4}\right) = \lambda_1(0,0) + \lambda_2(0,1) + \lambda_3(1,0)$ where $\lambda_1 + \lambda_2 + \lambda_3 = 1, \lambda_i \geq 0$

$\Rightarrow \quad \left(\dfrac{1}{2}, \dfrac{1}{4}\right) = (\lambda_3, \lambda_2)$

$\Rightarrow \quad \lambda_2 = 1/4$ and $\lambda_3 = 1/2$

Now, $\lambda_1 = 1 - \lambda_2 - \lambda_3 = = 1 - \dfrac{1}{4} - \dfrac{1}{2} = \dfrac{1}{4}$

Hence, $\left(\dfrac{1}{2}, \dfrac{1}{4}\right) = \dfrac{1}{4}(0,0) + \dfrac{1}{4}(0,1) + \dfrac{1}{2}(1,0)$

EXAMPLE 13. *Find the extreme points of the polygonal convex set X determined by the system.*

$$2x_1 + x_2 + 9 \geq 0 \; ; \; -x_1 + 3x_2 + 6 \geq 0, \; x_1 + x_2 \leq 0, \; x_1 + 2x_2 - 3 \leq 0$$

SOLUTION. Draw the graph of the following given equations

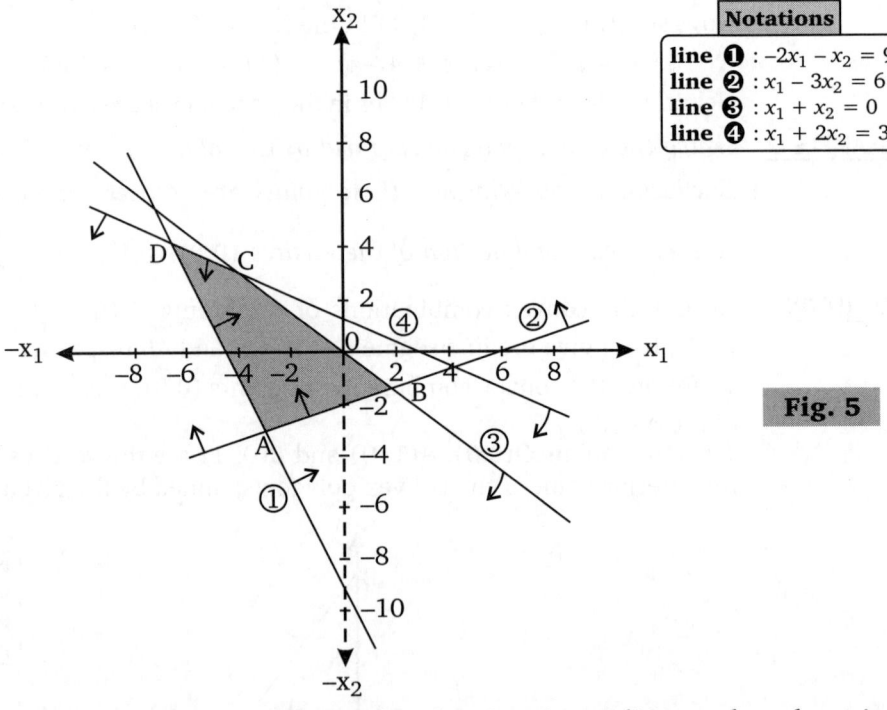

Notations
line ❶ : $-2x_1 - x_2 = 9$
line ❷ : $x_1 - 3x_2 = 6$
line ❸ : $x_1 + x_2 = 0$
line ❹ : $x_1 + 2x_2 = 3$

Fig. 5

From the above graph, we observe that the given inequalities enclose the points within and on the edge of quadrilateral $ABCD$ and it is a convex set.

The corners $A(-3, -3)$, $B\left(\dfrac{3}{2}, \dfrac{-3}{2}\right)$, $C(-3, 3)$ and $D(-7, 5)$ are the extreme points of this polygonal convex set.

EXAMPLE 14. *Find the convex hull of the set* $A = \{(x_1, x_2) : x_1^2 + x_2^2 = 1\}$

SOLUTION.   Let $\boldsymbol{x} = (x_1, x_2)$ and $\boldsymbol{y} = (y_1, y_2)$ be any two points of A

Then by definition of A, we can write

$$x_1^2 + x_2^2 = 1 \text{ and } y_1^2 + y_2^2 = 1 \qquad \qquad ...(1)$$

Now, let $\boldsymbol{z}$ be the convex combination of $\boldsymbol{x}$ and $\boldsymbol{y}$

Then $\boldsymbol{z} = \lambda \boldsymbol{x} + (1 - \lambda)\boldsymbol{y} = \lambda(x_1, x_2) + (1 - \lambda)(y_1, y_2)$; $0 \leq \lambda \leq 1$

$\Rightarrow \quad \boldsymbol{z} = (z_1, z_2) = (\lambda x_1 + (1 - \lambda)y_1, \lambda x_2 + (1 - \lambda)y_2)$

Therefore, $z_1 = \lambda x_1 + (1 - \lambda)y_1$, $z_2 = \lambda x_2 + (1 - \lambda)y_2$

Now,

$$z_1^2 + z_2^2 = (\lambda x_1 + (1 - \lambda)y_1)^2 + (\lambda x_2 + (1 - \lambda)y_2)^2$$

$$= \lambda^2(x_1^2 + x_2^2) + (1 - \lambda)^2(y_1^2 + y_2^2) + 2\lambda(1 - \lambda)(x_1 y_1 + x_2 y_2) \qquad ...(2)$$

We have,

$$(x_1^2 + x_2^2)(y_1^2 + y_2^2) - (x_1 y_1 + x_2 y_2)^2 = (x_1 y_2 - x_2 y_1)^2 \geq 0$$

$$\Rightarrow \quad (x_1 y_1 + x_2 y_2)^2 \leq (x_1^2 + x_2^2)(y_1^2 + y_2^2) = 1 \cdot 1$$

$$\Rightarrow \quad x_1 y_1 + x_2 y_2 \leq 1 \qquad \qquad ...(3)$$

Using (1) and (3) in (2) we get

$$z_1^2 + z_2^2 \leq \lambda^2 \cdot 1 + (1 - \lambda)^2 \cdot 1 + 2\lambda(1 - \lambda) \cdot 1$$

$$\Rightarrow \quad z_1^2 + z_2^2 \leq 1$$

$$\Rightarrow \quad \boldsymbol{z} = (z_1, z_2) \in S = \{(x_1, x_2) : x_1^2 + x_2^2 \leq 1\}$$

Hence, S is the set, containing the convex combinations of the elements of A.

 Exercise-1.2

1. Show that the set $\{(x_1, x_2) : x_1 \geq 2, x_1 \leq 3)\}$ is a convex set.

2. Let A be an $m \times n$ matrix and $\boldsymbol{b}$, an m-vector, then show that $\{x \in R^n : A\boldsymbol{x} \leq \boldsymbol{b}\}$ is a convex set.

3. Show that the following sets are not convex.
   (i) $\{(x_1, x_2) : x_1 x_2 \leq 1, x_1 \geq 0, x_2 \geq 0\}$
   (ii) $\{(x_1, x_2) : x_2 - 3 \geq -x_1^2, x_1 \geq 0, x_2 \geq 0\}$

4. Show that the set $\{(x_1, x_2) : x_1^2 + x_2^2 \leq 4\}$ is a convex set.   [MEERUT–2011]

5. Show that the following sets are convex
   (i) $S_1 = \left\{(x_1, x_2) : \dfrac{x_1^2}{4} + \dfrac{x_2^2}{9} \leq 1\right\}$
   (ii) $S_2 = \{(x_1, x_2) : x_1^2 + x_2^2 \leq 1, x_1 + x_2 \geq 1\}$

6. If $x_1, x_2 \in S$ implies $1/2(x_1 + x_2) \in S$. Show that S is convex.

7. Show that the set $S = \{x : |x| = 1, x \in E^n\}$ is not convex.

8. Show that union of two convex sets is not necessarily convex.

9. Show that the vector $[7, 0]$ is a convex combination of the vectors $[6, 3]$, $[9, -6]$, $[1, 2]$ and $[1, -1]$.

10. Show that the vector $[2, 1]$ can not be expressed as a convex combinations of $[1, 1]$ and $[-1, 2]$.

11. Find the extreme points of the set $\{(x, y) : |x| \leq 1, |y| \leq 1\}$.

12. Find the extreme points of the set $\{(x, y) : x^2 + y^2 \leq 25\}$.

ANSWERS

11. $(1, 1), (-1, 1), (-1, -1)$ and $(1, -1)$ **12.** Every point on the circumference is an extreme point.

1.24 FEASIBLE AND BASIC FEASIBLE SOLUTIONS

Definition 1. *For a system* $AX = B$ *of m equations in n unknown and if* $n > m$ *and* $Rank(A) = Rank(A|B) = m$ *then in order to solve the system of given equations, we set* $(n - m)$ *variables to zero. Thus, a solution to resulting system of equations is said to be basic solution provided, the determinant of the coefficient of the remaining m variables is not zero.*

Definition 2. *If all the basic variables are non-negative then a basic solution is called feasible.*

Definition 3. *If at least one of the basic variables is negative then a basic solution is called infeasible.*

RELATED THEOREMS

THEOREM 1.   *The set of all feasible solutions (if not empty) of a LPP is a convex set.*

PROOF. Let X be the set of all feasible solutions of a *LPP given by*

$$AX = B, X \geq 0 \qquad \qquad \qquad ...(1)$$

Now, we have the following cases:

Case I.

If the set X has only one element, then X is convex. In this case, theorem is true.

Case II.

If the set X has at least two elements

If $\boldsymbol{x}_1, \boldsymbol{x}_2 \in X$ then

$$A\boldsymbol{x}_1 = \boldsymbol{b}, \; x_1 \geq 0 \; \text{ and } \; A\boldsymbol{x}_2 = \boldsymbol{b}, \; x_2 \geq 0$$

Let $\qquad\qquad\qquad \boldsymbol{x}_3 = \lambda\boldsymbol{x}_1 + (1-\lambda)\boldsymbol{x}_2 : 0 \leq \lambda \leq 1$

Then $\qquad\qquad\qquad A\boldsymbol{x}_3 = A\lambda\boldsymbol{x}_1 + (1-\lambda)A\boldsymbol{x}_2$

$$= \lambda\boldsymbol{b} + (1-\lambda)\boldsymbol{b} = \boldsymbol{b}$$

Further, since $\boldsymbol{x}_1 \geq 0, \; \boldsymbol{x}_2 \geq 0, \; 1 - \lambda \geq 0$ $\qquad\qquad\qquad\qquad (\because \; 0 \leq \lambda \leq 1)$

So, $\qquad\qquad\qquad \boldsymbol{x}_3 = \lambda\boldsymbol{x}_1 + (1-\lambda)\boldsymbol{x}_2 \geq 0$

$\Rightarrow \; \boldsymbol{x}_3$ satisfies (1).

Therefore, $\boldsymbol{x}_3 = \lambda\boldsymbol{x}_1 + (1-\lambda)\boldsymbol{x}_2$ is also a feasible solution and hence belongs to X.

$\Rightarrow$ Convex combination of any two points $\boldsymbol{x}_1$ and $\boldsymbol{x}_2$ in X belongs to X.

Hence, X is a convex set.

THEOREM 2.   *Every basic feasible solution of the system* $A\boldsymbol{x} = B, \; \boldsymbol{x} \geq 0$ *is an extreme point of the convex set of feasible solutions and conversly.*

PROOF. Necessary Part: We have to prove that every B.F.S is an extreme point of the convex set of all feasible solutions.

Let us suppose $\boldsymbol{x}$ be a B.F.S. of $A\boldsymbol{x} = \boldsymbol{b}$

Also let $\boldsymbol{x}_B$ and B be the vector of m basic variables and the matrix of vectors associated to basic variables in the B.F.S. respectively, then $\boldsymbol{x} = [\boldsymbol{x}_B, \boldsymbol{O}]$ $\qquad\qquad ...(1)$

where $\boldsymbol{O}$ is a null vector of $(n\text{-}m)$ components

Also, $A\boldsymbol{x} = \boldsymbol{b} \; \Rightarrow \; B \cdot \boldsymbol{x}_B = \boldsymbol{b}$ $\qquad\qquad\qquad\qquad\qquad ...(2)$

We want to prove that $\boldsymbol{x}$ is an extreme point. Let if possible $\boldsymbol{x}$ is not an extreme point.

Since $\boldsymbol{x}$ is not an extreme point, then there exist two distinct points $\boldsymbol{x}_1$ and $\boldsymbol{x}_2$ of X

(the convex set of all feasible solutions of $Ax = b$) such that $x = \lambda x_1 + (1-\lambda)x_2$, $0 < \lambda < 1$ ...(3)

But we can write,

$$x_1 = [u_1, v_1] \text{ and } x_2 = [u_2, v_2]$$ ...(4)

where u_1 and u_2 are vectors of m components of x_1 and x_2 respectively and v_1, v_2 are $(n-m)$ components vectors.

Now using (1) and (4) in (3) we get

$$[x_B, 0] = \lambda[u_1, v_1] + (1-\lambda)[u_2, v_2], 0 < \lambda < 1$$
$$= [\lambda u_1 + (1-\lambda)u_2, \lambda v_1 + (1-\lambda)v_2]$$
$$\Rightarrow \qquad 0 = \lambda v_1 + (1-\lambda)v_2, \ 0 < \lambda < 1$$ ...(5)

Clearly, $1 > \lambda > 0$, $1 - \lambda > 0$ and the components of v_1 and v_2 are greater than equal to 0. So, (5) is satisfied only when $v_1 = 0$ and $v_2 = 0$

So, $$x_1 = [u_1, 0], x_2 = [u_2, 0]$$

Also, $x_1 \in X, x_2 \in X$, then from (2) we have

$$A x_1 = B u_1 = b$$
and $$A x_2 = B u_2 = b$$
$$\Rightarrow \qquad x_B = u_1 = u_2$$

So, $x = x_1 = x_2$, which contradict the fact that $x_1 \neq x_2$

$\Rightarrow x$ cannot be expressed as a convex combinations of any two distinct points in the set of all feasible solutions.

Hence, x is an extreme point.

Conversly, let us suppose that $x = (x_1, x_2, ..., x_n)$ be an extreme point. We have to prove that x is a BFS.

Here, it is sufficient to prove that the vector associated with the positive elements of x are linearly independent.

Let us suppose p-components in x are non-zero and $(n-p)$ components are zero.

We can assume these components as the first p-components of x.

So, $$\sum_{i=1}^{p} \alpha_i x_i = b, \ x_i > 0, \ i = 1, 2, ..., p$$ ...(6)

where α_i is the column vector in A associated to the ith variable in x

Let if possible the column vectors $\alpha_1, \alpha_2, ..., \alpha_p$ of matrix be linearly dependent. Then by definition, there exist some scalars λ_i $(i = 1, 2, ..., p)$ with at least one of them non-zero such that $$\sum_{i=1}^{p} \lambda_i \alpha_i = 0$$ ...(7)

Now, from (6) and (7), for some arbitrary $\delta > 0$ we have

$$\sum_{i=1}^{p} x_i \alpha_i \pm \delta \sum_{i=1}^{p} \lambda_i \alpha_i = b$$

$$\Rightarrow \qquad \sum_{i=1}^{p} (x_i \pm \delta \lambda_i)\alpha_i = b$$

Then, clearly, two points

$$x_1^* = [x_1 + \delta\lambda_1, x_2 + \delta\lambda_2,..., x_p + \delta\lambda_p, 0, 0, ..., 0 \ (n-p) \text{ in numbers}]$$

and $\quad x_2^* = [x_1 - \delta\lambda_1, x_2 - \delta\lambda_2,...,x_p - \delta\lambda_p, 0, 0, ..., 0]$

satisfy $\quad A\boldsymbol{x} = \boldsymbol{b}$

Further since, $x_i > 0$ therefore taking δ such that

$$0 < \delta < \min_i \left\{ \frac{x_i}{|\lambda_i|} \right\}, \ \lambda_i \neq 0, \ i = 1, 2,...,p$$

$\Rightarrow$ p component of $\boldsymbol{x}_1^*$ and $\boldsymbol{x}_2^*$ are always positive but the remaining components of $\boldsymbol{x}_1^*$ and $\boldsymbol{x}_2^*$ are zero.

$\Rightarrow$ $\boldsymbol{x}_1^*$ and $\boldsymbol{x}_2^*$ are feasible solutions different from $\boldsymbol{x}$.

Further, $\quad \boldsymbol{x}_1^* + \boldsymbol{x}_2^* = 2[x_1, x_2,..., x_p, 0, 0, 0, ..., 0]$

$\Rightarrow \qquad \frac{1}{2}\boldsymbol{x}_1^* + \frac{1}{2}\boldsymbol{x}_2^* = [x_1, x_2,..., x_p, 0, 0, ..., 0] = \boldsymbol{x}$

$\Rightarrow \qquad \boldsymbol{x} = \lambda\boldsymbol{x}_1^* + (1-\lambda)\boldsymbol{x}_2^* \qquad (\lambda = \frac{1}{2})$

$\Rightarrow$ $\boldsymbol{x}$ can be expressed as a convex combination of two distinct feasible solutions $\boldsymbol{x}_1^*$ and $\boldsymbol{x}_2^*$, which is a contradiction, because $\boldsymbol{x}$ is an extreme point.

$\Rightarrow \alpha_1, \alpha_2, ...,\alpha_p$ are linearly independent.

$\Rightarrow \alpha_1, \alpha_2, ...,\alpha_p$ can not be more than m

$\Rightarrow$ extreme point $\boldsymbol{x}$ will have atmost m non-zero variables $i.e.$, at least $(n-m)$ variables will be zero.

$\Rightarrow \boldsymbol{x}$ is a BFS.

Hence, every extreme point of the convex set of feasible solution is a BFS.

THEOREM 3. *The extreme point of the convex set of feasible solutions are finite in number.*

PROOF. Using Theorem 2 we have that there is only one extreme point for a given BFS and conversly.

$\Rightarrow$ there is one-to-one correspondance between the extreme points and the BFS in the absence of degeneracy.

Also, in case of degeneracy, corresponding to an extreme point with the number of non-zero variables less than m, we can form more than one degenerate BFS.

Hence, the number of extreme points of the feasible region is finite and it can not exceed the number of its basic feasible solution.

☛ **REMARKS**

- An extreme point can have atmost m positive $x_i's$ where m is the no. of constraints.
- In an extreme point, vectors associated to the positive $x_i's$ are linearly independent.

THEOREM 4. *If the convex set of the feasible solutions of $A\boldsymbol{x} = \boldsymbol{b}$, $\boldsymbol{x} \geq 0$ is a convex polyhedron then at least one of the extreme points gives an optimal solution.*

PROOF. We know that the extreme points of the convex set of feasible solutions of $A\boldsymbol{x} = \boldsymbol{b}$, $\boldsymbol{x} \geq 0$ are finite.

Thus, we suppose that $\boldsymbol{x}_1, \boldsymbol{x}_2, ..., \boldsymbol{x}_k$ are the extreme points of the set X of all feasible solutions of $A\boldsymbol{x} = \boldsymbol{b}$, $\boldsymbol{x} \geq 0$

Further, let z be the objective function such that $z = \boldsymbol{cx}$

We have to be maximized z.

If $\boldsymbol{x}^* \in X$ is the optimal solution, then

$$\max z = \boldsymbol{c}\,\boldsymbol{x}^*$$

If $\boldsymbol{x}^*$ is the extreme point, then result is obvious.

If $\boldsymbol{x}^*$ is not an extreme point in X, then $\boldsymbol{x}^*$ can be expressed as a convex combinations of the extreme point of X.

i.e.,

$$\boldsymbol{x}^* = \lambda_1\,\boldsymbol{x}_1 + \lambda_2\,\boldsymbol{x}_2 + \ldots + \lambda_k \boldsymbol{x}_k$$

$$= \sum_{i=1}^{n} \lambda_i \boldsymbol{x}_i, \ \lambda_i \geq 0 \text{ and } \Sigma\lambda_i = 1$$

$\Rightarrow$

$$z^* = \boldsymbol{c}\boldsymbol{x}^* = \boldsymbol{c}(\lambda_1\,\boldsymbol{x}_1 + \lambda_2\,\boldsymbol{x}_2 + \ldots + \lambda_k\,\boldsymbol{x}_k)$$

$$= (\lambda_1\,\boldsymbol{c}\,\boldsymbol{x}_1 + \lambda_2\,\boldsymbol{c}\,\boldsymbol{x}_2 + \ldots + \lambda_k\,\boldsymbol{c}\,\boldsymbol{x}_k)$$

Suppose maximum of $c\boldsymbol{x}_i$ is $c\boldsymbol{x}_p$, then

$$z^* \leq (\lambda_1 + \lambda_2 + \ldots + \lambda_k) \cdot \boldsymbol{c}\boldsymbol{x}_p$$

$\Rightarrow$

$$z^* \leq c\boldsymbol{x}_p$$

Since, z^* is the maximum value of z, so

$$z^* = \boldsymbol{c}\boldsymbol{x}_p$$

$\Rightarrow$

$$c\boldsymbol{x}^* = \boldsymbol{c}\boldsymbol{x}_p$$

$\Rightarrow$

$$\boldsymbol{x}^* = \boldsymbol{x}_p, \text{ one of the extreme point}$$

which shows that the optimal solution is attained at the extreme point.

THEOREM 5. *If the objective function of a LPP assumes its optimal value at more than one extreme point, then every convex combinations of these extreme points gives the optimal value of the objective function.*

PROOF. Let us consider an LPP as given below.

$$\max \cdot z = \boldsymbol{c}\,\boldsymbol{x}$$

such that $\quad A\,\boldsymbol{x} = b, \boldsymbol{x} \geq 0$

Let $\boldsymbol{x}_1, \boldsymbol{x}_2, \ldots, \boldsymbol{x}_k$ be the extreme points of the feasible region. If the objective function z assume its optimal value z^* at the extreme points $\boldsymbol{x}_1, \boldsymbol{x}_2,\ldots,\boldsymbol{x}_p$ $(p \leq k)$

Then $\quad z^* = \boldsymbol{c}\,\boldsymbol{x}_1 = \boldsymbol{c}\,\boldsymbol{x}_2 = \ldots = \boldsymbol{c}\,\boldsymbol{x}_p$

If $\boldsymbol{x}_0$ is the convex combination of the extreme points $\boldsymbol{x}_1, \boldsymbol{x}_2, \ldots, \boldsymbol{x}_p$

Then we have

$$\boldsymbol{x}_0 = \lambda_1\,\boldsymbol{x}_1 + \lambda_2\,\boldsymbol{x}_2 + \ldots + \lambda_p\,\boldsymbol{x}_p \qquad \lambda_i \geq 0, \Sigma\lambda_i = 1$$

Thus,

$$\boldsymbol{c}\boldsymbol{x}_0 = \boldsymbol{c}(\lambda_1\,\boldsymbol{x}_1 + \lambda_2\,\boldsymbol{x}_2 + \ldots + \lambda_p\,\boldsymbol{x}_p)$$
$$= \lambda_1\,\boldsymbol{c}\,\boldsymbol{x}_1 + \lambda_2\,\boldsymbol{c}\,\boldsymbol{x}_2 + \ldots + \lambda_p\,\boldsymbol{c}\boldsymbol{x}_p$$
$$= \lambda_1 z^* + \lambda_2 z^* + \ldots + \lambda_p z^*$$
$$= (\lambda_1 + \lambda_2 + \ldots + \lambda_p) z^*$$
$$= \Sigma\,\lambda_i \cdot z^*$$
$$= z^* \qquad\qquad\qquad (\because \Sigma\lambda_i = 1)$$

Hence, the optimal value z^* is also attained at $\boldsymbol{x}_0$ which is the convex combination of the extreme points at which optimal value occur.

Solved Examples

EXAMPLE 1. *Show that the feasible solution $x_1 = 1$, $x_2 = 0$, $x_3 = 1$, $z = 6$ to the system*

$$x_1 + x_2 + x_3 = 2$$
$$x_1 - x_2 + x_3 = 2$$
$$2x_1 + 3x_2 + 4x_3 = z \text{ (minimized)}, x_i > 0 \text{ is not basic}$$

SOLUTION. Here the objective funtion is given by,

minimized $z = 2x_1 + 3x_2 + 4x_3$

Here, we observe that first two equations in three variables x_1, x_2, x_3, only one variable can be assigned.

The given feasible solution is $x_1 = 1, x_2 = 0, x_3 = 1$ in which the variable x_1 and x_3 are non-zero, therefore, we shall take the vectors α_1 and α_2 associated to these variables, then

$$\alpha_1 = \text{column vector corresponding to } x_1 = \begin{bmatrix} 1 \\ 1 \end{bmatrix}$$

$$\alpha_2 = \text{column vector corresponding to } x_2 = \begin{bmatrix} 1 \\ 1 \end{bmatrix}$$

$\Rightarrow \qquad \alpha_1 = \alpha_2$

$\Rightarrow \qquad 1 \cdot \alpha_1 + (-1)\alpha_2 = 0$

$\Rightarrow \qquad \exists$ two scalars $\lambda_1 = 1, \lambda_2 = -1$ such that $\lambda_1\alpha_1 + \lambda_2\alpha_2 = 0$

$\Rightarrow \qquad \alpha_1, \alpha_2$ are linearly dependent

Hence, the given feasible solution is not basic.

EXAMPLE 2. *Find all the basic feasible solution for the system of equations*

$$2x_1 + 6x_2 + 2x_3 + x_4 = 3$$
$$6x_1 + 4x_2 + 4x_3 + 6x_4 = 2$$

and determine the associated general convex combinations of the extreme points solutions. [MEERUT–2007; KANPUR–2010; RAJASTHAN–2012]

SOLUTION. We can write the given system of equations as

$$\begin{bmatrix} 2 & 6 & 2 & 1 \\ 6 & 4 & 4 & 6 \end{bmatrix} \begin{bmatrix} x_1 \\ x_2 \\ x_3 \\ x_4 \end{bmatrix} = \begin{bmatrix} 3 \\ 2 \end{bmatrix}$$

$$\Rightarrow \qquad AX = B$$

where $A = \begin{bmatrix} 2 & 6 & 2 & 1 \\ 6 & 4 & 4 & 6 \end{bmatrix}$, $X = \begin{bmatrix} x_1 \\ x_2 \\ x_3 \\ x_4 \end{bmatrix}$, $B = \begin{bmatrix} 3 \\ 2 \end{bmatrix}$

If α_1, α_2, α_3 and α_4 are the column vectors in A then we have

$$\alpha_1 = \begin{bmatrix} 2 \\ 6 \end{bmatrix} \quad \alpha_2 = \begin{bmatrix} 6 \\ 4 \end{bmatrix} \quad \alpha_3 = \begin{bmatrix} 2 \\ 4 \end{bmatrix} \quad \alpha_4 = \begin{bmatrix} 1 \\ 6 \end{bmatrix}$$

$\Rightarrow \qquad A = [\alpha_1 \quad \alpha_2 \quad \alpha_3 \quad \alpha_4]$

Here, $\qquad n = $ number of unknowns $= 4$

$\qquad m = $ number of equations $= 2$

$\Rightarrow$ There can be atmost $^4C_2 = \dfrac{4!}{2!\cdot 2!} = 6$ feasible solutions

Now, six set of two vectors out of α_1, α_2, α_3, α_4 are given below:

$$|B_1| = \begin{vmatrix} 2 & 6 \\ 6 & 4 \end{vmatrix} = -28 \neq 0 \;;\; |B_2| = \begin{vmatrix} 2 & 2 \\ 6 & 4 \end{vmatrix} = -4 \neq 0$$

$$|B_3| = \begin{vmatrix} 2 & 1 \\ 6 & 6 \end{vmatrix} = 6 \neq 0 \;;\; |B_4| = \begin{vmatrix} 6 & 2 \\ 4 & 4 \end{vmatrix} = 16 \neq 0$$

$$|B_5| = \begin{vmatrix} 6 & 1 \\ 4 & 6 \end{vmatrix} = 32 \neq 0 \;;\; |B_6| = \begin{vmatrix} 2 & 1 \\ 4 & 6 \end{vmatrix} = 8 \neq 0$$

$\Rightarrow$ all these set of vectors are linearly independent. Hence, all the basic solutions exist.

If $x_{B_i} : i = 1, 2, \ldots, 6$ are the vectors of corresponding basic variables respectively, then the given system of equations reduces in the following forms.

$$B_1 x_1 = b, \qquad B_2 x_2 = b, \qquad B_3 x_3 = b$$
$$B_4 x_4 = b, \qquad B_5 x_5 = b, \qquad B_6 x_6 = b$$

which implies

$$x_{B_1} = B_1^{-1} b = -\frac{1}{28} \begin{bmatrix} 4 & 6 \\ -6 & 2 \end{bmatrix} \begin{bmatrix} 3 \\ 2 \end{bmatrix} = \begin{bmatrix} 0 \\ \frac{1}{2} \end{bmatrix}$$

$$x_{B_2} = B_2^{-1} b = -\frac{1}{4} \begin{bmatrix} 4 & -2 \\ -6 & 2 \end{bmatrix} \begin{bmatrix} 3 \\ 2 \end{bmatrix} = \begin{bmatrix} -2 \\ \frac{7}{2} \end{bmatrix}$$

$$x_{B_3} = B_3^{-1} b = \frac{1}{6} \begin{bmatrix} 6 & -1 \\ -6 & 2 \end{bmatrix} \begin{bmatrix} 3 \\ 2 \end{bmatrix} = \begin{bmatrix} \frac{8}{3} \\ \frac{-7}{3} \end{bmatrix}$$

$$x_{B_4} = B_4^{-1} b = \frac{1}{16} \begin{bmatrix} 4 & -2 \\ -4 & 6 \end{bmatrix} \begin{bmatrix} 3 \\ 2 \end{bmatrix} = \begin{bmatrix} \frac{1}{2} \\ 0 \end{bmatrix}$$

$$x_{B_5} = B_5^{-1} b = \frac{1}{32} \begin{bmatrix} 6 & -1 \\ -4 & 6 \end{bmatrix} \begin{bmatrix} 3 \\ 2 \end{bmatrix} = \begin{bmatrix} \frac{1}{2} \\ 0 \end{bmatrix}$$

and $\qquad x_{B_6} = B_6^{-1}\boldsymbol{b} = \dfrac{1}{8}\begin{bmatrix} 6 & -1 \\ -4 & 2 \end{bmatrix}\begin{bmatrix} 3 \\ 2 \end{bmatrix} = \begin{bmatrix} 2 \\ -1 \end{bmatrix}$

Now, we will find the basic solutions

In the basic matrix B_1, basic vectors are α_1 and α_2 and

$$x_{B_1} = \begin{bmatrix} x_1 \\ x_2 \end{bmatrix} = \begin{bmatrix} 0 \\ \dfrac{1}{2} \end{bmatrix}$$

$\Rightarrow \qquad x_1 = 0,\ x_2 = 1/2$

These two variables are the basic variables and the remaining $x_3,\ x_4$ are non-basic variables. The non-basic variables are zero.

Therefore, the basic solution associated to the basis B_1 is given by $(0,\ 1/2,\ 0,\ 0)$

In a similar way we can write all other basic solutions as follows:

$$\left(-2, 0, \frac{7}{2}, 0\right), \left(\frac{8}{3}, 0, 0, \frac{-7}{3}\right), \left(0, \frac{1}{2}, 0, 0\right), \left(0, \frac{1}{2}, 0, 0\right) \text{ and } (0, 0, 2, -1)$$

But out of these basic solutions, the BFS are $\left(0, \dfrac{1}{2}, 0, 0\right), \left(0, \dfrac{1}{2}, 0, 0\right), \left(0, \dfrac{1}{2}, 0, 0\right)$

Clearly, the extreme points are $\boldsymbol{x}_1 = \left(0, \dfrac{1}{2}, 0, 0\right),\ \ \boldsymbol{x}_2 = \left(0, \dfrac{1}{2}, 0, 0\right),\ \ \boldsymbol{x}_3 = \left(0, \dfrac{1}{2}, 0, 0\right)$

$\Rightarrow$ all the extreme points are same.

Hence, there is unique extreme point solution.

 Exercise-1.3

1. Find all the basic solutions for the following system of linear equations
   $$x_1 + 2x_2 + x_3 = 4$$
   $$2x_1 + x_2 + 5x_3 = 4$$

2. Find all the basic solutions of the following system of linear equations
   $$x_1 + x_2 + x_3 = 4$$
   $$2x_1 + 5x_2 - 2x_3 = 0$$

3. Show that the basic solution $x_1 = 1,\ x_2 = 1/2,$ $x_3 = x_4 = x_5 = 0$ of the equations
   $$x_1 + 2x_2 + x_3 + x_4 = 2$$
   $$x_1 + 2x_2 + 1/2\ x_3 + x_5 = 2$$
   is not basic.

4. Find a basic feasible solution of the system of the equations
   $$x_1 + 2x_3 = 3$$
   $$x_2 + x_3 = 4 \text{ and } x_1, x_2, x_3 \geq 0$$

5. Find all basic feasible solutions of the equations:
   $$2x_1 + x_2 + 3x_3 = 3$$
   $$x_1 + 2x_2 + x_3 = 3 \text{ and } x_1, x_2, x_3 \geq 0$$

6. Show that if $x_1, x_2, \dots x_k$ are k different optimal basic feasible solutions to an LPP then any convex combinations of $x_1, x_2, \dots, x_k$ is also an optimal solution.

ANSWERS

1. $(2,1,0),\ (5,0,-1),\ \left(0, \dfrac{5}{3}, \dfrac{2}{3}\right)$

2. $\left(\dfrac{17}{3}, \dfrac{-5}{3}, 0\right), \left(0, \dfrac{11}{7}, \dfrac{17}{7}\right), \left(\dfrac{11}{4}, 0, \dfrac{5}{4}\right)$

4. $x_1 = 1,\ x_2 = 4,\ x_3 = 0$

5. $(1,1,0),\ \left(0, \dfrac{6}{5}, \dfrac{3}{5}\right)$

Glossary

- **Non-degenerate Basic solution:** If none of the basic variable is zero. Then, basic solution is called non-degenerate.
- **Degenerate Basic solution:** If at least one of the basic variable is zero, then a basic solution is called degenerate.
- **Feasible Basic solution:** If all the basic variables are non-negative, then a basic solution is called feasible.
- **Hypersphere:** In n-dimensional space, a hypersphere, with centre a and radius $r(>0)$ is the set of points
  $$X = \{x : |x - a| = r\}$$
  The equation of hypersphere in E^n (or R^n) is given by $\Sigma(x_i - a_i)^2 = r^2$.
- **Hyperplane:** It is defined as the set of points $(x_1, x_2, ..., x_n)$ satisfying $c_1x_1 + c_2x_2 + ... + c_nx_n = z$, (not all $c_i = 0$) for prescribed values of $c_1, c_2, ..., c_n$ and z.
- **Convex set:** A set of points is said to be convex if for any two points in the set, the line segment joining these points is also in the set, *i.e.*, a set is said to be convex if convex combination of any two points in the set is also

in the set.
- **Convex Hull:** The set of all convex combinations of sets of points from the set X of points is called convex hull, *i.e.*, the intersection of all convex sets containing X in n-dimensional space is called the convex hull of X. Hence, the convex hull of a set $X \subseteq E^n$ is the smallest convex set containing X.
- **Convex function:** A function $f(x)$ is said to be strictly convex at x if for any two other distinct points x_1 and x_2
  $$f\{\lambda x_1 + (1 - \lambda)x_2\} < \lambda f(x_1) + (1 - \lambda) f(x_2),$$
  $$0 < \lambda < 1$$
- **Convex Polyhedron:** The set of all convex combinations of finite number of points is said to be the convex polyhedron generated by these points.
- **Extreme Point:** A point x in a convex set C is an extreme point of C if it does not lie on the line segment of any two points, different from x in the set, *i.e.*, it can not be expressed as a convex combinations of any two distinct points x_1 and x_2 in C.

REVIEW QUESTIONS

1. What do you mean by an extreme point of a convex set?
2. Write a short note on convex set and their applications to linear programming problem.
3. Obtain the convex hull of the boundary of a circle.
4. Prove that the convex hull of a finite number of points is a convex set.
5. Define: Hyperplane, Convex set
6. What is meant by convex polyhedron.
7. Explain the procedure of generating extreme points solutions to a linear programming problem pointing out the assumption made. if any?

MULTIPLE CHOICE QUESTIONS (CHOOSE THE MOST APPROPRIATE ONE)

1. The number of vertices of any non empty closed bounded convex set can not be:
   (a) finite
   (b) not finite
   (c) infinite
   (d) None of these
2. The closed half spaces in E_n or E^n is a:
   (a) open convex set
   (b) unbounded convex set
   (c) closed convex set
   (d) no convex set
3. The set of all feasible solution (if not empty) of a L.P.P. is a:
   (a) non convex set
   (b) poly convex set
   (c) convex set
   (d) none of these
4. The union of two convex sets may or may not be a:
   (a) Non convex set
   (b) Convex set
   (c) Poly convex set
   (d) None of these
5. Every extreme point of a convex set is:
   (a) boundary value of the set
   (b) boundary point of the set
   (c) both (a) and (b)
   (d) none of these
6. A hyper plane is:
   (a) convex
   (b) feasible
   (c) concave
   (d) none of these
7. Let S and T are two convex sets in E^n, then $S + T$, $S - T$ and $\alpha S + \beta T$, where α and β are scalars, are called:
   (a) non convex
   (b) convex sets
   (c) convex point
   (d) none of these

8. Convex hull of set of points on the circle is:
   (a) whole circle   (b) half circle
   (c) extreme point   (d) none of these

9. A hyper plane is:
   (a) non convex   (b) convex
   (c) convex point   (d) none of these

10. In a L.P.P. the set S is polytope when S made by ... of constraints:
    (a) finite number   (b) infinite number
    (c) no number   (d) none of these

11. The vertices of the polygons when are convex sets are called:
    (a) extreme point   (b) feasible
    (c) convex   (d) none of these

12. The intersection of any finite number of convex sets is also a:
    (a) finite set   (b) convex set
    (c) infinite set   (d) non convex set

13. A set S be non empty, then it has at least:
    (a) one vertex   (b) two vertex
    (c) no vertex   (d) none of these

14. The intersection of a finite number of closed half spaces is called:
    (a) monotope   (b) polytope
    (c) nonetope   (d) polygon

15. The intersection of two convex sets is also a:
    (a) convex set   (b) non convex
    (c) extreme point   (d) none of these

16. Any points on line segment joint two points in R^n can be expressed as a convex combination of:
    (a) two points   (b) one point
    (c) variable   (d) none of these

17. A set S be closed convex bounded above/bounded below, then S has at least:
    (a) one vertex   (b) two vertex
    (c) no vertex   (d) none of these

18. Set of all feasible solution of L.P.P. is:
    (a) convex   (b) concave
    (c) (a) and (b) both   (d) none of these

19. Convex linear combination of the points x_1, x_2 is given by $x_p = \lambda_1 x_1 + \lambda_2 x_2$

   (a) $(\lambda_1 + \lambda_2)$ (b) $(\lambda_1 - \lambda_2)$
   (c) $(\lambda_2 - \lambda_1)$ (d) none of these

20. The arbitrary intersection of convex sets is also:
    (a) a convex set   (b) a non convex set
    (c) a finite set   (d) none of these

21. A point X is called a convex linear combination of points x_1 and x_2 if there exist a λ, $0 \le \lambda \le 1$ such that:
    (a) $X = \lambda x_1 + (1 - \lambda)x_2$
    (b) $X = \lambda x_1 + (\lambda - 1)x_2$
    (c) $X = \lambda x_1 - (1 - \lambda)x_2$
    (d) $X = \lambda x_1 + \lambda x_2$

22. The set of all convex linear combination of finite number of points is a:
    (a) c.l.c   (b) no c.l.c
    (c) non convex   (d) convex set

23. The set of all convex combination of a finite number of points is called the .... which generated by these points:
    (a) convex polyhedron
    (b) convex set
    (c) non convex set
    (d) none of these

24. A set C in n dimensional space is said to be convex if every point on the line segment joining any two distinct points of C lie in:
    (a) R
    (b) C
    (c) neither (a) nor (c)
    (d) none of these

25. Let S be a closed set and x is an extreme point of S then $S - \{x\}$ is:
    (a) convex
    (b) concave
    (c) may or may not be convex
    (d) none of these

26. A simplex in n dimension is a convex polyhedron having exactly:
    (a) n vertices   (b) $n + 1$ vertices
    (c) $n - 1$ vertices   (d) none of these

ANSWERS

1. (c)	**2.** (d)	**3.** (c)	**4.** (b)	**5.** (b)	**6.** (a)	**7.** (b)	**8.** (b)	**9.** (b)
10. (a)	**11.** (a)	**12.** (b)	**13.** (a)	**14.** (b)	**15.** (a)	**16.** (a)	**17.** (a)	**18.** (a)
19. (a)	**20.** (a)	**21.** (a)	**22.** (d)	**23.** (a)	**24.** (b)	**25.** (c)	**26.** (b)	

Linear Programming Problems: Formulation and Graphical solutions

2

Linear programming is related to the most advanced mathematics, *i.e.*, operation research. The word 'linear' is used to describe the relationship among two or more variables which are directly proportional while the word 'programming' means planning of activities in such a manner that achieves some optimal results with some restricted resources. Thus, linear programming (L.P.) is one of the most important optimization techniques in the field of operation research.

According to the obtained fact during the world war-II (1939-1945), there was a need of such an effective management that can arrange the limited military resources and the army in the most successful way. There was the need of such methods, which when adopted, gives the maximum profit and to minimize the cost. This was the birth of linear programming.

Mathematically in various situation, the problems are seen in which the number of relations is not equal to the number of variables and many of the relations are in the form of inequality ($\leq$ or $\geq$) to optimize (maximize or minimize) a linear function of the variables subject to such conditions. Therefore, we can defined the linear programming (LP) and linear programming problems (LPP) as follows:

(i) **Linear Programming (LP):** Linear programming is the most general technique, that is used for the optimization (maximization or minimization) of a function to obtain the maximum or minimum value under the certain conditions.

(ii) **Linear Programming Problems (LPP):** Linear programming problem is used to optimize a given linear function of variables, called the objective function which is subject to a certain set of linear equations or inequations (called the constraints or restrictions).

☛ **Remark**
- A mathematical programming is said to be linear if both the objective function and the constraints are linear in decision variables.

2.2 BASIC TERMINOLOGY OF LINEAR PROGRAMMING

(i) **Objective function:** The function which is optimized (maximized or minimized) is known as objective function.

(ii) **Constraints:** The system of linear equations (inequations) under which the objective function is to be optimized are known as constraints.

(iii) **Decision Variables:** The decision variables refers to any activity that is competing with other activities for limited resources. In LPP, the relationship among decision variables should be linear.

2.3 BASIC REQUIREMENTS OF LPP

To define any LPP, there are some basic requirements which are given below:
(i) The relationship between the decision variables should be linear.
(ii) A well defined objective function must be stated which may be either to maximize or to minimize.
(iii) There must be some limitations in the form of constraints which are to be allocated among various activities.
(iv) There must be alternative course of action to choose the best from.
(v) All decision variables must assumes non-negative values.
(vi) Each element of the problem is capable of being quantified by means of measurement.

☞ Remarks
- In LPP, both the objective function and constraints must be expressed in terms of linear equations or inequalities which can be graphically represented by a straight line.
- Every linear programming problem has two important features in common, an objective function to be maximized or minimized and constraints or restrictions.
- The number of activities involved in a problem should be finite, leads to a finite number of constraints in the problem.

2.4 BASIC ASSUMPTIONS OF LP MODEL

The major assumption of LP model are given as follows:
(i) **Certainty:** All model parameters such as availability of resources, profit contribution of a unit of decision variables and consumption of resources by a unit of decision variable must be known and may be constant i.e., the coefficient in the objective function and constraints are completely known and do not change during the period being studied.
(ii) **Linearity:** Both the objective function and constraints must be expressed in terms of linear equations or inequations.
(iii) **Additivity:** The value of the objective function and the total amount of each resource used must be equal to the sum of the respective individual contributions by decision variables. In other words, interaction among the activities of the resources does not exists.
(iv) **Divisibility:** Divisibility implies that solution values of decision variables and resources can take any non-negative values, *i.e.*, fractional values of the decision variables are permitted. This, however is not always desirable.
(v) **Non-Negativity:** It is assumed that variable will take only non-negative values.
(vi) **Finiteness:** An optimal solution can not be evaluated in the situations where there are infinite number of alternative activities and resource restrictions.

2.5 ADVANTAGE OF LINEAR PROGRAMMING

Following are the main advantages of LP:
(i) LP helps in attaining the optimal use of productive factors (resources). It also indicate how a decision maker can employ his productive factors effectively by selecting and

distributing these resources (factors).

(ii) LP techniques improve the quality of decision. The decision making approach of the users of this technique become more objective and less subjective.

(iii) LP techniques provide possible and practical solutions, since there might be other constraints operating outside the problem which must be taken into account.

(iv) The most significant advantage of LP technique is the highlighting of bottlenecks in the production process.

(v) LP also helps in re-evaluation of a basic plan for changing conditions. Plan can be laid for several sets of conditions to find out how to best prepare for possible future changes.

(vi) If conditions change; when the plan is partly carried out, they can be determined so as to adjust the remainder of the plans for best results.

(vii) It also contributes the development of executives through the technique of model building and corresponding interpretation.

2.6 LIMITATIONS OF LINEAR PROGRAMMING

(i) Generally, neither the objective functions nor the constraints in reality are linear related to the variable, while in LPP, all relationship among decision variables are linear.

(ii) In linear programming problem, fractional values are permitted for the decision variable. However many decision problems requires that the decision variables should be obtained in non-fractional values.

(iii) Parameters appearing in the LPP, are assumed to be constant but in real life situations they are frequently neither known nor constant.

(iv) LPP deals with only single objective function, but in real life situations, we may come across conflicting multi-objective problem.

(v) The coefficient of the basic variables can not be determined with certainity, however they can be stated only with probability.

(vi) The LPP does not take into consideration the effect of time and uncertainity.

(vii) In LPP, coefficient in the objective function and the constraints equation must be completely known and they should not change during the study period, but practically, it may not be possible to state all coefficients in the objective function and constraints with certainity.

(viii) The approximation which must be made in the complex relationship between constraints and decision variables, to reduce the problem to meaningful dimensions frequently place the final results in some doubt.

2.7 APPLICATION AREAS OF LINEAR PROGRAMMING

1. **Production Scheduling:** In production scheduling problems the technique may often be used in situations where several products can be made on each of several different machine, the problem is to decide on a programme which will maximize the profit or minimize the cost.

2. **Product Mix Problems:** In product mix problems, one make selection of the optimal product mix to make best use of machine and man hour available while maximize the firm profit.

3. **Transportation Problem:** In transportation problem, we determined the distribution system that will minimize total shipping cost from several warehouses to various market locations.

4. **Travelling Salesman Problem:** Here, we have to decide the shortest route for a salesman starting from a given city, visiting each of the specified cities and then returning to the original city of departure.

5. **Blending Problems:** Blending problem arise when a product can be made from a variety of available raw materials of various composition. In each problem we determine optimal amount of various raw materials, to be used in producing a set of products while determining the optimum quantity of each product to be produced.

6. **Investment Problems:** In investment problems, we find the amount that should be invested in a number of fixed income securities in order to maximize the return on investment.

7. **Diet Problems:** In such type of problems, we determine the amount of different feed ingradient combinations used in desired diet that will satisfy stated nutritional requirements at a minimum cost level.

8. **Media Selection Problems:** In media selection problems, we find the optimum allocation of advertisement in different effective media mix in order to maximize the audience exposure.

9. **Manufacturing Problems:** In manufacturing problems, we find the number of items of each type that should be manufactured so as to maximize the profit subject to production restrictions imposed by limitations on the use of machinery and labour.

10. **Assembly Problems:** In assembly problems, we have to determine the best combinations of basic components to produce goods according to certain specifications.

11. **Job-assigning Problems:** In such problems, we have to assign the job to the workers for maximum effectiveness and optimum results subject to the restrictions of wages and costs.

2.8 STANDARD FORM OF LINEAR PROGRAMMING PROBLEMS

In linear programming problem, we have to determine the values of n decision variables $x_1, x_2, ..., x_n$ such that the linear objective function of these variables assumes an optimal (maximum or minimum) values, when these variables are subject to the set of m linear constraints.

The general LPP with n decision variables and m constraints is stated as follows:

Optimize (maximize or minimize), $Z = c_1x_1 + c_2x_2 + ... + c_nx_n$ *(objective function) subject to the constraints*

$$\left.\begin{array}{l} a_{11}x_1 + a_{12}x_2 + ... + a_{1n}x_n (\leq, =, \geq) b_1 \\ a_{21}x_1 + a_{22}x_2 + ... + a_{2n}x_n (\leq, =, \geq) b_2 \\ \quad\vdots \qquad\qquad\qquad \vdots \\ a_{m1}x_1 + a_{m2}x_2 + ... + a_{mn}x_n (\leq, =, \geq) b_m \end{array}\right] \text{constraints}$$

such that $x_1, x_2, ..., x_n \geq 0$ *(non-negative restrictions)* where

(i) $x_1, x_2, ..., x_n$ are decision variables.

(ii) $c_1, c_2, ..., c_n$ are cost or profit coefficients.

(iii) $a_{ij} : i = 1, 2, ... m, j = 1, 2, ..., n$ are input output coefficients.

(iv) $b_1, b_2, ..., b_n$ represent requirements or availability of m constraints.

(v) The expression $\leq, =, \geq$ means that each constraints may assume only one of three possible forms.

(vi) The restriction, $x_i \geq 0$ implies that x_j's must be non-negative.

In compact form the general LPP can be written as follows:

$$\text{optimize (maximize or minimize) } z = \sum_{j=1}^{n} c_j x_j \qquad \text{(objective function)}$$

$$\text{subject to } \sum_{j=1}^{n} a_{ij} x_j (\leq, =, \geq) b_i : i = 1, 2, ..., m \qquad \text{(constaints)}$$

$$\text{such that } x_j \geq 0, j = 1, 2, ..., n \qquad \text{(non-negative restrictions)}$$

☞ **Remarks**

- The input-output coefficients a_{ij} are also known as structural coefficients or technological coefficients which represent the exchange coefficients of the jth decision variable in the ith constraints.

- The constraints are generally expressed by "less than or equals ($\leq$)" sign in maximization problem. where as in case of minimization problem, it can be expressed by 'greater than or equal to ($\geq$) sign.

2.9 MATRIX FORM OF LPP

In matrix notation the general form of LPP can be written as follows:

$$\text{Optimize (Maximize or minimize) } Z = (c_1, c_2, ..., c_n) \begin{bmatrix} x_1 \\ x_2 \\ \vdots \\ x_n \end{bmatrix}$$

$$\text{subject to the constraints } \begin{bmatrix} a_{11} & a_{12} & ... & a_{1n} \\ a_{21} & a_{22} & ... & a_{2n} \\ \vdots & & & \\ a_{m1} & a_{m2} & ... & a_{mn} \end{bmatrix} \begin{bmatrix} x_1 \\ x_2 \\ \vdots \\ x_n \end{bmatrix} (\leq, =, \geq) \begin{bmatrix} b_1 \\ b_2 \\ \vdots \\ b_m \end{bmatrix}$$

$$\text{such that } x_j \geq 0 \; \forall \, j = 1, 2, ..., n$$

☞ **Remark**

- In a general LPP, it is assumed that the number of rows of coefficient matrix is less than its number of columns.

2.10 MATHEMATICAL FORMULATION OF LINEAR PROGRAMMING PROBLEM

In order to formulate of LPP as a mathematical model we use the following procedure:

WORKING PROCEDURE

STEP 1. Identify the unknown decision variables and assign symbols $x_1, x_2, \ldots$ etc to them.

STEP 2. Identify all restrictions or constraints in terms of requirement and availability of each resources and express them as linear inequations (inequalities) of decision variables.

STEP 3. Identify whether the objective function is to be maximized or minimized and then express it as a linear function of decision variables and then convert it into a linear mathematical expression in terms of decision variables multiplied by their profit or cost contributions.

☞ **Remark**

- Before the formulation of linear programming problem as a mathematical model, find the key decision to be made from the study of the problem. In this connection looking for variable helps considerably.

2.11 EXTRA VARIABLE NEEDED:

1. **Slack Variables:** If a constraint has $\leq$ sign, then in order to make it an equality, we have to add something positive to the LHS.

   Definition. *The non-negative variables which are added to LHS of the constraints to convert them into equalities are known as slack variables.*

2. **Surplus Variables:** If a constraint has $\geq$ sign, then in order to make it an equality, we have to subtract something positive from its LHS.

   Definition. *The non-negative variables which are subtracted from the LHS of the constraints to convert them into equalities are called the surplus variables.*

3. **Artificial Variables:** The artificial variables are introduced for the limited purpose of obtaining an initial solution when constraints of the type $\geq$ or $=$.

   Definition. *When we use surplus variables to convert inequlities into equations, then to obtain basic matrix as identity matrix, we used artificial variable in each constraints.*

   The summary of the extra variables to be added in the given LPP to convert it into standard form is given in the following table:

Type of constraints	Extra variable	Operation	Coefficient of extra variables in the objective functions	
			Max. z	Min. z
$\leq$	slack variable	added	0	0
$\geq$	surplus variable	subtracted	0	0
	artificial variable	added	$-M$	$+M$
$=$	Artificial variable	added	$-M$	$+M$

☞ **Remarks**

- Surplus variables are also known as negative slack variables.
- Surplus and slack variables carry a zero coefficient in the objective function.

Solved Examples

EXAMPLE 1. **(Diet Problem)** *The objective of a diet problem is to ascertain the quantities of certain foods that should be eaten to meet certain nutritional requirement at a minimum cost. The consideration is limited to milk, green vegetables and eggs and to vitamins A, B, C. The number of milligrams of each of these vitamin contained within a unit of each food and their daily minimum requirement along with the cost of each food is given as below:*

Vitamin	Litre of milk	Vegetables (in kg)	Eggs (Dozen)	Minimum daily requirement
A	1	1	10	1 mg
B	100	10	10	50 mg
C	10	100	10	10 mg
Cost in ₹	20	10	8	

Formulate a linear programming problem for this diet problem.

[MEERUT–2004, 10, GARHWAL–2008]

SOLUTION. Let the diet contain x_1 litres of milk, x_2 kg of vegetables and x_3 dozens of eggs

Then total cost (Z) per day in rupees is

$$Z = 20x_1 + 10x_2 + 8x_3 \qquad \text{...(1) (objective function)}$$

Further, total amount of vitamin A in daily diet is

$$(x_1 + x_2 + 10x_3)\text{mg}$$

According to question

$$x_1 + x_2 + 10x_3 \geq 1 \qquad \text{...(2)}$$

Similarly for vitamin B and C, we must have

$$100x_1 + 10x_2 + 10x_3 \geq 50 \qquad \text{...(3)}$$

and $\qquad 10x_1 + 100x_2 + 10x_3 \geq 10 \qquad \text{...(4)}$ (Constraints)

Also, the quantities of different food items to be consumed can not be negative,

so $\qquad x_1 \geq 0, x_2 \geq 0, x_3 \geq 0$ (Non-negative restrictions)

Hence, the mathematical model of given LPP is given below:

Minimize $\qquad Z = 20x_1 + 10x_2 + 8x_3$ (*objective function*)

subject to $\qquad x_1 + x_2 + 10x_3 \geq 1$

$$100x_1 + 10x_2 + 10x_3 \geq 50 \qquad (constraints)$$

$$10x_1 + 100x_2 + 10x_3 \geq 10$$

such that $\qquad x_1 \geq 0, x_2 \geq 0, x_3 \geq 0$ (*Non-negative restrictions*)

EXAMPLE 2. *A furniture dealer deals in two items viz, tables and chairs. He has ₹ 10,000 to invest and a space to store almost 60 pieces. A table costs him ₹ 500 and a chair of ₹ 200. He can sell a table at profit of ₹ 50 and a chair at a profit of ₹ 15. Assume that he can sell all the items that he buys. Formulate the problem as an LPP, so that he can maximize the profit.*

SOLUTION. Let x_1 be the number of tables and x_2 be the chairs

Then, the profit on x_1 tables = ₹ $50x_1$

and, the profit on x_2 chairs = ₹ $15x_2$

$\Rightarrow$ total profit $Z = 50x_1 + 15x_2$ ...(1)

Here, we have to maximize the profit Z.

Now, the cost of x_1 tables = $500x_1$

and the cost of x_2 chairs = $200x_2$

$\Rightarrow$ total cost of x_1 tables and x_2 chairs = $500x_1 + 200x_2$

Since, the dealer invest ₹ 10,000, so

$$500x_1 + 200x_2 \leq 10000 \qquad ...(2)$$

Also, total items = $x_1 + x_2$

which can be atmost 60 therefore

$$x_1 + x_2 \leq 60 \qquad ...(3)$$

Further, since number of tables and chairs cannot be negative

Therefore, $x_1 \geq 0, x_2 \geq 0$

Thus, the mathematical model of the given LPP is as follows:

Maximize $Z = 50x_1 + 15x_2$ (*objective function*)

subject to $\left.\begin{array}{l} 5x_1 + 2x_2 \leq 100 \\ x_1 + x_2 \leq 60 \end{array}\right]$ (*constraints*)

such that $x_1 \geq 0, x_2 \geq 0$ (*non-negative restrictions*)

EXAMPLE 3.   *A goldsmith manufactures necklaces and bracelets. The total number of necklaces and bracelets that he can handel per day is atmost 24. It takes one hour to make a bracelets and half an hour to make a necklace. It is assumed that he can work for a maximum of 16 hours a day. Further the profit on a bracelet is ₹300 and the profit on a necklace is ₹100. Formulate this problem as an LPP so as to maximize the profit.*

SOLUTION.   Let us suppose

Total number of necklaces manufactured = x_1

and Total number of bracelets manufactured = x_2

Since the profit on a necklace is ₹100 $\Rightarrow$ Profit on x_1 necklaces = $100x_1$

Similarly, the profit on a bracelet is ₹300 $\Rightarrow$ Profit on x_2 bracelets = $300x_2$

$\therefore$ Total profit $Z = 100x_1 + 300x_2$ ...(1)

To maximize the profit, we have to maximize Z.

Further, it takes one hour to make one bracelet,

$\therefore$ Total time required to make x_2 bracelets = $1 \cdot x_2 = x_2$ hours

Also, it takes half an hour to make one necklace,

$\therefore$ Total time required to make x_1 necklace = $(1/2)x_1$

So, total time required to make x_1 necklaces and x_2 bracelets $= \dfrac{x_1}{2} + x_2$

Here, total time available per day is 16 hours, so we can write $\dfrac{x_1}{2} + x_2 \leq 16$

$\Rightarrow$ $x_1 + 2x_2 \leq 32$ ...(2)

It is also given that the total number of necklaces and bracelets that the goldsmith can manufacture in a day is atmost 24,

So, $x_1 + x_2 \leq 24$ ...(3)

Also, the number of necklaces and bracelets manufactured can not be negative,

So, $x_1 \geq 0, x_2 \geq 0$

Hence, the mathematical model of the LPP is given by

$$\begin{array}{lll} \text{Maximize} & Z = 100x_1 + 300x_2 & \textit{(objective function)} \\ \text{subject to} & \left.\begin{array}{l} x_1 + 2x_2 \leq 32 \\ x_1 + x_2 \leq 24 \end{array}\right] & \textit{(constraints)} \\ \text{such that} & x_1 \geq 0, x_2 \geq 0 & \textit{(non-negative restrictions)} \end{array}$$

EXAMPLE 4. *A dietician decides a certain minimum intake of vitamins A, B and C for a family. The minimum daily needs of vitamins A, B, C are 30, 20, 16 units respectively. For the supply of these, the dietician depends on two types of foods X and Y. The first one gives 7, 5, 2 units per gram of vitamin A, B, C respectively.*

The first food costs ₹2 per gram and second ₹1 per gram. How many grams of each food stuff should the family buy everyday to keep the food expense at a minimum? Formulate a linear programming problem for this problem.

SOLUTION. It is often convenient to construct the following table after understanding the problem carefully.

Foods	Vitamin A	Vitamin B	Vitamin C	Cost per gram in ₹
$X(x_1)$	7	5	2	2
$Y(x_2)$	2	4	8	1
Minimum daily needs	30	20	16	

Clearly, $Z = 2x_1 + x_2$

Since, we have to keep the expenses minimum,

So we have to minimize $Z = 2x_1 + x_2$ ...(1)

Also, we observe that

$$\begin{array}{ll} 7x_1 + 2x_2 \geq 30 & \text{...(2)} \\ 5x_1 + 4x_2 \geq 20 & \text{...(3)} \\ 2x_1 + 8x_2 \geq 16 & \text{...(4)} \end{array}$$

and $x_1 \geq 0, x_2 \geq 0$

Hence, the mathematical model of the LPP is given below

$$\begin{array}{lll} \text{Minimize } Z = 2x_1 + x_2 & \textit{(objective function)} \\ \text{Subject to } \left.\begin{array}{l} 7x_1 + 2x_2 \geq 30 \\ 5x_1 + 4x_2 \geq 20 \\ 2x_1 + 8x_2 \geq 16 \end{array}\right] & \textit{(constraints)} \end{array}$$

such that $x_1 \geq 0, x_2 \geq 0$ *(non-negative restrictions)*

EXAMPLE 5. *A toy company manufactures two type of dolls; an ordinary doll A and a deluxe doll B. Each type of doll B takes twice as long to produce as one of the type A. It is given that the company would have time to make a maximum of 2000 dolls per day if it produces only the ordinary version. The supply of plastic is sufficient to* *produce 1500 dolls per day (both A and B combined). The deluxe version requires a fancy dress of which there are only 600 pieces per day available. If the company makes profit of ₹30 and ₹50 per doll respectively on doll A and B, formulate the problem as an LPP to maximize the profit.*   [MEERUT 2007, 09, 11]

SOLUTION.   Let the no. of dolls of type $A = x_1$

and the no. of dolls of type $B = x_2$

If t hours are required to produce one doll of type A then the time required to produce one doll of type B will be $2t$ hours.

Now, since the time available per day is $2000t$ hours

Therefore, $x_1 \cdot t + x_2 \cdot 2t \leq 2000t$

$\Rightarrow$ $x_1 + 2x_2 \leq 2000$ ...(1)

Further, since the supply of plastic is sufficient to produce 1500 dolls per day (both types)

So, $x_1 + x_2 \leq 1500$ ...(2)

Again, the fancy dress for deluxe version of doll B is available for 600 pieces per day only

So, $x_2 \leq 600$ ...(3)

Also, total profit on both the types of dolls per day is

$$Z = 30x_1 + 50x_2$$ ...(4)

We have to maximize Z.

Also, no. of dolls can't be negative, *i.e.*, $x_1 \geq 0, x_2 \geq 0$

Hence, the mathematical model of LPP is given as below:

Maximize $Z = 30x_1 + 50x_2$ *(objective function)*

subject to $x_1 + 2x_2 \leq 2000$ ⎤

$x_1 + x_2 \leq 1500$ *(constraints)*

$x_2 \leq 600$ ⎦

s.t. $x_1 \geq 0, x_2 \geq 0$ *(non-negative restrictions)*

EXAMPLE 6.   *A furniture firm manufactures chairs and tables, each requiring the use of three machines A, B and C. Production of one chair requires 2 hours on machine A, 1 hour on machine B and 1 hour on machine C. Each table requires 1 hour each on machine A and B and 3 hours on machine C. The profit realized by selling one chair is ₹300 while for a table the figure is ₹600. The total time available per week on machine A is 70 hours, on machine B is 40 hours and on machine C is 90 hours. How many chairs and tables should be made per week to maximize the profit? Formulate a mathematical model for the problem.*   [MEERUT–2008, 11, 13]

SOLUTION.   Let the number of chairs manufactured by firm $= x_1$

and the number of tables manufactured by firm $= x_2$

It is often convenient to construct the table after understanding the problem carefully.

	Time required per piece on machine			Profit (in ₹)
	A	**B**	**C**	
Chair	2	1	1	300
Table	1	1	3	600
Time available per week in hours	70	40	90	

As per given, for the manufactures of x_1 chairs and x_2 tables the time required on machines A, B, C are $2x_1 + x_2$; $x_1 + x_2$ and $x_1 + 3x_2$ hours respectively.

While the time available on these machines are 70, 40 and 90 hours respectively.

Also, the profit is given by

$$Z = 300x_1 + 600x_2$$

Hence, the mathematical model of LPP is given by

Maximize $\quad\quad Z = 300x_1 + 600x_2 \quad\quad\quad\quad$ (objective function)

subject to $\quad 2x_1 + x_2 \le 70$

$\quad\quad\quad\quad x_1 + x_2 \le 40 \quad\quad\quad\quad\quad\quad\quad\quad$ (constraints)

$\quad\quad\quad\quad x_1 + 3x_2 \le 90$

s.t. $\quad\quad x_1 \ge 0; x_2 \ge 0 \quad\quad\quad\quad$ (non-negative restrictions)

EXAMPLE 7. *A factory produces two products A and B. Each of the product A requires 2 hours of moulding, 3 hours of grinding and 4 hours for polishing and each of the product B requires 4 hours for moulding, 2 hours for grinding and 2 hours for polishing. The moulding machine can work for 20 hours grinding machine for 24 hours and the polishing machine available for 13 hours. The profit is ₹50 per unit of A and ₹30 per unit of B. Assuming that the factory can sell all that it produces, formulate the problem as an LPP to maximize the profit.*

SOLUTION. Let total units of product A produced $= x_1$

total units of product B produced $= x_2$

After carefully understanding the problem, the given information can be systematically arranged in the form of following table

Product	Moulding time in hours	Grinding time in hours	Polishing time in hours	Profit on one unit (in ₹)
A	2	3	4	50
B	4	2	2	30
Time availability in hours	20	24	13	

Using the above information, the mathematical model of LPP is given as below:

Maximize $\quad\quad Z = 50x_1 + 30x_2 \quad\quad\quad\quad$ (objective function)

subject to $\quad 2x_1 + 4x_2 \le 20$

$\quad\quad\quad\quad 3x_1 + 2x_2 \le 24 \quad\quad\quad\quad\quad\quad\quad$ (constraints)

$\quad\quad\quad\quad 4x_1 + 2x_2 \le 13$

s.t. $\quad\quad x_1 \ge 0, x_2 \ge 0 \quad\quad\quad\quad$ (non-negative restrictions)

EXAMPLE 8. *A diet is to contain at least 4000 units of carbohydrates 500 units of fats and 300 units of Proteins. Two foods A and B are available. Food A cost ₹20 per unit* *and food B cost ₹40 per unit. A unit of food A contains 10 units of carbohydrate, 20 units of fat and 15 units of protein. A unit of food B contains 25 units of carbohydrate, 10 units of fat and 20 units of protein. Formulate the problem as a LPP so as to find the minimized costs for a diet that consist of a mixture of these two foods and also meets the minimum nutritions requirements.*

SOLUTION.   Let a diet consist of x_1 units of food A and x_2 units of food B. Then after carefully understanding the problem, the given information can be systematically arranged in the following table.

Food	Carbohydrates	Fat	Protein	Cost in ₹
A	10	20	15	20
B	25	10	20	40
Requirement (least)	4000	500	300	

Clearly, in a diet, carbohydrates, fats and proteins are $10x_1 + 25x_2$; $20x_1 + 10x_2$; $15x_1 + 20x_2$ units, while they are required at least 4000, 500 and 300 units respectively.

Also, the profit $Z = 20x_1 + 40x_2$, has to be minimized.

Hence, the required mathematical model of LPP is given as follows:

Minimize $Z = 2x_1 + 4x_2$ (*objective function*)

subjecte to $\left. \begin{array}{l} 10x_1 + 25x_2 \geq 4000 \\ 20x_1 + 10x_2 \geq 500 \\ 15x_1 + 20x_2 \geq 300 \end{array} \right]$ (*constraints*)

s.t. $x_1 \geq 0, x_2 \geq 0$ (*non-negative restrictions*)

 **Exercise-2.1**

1. According to the medical experts it is necessary for an adult to consume at least 75 grams of proteins, 85 grams of fat and 300 grams of carbohydrates daily. The following table gives the analysis of the food items readily available in the market with their respective costs.

Food Type	Food value (in gms) per 100 gms			Cost in ₹ (per kg)
	Proteins	Fats	Carbohyd-rates	
A	18	15	—	30
B	16	4	7	40
C	4	20	2.5	20
D	5	8	40.0	15
Minimum daily require-ments	75	85	300	

Formulate a linear programming problem for an optimum diet.

2. A firm manufactures two types of product A and B and sells them at a profit of ₹ 20 on type A and ₹ 30 on type B. Each product is processed on two machines E and F. Type A requires one minute of processing time on E and two minutes on F, type B requires one minute while machine F is available for 10 hours during any working day. Formulate the problem as a linear programming problem.

3. A firm can produce two products A and B during a given period of time. Each of these products requires four different operations viz, Grinding, Turning, Assembling and Testing. The requirement in hours per unit of manufacturing of these products is given below:

Products	A	B
Grinding	1	2
Turning	3	1
Assembling	4	3
Testing	5	4

The available capacities of these operations in hours for the given time period are: 30 for grinding, 60 for turning, 200 for assembling and 200 for testing. Profit on each unit of A is ₹30 and that for each unit of B is ₹20. Formulate the problem as a linear programming model to maximize the profit assuming that the firm can sell all the items that it produces at the prevailing market price.

4. A resourceful home decorator manufactures two types of lamps say A and B. Both lamps go through two technicians, first a cutter, second a finisher. Lamp A requires 2 hours of the cutter's time and 1 hour of the finisher's time. Lamp B requires 1 hour of cutters and 2 hours of finisher's time. The cutter has 104 hours and finishers has 76 hours of time available each month. Profit on the lamp A is ₹ 60 and on the lamp B is ₹ 110. Assuming that each can sell all that he produces, how many of each type of lamps should be manufacture per month to obtain the best returns? Formulate a LPP for this problem.

5. The manager of an oil refinery must decide on the optimal mix of 2 possible blending process of which the input and output per production run are as follows:

Process	Input		Output	
	Crude A	Crude B	Gasoline X	Gasoline Y
1	6	4	6	9
2	5	6	5	5

The maximum amount available of crudes A and B are 500 units and 400 units respectively. Market demand shows that at least 300 units of gasoline X and 260 units of gasoline Y must be produced. The profit per production run from process 1 and 2 are ₹400 and ₹500 respectively. Formulate the LPP for maximizing the profit.

6. A factory produces two products A and B. To manufacture one unit of product A, a machine has to work for $1\frac{1}{2}$ hours and a craftsman has to work for 2 hours. To manufacture one unit of product B, the machine has to work for 3 hours and the craftsman for one hour. In a week, the factory can avail of 80 hours of machine time and 70 hours of craftsman time. The profit on the sale of each unit of A and

B is of ₹ 100 and ₹ 80 respectively. If the manufacture can sell all the items produced; formulate the problem as a LPP.

7. A firm manufactures 3 products A, B and C. The profits are ₹30, ₹20 and ₹40 respectively. The firm has 2 machines and below the required processing time in minutes for each machine on each product.

Machine	Products		
	A	B	C
G	4	3	5
H	2	2	4

Machines G and H have 2000 and 2500 machines minutes respectively. The firm must manufactures 100 A's, 200 B's and 50 C's but not more than 150 A's. Setup a linear programming problem to maximize profit.

8. A manufacturer produces three models I, II and III of a certain product. He used two types of raw material A and B of which 4000 and 6000 units respectively are available. The raw material requirements per unit of the three models are given below:

Raw material	Requirement per unit of given model		
	I	II	III
A	2	3	5
B	4	2	7

The labour time for each unit of model I is twice that of model II and three times that of model III. The entire labour force of the factory can produce the equivalent of 2500 units of model I. A market survey indicates the minimum demand of the three models are 500, 500 and 345 units respectively. Formulate the problem as a LPP in order to determine the number of units of each product which will maximize profit.

9. A complete unit of a certain product consist of four units of component A and three units of component B. The two components (A and B) are manufactured from two different raw materials of which 100 units and 200 units respectively are available. Three departments are engaged in the production process with each department using a different method for manufacturing the components. The following table gives the raw material requirements per production

run and resulting units of each component. The objective is to determine the number of production runs for each department which will maximize the total number of component units of the final product.

Depart-ment	Input per run (units)		Output per run (units)	
	Raw Material		Components	
	I	II	A	B
1	7	5	6	4
2	4	8	5	8
3	2	7	7	3

Formulate a linear programming model to the above problem.

10. The owner of the Metro sports wishes to determine how many advertisement to place in the selected three monthly magazines A, B and C. His objective is to advertise in such a way that total exposure to principal buyers of expensive sports good is maximized. Percentage of readers for each magzine are known. Exposure in any particular magazine is the number of advertisements placed multiplied by the number of principal buyers. The following data may be used

Exposure Category	Magazine		
	A	B	C
Reader (in lakhs)	1	0.6	0.4
Principal buyers	10%	15%	7%
Cost per advertisement (in ₹)	5000	4500	4250

The budgeted amount is atmost ₹ 1 lakh for the advertisement. The owner has already

decided that magazine A should have no more than 6 advertisement and that B and C each have at least two advertisement. Formulate the LPP model.

11. A city hospital has the following minimal daily requirement for nurses

Period	Clock Time (24 hr. day)	Minimum no. of nurses required
1	6 AM – 10 AM	2
2	10 AM – 2 PM	7
3	2 PM – 6 PM	15
4	6 PM – 10 PM	8
5	10 PM – 2 AM	20
6	2 AM – 6 AM	6

Nurses report to the hospital at the begining of each period and work for 8 consecutive hours. The hospital wants to determine the minimum number of nurses available for each period. Formulate this LPP.

[MEERUT–2015]

12. A tyre factory produces three types of tyres T_1, T_2, T_3. Three different type of chemicals say C_1, C_2, C_3 are required for production. One T_1 tyre needs 2 units of C_1, 3 units of C_3, one T_2 tyre needs 3 units of C_1, 2 units of C_2 and 2 units of C_3 and one T_3 tyre needs 5 units of C_2 and 4 units of C_3. The factory has only a stock of 20 units of C_1, 25 units of C_2 and 30 units of C_3. Further the profit from the sale of one tyre T_1 is ₹60, one tyre T_2 is ₹100 and one tyre of type T_3 is ₹80. Assuming that the factory can sell all that its produces. Formulate a linear programming to maximize the profit.

ANSWERS

1. minimize $Z = 30x_1 + 40x_2 + 20x_3 + 15x_4$
   subject to the constraints
   $180x_1 + 160x_2 + 40x_3 + 50x_4 \geq 75$
   $150x_1 + 40x_2 + 200x_3 + 80x_4 \geq 85$
   $70 x_2 + 25x_3 + 400x_4 \geq 300$
   and $x_1 \geq 0; x_2 \geq 0; x_3 \geq 0; x_4 \geq 0$

3. maximize $Z = 30x_1 + 20x_2$
   subject to the constraints
   $x_1 + 2x_2 \leq 30$
   $3x_1 + x_2 \leq 60$
   $4x_1 + 3x_2 \leq 200$
   $5x_1 + 4x_2 \leq 200$
   and $x_1 \geq 0; x_2 \geq 0$

2. maximize $Z = 20x_1 + 30x_2$
   subject to the constraints
   $x_1 + x_2 \leq 400$
   $2x_1 + x_2 \leq 600$
   and $x_1 \geq 0; x_2 \geq 0$

4. maximize $Z = 60x_1 + 110x_2$
   subject to the constraints
   $2x_1 + x_2 \leq 104$
   $x_1 + 2x_2 \leq 76$
   and $x_1 \geq 0; x_2 \geq 0$

5. maximize $Z = 400x_1 + 50x_2$
subject to the constraints
$6x_1 + 5x_2 \leq 500$; $4x_1 + 6x_2 \leq 400$
$6x_1 + 5x_2 \geq 300$; $9x_1 + 5x_2 \geq 260$
and $x_1 \geq 0$; $x_2 \geq 0$

6. maximize $Z = 100x_1 + 80x_2$
subject to the constraints
$1.5x_1 + 3x_2 \leq 80$
$2x_1 + x_2 \leq 70$
and $x_1 \geq 0$; $x_2 \geq 0$

7. maximize $Z = 30x_1 + 20x_2 + 40x_3$
subject to the constraints
$4x_1 + 3x_2 + 5x_3 \leq 2000$
$2x_1 + 2x_2 + 4x_3 \leq 2500$
$100 \leq x_1 \leq 150$; $0 \leq x_2 \leq 200$ and $0 \leq x_3 \leq 50$

8. maximize $Z = 60x_1 + 40x_2 + 100x_3$
subject to the constraints
$2x_1 + 3x_2 + 5x_3 \leq 4000$
$4x_1 + 2x_2 + 7x_3 \leq 6000$
$6x_1 + 3x_2 + 2x_3 \leq 15000$
$2x_1 = 3x_2$, $5x_2 = 2x_3$
and $x_1 \geq 500$; $x_2 \geq 500$; $x_3 \geq 375$

9. maximize $Z = V$, where $V = \min\left[\dfrac{1}{4}(6x_1 + 5x_2 + 7x_3), \dfrac{1}{3}(4x_1 + 8x_2 + 3x_3)\right]$

subject to the constraints
$6x_1 + 5x_2 + 7x_3 - 4V \geq 0$; $4x_1 + 8x_2 + 3x_3 - 3V \geq 0$
$7x_1 + 4x_2 + 2x_3 \leq 100$; $5x_1 + 8x_2 + 7x_3 \leq 200$
and $x_1, x_2, x_3, V \geq 0$

10. maximize $Z = 10000x_1 + 9000x_2 + 2800x_3$
subject to the constraints
$5000x_1 + 4500x_2 + 4250x_3 \leq 100000$
$x_1 \leq 6, x_2 \geq 2, x_3 \geq 2$
and $x_1, x_2, x_3 \geq 0$

11. minimize $Z = x_1 + x_2 + x_3 + x_4 + x_5 + x_6$
subject to the constraints
$x_1 + x_2 \geq 7$; $x_2 + x_3 \geq 15$; $x_3 + x_4 \geq 8$;
$x_4 + x_5 \geq 20$; $x_5 + x_6 \geq 6$; $x_6 + x_1 \geq 2$
and $x_i \geq 0 \ \forall \ i = 1, 2, ..., 6$

12. maximize $Z = 60x_1 + 100x_2 + 80x_3$
subject to the constraints
$2x_1 + 3x_2 \leq 20$
$2x_1 + 5x_3 \leq 25$
$3x_1 + 2x_2 + 4x_3 \leq 30$
and $x_1 \geq 0, x_2 \geq 0, x_3 \geq 0$

2.12 SOLUTION OF LINEAR PROGRAMMING PROBLEM

In this section we shall discuss solution method of LPP which are given below:
(1) ISO-Profit or ISO-Cost method
(2) Corner point or Extreme point solution method
Before discussing these method, let us define the following terms:

2.12.1 Solution

A set of values of decision variables, x_i which satisfy all the constraints of a linear programming problem is called 'solution' of that problem.

2.12.2 Feasible Solution

Any solution of LPP which also satisfy the non-negative restrictions is called feasible solution, *i.e.*, the set of values of decision variables which satisfy all the constraints and non-negative restriction of an LPP is called feasible solution.

2.12.3 Infeasible Solution

Any solution of an LPP which does not satisfy the non-negative restrictions, *i.e.*, if all the values of decision variables in any solution are negative, the solution is called infeasible solution.

2.12.4 Basic Solution

For a set of m simultaneous equations in n-variables $(n > m)$, a solution obtained by setting $(n - m)$ variables equal to zero and solving for remaining m equations in m variables is called a basic solution.

☛ **Remark**
- The $(n - m)$ variables whose value did not appear in the above solution are called non-basic variables and the remaining m variables are known as basic variables.

2.12.5 Basic Feasible Solution

A feasible solution to a general LPP which is also basic solution, *i.e.*, all basic variables assume non-negative values is called basic feasible solution.

Generally, basic feasible solutions are of two types:

(i) **Degenerate Basic Feasible solution:** A basic solution to the system of equations is called degenerate if one or more of the basic variables become equal to zero.

(ii) **Non-degenerate Basic Feasible solution:** A basic solution is called non-degenerate if values of m basic variables are non-zero and positive.

2.12.6 Optimal Basic Feasible Solution

Any basic feasible solution which optimize (maximize or minimize) the objective function of a general LPP is called optimal basic feasible solution.

2.12.7 Unbounded Solution

A solution which can increase or decrease the value of the objective function of an LPP indefinitely is said to be an unbounded solution.

2.12.8 Feasible Region

The common region formed by all the constraints and non-negative restrictions of an LPP is called feasible region.

2.12.9 Infeasible Region

The region common to all constraints in which all the decision variables are negative is called infeasible region.

2.12.10 Convex Region

If the line segment joining any two arbitrary points of the region lies entirely within the region, then this region is said to be convex region.

2.13 GRAPHICAL SOLUTION METHODS OF AN LPP

To obtain the solution of an LPP by graphical method, we shall recall the following important results:

(1) An optimum solution of an LPP, if it exists, occurs at one of the extreme points, i.e., corner points of the convex polygon of the set of feasible solutions (Fundamental extreme point theorem)

(2) If the optimum solution occurs at more than one extreme point the value of the objective function will be the same for all Convex Combinations of these extreme points.

(3) The collection of all feasible solutions to an LPP constitutes a convex set whose extreme points correspond to the Basic feasible solution.

(4) There are a finite number of basic feasible solutions within the feasible solution space.

(5) If the convex set of all feasible solutions of the system of simultaneous equations $AX = B, X \geq 0$ is a convex polyhedron, then at least one of the extreme points gives an optimal solution.

(6) Each corner point of the feasible region falls at the intersection of two constraints equalities.

(7) The extreme point of the convex set give the basic feasible solution to an LPP.

WORKING PROCEDURE

STEP 1.	**Formulate the given problem:** If the given problem is not in mathematical form, then formulate the problem in terms of a series of mathematical constraints and an objective function.
STEP 2.	**Plotting the constraints:** Each inequality in the constraints equation be written as inequality. Give any arbitrary value to one variable and get the value of other variable by solving the equation. Similarly, give another arbitrary value to the variable and find the corresponding value of the other variable. Plot these two set of values. Connect these points by a straight line. Do the same exercise for each constraints and get as many straight line as there are equations. Here each straight line representing one constraint.
STEP 3.	**Identify the Feasible Region:** We have to identify the area which satisfy all the constraints simultaneously. For "greater than (>)" constraints the feasible region will be the area which lie above the constraints lines. For "less than (<)" constraints this area is generally the region below these lines. On ≥ or ≤ constraints, the feasible region includes the points on the constraints line also.
STEP 4.	Select one of the following two techniques of solutions.

(A) Corner Point Method

Since, we know that an optimal solution of an LPP always lie at one of the corner points of the feasible solution space. Thus, first we determined the coordinates of all corner points of the feasible region and then compute the value of the objective function at these points and then compared.

The steps of the solutions will be clear by the following working procedure.

WORKING PROCEDURE

STEP 1.	Identify each of the corner point of the feasible region either by inspection or method of simultaneous equations.
STEP 2.	Determine the coordinate of each extreme point of the feasible region.
STEP 3.	Compute the value of objective function at each extreme point and then compare.
STEP 4.	Identify the extreme point that gives optimal value of the objective function.

The point where the objective function attains its optimum value, gives the optimal value of the LPP.

☞ **Remark**

- If two vertices of the Convex polygon gives the same optimum value of the objective function, then all points on the line segment joining these two vertices will give the optimum value of the objective function. In this case the LPP is said to have infinite number of optimum solutions

> To determine which side of a constraint is in the feasible region, examine whether the origin (0, 0) satisfies the constraints. If it does, then all points on and below the constraint towards the origin are feasible points. If it does not, then all points on and above the constraint away from the origin are feasible points.

(B) ISO-Profit or ISO-Cost Method

In this method, the optimal solution is found by using the slope of the objective function. An iso-profit line is a collection of points which designate with same value of the objective function. By assigning different values to Z, we get different profit lines. The steps of the solution by this method are as follows:

WORKING PROCEDURE

STEP 1. Identify the feasible region and its extreme points

STEP 2. Assign a constant value say Z_1 to the objective function and draw the corresponding line of objective function (called iso-profit line).

STEP 3. Assign another constant value say Z_2 to the objective function and draw the corresponding line of the objective function.

STEP 4. If $Z_1 > Z_2$ and objective function Z is to be maximized then we move the line corresponding to Z_1 to the line corresponding to Z_2 parallel to itself as farthest point within the feasible region is touched by this line and any further displacement of this line takes it out of the feasible region. The coordinates of the farthest point so obtained will give the maximum value of Z.

But if we have to minimize the objective function Z, we have the line corresponding to Z_2 to the line corresponding to Z_1 and find the nearest point in the same way as we had find the farthest point. Then, coordinate of the nearest point will give minimum value of the objective function.

Since every point in the feasible region satisfies all the constraints of an LPP and there are infinitely many points so it is not easy to anyone to find a point that gives a maximum or minimum value of the objective function. To handle this situation, following results seems very useful.

Let R be the feasible region for an LPP and let $Z = ax + by$ be the objective function

(i) When Z has an optimal value, where x and y are subject to constraints describe by the line or inequation, this optimal value must occur at a corner point of the feasible region.

(ii) If R is bounded, then the objective function Z has both a maximum or minimum value at that point on R.

☛ **Remark**

- The iso profit line method is also applicable if the problem has more constraints.

 Solved Examples

Type 1. Based on Finding the solution set of Simultaneous Linear Inequations

EXAMPLE 1. *Draw the graph of the solution set of the inequations $2x + 3x \geq 6$, $x + 4y \leq 4$, $x \geq 0$ and $y \geq 0$.*

SOLUTION. The corresponding equations of given inequations are:

$$2x + 3y = 6, x + 4y = 4, x = 0, y = 0$$

Region represented by $2x + 3y \geq 6$. The line $2x + 3y \geq 6$, meets x-axis at $A(3, 0)$ (put $y = 0$) and y-axis at $B(0, 2)$ (put $x = 0$ in it). We find that $(0, 0)$ does not satisfy the inequation $2x + 3y \geq 6$, the portion not containing the origin represented by the given inequation represents the solution set.

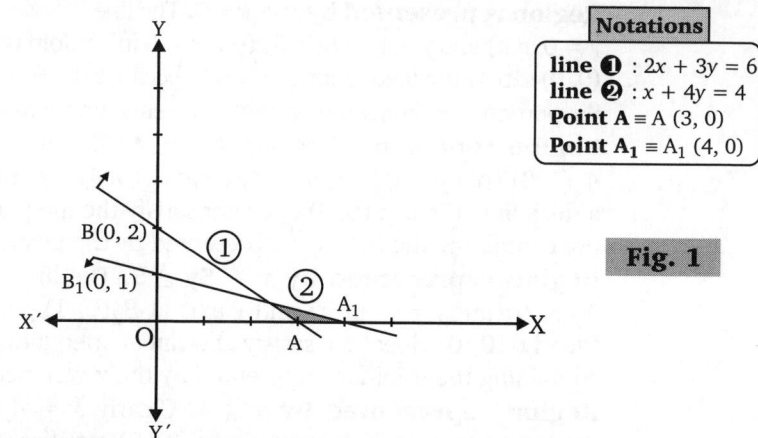

Notations

line ❶ : $2x + 3y = 6$
line ❷ : $x + 4y = 4$
Point $A \equiv A(3, 0)$
Point $A_1 \equiv A_1(4, 0)$

Fig. 1

Region represented by $x + 4y \leq 4$. The line $x + 4y = 4$ meets x-axis at $A_1(4, 0)$ (put $y = 0$ in it) and y-axis at $B_1(0, 1)$ (put $x = 0$ in it). We find that $(0, 0)$ satisfy the inequation $x + 4y \leq 4$. So, the portion containing the origin represented by the given inequation represents the solution set.

Region represented by $x \geq 0, y \geq 0$, represent the first quadrant.

Hence, the shaded region given in the figure represents the solution set of the given linear inequations.

EXAMPLE 2. *Draw the graph of the solution set of the linear inequations.*

$$x + y \leq 5, \, 4x + y \geq 4, \, x + 5y \geq 5, \, x \leq 4, \, y \leq 3.$$

SOLUTION. The corresponding linear equations of the given linear inequations are $x + y = 5$, $4x + y = 4$, $x + 5y = 5$, $x = 4$, $y = 3$.

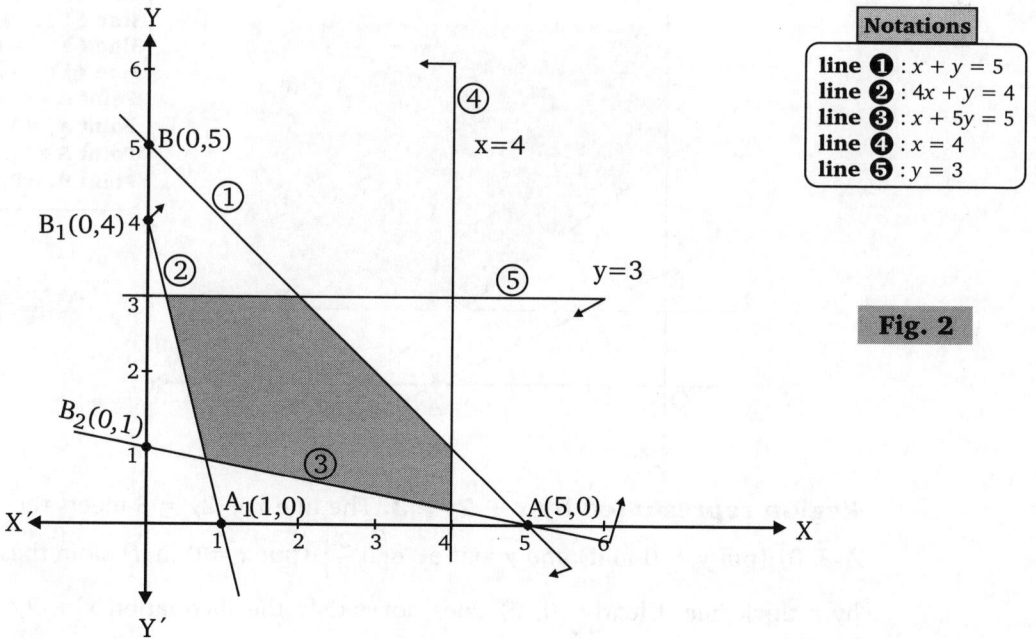

Notations

line ❶ : $x + y = 5$
line ❷ : $4x + y = 4$
line ❸ : $x + 5y = 5$
line ❹ : $x = 4$
line ❺ : $y = 3$

Fig. 2

Region represented by $x + y \leq 5$. The line $x + y = 5$ meets the x-axis at $A(5, 0)$ (put $y = 0$ in it) and y-axis at $B(0, 5)$ (put $x = 0$ in it). Join these points by a thick line. Clearly, $(0, 0)$ satisfy the inequation $x + y \leq 5$, because $0 \leq 5$.

So, portion containing the oritgin represents the solution set of the inequation $x + y \leq 5$.

Region represented by $4x + y \geq 4$. The line $4x + y = 4$ meets the x-axis at $A_1(1, 0)$ (put $y = 0$ in it) and y-axis at $B_1(0, 4)$) (put $x = 0$ in it). Join these points by a thick line. Clearly, $(0, 0)$ does not satisfy the inequation $4x + y \geq 4$. So, the portion not containing the origin is represented by the given inequation.

Region represented by $x + 5y \geq 5$. The line $x + 5y = 5$ meets the x-axis at $A_2(5, 0)$ (put $y = 0$ in it) and y-axis at $B_2(0, 1)$. Join these points by a thick line. Clearly, $(0, 0)$ does not satisfy the linear inequation $x + 5y \geq 5$. So, portion not containing the origin is represented by the given inequation.

Region represented by $x \leq 4$. Clearly, $x = 4$ is a line parallel to y-axis at a distance 4 from it to its right. Since, $(0, 0)$ satisfies the inequation $x \leq 4$. So, portion lying to the left of $x = 4$ is the shaded region.

Region represented by $y \leq 3$. Clearly $y = 3$ is a line parallel to x-axis at a distance 3 from it. Since, $(0, 0)$ satisfies the inequation $y \leq 3$. So, portion lying below $y = 3$ is the shaded region.

Hence, the shaded region given in figure represents the solution set of the given linear inequations.

EXAMPLE 3.   *Solve graphically the following system of inequations:*

$$x + 2y \geq 3, \ 3x + 4y \geq 12, \ x \geq 0, \ y \geq 1$$

SOLUTION.   The corresponding linear equations of the given linear inequations are:

$$x + 2y = 3, \ 3x + 4y = 12, \ x = 0, \ y = 1$$

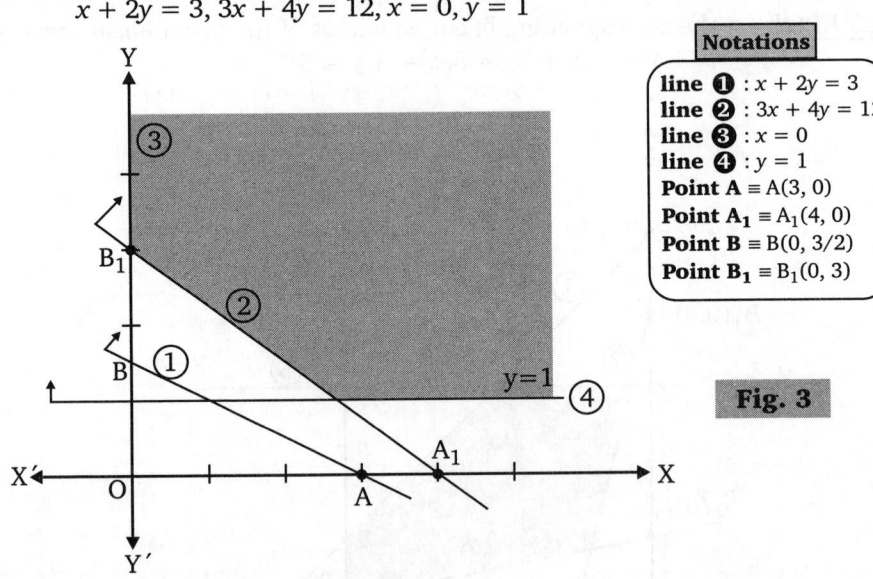

Notations
line ❶ : $x + 2y = 3$
line ❷ : $3x + 4y = 12$
line ❸ : $x = 0$
line ❹ : $y = 1$
Point A $\equiv A(3, 0)$
Point A_1 $\equiv A_1(4, 0)$
Point B $\equiv B(0, 3/2)$
Point B_1 $\equiv B_1(0, 3)$

Fig. 3

Region represented by $x + 2y \geq 3$. The line $x + 2y = 3$ meets the x-axis at $A(3, 0)$ (put $y = 0$ in it) and y-axis at $B\left(0, \dfrac{3}{2}\right)$ (put $x = 0$ in it). Join these points by a thick line. Clearly $(0, 0)$ does not satisfy the inequation $x + 2y \geq 3$. So, portion not containing the origin is represented by the given inequation.

Region represented by the $3x + 4y \geq 12$. The line $3x + 4y = 12$ meets the x-axis at $A_1(4, 0)$ (put $y = 0$ it it) and y-axis at $B_1(0, 3)$ (put $x = 0$ in it). Join these points by a thick line. Clearly, $(0, 0)$ does not satisfy the given inequation. So, porion not containing the origin is represented by the inequation $3x + 4y \geq 12$.

Region represented by $x \geq 0$. Clearly, $x \geq 0$ represents the right y-axis including y-axis.

Region represented by $y \geq 1$. Clearly, $y = 1$ is a line parallel to x-axis at a distance of 1 unit from it. Since, $(0, 0)$ does not satisfy $y \geq 1$. So, portion not containing the origin is represented by the given inequation.

Hence, the shaded region represents the solution set of the given inequations.

EXAMPLE 4.   *Solve the following system of inequations graphically.*

$$2x + y - 3 \geq 0$$
$$x - 2y + 1 \leq 0 \qquad (x \geq 0, y \geq 0).$$

SOLUTION.   The corresponding linear equations of the given linear inequations are:
$$2x + y - 3 = 0, x - 2y + 1 = 0, x = 0, y = 0.$$

Region represented by $2x + y - 3 \geq 0$. The line $2x + y - 3 = 0$ meets the x-axis at $A\left(\frac{3}{2}, 0\right)$ (put $y = 0$ in it) and y-axis at $B(0, 3)$ (put $x = 0$ in it). Join these points by a thick line. Clearly, $(0, 0)$ does not satisfy the inequation $2x + y - 3 \geq 0$. So, portion not containing the origin is represented by the given inequation $2x + y - 3 \geq 0$.

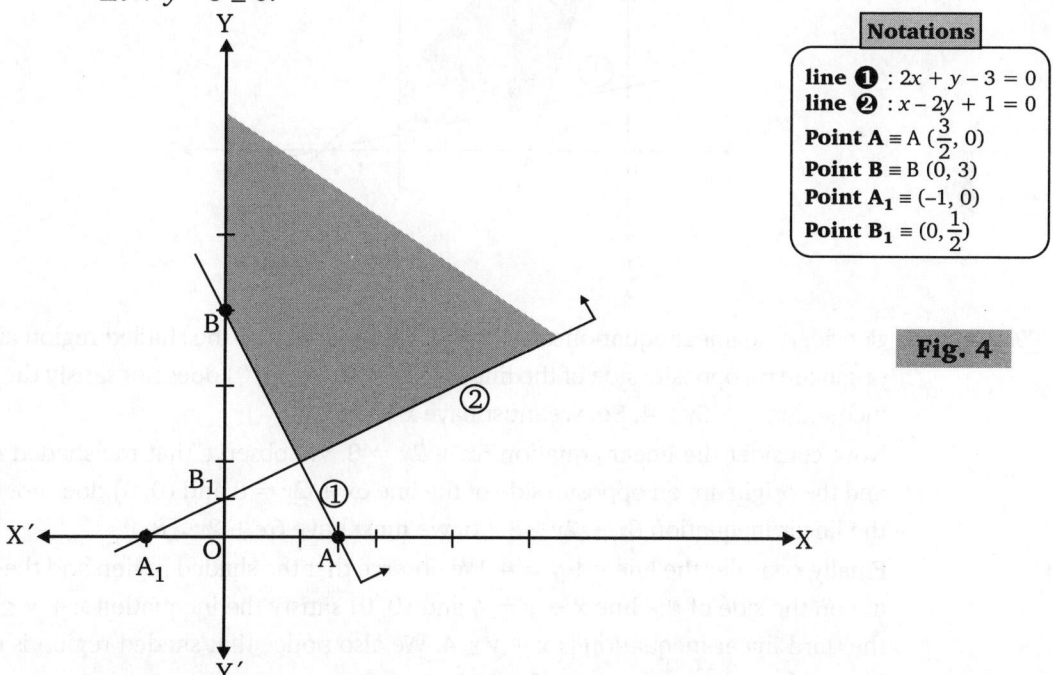

Notations

line ❶ : $2x + y - 3 = 0$
line ❷ : $x - 2y + 1 = 0$
Point A ≡ A $\left(\frac{3}{2}, 0\right)$
Point B ≡ B $(0, 3)$
Point A_1 ≡ $(-1, 0)$
Point B_1 ≡ $\left(0, \frac{1}{2}\right)$

Fig. 4

Region represented by $x - 2y + 1 \leq 0$. The line $x - 2y + 1 = 0$ meets the x-axis at $A_1(-1, 0)$ and y-axis at $B_1\left(0, \frac{1}{2}\right)$. Join these points by a thick line. Clearly,

(0, 0) does not satisfy the inequation $x - 2y + 1 \le 0$. So, portion not containing the origin is represented by the given inequation $x - 2y + 1 \le 0$.

Region represented by $x \ge 0, y \ge 0$. Clearly, $x \ge 0, y \ge 0$ represents the first quadrant.

Hence, the shaded region given in figure represents the solution set of the given linear inequations.

Type 2. Based on Finding the Linear Inequations, when their Solution Set is given

EXAMPLE 1. *Find the linear inequations for which the solution set is the shaded region given in figure.*

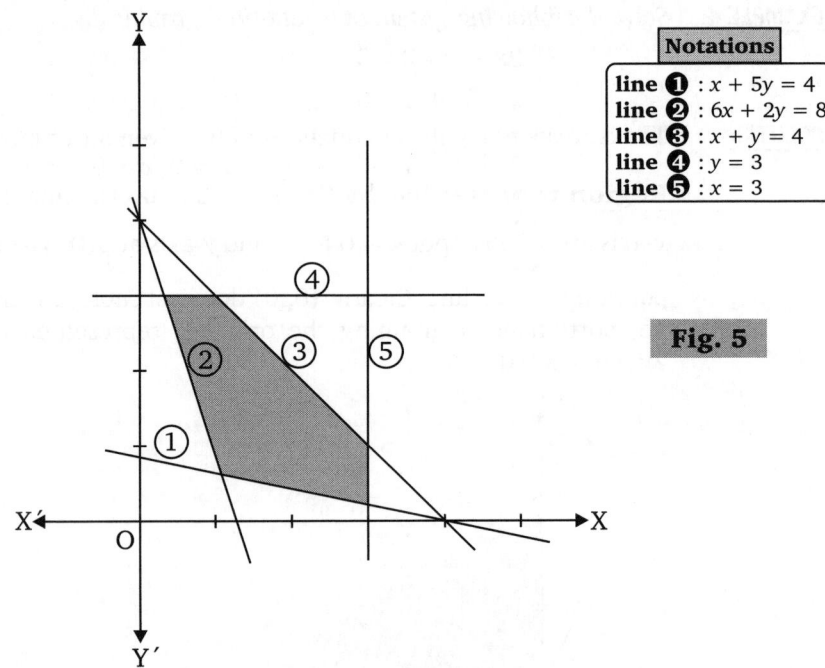

Notations

line ❶ : $x + 5y = 4$
line ❷ : $6x + 2y = 8$
line ❸ : $x + y = 4$
line ❹ : $y = 3$
line ❺ : $x = 3$

Fig. 5

SOLUTION. Consider the linear equation $x + 5y = 4$. We observe that the shaded region and the origin are on opposite side of the line $x + 5y = 4$ and (0, 0) does not satisfy the linear inequation $x + 5y \ge 4$. So, we must have $x + 5y \ge 4$.

Now, consider the linear equation $6x + 2y = 8$. We observe that the shaded region and the origin are on opposite side of the line $6x + 2y = 8$ and (0, 0) does not satisfy the linear inequation $6x + 2y \ge 8$. So, we must have $6x + 2y \ge 8$.

Finally, consider the line $x + y = 4$. We observe that the shaded region and the origin are on the side of the line $x + y = 4$ and (0, 0) satisfy the inequation $x + y \le 4$. So, the third linear inequation is $x + y \le 4$. We also notice that shaded region is on the same side of the origin so, $y \le 3$ and also $x \le 3$.

Thus, the linear inequations corresponding to the given solution set are:

$$x + 5y \ge 4, \ 6x + 2y \ge 8, \ x + y \le 4, x \le 3, y \le 3, x \ge 0, y \ge 0$$

EXAMPLE 2. *Find the linear inequations, for which the shaded region in the figure is the solution set.*

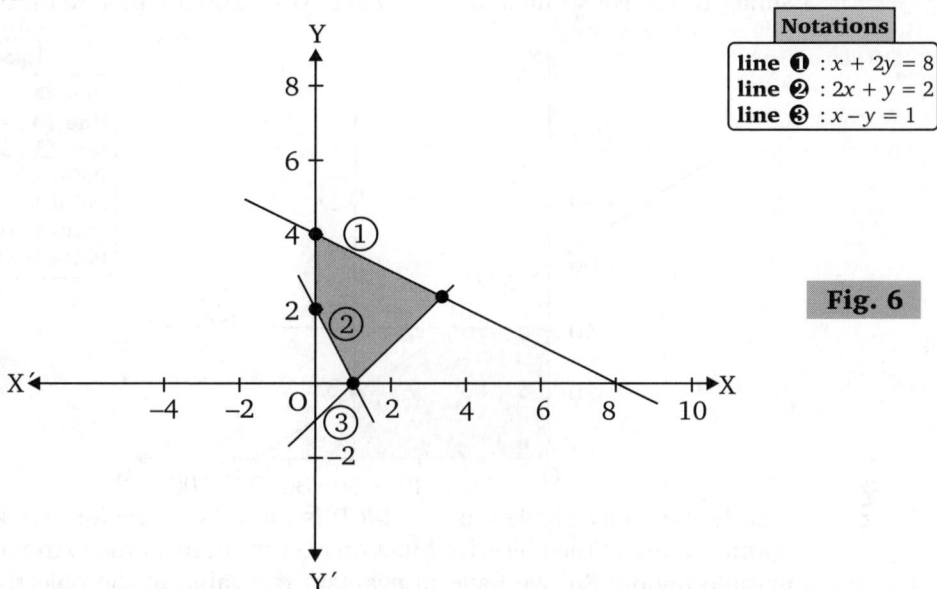

Notations

line ❶ : $x + 2y = 8$
line ❷ : $2x + y = 2$
line ❸ : $x - y = 1$

Fig. 6

SOLUTION. Consider the linear equation $x + 2y = 8$. We observe that the shaded region and the origin are on opposite side of the line $x + 2y = 8$ and $(0, 0)$ satisfy the linear inequation $x + 2y \le 8$. So, we must have one of the linear inequations as $x + 2y \le 8$. Now, consider the linear equation $2x + y = 2$. We find that shaded region and the origin are on opposite side of the line $2x + y = 2$ and $(0, 0)$ does not satisfy the inequation $2x + y \ge 2$. So, the second inequation is $2x + y \ge 2$. Finally, consider the line $x - y = 1$. We observe that the shaded region and the origin are on the side of the line $x - y = 1$ and $(0, 0)$ satisfy the inequation $x - y \le 1$. So, the third inequation is $x - y \le 1$. We also observe that the shaded region is above x-axis and is on the right side of y-axis. So, we have $x \le 0, y > 0$.

Thus, the linear inequations are

$$x + 2y \le 8, 2x + y \ge 2, x - y \le 1, x \le 0, y > 0$$

Type 3. Based on the Corner Point Method: Maximization Problems

EXAMPLE 1. *Solve graphically, the following LPP*

$$\text{Maximize } Z = 15x_1 + 10x_2$$

subject to the constraints

$$4x_1 + 6x_2 \le 360$$
$$3x_1 + 0x_2 \le 180$$
$$0x_1 + 5x_2 \le 200$$

and $x_1, x_2 \ge 0$

SOLUTION. Converting the given inequalities into equations. Now find any two points that satisfy the equation and then drawing the straight line through these two points. When $x_1 = 0$ we get $6x_2 = 360 \Rightarrow x_2 = 60$. Similarly, when $x_2 = 0, 4x_1 = 360$, *i.e.*, $x_1 = 90$. These two points are then connected by the straight line. It is also clear that any point below the line satisfy $4x_1 + 6x_2 \leq 360$.

Similarly, the constraints $3x_1 \leq 180$ and $5x_2 \leq 200$ are plotted on the graph.

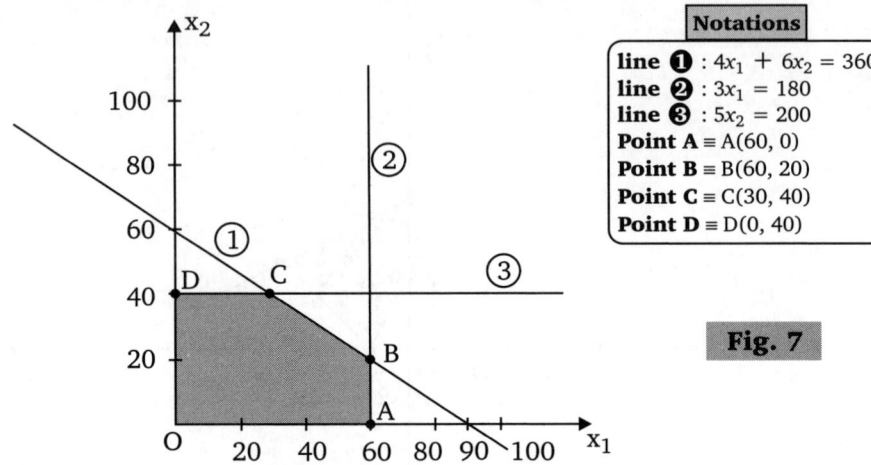

Notations

line ❶ : $4x_1 + 6x_2 = 360$
line ❷ : $3x_1 = 180$
line ❸ : $5x_2 = 200$
Point A $\equiv A(60, 0)$
Point B $\equiv B(60, 20)$
Point C $\equiv C(30, 40)$
Point D $\equiv D(0, 40)$

Fig. 7

Clearly the above shaded area $OABCD$ is the feasible region. We know that the optimal value of the objective function occurs at one of the extreme points of the feasible region. So, we have to evaluate the value of the objective function at each extreme point of the feasible region as shown below:

Extreme point	Coordinates (x_1, x_2)	Value of the objective function $Z = 15x_1 + 10x_2$
O	(0, 0)	$15 \times 0 + 10 \times 0 = 0$
A	(60, 0)	$15 \times 60 + 10 \times 0 = 900$
B	(60, 20)	$15 \times 60 + 10 \times 20 = 1100$
C	(30, 40)	$15 \times 30 + 10 \times 40 = 850$
D	(0, 40)	$15 \times 0 + 10 \times 40 = 400$

It is clear from the above table that Z is maximum at $B(60, 20)$ and its maximum value is 1100. Hence, optimal solution to the given LPP is $x_1 = 60, x_2 = 20$ and max $Z = 1100$.

EXAMPLE 2. *Solve the following LPP by graphical method*

Maximize $Z = 4x_1 + x_2$

subject to the constraints

$x_1 + x_2 \leq 50$

$3x_1 + x_2 \leq 90$

$x_1, x_2 \geq 0$

SOLUTION.STEP 1. First consider the constraints as equations, *i.e.,*

$x_1 + x_2 = 50$

$3x_1 + x_2 = 90$

STEP 2. In order to draw $x_1 + x_2 = 50$, put $x_2 = 0$ we get $x_1 = 50$ and by putting $x_1 = 0$, we get $x_2 = 50$

$\Rightarrow$ The line $x_1 + x_2 = 50$ passes through the point (50, 0) and (0, 50).

Now, plot the points (50, 0) and (0, 50) on coordinate axes and joining them by a straight line.

Similarly, we draw other line $3x_1 + x_2 = 90$.

STEP 3. Since, $x_1 + x_2 \leq 50$ and $3x_1 + x_2 \leq 90$ therefore, the region below the line $x_1 + x_2 = 50$ in the positive quadrant is the possible region and also the region below the line $3x_1 + x_2 = 90$ in the positive quadrant is the possible region. So the common region denoted by the shaded region $OABC$ is the feasible region.

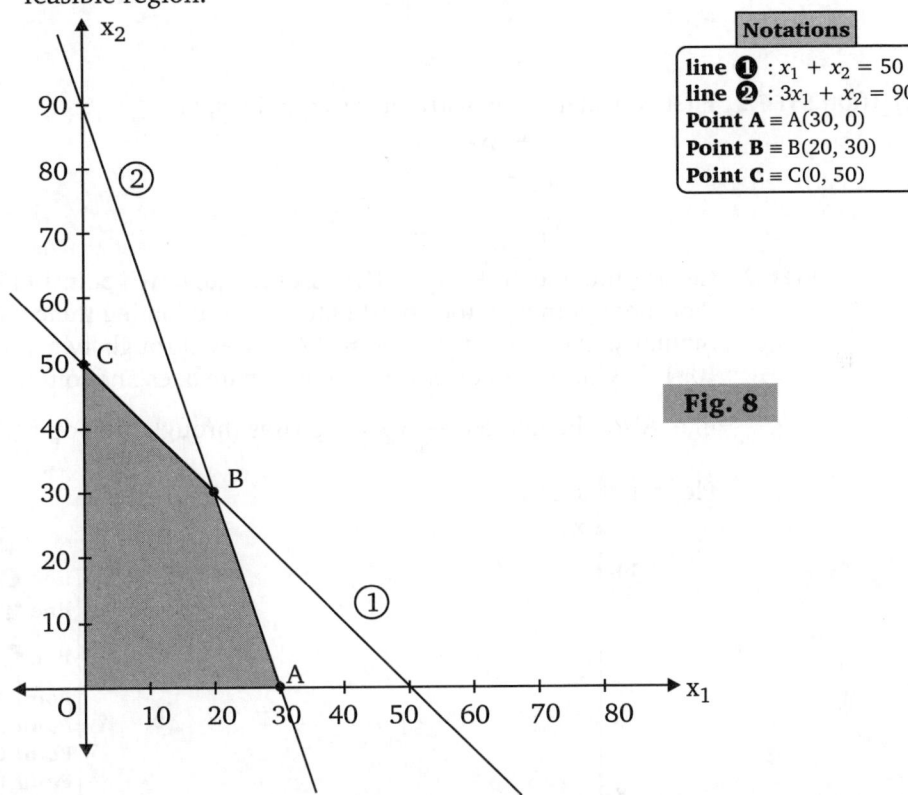

Notations

line ❶ : $x_1 + x_2 = 50$
line ❷ : $3x_1 + x_2 = 90$
Point A $\equiv$ A(30, 0)
Point B $\equiv$ B(20, 30)
Point C $\equiv$ C(0, 50)

Fig. 8

Clearly, the above shaded area $OABC$ is the feasible region. We know that the optimal value of the objective function occurs at one of the extreme point of the feasible region. So, we have to evaluate the value of the objective function at each extreme point of the feasible region as shown below:

Extreme point	Coordinates (x_1, x_2)	Value of the objective function $Z = 4x_1 + x_2$
O	(0, 0)	$4 \times 0 + 0 = 0$
A	(30, 0)	$4 \times 30 + 0 = 120$
B	(20, 30)	$4 \times 20 + 30 = 110$
C	(0, 50)	$4 \times 0 + 50 = 50$

It is clear from the above table that Z is maximum at $A(30, 0)$ and its maximum value is 120.

Hence, the optimal solution is given by $x_1 = 30, x_2 = 0$ and optimal value of Z is 120.

EXAMPLE 3. *Solve the following LPP by corner point method*

$$Maximize\ Z = 60x_1 + 40x_2$$

subject to the constraints

$$x_1 + 2x_2 \le 12$$
$$2x_1 + x_2 \le 12$$
$$x_1 + \frac{5}{4}x_2 \ge 5$$
$$x_1, x_2 \ge 0$$

SOLUTION. STEP 1. First consider the constraints as equations, *i.e.*,

$$x_1 + 2x_2 = 12$$
$$2x_1 + x_2 = 12$$
$$x_1 + \frac{5}{4}x_2 = 5$$

STEP 2. Clearly, the line $x_1 + 2x_2 = 12$ passes through two points (12, 0) and (0, 6). Plot these points on the coordinate axes and joining them by a straight line. Similarly, the line $2x_1 + x_2 = 12$ passes through two points (6, 0) and (0, 12). Plot these points on the coordinate axes and join them by a straight line. Also, the line $x_1 + \frac{5}{4}x_2 = 5$ passes through the point (5, 0) and (0, 4).

Plot this line also.

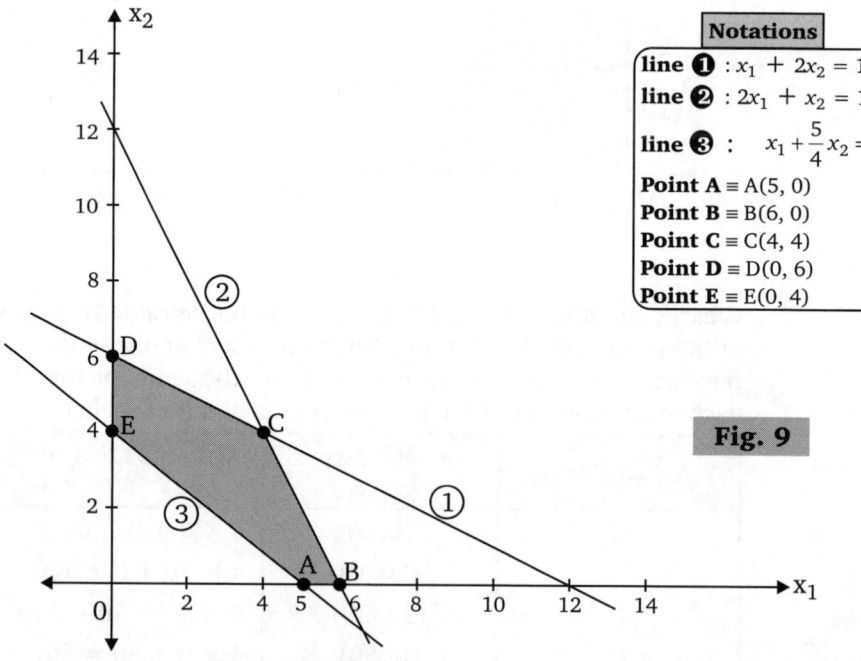

Notations

line ❶ : $x_1 + 2x_2 = 12$
line ❷ : $2x_1 + x_2 = 12$
line ❸ : $x_1 + \frac{5}{4}x_2 = 5$
Point **A** ≡ A(5, 0)
Point **B** ≡ B(6, 0)
Point **C** ≡ C(4, 4)
Point **D** ≡ D(0, 6)
Point **E** ≡ E(0, 4)

Fig. 9

Clearly, the above shaded area *ABCDE* is the feasible region and optimal value of the objective function at one of the extreme points of the feasible region. So, we have to evaluate the value of the objective function at each extreme point of the feasible region as shown below:

Extreme point	Coordinates (x_1, x_2)	Value of the objective function $Z = 60x_1 + 40x_2$
A	(5, 0)	$60 \times 5 + 40 \times 0 = 300$
B	(6, 0)	$60 \times 6 + 40 \times 0 = 360$
C	(4, 4)	$60 \times 4 + 40 \times 4 = 400$
D	(0, 6)	$60 \times 0 + 40 \times 6 = 240$
E	(0, 4)	$60 \times 0 + 40 \times 4 = 160$

It is clear from the above table that Z is maximum at $C(4, 4)$. Hence, the optimal solution is given by $x_1 = 4$, $x_2 = 4$ and optimal value of Z is 400.

Type 4. Based on Corner Point Method: Minimization Problem:

EXAMPLE 4. *Solve the following LPP by corner point method:*

$$\text{Minimize } Z = 4x_1 + 3x_2$$
subject to the constraints
$$200x_1 + 100x_2 \geq 4000$$
$$x_1 + 2x_2 \geq 50$$
$$40x_1 + 40x_2 \geq 1400$$
$$x_1, x_2 \geq 0$$

SOLUTION. STEP 1. Consider the constraints as equations, *i.e.*,
$$200x_1 + 100x_2 = 4000$$
$$x_1 + 2x_2 = 50$$
$$40x_1 + 40x_2 = 1400$$

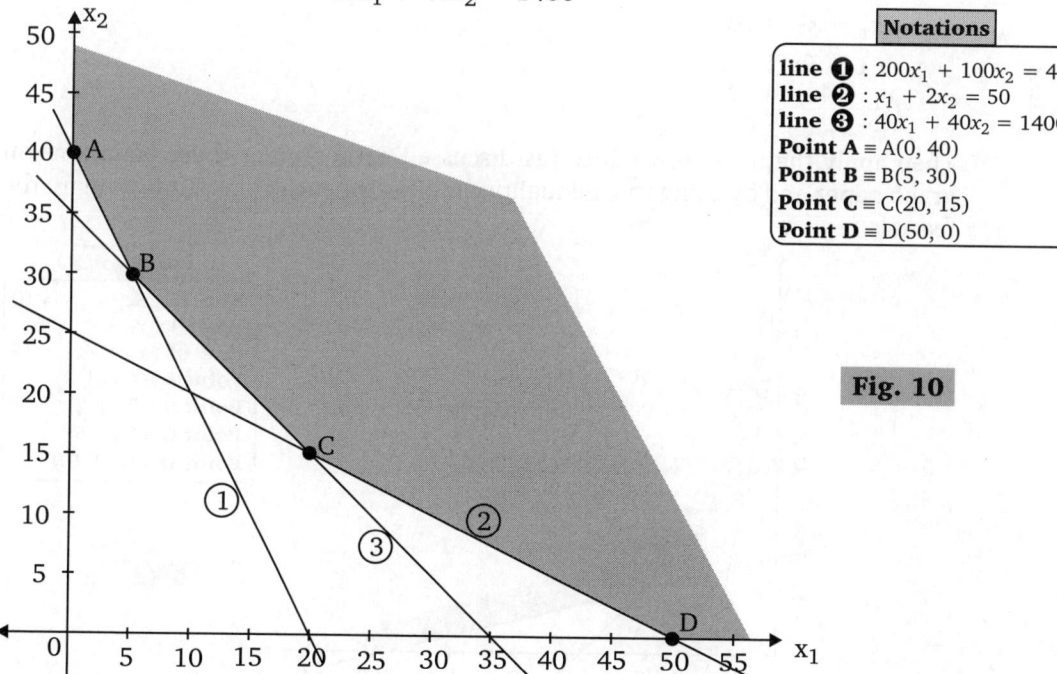

Notations

line ❶ : $200x_1 + 100x_2 = 4000$
line ❷ : $x_1 + 2x_2 = 50$
line ❸ : $40x_1 + 40x_2 = 1400$
Point A $\equiv$ A(0, 40)
Point B $\equiv$ B(5, 30)
Point C $\equiv$ C(20, 15)
Point D $\equiv$ D(50, 0)

Fig. 10

STEP 2. The line $200x_1 + 100x_2 = 4000$ passes through the point $(20, 0)$ and $(0, 40)$ and the line $x_1 + 2x_2 = 50$ passes through the point $(50, 0)$ and $(0, 25)$ while the line $40x_1 + 40x_2 = 1400$ passes through the points $(35, 0)$ and $(0, 35)$. Plot these points on the coordinate axes and join them by a straight line as shown in the adjoining graph.

Clearly, the above shaded area in the positive quadrant is the feasible region. Now, we have to evaluate the value of the objective function at each extreme point of the feasible region.

Extreme point	Coordinates (x_1, x_2)	Value of the objective function $Z = 4x_1 + 3x_2$
A	(0, 40)	$4 \times 0 + 3 \times 40 = 120$
B	(5, 30)	$4 \times 5 + 3 \times 30 = 110$
C	(20, 15)	$4 \times 20 + 3 \times 15 = 125$
D	(50, 0)	$4 \times 50 + 3 \times 0 = 200$

Clearly the point B give the minimum value of the objective function. Hence, the optimal solution of the given LPP is $x_1 = 5$ and $x_2 = 30$ and optimal value of $Z = 110$.

EXAMPLE 5. *Solve the following LPP by corner-point method.*

$$\text{Minimize } Z = 3x_1 + 2x_2$$
subject to the constraints
$$5x_1 + x_2 \geq 10, \ x_1 + x_2 \geq 6,$$
$$x_1 + 4x_2 \geq 12, \ x_1, x_2 \geq 0$$

SOLUTION. Consider the constraints as equations
$$5x_1 + x_2 = 10$$
$$x_1 + x_2 = 6$$
$$x_1 + 4x_2 = 12$$
$$x_1, x_2 \geq 0$$

Then apply the usual procedure (as discussed earlier). Plot these equations on graph paper and by using the inequality condition of each constraints to mark the feasible region.

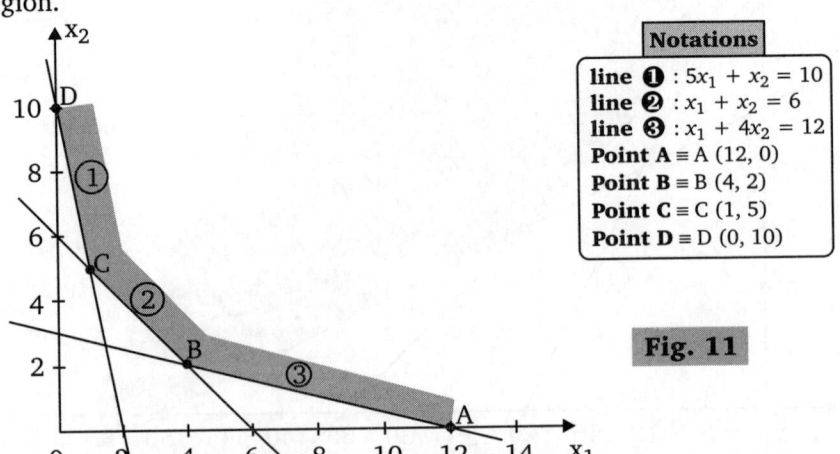

Notations

line ❶ : $5x_1 + x_2 = 10$
line ❷ : $x_1 + x_2 = 6$
line ❸ : $x_1 + 4x_2 = 12$
Point A ≡ A (12, 0)
Point B ≡ B (4, 2)
Point C ≡ C (1, 5)
Point D ≡ D (0, 10)

Fig. 11

Clearly, the above shaded region is the feasible region. Now, we have to evaluate the value of the objective function at each point (extreme) of the feasible region.

Extreme point	Coordinates (x_1, x_2)	Value of the objective function $Z = 3x_1 + 2x_2$
A	(12, 0)	$3 \times 12 + 2 \times 0 = 36$
B	(4, 2)	$3 \times 4 + 2 \times 2 = 16$
C	(1, 5)	$3 \times 1 + 2 \times 5 = 13$
D	(0, 10)	$3 \times 0 + 2 \times 10 = 20$

Clearly, the minimum value of the objective function $Z = 13$ occurs at the point $C(1, 5)$. Hence, the optimal solution to the given LPP is $x_1 = 1$, $x_2 = 5$ and minimum $Z = 13$.

EXAMPLE 6.   *Solve the following LPP by graphical method.*

$$\text{Minimize } Z = -x_1 + 2x_2$$

subject to the constraints

$$-x_1 + 3x_2 \le 10$$
$$x_1 + x_2 \le 6$$
$$x_1 - x_2 \le 2$$
$$x_1, x_2 \ge 0$$

SOLUTION.   First consider the constraints as equations and then plot each constraints on a graph paper. Use the inequality condition of each constraints to mark the feasible region as given below:

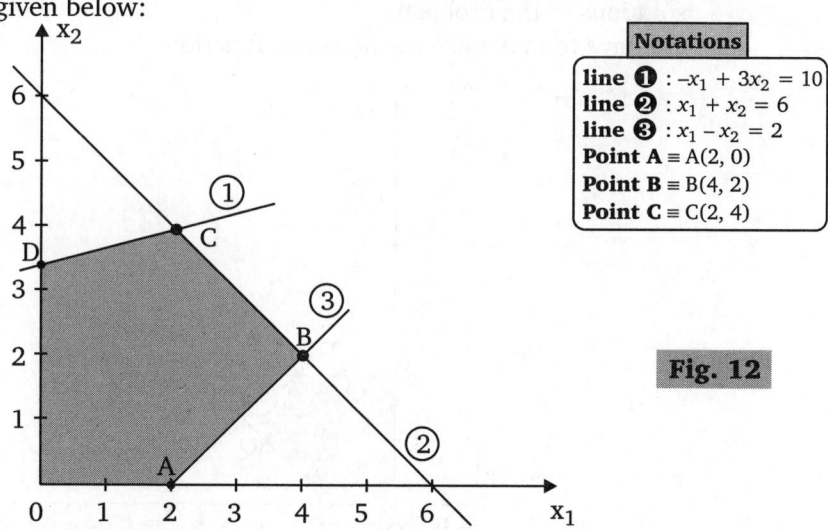

Notations

line ❶ : $-x_1 + 3x_2 = 10$
line ❷ : $x_1 + x_2 = 6$
line ❸ : $x_1 - x_2 = 2$
Point A ≡ A(2, 0)
Point B ≡ B(4, 2)
Point C ≡ C(2, 4)

Fig. 12

We observe that the area below the lines $-x_1 + 3x_2 = 10$ and $x_1 - x_2 = 2$ is not desirable, due to the reason that value of x_1 and x_2 are desired to be negative (*i.e.*, $x_1 \ge 0, x_2 \ge 0$)

Now, we have to calculate the value of the objective function at each extreme point of the above feasible region.

Extreme point	Coordinates (x_1, x_2)	Value of the objective function $Z = -x_1 + 2x_2$
O	$(0, 0)$	$-1 \times 0 + 2 \times 0 = 0$
A	$(2, 0)$	$-1 \times 2 + 2 \times 0 = -2$
B	$(4, 2)$	$-1 \times 4 + 2 \times 2 = 0$
C	$(2, 4)$	$-1 \times 2 + 2 \times 4 = 6$
D	$\left(0, \dfrac{10}{3}\right)$	$-1 \times 0 + 2 \times \dfrac{10}{3} = \dfrac{20}{3}$

Clearly, the minimum value of the objective function $Z = -2$ occur at the extreme point $A(2, 0)$. Hence, the optimal solution of the given LPP is : $x_1 = 2, x_2 = 0$ and min $Z = -2$.

Type 5. Based on ISO-Profit Method: Maximization Problem

EXAMPLE 1. *Using, iso-profit method, find the maximum value of the function $Z = 2x_1 + 3x_2$ subject to the constraints*

$$2x_1 + x_2 \leq 5$$
$$x_1 - x_2 \leq 1$$
$$x_2 \leq 2$$

and $\qquad x_1 \geq 0, x_2 \geq 0$

SOLUTION. Converting the given condition into equations and drawing these lines, the permissible region is $OPQRSO$ (shaded region) which is the set of all feasible solutions of the problem.

We have to maximize the objective function

$$Z = 2x_1 + 3x_2$$

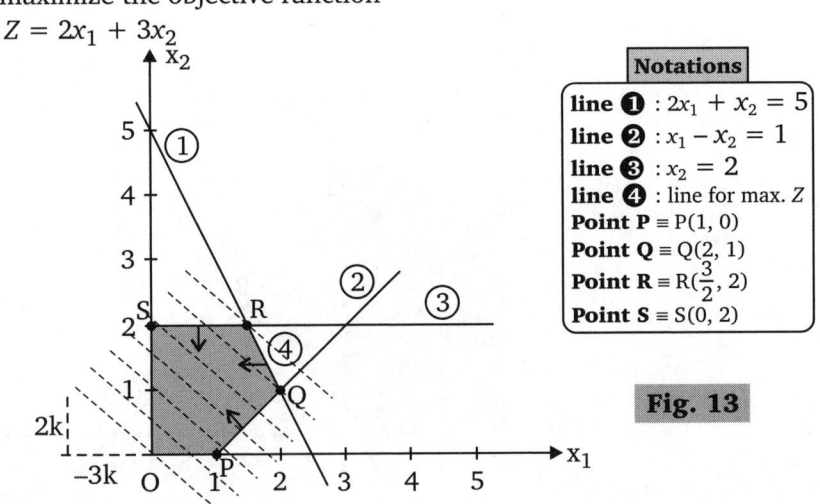

Notations

line ❶ : $2x_1 + x_2 = 5$
line ❷ : $x_1 - x_2 = 1$
line ❸ : $x_2 = 2$
line ❹ : line for max. Z
Point P ≡ P(1, 0)
Point Q ≡ Q(2, 1)
Point R ≡ R($\frac{3}{2}$, 2)
Point S ≡ S(0, 2)

Fig. 13

Taking $Z = 0$, we get $\dfrac{x_1}{x_2} = -\dfrac{3}{2}$

The line corresponding to $Z = 0$ is shown by dotted line through the origin O which is parallel to the iso-profit line. Now draw dotted parallel lines away from the origin. The farthest line from the origin passes through $R\left(\dfrac{3}{2}, 2\right)$.

Since, the farthest line from the origin O passes through the vertex $R\left(\dfrac{3}{2}, 2\right)$.

Hence the required optimal solution is given by $x_1 = \dfrac{3}{2}, x_2 = 2$ and

$$\text{max. } Z = 2 \times \dfrac{3}{2} + 3 \times 2 = 9.$$

EXAMPLE 2. *Solve the following L.P.P. graphically using Iso-profit method.*

$$\text{Maximize } Z = 2x_1 + x_2$$

Subject to the constraints

$$5x_1 + 10x_2 \le 50$$
$$x_1 + x_2 \ge 1$$
$$x_2 \le 4$$
$$x_1 - x_2 \le 0$$

and $\quad x_1 \ge 0, x_2 \ge 0$

SOLUTION. To solve the above L.P.P. we proceed as follows:

STEP 1. Making all the constraints as equations:

$$5x_1 + 10x_2 = 50$$
$$x_1 + x_2 = 1$$
$$x_2 = 4$$
$$x_1 - x_2 = 0$$

STEP 2. Draw these lines on the $x_1 x_2$ - plane as follows:

The line $5x_1 + 10x_2 = 50$ passes through the two points (10, 0) and (0, 5). Plot these points on the co-ordinate axes i.e., x_1- axis and x_2- axis respectively and join them by a straight line.

The line $x_1 + x_2 = 1$ passes through the points (1, 0) and (0, 1). Plot these point on the x_1- axis and x_2- axis respectively and join them by a straight line.

Similarly, we draw the lines $x_2 = 4$ *and* $x_1 - x_2 = 0$.

STEP 3. Since $5x_1 + 10x_2 \le 50, x_1 + x_2 \ge 1, x_2 \le 4, x_1 - x_2 \le 0$. So the regions below the lines $5x_1 + 10x_2 = 50$ in the positive regions, and the regions above the line $x_1 + x_2 = 1$ and $x_1 - x_2 = 0$ in the positive quadrant is the possible region.

Now the region common to all above regions is the feasible region, which is as shown in following figure by the shaded region *ABCDE*.

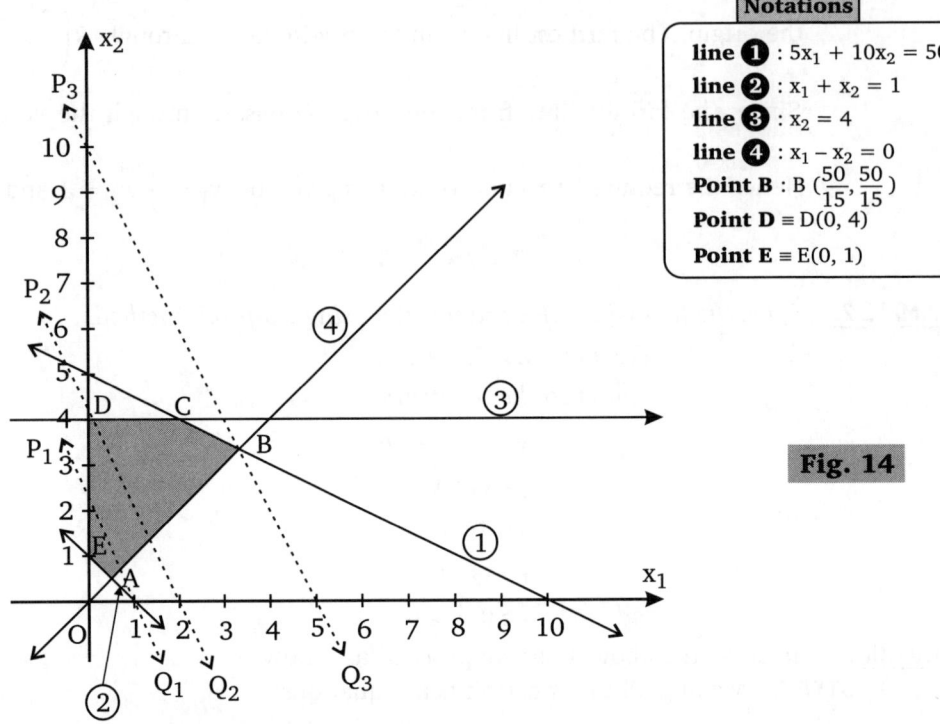

Notations

line ❶ : $5x_1 + 10x_2 = 50$
line ❷ : $x_1 + x_2 = 1$
line ❸ : $x_2 = 4$
line ❹ : $x_1 - x_2 = 0$
Point B : B $(\frac{50}{15}, \frac{50}{15})$
Point D ≡ D(0, 4)
Point E ≡ E(0, 1)

Fig. 14

STEP 4. Iso-profit Method:

(a) Assign a constant value $Z_1 = 2$ (L.C.M. of 2 and 1) to the objective function Z, we get a line

$$2x_1 + x_2 = 2 \qquad \qquad ...(1)$$

Now draw this line (1) which is as shown in figure by P_1Q_1.

(b) Again assign a constant value $Z_2 = 4$ to the objective function Z, we get a line

$$2x_1 + x_2 = 4 \qquad \qquad ...(2)$$

Now draw this line (2) which is shown in figure by P_2Q_2, clearly both the lines given by eqns. (1) and (2) and parallel to each other.

Since $Z_1 < Z_2$ and the L.P.P. is of maximization, so we move the line P_1Q_1

from P_1Q_1 to P_2Q_2 parallel to itself as far as possible, until the farthest point

$B\left(\frac{50}{15}, \frac{50}{15}\right)$within the feasible region *ABCDE* is touched by the line. Any

further displacement of the line takes it out of the feasible region *ABCDE*.

Thus, the point $B\left(\frac{50}{15}, \frac{50}{15}\right)$obtained on solving the lines $x_1 - x_2 = 0$ and

$5x_1 + 10x_2 = 50$, gives the maximum value of the objective function Z, and

the line P_3Q_3 is the line of maximum objective function.

Hence,   the optimal solution is $x_1 = \frac{50}{15}, x_2 = \frac{50}{15}$

and   the optimal value of $Z = 2 \times \frac{50}{15} + \frac{50}{15} = 10$

EXAMPLE 3. *Solve the following L.P.P. by graphical method:*

$$Maximize \ Z = 3x_1 + 5x_2$$
subject to the constraints

$$x_1 + x_2 \le 1500$$
$$x_1 + 2x_2 \le 2000$$
$$x_2 \le 600$$

and $\quad x_1 \ge 0, x_2 \ge 0$

SOLUTION. STEP 1. Making the given constraints as equations:

$$x_1 + x_2 = 1500$$
$$x_1 + 2x_2 = 2000$$
$$x_2 = 600$$

STEP 2. The line $x_1 + x_2 = 1500$ passes through the two points (1500, 0) and (0, 1500). Plot the points (1500, 0) and (0, 1500) on x_1- axis and x_2- axis respectively and joining them by a straight line.

The line $x_1 + 2x_2 = 2000$ passes through the points (2000, 0) and (0, 1000). Plot these point on x_1- axis and x_2- axis and joining them by a straight line. The line $x_2 = 600$ is parallel to the x_1- axis and passing through the point (0, 600).

STEP 3. Since we have,

$$x_1 + x_2 \le 1500, x_1 + 2x_2 \le 2000, x_2 \le 600 \text{ and } x_1 \ge 0, x_2 \ge 0$$

Therefore, the region below the line $x_1 + x_2 = 1500$ in the positive quadrant is the possible region, the region below the line $x_1 + 2x_2 = 2000$ in the possible region, the region below the line $x_2 = 600$ in the positive quadrant is the possible region.

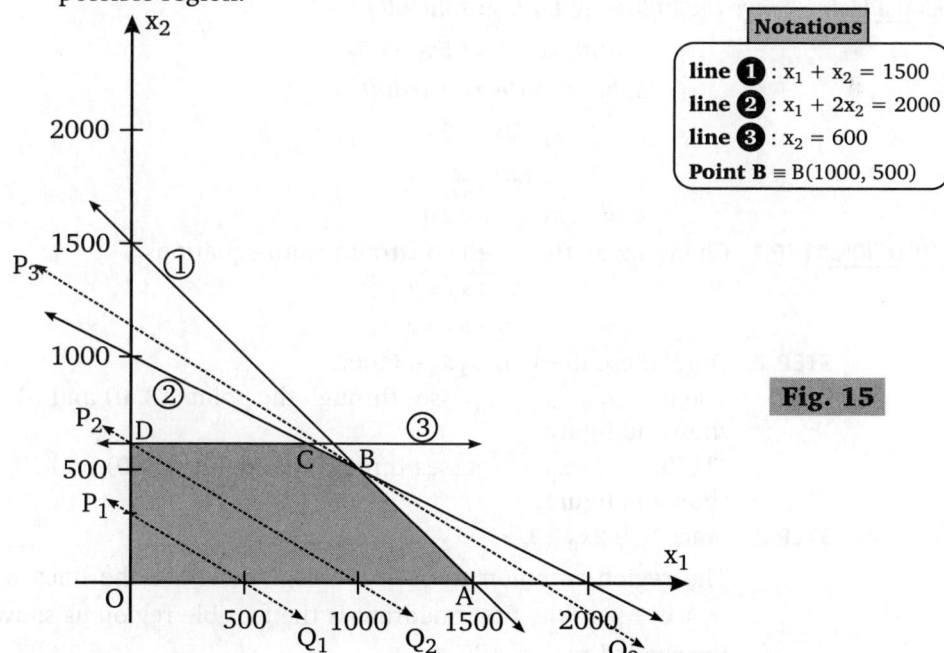

Notations

line ① : $x_1 + x_2 = 1500$
line ② : $x_1 + 2x_2 = 2000$
line ③ : $x_2 = 600$
Point B $\equiv$ B(1000, 500)

Fig. 15

Thus, the region common to above three regions in the positive quadrant is the feasible region, which is as shown in figure by the shaded region *OABCD*.

STEP 4. Iso-profit Method:

(a) Assign a constant $Z_1 = 1500$ to the objective function Z, we get a line
$$3x_1 + 5x_2 = 1500 \qquad \qquad ...(1)$$

Now, draw this line (1), which is as shown in figure by P_1Q_1.

(b) Again assign a constant value $Z_1 = 3000$ to the objective function Z, we get a line
$$3x_1 + 5x_2 = 3000 \qquad \qquad ...(2)$$

Now draw this line (2), which is as shown in figure by P_2Q_2.

Clearly, both the lines P_1Q_1 and P_2Q_2 are parallel to each other.

Since $Z_1 < Z_2$ and the L.P.P is of maximization, so we move the line P_1Q_1 from P_1Q_1 to P_2Q_2 parallel to itself as far as possible, until the farthest point B (1000, 500) within the feasible region *OABCD* is touched by the line. Any further displacement of the line takes it out of the feasible region *OABCD*.

Thus, the point B (1000, 500) is obtained on solving the equations $x_1 + x_2 = 1500$ and $x_1 + 2x_2 = 2000$, gives the maximum value of the objective function Z, and the line P_3Q_3 is the line of maximum objective function.

Hence, the optimal solution is
$$x_1 = 1000, x_2 = 500$$

and the optimal value of $Z = 3 \times 1000 + 5 \times 500 = 5500$

Type 6. Based on ISO-Profit Method: Minimization Problem

EXAMPLE 1. *Solve the following L.P.P. graphically.*

$$\text{Minimize: } Z = 1.5x_1 + 2.5x_2$$

subject to the constraints

$$x_1 + 3x_2 \geq 3$$
$$x_1 + x_2 \geq 2$$
$$and \quad x_1 \geq 0, x_2 \geq 0$$

SOLUTION. STEP 1. Changing all the given constraints into equations:
$$x_1 + 3x_2 = 3$$
$$x_1 + x_2 = 2$$

STEP 2. Draw these lines on $x_1 x_2$-plane.

The line $x_1 + 3x_2 = 3$ passes through the points (3, 0) and (0, 1), which is as shown in figure.

The line $x_1 + x_2 = 2$ passes through the points (2, 0) and (0, 2) which is as shown in figure.

STEP 3. Since $x_1 + 3x_2 \geq 3$, $x_1 + x_2 \geq 2$ and $x_1 \geq 0, x_2 \geq 0$

The region common to both the regions above the lines $x_1 + 3x_2 = 3$ and $x_1 + x_2 = 2$ in the first quadrant is the feasible region as shown in figure by the shaded region *ABC*.

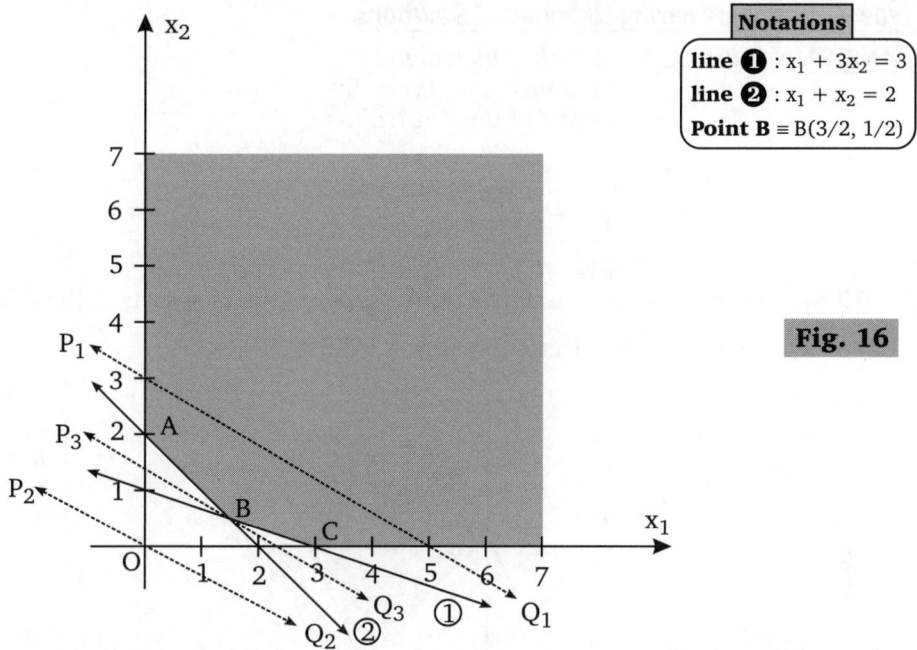

Notations
line **1** : $x_1 + 3x_2 = 3$
line **2** : $x_1 + x_2 = 2$
Point **B** $\equiv$ B(3/2, 1/2)

Fig. 16

STEP 4. Iso-profit Method:

(a) Assign a constant value $Z_1 = 7.5$ (LCM of 1.5 and 2.5) to the objective function Z, we get a line

$$1.5x_1 + 2.5x_2 = 7.5 \qquad \ldots(1)$$

Now draw this line (1) on $x_1 x_2$-plane, which is as shown in figure by $P_1 Q_1$.

(b) Again assign a constant value $Z_2 = 10$ to the objective function Z, we get a line

$$1.5x_1 + 2.5x_2 = 10 \qquad \ldots(2)$$

Now draw this line (2) on $x_1 x_2$-plane which is shown in Fig. 1.16 by $P_2 Q_2$.

Since, $Z_2 > Z_1$ so we move the line $P_1 Q_1$ from $P_1 Q_1$ to $P_2 Q_2$ parallel to itself, until a point B of the feasible region ABC is touched by the line. The point B is the intersection of the lines $x_1 + x_2 = 2$ and $x_1 + 3x_2 = 3$, so on solving these

equations, we obtain the co-ordinates of B as $\left(\dfrac{3}{2}, \dfrac{1}{2}\right)$.

Thus, the point $B\left(\dfrac{3}{2}, \dfrac{1}{2}\right)$ will give the minimum value of Z.

Hence, the optimum solution of L.P.P. is $x_1 = \dfrac{3}{2}$, $x_2 = \dfrac{1}{2}$ and the optimum value of

objective function $Z = 1.5 \times \dfrac{3}{2} + 2.5 \times \dfrac{1}{2} = 3.5$

Type 7. Problems having Unbounded Solutions

EXAMPLE 1. *Solve graphically the following LPP*

$$\text{Maximize } Z = 4x_1 + 5x_2$$

subject to the constraints

$$x_1 + x_2 \geq 1,$$
$$-2x_1 + x_2 \leq 1,$$
$$4x_1 - 2x_2 \leq 1,$$

and $\qquad x_1, x_2 \geq 0$

SOLUTION. Proceeding as usual, the permissible region is shown in the following graph by shaded region. Here, $Z = 4x_1 + 5x_2 \Rightarrow \dfrac{x_1}{x_2} = -\dfrac{5}{4}$

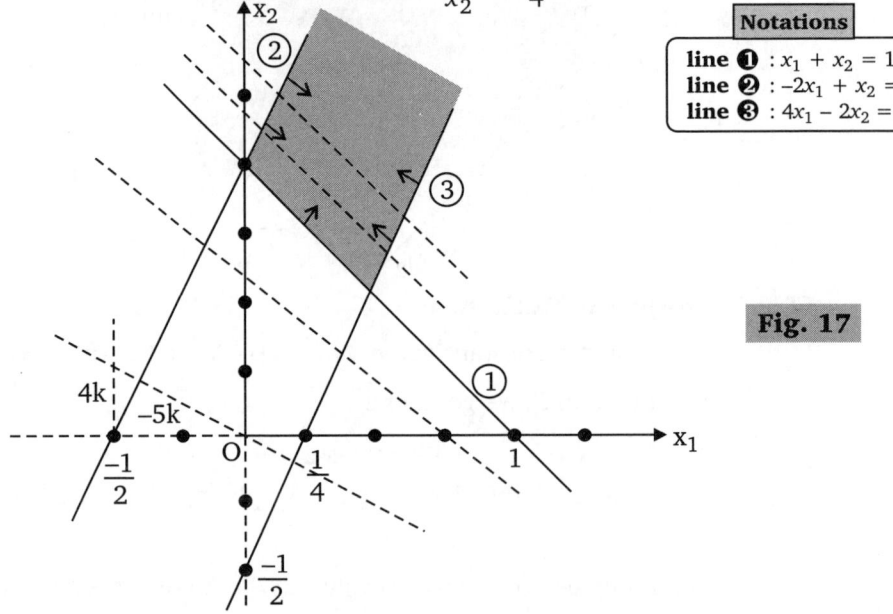

Notations

line ❶ : $x_1 + x_2 = 1$
line ❷ : $-2x_1 + x_2 = 1$
line ❸ : $4x_1 - 2x_2 = 1$

Fig. 17

From the above figure, it is clear that dotted line through the origin representing $Z = 0$ can be moved parallel to itself in the direction of Z increasing and still has some point in the permissible region. Also, $Z = 4x_1 + 5x_2 = 0 \Rightarrow \dfrac{x_1}{x_2} = -\dfrac{5}{4}$

Hence, Z can be made arbitrarily large and so the problem has no finite maximum value of Z. Thus problem has unbounded solution.

EXAMPLE 2. *Solve the following LPP graphically*

$$\text{Maximize } Z = 0.75x_1 + x_2$$

subject to the constraints

$$x_1 - x_2 \geq 0, \, -0.5x_1 + x_2 \leq 1 \text{ and } x_1, x_2 \geq 0$$

SOLUTION. Proceeding as usual, we get the permissible region which shaded in the following figure. Hence, $Z = 0 \Rightarrow \dfrac{x_1}{x_2} = -\dfrac{4}{3}$

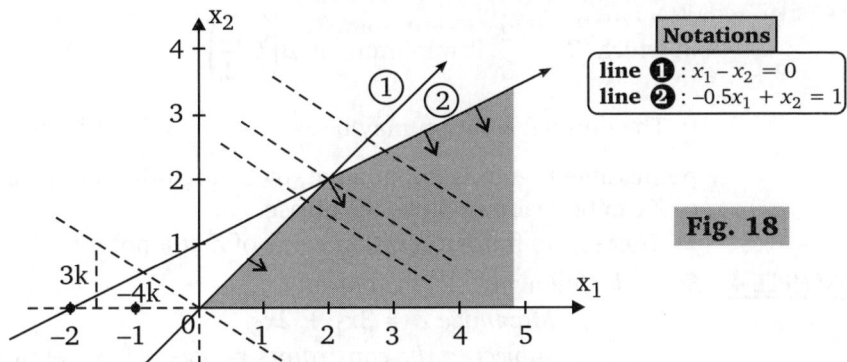

Notations

line **①** : $x_1 - x_2 = 0$
line **②** : $-0.5x_1 + x_2 = 1$

Fig. 18

From the above graph, it is clear that the dotted line passes through the origin representing $Z = 0$ can be moved parallel to itself in the direction of Z increasing and will have some points in the feasible region. Therefore, Z can be made arbitrarily large and hence the problem has no finite maximum value of Z. Thus the problem has unbounded solution.

EXAMPLE 3. *Solve the following LPP graphically:*

$$Minimize\ Z = 5x_1 - 2x_2$$

subject to the constraints

$$2x_1 + 3x_2 \geq 1$$

and $\quad x_1, x_2 \geq 0$

SOLUTION. Proceeding as usual, we get the permissible region, which is shaded in the figure given below.

Also, $Z = 0 \Rightarrow \dfrac{x_1}{x_2} = \dfrac{2}{5}$

The permissible region is the set of all the constraints and the non-negative restrictions, is unbounded. Solving simultaneously the equations of the corresponding intersecting lines we get two vertices $A\left(\dfrac{1}{2}, 0\right)$ and $B\left(0, \dfrac{1}{3}\right)$.

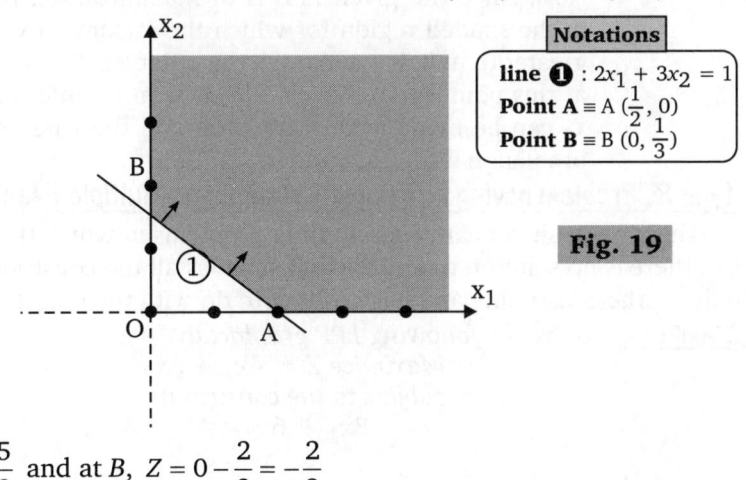

Notations

line **①** : $2x_1 + 3x_2 = 1$
Point **A** $\equiv$ A $\left(\dfrac{1}{2}, 0\right)$
Point **B** $\equiv$ B $\left(0, \dfrac{1}{3}\right)$

Fig. 19

At A, $Z = \dfrac{5}{2} - 0 = \dfrac{5}{2}$ and at B, $Z = 0 - \dfrac{2}{3} = -\dfrac{2}{3}$

in which $Z = -\dfrac{2}{3}$ is minimum at $B\left(0, \dfrac{1}{3}\right)$

$\therefore$ The optimal solution may be $x_1 = 0$, $x_2 = \dfrac{1}{3}$ and min $Z = -\dfrac{2}{3}$. But, since the permissible region is unbounded, by taking other points in the region, the value of Z can be made small as we please.

$\Rightarrow$ There is no finite minimum value of Z at a point in the permissible region.

EXAMPLE 4. *Solve the following LPP graphically*

$$\text{Maximize } Z = 3x_1 + 2x_2$$
$$\text{subject to the constraints } x_1 - x_2 \geq 1, x_1 + x_2 \geq 3$$
$$\text{and } x_1, x_2 \geq 0$$

SOLUTION. Apply the usual procedure, we plot the constraints on the graph and permissible region is shaded and is bounded from below and unbounded from above.

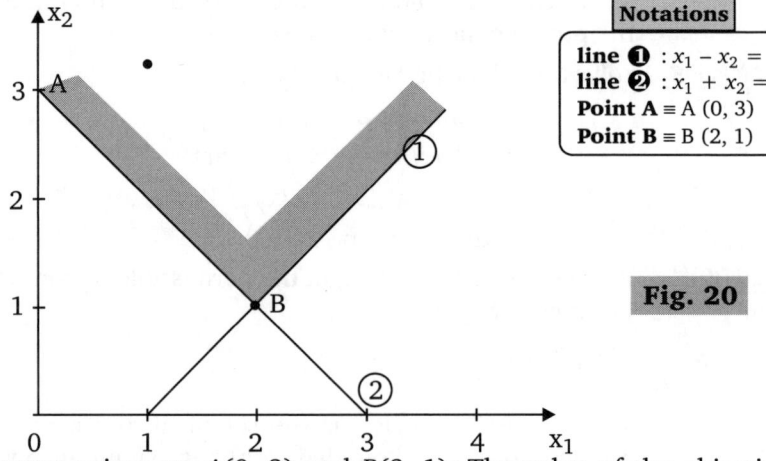

Notations
line ❶ : $x_1 - x_2 = 1$
line ❷ : $x_1 + x_2 = 3$
Point A $\equiv$ A $(0, 3)$
Point B $\equiv$ B $(2, 1)$

Fig. 20

Clearly, two corner points are $A(0, 3)$ and $B(2, 1)$. The value of the objective function at these points is 6 and 8 respectively.

Now, since the given LPP is of maximization, there exists a number of points in the shaded region for which the maximum value of the objective function is greater than 8. For example, the point $(2, 3)$ lies in the region and function value at this point is 12, which is clearly more than 8. So, both the variables x_1 and x_2 can be made arbitrarily large and the value of Z also increases. Hence, the problem has an unbounded solution.

Type 8. Problem having Infeasible Solutions or Multiple Feasible Region

There are some linear programming problems in which the constraints are not consistent, *i.e.*, there is no solution to an LPP that satisfies all the constraints. Here, infeasibility depends only on the constraints and has nothing to do with the objective function.

EXAMPLE 1. *Solve the following LPP graphically:*

$$\text{Maximize } Z = 4x_1 + 3x_2$$
$$\text{subject to the constraints}$$
$$8x_1 + 6x_2 \leq 48$$
$$x_1 \leq 6$$
$$\text{and} \quad x_1, x_2 \geq 0$$

SOLUTION. Draw the converted line on x_1x_2-plane by following the usual procedure. Clearly, the line $8x_1 + 6x_2 = 48$ passes through the point $(6, 0)$ and $(0, 8)$ and the line $x_1 = 6$ is parallel to x_2-axis and passes through the point $(6, 0)$. Also, the region below the line $8x_1 + 6x_2 = 48$ in the positive quadrant is the permissible region and the region left to the $x_1 = 6$ in the positive quadrant is the permissible region. The common to both regions is the feasible region which is shown in following figure by the shaded region OAB.

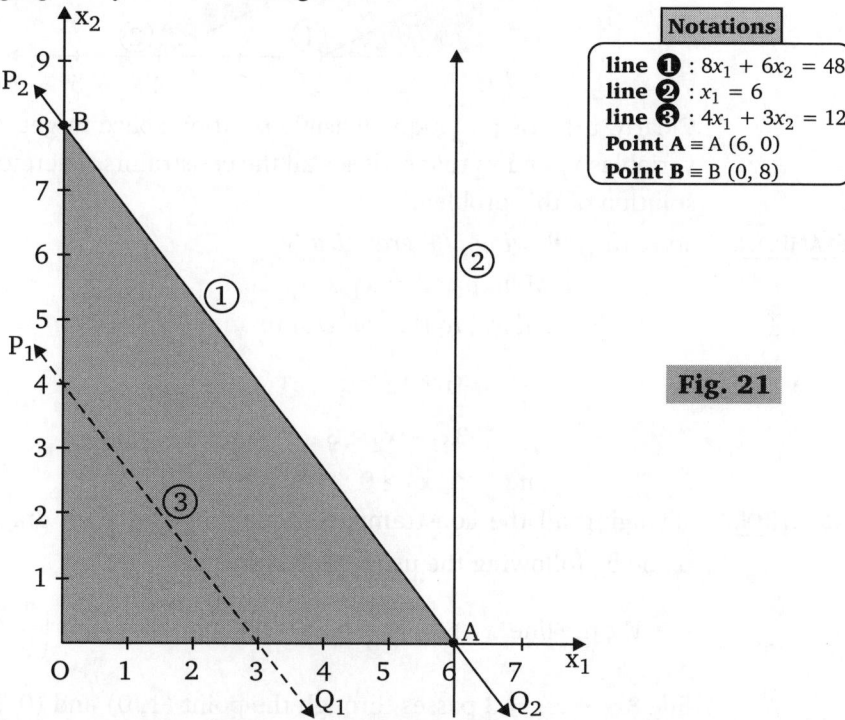

Notations
line ❶ : $8x_1 + 6x_2 = 48$
line ❷ : $x_1 = 6$
line ❸ : $4x_1 + 3x_2 = 12$
Point A ≡ A $(6, 0)$
Point B ≡ B $(0, 8)$

Fig. 21

Now, assign a constant value $Z_1 = 12$ to the objective function Z, we get a line $4x_1 + 3x_2 = 12$, which is shown in the above figure by P_1Q_1. Similarly, assign a constant value $Z_2 = 24$ to the objective function Z we may get a line $4x_1 + 3x_2 = 24$, shown in the above figure by the line P_2Q_2.

Now, since $Z_2 < Z_1$ and LPP is of maximization, thus we move P_1Q_1 towards P_2Q_2 parallel to itself, untill the further point of feasible region is bounded by the line. But the iso-profit line lies along the constraint $8x_1 + 6x_2 \leq 48$. Hence, the given LPP has multiple feasible solutions.

EXAMPLE 2. *Solve the following LPP graphically:*
Maximize $Z = 6x_1 - 4x_2$
subject to the constraints
$2x_1 + 4x_2 \leq 4$, $4x_1 + 8x_2 \geq 16$
and $x_1, x_2 \geq 0$

SOLUTION. Following the usual procedure, the constraints are plotted on the graph as follows:

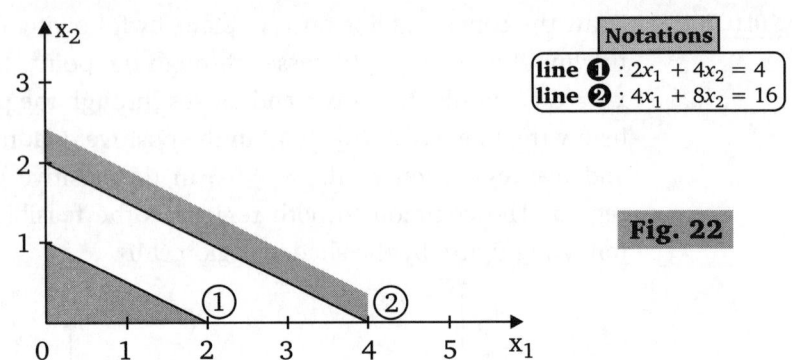

Fig. 22

Clearly, there is no unique feasible solution space to get unique set of values of variables x_1 and x_2 that satisfies all the constraints. Therefore, there is no feasible solution of this problem.

EXAMPLE 3. *Solve the following LPP graphically*

$$\text{Minimize } Z = x_1 + x_2$$

subject to the constraints

$$x_1 + x_2 \le \frac{1}{2}$$

$$3x_1 + x_2 \ge 3$$

and $x_1, x_2 \ge 0$

SOLUTION. Changing all the constraints to equations and then draw these lines on x_1x_2-plane by following the usual procedure.

Clearly, the line $x_1 + x_2 = \frac{1}{2}$ passes through the point $\left(\frac{1}{2}, 0\right)$ and $\left(0, \frac{1}{2}\right)$ and the line $3x_1 + x_2 = 3$ passes through the point $(1, 0)$ and $(0, 3)$.

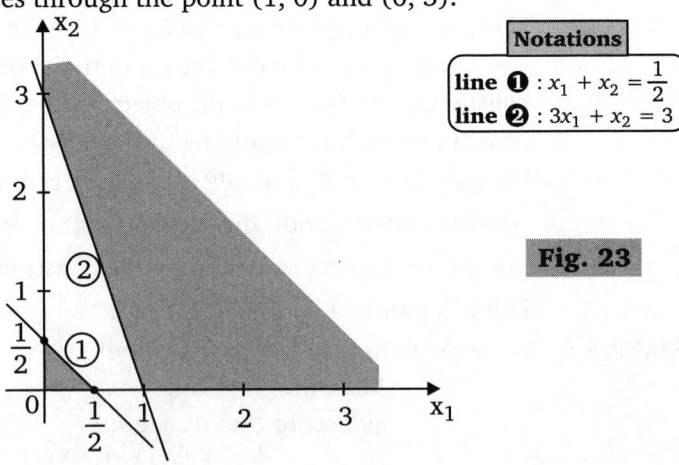

Fig. 23

Now, since we have $x_1 + x_2 \le \frac{1}{2}, 3x_1 + x_2 \ge 3$ and $x_1 \ge 0, x_2 \ge 0$, the region below the line $x_1 + x_2 = 1$ is the permissible region and the region above the line

$3x_1 + x_2 = 3$ is the permissible region in the positive quadrant. The solution spaces are shaded in the above figure, which shows that there is no point satisfying both the constraints. Hence the given LPP has infeasible solution.

EXAMPLE 4. *Solve the following LPP graphically*

$$\text{Minimize } Z = 5x_1 + 2x_2$$

subject to the constraints

$$x_1 + x_2 \le 2$$

$$3x_1 + 3x_2 \ge 12$$

and $\qquad x_1, x_2 \ge 0$

SOLUTION. Changing all the constraints to equations and draw them on $x_1 x_2$-plane. Clearly the line $x_1 + x_2 = 2$ passes through the point (2, 0) and (0, 2) and the line $3x_1 + 3x_2 = 12$ passes through the point (4, 0) and (0, 4).

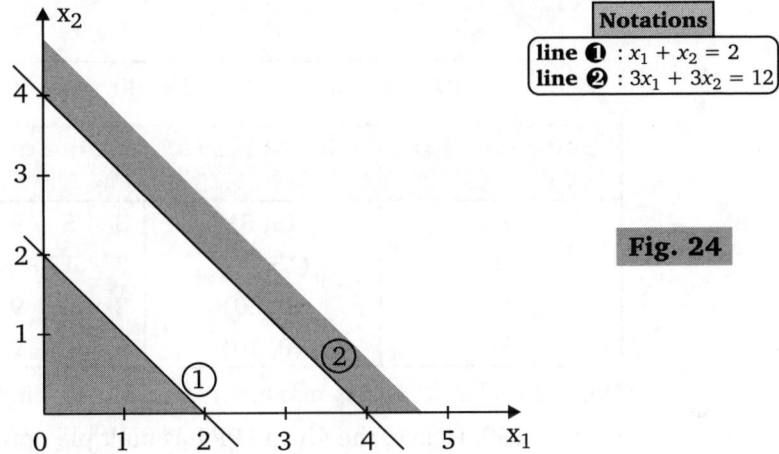

Notations
line ❶ : $x_1 + x_2 = 2$
line ❷ : $3x_1 + 3x_2 = 12$

Fig. 24

Clearly, the region below the line $x_1 + x_2 = 2$ and above the line $3x_1 + 3x_2 = 12$ is the permissible region. The solution spaces are shaded in the above figure. Clearly, there is no common part to both the region, so there is no point satisfying both the constraints. Hence, the given LPP has infeasible solutions.

EXAMPLE 5. *Solve the following LPP graphically*

$$\text{Minimize } Z = 3x_1 + 9x_2$$

subject to the constraints

$$x_1 + 3x_2 \le 60,$$

$$x_1 + x_2 \ge 10, x_1 \le x_2$$

and $\qquad x_1, x_2 \ge 0$

SOLUTION. Converting the given constraints into equations, then plot these lines on $x_1 x_2$-plane as shown below:

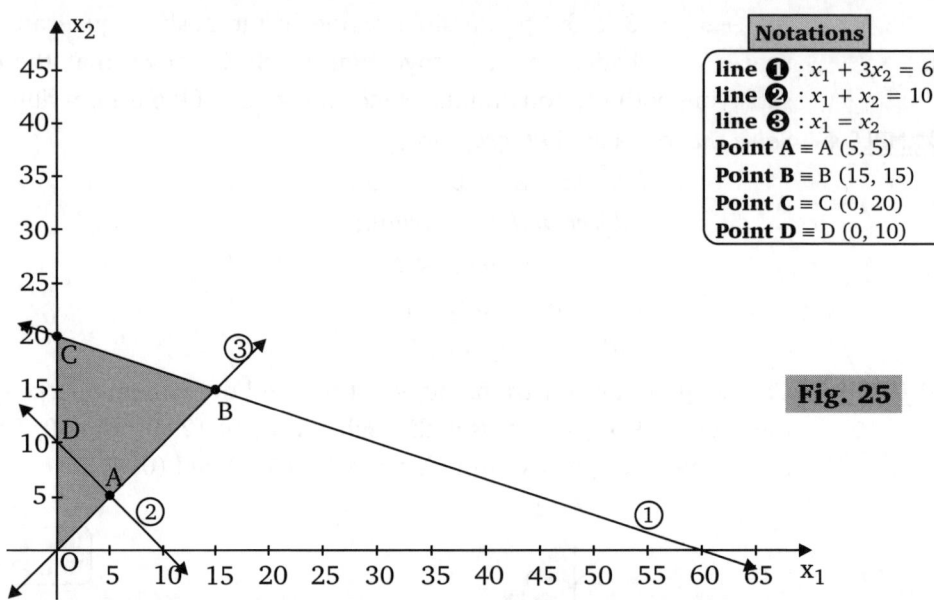

Fig. 25

Extreme point	Coordinates (x_1, x_2)	Value of the objective function $Z = 3x_1 + 9x_2$
A	(5, 5)	$3 \times 5 + 9 \times 5 = 60$
B	(15, 15)	$3 \times 15 + 9 \times 15 = 180$
C	(0, 20)	$3 \times 0 + 9 \times 20 = 180$
D	(0, 10)	$3 \times 0 + 9 \times 10 = 90$

Here, it is clear that Z is maximum at $B(15, 15)$ and $C(0, 20)$ and its maximum value is 180. Hence, the given LPP has multiple optimal solutions at B and C.

Type 9. More Problems on Mixed Constraints

A redundant constraint is one that does not affect the feasible solution space and thus redundancy of any constraint does not cause any difficulty in solving LPP graphically.

EXAMPLE 1. *Solve the following LPP graphically*

$$\text{Minimize } Z = x_1 + x_2$$
subject to the constraints
$$5x_1 + 10x_2 \leq 50$$
$$x_1 + x_2 \geq 1$$
$$x_2 \geq 4$$
and $$x_1, x_2 \geq 0$$

SOLUTION. Converting the given constraints into equations and drawing these lines, we have the following figure. Here, $Z = x_1 + x_2 = 0 \Rightarrow \dfrac{x_1}{x_2} = -\dfrac{1}{1}$.

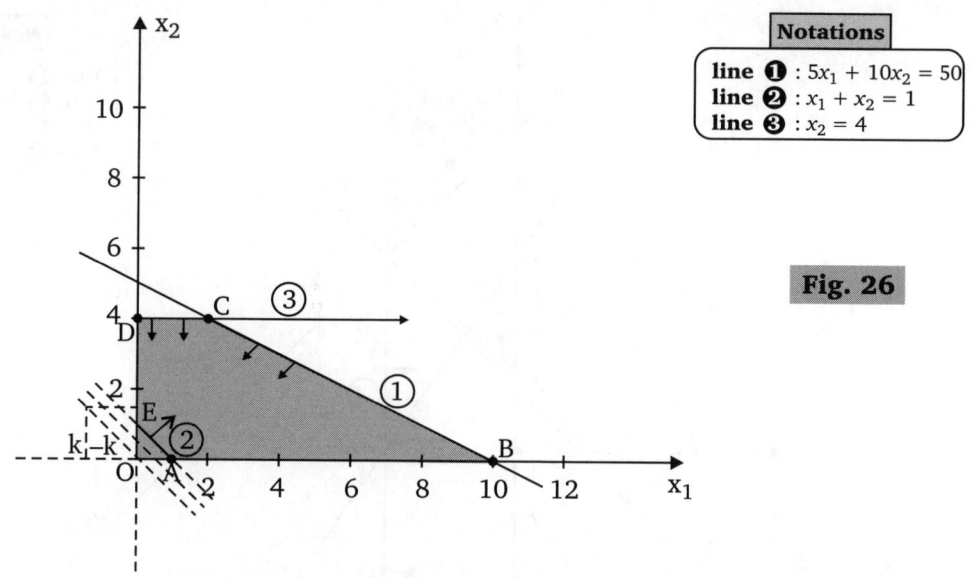

Notations
line ❶ : $5x_1 + 10x_2 = 50$
line ❷ : $x_1 + x_2 = 1$
line ❸ : $x_2 = 4$

Fig. 26

Clearly, the permissible region is *ABCDEOA* (shaded region) which is the set of all feasible solutions of the given LPP.

Taking $Z = x_1 + x_2 = 0 \Rightarrow \dfrac{x_1}{x_2} = -\dfrac{1}{1}$

The line corresponding to $Z = 0$ is shown by dotted line through the origin O. Drawing lines parallel to this line, we see that a parallel dotted line passing through the vertices $A(1, 0)$ and $E(0, 1)$ of the convex polygon, which is the nearest to the origin O.

$\Rightarrow$ Every point of the line segment AE gives the optimal solution of the problem. Hence, the given LPP has infinite number of solutions.

EXAMPLE 2. *Solve the following LPP graphically*

$$\text{Maximize } Z = x_1 + \frac{1}{2}x_2$$

subject to the constraints

$$3x_1 + 2x_2 \le 12$$
$$x_1 \le 2$$
$$x_1 + x_2 \ge 8$$
$$-x_1 + x_2 \ge 4$$

and $\qquad x_1, x_2 \ge 0$

SOLUTION. Proceeding stepwise as usual, we observe that there is no permissible region of the problem, whose points satisfy all the constraints, i.e., there is no point which will satisfy all the constraints.

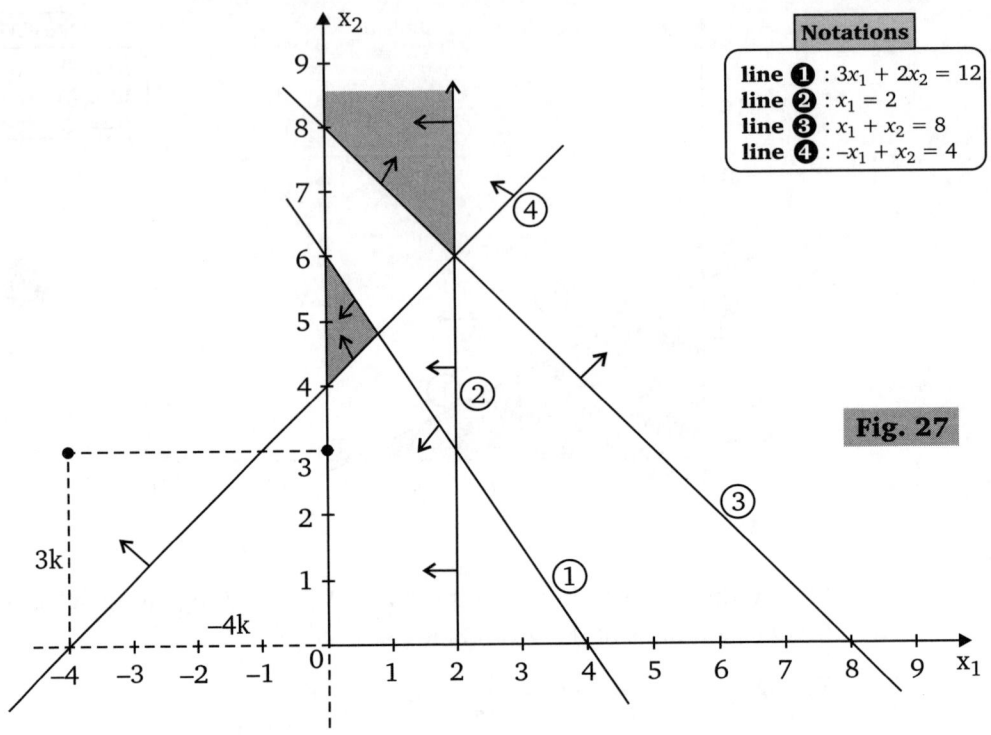

Fig. 27

EXAMPLE 3. *Solve the following LPP graphically*

$$\text{Maximize } Z = 7x_1 + 3x_2$$

subject to the constraints

$$x_1 + 2x_2 \geq 3, \ x_1 + x_2 \leq 4, \ x_1 \leq 5/2, \ x_2 \leq 3/2 \ and \ x_1, x_2 \geq 0$$

SOLUTION. Converting the given constraints into equations and then plot them by following the usual procedure and then use the inequality conditions to mark the feasible region as shown in the following figure.

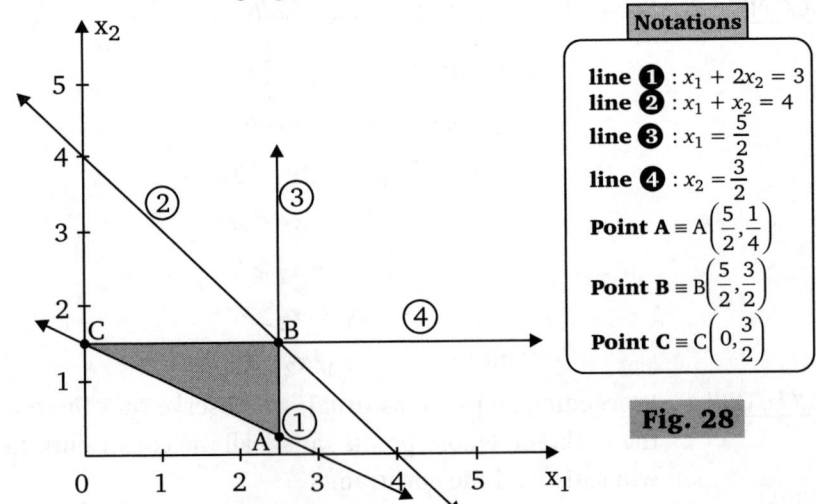

Fig. 28

Here, the extreme points of the feasible region are A, B and C. Now, we compute the value of the objective function at each of these extreme points.

Extreme point	Coordinates (x_1, x_2)	Value of the objective function $Z = 7x_1 + 3x_2$
A	$\left(\dfrac{5}{2}, \dfrac{1}{4}\right)$	$7 \times \dfrac{5}{2} + 3 \times \dfrac{1}{4} = \dfrac{73}{4}$
B	$\left(\dfrac{5}{2}, \dfrac{3}{2}\right)$	$7 \times \dfrac{5}{2} + 3 \times \dfrac{3}{2} = 22$
C	$\left(0, \dfrac{3}{2}\right)$	$0 \times 7 + 3 \times \dfrac{3}{2} = \dfrac{9}{2}$

Clearly, the maximum value of the objective function $Z = 22$ occurs at the extreme point $B\left(\dfrac{5}{2}, \dfrac{3}{2}\right)$. Hence, the optimal solution to the given LPP is $x_1 = \dfrac{5}{2}$, $x_2 = \dfrac{3}{2}$ and max $Z = 22$.

EXAMPLE 4. *Solve the following LPP graphically*

$$\text{minimize } Z = 5x_1 + 6x_2$$

subject to the constraints

$$x_1 + x_2 \geq 50, \ x_1 + 2x_2 \leq 40$$
$$3x_1 + 4x_2 \leq 100$$

and $\quad x_1 \geq 0; x_2 \geq 0$

SOLUTION. Converting the given constraints into equation and then plot them by following procedure and then use the inequality conditions to mark the feasible region as shown in the following figure.

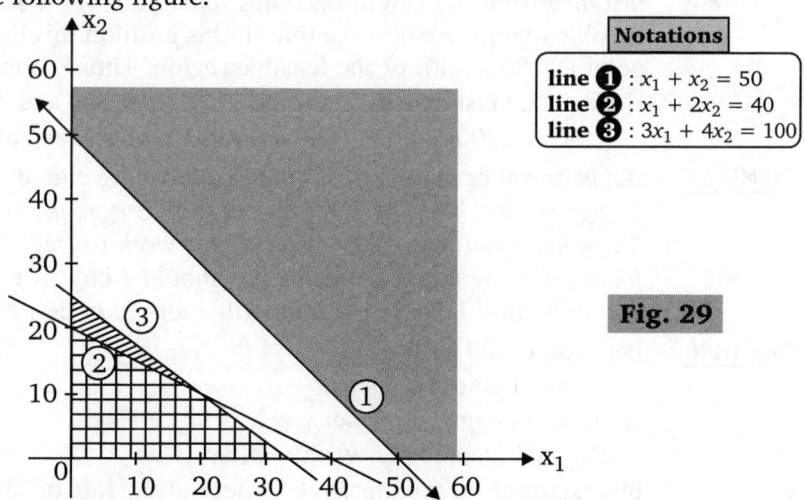

Notations
line ❶ : $x_1 + x_2 = 50$
line ❷ : $x_1 + 2x_2 = 40$
line ❸ : $3x_1 + 4x_2 = 100$

Fig. 29

Here, we observe that there exists no value of x_1 and x_2 simultaneously satisfy all the constraints and the non-negative restrictions. Hence, the given problem does not have any feasible solution.

Type 10. Value Based Questions

EXAMPLE 1. *A toy company manufactures two types of dolls, a basic version doll A and a deluxe version doll B. Each doll of type B takes twice as long to produces as one of type A and the company would have to make a maximum 2000 per day if it produces 1500 dolls per day (both A and B combined). The deluxe version requires a fency dress of which there are only 600 pieces per day available. If the company makes a profit of ₹30 and 50 per doll respectively on doll A and B, how many of each should be produced per day in order to maximize the profit.*

SOLUTION. We have already formed the LPP (in solved example 5 of Formulation of an LPP) which is given as below

Maximize $Z = 30x_1 + 50x_2$

subject to the constraints

$$x_1 + 2x_2 \leq 2000$$
$$x_1 + x_2 \leq 1500$$
$$x_2 \leq 600$$

and $x_1, x_2 \geq 0$

Now proceeding as usual, by converting the inequalities into equations and plotting them on the graph, as follows:

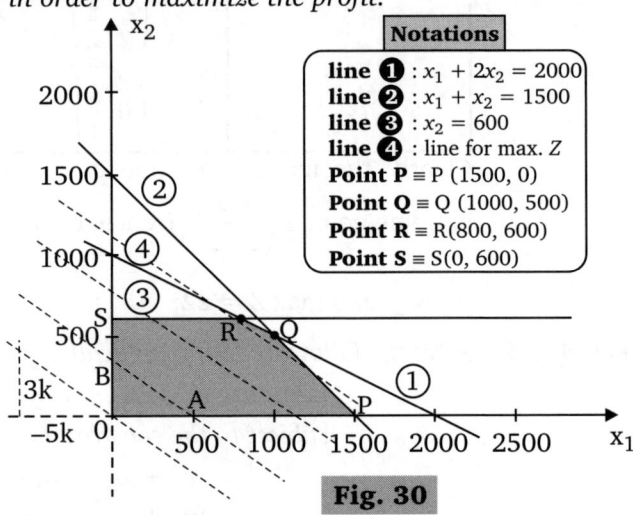

Notations

line ❶ : $x_1 + 2x_2 = 2000$
line ❷ : $x_1 + x_2 = 1500$
line ❸ : $x_2 = 600$
line ❹ : line for max. Z
Point P ≡ P (1500, 0)
Point Q ≡ Q (1000, 500)
Point R ≡ R(800, 600)
Point S ≡ S(0, 600)

Fig. 30

To find the maximum value of the objective function Z, we draw the dotted line $Z = 3x_1 + 5x_2 = 0$ passing through the origin which is parallel to the iso-profit line. We now move this line parallel to itself away from the origin so that its distance from the origin becomes maximum yet it has atleast one point in the feasible region. We observe that in this position this line passes through only one point Q(1000, 500) of the feasible region. Thus Z is maximum at Q.

Hence, Z is maximum for $x_1 = 1000$ and $x_2 = 500$ and the maximum value of Z is

$$30 \times 1000 + 50 \times 500 = 30000 + 25000 = 55000$$

EXAMPLE 2. *Old hens can be bought at ₹20 each and young ones at ₹50 each. The old hens lay 3 eggs per week and the young ones lay 5 eggs per week, each egg being worth ₹3. A hen (young or old) cost ₹10 per week to feed. I have only ₹800 to spends for hens. How many of each kind should I buy to give me a maximum profit, assuming that I can not accomodate more than 20 hens.*

SOLUTION. Let   No. of old hens = x_1

No. of young hens = x_2

Total no. of eggs I have per week = $3x_1 + 5x_2$

∴ My total income per week = $3(3x_1 + 5x_2)$

But expanses for feeding $x_1 + x_2$ hens at the rate of ₹10 per hen is = $10(x_1 + x_2)$

∴   Total profit $Z = 3(3x_1 + 5x_2) - 10(x_1 + x_2)$

$$= -x_1 + 5x_2$$

Also, $20x_1 + 50x_2 \leq 800$

and $\qquad x_1 + x_2 \leq 20$

$\therefore$ The mathematical model of an LPP is given by

$$\text{Max. } Z = -x_1 + 5x_2$$

subject to the constraints

$$20x_1 + 50x_2 \leq 800 \Rightarrow 2x_1 + 5x_2 \leq 80$$
$$x_1 + x_2 \leq 20$$
$$x_1, x_2 \geq 0$$

Following the usual procedure, we draw the following graph:

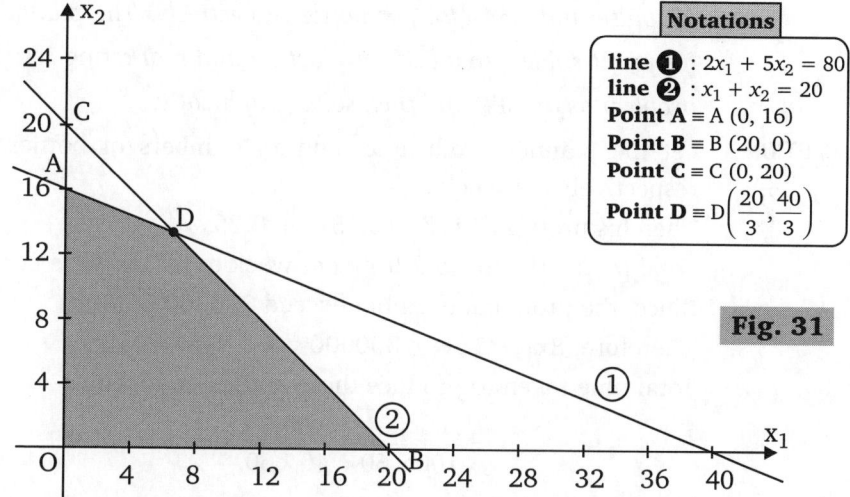

Notations

line ❶ : $2x_1 + 5x_2 = 80$
line ❷ : $x_1 + x_2 = 20$
Point A $\equiv$ A (0, 16)
Point B $\equiv$ B (20, 0)
Point C $\equiv$ C (0, 20)
Point D $\equiv$ D$\left(\dfrac{20}{3}, \dfrac{40}{3}\right)$

Fig. 31

Solving simultaneously, the equations of the corresponding intersecting lines of the feasible region, we get the coordinates of the vertices of the feasible region is

$O(0, 0), A(0, 16), B(20, 0)$ and $D = \left(\dfrac{20}{3}, \dfrac{40}{3}\right)$.

The value of the objective function at these corner points as given below:

Extreme point	Coordinates (x_1, x_2)	Value of the objective function $Z = -x_1 + 5x_2$
O	(0, 0)	$-0 + 5 \times 0 = 0$
A	(0, 16)	$-1 \times 0 + 5 \times 16 = 80$ (max.)
B	(20, 0)	$-1 \times 20 + 0 \times 5 = -20$
D	$\left(\dfrac{20}{3}, \dfrac{40}{3}\right)$	$-1 \times \dfrac{20}{3} + 5 \times \dfrac{40}{3} = 60$

Hence, the maximum value of Z is 80 at the point

$$x_1 = \frac{20}{3}, x_2 = \frac{40}{3}$$

EXAMPLE 3. *A soft drink plant has two bottling machines A and B. It produces and sells 8 ounce and 16 ounce bottles. The following data is available:*

Machine	8 Ounce	16 Ounce
A	100/minute	40/minute
B	60/minute	75/minute

The machines can be run 8 hours per day, 5 days per week. Weekly production of the drinks cannot exceed 300000 ounces and the market can absorb 25000 eight ounce bottles and 7000 sixteen ounce bottles per week. Profit on these bottles is 15 paise and 25 paise per bottle respectively. The planner wishes to maximize his profit subject to all the production and marketing restrictions. Formulate the problem as an LPP and then solve graphically.

SOLUTION. Let the planner produce x_1 and x_2 numbers of bottles of 8 and 16 ounces respectively per week.

Then his profit in ₹ is $Z = 0.15x_1 + 0.25x_2$

Total production of soft drink per week to fill up these bottles $= 8x_1 + 16x_2$

Since, the production cannot exceed 300000 ounces.

Therefore, $8x_1 + 16x_2 \le 300000$

Total time taken to produce these bottles on machine A

$$= \frac{x_1}{100 \times 60} + \frac{x_2}{40 \times 60} \text{ hours}$$

On machine $B = \frac{x_1}{60 \times 60} + \frac{x_2}{75 \times 60}$ hours

But machines can run for $8 \times 5 = 40$ hours per week

$\therefore \quad \dfrac{x_1}{100 \times 60} + \dfrac{x_2}{40 \times 60} \le 40$ and $\dfrac{x_1}{60 \times 60} + \dfrac{x_2}{75 \times 60} \le 40$

or $\quad 2x_1 + 5x_2 \le 480000$ and $5x_1 + 4x_2 \le 720000$

Also, as per given $x_1 \le 25000$ and $x_2 \le 7000$

$\qquad x_1, x_2 \ge 0$

Hence, the mathematical model of the given LPP as follows:

$\qquad$ Max $Z = 0.15x_1 + 0.25x_2$

subject to the constraints

$\qquad 8x_1 + 16x_2 \le 300000, \; 2x_1 + 5x_2 \le 480000$

$\qquad 5x_1 + 4x_2 \le 720000, \; x_1 \le 25000, \; x_2 \le 7000$ and $x_1, x_2 \ge 0$

Following the usual procedure, we draw the graph as follows:

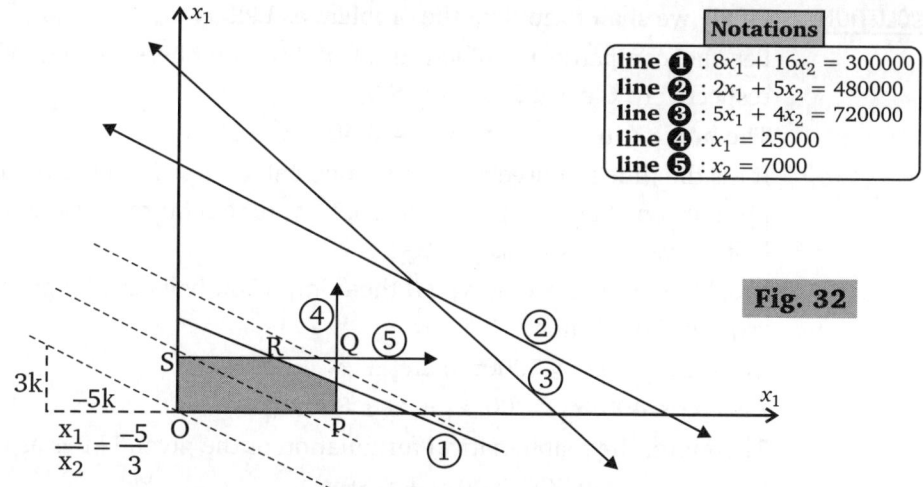

Notations
line ❶ : $8x_1 + 16x_2 = 300000$
line ❷ : $2x_1 + 5x_2 = 480000$
line ❸ : $5x_1 + 4x_2 = 720000$
line ❹ : $x_1 = 25000$
line ❺ : $x_2 = 7000$

Fig. 32

Solving simultaneously, the equations of the corresponding intersecting lines, the coordinates of the vertices of the convex polygon are $O(0, 0)$, $P(25000, 0)$, $Q(25000, 6250)$, $R(23500, 7000)$, $S(0, 7000)$.

Now, we draw dotted lines through the origin corresponding to $Z = 0$. Drawing lines parallel to this line away from the origin we see that the farthest line for maximum Z in the permissible region passes through the vertex $Q(25000, 6250)$ of the convex polygon which is the point of intersection of the lines

$$8x_1 + 16x_2 = 300000 \text{ and } x_1 = 25000.$$

Hence, the optimal solution is given by

$$x_1 = 25000, x_2 = 6250$$

and max. $Z = 0.15 \times 25000 + 0.25 \times 6250 = ₹ 5312.50$

We can also solve this LPP by corner point method as follows:

Extreme point	Coordinates (x_1, x_2)	Value of the objective function $Z = 0.15x_1 + 0.25x_2$
O	(0, 0)	$0.15 \times 0 + 0.25 \times 0 = 0$
P	(25000, 0)	$0.15 \times 25000 + 0.25 \times 0 = 3750$
Q	(25000, 6250)	$0.15 \times 25000 + 0.25 \times 6250 = 5312.50$ (max)
R	(23500, 7000)	$0.15 \times 23500 + 0.25 \times 7000 = 5275$
S	(0, 7000)	$0 + 0.25 \times 7000 = 1750$

Therefore, $Z = ₹ 5312.50$ and is maximum when $x_1 = 25000$ and $x_2 = 6250$.

EXAMPLE 4. *A company produces two types of leather belts say type A and B. Belt A is of superior quality and B is of lower quality. Profits on the two types of belts are 40 and 30 paise per belt respectively. Each belt of type A requires twice as much time as required by a belt of type B. If all belts were of type B, the company would produces 1000 belts per day. But the supply of leather is sufficient only for 800 per day. Belt A requires a fancy buckle and 400 fancy buckles are available for this per day. For belt of type B, only 700 buckles are available per day. How should the company manufacture the two types of belts in order to have maximum overall profit.*

SOLUTION. Firstly we shall formulate the problem as LPP.

Let the company manufacture x_1 and x_2 numbers of belt of type A and B respectively. Clearly, $x_1 \geq 0$, $x_2 \geq 0$.

The profit Z in ₹ is given by $Z = 0.40x_1 + 0.30x_2$

If t is the time required to produce one belt of type B and so the time required to produce one belt of type A is $2t$. So, total time required for the production of the belts of two types $= 2tx_1 + tx_2$

If belts produced are of type B then only 1000 belts can be produced and it will requires $1000t$ time.

Thus, $2tx_1 + tx_2 \leq 1000t \Rightarrow 2x_1 + x_2 \leq 1000$

Also, $x_1 \leq 400$, $x_2 \leq 700$, $x_1 + x_2 \leq 800$

Therefore, the mathematical formulation of the given LPP is as follows:

$$\text{Max } Z = 0.40x_1 + 0.30x_2$$

subject to the constraints $2x_1 + x_2 \leq 1000$, $x_1 + x_2 \leq 800$, $x_1 \leq 400$, $x_2 \leq 700$ and $x_1, x_2 \geq 0$

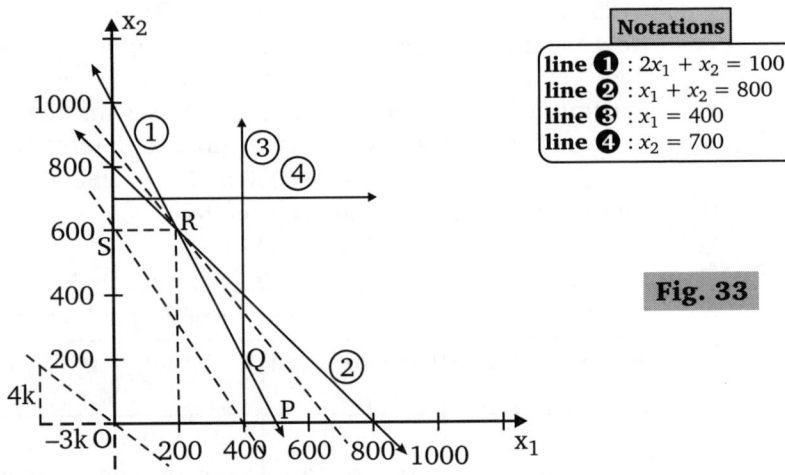

Notations
line ❶ : $2x_1 + x_2 = 1000$
line ❷ : $x_1 + x_2 = 800$
line ❸ : $x_1 = 400$
line ❹ : $x_2 = 700$

Fig. 33

To solve the above LPP, proceeding stepwise the permissible region is $OPQRSO$ and the optimal value of Z occurs at the corner $R(200, 600)$ of the convex polygon, which is clearly the point of the intersection of the lines.

$$2x_1 + x_2 = 1000$$
$$x_1 + x_2 = 800$$

Therefore, $x_1 = 200$, $x_2 = 600$

and max. $Z = 0.40 \times 200 + 0.30 \times 600 = ₹ 260$

Hence, company should produce 200 belts of type A and 600 belts of type B and the maximum profit of the company will be ₹260.

Exercise-2.2

Solve the following LPP by graphical method (Qus 1-14)

1. Max. $Z = x_1 + x_2$

subject to the constraints

$$x_1 + 2x_2 \le 2000$$
$$x_1 + x_2 \le 1500$$
$$x_2 \le 600$$

and $\quad x_1, x_2 \ge 0$

2. Minimize $Z = 20x_1 + 10x_2$

subject to the constraints

$$x_1 + x_2 \le 40$$
$$3x_1 + x_2 \ge 30$$
$$4x_1 + 3x_2 \ge 60$$

and $\quad x_1, x_2 \ge 0$

3. Maximize $Z = 8x_1 + 7x_2$

subject to the constraints

$$3x_1 + x_2 \le 66000$$
$$x_1 + x_2 \le 45000$$
$$x_1 \le 20000$$
$$x_2 \le 40000$$

and $\quad x_1, x_2 \ge 0$

4. Maximize $Z = 6x_1 + 11x_2$

subject to the constraints

$$2x_1 + x_2 \le 104$$
$$x_1 + 2x_2 \le 76$$

and $\quad x_1, x_2 \ge 0$

5. Minimize $Z = 3x_1 + 5x_2$

subject to the constraints

$$-3x_1 + 4x_2 \le 12$$
$$2x_1 - x_2 \ge -2$$
$$2x_1 + 3x_2 \ge 12$$
$$x_1 \le 4$$
$$x_2 \ge 2$$

and $\quad x_1, x_2 \ge 0$

6. Maximize $Z = 3x_1 + 4x_2$

subject to the constraints

$$x_1 - x_2 \le -1$$
$$-x_1 + x_2 \le 0$$

and $\quad x_1, x_2 \ge 0$

7. Maximize $Z = 6x_1 - 2x_2$

subject to the constraints

$$2x_1 - x_2 \le 2$$
$$x_1 \le 3$$

and $\quad x_1, x_2 \ge 0$

8. Maximize $Z = 3x_1 + 4x_2$

subject to the constraints

$$5x_1 + 4x_2 \le 200$$
$$3x_1 + 5x_2 \le 150$$
$$5x_1 + 4x_2 \ge 100$$
$$8x_1 + 4x_2 \ge 80$$

and $\quad x_1, x_2 \ge 0$

9. Minimize $Z = 2x_1 - 10x_2$

subject to the constraints

$$x_1 - x_2 \ge 0$$
$$x_1 - 5x_2 \ge -5$$

and $\quad x_1, x_2 \ge 0$

10. Maximize $Z = 7x_1 + 3x_2$

subject to the constraints

$$x_1 + 2x_2 \le 3$$
$$x_1 + x_2 \le 4$$

$$0 \le x_1 \le \frac{5}{2}$$

$$0 \le x_2 \le \frac{3}{2}$$

11. Maximize $Z = 5x_1 + 3x_2$

subject to the constraints

$$3x_1 + 5x_2 \le 15$$
$$5x_1 + 2x_2 \le 10$$

and $\quad x_1, x_2 \ge 0$

12. Maximize $Z = 2x_1 + x_2$

subject to the constraints

$$x_1 + 2x_2 \le 10$$
$$x_1 + x_2 \le 6$$
$$x_1 - x_2 \le 2$$
$$x_1 - 2x_2 \le 1$$

and $\quad x_1, x_2 \ge 0$

13. Maximize $Z = 3x_1 + 2x_2$

subject to the constraints

$$x_1 - x_2 \le 1$$
$$x_1 + x_2 \ge 3$$

and $\quad x_1, x_2 \ge 0$

14. Maximize $Z = -3x_1 + 2x_2$

subject to the constraints

$$x_1 \le 3$$
$$x_1 - x_2 \le 0$$

and $\quad x_1, x_2 \ge 0$

15. A dietician mixes two types of food in such a way that the vitamin contents of the mixture contain at least 8 units of vitamin A and 10 units of vitamin B. Food X contains 2 units/Kg of vitamin A and 1 unit/Kg of Vitamin C while Food Y contains 1 unit/Kg of vitamin A and 2 units/Kg of vitamin C. One Kg of Food X costs ₹5 whereas one Kg of food Y costs

₹7. Determine the minimum cost of such a mixture.

16. A farm is engaged in breeding hens. In view of the need to ensure certain nutrients (say x_1, x_2, x_3), it is necessary to buy two types of food say A and B. One unit of food A contains 36 units of x_1, 3 units of x_2 and 20 units of x_3. One unit of food B contains 6 units of x_1, 12 units of x_2 and 10 units of x_3. The minimum daily requirement of x_1, x_2 and x_3 is 108, 36 and 100 units respectively. The cost of food A is ₹20 per unit whereas food B costs ₹40 per unit. Find the minimum food cost so as to meet the minimum daily requirement of nutrients.

17. Two manager of courier company wishes to hire extra helpers during the Christmas season, because of a large increase in the volume of mail handling and number of temporary helpers must not exceed 10. According to the past experience, a man can handle 300 letters and 80 packages per day and a woman can handle 400 letters and 50 packages per day. It is believed that daily volume of the extra mail and packages will be no less than 3400 and 680 respectively. A man receives ₹25 per day and a woman receives ₹22 a day. How many men and women helpers should be hired to keep the pay roll at a minimum.

18. The ABC electric appliances company produces two products refrigerators and coolers. Production takes place in two separate departments. Refrigerators are produced in department-I and coolers produced in department-II. The company's two products are produced and sold on a weekly basis. The weekly production can not exceed 25 refrigerators in department-I and 35 coolers in department-II, because of the limited available facilities in these two departments. A refrigerator requires 2 man-week of labour, while a cooler requires 1 man-week of labour. A refrigerator contributes a profit of ₹60 and a cooler contributes a profit of ₹40. How many units of refrigerators and coolers should the company produces to realize maximum profit.

19. An automobile manufacture makes automobiles and trucks in a factory that is divided into two shops. Shop A, which performs the basic assembly operation, must work 5 man-days on each truck but only 2 man-days on each automobile. Shop B, which performs finishing operation must work 3 man-days for each automobile or truck that it produces. Because of men and machine limitations, shop A has 180 man-days per week available while shop B has 135 man-days per week. If the manufacturer makes a profit of ₹300 on each truck and ₹200 on each automobile, how many of each should he produce to maximize the profit.

20. A person requires 10, 12 and 12 units of chemicals A, B and C respectively for his garden. A liquid product contains 5, 2 and 1 units of A, B and C respectively per jar. A dry product contains 1, 2 and 4 units of A, B and C per carton. If the liquid product sells ₹3 per jar and the dry product sells ₹2 per carton. How many of each should be purchased to minimize the cost and meet the requirements.

Answers

1. $x_1 = 1000, x_2 = 500$, max $Z = 1500$

2. $x_1 = 6, x_2 = 12$, min. $Z = 240$

3. $x_1 = 10500, x_2 = 34500$, max $Z = 325000$

4. $x_1 = 44, x_2 = 16$, max $Z = 440$

5. $x_1 = 3, x_2 = 2$, min$(Z) = 19$ 6. No solution

7. $x_1 = 3, x_2 = 4$, max$(Z) = 10$

8. $x_1 = \dfrac{400}{13}, x_2 = \dfrac{150}{13}$, max$(Z) = \dfrac{1800}{13}$

9. $x_1 = \dfrac{5}{4}, x_2 = \dfrac{5}{4}$, min$(Z) = -10$

10. $x_1 = \dfrac{5}{2}, x_2 = \dfrac{1}{4}$, max $Z = \dfrac{73}{4}$

11. $x_1 = \dfrac{20}{19}, x_2 = \dfrac{45}{19}$, max $Z = \dfrac{235}{19}$

12. $x_1 = 4, x_2 = 2$, max $Z = 10$ 13. Unbounded solution   14. Unbounded solution

15. ₹38   16. ₹160   17. men = 6, women = 4, ₹238

18. Refrigerators $= \dfrac{25}{2}$, Coolers = 35, ₹2150 per week

19. Automobiles = 15, Trucks = 30, maximum profit = ₹12000

20. 200 belts of type A, 600 belts of type B, maximum profit = ₹260

Glossary

- **Objective Function:** The function which is to be optimized (maximized or minimized) is called an objective function.
- **Linear Programming:** It is an optimization technique used in decision making in business and everyday life for obtaining the maximum or minimum values as required of a linear expression subject to satisfying certain number of given linear restrictions.
- **Constraints:** The system of linear inequations (or euqations) under which the objective function is to be maximized are called the constraints.
- **Linearity:** Linearity implies that the products of variables such as xy, x^3y^2, x^3 etc and combinations of variables of the type $ax + b + \log y$ etc are not allowed.
- **Feasible solution:** Any solution that also satisfies the non-negative restrictions of the general LPP is said to be a feasible region.

- **Basic solution:** For a set of m simultaneous equations in n variables ($n > m$), a solution obtained by setting ($n - m$) of the variables equal to zero and solving the remaining m equations in m unknowns is called a basic solution.
- **Degenerate solution:** A basic solution to the system of equations is called degenerate if one or more of the basic variables became equal to zero.
- **Infeasible solution:** Infeasibility is a condition that arises when there is no solution to an LPP that satisfies all the constraints.
- **Unboundedness:** The condition when the objective of a linear programming problem can be made infinitely large without violating any of the constraints.
- **Iso-Profit line:** A straight line representing all non-negative combinations of x_1 and x_2 for a particular profit level.

REVIEW QUESTIONS

1. What is linear programming problem?
2. Define constraints, objective function and non-negative restrictions.
3. Define feasible and infeasible solution of an LPP.
4. Define optimal feasible solution and unbounded solution of an LPP.
5. Explain the graphical method of solving an LPP.
6. Define the term 'Feasible region'.
7. Define iso-profit and iso-cost lines.
8. Explain the procedure for generating extreme points solutions to an LPP.
9. How the unbounded solution be recognized in the graphical method.

MULTIPLE CHOICE QUESTIONS (CHOOSE THE MOST APPROPRIATE ONE)

1. For the inequality $4x + 3y > 24$, the point of intersection are:
   (a) $(0, 0), (0, 0)$ (b) $(0, 0), (0, 6)$
   (c) $(0, 8), (6, 0)$ (d) None of these

2. The set of values of the variable $x_1, x_2, ..., x_n$ satisfying the constraints and non negative restriction of a L.P.P. is called:
   (a) bounded Solution (b) feasible Solution
   (c) no Solution   (d) none of these

3. The positive variables which are subtracted from the left hand side of the constraints to convert them into equalities are called the:
   (a) slack variables   (b) surplus variable
   (c) non zero variable (d) none of the above

4. In a L.P.P. if the standard primal problem is of minimization, all the constraints involve the sign:

   (a) $\geq$ (b) $\leq$
   (c) $=$ (d) None of these

5. If the value of the objective function, Z can be increased or decreased infinitely. Such solution are called:
   (a) bounded Solution
   (b) unbounded Solution
   (c) optimal Solution
   (d) none of these

6. The system of linear inequations in an L.P.P. under which the given function is to be optimized are called as:
   (a) equations   (b) inequalities
   (c) constraints   (d) none of these

7. How many number of variables at least must vanish for a feasible solution to be a Basic feasible solution?

(a) n (b) m

(c) $n - m$ (d) $n - 1$

8. By the graphical method the solution of L.P.P. given by max $Z = 3x_1 + 2x_2$

   s.t. $x_1 - x_2 \leq 1$

   $x_1 + x_2 \geq 3$ and $x_1, x_2 \geq 0$ is:

   (a) unbounded solution

   (b) bounded solution

   (c) finite solution

   (d) None of these

9. For a L.P.P. min $Z = 2x_1 + 3x_2 + 4x_3$

   s.t. $x_1 + x_2 + x_3 = 2$

   $x_1 - x_2 + x_3 = 2$ and $x_1, x_2, x_3 \geq 0$

   the feasible solution is $x_1 = 1, x_2 = 0, x_3 = 1$ is:

   (a) a basic solution

   (b) not a basic solution

   (c) finite solution

   (d) None of these

10. How many variables in L.P.P. can be solved easily graphically?

    (a) two   (b) three

    (c) four   (d) None of these

11. A B.F.S. of a L.P.P. is said to be degenerate B.F.S. if:

    (a) all basic variables are zero

    (b) at least one of the basic variable is zero

    (c) none of the basic variables is zero

    (d) none of these

12. In graphical solution of solving L.P. problem to convert inequalities into equations:

    (a) use slack variables

    (b) use surplus variables

    (c) use artificial variables

    (d) simply assume them to be equations

13. Given a system of m simultaneous linear equations in n unknowns ($m < n$) the number of basic variables will be:

    (a) m (b) n

    (c) $n - m$ (d) $n + m$

14. For solution of a L.P.P. the Iso-Cost method is:

    (a) graphical Method

    (b) an analytical method

    (c) simplex method

    (d) none of these

15. In a L.P.P. :

    (a) objective function of constraints and variables are all linear

    (b) only constraints are linear

    (c) only objective function are linear

    (d) only variable are linear

16. In a L.P.P. of the standard primal problem of maximization all the constraints involve the sign:

    (a) $\geq$ (b) $\leq$

    (c) $=$ (d) None of these

17. Linear inequation in L.P.P. are called:

    (a) constraints   (b) F.S.

    (c) two variables   (d) none of these

18. If a L.P.P. have no feasible region then we say that the problem has:

    (a) unbounded solution

    (b) no solution

    (c) bounded solution

    (d) none of these

19. The Iso-Cost line represents:

    (a) a boundary of the feasible region

    (b) an infinite number of solution all of which yield the same profit

    (c) an infinite number of solution all of which yield the same cost

    (d) none of these

20. Solving a L.P.P. graphically, the area bounded by equation of the constraints called:

    (a) unbounded region

    (b) infeasible region

    (c) feasible region

    (d) none of these

21. Set of values of the variables $x_1, x_2, ..., x_n$ satisfying constraints and Non-negative restrictions is a:

    (a) feasible solution   (b) ^nC_m

    (c) vector   (d) None of these

22. The graphical method used in L.P.P. for:

    (a) linear equations

    (b) constraint equations

    (c) objective function equations

    (d) all of these

23. Which of the following is an assumption of an L.P. model:

    (a) additivity   (b) proportionality

    (c) divisibility   (d) all of these

24. Necessary and sufficient condition for existence and non-degeneracy of all the basic solutions of $AX = B$, is that every set of m columns of (A/B) is:

    (a) linearly independent

(b) F.S.

(c) vector

(d) none of these

25. If a non-redundant constraint is removed from a L.P.P. then:

(a) feasible region will become larger

(b) feasible region will become smaller

(c) solution will become infeasible

(d) none of these

26. The variables are to be determined in L.P.P. are always:

(a) negative   (b) non negative

(c) positive   (d) non positive

27. Solving a L.P.P. in feasibility may be removed by:

(a) adding another variable

(b) adding another constraint

(c) removing a variable

(d) removing constraint

28. A constraint in a L.P.P. become redundant because:

(a) the constraint is not satisfied by the solution values

(b) two iso profit lines may be parallel to each other

(c) the solution is unbounded

(d) none of these

29. If a L.P.P. have B.F.S, if we drop one basis vector and introduce a Non basis vector in basis, then new solution obtained is:

(a) a B.F.S.   (b) a non B.F.S

(c) no solution exist   (d) none of these

30. If an iso-profit line yielding the optimal solution coincides with a constraint line, then:

(a) the solution is unbounded

(b) the solution is feasible

(c) the constraint which coincides is redundant

(d) None of these

31. Which of the following statement is true with respect to the optimal solution of an L.P. problem:

(a) every L.P. problem has an optimal solution.

(b) optimal solution of an L.P. problem always occurs at an extreme points.

(c) at optimal solution a resources are used completely.

(d) if an optimal solution exists, there will always be at least one at a corner

32. A feasible solution to an LP problem:

(a) must satisfy all the problems constraints simultaneously

(b) need not satisfy all of the constraints

(c) only some of them must be a corner point of the feasible region

(d) must optimize the value of the objective function

33. Non-Negativity condition is an important component of LP model because:

(a) variables are interrelated in terms of limited resources

(b) variable values should remain under the control of decision maker

(c) value of variables make sense and correspond to real word problems

(d) none of the above

34. Constraints in an LP model represents:

(a) limitations

(b) requirements

(c) balancing limitations and requirements

(d) all of the above

35. The distinguishing feature of an LP model is:

(a) it has single objective function and constraints

(b) value of decision variables is non-negative

(c) relationship among all variables is linear

(d) all of the above

36. A variable which has no physical meaning but is used to obtain an initial basic feasible solution to the linear programming problem is referred to as:

(a) basic variable   (b) non-basic variable

(c) artificial variable (d) basis

37. A feasible solution to the linear programming problem should:

(a) satisfy the problem constraints

(b) optimize the objective function

(c) satisfy the problem constraints and non-negativity restrictions

(d) satisfy the non-negativity restrictions

38. The solution space (region) of an LP problem is unbounded due to:

(a) objective function is unbalanced

(b) an incorrect formulation of the LP model

(c) neither (a) nor (b)

(d) both (a) and (b)

39. Constraints in LP Problem are called active if they:
(a) represent optimal solution
(b) at optimality do not consume all the available resources
(c) both (a) and (b)
(d) none of the above

40. If two constraints do not intersect in the positive quadrant of the graph, then:
(a) one of the constraint is redundant
(b) the region is unbounded
(c) the problem is infeasible
(d) none of the above

41. A constraint in an LP model becomes redundant because:
(a) two iso-profit lines may be parallel to each other
(b) the solution is unbounded
(c) this constraint is not satisfied by the solution value
(d) none of the above

42. While plotting constraints on a graph paper, terminal points on both the axes are connected by a straight line because:
(a) the constraints are linear equations or inequalities
(b) the resources are limited in supply
(c) the objective function is a linear function
(d) all of the above

43. If there is no feasible solution in a L.P.P. then we say that the problem has:
(a) infinite solution
(b) no solution
(c) unbounded solution
(d) none of these

44. An iso-profit line represents:
(a) an infinite number of solutions all of which yield the same profit
(b) an infinite number of optimum solutions
(c) an infinite number of solutions all of which yield the same cost
(d) a boundary of the feasible region

45. In a L.P.P.:
(a) number of BFS ≤ Number of vertices
(b) number of BFS = Number of vertices
(c) number of BFS ≥ Number of vertices
(d) none of these

Answers

1. (c)	**2.** (b)	**3.** (b)	**4.** (a)	**5.** (b)	**6.** (c)	**7.** (c)	**8.** (a)	**9.** (b)
10. (a)	**11.** (b)	**12.** (d)	**13.** (c)	**14.** (a)	**15.** (a)	**16.** (b)	**17.** (a)	**18.** (b)
19. (c)	**20.** (c)	**21.** (a)	**22.** (d)	**23.** (d)	**24.** (a)	**25.** (a)	**26.** (b)	**27.** (d)
28. (d)	**29.** (a)	**30.** (c)	**31.** (b)	**32.** (b)	**33.** (c)	**34.** (d)	**35.** (c)	**36.** (b)
37. (c)	**38.** (c)	**39.** (b)	**40.** (a)	**41.** (b)	**42.** (b)	**43.** (b)	**44.** (d)	**45.** (c)

Simplex Method

3.1 INTRODUCTION

The 'simplex' is an important term in Mathematics that represents an object in n-dimensional space containing $(n + 1)$ points. Obviously, in one dimension, a simplex is a line segment connecting two points, in two dimensions, it is a triangle formed by joining these points in three dimensions, it is a four sided pyramid having four corners. In this chapter, we shall discuss a procedure called the 'simplex method' for solving an LP model, which was developed by G.B. Dantzig in 1947. The simplex method examines the extreme points in a systematic manner repeating the same set of steps of the algorithm until an optimal solution is reached. Due to this reason, it is called an iterative method.

The simplex method is an algebraic procedure that start at a feasible extreme point of the simplex or convex set; normally the origin and systematically moves from one feasible extreme point to another until an optimal extreme point is located.

3.2 TERMINOLOGY AND NOTATIONS

The general form of an LPP is given as below:

$$\text{Max. } Z = c \cdot x$$

subject to $Ax = b, x \geq 0$

where
$$A = [a_{ij}]_{m \times N}, x = (x_1, x_2, ..., x_n, ..., x_N)_{N \times 1}$$
$$c = (c_1, c_2, ..., c_n, 0, 0, ..., 0)_{1 \times N}$$

and
$$b = [b_1, b_2, ..., b_m]_{m \times 1}$$

Now, different symbolic representation is given in the following table:

S.No.	Name	Representation	Description
1.	Basic matrix	$B = \{\beta_1, \beta_2, ..., \beta_m\}$	A non-singular submatrix B of order $m \times n$ whose column vectors are m no. of linearly independent columns selected from A.
2.	Basic Variables	$x_{B_1}, x_{B_2} ... x_{B_m}$	The variable corresponding to β_1, β_2, ..., β_m are called basic variables.
3.	Basic Feasible Solution	$x_B = B^{-1} \cdot b$	$x_B = [x_{B_1}, x_{B_2}, ... x_{B_m}] = B^{-1} \cdot b$
4.	Non-basic variables	$x_{B_{m_i}} \; i > m$	The variables, other than basic, are called non-basic variables
5.	Coefficient of basic variables	$C_{B_i} : i = 1$ to m	Corresponding to any x_B, C_B will represent the row vector containing the constants $C_{B_1}, C_{B_2} ... C_{B_m}$

General Representation: We shall represent column vectors by [ ] without using transpose symbol and row vector by ( ).

3.3 FUNDAMENTAL THEOREM OF LINEAR PROGRAMMING

If a LPP *max. Z = cx*

subject to Ax = b, x ≥ 0

where A is an m × N (N = m + n) matrix of coefficients given by A = (α_1, α_2, ..., α_N) has an optimal feasible solution then at least one basic feasible solution must be optimal.

[MEERUT–2009, 10, 11, 14, 16; KANPUR–2010, 12; BANARAS–2014; DELHI–2010, 12]

Proof. Let us suppose that $x^* = [x_1, x_2, ..., x_n]$ be an optimal feasible solution of the given LPP and $Z^* = \sum\limits_{i=1}^{N} c_i x_i$ be the corresponding optimum value of the objective function.

If $k(k \leq N)$ variables in x^* are non-zero and the remaining $N - k$ are zero such that

$$x^* = [x_1, x_2, ..., x_k, 0, 0, ..., 0]$$

Therefore, $\sum\limits_{i=1}^{k} x_i \alpha_i = b$ ...(1)

and $Z^* = \sum\limits_{i=1}^{k} c_i x_i$ ...(2)

Now we have the following two cases:

Case-I: If α_1, α_2, ..., α_k are linearly independent

Since, we know that any feasible solution for which the vectors $\alpha_i : i = 1, 2, ..., k$ associated with non-zero variables x_i ($i = 1, ..., k$) are linearly independent is called a basic feasible solution.

⇒ x^* is a basic feasible solution which is also optimal.

Here, it is also clear that x^* is degenerate if $k < m$ and non-degenerate if $k = m$.

Case-II: If α_1, α_2, ..., α_k are linearly dependent and $k > m$.

Let α_1, α_2, ..., α_k are linearly dependent. Therefore, we can write

$\lambda_1 \alpha_1 + \lambda_2 \alpha_2 + ... \lambda_k \alpha_k = 0$ for scalars $\lambda_1, \lambda_2 ..., \lambda_k$ such that at least one $\lambda_i \neq 0$.

⇒ $\sum\limits_{i=1}^{k} \lambda_i \alpha_i = 0$ ...(3)

Now, assume that at least one λ_i is positive (because if none is positive then we can multiply (3) by –1 and get positive λ_i)

Further suppose $V = \max\limits_{1 \leq i \leq k} \left(\dfrac{\lambda_i}{x_i} \right)$ ...(4)

Clearly, V is positive ($\because x_i > 0 \ \forall i$ and $\lambda_i > 0$ for at least one i)

So, multiply (3) by $\dfrac{1}{V}$ and then subtracting from (1) we get

$$\sum\limits_{i=1}^{k} \left(x_i - \frac{\lambda_i}{V} \right) \alpha_i = b$$ ...(5)

⇒ $x' = \left[x_1 - \dfrac{\lambda_1}{V}, x_2 - \dfrac{\lambda_2}{V}, ..., x_k - \dfrac{\lambda_k}{V}, 0, 0, ..., 0 \right]$ is the new solution of $Ax = b$.

From (4), for $i = 1, 2, ..., k$, we have

$$V \geq \frac{\lambda_i}{x_i} \text{ or } x_i \geq \frac{\lambda_i}{V} \qquad\qquad (\because x_i > 0, V > 0)$$

$$\Rightarrow \quad x_i - \frac{\lambda_i}{V} \geq 0$$

$\Rightarrow$ all the components of x' are non-negative.

$\Rightarrow$ x' is a feasible solution of the given LPP.

Further, $V = \dfrac{\lambda_i}{x_i}$, for at least one i $(i = 1, 2, ..., k)$

$$\Rightarrow \quad V - \frac{\lambda_i}{x_i} = 0 \text{ for at least one } i.$$

$\Rightarrow$ x' can not have more than $k - 1$ non-zero variables.

Proceeding in this way, we find a new feasible solution from the given optimal feasible solution which contains less number of non-zero variables. This solution is BFS if the column vector associated to non-zero variables in this new solution are linearly independent. But if these associated vectors are not linearly independent, we shall repeat the whole process same as above. Therefore, continuing in this way for finite number of times we will find a solution in which columns corresponding to positive variables are linearly independent.

$\Rightarrow$ a BFS of the system under consideration exist.

Optimality of x^*

Let Z' be the new value of the objective function corresponding to the new solution x'. Then

$$Z' = \sum_{i=1}^{k} c_i \left(x_i - \frac{\lambda_i}{V} \right) = \sum_{i=1}^{k} c_i x_i - \frac{1}{V} \sum_{i=1}^{k} c_i \lambda_i$$

$$\Rightarrow \quad Z' = Z^* - \frac{1}{V} \sum_{i=1}^{k} c_i \lambda_i \qquad \text{(From (2))} \qquad\qquad ...(6)$$

For optimality, we must have $Z' = Z^*$, i.e., x' will be optimal if $\sum\limits_{i=1}^{k} c_i \lambda_i = 0$ ...(7)

Let if possible $\sum\limits_{i=1}^{k} c_i \lambda_i \neq 0$. Then either $\sum\limits_{i=1}^{k} c_i \lambda_i < 0$ or $\sum\limits_{i=1}^{k} c_i \lambda_i > 0$

In either of these two cases, we can find a real number r such that

$$r \cdot \sum_{i=1}^{k} c_i \lambda_i > 0$$

$$\Rightarrow \qquad\qquad \sum_{i=1}^{k} c_i (r \lambda_i) > 0$$

$$\Rightarrow \qquad \sum_{i=1}^{k} c_i (r \lambda_i) + \sum_{i=1}^{k} c_i x_i > \sum_{i=1}^{k} c_i x_i$$

$$\Rightarrow \qquad\qquad \sum_{i=1}^{k} c_i (x_i + r \lambda_i) > Z^* \qquad \text{(From 2)} \qquad\qquad ...(8)$$

Now, multiplying (3) by r and then adding to (1) we get

$$\sum_{i=1}^{k} x_i \alpha_i + \sum_{i=1}^{k} r\lambda_i \alpha_i = \boldsymbol{b}$$

$$\Rightarrow \qquad \sum_{i=1}^{k} (x_i + r\lambda_i)\alpha_i = \boldsymbol{b}$$

$$\Rightarrow \quad [x_1 + r\lambda_1, x_2 + r\lambda_2, \ldots, x_K + r\lambda_K, \underbrace{0, 0, \ldots, 0}_{(N-k)}] \qquad \ldots(A)$$

is also a solution of $A\boldsymbol{x} = \boldsymbol{b} \; \forall r$.

Let us choose r such that

$$x_i + r\lambda_i \geq 0, \; i = 1, 2, \ldots, k$$

$$\Rightarrow \qquad\qquad r\lambda_i \geq -x_i$$

$$\Rightarrow \qquad\qquad r \geq -\frac{x_i}{\lambda_i} \text{ if } \lambda_i > 0$$

$$\Rightarrow \qquad\qquad r \leq -\frac{x_i}{\lambda_i} \text{ if } \lambda_i < 0$$

and r is unrestricted if $\lambda_i = 0$.

So, $x_i + r\lambda_i \geq 0$ for $i = 1, 2, \ldots, k$ if we select r such that

$$\max_{(\lambda_i > 0)}\left(-\frac{x_i}{\lambda_i}\right) \leq r < \min_{(\lambda_i < 0)}\left(-\frac{x_i}{\lambda_i}\right) \qquad \ldots(9)$$

$$\Rightarrow \qquad \max_{i \atop (\lambda_i > 0)}\left(-\frac{x_i}{\lambda_i}\right) < 0 \text{ and } \min_{i \atop (\lambda_i < 0)}\left(-\frac{x_i}{\lambda_i}\right) > 0$$

It is clear that the interval given by (9) is non-empty.

$\Rightarrow$ r lies in the non-empty interval given by (9).

Thus, an infinite no. of solutions given by (A) satisfy the non-negative restriction also.

Finally, from (8) we find that $\sum_{i=1}^{k} c_i(x_i + r\lambda_i)$ gives the value of objective function which is greater than the greatest value Z^* of objective function, which is not possible.

So, we must have $\sum_{i=1}^{k} c_i \lambda_i = 0$

$$\Rightarrow Z' = Z^*$$

Hence, $\boldsymbol{x}' = \left[x_1 - \dfrac{\lambda_1}{V}, x_2 - \dfrac{\lambda_2}{V}, \ldots, x_k - \dfrac{\lambda_k}{V}, \underbrace{0, 0, \ldots, 0}_{(N-k)}\right]$ is also an optimal solution.

3.4 REDUCTION OF FEASIBLE SOLUTION TO BASIC FEASIBLE SOLUTION

THEOREM 1. *If a linear programming problem*

$$\text{Max. } Z = \boldsymbol{Cx}$$

subject to $A\boldsymbol{x} = \boldsymbol{b}, \boldsymbol{x} \geq 0$

where $A = (\alpha_1, \alpha_2, \ldots, \alpha_N)$ is the coefficient matrix of order $m \times N$ ($N = m + n$) has at least one feasible solution then it has one basic feasible solution also.

[MEERUT–2005, 08]

PROOF. Let $x^* = (x_1, x_2, ..., x_N) : x_i \geq 0$ ...(1)

be an arbitrary feasible solution of given LPP

Suppose that k ($k \leq N$) variables in x^* have positive values and the remaining $N - k$ variables are zero, which can be written as follows:

$$x^* = [x_1, x_2, ..., x_k, \underbrace{0, 0, ..., 0}_{(N-k)}]$$

Also, $\sum_{i=1}^{k} x_i \alpha_i = b$ ...(2)

Now, we have the following cases:

CASE I. If $\alpha_1, \alpha_2, ..., \alpha_k$ are linearly independent:

We know that any feasible solution for which the vectors $\alpha_i : i = 1, 2, ..., k$ associated with non-zero variables $x_i : i = 1, 2, ..., k$ are linearly independent is called a BFS. Therefore, clearly x^* is a BFS which is also optimal.

CASE II. If $\alpha_1, \alpha_2, ..., \alpha_k$ are linearly dependent and $k > m$

Let $\alpha_1, \alpha_2, ..., \alpha_k$ be the linearly dependent. Then by definition $\exists$ scalars $\lambda_1, \lambda_2, ..., \lambda_k$ such that

$$\lambda_1 \alpha_1 + \lambda_2 \alpha_2 + ... + \lambda_k \alpha_k = 0, \text{ at least one } \lambda_i \neq 0$$

$$\Rightarrow \sum_{i=1}^{k} \lambda_i \alpha_i = 0 \text{ for at least one } \lambda_i \neq 0 \qquad ...(3)$$

Without loss of any generality, we may assume that at least one λ_i is positive ($\because$ if none is positive then we can multiply (3) by -1 and get positive λ_i)

Further, suppose that

$$V = \max_{1 \leq i \leq k} \left(\frac{\lambda_i}{x_i} \right) \qquad ...(4)$$

Clearly, since $x_i > 0 \; \forall \; i$ and at least one λ_i is positive therefore, V is positive.

Now multiplying (3) by $\frac{1}{V}$ and then subtracting from (2), we get

$$\sum_{i=1}^{k} \left(x_i - \frac{\lambda_i}{V} \right) \alpha_i = b \qquad ...(5)$$

Here, (5) gives a new solution of $Ax = b$ which is given by

$$x' = \left[x_1 - \frac{\lambda_1}{V}, x_2 - \frac{\lambda_2}{V}, ..., x_k - \frac{\lambda_k}{V}, 0, 0, ..., 0 \right]$$

Also, from (4) for $i = 1, 2, ..., k$ we get

$$V \geq \frac{\lambda_i}{x_i} \text{ or } x_i \geq \frac{\lambda_i}{V} \qquad (\because x_i, V \geq 0)$$

$$\Rightarrow \qquad x_i - \frac{\lambda_i}{V} \geq 0$$

$\Rightarrow$ All the components of x' are non-negative.

$\Rightarrow x'$ is a feasible solution.

Also, $V = \frac{\lambda_i}{x_i}$, for at least one i, $1 \leq i \leq k$

$\Rightarrow$ For this value of i, $1 \leq i \leq k, V - \dfrac{\lambda_i}{x_i} = 0$

$\Rightarrow$ New feasible solution $\mathbf{x}'$ cannot have more than $k - 1$ non-zero variables.

Proceeding in the same way, as above, we have derived a new feasible solution from the given optimal feasible solution which contains less number of non-zero variables. Further, this solution is BFS if the column vectors associated to non-zero variables in this new solution are linearly independent. If these associated vectors are not linearly independent we shall repeat the whole reduction process as above.

Hence, continuing in this way for a finite number of times we shall derive a solution in which column corresponding to positive variables are linearly independent, i.e., we will get a basic feasible solution of the given system.

 Solved Examples

EXAMPLE 1. If $x_1 = 2$, $x_2 = 4$ and $x_3 = 1$ be a feasible solution (FS) to the system of equations.

$$2x_1 - x_2 + 2x_3 = 2$$
$$x_1 + 4x_2 = 18$$

Then find two basic feasible solutions. (BFS).   [GORAKHPUR–2007]

SOLUTION. We can write the given system of equations in matrix form as

$$\begin{bmatrix} 2 & -1 & 2 \\ 1 & 4 & 0 \end{bmatrix} \begin{bmatrix} x_1 \\ x_2 \\ x_3 \end{bmatrix} = \begin{bmatrix} 2 \\ 18 \end{bmatrix}$$

which can be written as

$$\alpha_1 x_1 + \alpha_2 x_2 + \alpha_3 x_3 = \mathbf{b} \qquad \ldots(1)$$

where $\alpha_1 = \begin{bmatrix} 2 \\ 1 \end{bmatrix}, \alpha_2 = \begin{bmatrix} -1 \\ 4 \end{bmatrix}, \alpha_3 = \begin{bmatrix} 2 \\ 0 \end{bmatrix}$ and $\mathbf{b} = \begin{bmatrix} 2 \\ 18 \end{bmatrix}$

The solution is given by

$$x_1 = 2, x_2 = 4 \text{ and } x_3 = 1$$

$\Rightarrow \quad 2\alpha_1 + 4\alpha_2 + \alpha_3 = \mathbf{b} \qquad \ldots(2)$

Clearly, we have $m = 2$, $n = 3$, therefore, the BFS have at least $n - m = 3 - 2 = 1$ zero variable.

$\Rightarrow$ BFS can not have more than two non-zero variables.

$\Rightarrow$ given FS is not BFS.

Now, we have to reduce this FS to BFS, i.e., we have to make at least one variable equal to 0.

Let α_1, α_2, α_3 be linearly dependent

$\Rightarrow$ $\exists$ a linear relation between them.

Let $\alpha_1 = a\alpha_2 + b\alpha_3$

$\Rightarrow \quad \begin{bmatrix} 2 \\ 1 \end{bmatrix} = a\begin{bmatrix} -1 \\ 4 \end{bmatrix} + b\begin{bmatrix} 2 \\ 0 \end{bmatrix} = \begin{bmatrix} -a + 2b \\ 4a \end{bmatrix}$

$\Rightarrow \quad -a + 2b = 2, \ 4a = 1$

$\Rightarrow \quad a = 1/4, b = 9/8$

$\therefore \quad \alpha_1 = \dfrac{1}{4}\alpha_2 + \dfrac{9}{8}\alpha_3 \quad \Rightarrow \quad 8\alpha_1 - 2\alpha_2 - 9\alpha_3 = \mathbf{0}$ ...(3)

$\Rightarrow \quad \sum\limits_{i=1}^{3} \lambda_i \alpha_i = \mathbf{0}$ where $\lambda_1 = 8, \lambda_2 = -2, \lambda_3 = -9$

Now, let $\quad V = \max\left(\dfrac{\lambda_i}{x_i}\right) = \max\left(\dfrac{\lambda_1}{x_1}, \dfrac{\lambda_2}{x_2}, \dfrac{\lambda_3}{x_3}\right)$

$= \max\left(\dfrac{8}{2}, -\dfrac{2}{4}, \dfrac{-9}{1}\right) = \dfrac{8}{2} = \dfrac{\lambda_1}{x_1}$

So, x_1 should be zero for which the vector α_1 should be eliminated between (2) and (3), such that

$$\frac{2(2\alpha_2 + 9\alpha_3)}{8} + 4\alpha_2 + \alpha_3 = \mathbf{b}$$

$\Rightarrow \quad \dfrac{9}{2}\alpha_2 + \dfrac{13}{4}\alpha_3 = \mathbf{b}$

$\therefore$ The new feasible solution is $x_1 = 0$ (non-basic), $x_2 = \dfrac{9}{2}, x_3 = \dfrac{13}{4}$ (basic)

Now, the column vectors α_2, α_3 corresponding to basic variables in this FS are linearly independent.

Since, $\quad |(\alpha_2, \alpha_3)| = \begin{vmatrix} -1 & 2 \\ 4 & 0 \end{vmatrix} = -8 \neq 0$

Therefore, the feasible solution given by $x_1 = 0, x_2 = \dfrac{9}{2}, x_3 = \dfrac{13}{4}$ is a BFS.

Now, we have to find another BFS

We can rewrite equation (3) as

$-8\alpha_1 + 2\alpha_2 + 9\alpha_3 = 0$

$\Rightarrow \quad \sum\limits_{i=1}^{3} \lambda_i \alpha_i = \mathbf{0}$ where $\lambda_1 = -8, \lambda_2 = 2, \lambda_3 = 9$

So, $\quad V = \max\limits_{1 \leq i \leq 3}\left(\dfrac{\lambda_i}{x_i}\right) = \max\left(\dfrac{\lambda_1}{x_1}, \dfrac{\lambda_2}{x_2}, \dfrac{\lambda_3}{x_3}\right) = \max\left(\dfrac{-8}{2}, \dfrac{2}{4}, \dfrac{9}{1}\right) = \dfrac{9}{1} = \dfrac{\lambda_3}{x_3}$

Therefore, x_3 should be zero for which α_3 should be eliminated between (2) and (4) such that

$2\alpha_1 + 4\alpha_2 + (8\alpha_1 - 2\alpha_2)/9 = \mathbf{b}$

or $\quad \left(\dfrac{26}{9}\right)\alpha_1 + \left(\dfrac{34}{9}\right)\alpha_2 = \mathbf{b}$

So, feasible solution is given by

$$x_1 = \frac{26}{9}, x_2 = \frac{34}{9} \text{ (basic) and } x_3 = 0 \text{ (non-basic)}$$

This solution is also BFS, since the vectors α_1, α_2 corresponding to the basic variables x_1 and x_2 are linearly independent as $|(\alpha_1, \alpha_2)| = \begin{vmatrix} 2 & -1 \\ 1 & 4 \end{vmatrix} = 9 \neq 0$.

Hence, two different BFS of the given system are

$$x_1 = 0, x_2 = \frac{9}{2}, x_3 = \frac{13}{4} \text{ and } x_1 = \frac{26}{9}, x_2 = \frac{34}{9}, x_3 = 0.$$

EXAMPLE 2. $(1, 1, 1)$ *is a feasible solution to the system of equations*

$$x_1 + x_2 + 2x_3 = 4$$
$$2x_1 - x_2 + x_3 = 2$$

Reduce the Feasible Solution to Basic Feasible Solution.

SOLUTION. We can write the given system of equation in matrix form, as follows:

$$\begin{bmatrix} 1 & 1 & 2 \\ 2 & -1 & 1 \end{bmatrix} \begin{bmatrix} x_1 \\ x_2 \\ x_3 \end{bmatrix} = \begin{bmatrix} 4 \\ 2 \end{bmatrix}$$

or $\qquad \alpha_1 x_1 + \alpha_2 x_2 + \alpha_3 x_3 = \boldsymbol{b}$ $\qquad$ ...(1)

where $\qquad \alpha_1 = \begin{bmatrix} 1 \\ 2 \end{bmatrix}, \alpha_2 = \begin{bmatrix} 1 \\ -1 \end{bmatrix}, \alpha_3 = \begin{bmatrix} 2 \\ 1 \end{bmatrix}$ and $\boldsymbol{b} = \begin{bmatrix} 4 \\ 2 \end{bmatrix}$

Here, $m = 2, n = 3 \Rightarrow n > m$

$\Rightarrow$ BFS will have at least $n - m = 3 - 2 = 1$ variable whose value is zero.

$\Rightarrow$ BFS cannot have more than two non-zero variables.

$\Rightarrow$ Given feasible solution $x_1 = 1, x_2 = 1, x_3 = 1$ is not BFS.

Now substituting the given FS in (1) we get

$$1 \cdot \alpha_1 + 1 \cdot \alpha_2 + 1 \cdot \alpha_3 = \boldsymbol{b}$$

$\Rightarrow \qquad \alpha_1 + \alpha_2 + \alpha_3 = \boldsymbol{b}$ $\qquad$ ...(2)

To reduce this FS to BFS, we have to make at least one variable zero, as follows:

Let $\alpha_1, \alpha_2, \alpha_3$ be linearly dependent. Then let

$$\alpha_1 = a\alpha_2 + b\alpha_3$$

$\Rightarrow \qquad \begin{bmatrix} 1 \\ 2 \end{bmatrix} = a \begin{bmatrix} 1 \\ -1 \end{bmatrix} + b \begin{bmatrix} 2 \\ 1 \end{bmatrix} = \begin{bmatrix} a + 2b \\ -a + b \end{bmatrix}$

$\Rightarrow \qquad a + 2b = 1$ and $-a + b = 2$

$\Rightarrow \qquad a = -1, b = 1$

$\therefore \qquad \alpha_1 = -1\alpha_2 + 1 \cdot \alpha_3$ or $\alpha_1 + \alpha_2 - \alpha_3 = 0$ $\qquad$ ...(3)

$\Rightarrow \displaystyle\sum_{i=1}^{3} \lambda_i \alpha_i = 0$, where $\lambda_1 = 1, \lambda_2 = 1, \lambda_3 = -1$

Further, let $\qquad V = \max_{1 \le i \le 3} \left(\dfrac{\lambda_i}{x_i} \right) = \max \left(\dfrac{\lambda_1}{x_1}, \dfrac{\lambda_2}{x_2}, \dfrac{\lambda_3}{x_3} \right)$

$$= \max \left(\frac{1}{1}, \frac{1}{1}, \frac{-1}{1} \right) = \frac{1}{1} = \frac{\lambda_1}{x_1} \text{ or } \frac{\lambda_2}{x_2}$$

$\Rightarrow$ Either x_1 or x_2 should be zero for which α_1 or α_2 should be eliminated.

Eliminate α_1 between (2) and (3), we get

$$(-\alpha_2 + \alpha_3) + \alpha_2 + \alpha_3 = \boldsymbol{b}$$

or $\qquad\qquad 0 \cdot \alpha_2 + 2\alpha_3 = \boldsymbol{b}$

$\Rightarrow$ The new FS is $x_1 = 0$ (non-basic), $x_2 = 0, x_3 = 2$ (basic) which is also BFS, since the vectors α_2, α_3 (basic) corresponding to basic variables x_2, x_3 are linearly independent as

$$|(\alpha_2, \alpha_3)| = \begin{vmatrix} 1 & 2 \\ -1 & 1 \end{vmatrix} = 3 \ne 0$$

Hence, the new BFS is $x_1 = 0, x_2 = 0, x_3 = 2$, i.e., $(0, 0, 2)$
Now we have to find another BFS. Eq^n (3) can be written as
$$-\alpha_1 - \alpha_2 + \alpha_3 = 0$$
Then $\lambda_1 = -1, \lambda_2 = -1, \lambda_3 = 1$

In this case $V = \max\limits_{1 \le i \le 3}\left(\dfrac{\lambda_i}{x_i}\right) = \dfrac{\lambda_3}{x_3}$ so variable x_3 should be zero for which the

vector α_3 should be eliminated between (2) and (3).
Eliminate α_3 between (2) and (3) we get
$$2\alpha_1 + 2\alpha_2 = \boldsymbol{b}$$
$\Rightarrow$ Another FS is $x_1 = 2, x_2 = 2$ (basic) and $x_3 = 0$ (non-basic)
Since, the column vector α_1, α_2 corresponding to these basic variables x_1, x_2 are linearly independent as
$$|(\alpha_1, \alpha_2)| = \begin{vmatrix} 1 & 1 \\ 2 & -1 \end{vmatrix} = -3 \ne 0$$
Hence, the FS $x_1 = 2, x_2 = 2, x_3 = 0$, i.e., $(2, 2, 0)$ is another BFS of the given system of equations.

EXAMPLE 3. *If $(2, 3, 1)$ is a feasible solution of the linear programming problem*
$$\text{Max. } Z = x_1 + 2x_2 + 4x_3$$
$$\text{s.t. } 2x_1 + x_2 + 4x_3 = 11$$
$$3x_1 + x_2 + 5x_3 = 14$$
$$x_1, x_2, x_3 \ge 0$$
Then find a basic feasible solution from the given feasible solution.

[MEERUT–1994, 95, 2006, 10]

SOLUTION. We can write the given LPP as
Max. $Z = x_1 + 2x_2 + 4x_3$
such that
$$A\boldsymbol{x} = \boldsymbol{b}$$
or
$$\begin{bmatrix} 2 & 1 & 4 \\ 3 & 1 & 5 \end{bmatrix} \begin{bmatrix} x_1 \\ x_2 \\ x_3 \end{bmatrix} = \begin{bmatrix} 11 \\ 14 \end{bmatrix}$$
or $\alpha_1 x_1 + \alpha_2 x_2 + \alpha_3 x_3 = \boldsymbol{b}$ ...(1)
where $\alpha_1 = \begin{bmatrix} 2 \\ 3 \end{bmatrix}, \alpha_2 = \begin{bmatrix} 1 \\ 1 \end{bmatrix}, \alpha_3 = \begin{bmatrix} 4 \\ 5 \end{bmatrix}$ and $\boldsymbol{b} = \begin{bmatrix} 11 \\ 14 \end{bmatrix}$

Since, $(2, 3, 1)$ is a feasible solution of the given LPP, therefore from (1)
$$2\alpha_1 + 3\alpha_2 + \alpha_3 = \boldsymbol{b} \qquad\qquad ...(2)$$
Clearly, the vectors $\alpha_1, \alpha_2, \alpha_3$ associated with non-zero variables x_1, x_2, x_3 will be linearly dependent if one of these vectors can be expressed as the linear combination of the remaining two vectors.
Let us suppose $\alpha_1 = a\alpha_2 + b\alpha_3$ ...(3)
$$\Rightarrow \qquad \begin{bmatrix} 2 \\ 3 \end{bmatrix} = a\begin{bmatrix} 1 \\ 1 \end{bmatrix} + b\begin{bmatrix} 4 \\ 5 \end{bmatrix} = \begin{bmatrix} a + 4b \\ a + 5b \end{bmatrix}$$
$$\Rightarrow \qquad\qquad a + 4b = 2, a + 5b = 3$$
$$\Rightarrow \qquad\qquad\qquad a = -2, b = 1$$

Using these values in (3) we get
$$\alpha_1 = -2\alpha_2 + \alpha_3 \text{ or } \alpha_1 + 2\alpha_2 - \alpha_3 = 0 \qquad \dots(4)$$

or $\quad \sum_{i=1}^{3} \lambda_i \alpha_i = 0$ which gives $\lambda_1 = 1, \lambda_2 = 2, \lambda_3 = -1$

Let
$$V = \max_{1 \le i \le 3} \left(\frac{\lambda_i}{x_i} \right) = \max \left(\frac{\lambda_1}{x_1}, \frac{\lambda_2}{x_2}, \frac{\lambda_3}{x_3} \right) = \max \left(\frac{1}{2}, \frac{2}{3}, \frac{-1}{1} \right) = \frac{2}{3} = \frac{\lambda_2}{x_2}$$

$\Rightarrow \quad x_2$ should be 0, for which we should eliminate α_2 between (2) and (4)

Now, eliminating α_2 between (2) and (4) we get
$$2\alpha_1 - 3(\alpha_1 - \alpha_3)/2 + \alpha_3 = \boldsymbol{b}$$
$$\Rightarrow \qquad \frac{1}{2}\alpha_1 + \frac{5}{2}\alpha_3 = \boldsymbol{b}$$

$\therefore \quad$ The new FS is given by $x_1 = \dfrac{1}{2}, x_2 = 0, x_3 = \dfrac{5}{2}$

Now, $\quad |(\alpha_1, \alpha_3)| = \begin{vmatrix} 2 & 4 \\ 3 & 5 \end{vmatrix} = -2 \ne 0$

$\Rightarrow \quad \alpha_1$ and α_3 are linearly independent.

Hence, the basic feasible solution obtained from the given Feasible solution is
$$x_1 = \frac{1}{2}, x_2 = 0, x_3 = \frac{5}{2}.$$

EXAMPLE 4. *If (1, 1, 1, 0) is a feasible solution to the system of equations*
$$x_1 + 2x_2 + 4x_3 + x_4 = 7$$
$$2x_1 - x_2 + 3x_3 - 2x_4 = 4$$

Reduce the feasible solution to two different BFS. [MEERUT–2008]

SOLUTION. The given system of equations can be written as

$$\begin{bmatrix} 1 & 2 & 4 & 1 \\ 2 & -1 & 3 & -2 \end{bmatrix} \begin{bmatrix} x_1 \\ x_2 \\ x_3 \\ x_4 \end{bmatrix} = \begin{bmatrix} 7 \\ 4 \end{bmatrix}$$

or $\alpha_1 x_1 + \alpha_2 x_2 + \alpha_3 x_3 + \alpha_4 x_4 = \boldsymbol{b}$ $\qquad \dots(1)$

where $\quad \alpha_1 = \begin{bmatrix} 1 \\ 2 \end{bmatrix}, \alpha_2 = \begin{bmatrix} 2 \\ -1 \end{bmatrix}, \alpha_3 = \begin{bmatrix} 4 \\ 3 \end{bmatrix}, \alpha_4 = \begin{bmatrix} 1 \\ -2 \end{bmatrix}$ and $\boldsymbol{b} = \begin{bmatrix} 7 \\ 4 \end{bmatrix}$

Since, $x_1 = 1, x_2 = 1, x_3 = 1, x_4 = 0$ is the solution of (1) therefore,
$$\alpha_1 + \alpha_2 + \alpha_3 = \boldsymbol{b} \qquad \dots(2)$$

Also here $m = 2, n = 4 \quad \Rightarrow \quad m < n$

Therefore, BFS will have at least $n - m = 4 - 2 = 2$ zero variables, *i.e.*, BFS cannot have more than 2 non-zero variables. So, given FS is not BFS.

Now, to reduce FS to BFS, we have to make at least one variable zero.

We know that the column vectors $\alpha_1, \alpha_2, \alpha_3$ corresponding to non-zero variables x_1, x_2, x_3 are linearly dependent if one of them can be expressed as a linear combination of others.

So, let $\qquad \alpha_1 = a\alpha_2 + b\alpha_3$

$$\Rightarrow \qquad \begin{bmatrix} 1 \\ 2 \end{bmatrix} = a \begin{bmatrix} 2 \\ -1 \end{bmatrix} + b \begin{bmatrix} 4 \\ 3 \end{bmatrix} \qquad \Rightarrow \qquad a = -1/2, \ b = 1/2$$

So, $\alpha_1 = \left(-\dfrac{1}{2}\right)\alpha_2 + \left(\dfrac{1}{2}\right)\alpha_3$ or $2\alpha_1 + \alpha_2 - \alpha_3 = 0$ ...(3)

or $\displaystyle\sum_{i=1}^{3} \lambda_i \alpha_i = 0$ where $\lambda_1 = 2, \ \lambda_2 = 1, \ \lambda_3 = -1$

Now,
$$V = \max\left(\frac{\lambda_i}{x_i}\right) = \max\left(\frac{\lambda_1}{x_1}, \frac{\lambda_2}{x_2}, \frac{\lambda_3}{x_3}\right)$$

$$= \max\left(\frac{2}{1}, \frac{1}{1}, \frac{-1}{1}\right) = \frac{2}{1} = \frac{\lambda_1}{x_1}$$

$\Rightarrow x_1$ should be zero for which we have to eliminate α_1 between (2) and (3) such that

$$\frac{(-\alpha_2 + \alpha_3)}{2} + \alpha_2 + \alpha_3 = \boldsymbol{b}$$

$$\Rightarrow \qquad \frac{1}{2}\alpha_2 + \frac{3}{2}\alpha_3 = \boldsymbol{b}$$

$$\Rightarrow \qquad x_1 = 0 \text{ (non-basic)}, \ x_2 = \frac{1}{2}, x_3 = \frac{3}{2} \text{ (basic)}$$

Now, since $|(\alpha_2, \alpha_3)| = \begin{vmatrix} 2 & 4 \\ -1 & 3 \end{vmatrix} = 10 \neq 0$

$\Rightarrow$ Column vectors α_2, α_3 corresponding to basic variables x_2, x_3 are linearly independent.

$\Rightarrow$ This feasible solution is BFS.

Further, we have to obtain another BFS. Rewrite (3) as

$$-2\alpha_1 - \alpha_2 + \alpha_3 = 0$$

$$\Rightarrow \qquad \sum_{i=1}^{3} \lambda_i \alpha_i = 0, \text{ where } \lambda_1 = -2, \ \lambda_2 = -1, \ \lambda_3 = 1$$

$$\therefore \qquad V = \max\left(\frac{\lambda_1}{x_1}, \frac{\lambda_2}{x_2}, \frac{\lambda_3}{x_3}\right) = \max\left(\frac{-2}{1}, \frac{-1}{1}, \frac{1}{1}\right) = \frac{1}{1} = \frac{\lambda_3}{x_3}$$

$\Rightarrow$ x_3 should be zero.

For which eliminate α_3 between (2) and (4) such that

$$\alpha_1 + \alpha_2 + (2\alpha_1 + \alpha_2) = 0 \quad \Rightarrow \quad 3\alpha_1 + 2\alpha_2 = 0$$

$$\Rightarrow \quad x_1 = 3, \ x_2 = 2 \text{ (basic)}, \ x_3 = 0, \ x_4 = 0 \text{ (non-basic)}$$

Also, $|(\alpha_1, \alpha_2)| = \begin{vmatrix} 1 & 2 \\ 2 & -1 \end{vmatrix} = -5 \neq 0$

$\Rightarrow$ Column vectors α_1, α_2 corresponding to basic variables x_1, x_2 are linearly independent.

$\Rightarrow$ This solution is BFS.

Hence, the given FS reduces to the following two BF solutions.

$$x_1 = 0, x_2 = \frac{1}{2}, x_3 = \frac{3}{2}, x_4 = 0 \text{ and } x_1 = 3, x_2 = 2, x_3 = 0, x_4 = 0.$$

EXAMPLE 5. *If* $(1, 2, 1, 3)$ *is a feasible solution of the set of equations*

$$5x_1 - 4x_2 + 3x_3 + x_4 = 3$$
$$2x_1 + x_2 + 5x_3 - 3x_4 = 0$$
$$x_1 + 6x_2 - 4x_3 + 2x_4 = 15$$
$$x_1, x_2, x_3, x_4 \geq 0$$

Find a corresponding BFS. [MEERUT–2005, 14; GORAKHPUR–2011, 17]

SOLUTION. We can write

$$A\boldsymbol{x} = \boldsymbol{b}$$

$$\Rightarrow \quad \begin{bmatrix} 5 & -4 & 3 & 1 \\ 2 & 1 & 5 & -3 \\ 1 & 6 & -4 & 2 \end{bmatrix} \begin{bmatrix} x_1 \\ x_2 \\ x_3 \\ x_4 \end{bmatrix} = \begin{bmatrix} 3 \\ 0 \\ 15 \end{bmatrix}$$

or $\quad \alpha_1 x_1 + \alpha_2 x_2 + \alpha_3 x_3 + \alpha_4 x_4 = \boldsymbol{b}$ ...(1)

where $\alpha_1 = \begin{bmatrix} 5 \\ 2 \\ 1 \end{bmatrix}, \alpha_2 = \begin{bmatrix} -4 \\ 1 \\ 6 \end{bmatrix}, \alpha_3 = \begin{bmatrix} 3 \\ 5 \\ -4 \end{bmatrix}, \alpha_4 = \begin{bmatrix} 1 \\ -3 \\ 2 \end{bmatrix}$ and $\boldsymbol{b} = \begin{bmatrix} 3 \\ 0 \\ 15 \end{bmatrix}$

Since $(1, 2, 1, 3)$ is a solution of (1) therefore,

$$\alpha_1 + 2\alpha_2 + \alpha_3 + 3\alpha_4 = \boldsymbol{b}$$ ...(2)

The vectors $\alpha_1, \alpha_2, \alpha_3, \alpha_4$ associated with non-zero variables x_1, x_2, x_3, x_4 will be linearly dependent if one of these vectors can be expressed as the linear combination of the remaining three vectors.

Let $\qquad\qquad \alpha_1 = a\alpha_2 + b\alpha_3 + c\alpha_4$ ...(3)

$$\Rightarrow \qquad \begin{bmatrix} 5 \\ 2 \\ 1 \end{bmatrix} = a \begin{bmatrix} -4 \\ 1 \\ 6 \end{bmatrix} + b \begin{bmatrix} 3 \\ 5 \\ -4 \end{bmatrix} + c \begin{bmatrix} 1 \\ -3 \\ 2 \end{bmatrix}$$

$$\Rightarrow \qquad \begin{bmatrix} -4a + 3b + c \\ a + 5b - 3c \\ 6a - 4b + 2c \end{bmatrix} = \begin{bmatrix} 5 \\ 2 \\ 1 \end{bmatrix}$$

$\Rightarrow \quad -4a + 3b + c = 5; \ a + 5b - 3c = 2$ and $6a - 4b + 2c = 1$

On solving we get

$$a = \frac{22}{43}, b = \frac{139}{86}, c = \frac{189}{86}$$

Putting all these values in (3) we get

$$\frac{22}{43}\alpha_2 + \frac{139}{86}\alpha_3 + \frac{189}{86}\alpha_4 = \alpha_1$$

$\Rightarrow \quad 86\alpha_1 - 44\alpha_2 - 139\alpha_3 - 189\alpha_4 = 0$ ...(4)

$\Rightarrow \quad \sum\limits_{i=1}^{4} \lambda_i \alpha_i = 0$ where $\lambda_1 = 86, \lambda_2 = -44, \lambda_3 = -139, \lambda_4 = -189$

Now, let $\qquad\qquad V = \max\limits_{1 \leq i \leq 4} \left(\frac{\lambda_i}{x_i} \right) = \max \left(\frac{\lambda_1}{x_1}, \frac{\lambda_2}{x_2}, \frac{\lambda_3}{x_3}, \frac{\lambda_4}{x_4} \right)$

$$= \max\left(\frac{86}{1}, \frac{-44}{2}, \frac{-139}{1}, \frac{-189}{3}\right) = \frac{86}{1} = \frac{\lambda_1}{x_1}$$

$\Rightarrow$ x_1 should be zero.

So, eliminate α_1 between (2) and (4) we get

$$\frac{1}{86}(44\alpha_2 + 139\alpha_3 + 189\alpha_4) + 2\alpha_2 + \alpha_3 + 3\alpha_4 = \boldsymbol{b}$$

$$\Rightarrow \quad 0 \cdot \alpha_1 + \frac{216}{86}\alpha_2 + \frac{225}{86}\alpha_3 + \frac{447}{86}\alpha_4 = \boldsymbol{b}$$

$$\Rightarrow \quad x_1 = 0, x_2 = \frac{216}{86}, x_3 = \frac{225}{86}, x_4 = \frac{447}{86} \text{ is the new BFS.}$$

Also, $|(\alpha_2, \alpha_3, \alpha_4)| = \begin{vmatrix} -4 & 3 & 1 \\ 1 & 5 & -3 \\ 6 & -4 & 2 \end{vmatrix} \neq 0$

$\Rightarrow$ The vectors corresponding to non-zero basic variables are linearly independent. Hence, the new FS is BFS.

EXAMPLE 6. *If* (2, 1, 2, 2, 1) *is a feasible solution of the set of equations*

$$2x_1 - 3x_2 + 4x_3 + 6x_4 = 21$$

$$x_1 + 2x_2 + 3x_3 - 3x_4 + 5x_5 = 9$$

Then reduce it to two different BFS.

SOLUTION. We can write the given system of equations as

$$\begin{bmatrix} 2 & -3 & 4 & 6 & 0 \\ 1 & 2 & 3 & -3 & 5 \end{bmatrix} \begin{bmatrix} x_1 \\ x_2 \\ x_3 \\ x_4 \\ x_5 \end{bmatrix} = \begin{bmatrix} 21 \\ 9 \end{bmatrix}$$

or $\alpha_1 x_1 + \alpha_2 x_2 + \alpha_3 x_3 + \alpha_4 x_4 + \alpha_5 x_5 = \boldsymbol{b}$ ...(1)

where $\alpha_1 = \begin{bmatrix} 2 \\ 1 \end{bmatrix}, \alpha_2 = \begin{bmatrix} -3 \\ 2 \end{bmatrix}, \alpha_3 = \begin{bmatrix} 4 \\ 3 \end{bmatrix}, \alpha_4 = \begin{bmatrix} 6 \\ -3 \end{bmatrix}, \alpha_5 = \begin{bmatrix} 0 \\ 5 \end{bmatrix}$ and $\boldsymbol{b} = \begin{bmatrix} 21 \\ 9 \end{bmatrix}$

The feasible solution is given by $x_1 = 2, x_2 = 1, x_3 = 2, x_4 = 2$ and $x_5 = 1$

Then from (1)

$$2\alpha_1 + \alpha_2 + 2\alpha_3 + 2\alpha_4 + \alpha_5 = \boldsymbol{b} \qquad \text{...(2)}$$

Also, here $m = 2, n = 5$, *i.e.*, $m < n$ therefore BFS will have at least $n-m=5-2=3$ zero variables, *i.e.*, it can not have more than two non-zero variables. So, given feasible solution is not basic. To reduce FS to BFS we proceed as follows:

(1) To Reduce one of the variables to zero value.

Since, $\alpha_1, \alpha_2, \alpha_3, \alpha_4$ and α_5 are linearly dependent therefore, by inspection we have

$$2\alpha_1 + 2\alpha_2 - \alpha_3 + \alpha_4 + 0 \cdot \alpha_5 = 0 \qquad \text{...(3)}$$

$$\Rightarrow \quad \sum_{i=1}^{5} \lambda_i \alpha_i = 0 \text{ where } \lambda_1 = 2, \lambda_2 = 2, \lambda_3 = -1, \lambda_4 = 1, \lambda_5 = 0$$

Further, let $V = \max_{1 \le i \le 5}\left(\dfrac{\lambda_i}{x_i}\right) = \max\left(\dfrac{\lambda_1}{x_1},\dfrac{\lambda_2}{x_2},\dfrac{\lambda_3}{x_3},\dfrac{\lambda_4}{x_4},\dfrac{\lambda_5}{x_5}\right)$

$$= \max\left(\dfrac{2}{2},\dfrac{2}{1},\dfrac{-1}{2},\dfrac{1}{2},\dfrac{0}{1}\right) = \dfrac{2}{1} = \dfrac{\lambda_2}{x_2}$$

$\Rightarrow$ x_2 should be zero.

$\Rightarrow$ We have to eliminate α_2 between 2 and 3 as follows:

$$2\alpha_1 + \left(-\alpha_1 + \dfrac{1}{2}\alpha_3 - \dfrac{1}{2}\alpha_4\right) + 2\alpha_3 + 2\alpha_4 + \alpha_5 = \boldsymbol{b}$$

$\Rightarrow$ $\alpha_1 + \dfrac{5}{2}\alpha_3 + \dfrac{3}{2}\alpha_4 + \alpha_5 = \boldsymbol{b}$ ...(4)

$\Rightarrow x_1 = 1, x_2 = 0, x_3 = \dfrac{5}{2}, x_4 = \dfrac{3}{2}$ and $x_5 = 1$ is the another feasible solution of the problem, but it is again not a BFS, as it contains more than two non-zero variables.

(2) To Reduce other variable to Zero value.

Since, $\alpha_1, \alpha_3, \alpha_4, \alpha_5$ are linearly dependent, so by inspection we have

$$0 \cdot \alpha_1 - 3 \cdot \alpha_3 + 2 \cdot \alpha_4 + 3\alpha_5 = \boldsymbol{0}$$ ...(5)

$\Rightarrow$ $\lambda_1 = 0, \lambda_3 = -3, \lambda_4 = 2, \lambda_5 = 3$

Further, let $V = \max\left(\dfrac{\lambda_1}{x_1},\dfrac{\lambda_3}{x_3},\dfrac{\lambda_4}{x_4},\dfrac{\lambda_5}{x_5}\right)$

$$= \max\left(\dfrac{0}{1},\dfrac{-3}{(5/2)},\dfrac{2}{(3/2)},\dfrac{3}{1}\right) = \dfrac{3}{1} = \dfrac{\lambda_5}{x_5}$$

So, x_5 should be zero, i.e., we have to eliminate α_5 between (4) and (5) as follows:

$$\alpha_1 + \dfrac{5}{2}\alpha_3 + \dfrac{3}{2}\alpha_4 + \left(\alpha_3 - \dfrac{2}{3}\alpha_4\right) = \boldsymbol{b}$$

$\Rightarrow$ $\alpha_1 + \dfrac{7}{2}\alpha_3 + \dfrac{5}{6}\alpha_4 = \boldsymbol{b}$ ...(6)

$\Rightarrow x_1 = 1, x_2 = 0, x_3 = 7/2, x_4 = 5/6, x_5 = 0$ is another FS.
It is also not a BFS as it contains more than two non-zero variables.

(3) To Reduce one more variable to zero.

Since, α_1, α_3 and α_4 are linearly dependent.

$\Rightarrow$ $\alpha_1 = a\alpha_3 + b\alpha_4$

$\Rightarrow$ $\begin{pmatrix}2\\1\end{pmatrix} = a\begin{pmatrix}4\\3\end{pmatrix} + b\begin{pmatrix}6\\-3\end{pmatrix} = \begin{pmatrix}4a+6b\\3a-3b\end{pmatrix}$

$\Rightarrow$ $4a + 6b = 2, 3a - 3b = 1$

$\Rightarrow$ $a = \dfrac{2}{5}, b = \dfrac{1}{15}$

So, $\alpha_1 = \dfrac{2}{5}\alpha_3 + \dfrac{1}{15}\alpha_4 = 0$

$\Rightarrow$ $15\alpha_1 - 6\alpha_3 - \alpha_4 = 0$ ...(7)

$\Rightarrow$ $\lambda_1 = 15, \lambda_3 = -6, \lambda_4 = -1$

Further let $V = \max\limits_{1 \le i \le 4}\left\{\dfrac{\lambda_i}{x_i}\right\} = \max\left\{\dfrac{\lambda_1}{x_1}, \dfrac{\lambda_3}{x_3}, \dfrac{\lambda_4}{x_4}\right\}$

$$= \max\left\{\dfrac{15}{1}, \dfrac{-6}{(7/2)}, \dfrac{-1}{(5/6)}\right\} = \dfrac{15}{1} = \dfrac{\lambda_1}{x_1}$$

$\Rightarrow$ x_1 should be zero, *i.e.*, α_1 should be eliminated from (6) and (7) as follows:

$$\frac{2}{5}\alpha_3 + \left(\frac{1}{15}\right)\alpha_4 + \frac{7}{2}\alpha_3 + \frac{5}{6}\alpha_4 = \boldsymbol{b}$$

$\Rightarrow$ $\qquad \dfrac{39}{10}\alpha_3 + \dfrac{9}{10}\alpha_4 = \boldsymbol{b}$

$\Rightarrow$ $\qquad x_1 = 0, x_2 = 0, x_3 = \dfrac{39}{10}, x_4 = \dfrac{9}{10}, x_5 = 0$

which is a feasible solution with two non-zero variables.

Since, $|(\alpha_3, \alpha_4)| = \begin{vmatrix} 4 & 6 \\ 3 & -3 \end{vmatrix} = -30 \ne 0$

$\Rightarrow$ $\qquad \alpha_3, \alpha_4$ are linearly independent.
$\Rightarrow$ $\qquad$ Feasible solution is BFS.
Now, to get another BFS, rewrite (7) as
$\qquad -15\alpha_1 + 6\alpha_3 + \alpha_4 = 0$
Then $\lambda_1 = -15, \lambda_3 = 6, \lambda_4 = 1$

$\Rightarrow$ $\qquad \max\left\{\dfrac{\lambda_1}{x_1}, \dfrac{\lambda_3}{x_3}, \dfrac{\lambda_4}{x_4}\right\} = \max\left\{\dfrac{-15}{1}, \dfrac{6}{(7/2)}, \dfrac{1}{(5/6)}\right\} = \dfrac{12}{7} = \dfrac{\lambda_3}{x_3}$

$\Rightarrow$ x_3 should be zero, *i.e.*, α_3 should be eliminated between (6) and (7) as follows:

$$\frac{39}{4}\alpha_1 + \frac{1}{4}\alpha_4 = \boldsymbol{b}$$

$\therefore$ Feasible solution is $x_1 = \dfrac{39}{4}, x_2 = 0, x_3 = 0, x_4 = \dfrac{1}{4}, x_5 = 0$ which is also BFS,

because column vectors α_1, α_4 corresponding to x_1 and x_4 are linearly independent.

Exercise-3.1

1. If (1, 0, 1) is the feasible solution of the LPP given by
   min. $Z = 2x_1 + 3x_2 + 4x_3$
   subject to
   $x_1 + x_2 + x_3 = 2$
   $x_1 - x_2 + x_3 = 0$
   $x_1, x_2, x_3 \ge 0$
   Then, show that the given feasible solution is not basic.

2. If (2, 4, 5) is the feasible solution of the system of equations
   $2x_1 - x_2 + 2x_3 = 10$
   $x_1 + 4x_2 = 18$
   $x_1, x_2, x_3 \ge 0$
   Reduce this FS to BFS.

3. If (2, 13) is a feasible solution of the set of equations
   $4x_1 + 2x_2 - 3x_3 = 1$
   $6x_1 + 4x_2 - 5x_3 = 1$
   Reduce this FS to BFS.

4. If (2, 1, 3, 2, 1) is a Feasible solution of the system of equations
   $2x_1 - 3x_2 + 4x_3 + 6x_4 = 25$
   $x_1 + 2x_2 + 3x_3 - 3x_4 + 5x_5 = 12$
   Reduce FS to a BFS of the system.

3.5 TO FIND THE IMPROVED BASIC FEASIBLE SOLUTION (BFS) FROM A GIVEN BFS

THEOREM I. Let $\boldsymbol{x}_B = B^{-1}\boldsymbol{b}$ be a basic feasible solution of a LPP with $Z = \boldsymbol{C}_B\boldsymbol{x}_B$ as the value of the objective function. If for any column α_j in A, but not in B the condition $c_j - Z_j > 0$ or < 0 holds and if at least one $y_{ij} > 0$, $i = 1, 2, ..., m$, then it is possible to find a new BFS by replacing one of the columns in B by α_j and if Z' is the new value of the objective function then $Z' \geq Z$. Further, if the initial BFS $\boldsymbol{x}_B = B^{-1}\boldsymbol{b}$ is non-degenerate then $Z' > Z$.

PROOF. Let us consider an LPP given by

$$\text{max. } Z = \boldsymbol{Cx} \text{ such that } A\boldsymbol{x} = \boldsymbol{b}, x \geq 0$$

where $A = (\alpha_1, \alpha_2, ..., \alpha_N)$

$$N = m + n$$

and basic matrix $B = (\beta_1, \beta_2, ..., \beta_m)$

Suppose that $x_B = [x_{B_1}, x_{B_2}, ..., x_{B_m}]$ be a basic feasible solution of the given LPP.

Now, since the vectors $\beta_1, \beta_2, ..., \beta_m$ are in the basis of A, so by definition of basis, the given vectors can be expressed as

$$\alpha_j = \sum_{i=1}^{m} y_{ij}\beta_i = y_{1j}\beta_1 + y_{2j}\beta_2 + ... + y_{mj}\beta_m \qquad ...(1)$$

Here, if $y_{rj} \neq 0$ then α_j can replace β_r in B and B is still a basis matrix.

Now, let $y_{rj} \neq 0$ then from (1) we can write

$$\beta_r = \frac{1}{y_{rj}}\alpha_j - \frac{y_{1j}}{y_{rj}}\beta_1 - ... - \frac{y_{(r-1)j}}{y_{rj}}\beta_{r-1} - \frac{y_{(r+1)j}}{y_{rj}}\beta_{r+1} - ... - \frac{y_{mj}}{y_{rj}}\beta_m$$

$$\Rightarrow \qquad \beta_r = \frac{1}{y_{rj}}\alpha_j - \sum_{\substack{i=1 \\ i \neq r}}^{m} \frac{y_{ij}}{y_{rj}}\beta_i \qquad ...(2)$$

Consider $B \cdot \boldsymbol{x}_B = \boldsymbol{b}$

$$\Rightarrow \quad (\beta_1, \beta_2, ..., \beta_r, ..., \beta_m)[x_{B_1}, x_{B_2}, ..., x_{B_r}, ..., x_{B_m}] = \boldsymbol{b}$$

$$\Rightarrow \qquad \beta_1 x_{B_1} + \beta_2 x_{B_2} + ... + \beta_r x_{B_r} + ... + \beta_m x_{B_m} = \boldsymbol{b}$$

$$\Rightarrow \qquad \sum_{\substack{i=1 \\ i \neq r}}^{m} \beta_i x_{B_i} + \beta_r x_{B_r} = \boldsymbol{b} \qquad ...(3)$$

Using (2) in (3) we get

$$\sum_{\substack{i=1 \\ i \neq r}}^{m} \beta_i \left[x_{B_i} - \frac{y_{ij}}{y_{rj}} x_{B_r} \right] + \frac{x_{B_r}}{y_{rj}} \alpha_j = \boldsymbol{b} \qquad ...(4)$$

$$\Rightarrow \qquad \sum_{\substack{i=1 \\ i \neq r}}^{m} x'_{B_i} \beta_i + x'_{B_r} \alpha_j = \boldsymbol{b} \qquad ...(5)$$

where $x'_B = x_{B_i} - x_{B_r} \dfrac{y_{ij}}{y_{rj}} ; i = 1,2,...,m, i \neq r$

and $\quad x'_{B_r} = \dfrac{x_{B_r}}{y_{rj}} , i = r$

$\quad\quad\quad\quad\quad\quad\quad\quad\quad\quad\quad\quad$ ...(6)

On comparing (3) and (5), the new basic solution of $A\boldsymbol{x} = \boldsymbol{b}$ is given by

$$\boldsymbol{x}'_B = [x'_{B_i}, x'_{B_r}] \; i = 1,2,...,m, i \neq r$$

The above basic solution will be feasible if

$x_{B_i} - \dfrac{y_{ij}}{y_{rj}} x_{B_r} \geq 0, i = 1,2,...,m, i \neq r$

and $\quad\quad \dfrac{x_{B_r}}{y_{rj}} \geq 0, i = r$

$\quad\quad\quad\quad\quad\quad\quad\quad\quad\quad\quad\quad$ ...(7)

Now, since $\boldsymbol{x}_B$ is the IBFS, we have

$$x_{B_r} \geq 0 : r = 1,2,...,m$$

Then (7) holds only if $y_{rj} > 0$ and $y_{ij} \leq 0$, $i = 1, 2, ..., m, i \neq r$
If $y_{rj} > 0$ and $y_{ij} > 0$ then (7) is satisfied only if

$$\dfrac{x_{B_i}}{y_{ij}} - \dfrac{x_{B_r}}{y_{rj}} \geq 0 \Rightarrow \dfrac{x_{B_r}}{y_{rj}} \leq \dfrac{x_{B_i}}{y_{ij}}$$

$\Rightarrow \quad\quad \dfrac{x_{B_r}}{y_{rj}} = \min_i \left\{ \dfrac{x_{B_i}}{y_{ij}}, y_{ij} > 0 \right\} = V$

$\quad\quad\quad\quad\quad\quad\quad\quad\quad\quad\quad\quad$ ...(8)

Thus, we conclude that

A new basic feasible solution can be obtained from the initial basic feasible solution by removing the column vector β_r of the basis matrix B by α_j is r is to be selected such that

$$V = \dfrac{x_{B_r}}{y_{rj}} = \min_i \left\{ \dfrac{x_{B_i}}{y_{ij}}, y_{ij} > 0 \right\}$$

Now, it remains to prove that $Z' \geq Z$.

Clearly, the value of the objective function for the IBFS x_B is given by

$$Z = C_B \boldsymbol{x}_B$$
$$= (C_{B_1} C_{B_2} ... C_{B_m})[x_{B_1}, x_{B_2}, ..., x_{B_m}]$$
$$= \sum_{i=1}^{m} C_{B_i} x_{B_i}$$

Now, corresponding to new BFS, $\boldsymbol{x}'_B$, the value of the objective function is Z', therefore, we have

$$Z' = \sum_{i=1}^{m} C'_{B_i} x'_{B_i}$$

$\quad\quad\quad\quad\quad\quad\quad\quad\quad\quad\quad\quad$ ...(9)

Here, C'_{B_i} are the coefficients of the basic variables $x'_{B_i} : i = 1,2,...,m$ in the objective function.

Clearly, $\quad C'_{B_i} = C_{B_i} : i = 1,2,...,m, i \neq r$

and $\qquad C'_{B_r} = c_j$

$\Rightarrow \qquad Z' = \sum_{\substack{i=1 \\ i \neq r}}^{m} C_{B_i} x'_{B_i} + c_j C'_{B_r}$ ...(10)

Using (6) in (10) we get

$$Z' = \sum_{\substack{i=1 \\ i \neq r}}^{m} C_{B_i}\left(x_{B_i} - x_{B_r}\frac{y_{ij}}{y_{rj}}\right) + c_j\frac{x_{B_r}}{y_{rj}}$$

$$= \sum_{i=1}^{m} C_{B_i}\left(x_{B_i} - x_{B_r}\frac{y_{ij}}{y_{rj}}\right) + c_j\frac{x_{B_r}}{y_{rj}}$$

Since, $C_{B_i}\left(x_{B_i} - x_{B_r}\dfrac{y_{ij}}{y_{rj}}\right) = 0$ when $i = r$, therefore,

$$Z' = \sum_{i=1}^{m} C_{B_i} x_{B_i} - \frac{x_{B_r}}{y_{rj}}\sum_{i=1}^{m} C_{B_i} y_{ij} + \frac{x_{B_r}}{y_{rj}}c_j$$

$$= Z + \frac{x_{B_r}}{y_{rj}}(c_j - z_j) \text{ where } Z_j = \sum_{i=1}^{m} C_{B_i} y_{ij}$$

$$= Z + V(c_j - z_j) \qquad \qquad ...(11)$$

Equation (11) clearly shows that $Z' \geq Z$ if $V[c_j - z_j] \geq 0$

$\because \qquad V \geq 0$ therefore, $Z' \geq Z$ only if $c_j - z_j \geq 0$

So, we conclude that

> By choosing the vector α_j for which $c_j - z_j > 0$ and at least one $y_{ij} > 0$ we obtain a new improved value of the objective function.

☞ REMARKS
- If $V = 0$ then initial basic feasible solution is degenerate and hence new BFS is also degenerate.
- If the initial Basic Feasible solution is non-degenerate then $V > 0$ and hence $Z' > Z$.

3.6 CONDITIONS FOR THE EXISTENCE OF UNBOUNDED SOLUTIONS

We have already discussed in previous article that for any column α_j in A (not in B) the condition $c_j - z_j > 0$ holds and if at least one $y_{ij} > 0$, $i = 1, 2, ..., m$ then it is possible to obtain an improved BFS. But if for at least one α_j, all $y_{ij} \leq 0$, then we get an unbounded solution.

THEOREM 1. *If for any basic feasible solution $x_B = B^{-1}b$ to $Ax = b$ there is some column α_j in A but not in B for which $c_j - z_j > 0$ and $y_{ij} < 0$, $i = 1, 2, ..., m$ then if the objective function is to be maximized, then problem has an unbounded solution.*

PROOF. Let $\qquad$ max. $Z = Cx$

such that $Ax = b, x \geq 0$

where $A = (\alpha_1, \alpha_2, ..., \alpha_N)$ $(N = m + n)$

and $B = (\beta_1, \beta_2, ..., \beta_m)$

be the given LPP.

Further, suppose that $x_B = [x_{B_1}, x_{B_2}, ..., x_{B_m}]$ be a BFS of the given LPP. Then

$$Bx_B = b \text{ or } \sum_{i=1}^{m} x_{B_i}\beta_i = b \qquad \qquad ...(1)$$

and
$$Z = \boldsymbol{C}_B \boldsymbol{x}_B = \sum_{i=1}^{m} C_{B_i} x_{B_i} \qquad \qquad ...(2)$$

Equation (1) can be written as

$$\sum_{i=1}^{m} x_{B_i} \beta_i - \lambda \alpha_j + \lambda \alpha_j = \boldsymbol{b} \text{ for some scalars } \lambda. \qquad \qquad ...(3)$$

Suppose $\alpha_j \in A$ such that $\alpha_j \notin B$

Now, since the vectors $\beta_1, \beta_2, ..., \beta_m$ are in the basis of A, some α_j can be express as the linear combination of β's.

So,
$$\alpha_j = \sum_{i=1}^{m} y_{ij} \beta_i$$

$$\Rightarrow \qquad -\lambda \alpha_j = -\lambda \sum_{i=1}^{m} y_{ij} \beta_i$$

Using the above equation in (3) we get

$$\sum_{i=1}^{m} x_{B_i} \beta_i - \lambda \sum_{i=1}^{m} y_{ij} \beta_i + \lambda \alpha_j = \boldsymbol{b}$$

$$\Rightarrow \qquad \sum_{i=1}^{m} (x_{B_i} - \lambda y_{ij}) \beta_i + \lambda \alpha_j = \boldsymbol{b}$$

which gives a new solution given by
$$\boldsymbol{x}'_B = [x'_{B_1}, x'_{B_2}, ..., x'_{B_m}, \lambda] \qquad \qquad ...(4)$$

where $x'_{B_i} = x_{B_i} - \lambda y_{ij} : i = 1, 2, ..., m$

When $\lambda > 0$, $x_{B_i} - \lambda y_{ij} \geq 0 (\because y_{ij} \leq 0)$. Then from (4) we get a Feasible solution in which the number of positive variables is less than or equal to $(m + 1)$. It may be less than $(m + 1)$ because $x_{B_i} - \lambda y_{ij}$ may be zero for some i. If the number of positive variables in this solution is equal to $(m + 1)$ then this solution will be non-basic feasible solution.

Further, if Z' is the new value of the objective function corresponding to new solution then

$$Z' = \sum_{i=1}^{m} C_{B_i} (x_{B_i} - \lambda y_{ij}) + c_j \lambda$$

$$\Rightarrow \qquad Z' = \sum_{i=1}^{m} C_{B_i} x_{B_i} + \lambda (c_j - C_{B_i} y_{ij}) = Z + \lambda (c_j - z_j)$$

Since, $c_j - z_j > 0 \Rightarrow Z'$ can be made as large as we want by giving sufficiently large values to λ. Hence the given LPP has an unbounded solution.

☛ REMARKS
- An LPP has an unbounded solution if the value of the objective function can be increased or decreased arbitrarily.
- If for some α_j, $z_j - c_j > 0$ and $y_{ij} \leq 0 : i = 1, 2, ..., m$ then the LPP has an unbounded solution if the objective function is to be minimized.

3.7 CONDITION FOR IMPROVED BASIC FEASIBLE SOLUTION TO BECOME OPTIMAL

Let us suppose $\boldsymbol{x}_B = B^{-1} \boldsymbol{b}$ be the basic feasible solution of an LPP given by $Z = \boldsymbol{Cx}$ such that $A\boldsymbol{x} = \boldsymbol{b}, \boldsymbol{x} \geq 0$.

Let $Z^* = C_B x_B$ be the value of the objective function at any iteration of simplex method. If $c_j - z_j \leq 0$ for every column α_j in A but not in B, then Z^* is the optimum value of the objective function Z and x_B is an optimal basic feasible solution.

3.8 ALTERNATIVE OPTIMAL SOLUTIONS

An LPP is said to have an alternative optimal solution if the set of variables giving the optimal value of the objective function is not unique.

CONDITION FOR ALTERNATIVE OPTIMUM SOLUTION

Let us suppose there exists an optimal basic feasible solution to the given LPP and

(i) if for some α_j in A but not in B, $c_j - z_j = 0$ and $y_{ij} \leq 0$ for all $i = 1, 2, ..., m$ then a non-basic alternative optimal solution will exist.

(ii) if for some α_j in A but not in B, $c_j - z_j = 0$ and $y_{ij} > 0$ for at least one i, then an alternative basic optimal solution will exist.

3.9 INCONSISTENCY AND REDUNDANCY

Definition 1. *If the system of equations has more than enough number of constraints equations, i.e., it has more constraints equations than no. of variables. Then it is called redundancy.*

i.e., $\qquad r(A) = r(Ab) = k \leq n < m$

In this case, there will be $(m - k)$ redundant equations.

Definition 2. *The set of linear equations is said to be inconsistent if $r(A) \neq r(Ab)$.*

☛ REMARK

- When we solve an LPP by simplex method we should have $r(A) = r(Ab)$, i.e., the constraints equations after introducing the slack and artificial variables should be consistent.

If the system $Ax = b$ involves artificial variables, then we have the following cases:

(i) If the basis B consist no artificial vector and optimality condition is satisfied, then the current solution is a Basic Feasible Solution.

(ii) If one or more artificial variable appear in B at a zero level, i.e., the value of the artificial variables corresponding to aritificial vectors in B are zero and the optimality condition is satisfied, then the system is consistent. But if $y_{ij} = 0 \; \forall j$ and $x_{B_r} = 0$ and r corresponding to the row containing an artificial vector, then r^{th} equation is redundant.

(iii) If at least one artificial variable appears in the basis B at a positive level, i.e., the value of at least one artificial variable corresponding to artificial vector in B is non-zero and the optimality condition is satisfied then there exists no feasible solution.

3.10 PROCEDURE TO OBTAIN INITIAL BASIC FEASIBLE SOLUTION

CASE I. WHEN ALL ORIGINAL CONSTRAINTS HAVE ≤ SIGN

Firstly convert all the constraints into equations by introducing slack variables s_i, as given below:

$$a_{11}x_1 + a_{12}x_2 + ... + a_{1n}x_n + 1 \cdot s_1 = b_1$$
$$a_{21}x_1 + a_{22}x_2 + ... + a_{2n}x_n + 1 \cdot s_2 = b_2$$
$$\vdots$$
$$\vdots$$
$$a_{m1}x_1 + a_{m2}x_2 + ... + a_{mn}x_n + 1 \cdot s_m = b_m$$

where each s_i is a slack variable.

The above system of equations can be written in matrix form as follows:

$$\begin{bmatrix} a_{11} & a_{12} & \cdots & a_{1n} & 1 & 0 & \cdots & 0 \\ a_{21} & a_{22} & \cdots & a_{2n} & 0 & 1 & \cdots & 0 \\ \vdots & & & & & & & \\ a_{m1} & a_{m2} & \cdots & a_{mn} & 0 & 0 & \cdots & 1 \end{bmatrix} \begin{bmatrix} x_1 \\ x_2 \\ \vdots \\ x_n \\ s_1 \\ s_2 \\ \vdots \\ s_m \end{bmatrix} = \begin{bmatrix} b_1 \\ b_2 \\ \vdots \\ b_m \end{bmatrix}$$

Let $B = I_m$, a unit matrix of order $m \times m$. Then initial basis matrix is given by

$$x_B = B^{-1}\boldsymbol{b} = I_m \cdot \boldsymbol{b} = \boldsymbol{b} \geq 0$$

Hence, the initial basic feasible solution (IBFS) is given by

$$s_1 = x_{B_1} = b_1, s_2 = x_{B_2} = b_2, \ldots, s_m = x_{B_m} = b_m$$

WORKING PROCEDURE

> The initial basic feasible solution can be obtained by writing all the non-basic variables x_1, x_2, ..., x_n equal to zero and solving the equations for remaining variables s_1, s_2, ..., s_m (i.e., slack variables).

CASE-2 WHEN ALL ORIGINAL CONSTRAINTS HAVE $\geq$ SIGN

Firstly, convert all the constraints into equations by introducing surplus variables s_i, as given below:

$$a_{11}x_1 + a_{12}x_2 + \ldots + a_{1n}x_n - s_1 = b_1$$
$$a_{21}x_1 + a_{22}x_2 + \ldots + a_{2n}x_n - s_2 = b_2$$
$$\vdots$$
$$a_{m1}x_1 + a_{m2}x_2 + \ldots + a_{mn}x_n - s_m = b_m$$

where each s_i is a surplus variable.

The above system of equations can be written in matrix form as follows:

$$\begin{bmatrix} a_{11} & a_{12} & \cdots & a_{1n} & -1 & 0 & \cdots & 0 \\ a_{21} & a_{22} & \cdots & a_{2n} & 0 & -1 & \cdots & 0 \\ \vdots & & & & & & & \\ a_{m1} & a_{m2} & \cdots & a_{mn} & 0 & 0 & \cdots & -1 \end{bmatrix} \begin{bmatrix} x_1 \\ x_2 \\ \vdots \\ x_n \\ s_1 \\ s_2 \\ \vdots \\ s_m \end{bmatrix} = \begin{bmatrix} b_1 \\ b_2 \\ \vdots \\ b_m \end{bmatrix}$$

If we take the initial basis matrix $B = -I_m$. Then

$$x_B = B^{-1}\boldsymbol{b} = -I_m \cdot \boldsymbol{b} = -\boldsymbol{b} \leq 0$$

which is not a basic feasible solution (BFS).

To counter this difficulty introduce artificial variables in each constraints as given below:

$$a_{11}x_1 + a_{12}x_2 + \ldots + a_{1n}x_n - s_1 + a_1 = b_1$$
$$a_{21}x_1 + a_{22}x_2 + \ldots + a_{2n}x_n - s_2 + a_2 = b_2$$
$$\vdots$$
$$\vdots$$
$$a_{m1}x_1 + a_{m2}x_2 + \ldots + a_{mn}x_n - s_m + a_m = b_m$$

which can be written in the matrix form as follows:

$$\begin{bmatrix} a_{11} & a_{12} & \ldots & a_{1n} & -1 & 0 & \ldots & 0 & 1 & 0 & \ldots & 0 \\ a_{21} & a_{22} & \ldots & a_{2n} & 0 & -1 & \ldots & 0 & 0 & 1 & \ldots & 0 \\ \vdots & & & & & & & & & & & \\ a_{m1} & a_{m2} & \ldots & a_{mn} & 0 & 0 & \ldots & -1 & 0 & 0 & \ldots & 1 \end{bmatrix} \begin{bmatrix} x_1 \\ x_2 \\ \vdots \\ x_n \\ s_1 \\ s_2 \\ \vdots \\ s_m \\ a_1 \\ a_2 \\ \vdots \\ a_m \end{bmatrix} = \begin{bmatrix} b_1 \\ b_2 \\ \vdots \\ b_m \end{bmatrix}$$

Now, taking the basis matrix $B = I_m$, therefore, $x_B = B^{-1}b = I_m b = b \geq 0$ which is the required basic feasible solution.

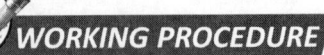 **WORKING PROCEDURE**

The basic feasible solution is $a_1 = x_{B_1} = b_1, a_2 = x_{B_2} = b_2, \ldots, a_m = x_{B_m} = b_m$ can be obtained by writing all the non-basic variables $x_1, x_2, \ldots, x_n, s_1, s_2, \ldots, s_m$ equal to zero and solving the equations for the remaining basic variables (artificial variables) $a_1, a_2, \ldots, a_m$.

CASE -3 WHEN ALL ORIGINAL CONSTRAINTS HAVE '$\leq$', '$\geq$' AND '$=$' SIGN

In such type of cases, we convert the constraints into equations by introducing and inserting slack, surplus and artificial variables. The basis matrix $B = I_m$ is obtained by introducing unit column vectors corresponding to the slack and artificial variables.

WORKING PROCEDURE

To find the initial basic feasible solution of the given LPP, put all the non-basic variables equal to zero and solve for remaining basic variables. Then initial basic feasible solution is $x_B = b \geq 0$.

If identity is present in coefficient matrix A without introducing the artificial variables, then there is no need to introduce the artificial variables.

3.11 SIMPLEX ALGORITHM

To find the optimal solution of the given linear programming problem, we use the following steps:

STEP 1. **General Steps:**

    (i) If the given problem is of minimization, first convert it into the maximization problem by multiplying both sides of the objective function by -1 and put $-z = z^*$. Remember that if V is the maximum value of z^* then $-V$ will be the minimum value of z.

    (ii) The RHS of each of the constraints should be non-negative. If there is any constraints for which b_i is negative then multiply this constraints by -1 to convert it into positive values.

    (iii) Replace each unrestricted variable (if any) with the difference of two non-negative variables, replace each non-positive variables with a new non-negative variable whose value is the negative of the original variable.

    (iv) Express the given LPP into standard form by adding additional variables to the left side of each constraints and assign a zero cost coefficients to these in the objective function.

STEP 2. Convert the given inequalities of constraints into equations. For this we introduce, slack, surplus or artificial variables (whichever is required). The coefficients of slack or surplus variables in the objective function is zero. Note that, if artificial variables are introduced then we use two-phase or Big-M method.

STEP 3. **Construction of Initial Simplex Table:**

    Construct the initial simplex table as given below:

		$c_j \rightarrow$	c_1	c_2	$\cdots$	c_n	0	0	$\cdots$	0
Basic Variables B.V.	**Coefficients of basic variables C_B**	**Value of basic variables $b\,(=x_B)$**			Variables					
			x_1	x_2	$\cdots$	x_n	s_1	s_2	$\cdots$	s_m
s_1	C_{B_1}	$x_{B_1} = b_1$	a_{11}	a_{12}	$\cdots$	a_{1n}	1	0	$\cdots$	0
s_2	C_{B_2}	$x_{B_2} = b_2$	a_{21}	a_{22}	$\cdots$	a_{2n}	0	1	$\cdots$	0
$\vdots$	$\vdots$	$\vdots$	$\vdots$							
s_m	C_{B_m}	$x_{B_m} = b_m$	a_{m1}	a_{m2}	$\cdots$	a_{mn}	0	0	$\cdots$	1
$Z = \Sigma C_{B_i} x_{B_i}$		$z_j = \Sigma C_{B_i} x_j$	0	0	$\cdots$	0	0	0	$\cdots$	0
		$c_j - z_j$	$c_1 - z_1$	$c_2 - z_2$	$\cdots$	$c_n - z_n$	0	0	$\cdots$	0

In the above table we observe that

    (i) The first row provide coefficients of the current basic variables in the objective function. Column headed by x_B represents the current values of the corresponding variables in the basis.

    (ii) The second row indicate the coefficient c_j of variables in the objective function which remains the same in successive simplex tables. These values are used to determine the variables to be entered into the basis matrix B.

    (iii) The basic matrix (identity matrix) represents the coefficients of slack variables which have been added to the constraints. Each column of the identity matrix also represents a basic variables to be listed in Column B.

    (iv) The value of z_j represent the amount by which the value of the objective function z should be decreased (or increased) if one unit of given variable is added to the new solution.

Each of the value in the $c_j - z_j$ row represents the net amount of increase (or decrease) in the objective function that would occur when one unit of the variables represented by the column head is introduced into the solution.

STEP 4. Test for optimality:

Calculate $\Delta_j = c_j - z_j$ for all non-basic variables. Then we have the following cases:

(i) If $\Delta_j \leq 0 \ \forall \ j$, then solution under consideration is optimal.

(ii) Alternative optimal solution will exist if any Δ_j, for non-basic variables is zero.

(iii) If $\Delta_j < 0 \ \forall \ j$, corresponding to non-basic variables, then unique optimal solution exists.

(iv) If $\Delta_j > 0$ for any j, *i.e.*, if at least one Δ_j is positive then solution is not optimal and it indicates that an improvement in the value of the objective function Z is possible.

(v) If corresponding to maximum positive Δ_j, all elements of the column of basic variables are negative or zero, then solution is unbounded.

STEP 5. Selection of incoming and outgoing vector:

(i) The incoming vector α_k is selected corresponding to the largest positive value of Δ_j. But if maximum value of Δ_j occur for more than one α_j then we may select any of these vectors as incoming vector.

(ii) The outgoing vector β_r is selected corresponding to that value of r for which

$$\frac{x_{B_r}}{y_{rk}} = \min_i \left\{ \frac{x_{B_i}}{y_{i_k}}, y_{i_k} > 0 \right\}$$

where α_k is the incoming vector.

If the minimum value is not unique then more than one variable will vanish in the next solution and a degenerate basic feasible solution exists.

Here, it must be noted that the division by negative or zero is not permitted.

STEP 6. Test for Feasibility. If α_k is incoming vector and β_r is the outgoing vector then the element y_{rk} which lies at the intersection of minimum ratio arrow ($\rightarrow$) and incoming vector arrow ($\uparrow$) is called the pivot element. Put this element in the box $\square$.

STEP 7. Finding the New Solution. Use the following steps:

(i) If the key element is 1, then row remains unchanged in the next simplex table.

(ii) If the key element is not 1 then divide each element of the row by the key element.

(iii) The new values of the element in the remaining rows for the new simplex table can be obtained by performing elementary row operations on all rows so that all the elements except the kew element in the key column are zero.

Then, new entries in C_B and x_B columns are updated in the new simplex table of the current solution.

Therefore, we get improved basic feasible solution.

STEP 8. Test the improved BFS for optimality.

Solved Examples

EXAMPLE 1. *Using simplex method, solve the following LPP:*

$$Max.\ Z = 16x_1 + 17x_2 + 10x_3$$

subject to the constraints

$$x_1 + x_2 + 4x_3 \leq 2000$$
$$2x_1 + x_2 + x_3 \leq 3600$$
$$x_1 + 2x_2 + 2x_3 \leq 2400$$
$$x_1 \leq 30$$

and $x_1, x_2, x_3 \geq 0$

SOLUTION. Clearly, the given LPP is in standard form.

Now introducing slack variables s_1, s_2, s_3 and s_4 to convert the given inequalities to equations. Then LP model can be written as

$Max.\ Z = 16x_1 + 17x_2 + 10x_3 + 0s_1 + 0s_2 + 0s_3 + 0s_4$
subject to the constraints

$$x_1 + x_2 + 4x_3 + s_1 = 2000$$
$$2x_1 + x_2 + x_3 + s_2 = 3600$$
$$x_1 + 2x_2 + 2x_3 + s_3 = 2400$$
$$x_1 + s_4 = 30$$

and $x_1, x_2, x_3, s_1, s_2, s_3, s_4 \geq 0$

Now we prepare the initial simplex table as follows:

Simplex Table-1

B.V.	C_B	x_B	$c_j \rightarrow$ 16 x_1	17 x_2	10 x_3	0 s_1	0 s_2	0 s_3	0 s_4	Min Ratio x_B/x_2
s_1	0	2000	1	1	4	1	0	0	0	2000/1 = 2000
s_2	0	3600	2	1	1	0	1	0	0	3600/1 = 3600
s_3	0	2400	1	②	2	0	0	1	0	2400/2 = 1200 → (min)
s_4	0	30	1	0	0	0	0	0	1	—
$\Delta_j = c_j - z_j$			16	17 ↑	10	0	0	0	0	

In the above simplex table, $c_2 - z_2 = 17$ in x_2-column is the largest positive value. Now to obtain a new improved solution, by entering variable x_2 into the basis and removing variable s_3 from the basis, use the following steps:

$$R_3(\text{new}) \rightarrow R_3\ (\text{old}) \div 2$$
$$R_1\ (\text{new}) \rightarrow R_1\ (\text{old}) - R_3\ (\text{new})$$
$$R_2\ (\text{new}) \rightarrow R_2\ (\text{old}) - R_3\ (\text{new})$$

Then we have the following simplex table:

Simplex Table-2

	$c_j \rightarrow$		16	17	10	0	0	0	0	Min Ratio
B.V.	C_B	x_B	x_1	x_2	x_3	s_1	s_2	s_3	s_4	x_B/x_1
s_1	0	800	1/2	0	3	1	0	−1/2	0	800/(1/2) = 1600
s_2	0	2400	3/2	0	0	0	1	−1/2	0	2400/(3/2) = 1600
x_2	17	1200	1/2	1	1	0	0	1/2	0	1200/(1/2) = 2400
s_4	0	30	①	0	0	0	0	0	1	30/1 = 30 → (min)
$\Delta_j = c_j - z_j$			15/2 ↑	0	−7	0	0	−17/2	0	

The solution obtained in the above table is not optimal because $\Delta_j > 0$ in x_1-column. To improved this solution, applying the following row operations to get a new improved solution by entering variable x_1 into the basis and removing the variable s_4 from the basis.

$$R_4 \text{ (new)} \rightarrow R_4 \text{ (old)} \div \text{ (Key element)}$$

$$R_1 \text{ (new)} \rightarrow R_1 \text{ (old)} - \frac{1}{2} R_4 \text{ (new)}$$

$$R_2 \text{ (new)} \rightarrow R_2 \text{ (old)} - \frac{3}{2} R_4 \text{ (new)}$$

$$R_3 \text{ (new)} \rightarrow R_3 \text{ (old)} - \frac{1}{2} R_4 \text{ (new)}$$

Then we have the following simplex table:

Simplex Table-3

	$c_j \rightarrow$		16	17	10	0	0	0	0
B.V.	C_B	x_B	x_1	x_2	x_3	s_1	s_2	s_3	s_4
s_1	0	785	0	0	3	1	0	−1/2	−1/2
s_2	0	2355	0	0	0	0	1	−1/2	−3/2
x_2	17	1185	0	1	1	0	0	1/2	−1/2
x_1	16	30	1	0	0	0	0	0	1
$\Delta_j = c_j - z_j$			0	0	−7	0	0	−17/2	−15/2

Clearly, in the above table all $\Delta_j < 0$ corresponding to non-basic variables column. Hence, the solution is optimal and given by

$$x_1 = 30, x_2 = 1185, x_3 = 0$$

and max. $Z = 0 \times 785 + 0 \times 2355 + 17 \times 1185 + 16 \times 30 = 20625$

EXAMPLE 2. *Solve the following LPP by simplex method*

$$\text{Max. } Z = 8x_1 + 11x_2$$

subject to the constraints
$$3x_1 + x_2 \le 7$$
$$x_1 + 3x_2 \le 8$$
and $\qquad x_1, x_2 \ge 0$

SOLUTION. The given maximization problem is in standard form with all $b_i's$ are non-negative.

Now introduce the slack variables s_1 and s_2 to convert the given inequalities into equations. Then given problem becomes

max. $Z = 8x_1 + 11x_2 + 0s_1 + 0s_2$

s.t. $\qquad 3x_1 + x_2 + s_1 = 7$
$$x_1 + 3x_2 + s_2 = 8$$
and $\qquad x_1, x_2, s_1, s_2 \ge 0$

Then apply the simplex method, we construct the following table:

Sinplex table

B.V.	C_B	x_B	$c_j \rightarrow$ 8 x_1	11 x_2	0 s_1	0 s_2	Min Ratio x_B/x_2
s_1	0	7	3	1	1	0	7/1 = 7
s_2	0	8	1	③	0	1	8/3 (min) →
$Z = C_B x_B = 0$		$\Delta_j =$	8	11	0	0	x_B/x_1
				↑		↓	
s_1	0	13/3	⑧/③	0	1	−1/3	13/8 (min) →
x_2	11	8/3	1/3	1	0	1/3	8/1 = 8
$Z = C_B x_B = 88/3$		$\Delta_j =$	13/3	0	0	−11/3	
			↑		↓		
x_1	8	13/8	1	0	3/8	−1/8	
x_2	11	17/8	0	1	−1/8	3/8	
$Z = C_B x_B = 291/8$		$\Delta_j =$	0	0	−13/8	−25/8	

Clearly, in the last table all $\Delta_j \le 0$. Thus, solution is optimal and is given by

$$x_1 = \frac{13}{8}, x_2 = \frac{17}{8} \text{ and max } Z = \frac{291}{8}$$

EXAMPLE 3. *Solve the following LPP by simplex method*

$\qquad$ *max.* $Z = 3x_1 + 2x_2$

$\qquad$ *subject to the constraints*

$$x_1 + x_2 \le 4$$
$$x_1 - x_2 \le 2$$
and $\quad x_1, x_2 \ge 0$

SOLUTION. The given LPP is in standard form. Introduce slack variables s_1 and s_2 to convert the given inequalities into equations. The given LPP becomes

Max. $Z = 3x_1 + 2x_2 + 0s_1 + 0s_2$

s.t.
$$x_1 + x_2 + s_1 = 4$$
$$x_1 - x_2 + s_2 = 2$$

and
$$x_1, x_2, s_1, s_2 \geq 0$$

Then solution to the problem using simplex method is given in the following table.

Simplex table

B.V.	C_B	x_B	$c_j \to$ 3 x_1	2 x_2	0 s_1	0 s_2	Min Ratio x_B/x_1
s_1	0	4	1	1	1	0	4/1 = 4
s_2	0	2	①	−1	0	1	2/1 = 2 (min)→
$Z = C_B x_B = 0$		$\Delta_j =$	3 ↑	2	0	0 ↓	
s_1	0	2	0	②	1	−1	2/2 = 1 (min) →
x_1	3	2	1	−1	0	1	—
$Z = C_B x_B = 6$		$\Delta_j =$	0	5 ↑	0	−3 ↓	
x_2	2	1	0	1	1/2	−1/2	
x_1	3	3	1	0	1/2	1/2	
$Z = C_B x_B = 11$		$\Delta_j =$	0	0	−5/2	−1/2	

We observe that in the last table all $\Delta_j < 0$.

$\Rightarrow$ Solution is optimal.

Hence, solution is given by

$$x_1 = 3, x_2 = 1 \text{ and max } Z = 11.$$

EXAMPLE 4. *Solve the following LPP by simplex method*

$$\text{Max. } Z = 3x_1 + 2x_2 - 2x_3$$

subject to the constraints

$$x_1 + 2x_2 + 2x_3 \leq 10$$
$$2x_1 + 4x_2 + 3x_3 \leq 15$$

and $\qquad x_1, x_2, x_3 \geq 0$

SOLUTION. Clearly, the given maximization problem is in standard form in which all $b_i's$ are non-negative.

Now, introduce the slack variables s_1 and s_2 to convert the given inequalities into equations such that

$$\text{max. } Z = 3x_1 + 2x_2 - 2x_3 + 0s_1 + 0s_2$$

s.t.

$$x_1 + 2x_2 + 2x_3 + s_1 = 10$$
$$2x_1 + 4x_2 + 3x_3 + s_2 = 15$$

and $\qquad x_1, x_2, x_3, s_1, s_2 \geq 0$

Now by applying the simplex method, we have the following simplex table.

Simplex table

B.V.	C_B	x_B	x_1	x_2	x_3	s_1	s_2	Min Ratio x_B/x_1
$c_j \rightarrow$			3	2	–2	0	0	
s_1	0	10	1	2	2	1	0	10/1 = 10
s_2	0	15	②	4	3	0	1	15/2 = (min)→
$Z = C_B x_B = 0, \Delta_j =$			3 ↑	2	–2	0	0 ↓	
s_1	0	5/2	0	0	1/2	0	–1/2	
x_1	3	15/2	1	2	3/2	0	1/2	
$Z = C_B x_B = \dfrac{45}{2}, \Delta_j =$			0	– 4	–13/2	0	–3/2	

In the last row of the above table all $\Delta_j < 0$. So, the obtained solution is optimal, which is given by

$$x_1 = \frac{15}{2}, x_2 = 0, x_3 = 0 \text{ and max. } Z = \frac{45}{2}$$

EXAMPLE 5. *Solve the following LPP by simplex method.*

$$max. \ Z = 3x_1 + 5x_2$$

subject to the constraints

$$3x_1 + 2x_2 \le 18$$
$$x_1 \le 4$$
$$x_2 \le 6$$
and $\qquad x_1, x_2 \ge 0$

SOLUTION. Clearly, the given problem of maximization is in standard form such that all $b_i's \ge 0$.

Converting the inequalities into equations by introducing slack variables s_1, s_2, s_3 as follows:

$$max. \ Z = 3x_1 + 5x_2 + 0s_1 + 0s_2 + 0s_3$$
s.t. $\qquad 3x_1 + 2x_2 + s_1 = 18$
$$x_1 + s_2 = 4$$
$$x_2 + s_3 = 6$$
and $\qquad x_1, x_2, s_1, s_2, s_3 \ge 0$

Then we have the following simplex table

Simplex table

B.V.	C_B	x_B	x_1	x_2	s_1	s_2	s_3	Min Ratio
$c_j \rightarrow$			3	5	0	0	0	
s_1	0	18	3	2	1	0	0	18/2 = 9
s_2	0	4	1	0	0	1	0	—
s_3	0	6	0	①	0	0	1	6/1 = 6 (min)→
$Z = C_B x_B = 0, \Delta_j =$			3 ↑	5	0	0 ↓	0	

s_1	0	6	③	0	1	0	–2	6/3 = 2 (min) →
s_2	0	4	1	0	0	1	0	4/1 = 4
x_2	5	6	0	1	0	0	1	—
$Z = C_B x_B = 30, \Delta_j =$		3 ↑	0	0 ↓	0	–5		
x_1	3	2	1	0	1/3	0	–2/3	
s_2	0	2	0	0	–1/3	1	2/3	
x_2	5	6	0	1	0	0	1	
$Z = C_B x_B = 36, \Delta_j =$		0	0	–1	0	–3		

In the last row of the above table all $\Delta_j < 0$, so the obtained solution is optimal and is given by $x_1 = 2$, $x_2 = 6$ and max. $Z = 36$

EXAMPLE 6. *Solve the following LPP by simplex method*

$$\text{max. } Z = 3x_1 + 2x_2$$
subject to the constraints
$$2x_1 + x_2 \le 40, \; x_1 + x_2 \le 24, \; 2x_1 + 3x_2 \le 60$$
and $\quad x_1, x_2 \ge 0$

SOLUTION. Since, the given problem is in standard form. Converting the given inequalities into equations by introducing slack variables s_1, s_2, s_3 such that

$$\text{max. } Z = 3x_1 + 2x_2 + 0s_1 + 0s_2 + 0s_3$$
s.t. $\quad 2x_1 + x_2 + s_1 = 40, \; x_1 + x_2 + s_2 = 24, \; 2x_1 + 3x_2 + s_3 = 60$
and $\quad x_1, x_2, s_1, s_2, s_3 \ge 0$

Now applying simplex method, we get the following table:

Simplex table

B.V.	C_B	x_B	x_1	x_2	s_1	s_2	s_3	Min Ratio
$c_j \to$			3	4	0	0	0	
s_1	0	9	1	③	1	0	0	9/3 = 3 (min)→
s_2	0	8	2	–1	0	1	0	—
s_3	0	5	1	1	0	0	1	5/1 = 5
$Z = C_B x_B = 0, \Delta_j \to$			3	4 ↑	0 ↓	0	0	
x_2	4	3	1/3	1	1/3	0	0	9
s_2	0	11	7/3	0	1/3	1	0	33/7
s_3	0	2	②/3	0	–1/3	0	1	3 (min) →
$Z = C_B x_B = 12, \Delta_j \to$			5/3 ↑	0	– 4/3	0	0 ↓	
x_2	4	2	0	1	1/2	0	–1/2	
s_2	0	4	0	0	3/2	1	–7/2	
x_1	3	3	1	0	–1/2	0	3/2	
$Z = C_B x_B = 17, \Delta_j \to$			0	0	–1/2	0	–5/2	

In the last row of the above table all $\Delta_j \le 0$. Hence, the obtained solution is optimal which is given by

$$x_1 = 3, x_2 = 2 \text{ and max. } Z = 17$$

EXAMPLE 7. *Solve the following LPP by simplex method*

$$\text{max. } Z = 2x_1 + x_2$$

subject to the constraints

$$x_1 + 2x_2 \le 10, x_1 + x_2 \le 6$$
$$x_1 - x_2 \le 2, x_1 - 2x_2 \le 1$$

and $x_1, x_2 \ge 0$

SOLUTION. Clearly, the given LPP is in standard form.

Now, introduce slack variables s_1, s_2, s_3 and s_4 to convert the given inequalities into equations such that

$$\text{max. } Z = 2x_1 + x_2 + 0s_1 + 0s_2 + 0s_3 + 0s_4$$

s.t. $\quad x_1 + 2x_2 + s_1 = 10, x_1 + x_2 + s_2 = 6$
$$x_1 - x_2 + s_3 = 2, x_1 - 2x_2 + s_4 = 1$$

and $\quad x_1, x_2, s_1, s_2, s_3, s_4 \ge 0$

Now we have the following simplex table

Simplex table

B.V.	C_B	x_B	x_1	x_2	s_1	s_2	s_3	s_4	Min Ratio
	$c_j \to$		2	1	0	0	0	0	
s_1	0	10	1	2	1	0	0	0	10/1 = 10
s_2	0	6	1	1	0	1	0	0	6/1 = 6
s_3	0	2	1	−1	0	0	1	0	2/1 = 2
s_4	0	1	①	−2	0	0	0	1	1/1 = 1 (min)→
$Z = 0,$		$\Delta_j \to$	2 ↑	1	0	0	0	0 ↓	
s_1	0	9	0	4	1	0	0	−1	9/4
s_2	0	5	0	3	0	1	0	−1	5/3
s_3	0	1	0	①	0	0	1	−1	1/1 = 1 (min)→
x_1	2	1	1	−2	0	0	0	1	—
$Z = 2,$		$\Delta_j \to$	0	5 ↑	0	0	0 ↓	−2	
s_1	0	5	0	0	1	0	− 4	3	5/3
s_2	0	2	0	0	0	1	−3	②	2/2 = 1 (min)→
x_2	1	1	0	1	0	0	1	−1	—
x_1	2	3	1	0	0	0	2	−1	—
$Z = 8,$		$\Delta_j \to$	0	0	0	0	−5 ↓	3 ↑	

s_1	0	2	0	0	1	–3/2	1/2	0
s_4	0	1	0	0	0	1/2	–3/2	1
x_2	1	2	0	1	0	1/2	–1/2	0
x_1	2	4	1	0	0	1/2	1/2	0
$Z = C_B x_B = 10, \Delta_j \rightarrow$	0	0	0	–3/2	–1/2	0

In the last row of the above table all $\Delta_j \leq 0$. So, obtained solution is optimal and is given by $x_1 = 4$, $x_2 = 2$ and max. $Z = 10$

EXAMPLE 8. *Solve the following LPP by simplex method*

$$max.\ Z = 3x_1 + 5x_2 + 4x_3$$
subject to the constraints
$$2x_1 + 3x_2 \leq 8,\ 2x_1 + 5x_3 \leq 10,\ 3x_1 + 2x_2 + 4x_3 \leq 15$$
and $x_1, x_2, x_3 \geq 0$

[MEERUT–2007, 11, 14; KANPUR–2010, AVADH–2012; ALLAHBAD–2014]

SOLUTION. Clearly, the LPP is given in standard form.

Now, to convert the given inequalities into equations, we introduce the slack variables s_1, s_2, s_3 such that

max. $Z = 3x_1 + 5x_2 + 4x_3 + 0s_1 + 0s_2 + 0s_3$

s.t. $2x_1 + 3x_2 + s_1 = 8,\ 2x_1 + 5x_3 + s_2 = 10,\ 3x_1 + 2x_2 + 4x_3 + s_3 = 15$

and $x_1, x_2, x_3, s_1, s_2, s_3 \geq 0$

Now applying the usual procedure, we have the following simplex table.

Simplex table

B.V.	C_B	x_B	$c_j \rightarrow$ 3 x_1	5 x_2	4 x_3	0 s_1	0 s_2	0 s_3	Min Ratio
s_1	0	8	2	③	0	1	0	0	$\dfrac{8}{3}$ (min)→
s_2	0	10	0	2	5	0	1	0	10/2 = 5
s_3	0	15	3	2	4	0	0	1	15/2
$Z = C_B x_B = 0,$ $\Delta_j \rightarrow$			3 ↑	5 ↓	4	0	0	0	
x_2	5	$\dfrac{8}{3}$	$\dfrac{2}{3}$	1	0	1/3	0	0	—
s_2	0	$\dfrac{14}{3}$	$-\dfrac{4}{3}$	0	⑤	–2/3	1	0	$\dfrac{14}{15}$ (min)→
s_3	0	$\dfrac{29}{3}$	$\dfrac{5}{3}$	0	4	–2/3	0	1	$\dfrac{29}{12}$
$Z = C_B x_B = \dfrac{40}{3},$ $\Delta_j \rightarrow$			$-\dfrac{1}{3}$	0	4 ↑	–5/3	0 ↓	0	

	C_B	x_B							ratio
x_2	5	$\dfrac{8}{3}$	$\dfrac{2}{3}$	1	0	1/3	0	0	1
x_3	4	$\dfrac{14}{15}$	$-\dfrac{4}{15}$	0	1	$-2/15$	1/5	0	—
s_3	0	$\dfrac{89}{15}$	$\boxed{\dfrac{41}{15}}$	0	0	$-2/15$	$-4/5$	1	$\dfrac{89}{41}$ (min)→
$Z = C_B x_B = \dfrac{256}{45}, \Delta_j \rightarrow$			$\dfrac{11}{15}$ $\uparrow$	0	0	$-17/15$	$-4/5$ $\downarrow$	0	
x_2	5	$\dfrac{50}{41}$	0	1	0	$-15/41$	8/41	$-10/41$	
x_3	4	$\dfrac{62}{41}$	0	0	1	$-6/41$	5/41	4/41	
x_1	3	$\dfrac{89}{41}$	1	0	0	$-2/41$	$-12/41$	15/41	
$Z = C_B x_B = \dfrac{765}{41}, \Delta_j \rightarrow$			0	0	0	$-45/41$	$-14/41$	$-11/41$	

Clearly, in the last row of the above table, all $\Delta_j \leq 0$.

Hence, the optimal solution is given by

$$x_1 = \frac{89}{41}, x_2 = \frac{50}{41}, x_3 = \frac{62}{41} \text{ and max. } Z = \frac{765}{41}$$

EXAMPLE 9. *Solve the following LPP by simplex method*

$$\text{max. } Z = 3x_1 + 2x_2 + 5x_3$$

subject to the constraints

$$x_1 + 2x_2 + x_3 \leq 430$$
$$3x_1 + 2x_3 \leq 460$$
$$x_1 + 4x_2 \leq 420$$

and $\quad x_1, x_2, x_3 \geq 0$

[GARHWAL–2013; MEERUT–2008, 12; GORAKHPUR–2010, 11]

SOLUTION. Clearly, the given LPP is in standard form. Now, introduce slack variables s_1, s_2 and s_3 to convert the given inequalities into equations such that

$$\text{max. } Z = 3x_1 + 2x_2 + 5x_3 + 0s_1 + 0s_2 + 0s_3$$

s.t. $\qquad x_1 + 2x_2 + x_3 + s_1 = 430$

$$3x_1 + 2x_3 + s_2 = 460$$

$$x_1 + 4x_2 + s_3 = 420$$

and $\qquad x_1, x_2, x_3, s_1, s_2, s_3 \geq 0$

Simplex Table

B.V.	C_B	x_B	x_1	x_2	x_3	s_1	s_2	s_3	Min Ratio
		$c_j \rightarrow$	3	2	5	0	0	0	
s_1	0	430	1	2	1	1	0	0	430
s_2	0	460	3	0	②	0	1	0	230 (min)→
s_3	0	420	1	4	0	0	0	1	—
$Z = 0,$		$\Delta_j \rightarrow$	3	2	5	0	0	0	
				↑			↓		
s_1	0	200	$-1/2$	②	0	1	$-1/2$	0	100 (min)→
x_3	5	230	$3/2$	0	1	0	$1/2$	0	—
s_3	0	420	1	4	0	0	0	1	420/4
$Z = 1150,$		$\Delta_j \rightarrow$	$-9/2$	2	0	0	$-5/2$	0	
				↑			↓		
x_2	2	100	$-1/4$	1	0	$1/2$	$-1/4$	0	
x_3	5	230	$3/2$	0	1	0	$1/2$	0	
s_3	0	20	2	0	0	-2	1	1	
$Z = 1350,$		$\Delta_j \rightarrow$	-4	0	0	-1	-2	0	

Clearly in the last row of the above table all $\Delta_j \leq 0$.

$\Rightarrow$ Solution is optimal.

Hence, optimal solution is given by

$x_1 = 0$, $x_2 = 100$, $x_3 = 230$ and max. $Z = 1350$.

EXAMPLE 10. *Solve the following LPP by simplex method*

$$max. \ Z = 2x_1 + 4x_2 + 3x_3$$

subject to the constraints

$$3x_1 + 4x_2 + 2x_3 \leq 60$$
$$2x_1 + x_2 + 2x_3 \leq 40$$
$$x_1 + 3x_2 + 2x_3 \leq 80$$

and $x_1, x_2, x_3 \geq 0$

SOLUTION. Clearly, the given LPP is in standard form. Introduce slack variables s_1, s_2 and s_3 to convert the given inequalities into equations. Then we get

max. $Z = 2x_1 + 4x_2 + 3x_3 + 0s_1 + 0s_2 + 0s_3$

s.t. $3x_1 + 4x_2 + 2x_3 + s_1 = 60$

$2x_1 + x_2 + 2x_3 + s_2 = 40$

$x_1 + 3x_2 + 2x_3 + s_3 = 80$

and $x_1, x_2, x_3, s_1, s_2, s_3 \geq 0$

Now, apply the simplex method, we have the following simplex table:

Simplex table

B.V.	C_B	x_B	x_1	x_2	x_3	s_1	s_2	s_3	Min Ratio
$c_j \rightarrow$			2	4	3	0	0	0	
s_1	0	60	3	④	2	1	0	0	60/4 = 15 (min)→
s_2	0	40	2	1	2	0	1	0	40/1 = 40
s_3	0	80	1	3	2	0	0	1	80/3
$Z = 0$,		$\Delta_j \rightarrow$	2	4	3	0	0	0	
			↑	↓					
x_2	4	15	3/4	1	1/2	1/4	0	0	—
s_2	0	25	5/4	0	③/②	–1/4	1	0	25/(3/2)=(50/3)(min)→
s_3	0	35	–5/4	0	1/2	–3/4	0	1	—
$Z = 60$,		$\Delta_j \rightarrow$	–1	0	1	–1	0	0	
					↑	↓			
x_2	4	20/3	1/6	1	0	1/3	–1/3	0	
x_3	3	50/3	5/6	0	1	–1/6	2/3	0	
s_3	0	80/3	–5/3	0	0	–1/12	–1/3	1	
$Z = C_B x_B = 230/3$, $\Delta_j \rightarrow$			–11/6	0	0	–5/6	–2/3	0	

Clearly in the last row of the above table all $\Delta_j \leq 0$.

⇒ Solution is optimal.

Hence, optimal solution is given by

$$x_1 = 0, x_2 = \frac{20}{3}, x_3 = \frac{50}{3} \text{ and max. } Z = \frac{230}{3}$$

Exercise-3.2

Solve the following LPP by simplex method:

1. max. $Z = 40x_1 + 35x_2$
   s.t. $2x_1 + 3x_2 \leq 60$
   $4x_1 + 3x_2 \leq 96$
   $x_1, x_2 \geq 0$

2. max. $Z = 7x_1 + 5x_2$
   s.t. $-x_1 - 2x_2 \geq -6$
   $4x_1 + 3x_2 \leq 12$
   $x_1, x_2 \geq 0$

3. max. $Z = 5x_1 + 3x_2$
   s.t. $3x_1 + 5x_2 \leq 15$
   $5x_1 + 2x_2 \leq 10$
   $x_1, x_2 \geq 0$

4. max. $Z = 7x_1 + x_2 + 2x_3$
   s.t. $x_1 + x_2 - 2x_3 \leq 10$
   $4x_1 + x_2 + x_3 \leq 20$
   $x_1, x_2, x_3 \geq 0$

5. max. $Z = 5x_1 + 7x_2$
   s.t. $x_1 + x_2 \leq 4$
   $3x_1 - 8x_2 \leq 24$
   $10x_1 + 7x_2 \leq 35$
   $x_1, x_2 \geq 0$

6. max. $Z = 3x_1 + 2x_2$
   s.t. $2x_1 + x_2 \leq 40$
   $x_1 + x_2 \leq 24$
   $2x_1 + 3x_2 \leq 60$
   $x_1, x_2 \geq 0$

7. max. $Z = 2x_1 + 4x_2$
   s.t. $2x_1 + 3x_2 \leq 48$
   $x_1 + 3x_2 \leq 42$
   $x_1 + x_2 \leq 21$
   $x_1, x_2 \geq 0$

8. max. $Z = 5x_1 + 10x_2 + 8x_3$
   s.t. $3x_1 + 5x_2 + 2x_3 \leq 60$
   $4x_1 + 4x_2 + 4x_3 \leq 72$
   $2x_1 + 4x_2 + 5x_3 \leq 100$
   $x_1, x_2, x_3 \geq 0$ [MEERUT–2009, 12]

9. max. $Z = x_1 - x_2 + 3x_3$

s.t. $\quad x_1 + x_2 + x_3 \le 10$

$\qquad 2x_1 - x_3 \le 2$

$\qquad 2x_1 - 2x_2 + 3x_3 \le 0$

$\qquad x_1, x_2, x_3 \ge 0$

10. max. $Z = 2x_1 + 5x_2 + 7x_3$

s.t. $3x_1 + 2x_2 + 4x_3 \le 100$

$\qquad x_1 + 4x_2 + 2x_3 \le 100$

$\qquad x_1 + x_2 + 3x_3 \le 100$

$\qquad x_1, x_2, x_3 \ge 0$

[MEERUT–2009, 11; KANPUR–2009]

11. max. $Z = 4x_1 + 5x_2 + 9x_3 + 11x_4$

s.t. $x_1 + x_2 + x_3 + x_4 \le 15$

$\qquad 7x_1 + 5x_2 + 3x_3 + 2x_4 \le 120$

$\qquad 3x_1 + 5x_2 + 10x_3 + 15x_4 \le 100$

$\qquad x_1, x_2, x_3, x_4 \ge 0$

ANSWERS

1. $x_1 = 8, x_2 = 18$, max. $Z = 1000$

2. $x_1 = 3, x_2 = 0$, max. $Z = 21$

3. $x_1 = \dfrac{20}{19}, x_2 = \dfrac{45}{19}$, max. $Z = \dfrac{235}{19}$

4. $x_1 = x_2 = 0, x_3 = 20$, max. $Z = 40$

5. $x_1 = 0, x_2 = 4$, max. $Z = 28$

6. $x_1 = 16, x_2 = 8$, max. $Z = 64$

7. $x_1 = 6, x_2 = 12$, max. $Z = 60$

8. $x_1 = 0, x_2 = 8, x_3 = 10$, max. $Z = 160$

9. $x_1 = 0, x_2 = 6, x_3 = 4$, max. $Z = 6$

10. $x_1 = 0, x_2 = \dfrac{50}{3}, x_3 = \dfrac{50}{3}$, max. $Z = 200$

11. $x_1 = \dfrac{50}{7}, x_2 = 0, x_3 = \dfrac{55}{7}$ max. $Z = \dfrac{695}{7}$

3.12 SIMPLEX METHOD : CASE OF MINIMIZATION

In case of minimization of an LPP, we have the following cases:

Case I. When the constraints are of the type $\le$, *i.e.*,

$$\sum_{j=1}^{n} a_{ij}x_j \le b_i, x_j \ge 0$$

But some right hand side constraints are negative ($b_i < 0$). Then, after adding the non-negative slack variables s_i, the initial solution so obtained will be $s_i = -b_i$ for some i. It is not the feasible solution.

Case II. When the constraints are of the $\ge$ type, *i.e.*,

$$\sum_{j=1}^{n} a_{ij}x_j \ge b_i, x_j \ge 0$$

Then convert the inequalities into equation's form, adding surplus variables such that

$$\sum_{j=1}^{n} a_{ij}x_j - s_i = b_i, x_j \ge 0, s_i \ge 0$$

Letting $x_j = 0$ we get an initial solution $-s_i = b_j$ or $s_j = -b_i$ which is also not a feasible solution. In this case we add artificial variable A_i to get an initial basic feasible solution.

> If the given problem is of minimization, we multiply the objective function by -1, *i.e.*,
> $$Z' = -Z = -c_i x_i.$$

3.13 ARTIFICIAL VARIABLE TECHNIQUE

There are following two methods for eliminating artificial variables from the solution:

(i) Big-M method

(ii) Two phase method

(i) Big-M method: The artificial variable like surplus and slack variables are imaginary variables which are introduced in the constraints with $\geq$ or $=$ sign. These are just introduced to form an identity matrix. These variables have no physical significance. So, these variables have to be removed by the simplex method to have the meaningful solution. Here a very large price say $-M$ has to be alloted to the artificial variable in the objective function. This method of solution by aritificial variables was given by Charnes and is hence known as Charnes Big-M method.

 Solved Examples

EXAMPLE I. *Solve the following LPP by Big-M method.*
$$\text{Min } Z = 5x_1 + 6x_2$$
subject to the constraints
$$2x_1 + 5x_2 \geq 1500,\ 3x_1 + x_2 \geq 1200$$
and $\quad x_1, x_2 \geq 0$

SOLUTION. The given problem is a minimization problem.
So, first convert it to maximization problem by using $Z' = Z$.
Then objective function becomes max. $Z' = -5x_1 - 6x_2$
Now, introduce the necessary surplus variables s_1, s_2 and artificial variables A_1 and A_2, then given problem becomes
$$\text{max } Z' = -5x_1 - 6x_2 + 0s_1 + 0s_2 - MA_1 - MA_2$$
s.t. $\quad 2x_1 + 5x_2 - s_1 + A_1 = 1500$
$\qquad\quad 3x_1 + x_2 - s_2 + A_2 = 1200$
and $\quad x_1, x_2, s_1, s_2, A_1, A_2 \geq 0$

Now apply the simplex method, we have the following simplex table.

Simplex table

B.V.	C_B	x_B	x_1	x_2	s_1	s_2	A_1	A_2	Min Ratio
	$c_j \rightarrow$		-5	-6	0	0	$-M$	$-M$	
A_1	$-M$	1500	2	⑤	-1	0	1	0	$1500/5 = 300$ (Min) $\rightarrow$
A_2	$-M$	1200	3	1	0	-1	0	1	$1200/1$
$Z' = C_B x_B,$ $\Delta_j \rightarrow$ $= -2700M$			$5(M-1)$	$6(M-1)$ $\uparrow$	$-M$	$-M$ $\downarrow$	0	0	
x_2	-6	300	2/5	1	$-1/5$	0	—	0	750
A_2	$-M$	900	⑬/5	0	1/5	-1	—	1	$4500/13$ (Min) $\rightarrow$
$Z' = -900M$ $\Delta_j \rightarrow$ $= -1800$			$(13/5)$ $(M-1)$ $\uparrow$	0	$(1/5)$ $(M-6)$	$-M$	—	0 $\downarrow$	
x_2	-6	2100/13	0	1	$-3/13$	2/13	—	—	
x_1	-5	4500/13	1	0	1/13	$-5/13$	—	—	
$Z' = C_B x_B,$ $\Delta_j \rightarrow$ $= -2700$			0	0	-1	-1			

Computation of Δ_j

For first table:

$\Delta_1 = c_1 - C_B x_1 = -5 - (-M, -M)(2, 3) = 5(1 + M)$

$\Delta_2 = c_2 - C_B x_2 = -6 - (-M, -M)(5, 1) = 6(M - 1)$

$\Delta_3 = c_3 - C_B s_1 = 0 - (-M, -M)(-1, 0) = -M$

$\Delta_4 = c_4 - C_B s_2 = 0 - (-M, -M)(0, -1) = -M$

For second table:

$\Delta_1 = c_1 - C_B x_1 = -5 - (-6, -M)\left(\dfrac{2}{5}, \dfrac{13}{5}\right) = \dfrac{13(M-1)}{5}$

$\Delta_2 = 0 = \Delta_5, \Delta_3 = c_3 - C_B s_1 = 0 - (-6, -M)\left(-\dfrac{1}{5}, \dfrac{1}{5}\right) = \dfrac{(M-6)}{5}$

$\Delta_4 = c_4 - C_B s_2 = 0 - (-6, -M)(0, -1) = -M$

For third table:

$\Delta_1 = 0 = \Delta_2, \Delta_3 = c_3 - C_B s_1 = 0 - (-6, -5)\left(-\dfrac{3}{13}, \dfrac{1}{13}\right) = -1$

$\Delta_4 = c_4 - C_B s_2 = 0 - (-6, -5)\left(\dfrac{2}{13}, \dfrac{-5}{13}\right) = -1$

Conclusion: In the last row of the third table all $\Delta_j \leq 0$ and no artificial variables appears in the basis. Thus, the solution is optimal and is given by

$$x_1 = \frac{4500}{13}, x_2 = \frac{2100}{13}, \min. Z = -\max. Z' = 2700$$

EXAMPLE 2. *Solve the following LPP by simplex method*

$$Max. Z = 2x_1 + 4x_2$$

subject to the constraints

$$2x_1 + x_2 \leq 18$$
$$3x_1 + 2x_2 \geq 30$$
$$x_1 + 2x_2 = 26$$

and $x_1, x_2 \geq 0$ [MEERUT–2007, 10, 12; BHOPAL–2014]

SOLUTION. Clearly, the problem under consideration is of maximization and all $b_i's$ are positive.

Now, introduce slack variable s_1, surplus variable s_2 and artificial variables A_1 and A_2. Then we get the following LPP

$$Max. Z = 2x_1 + 4x_2 + 0s_1 + 0s_2 - MA_1 - MA_2$$

s.t. $2x_1 + x_2 + s_1 = 18$

$3x_1 + 2x_2 - s_2 + A_1 = 30$

$x_1 + 2x_2 + A_2 = 26$

and $x_1, x_2, s_1, s_2, A_1, A_2 \geq 0$

Now apply the simplex method, we have the following simplex table.

Simplex table

B.V.	C_B	x_B	x_1	x_2	s_1	s_2	A_1	A_2	Min Ratio
		$c_j \to$	2	4	0	0	$-M$	$-M$	
s_1	0	18	2	1	1	0	0	0	18
A_1	$-M$	30	3	2	0	-1	1	0	15
A_2	$-M$	26	1	②	0	0	0	1	13 (Min) $\to$
$Z=C_B x_B=-56M, \Delta_j \to$			$(2+4M)$	$(4+4M)$ $\uparrow$	0	$-M$	0	0 $\downarrow$	
s_1	0	5	3/2	0	1	0	0	—	10/3
A_1	$-M$	4	②	0	0	-1	1	—	2 (Min) $\to$
x_2	4	13	1/2	1	0	0	0	—	26
$Z = 52 - 4M, \quad \Delta_j \to$			$2M$ $\uparrow$	0	0	$-M$	0	—	
s_1	0	2	0	0	1	3/4	—	—	
x_1	2	2	1	0	0	$-1/2$	—	—	
x_2	4	12	0	1	0	1/4	—	—	
$Z = 52 \qquad \Delta_j \to$			0	0	0	0			

Computation of Δ_j:

(i) For first table:

$$\Delta_1 = c_1 - C_B x_1 = 2 - (0, -M, -M)(2, 3, 1) = 2 + 4M$$
$$\Delta_2 = c_2 - C_B x_2 = 4 - (0, -M, -M)(1, 2, 2) = 4 + 4M$$
$$\Delta_3 = 0$$
$$\Delta_4 = c_4 - C_B s_2 = 0 - (0, -M, -M)(0, -1, 0) = -M$$

(ii) For second table:

$$\Delta_1 = c_1 - C_B x_1 = 2 - (0, -M, 4)\left(\frac{3}{2}, 2, \frac{1}{2}\right) = 2M$$

$$\Delta_2 = 0 = \Delta_3 = \Delta_5$$
$$\Delta_4 = c_4 - C_B s_2 = 0 - (0, -M, 4)(0, -1, 0) = -M$$

(iii) For third table:

$$\Delta_1 = 0 = \Delta_2 = \Delta_3$$
$$\Delta_4 = c_4 - C_B s_2 = 0 - (0, 2, 4)\left(\frac{3}{4}, \frac{-1}{2}, \frac{1}{4}\right) = 0$$

Conclusions.

In the last row of third table all $\Delta_j \leq 0$ and no artificial variable appears in the basis, so solution is optimal.

Hence, the optimal solution is given by

$$x_1 = 2, x_2 = 12 \text{ and max. } Z = 52$$

EXAMPLE 3. *Solve the following LPP by Big-M method*

$$\text{Max. } Z = -2x_1 - x_2$$

subject to the constraints

$$3x_1 + x_2 = 3$$
$$4x_1 + 3x_2 \geq 6$$
$$x_1 + 2x_2 \leq 4$$

and $x_1, x_2 \geq 0$

SOLUTION. The given maximization problem is in standard form. Now, introducing the surplus variables s_1 and s_2 and artificial variables A_1 and A_2, to convert the given inequalities into equations, we get,

$$\text{max. } Z = -2x_1 - x_2 + 0s_1 + 0s_2 - MA_1 - MA_2$$

s.t.
$$3x_1 + x_2 + A_1 = 3$$
$$4x_1 + 3x_2 - s_1 + A_2 = 6$$
$$x_1 + 2x_2 + s_2 = 4$$

Apply the Big-M method, we have the following simplex table:

Simplex table

B.V.	C_B	x_B	x_1 (-2)	x_2 (-1)	s_1 (0)	s_2 (0)	A_1 $(-M)$	A_2 $(-M)$	Min Ratio
A_1	$-M$	3	③	1	0	0	1	0	3/3 = 1 (Min) →
A_2	$-M$	6	4	3	-1	0	0	1	6/4
s_2	0	4	1	2	0	1	0	0	4/1
$Z=C_B x_B=-9M$, $\Delta_j \rightarrow$			$(-2+7M)$ ↑	$(-1+4M)$	$-M$	0 ↓	0	0	
x_1	-2	1	1	1/3	0	0	—	0	3
A_2	$-M$	2	0	⑤/3	-1	0	—	1	6/5 (Min) →
s_2	0	3	0	5/3	0	1	—	0	9/5
$Z=C_B x_B=-2-2M$, $\Delta_j \rightarrow$			0	$\left(\dfrac{-1+5M}{3}\right)$ ↑	$-M$	0	—	0 ↓	
x_1	-2	3/5	1	0	1/5	0	—	—	
x_2	-1	6/5	0	1	$-3/5$	0	—	—	
s_2	0	1	0	0	1	1	—	—	
$Z=C_B x_B=-12/5$, $\Delta_j \rightarrow$			0	0	$-1/5$	0			

Computation of Δ_j:
(i) For the first table:

$$\Delta_1 = c_1 - C_B x_1 = -2 - (-M, -M, 0)(3, 4, 1) = -2 + 7M$$
$$\Delta_2 = c_2 - C_B x_2 = -1 - (-M, -M, 0)(1, 3, 5) = -1 + 4M$$
$$\Delta_3 = c_3 - C_B s_1 = 0 - (-M, -M, 0)(0, -1, 0) = -M$$

(ii) For the second table:

$$\Delta_2 = c_2 - C_B x_2 = -1 - (-2, -M, 0)\left(\frac{1}{3}, \frac{5}{3}, \frac{5}{3}\right) = \frac{1}{3}(-1 + 5M)$$

$$\Delta_3 = c_3 - C_B s_1 = 0 - (-2, -M, 0)(0, -1, 0) = -M$$

(iii) For the third table:

$$\Delta_1 = \Delta_2 = \Delta_4 = 0$$

$$\Delta_3 = c_3 - C_B s_1 = 0 - (-2, -1, 0)\left(\frac{1}{5}, -\frac{3}{5}, 1\right) = -\frac{1}{5}$$

Conclusion. Since in the last row of the last table, all $\Delta_j \leq 0$ and no artificial variable appears in the basis, so solution is optimal.

Hence, optimal solution is given by $x_1 = \dfrac{3}{5}, x_2 = \dfrac{6}{5}, \max. Z = \dfrac{-12}{5}$

EXAMPLE 4. Solve the following LPP by Big-M method:

$$\text{Min. } Z = x_1 + x_2 + 3x_3$$

subject to the constraints

$$3x_1 + 2x_2 + x_3 \leq 3, \; 2x_1 + x_2 + 2x_3 \geq 3$$

and $\qquad x_1, x_2, x_3 \geq 0$

SOLUTION. Convert the given minimization problem to maximization such that

$$Z' = -Z = -x_1 - x_2 - 3x_3$$

Now introduce the slack variable s_1, surplus variable s_2 and artificial variable A_1. Then given problem reduces to

$$\text{Max. } Z' = -Z = -x_1 - x_2 - 3x_3 + 0x_4 + 0x_5 - MA_1$$

s.t. $\qquad 3x_1 + 2x_2 + x_3 + s_1 = 3$

$$2x_1 + x_2 + 2x_3 - s_2 + A_1 = 3$$

and $\qquad x_1, x_2, x_3, s_1, s_2, A_1 \geq 0$

Now, we have the following simplex tables:

Simplex table

B.V.	C_B	x_B	$c_j \to$ x_1 -1	x_2 -1	x_3 -3	s_1 0	s_2 0	A_1 $-M$	Min Ratio
s_2	0	3	③	2	1	1	0	0	$3/3 = 1$ (Min) $\to$
A_1	$-M$	3	2	1	2	0	-1	1	$3/2$
$Z' = C_B x_B,$ $= -2M$		$\Delta_j \to$	$(-1+2M)$ $\uparrow$	$(-1+M)$	$(-3+2M)$	0	$-M$ $\downarrow$	0	
x_1	-1	1	1	$2/3$	$1/3$	$1/3$	0	0	3
A_1	$-M$	1	0	$-1/3$	④/③	$-2/3$	-1	1	$3/4$ (Min) $\to$
$Z' = C_B x_B,$ $= -1 - M$		$\Delta_j \to$	0	$-\left(\dfrac{1+M}{3}\right)$	$\dfrac{4}{3}(-2+M)$ $\uparrow$	$\dfrac{1}{3}(1-2M)$	$-M$ $\downarrow$	0	
x_1	-1	$3/4$	1	$3/4$	0	$1/2$	$1/4$	—	
x_3	-3	$3/4$	0	$-1/4$	1	$-1/2$	$-3/4$	—	
$Z' = C_B x_B,$ $= -3$		$\Delta_j \to$	0	-1	0	-1	-2		

Computation of Δ_j:

(i) For the first table:

$$\Delta_1 = c_1 - C_B x_1 = -1 - (0, -M)(3, 2) = -1 + 2M$$
$$\Delta_2 = c_2 - C_B x_2 = -1 - (0, -M)(2, 1) = -1 + M$$
$$\Delta_3 = c_3 - C_B s_1 = -3 - (0, -M)(1, 2) = -3 + 2M$$
$$\Delta_4 = 0$$
$$\Delta_5 = c_5 - C_B s_2 = 0 - (0, -M)(0, -1) = -M$$

(ii) For the second table:

$$\Delta_1 = \Delta_6 = 0$$
$$\Delta_2 = c_2 - C_B x_2 = -1 - (-1, -M)\left(\frac{2}{3}, -\frac{1}{3}\right) = -\frac{(1+M)}{3}$$
$$\Delta_3 = c_3 - C_B x_3 = -3 - (-1, -M)\left(\frac{1}{3}, \frac{4}{3}\right) = \frac{4(-2+M)}{3}$$
$$\Delta_4 = c_4 - C_B s_1 = 0 - (-1, -M)\left(\frac{1}{3}, \frac{-2}{3}\right) = \frac{(1-2M)}{3}$$
$$\Delta_5 = c_5 - C_B s_2 = 0 - (-1, -M)(0, -1) = -M$$

(iii) For the third table:

$$\Delta_1 = \Delta_3 = 0$$
$$\Delta_2 = c_2 - C_B x_2 = -1 - (-1, -3)\left(\frac{3}{4}, -\frac{1}{4}\right) = -1$$
$$\Delta_4 = c_4 - C_B s_1 = 0 - (-1, -3)\left(\frac{1}{2}, -\frac{1}{2}\right) = -1$$
$$\Delta_5 = c_5 - C_B s_2 = 0 - (-1, -3)\left(\frac{1}{4}, \frac{-3}{4}\right) = -2$$

Conclusion. Since in the last row of the last table all $\Delta_j \leq 0$, so solution is optimal and hence optimal solution is given by

$$x_1 = \frac{3}{4}, x_2 = 0, x_3 = \frac{3}{4} \text{ and } \min.Z = -\max.Z' = 3$$

EXAMPLE 5. *Solve the following LPP by Big-M method*

$$Max.\ Z = x_1 + 2x_2 + 3x_3 - x_4$$

subject to the constraints

$$x_1 + 2x_2 + 3x_3 = 15$$
$$2x_1 + x_2 + 5x_3 = 20$$
$$x_1 + 2x_2 + x_3 + x_4 = 10$$

and $x_1, x_2, x_3, x_4 \geq 0$ [MEERUT–2009; GORAKHPUR–2008]

SOLUTION. The given maximization problem is in standard form. Also, constraints are equations. We observe that, to obtain a unit matrix of order 3, we need two more unit vectors as one unit vector is formed by the coefficient of x_4.

So, introducing two artificial variables A_1 and A_2 in the first two constraints, the given problem becomes

$$\text{Max. } Z = x_1 + 2x_2 + 3x_3 - x_4 - MA_1 - MA_2$$

$$\text{s.t.} \quad x_1 + 2x_2 + 3x_3 + A_1 = 15$$

$$2x_1 + x_2 + 5x_3 + A_2 = 20$$

$$x_1 + 2x_2 + x_3 + x_4 = 10$$

$$\text{and} \quad x_1, x_2, x_3, x_4, A_1, A_2 \geq 0$$

Now we have the following simplex table:

Simplex table

B.V.	C_B	x_B	x_1	x_2	x_3	x_4	A_1	A_2	Min Ratio
	$c_j \rightarrow$		1	2	3	-1	$-M$	$-M$	
A_1	$-M$	15	1	2	3	0	1	0	5
A_2	$-M$	20	2	1	⑤	0	0	1	4 (Min) →
x_4	-1	10	1	2	1	1	0	0	10
$Z = C_B x_B,$ $\Delta_j \rightarrow$ $= -35M - 10$			$(3M+2)$	$(3M+4)$	$(8M+4)$ $\uparrow$	0	0	0 $\downarrow$	
A_1	$-M$	3	$-1/5$	⑦/5	0	0	1	—	15/7(Min)→
x_3	3	4	2/5	1/5	1	0	0	—	20
x_4	-1	6	3/5	9/5	0	1	0	—	10/3
$Z = C_B x_B,$ $\Delta_j \rightarrow$ $= -3M + 6$			$\dfrac{(-M+2)}{5}$	$\dfrac{7M+16}{5}$ $\uparrow$	0	0	0 $\downarrow$	0	
x_2	2	15/7	$-1/7$	1	0	0	—	—	–ve
x_3	3	25/7	3/7	0	1	0	—	—	25/3
x_4	-1	15/7	⑥/7	0	0	1	—	—	5/2 (Min)→
$Z = 90/7,$ $\Delta_j \rightarrow$			6/7 $\uparrow$	0	0	0 $\downarrow$			
x_2	2	5/2	0	1	0	1/6			
x_3	3	5/2	0	0	1	$-1/2$			
x_1	1	5/2	1	0	0	7/6			
$Z = C_B x_B,$ $\Delta_j \rightarrow$ $= 15$			0	0	0	-1			

In the above table (last row) all $\Delta_j \leq 0$

Hence solution is optimal and is given by

$$x_1 = \frac{5}{2} = x_2 = x_3, x_4 = 0 \text{ and max. } Z = 15$$

EXAMPLE 6. *Solve the following LPP by Big-M method*

$$Max.\ Z = 4x_1 + 5x_2 - 3x_3$$

subject to the constraints

$$x_1 + x_2 + x_3 = 10$$
$$x_1 - x_2 \geq 1$$
$$2x_1 + 3x_2 + x_3 \leq 40$$

and $x_1, x_2, x_3 \geq 0$

SOLUTION. The given maximization problem is in standard form. So, introduce the surplus variable s_1, slack variable s_2 and artificial variable A_1 and A_2 in the following manner:

$$max.\ Z = 4x_1 + 5x_2 - 3x_3 + 0s_1 + 0s_2 - MA_1 - MA_2$$

s.t.
$$x_1 + x_2 + x_3 + A_1 = 10$$
$$x_1 - x_2 - s_1 + A_2 = 1$$
$$2x_1 + 3x_2 + x_3 + s_2 = 40$$

and $x_1, x_2, x_3, s_1, s_2, A_1, A_2 \geq 0$

Now we have the following simplex table:

Simplex table

B.V.	C_B	x_B	$c_j \rightarrow$ 4 x_1	5 x_2	–3 x_3	0 s_1	0 s_2	–M A_1	–M A_2	Min Ratio
A_1	–M	10	1	1	1	0	0	1	0	10/1 = 10
A_2	–M	1	①	–1	0	–1	0	0	1	1/1 = 1(Min)→
s_2	0	40	2	3	1	0	1	0	0	40/2 = 20
$Z = -21M,$	$\Delta_j \rightarrow$		(4 + 2M) ↑	5	(–3 + M)	–M	0	0	0 ↓	
A_1	–M	9	0	②	1	1	0	1	—	9/2(Min)→
x_1	4	1	1	–1	0	–1	0	0	—	Negative
s_2	0	38	0	5	1	2	1	0	—	38/5
$Z = -9M + 4, \Delta_j \rightarrow$			0	(9 + 2M) ↑	–3 + M	4 + M	0	0	↓	
x_2	5	9/2	0	1	1/2	1/2	0	—	—	
x_1	4	11/2	1	0	1/2	–1/2	0	—	—	
s_2	0	31/2	0	0	–3/2	–1/2	1	—	—	
$Z = C_B x_B,$ = 89/2	$\Delta_j \rightarrow$		0	0	–15/2	–1/2	0			

Computation of Δ_j:

(i) For the first table:

$$\Delta_1 = c_1 - C_B x_1 = 4 - (-M, -M, 0)(1, 1, 2) = 4 + 2M$$
$$\Delta_2 = c_2 - C_B x_2 = 5 - (-M, -M, 0)(1, -1, 3) = 5$$
$$\Delta_3 = c_3 - C_B x_3 = -3 - (-M, -M, 0)(1, 0, 1) = -3 + M$$

$$\Delta_4 = c_4 - C_B s_1 = 0 - (-M, -M, 0)(0, -1, 0) = -M$$
$$\Delta_5 = \Delta_6 = \Delta_7 = 0$$

(ii) For the second table:

$$\Delta_1 = 0, \Delta_2 = c_2 - C_B x_2 = 5 - (-M, 4, 0)(2, -1, 5) = 9 + 2M$$
$$\Delta_3 = c_3 - C_B x_3 = -3 - (-M, 4, 0)(1, 0, 1) = -3 + M$$
$$\Delta_4 = c_4 - C_B s_1 = 0 - (-M, 4, 0)(1, -1, 2) = 4 + M$$
$$\Delta_5 = \Delta_6 = 0$$

(iii) For the third table:

$$\Delta_1 = \Delta_2 = \Delta_5 = 0$$

$$\Delta_3 = C_3 - C_B x_3 = -3 - (5, 4, 0)\left(\frac{1}{2}, \frac{1}{2}, -\frac{3}{2}\right) = -\frac{15}{2}$$

$$\Delta_4 = C_4 - C_B s_1 = 0 - (5, 4, 0)\left(\frac{1}{2}, \frac{-1}{2}, \frac{-1}{2}\right) = \frac{-1}{2}$$

Conclusions: In the last row of last table all $\Delta_j \leq 0$, so solution is optimal. Hence, optimal solution is given by

$$x_1 = \frac{11}{2}, x_2 = \frac{9}{2}, x_3 = 0 \text{ and max. } Z = \frac{89}{2}$$

 Exercise-3.3

Solve the following LPP by Big-M method.

1. max. $Z = 4x_1 + 2x_2$
   s.t. $3x_1 + x_2 \leq 27$
   $x_1 + x_2 \geq 21$
   and $x_1, x_2 \geq 0$

2. min. $Z = 2x_1 + 3x_2$
   s.t. $x_1 + x_2 \geq 5$
   $x_1 + 2x_2 \geq 6$
   and $x_1, x_2 \geq 0$

3. max. $Z = 3x_1 - x_2$
   s.t. $2x_1 + x_2 \geq 2$

$x_1 + 3x_2 \leq 3$
$x_2 \leq 4$
and $x_1, x_2 \geq 0$

4. min. $Z = 0.60x_1 + 0.80x_2$
   s.t. $20x_1 + 30x_2 \geq 900$
   $40x_1 + 30x_2 \geq 1200$
   and $x_1, x_2 \geq 0$

5. min. $Z = 2x_1 + 8x_2$
   s.t. $5x_1 + 10x_2 = 150$
   $x_1 \leq 20$
   $x_2 \geq 14$
   and $x_1, x_2 \geq 0$

ANSWERS

1. $x_1 = 0, x_2 = 27$, max. $Z = 54$
2. $x_1 = 4, x_2 = 1$ and min. $Z = 11$
3. $x_1 = 3, x_2 = 0$, max. $Z = 9$
4. $x_1 = 15, x_2 = 20$, min. $Z = 25$
5. $x_1 = 2, x_2 = 14$, min. $Z = 116$

3.14 TWO PHASE METHOD
[KANPUR–2012]

This is an alternative of Big-M method. Using this method, we obtain the solution in two phases given as follows:

(1) In phase-I all the artificial variables are eliminated from the basis.

(2) In phase-II, we use the solution from phase-I as the initial basic feasible solution and then use the simplex method to obtain the optimal solution.

WORKING PROCEDURE

(A) For Phase-1

STEP 1. Convert the given LPP in the standard form.

STEP 2. Add the necessary artificial variables to the constraints as done in Big-M method to obtain an initial basic feasible solution.

STEP 3. Formulate an artificial objective function Z^* such that
$$Z^* = -A_1 - A_2 - \dots - A_n \ (= - \text{(sum of the artificial variables)})$$
by assigning -1 cost to each artificial variable A_i and zero cost to all other variables.

STEP 4. Maximize Z^* subject to the constraints of the original problem using the simplex method. Now, we have the following cases:

CASE-I. If max. $Z^* < 0$ and at least one artificial variable appears in the optimal basis at a positive level, then given LPP will not have any feasible solution and then we will not move to phase-II.

CASE-II. If max. $Z^* = 0$ and no artificial variables appears in the optimal basis then BFS is not obtained and in order to obtain optimal BFS, we move to phase-II.

CASE-III. If max. $Z^* = 0$ and at least one artificial variable appears in the optimal basis at zero level, then a feasible solution to the auxiliary LPP is also a feasible solution of the given LPP by setting all artificial variables to zero. Finally to obtain the basic feasible solution, remove all the artificial variable from the basis matrix.

(B) For Phase-2

STEP 1. Take the basic feasible solution, which was found at the end of Phase-1 as the starting BFS for the given LPP.

STEP 2. Apply simplex method to find the optimal basic feasible solution.

 Solved Examples

EXAMPLE 1. *Apply Two-phase method, solve the following LPP.*

$$\text{min. } Z = 40x_1 + 24x_2$$

subject to the constraints

$$20x_1 + 50x_2 \geq 4800$$
$$80x_1 + 50x_2 \geq 7200$$

and $\qquad x_1, x_2 \geq 0$ 　　　　　　　　　　　　　　[MEERUT–2008, 12]

SOLUTION. Firstly, convert the given minimization problem into maximization by taking the objective function as

$$Z^* = -Z = -40x_1 - 24x_2$$

Now, introduce the surplus variables s_1, s_2 and artificial variables A_1, A_2 such that

$$20x_1 + 50x_2 - s_1 + A_1 = 4800$$
$$80x_1 + 50x_2 - s_2 + A_2 = 7200$$

and $\qquad x_1, x_2, s_1, s_2, A_1, A_2 \geq 0$

PHASE-1. We assign the cost -1 to the artificial variable, and 0 to all other variables. Then the new objective function of auxiliary LPP becomes

$$\text{max. } Z^* = 0x_1 + 0x_2 + 0s_1 + 0s_2 - A_1 - A_2$$

subject to the above constraints.

Now by applying the simplex method in a usual manner, we have the following simplex table:

		$c_j \rightarrow$	0	0	0	0	-1	-1	
B.V.	C_B	x_B	x_1	x_2	s_1	s_2	A_1	A_2	Min Ratio
A_1	-1	4800	20	50	-1	0	1	0	4800/20=240
A_2	-1	7200	⑧⓪	50	0	-1	0	1	7200/80=90 (Min) →
$Z^* = -12000, \Delta_j \rightarrow$			100 ↑	100	-1	-1	0	0 ↓	
A_1	-1	3000	0	⑦⑤/②	-1	1/4	1	—	80 (Min)→
x_1	0	90	1	5/8	0	$-1/30$	0	—	144
$Z^* = -2910, \Delta_j \rightarrow$			0	75/2 ↑	0	1/4	0	↓	
x_2	0	80	0	1	$-2/75$	1/50			
x_1	0	40	1	0	1/60	$-1/60$			
$Z^* = 0,$	$\Delta_j \rightarrow$		0	0	0	0			

Clearly, in the last row of the above table all $\Delta_j \leq 0$. Also, no artificial variable appears in the basis. So, phase-1 is complete.

PHASE-2. Assign the actual costs to the original variables and cost zero to the surplus variables, the objective function becomes

$$\text{max. } Z^* = -40x_1 - 24x_2 + 0s_1 + 0s_2$$

Further, replace the c_j-row values in the final simplex table of phase-1 by the c_j-values of the original objective function and delete the artificial variables from the final simplex table of phase-1, we write the first simplex table of phase-2.

Now, again apply the simplex method to get the following simplex table.

		$c_j \rightarrow$	-40	-24	0	0	
B.V.	C_B	x_B	x_1	x_2	s_1	s_2	Min Ratio
x_2	-24	80	0	1	$-2/75$	1/50	$-$ve
x_1	-40	40	1	0	1/60	$-1/60$	2400 (Min) →
$Z^* = -3520, \Delta_j \rightarrow$			0 ↓	0	2/75	$-38/75$ ↑	
x_2	-24	144	8/5	1	0	$-1/50$	
s_1	0	2400	60	0	1	-1	
$Z^* = -3456, \Delta_j \rightarrow$			$-8/5$	0	0	$-12/25$	

Clearly in the last row of the above table, all $\Delta_j \le 0$.

Hence, the solution is optimal and is given by

$$x_1 = 0, x_2 = 144 \text{ and min. } Z = -Z^* = 3456.$$

EXAMPLE 2. *Solve the following LPP by Two-phase method.*

$$\text{max. } Z = 3x_1 - x_2$$

subject to the constraints

$$2x_1 + x_2 \ge 2$$
$$x_1 + 3x_2 \le 2$$
$$x_2 \le 4$$

and $\qquad x_1, x_2 \ge 0$

SOLUTION. Let us introduce the surplus variable s_1, slack variables s_2, s_3 and artificial variable A_1 such that the given constraints reduces to the following equations

$$2x_1 + x_2 - s_1 + A_1 = 2$$
$$x_1 + 3x_2 + s_2 = 2$$
$$x_2 + s_3 = 4$$

and $\qquad x_1, x_2, s_1, s_2, s_3, A_1 \ge 0$

PHASE-1. Now assigning the cost -1 to the artificial variable and cost 0 to all other variables, to get the new objective function of this auxiliary problem is given by

$$\text{max. } Z^* = 0x_1 + 0x_2 + 0s_1 + 0s_2 + 0s_3 - A_1$$

Now, apply the simplex method to get the following simplex table

		$c_j \rightarrow$	0	0	0	0	0	−1	
B.V.	C_B	x_B	x_1	x_2	s_1	s_2	s_3	A_1	Min Ratio
A_1	−1	2	②	1	−1	0	0	1	2/2 = 1 (Min) →
s_2	0	2	1	3	0	1	0	0	2/1
s_3	0	4	0	1	0	0	1	0	—
$Z^*=C_Bx_B=-2, \Delta_j \rightarrow$			2	1	−1	0	0	0	
			↑					↓	
x_1	0	1	1	1/2	−1/2	0	0	—	
s_2	0	1	0	5/2	1/2	1	0	—	
s_3	0	4	0	1	0	0	1	—	
$Z^*=C_Bx_B=0, \Delta_j \rightarrow$			0	0	0	0	0		

In the last row of the above table $\Delta_j \le 0$. Also, no artificial variable appears in the basis. So, we move to the phase-2.

PHASE-2. Assign the original costs to the original variables, cost 0 to slack and surplus variables, the objective function becomes

$$\text{max. } Z = 3x_1 - x_2 + 0s_1 + 0s_2 + 0s_3$$

Further, delete the artificial columns from the last simplex table of phase-1, we write the first simplex table of phase-2.

Now apply the simplex method in a usual manner to get the following simplex tables.

B.V.	C_B	x_B	x_1	x_2	s_1	s_2	s_3	Min Ratio
		$c_j \rightarrow$	3	−1	0	0	0	
x_1	3	1	1	1/2	−1/2	0	0	−ve
s_2	0	1	0	5/2	(1/2)	1	0	2 (Min) →
s_3	0	4	0	1	0	0	1	—
$Z^* = C_B x_B = 3$, $\Delta_j \rightarrow$			0	−5/2	3/2	0	0	
				↑	↓			
x_1	3	2	1	3	0	1	0	
s_1	0	2	0	5	1	2	0	
s_3	0	4	0	1	0	0	1	
$Z^* = C_B x_B = 6$, $\Delta_j \rightarrow$			0	−10	0	−3	0	

In the last row of the above table all $\Delta_j \leq 0$, so solution is optimal and is given by

$$x_1 = 2, x_2 = 0 \text{ and max. } Z = 6$$

EXAMPLE 3. *Solve the following LPP by two-phase method*

$$\text{min. } Z = x_1 - 2x_2 - 3x_3$$

subject to the constraints

$$-2x_1 + x_2 + 3x_3 = 2$$
$$2x_1 + 3x_2 + 4x_3 = 1$$

and $\qquad x_1, x_2, x_3 \geq 0$ $\qquad$ [MEERUT–2005, 09, 10; KANPUR–2012]

SOLUTION. Since the given problem is of minimization, so convert it into maximization problem such that $Z = -Z^*$.

$$\text{max. } Z^* = -x_1 + 2x_2 + 3x_3$$

Now, introduce the artificial variables A_1 and A_2 such that

$$-2x_1 + x_2 + 3x_3 + A_1 = 2$$
$$2x_1 + 3x_2 + 4x_3 + A_2 = 1$$

and $\qquad x_1, x_2, x_3, A_1, A_2 \geq 0$

PHASE-1. In a usual manner, assigning cost −1 to the artificial variables and cost 0 to all other variables, the new objective function of the auxiliary problem is:

$$\text{max. } Z^* = 0x_1 + 0x_2 + 0x_3 - A_1 - A_2$$

Now, we have the following simplex table:

$c_j \rightarrow$			0	0	0	−1	−1	
B.V.	C_B	x_B	x_1	x_2	x_3	A_1	A_2	Min Ratio
A_1	−1	2	−2	1	3	1	0	2/3
A_2	−1	1	2	3	④	0	1	1/4 (Min) →
$Z^* = -3,$		$\Delta_j \rightarrow$	0	4	7	0	0	
					↑		↓	
A_1	−1	5/4	−7/2	−5/4	0	1	−3/4	
x_3	0	1/4	1/2	3/4	1	0	1/4	
$Z^* = -5/4,$		$\Delta_j \rightarrow$	−7/4	−5/4	0	0	−3/4	

Clearly, all $\Delta_j's$ in the last row of above table are negative or zero, so an optimum basic feasible solution to the auxiliary problem has been attained. But on the other hand, the artificial variable A_1 appears in the basic solution at a positive level. Hence, the given LPP does not have any feasible solution.

 Exercise-3.4

Solve the following LPP by Two-phase method:

1. min. $Z = x_1 + x_2$

subject to the constraints

$$2x_1 + x_2 \geq 4$$
$$x_1 + 7x_2 \geq 7$$
and $$x_1, x_2 \geq 0$$

2. max. $Z = 5x_1 + 8x_2$

subject to the constraints

$$3x_1 + 2x_2 \geq 3$$

$$x_1 + 4x_2 \geq 4$$
$$x_1 + x_2 \leq 5$$
and $$x_1, x_2 \geq 0$$

3. max. $Z = 3x_1 + 2x_2 + x_3 + 4x_4$

subject to the constraints

$$4x_1 + 5x_2 + x_3 + 5x_4 = 5$$
$$2x_1 - 3x_2 - 4x_3 + 5x_4 = 7$$
$$x_1 + 4x_2 + 5x_3 - 4x_4 = 6$$
and $$x_1, x_2, x_3, x_4 \geq 0$$

ANSWERS

1. $x_1 = \dfrac{21}{13}, x_2 = \dfrac{10}{13}$ and min.$Z = \dfrac{31}{13}$ **2.** $x_1 = 0, x_2 = 5$ and max. $Z = 40$

3. No feasible solution exists.

3.15 SOME SPECIAL LINEAR PROGRAMMING PROBLEMS

(1) Problem of Unbounded Feasible Solution but Bounded Optimal Solution

EXAMPLE 1. Solve the following LPP

$$\text{max. } Z = 6x_1 - 2x_2$$

subject to the constraints

$$2x_1 - x_2 \leq 2$$
$$x_1 \leq 4$$
and $$x_1, x_2 \geq 0$$

SOLUTION. Introduce the slack variables s_1 and s_2 in the given LPP, then we have

$$\text{max. } Z = 6x_1 - 2x_2 + 0s_1 + 0s_2$$

s.t.
$$2x_1 - x_2 + s_1 = 2$$
$$x_1 + s_2 = 4$$

and
$$x_1, x_2, s_1, s_2 \geq 0$$

Now, apply the simplex method in a usual manner, we have the following simplex table:

Simplex table

	$c_j \rightarrow$		6	-2	0	0	
B.V.	C_B	x_B	x_1	x_2	s_1	s_2	Min Ratio
s_1	0	2	②	-1	1	0	1 (Min) $\rightarrow$
s_2	0	4	1	0	0	1	4
$Z = 0,$		$\Delta_j \rightarrow$	6 $\uparrow$	-2 $\downarrow$	0	0	
x_1	6	1	1	-1/2	1/2	0	-ve
s_2	0	3	0	⑴/2	-1/2	1	(min) $\rightarrow$
$Z = 6,$		$\Delta_j \rightarrow$	0	1 $\uparrow$	-3	0 $\downarrow$	
x_1	6	4	1	0	0	1	
x_2	-2	6	0	1	-1	2	
$Z = 12,$		$\Delta_j \rightarrow$	0	0	-2	-2	

Clearly, in the last row of the above table all $\Delta_j \leq 0$, so solution is optimal and is given by $x_1 = 4$, $x_2 = 6$ and max. $Z = 12$

Now, in the first simplex table, we note that the elements of column vector x_2 are negative or zero which indicate that the feasible region is not bounded.

Hence, the given LPP has unbounded feasible solution but not bounded optimal solution.

(2) PROBLEM OF UNBOUNDED SOLUTION

It is a very well known fact that, if in any situation there is at least one Δ_j greater than zero but there is no non-negative ratios or min ratios $\rightarrow \infty$ then we have unbounded solution. Unboundedness describes a LP problem that do not have finite solution.

EXAMPLE 1. *Solve the following LPP*
$$max.\ Z = 2x_1 + x_2$$
subject to the constraints
$$x_1 - x_2 \leq 10$$
$$2x_1 - x_2 \leq 40$$
and $x_1, x_2 \geq 0$

SOLUTION. Introducing slack variables s_1 and s_2 in the given LPP. Then we have
$$max.\ Z = 2x_1 + x_2 + 0s_1 + 0s_2$$
s.t.
$$x_1 - x_2 + s_1 = 10$$
$$2x_1 - x_2 + s_2 = 40$$
and
$$x_1, x_2, s_1, s_2 \geq 0$$

Then apply the simplex method in the usual manner, we have the following simplex table.

Simplex table

B.V.	C_B	x_B	$c_j \to$ 2 x_1	1 x_2	0 s_1	0 s_2	Min Ratio
s_1	0	10	1	−1	1	0	
s_2	0	40	2	−1	0	1	
$Z = C_B x_B = 0,$		$\Delta_j \to$	2	1	0	0	

Clearly in the last row of the above table, we observe that the value of $x_2 \notin B$ (the set of basic variable) and for x_2, we have $\Delta_j = 1 > 0$ and $x_{i2} \le 0$, $i = 1, 2$. Hence, the solution is unbounded.

EXAMPLE 2. Solve the following LPP:
$$max. \ Z = 10x_1 + 20x_2$$
subject to the constraints
$$2x_1 + 4x_2 \ge 16, \ x_1 + 5x_2 \ge 15$$
and
$$x_1, x_2 \ge 0$$

SOLUTION. Using surplus variables s_1, s_2 and artificial variables A_1, A_2, the given LPP becomes
$$max. \ Z = 10x_1 + 20x_2 + 0s_1 + 0s_2 - MA_1 - MA_2$$
subject to
$$2x_1 + 4x_2 - s_1 + A_1 = 16$$
$$x_1 + 5x_2 - s_2 + A_2 = 15$$
and
$$x_1, x_2, s_1, s_2, A_1, A_2 \ge 0$$

Now, apply the simplex method in a usual manner, we have the following simplex table:

Simplex table

B.V.	C_B	x_B	$c_j \to$ 10 x_1	20 x_2	0 s_1	0 s_2	−M A_1	−M A_2	Min Ratio
A_1	−M	16	2	4	−1	0	1	0	16/4
A_2	−M	15	1	⑤	0	−1	0	1	15/5(Min) →
$Z = -31M,$		$\Delta_j \to$	(3M + 10)	(9M + 20) ↑	−M	−M	0	0 ↓	
A_1	−M	4	⑥/⑤	0	−1	4/5	1	—	10/3(Min)→
x_1	20	3	1/5	1	0	−1/5	0	—	15
$Z = -4M + 60, \Delta_j \to$		$\frac{6}{5}(M + 5)$ ↑		0	−M	$\frac{4}{5}(M + 5)$	0	↓	—
x_1	10	10/3	1	0	−5/6	2/3	—	—	−ve
x_2	20	7/3	0	1	①/⑥	−1/3	—	—	14 (Min)→
$Z = 80,$		$\Delta_j \to$	0	0	5 ↓	0			
x_1	10	15	1	5	0	−1			
s_1	0	14	0	6	1	−2			
$Z = 150,$		$\Delta_j \to$	0	−30	0	10 ↑			

Clearly, in the last table, we observe that s_2 is the incoming vector but outgoing vector can not be found because all the elements in this column are negative.

Hence, the solution is unbounded.

(3) PROBLEM HAVING MORE THAN ONE OPTIMAL SOLUTIONS

An alternative optimal solutions can be obtained by considering the $c_j - z_j$ row of the simplex table. We know that for optimal solution of maximization problem all $\Delta_j \leq 0$. But if $\Delta_j = 0$ for some non-basic variables columns in the optimal simplex table, optimal solution may exist, because if a non-basic variable corresponding to which $\Delta_j = 0$ is entered into the basis, a new solution will be arrived at but value of the objective function remains the same.

EXAMPLE 1. *Solve the following LPP*

$$max. \ Z = 6x_1 + 4x_2$$
subject to the constraints
$$2x_1 + 3x_2 \leq 30$$
$$3x_1 + 2x_2 \leq 24$$
$$x_1 + x_2 \geq 3$$
and $x_1, x_2 \geq 0$

SOLUTION. Using slack variables s_1, s_2, surplus variable s_3 and artificial variable A_1 we have the following LPP.

$$max. \ Z = 6x_1 + 4x_2 + 0s_1 + 0s_2 + 0s_3 - MA_1$$
s.t.
$$2x_1 + 3x_2 + s_1 = 30$$
$$3x_1 + 2x_2 + s_2 = 24$$
$$x_1 + x_2 - s_3 + A_1 = 3$$
and $x_1, x_2, s_1, s_2, s_3, A_1 \geq 0$

Then apply the simplex method in a usual manner, we have the following simplex table.

B.V.	C_B	x_B	x_1	x_2	s_1	s_2	s_3	A_1	Min Ratio
		$c_j \rightarrow$	6	4	0	0	0	$-M$	
s_1	0	30	2	3	1	0	0	0	15
s_2	0	24	3	2	0	1	0	0	8
A_1	$-M$	3	①	1	0	0	-1	1	3 (Min) →
$Z = -3M$,		$\Delta_j \rightarrow$	$(6+M)$ ↑	$(4+M)$	0	0	$-M$	0 ↓	
s_1	0	24	0	1	1	0	2	—	12
s_2	0	15	0	-1	0	1	③	—	5 (Min) →
x_1	6	3	1	1	0	0	-1	—	—
$Z = 18$,		$\Delta_j \rightarrow$	0	-2	0	0 ↓	6 ↑		
s_1	0	14	0	⑤/3	1	$-2/3$	0	—	12/5 (Min) →
s_3	0	5	0	$-1/3$	0	1/3	1	—	—
x_1	6	8	1	2/3	0	1/3	0	—	24/2 = 12
$Z = 48$,		$\Delta_j \rightarrow$	0	0	0	-2	0		

We observe that in the last table all $\Delta_j \leq 0$, therefore, solution is optimal and is given by $x_1 = 8$, $x_2 = 0$, max. $Z = 48$.

Also, corresponding to non-basic variables x_2 in the last table $\Delta_2 = 0$ and x_2 is not in the basis B, so an alternative solution will exists. For alternative optimal solution we have the following simplex table.

Simplex table

		$c_j \rightarrow$	6	4	0	0	0	
B.V.	C_B	x_B	x_1	x_2	s_1	s_2	s_3	Min Ratio
x_2	4	42/5	0	1	3/5	−2/5	0	
s_3	0	39/5	0	0	1/5	1/5	1	
x_1	6	12/5	1	0	−2/5	3/5	0	
$Z = C_B x_B = 48$,		$\Delta_j \rightarrow$	0	0	0	−2	0	

Clearly all $\Delta_j \leq 0$ implies solution is optimal and is given by

$$x_1 = \frac{12}{5}, x_2 = \frac{42}{5} \text{ and max. } Z = 48.$$

☛ REMARK

- We know that the convex combination of BFS is also an optimal solution. Hence, if we obtain two alternative optimal solutions, then we can obtain any number of optimal solutions.

(4) LPP WITH UNRESTRICTED VARIABLES

Generally in LPP, we assume that all $x_j \geq 0$. But one or more variables can have either positive, negative or zero value, then these are called unrestricted variables.

For example, if x_r is unrestricted then we assume

$$x_r = x_r' - x_r''$$

where $x_r', x_r'' \geq 0$

EXAMPLE 1. *Solve the following LPP*

$$max. \; Z = x_1 + 3x_2$$

subject to the constraints

$$x_1 + x_2 \leq 2$$
$$-x_1 + x_2 \leq 4$$

and $x_2 \geq 0$, x_1 *is unrestricted.*

SOLUTION. Here x_1 is unrestricted. Assume $x_1 = x_1' - x_1''$ such that $x_1', x_1'' \geq 0$. Also, introducing slack variables s_1, s_2, the given LPP becomes

$$max. \; Z = (x_1' - x_1'') + 3x_2 + 0s_1 + 0s_2$$

s.t.
$$x_1' - x_1'' + x_2 + s_1 = 2$$
$$-x_1' + x_1'' + x_2 + s_2 = 4$$

and
$$x_1', x_1'', x_2, s_1, s_2 \geq 0$$

Now, apply the simplex method in a usual manner we have the following simplex table.

Simplex table

B.V.	C_B	x_B	$c_j \rightarrow$ x_1'	x_1''	x_2	s_1	s_2	Min Ratio
			1	−1	3	0	0	
s_1	0	2	1	−1	①	1	0	2 (Min) →
s_2	0	4	−1	1	1	0	1	4
$Z = 0,$		$\Delta_j \rightarrow$	1	−1	3 ↑	0 ↓	0	
x_2	3	2	1	−1	1	1	0	−2
s_2	0	2	−2	②	0	−1	1	1 (Min) →
$Z = 6,$		$\Delta_j \rightarrow$	−2	2 ↑	0	−3 ↓	0	
x_2	3	3	0	0	1	1/2	1/2	
x_1''	−1	1	−1	1	0	−1/2	1/2	
$Z = 8,$		$\Delta_j \rightarrow$	0	0	0	−2	−1	

Clearly, in the last row of the above table all $\Delta_j \le 0$, so solution is optimal and is given by

$$x_1' = 0, x_1'' = 1, x_2 = 3 \text{ and max. } Z = 8$$

$$\Rightarrow \quad x_1 = x_1' - x_1'' = -1, x_2 = 3 \text{ and max. } Z = 8$$

EXAMPLE 2. *Solve the following LPP*

$$max. \ Z = 2x_1 + 3x_2$$

subject to the constraints

$$-x_1 + 2x_2 \le 4$$
$$x_1 + x_2 \le 6$$
$$x_1 + 3x_2 \le 9$$

and x_1, x_2 *are unrestricted.*

SOLUTION. In the given LPP, x_1 and x_2 are unrestricted. So, making the transformation.

$$x_1 = x_1' - x_1'' \text{ and } x_2 = x_2' - x_2''$$

and using the slack variables s_1, s_2 and s_3. The given LPP reduces to

$$max. \ Z = 2x_1' - 2x_1'' + 3x_2' - 3x_2'' + 0 \cdot s_1 + 0 \cdot s_2 + 0 \cdot s_3$$

$$\text{s.t.} \quad -x_1' + x_1'' + 2x_2' - 2x_2'' + s_1 = 4$$

$$x_1' - x_1'' + x_2' - x_2'' + s_2 = 6$$

$$x_1' - x_1'' + 3x_2' - 3x_2'' + s_3 = 9$$

and $\qquad x_1', x_1'', x_2', x_2'', s_1, s_2, s_3 \ge 0$

Now apply simplex method in usual manner, we have the following simplex table

Simplex table

B.V.	C_B	x_B	$c_j \rightarrow$ 2 x_1'	-2 x_1''	3 x_2'	-3 x_2''	0 s_1	0 s_2	0 s_3	Min Ratio
s_1	0	4	-1	1	②	-2	1	0	0	2 (Min) $\rightarrow$
s_2	0	6	1	-1	1	-1	0	1	0	6
s_3	0	9	1	-1	3	-3	0	0	1	3
$Z=0$,		$\Delta_j \rightarrow$	2	-2	3 $\uparrow$	-3 $\downarrow$	0	0	0	
x_2'	3	2	$-1/2$	1/2	1	-1	1/2	0	0	$-$ve
s_2	0	4	3/2	$-3/2$	0	0	$-1/2$	1	0	8/3
s_3	0	3	⑤/2	$-5/2$	0	0	$-3/2$	0	1	6/5(Min) $\rightarrow$
$Z=6$,		$\Delta_j \rightarrow$	7/2 $\uparrow$	$-7/2$	0	0	$-3/2$	0	0 $\downarrow$	
x_2'	3	13/5	0	0	1	-1	1/5	0	1/5	13
s_2	0	11/5	0	0	0	0	2/5	1	$-3/5$	11/2(Min) $\rightarrow$
x_1'	2	6/5	1	-1	0	0	$-3/5$	0	2/5	$-$ve
$Z=51/5$,		$\Delta_j \rightarrow$	0	0	0	0	3/5 $\uparrow$	0 $\downarrow$	$-7/5$	
x_2'	3	3/2	0	0	1	-1	0	$-1/2$	1/2	
s_1	0	11/2	0	0	0	0	1	5/2	$-3/2$	
x_1'	2	9/2	1	-1	0	0	0	3/2	$-1/2$	
$Z=27/2$,		$\Delta_j \rightarrow$	0	0	0	0	0	$-3/2$	$-1/2$	

We observe that all $\Delta_j \leq 0$ (in the last row of the last table). Hence, solution is optimal and is given by

$$x_1' = \frac{9}{2}, x_1'' = 0, x_2' = \frac{3}{2}, x_2'' = 0, \max.Z = \frac{27}{2}$$

$$\Rightarrow \qquad x_1 = \frac{9}{2}, x_2 = \frac{3}{2} \text{ and max. } Z = \frac{27}{2}$$

(5) LPP WITH DEGENERACY: TIE FOR ENTERING AND LEAVING BASIC VARIABLES

Sometimes, at the stage of improving the solution during simplex algorithm, minimum ratio (x_B/x_k) is determined by the last column of simplex table to find the key row. But sometimes this ratio may not be unique.

(A) In order to break the tie the selection of key column (entering variable) can be made arbitrarily by keeping in mind the following points.

    (i) If there is a tie between two decision variable, then the selection can be made arbitrarily.

    (ii) If there is a tie between decision variable and slack (or surplus) variable, then select the decision variable to enter in the basis first.

    (iii) If there is a tie between two slack (or surplus) variables, then selection can be made arbitrarily.

(iv) If there is a tie between a slack and artificial variable, preference shall be given to the artificial variable.

(B) When there is a tie between two or more basic variables for leaving the basis, *i.e.*, the minimum ratio is not unique or values of one or more variables in x_B become equal to zero; then use the following points

(i) Divide the coefficient of slack variables in the simplex table where degeneracy is detected by the corresponding positive numbers of the key column in the row, starting from left to right.

(ii) The row which contains smallest ratio comparing from left to right columnwise becomes the key row.

EXAMPLE I. *Solve the following LPP by simplex method*

$$max. \ Z = 3x_1 + 9x_2$$

subject to the constraints

$$x_1 + 4x_2 \leq 8$$
$$x_1 + 2x_2 \leq 4$$

and $\quad x_1, x_2 \geq 0$

SOLUTION. Introducing slack variables, s_1 and s_2, the given LPP becomes

$$max. \ Z = 3x_1 + 9x_2 + 0s_1 + 0s_2$$

s.t. $\quad x_1 + 4x_2 + s_1 = 8$
$\quad x_1 + 2x_2 + s_2 = 4$

and $\quad x_1, x_2, s_1, s_2 \geq 0$

Now apply the simplex method in a usual manner, we have the following simplex table.

Simplex table

B.V.	C_B	x_B	$c_j \to$ x_1	9 x_2	0 s_1	0 s_2	Min Ratio
			3	9	0	0	
s_1	0	8	1	4	1	0	$\frac{8}{4} = 2$
s_2	0	4	1	②	0	1	$\frac{4}{2} = 2$
$Z = 0,$		$\Delta_j \to$	3	9 ↑	0	0	Tie
s_1	0	0	−1	0	1	−2	
x_2	9	2	1/2	1	0	1/2	
$Z = 18,$		$\Delta_j \to$	−3/2	0	0	−9/2	

Clearly all $\Delta_j \leq 0 \Rightarrow$ solution is optimal and is given by

$$x_1 = 0, x_2 = 2, Z = 18$$

Explanation of Tie: In the initial simplex table, both variables s_1 and s_2 are eligible to leave the basis as the minimum ratio (*i.e.*, 2) is same, *i.e.*, there is a tie between s_1 and s_2. To avoid this, we have used the following steps:

STEP 1. Write the coefficient of slack variables as given below.

Row	Key Column	Column	
		s_1	s_2
s_1	4	1	0
s_2	2	0	1

STEP 2. Dividing the coefficients by the corresponding elements of the key column, we get the following ratio.

Row	Key Column	Column	
		s_1	s_2
s_1	4	$\dfrac{1}{4} = \dfrac{1}{4}$	$\dfrac{0}{4} = 0$
s_2	2	$\dfrac{0}{2} = 0$	$\dfrac{1}{2} = \dfrac{1}{2}$

STEP 3. Comparing the ratio of step 2 from left to right, otherwise the minimum ratio occurs for the second row.

Hence, the variable s_2 is selected to leave the basis.

Miscellaneous Exercise

1. Apply Big-M method to solve the following LPP
   min. $Z = 4x_1 + 8x_2 + 3x_3$
   subject to the constraints
   $$x_1 + x_2 \geq 2$$
   $$2x_1 + x_3 \geq 5$$
   and $x_1, x_2, x_3 \geq 0$

2. Apply Big-M method to solve the following LPP
   max. $Z = -x_1 - x_2$
   subject to the constraints
   $$3x_1 + 2x_2 \geq 30$$
   $$-2x_1 + 3x_2 \leq -30$$
   $$x_1 + x_2 \leq 5$$
   and $x_1, x_2 \geq 0$

3. Solve the following LPP
   max. $Z = 6x_1 - 2x_2$
   subject to the constraints
   $$2x_1 - x_2 \leq 2$$
   $$x_1 \leq 4$$
   and $x_1, x_2 \geq 0$

4. Solve by simplex method the following LPP
   max. $Z = 3x_1 + 4x_2$
   subject to
   $$x_1 - x_2 \leq 1$$
   $$-x_1 + x_2 \leq 2$$
   and $x_1, x_2 \geq 0$

5. Solve the following LPP
   max. $Z = 6x_1 + 4x_2$
   subject to the constraints
   $$2x_1 + 3x_2 \leq 30$$
   $$3x_1 + 2x_2 \leq 24$$
   $$x_1 + x_2 \geq 3$$
   and $x_1, x_2 \geq 0$ [KANPUR–2011]

6. Solve the following LPP
   max. $Z = 4x_1 + 10x_2$
   subject to the constraints
   $$2x_1 + x_2 \leq 50$$
   $$2x_1 + 5x_2 \leq 100$$
   $$2x_1 + 3x_2 \leq 90$$
   and $x_1, x_2 \geq 0$
   Also find the alternative optimal solution if exist.

7. Solve the following LPP
   min. $Z = x_1 + x_2 + x_3$
   subject to the constraints
   $$x_1 - 3x_2 + 4x_3 = 5$$
   $$x_1 - 2x_2 \leq 3$$
   $$2x_2 - x_3 \geq 4$$
   and $x_1, x_2 \geq 0$, x_3 is unrestricted.

1. $x_1 = \dfrac{5}{2}, x_2 = 0, x_3 = 0, \min. Z = 10$ **2.** No feasible solution

3. $x_1 = 4, x_2 = 6, \max. Z = 12$ **4.** Unbounded solution

5. (I) $x_1 = 8, x_2 = 0, \max. Z = 48$ (II) $x_1 = \dfrac{12}{5}, x_2 = \dfrac{42}{5}, \max. Z = 48$

6. $x_1 = \dfrac{75}{4}, x_2 = \dfrac{25}{2}, \max. Z = 200$ **7.** $x_1 = 0, x_2 = \dfrac{21}{5}, x_3 = \dfrac{22}{5}, \min. Z = \dfrac{43}{5}$

3.16 SOLUTION OF SIMULTANEOUS LINEAR EQUATIONS BY SIMPLEX METHOD

Using simplex method, we can solve the system of simultaneous linear equations. For this, we consider a dummy objective function with zero cost to those variables involved in the equations with cost –1 to each artificial variable that is introduced. After that, apply the simplex method in a usual manner.

Solved Examples

EXAMPLE 1. *Using simplex method, solve the following system of linear equations:*
$$x_1 - x_3 + 4x_4 = 3, \ 2x_1 - x_2 = 3, \ 3x_1 - 2x_2 - x_4 = 1$$
and $\quad x_1, x_2, x_3, x_4 \geq 0$ [MEERUT–2001, 06, 07, 13; PUNJAB–2011; HIMACHAL–2010]

SOLUTION. Let us introduce a dummy objective function Z, with zero cost to each given variable and cost –1, to each artificial variable. Then we can write the LPP as
max. $Z = 0x_1 + 0x_2 + 0x_3 + 0x_4 - A_1 - A_2 - A_3$
subject to
$$x_1 - x_3 + 4x_4 + A_1 = 3, \ 2x_1 - x_2 + A_2 = 3, \ 3x_1 - 2x_2 - x_4 + A_3 = 1$$
and $x_1, x_2, x_3, x_4, A_1, A_2$ and $A_3 \geq 0$
Now, apply the simplex method in a usual manner, we have the following simplex table.

Simplex table

B.V.	C_B	x_B	x_1	x_2	x_3	x_4	A_1	A_2	A_3	Min Ratio
		$c_j \rightarrow$	0	0	0	0	–1	–1	–1	
A_1	–1	3	1	0	–1	4	1	0	0	3
A_2	–1	3	2	–1	0	0	0	1	0	3/2
A_3	–1	1	③	–2	0	–1	0	0	1	1/3 (Min) $\rightarrow$
$Z = 7,$	$\Delta_j \rightarrow$		6	–3	–1	3	0	0	0	
			$\uparrow$						$\downarrow$	
A_1	–1	8/3	0	2/3	–1	⑬/3	1	0	—	8/13 (Min) $\rightarrow$
A_2	–1	7/3	0	1/3	0	2/3	0	1	—	7/2
x_1	0	1/3	1	–2/3	0	–1/3	0	0	—	–ve
$Z = -5,$	$\Delta_j \rightarrow$		0	1	–1	5	0	0		
					$\uparrow$	$\downarrow$				

x_4	0	8/13	0	②/13	–3/13	1	0	8/2=4(min)→
A_2	–1	25/13	0	3/13	2/13	0	1	25/3
x_1	0	7/13	1	–8/13	–1/13	0	0	–ve
$Z=-25/13, \Delta_j \to$			1	3/13	2/13	0	0	
				↑	↓			
x_2	0	4	0	1	–3/2	13/2	0	–ve
A_2	–1	1	0	0	①/2	–3/2	1	2 (min) →
x_1	0	3	1	0	–1	4	0	–ve
$Z = -1, \quad \Delta_j \to$			0	0	1/2	–3/2	0	
					↑			
x_2	0	7	0	1	0	2		
x_3	0	2	0	0	1	–3		
x_1	0	5	1	0	0	1		
$Z = 0, \quad \Delta_j \to$			0	0	0	0		

Since, in the last row of the above table all $\Delta_j \le 0$. Hence, solution is optimal, and is given by $x_1 = 5, x_2 = 7, x_3 = 2, x_4 = 0$.

EXAMPLE 2. *Using simplex method, solve the following system of linear equations.*

$$3x_1 + 2x_2 = 4$$
$$4x_1 - x_2 = 6$$

SOLUTION. Clearly, to apply simplex method, all variables should be non-negative, so let us assume

$$x_1 = x_1' - x_1'', x_2 = x_2' - x_2'' \text{ such that } x_1', x_1'', x_2', x_2'' \ge 0$$

Now, introduce a dummy objective function Z with zero cost to each given variable and cost -1 to each artificial variables A_1 and A_2. Then we can write

$$\text{max.} Z = 0x_1' + 0x_1'' + 0x_2' + 0x_2'' - 1 \cdot A_1 - 1 \cdot A_2$$

s.t.
$$3x_1' - 3x_1'' + 2x_2' - 2x_2'' + A_1 = 4$$
$$4x_1' - 4x_1'' - x_2' + x_2'' + A_2 = 6$$

and
$$x_1', x_1'', x_2', x_2'', A_1, A_2 \ge 0$$

Now, apply the simplex method in a usual manner, we have the following simplex table.

Simplex table

B.V.	C_B	x_B	x_1'	x_1''	x_2'	x_2''	A_1	A_2	Min Ratio
		$c_j \rightarrow$	0	0	0	0	–1	–1	
A_1	–1	4	③	–3	2	–2	1	0	4/3 (Min) →
A_2	–1	6	4	– 4	–1	1	0	1	6/4
$Z = -10,$		$\Delta_j \rightarrow$	7	–7	1	–1	0	0	
			↑			↓			
x_1'	0	4/3	1	–1	2/3	–2/3	—	0	–ve
A_2	–1	2/3	0	0	–11/3	⑪/3	—	1	2/11 (min)→
$Z = -2,$		$\Delta_j \rightarrow$	0	0	–11/3	11/3		0	
						↑		↓	
x_1'	0	16/11	1	–1	0	0			
x_2''	0	2/11	0	0	–1	1			
$Z = 0,$		$\Delta_j \rightarrow$	0	0	0	0			

In the last row of the above table, we observe that all $\Delta_j \leq 0$.

$\Rightarrow$ solution is optimal and is given by

$$x_1' = \frac{16}{11}, x_1'' = 0, x_2' = 0, x_2'' = \frac{2}{11}$$

i.e.,
$$x_1 = x_1' - x_1'' = \frac{16}{11}, x_2 = x_2' - x_2'' = -\frac{2}{11}$$

3.17 INVERSE OF A MATRIX BY SIMPLEX METHOD

Let A be a non-singular matrix of order $n \times n$ with real entries.

Now, consider the system of equations
$$Ax = b, x \geq 0$$
where b is a dummy real matrix of order $n \times 1$.

Then, by introducing the artificial variable vector $A_i \geq 0$, form a dummy objective function Z with cost zero to the variable x and cost –1 to each artificial variable as
$$Z = 0 \cdot x - A_i$$
Then find the solution of the LPP given below by simplex method
$$\text{max. } Z = 0 \cdot x - A_i \cdot 1$$
subject to the constraints
$$Ax + 1 \cdot A_i = b$$
and
$$x, A_i \geq 0$$
Then use the following concepts

"When the columns of A becomes the column of unit matrix I_n, then the columns of the vectors which are in the initial basis matrix in the starting simplex table give the inverse of A."

☛ REMARK
- To find the dummy matrix, we assign any non-negative integer to each variable Z.

Solved Examples

EXAMPLE 1. *Find the inverse of the matrix* $\begin{bmatrix} 4 & 3 \\ 3 & 2 \end{bmatrix}$, *by simplex method.* [MEERUT–2008]

SOLUTION. Consider the system of equation

$$\begin{bmatrix} 4 & 3 \\ 3 & 2 \end{bmatrix}\begin{bmatrix} x_1 \\ x_2 \end{bmatrix} = \begin{bmatrix} b_1 \\ b_2 \end{bmatrix}$$

To find the dummy matrix $B = \begin{bmatrix} b_1 \\ b_2 \end{bmatrix}$, taking $x_1 = 1, x_2 = 1$, we get

$$B = \begin{bmatrix} 7 \\ 5 \end{bmatrix}$$

Now, introduce the artificial variables A_1, A_2 and the dummy objective function Z, we obtained the following form of LPP

max. $Z = 0 \cdot x_1 + 0x_2 - A_1 - A_2$

s.t. $4x_1 + 3x_2 + A_1 + 0A_2 = 7$

$3x_1 + 2x_2 + 0A_1 + A_2 = 5$

and $x_1, x_2, A_1, A_2 \geq 0$

Then apply the simplex method in a usual manner, we have the following simplex table.

Simplex table

B.V.	C_B	x_B	x_1	x_2	A_1	A_2	Min Ratio
$c_j \rightarrow$			0	0	−1	−1	
A_1	−1	7	4	3	1	0	7/4
A_2	−1	5	③	2	0	1	5/3 (Min) →
$Z = -12,$		$\Delta_j \rightarrow$	7 ↑	5	0	0	
A_1	−1	1/3	0	⟨1/3⟩	1	− 4/3	1 (Min) →
x_1	0	5/3	1	2/3	0	1/3	5/2
x_2	0	1	0	1	3	− 4	
x_1	0	1	1	0	−2	3	
$Z = 0,$		$\Delta_j \rightarrow$	0	0	−1	−1	

In the last row of the above table $\Delta_j \leq 0 \Rightarrow$ solution is optimal.

Also, the columns of the given matrix A are converted into the columns of unit matrix I. Hence, in the final iteration, we observe that $(-2, 3)$ is corresponding to x_1 and $(3, -4)$ is that of x_2.

Therefore, inverse of A is $\begin{bmatrix} -2 & 3 \\ 3 & -4 \end{bmatrix}$.

EXAMPLE 2. *Find the inverse of the following matrix by simplex method*

$$A = \begin{bmatrix} 1 & 2 & 3 \\ 2 & 4 & 5 \\ 3 & 5 & 6 \end{bmatrix}$$

SOLUTION.   Consider $Ax = B$

$$\Rightarrow \qquad \begin{bmatrix} 1 & 2 & 3 \\ 2 & 4 & 5 \\ 3 & 5 & 6 \end{bmatrix} \begin{bmatrix} x_1 \\ x_2 \\ x_3 \end{bmatrix} = \begin{bmatrix} b_1 \\ b_2 \\ b_3 \end{bmatrix}$$

To find the dummy matrix B, let us take $x_1 = 1, x_2 = 0, x_3 = 1$ we get
$$b_1 = 4, b_2 = 7, b_3 = 9$$

$$\Rightarrow \qquad B = \begin{bmatrix} 4 \\ 7 \\ 9 \end{bmatrix}$$

Further, introducing the artificial variables A_1, A_2, A_3 and the dummy objective function Z we get the following LPP

max. $Z = 0{\cdot}x_1 + 0{\cdot}x_2 + 0{\cdot}x_3 - A_1 - A_2 - A_3$
subject to the constraints
$$x_1 + 2x_2 + 3x_3 + A_1 + 0A_2 + 0A_3 = 4$$
$$2x_1 + 4x_2 + 5x_3 + 0{\cdot}A_1 + A_2 + 0{\cdot}A_3 = 7$$
$$3x_1 + 5x_2 + 6x_3 + 0{\cdot}A_1 + 0{\cdot}A_2 + A_3 = 9$$
and $x_1, x_2, x_3, A_1, A_2, A_3 \geq 0$

Now, apply the simplex method in a usual manner, we have the following simplex table.

Simplex table

B.V.	C_B	x_B	x_1	x_2	x_3	A_1	A_2	A_3	Min Ratio
	$c_j \to$		0	0	0	−1	−1	−1	
A_1	−1	4	1	2	③	1	0	0	4/3 (Min) $\to$
A_2	−1	7	2	4	5	0	1	0	7/5
A_3	−1	9	3	5	6	0	0	1	9/6
$Z = -20,$	$\Delta_j \to$		6	11	14 ↑	0	0	0	
x_3	0	4/3	1/3	2/3	1	1/3	0	0	2
A_2	−1	1/3	1/3	②/③	0	−5/3	1	0	1/2 (Min) $\to$
A_3	−1	1	1	1	0	−2	0	1	1
$Z = -4/3,$	$\Delta_j \to$		4/3	5/3 ↑	0	−14/3	0 ↓	0	
x_3	0	1	0	0	1	2	−1	0	—
x_2	0	1/2	1/2	1	0	−5/2	3/2	0	1/2
A_3	−1	1/2	①/②	0	0	1/2	−3/2	1	1/2 (Min) $\to$
$Z = -1/2,$	$\Delta_j \to$		1/2 ↑	0	0	−1/2	−5/2	0 ↓	
x_3	0	1	0	0	1	2	−1	0	
x_2	0	0	0	1	0	−3	3	−1	
x_1	0	1	1	0	0	1	−3	2	
$Z = 0,$	$\Delta_j \to$		0	0	0	−1	−1	−1	

We observe that in the last row of the above table all $\Delta_j \le 0$. So, optimality condition is satisfied. Also, the columns of the given matrix A are converted into the columns of the unit matrix I_3.

Also, in the final iteration, we see that $(1, -3, 2)$ corresponds to x_1, $(-3, 3, -1)$ is that of x_2 and $(2, -1, 0)$ is that of x_3.

Hence,
$$A^{-1} = \begin{bmatrix} 1 & -3 & 2 \\ -3 & 3 & -1 \\ 2 & -1 & 0 \end{bmatrix}$$

Exercise-3.5

1. Solve the following system of linear equations by simplex method:

   (i) $x_1 + x_2 = 1$, $2x_1 + x_2 = 3$

   (ii) $x_1 + x_2 + x_3 = 6$, $x_1 + 2x_2 + 3x_3 = 14$, $x_1 + 4x_2 + 9x_3 = 36$

2. Find the inverse of the following matrix by simple method:

   (i) $\begin{bmatrix} 1 & 2 \\ 3 & 2 \end{bmatrix}$

   (ii) $\begin{bmatrix} 3 & 2 \\ 4 & -1 \end{bmatrix}$ [GORAKHPUR–2007, 09]

   (iii) $\begin{bmatrix} 4 & 1 & 2 \\ 0 & 1 & 0 \\ 8 & 4 & 5 \end{bmatrix}$

ANSWERS

1. (i) 2, –1   (ii) 1, 3, 3

2. (i) $\dfrac{1}{4}\begin{bmatrix} -2 & 2 \\ 3 & -1 \end{bmatrix}$ (ii) $\dfrac{1}{11}\begin{bmatrix} 1 & 2 \\ 4 & -3 \end{bmatrix}$ (iii) $\dfrac{1}{4}\begin{bmatrix} 5 & 3 & -1 \\ 0 & 4 & 0 \\ -8 & -8 & 4 \end{bmatrix}$

📝 Glossary

- **Basic Matrix:** A non-singular submatrix B of order $m \times n$ whose column vectors are m no. of linearly independent columns selected from A.

- **Basic Variables:** The variable corresponding to $\beta_1, \beta_2, ..., \beta_m$ are called basic variables.

- **Non-Basic Variables:** The variables, other than basic, are called non-basic variables.

- **Coefficient of Basic Variables:** Corresponding to any x_B, C_B will represent the row vector containing the constants C_{B_1}, $C_{B_2} ... C_{B_m}$.

- **Alternative Optimal Solution:** An LPP is said to have an alternative optimal solution if the set of variables giving the optimal value of the objective function is not unique.

- **Redundancy:** If the system of equations has more than enough number of constraints, *i.e.*, it has more constraints equations than no. of variables. Then it is called redundancy.

  i.e., $r(A) = r(Ab) = k \leq n < m$

- **Inconsistency:** The set of linear equations is said to be inconsistent if $r(A) \neq r(Ab)$.

- **Big-*M* method:** The artificial variable like surplus and slack variables are imaginary variables which are introduced in the constraints with $\geq$ or $=$ sign. These are just introduced to form an identity matrix. These variables have no physical significance. So, these variables have to be removed by the simplex method to have the meaningful solution. Here a very large price say $-M$ has to be alloted to the artificial variable in the objective function. This method of solution by aritificial variables was given by Charnes and is hence known as Charnes Big-*M* method.

- **Two Phase Method:** This is an alternative of Big-M method. Using this method, we obtain the solution in two phases given as follows:

  (1) In phase-I all the artificial variables are eliminated from the basis.

  (2) In phase-II, we use the solution from phase-I as the initial basic feasible solution and then use the simplex method to obtain the optimal solution.

⚙ REVIEW QUESTIONS

1. Explain various steps involved in simplex method.
2. Explain various steps involved in Big-M method.
3. Explain various steps involved in two-phase method.
4. Define optimal feasible solution and basic feasible solution.
5. Define artificial variable and its need.
6. Explain the uses of artificial variables in linear programming problems.
7. Define slack and surplus variables in linear programming problem.
8. Give outlines of the simplex method in linear programming.

📖 MULTIPLE CHOICE QUESTIONS (CHOOSE THE MOST APPROPRIATE ONE)

1. A basic feasible solution is said to be optimal if $c_j - z_j$ ($\forall j$):
   (a) ≤ 0 (b) ≥ 0
   (c) 0 (d) None of these

2. If a LPP involves $\geq$ constraints, then IBFS is obtained by introducing:
   (a) decision variable   (b) artificial variable
   (c) two variables   (d) None of these

3. In Big-M method of the LPP, we take cost of slack surplus and artificial variables in the objective function respectively:
   (a) $0, 0, -M$ (b) $0, 0, M$
   (c) $-1, -1, M$ (d) None of these

4. When we solve simultaneous linear equations by simplex method we take cost of artificial variables in objective function and cost of the given variables as:
   (a) $-M, 0$ (b) $0, -1$
   (c) $-1, 0$ (d) None of these

5. To solve a LPP by simplex method all $b'_j s$ should be:
   (a) negative   (b) non-negative
   (c) positive   (d) None of these

6. Simplex method to solve LPP was developed by George Dantzig in:
   (a) 1945   (b) 1946
   (c) 1947   (d) None of these

7. Big-M method of LPP was developed by:
 (a) A. Carners   (b) G.Dantizig
 (c) Lagrange   (d) None of these

8. In two phase method, for solving a LPP in Phase-1 the new objective function will be max $Z' =$
 (a) – (sum of artificial variables)
 (b) sum of slack variables
 (c) – (sum of slack variables)
 (d) None of these

9. In Big-M method, M stands for:
 (a) very large positive price
 (b) very large negative price
 (c) very small negative price
 (d) None of these

10. In LPP, if in the final simplex table all $\Delta_j \leq 0$, then the optimal solution is:
 (a) unique   (b) not unique
 (c) does not exist (d) None of these

ANSWERS

1. (a)   **2.** (b)   **3.** (a)   **4.** (c)   **5.** (b)   **6.** (c)   **7.** (a)   **8.** (a)   **9.** (b)

10. (a)

Degeneracy in Linear Programming 4

4.1 INTRODUCTION

In previous chapter, we have discussed that if there are m equations in n variables $(n>m)$, a basic solution is obtained by solving for m variables in terms of the remaining $(n - m)$ variables and setting these $(n - m)$ variables equal to zero. Then remaining $(n - m)$ variables are called non-basic variables, whereas the m variables are known as basic. Further we have also discussed that, in a basic feasible solution, if all m variables are positive, then it is called a non-degenerate solution. On the other hand if at least one basic variable in a basic feasible solution is zero, then it is known as a degenerate basic solution.

4.2 DEGENERACY IN LINEAR PROGRAMMING

The phenomenon of obtaining a degenerate basic feasible solution in linear programming is called the degeneracy in linear programming.

4.3 THE NECESSARY AND SUFFICIENT CONDITION FOR THE EXISTENCE OF NON-DEGENERACY

(1) A necessary and sufficient condition for the existence of the non-degeneracy of the basic solution of a matrix equation $Ax = b$ is that every set of m columns of the augumented matrix $[A|b]$ is linearly independent.

(2) A necessary and sufficient condition for a given basic feasible solution $x_B = B^{-1}b$ to be non-degenerate is the linear independence of b and every set of $(m - 1)$ columns of B.

4.4 OCCURANCE OF DEGENERACY IN LINEAR PROGRAMMING

In a linear programming problem, the degeneracy may appear in the following two ways:

(1) The degeneracy appear in a LPP at the very first iteration when some component of vector b is zero.

(2) If none of the components of b is zero at any iteration and choice of outgoing vector at some iteration is not unique, then next solution is bound to be degenerate.

4.5 RESOLUTION OF DEGENERACY

Let x_k ($= \alpha_k$) be the incoming vector and $\min\left\{\dfrac{x_{B_i}}{x_{ik}}, x_{ik} > 0\right\}$ is not unique, i.e., same minimum ratio $\dfrac{x_{B_i}}{x_{ik}}, x_{ik} > 0$ occurs for more than one value of i.

i.e., the key element (hence the variable to leave the basis) is not uniquely determined or at the very first iteration, the value of one or more basic variables in the x_B column become equal to zero, this cause the problem of degeneracy.

However, if the minimum ratio is zero for two or more basic variables, degeneracy may result the simplex routine to cycle indefinitely. That is the solution which we have obtained in one iteration may repeat again after few iteration and therefore no optimum solution may be obtained under such circumstances. In such type of situation we use the following procedure.

WORKING PROCEDURE

STEP 1. Pick up the rows for which the minimum, non-negative ratio is same (tied). To be definite, suppose such rows are first, second, third etc. for example.

STEP 2. Rearrange the columns of the usual simplex table so that the columns forming the original unit matrix come first in proper order.

STEP 3. Find the minimum ratio.

$$\left[\frac{\text{elements of the first column of unit matrix}}{\text{Corresponding elements of Key column}} \right]$$

only for the rows for which minimum ratio was not unique. That is for the rows, first, second, third, etc. as picked in step 1.

Then observe that

(i) If the minimum is attained for the third row (say) then this row will determine the key element by intersecting the key column.

(ii) If this minimum is also not unique then go to the next step.

STEP 4. Now compute the minimum of the ratio

$$\left[\frac{\text{elements of second column of unit matrix}}{\text{Corresponding elements of Key column}} \right]$$

only for the rows for which minimum ratio was not unique in step 3. Then

(i) If this minimum ratio is unique for the first row (say) then this row will determine the key element by intersecting the key column.

(ii) If this minimum is still not unique then go to next step.

STEP 5. Compute the minimum of the ratio

$$\left[\frac{\text{elements of the third column of unit matrix}}{\text{Corresponding elements of Key column}} \right]$$

only for the rows for which minimum ratio was not unique in step 4. Then

(i) If this minimum ratio is unique for the third row (say) then this row will determine the key element by intersecting the key column.

(ii) If this minimum is still not unique, then go on repeating the above outlined procedure till the unique minimum ratio is obtained to resolve the degeneracy.

After the resolution of this tie, simplex method is applied to obtain the optimum solution.

Solved Examples

EXAMPLE I.   *Solve the following LPP*

$$max.\ Z = 2x_1 + 3x_2 + 10x_3$$

subject to the constraints

$$x_1 - 2x_3 = 0$$
$$x_2 + x_3 = 1$$

and $x_1, x_2, x_3 \geq 0$

SOLUTION.   We can write the given LPP as

$$max.\ Z = 2x_1 + 3x_2 + 10x_3$$

s.t.

$$\left.\begin{array}{l} x_1 + 0x_2 - 2x_3 = 0 \\ 0x_1 + x_2 - x_3 = 1 \end{array}\right] \qquad \qquad ...(1)$$

Clearly, the coefficients of x_1 and x_2 in (1) produce a unit matrix I_2 so that there is no need to introduce artificial variable.

We observe that by putting x_3 (non-basic variable) $= 0$ in (1) we get

$$x_1 = 0,\ x_2 = 1$$

$\Rightarrow$ $x_1 = 0, x_2 = 1, x_3 = 0$ is the starting basic feasible solution.

Now we have the following simplex table.

Simplex Table-1

B.V.	C_B	x_B	$c_j \rightarrow$ x_1	2 x_2	3	10 x_3	Min Ratio
x_1	2	0	1	0		−2	−ve
x_2	3	1	0	1		①	1 (Min) →
$Z = C_B x_B = 3,$		$\Delta_j \rightarrow$	0	0 ↓		11 ↑	

In the above table, using $\Delta_j = c_j - C_B x_j$, we have

$$\Delta_1 = c_1 - C_B x_1 = 2 - (2,3)\begin{pmatrix}1\\0\end{pmatrix} = 2 - 2 = 0$$

$$\Delta_2 = c_2 - C_B x_2 = 3 - (2,3)\begin{pmatrix}0\\1\end{pmatrix} = 3 - 3 = 0$$

$$\Delta_3 = c_3 - C_B x_3 = 10 - (2,3)\begin{pmatrix}-2\\1\end{pmatrix} = 10 - (-1) = 11$$

We observe that $\Delta_j \geq 0 \ \forall \ j$, so IBFS is not optimal.

Now, $\Delta_k = max\{\Delta_j\} = max\{\Delta_3\} = \Delta_3$

$\Rightarrow$ $k = 3$

$\Rightarrow x_3$ is the incoming vector.

Further, $\dfrac{x_{B_r}}{x_{rk}} = \min_i \left\{ \dfrac{x_{B_i}}{x_{ik}}, x_{ik} > 0 \right\}$

$\Rightarrow \qquad \dfrac{x_{B_r}}{x_{r3}} = \min_i \left\{ \dfrac{x_{B_i}}{x_{i3}}, x_{i3} > 0 \right\} = \min \left\{ \dfrac{1}{1} \right\} = \dfrac{x_{B_2}}{x_{23}}$

$\Rightarrow \qquad\qquad r = 2$

$\Rightarrow \quad x_2$ is the outgoing vector.

So, key element $= x_{23} = 1$

Now, entering vector x_3 in the basis matrix and remove x_2, the second simplex table is given as follows:

Simplex Table-2

B.V.	C_B	x_B	$c_j \rightarrow$ 2 x_1	3 x_2	10 x_3	Min Ratio
x_1	2	2	1	2	0	
x_3	10	1	0	1	1	
$Z = C_B x_B = 10, \ \Delta_j \rightarrow$			0	-11	0	

Here, in the last row of the above table, we observe that all $\Delta_j \leq 0$.

Hence, basic feasible solution $x_1 = 2$, $x_2 = 0$, $x_3 = 1$ is the non-degenerate optimal solution and the improved value of the objective function is max. $Z = 10$.

EXAMPLE 2. *Solve the following LPP*

$$\text{max. } Z = 2x_1 + 3x_2 + 10x_3$$

subject to the constraints

$$x_1 + 2x_3 = 0$$
$$x_2 + x_3 = 1$$

and $\qquad x_1, x_2, x_3 \geq 0$

SOLUTION. Clearly, this is the problem of degeneracy at the initial stage because in the first constraints, we have $b_1 = 0$.

Here, the given LPP can be written as

$$\text{max. } Z = 2x_1 + 3x_2 + 10x_3$$

s.t. $\qquad x_1 + 0x_2 + 2x_3 = 0$

$\qquad\qquad 0x_1 + x_2 + x_3 = 1$

and $\qquad\qquad x_1, x_2, x_3 \geq 0$

We observe that the coefficients of x_1 and x_2 in both constraints produce a unit matrix such that there is no need to introduce the artificial variables.

Also, by putting $x_1 = 0$ in the given constraints we get $x_1 = 0$, $x_2 = 1$.

$\Rightarrow \quad x_1 = 0$, $x_2 = 1$ and $x_3 = 0$ is the starting basic feasible solution.

Now, we have the following simplex table.

Simplex Table-1

B.V.	C_B	x_B	$c_j \to$ x_1 2	x_2 3	x_3 10	Min Ratio
x_1	2	0	1	0	②	0 (Min) →
x_2	3	1	0	1	1	1
$Z = 3$,		$\Delta_j \to$	0 ↓	0	3 ↑	

In the above table, by using $\Delta_j = c_j - C_B x_j$ we get

$$\Delta_1 = c_1 - C_B x_1 = 2 - 2 = 0$$
$$\Delta_2 = c_2 - C_B x_2 = 3 - 3 = 0$$
$$\Delta_3 = c_3 - C_B x_3 = 10 - 7 = 3$$

Clearly, $\Delta_j \geq 0$, so solution is not optimal.

Now, $\qquad \Delta_k = $ max. $\{\Delta_j\}$ for all non-basic variables.

$$= \text{max } \{\Delta_3\} = \Delta_3$$

$\Rightarrow \qquad\qquad k = 3$

$\Rightarrow \quad \Delta_3$ is the incoming vector, *i.e.*, x_3 is the incoming vector.

Also, $\qquad \dfrac{x_{B_r}}{x_{rk}} = \min\limits_{i}\left\{\dfrac{x_{B_i}}{x_{ik}}, x_{ik} > 0\right\}$

$\Rightarrow \qquad \dfrac{x_{B_r}}{x_{r3}} = \min\limits_{i}\left\{\dfrac{x_{B_i}}{x_{i3}}, x_{i3} > 0\right\} = \min\left\{\dfrac{0}{2}, \dfrac{1}{2}\right\} = \dfrac{0}{2} = \dfrac{x_{B_1}}{x_{13}}$

$\Rightarrow \qquad\qquad r = 1$

$\Rightarrow \quad x_1$ is the outgoing vector.

Now, entering vector x_3 in basis matrix and deleting x_1, the second simplex table is as follows:

Simplex Table-2

B.V.	C_B	x_B	$c_j \to$ x_1 2	x_2 3	x_3 10	Min Ratio
x_3	10	0	1/2	0	1	
x_2	3	1	$-1/2$	1	0	
$Z = C_B x_B = 3$,		$\Delta_j \to$	$-3/2$	0	0	

From the above table, we observe that all $\Delta_j \leq 0$. Hence, the BFS, $x_1 = 0$, $x_2 = 1$, $x_3 = 0$ is the degenerate BFS and max. $Z = 3$.

☛ **REMARK**

• In the above example, we observe that it is possible to move from one table to next without any harm due to degeneracy.

EXAMPLE 3. *Solve the following LPP :*

$$\text{max. } Z = 2x_1 + x_2$$

subject to the constraints

$$4x_1 + 3x_2 \leq 12$$
$$4x_1 + x_2 \leq 8 \qquad\qquad \text{[KANPUR–2008, 12]}$$

$$4x_1 - x_2 \leq 8$$

and $\qquad x_1, x_2 \geq 0$

SOLUTION.   Using the slack variables s_1, s_2 and s_3 in the given LPP, we get

max. $Z = 2x_1 + x_2 + 0s_1 + 0s_2 + 0s_3$

s.t. $\qquad 4x_1 + 3x_2 + s_1 + 0s_2 + 0s_3 = 12$

$\qquad\qquad 4x_1 + x_2 + 0s_1 + s_2 + 0s_3 = 8$

$\qquad\qquad 4x_1 - x_2 + 0s_1 + 0s_2 + s_3 = 8$

and $\qquad x_1, x_2, s_1, s_2, s_3 \geq 0$

Now apply the simplex method in a usual manner, we have the following simplex table.

Simplex Table-1

B.V.	C_B	x_B	$c_j \rightarrow$ $\overline{s_2}$ (x_1)	2 $\overline{s_3}$ (x_2)	1 x_1 (s_1)	0 x_2 (s_2)	0 $\overline{s_1}$ (s_3)	0 Min Ratio
s_1	0	12	4	3	1	0	0	3
s_2	0	8	$4(\overline{x}_{24})$	1	$0(\overline{x}_{21})$ $1(\overline{x}_{22})$		0	2
s_3	0	8	$4(\overline{x}_{34})$	-1	$0(\overline{x}_{31})$ $0(\overline{x}_{23})$		1	2
$Z = C_B x_B = 0,$		$\Delta_j \rightarrow$	2 $\uparrow$	1	0	0	0 $\downarrow$	

In the above table, by using $\Delta_j = c_j - C_B x_j$, we have

$$\Delta_1 = c_1 - C_B x_1 = 2 - (0,0,0)\begin{bmatrix} 4 \\ 4 \\ 4 \end{bmatrix} = 2$$

$$\Delta_2 = c_2 - C_B x_2 = 1 - (0,0,0)\begin{bmatrix} 3 \\ 1 \\ -1 \end{bmatrix} = 1$$

$$\Delta_3 = 0 = \Delta_4 = \Delta_5$$

$$\text{max. } \{\Delta_j\} = \Delta_1 = \alpha_k \quad \Rightarrow \quad k = 1$$

$\Rightarrow \quad x_1$ is the incoming vector.

Now, $\dfrac{x_{B_r}}{x_{r1}}$ is not unique and minimum occur at $i = 2, 3$.

$\Rightarrow$ degeneracy exists.

Now to select the outgoing vector, proceed as follows:

Firstly renumber the columns in the starting simplex table such that

$$s_1 = \overline{x}_1, s_2 = \overline{x}_2, s_3 = \overline{s}_1, x_1 = \overline{s}_2 \text{ and } x_2 = \overline{s}_3$$

Since, minimum ratio occurs at $i = 2$ and $i = 3$

$\Rightarrow \qquad\qquad I_1 = \{2, 3\}$

$\because \quad x_1(= \overline{s}_2)$ is the incoming vector, so we take $k = 4$. Now, for $i \in I_1$, we have

$$\min_{i \in I_1}\left\{\frac{\overline{x}_{i1}}{\overline{x}_{ik}}, x_{ik} > 0\right\} = \min_{i \in I_1}\left\{\frac{\overline{x}_{i1}}{\overline{x}_{14}}, \overline{x}_{14} > 0\right\} = \min\left\{\frac{\overline{x}_{21}}{\overline{x}_{24}}, \frac{\overline{x}_{31}}{\overline{x}_{34}}\right\} = \min\left\{\frac{0}{4}, \frac{0}{4}\right\} = \min\{0, 0\}$$

$\Rightarrow$ minimum ratio is not unique, because it occurs at $i = 2, 3$.

Let $I_2 = \{2, 3\} \subseteq I_1$. Now, for $i \in I_2$, we have

$$\min_{i \in I_2}\left\{\frac{\bar{x}_{12}}{\bar{x}_{14}}, \bar{x}_{14} > 0\right\} = \min\left\{\frac{\bar{x}_{22}}{\bar{x}_{24}}, \frac{\bar{x}_{32}}{\bar{x}_{34}}\right\} = \min\left\{\frac{1}{4}, \frac{0}{4}\right\} = 0 = \frac{\bar{x}_{32}}{\bar{x}_{34}}$$

$\Rightarrow$ minimum occurs at $i = 3$, so $\bar{s}_1(= s_3)$ is the outgoing vector.

Now, we have the following simplex table.

Simplex Table-2

B.V.	C_B	x_B	x_1	x_2	$s_1(x_3)$	$s_2(x_4)$	$s_3(x_5)$	Min Ratio
		$c_j \rightarrow$	2	1	0	0	0	
s_1	0	4	0	4	1	0	−1	4/4
s_2	0	0	0	②	0	1	−1	0/2 (Min) →
s_3	2	2	1	−1/4	0	0	1/4	−ve
$Z = C_B x_B = 4,$		$\Delta_j \rightarrow$	0	3/2	0	0	−1/2	
				↑		↓		

In the above table, using $\Delta_j = c_j - C_B x_j$ we have

$$\Delta_1 = 0, \Delta_2 = c_2 - C_B x_2 = 1 - (0, 0, 2)\begin{pmatrix} 4 \\ 2 \\ -1/4 \end{pmatrix} = 1 + \frac{1}{2} = \frac{3}{2}$$

$$\Delta_3 = 0, \Delta_4 = 0, \Delta_5 = c_5 - C_B x_5 = 0 - (0, 0, 2)\begin{pmatrix} -1 \\ -1 \\ 1/4 \end{pmatrix} = -\frac{1}{2}$$

$\Rightarrow$ all Δ_j are not negative.

For incoming vector $\alpha_k = \max\{\Delta_2, \Delta_5\} = \max\left\{\frac{3}{2}, -\frac{1}{2}\right\} = \frac{3}{2} = \Delta_2$

Also, $\dfrac{x_{B_r}}{x_{r2}} = \min_i\left\{\dfrac{x_{B_i}}{x_{i2}}, x_{i2} > 0\right\} = \min\left\{\dfrac{4}{4}, \dfrac{0}{2}\right\} = \dfrac{0}{2} = \dfrac{x_{B_2}}{x_{22}}$

$\Rightarrow \qquad r = 2$

$\Rightarrow \quad s_2$ is the outgoing vector.

$\Rightarrow \qquad$ key element $= x_{22} = 2$

So, the next simplex table is given as below.

Simplex Table-3

B.V.	C_B	x_B	x_1	x_2	s_1	s_2	s_3	Min Ratio
		$c_j \rightarrow$	2	1	0	0	0	
s_1	0	4	0	0	1	−2	①	4/1 (Min) →
x_2	1	0	0	1	0	1/2	−1/2	−ve
x_1	2	2	1	0	0	1/8	1/8	16
$Z = C_B x_B = 4,$		$\Delta_j \rightarrow$	0	0	0	−3/4	1/4	
					↓		↑	

In the above table, using $\Delta_j = c_j - x_B x_j$, we have

$$\Delta_1 = 0, \Delta_2 = 0, \Delta_3 = 0$$

$$\Delta_4 = c_4 - C_B x_4 = 0 - (0,1,2) \begin{pmatrix} -2 \\ 1/2 \\ 1/8 \end{pmatrix} = -\frac{3}{4}$$

$$\Delta_5 = c_5 - C_B x_5 = 0 - (0,1,2) \begin{pmatrix} 1 \\ -1/2 \\ 1/8 \end{pmatrix} = \frac{1}{4}$$

$\because$ Δ_j are non-negative, so solution is not optimal.

Now, for incoming vector, we have

$$x_k = \max(\Delta_j) = \max(\Delta_4, \Delta_5) = \max\left\{-\frac{3}{4}, \frac{1}{4}\right\} = \frac{1}{4} = \Delta_5$$

$\Rightarrow \qquad k = 5$

$\Rightarrow \quad x_5(s_3)$ is the incoming vector.

Now, for outgoing vector, we have

$$\frac{x_{B_r}}{x_{r5}} = \min_i\left\{\frac{x_{B_i}}{x_{i5}}, x_{i5} > 0\right\}$$

$$= \min_i\left\{\frac{4}{1}, \frac{2}{1/8}\right\} = \frac{4}{1} = \frac{x_{B_1}}{x_{15}}$$

$\Rightarrow \qquad r = 1$

$\Rightarrow \quad s_1$ is the outgoing vector.

$\Rightarrow \quad$ Key element = 1

Then proceed as usual, we have the following simplex table.

Simplex Table-4

B.V.	C_B	x_B	x_1	x_2	s_1	s_2	s_3	Min Ratio
		$c_j \rightarrow$	2	1	0	0	0	
s_3	0	4	0	0	1	-2	1	
x_2	1	2	0	1	1/2	$-1/2$	0	
x_1	2	3/2	1	0	$-1/8$	3/8	0	
$Z = C_B x_B = 5,$		$\Delta_j \rightarrow$	0	0	$-1/4$	$-1/2$	0	

We observe that in the above table all $\Delta_j \leq 0 \Rightarrow$ solution is optimal and is given by

$$x_1 = \frac{3}{2}, x_2 = 2 \text{ and max. } Z = 5$$

EXAMPLE 4.   *Solve the following LPP*

$$max. \ Z = \frac{3}{4}x_1 - 150x_2 + \frac{1}{50}x_3 - 6x_4$$

subject to the constraints

$$\frac{1}{4}x_1 - 60x_2 - \frac{1}{25}x_3 + 9x_4 \leq 0$$

$$\frac{1}{2}x_1 - 90x_2 - \frac{1}{50}x_3 + 3x_4 \leq 0$$

$$x_3 \leq 1$$

and $\qquad\qquad x_1, x_2, x_3, x_4 \geq 0$

SOLUTION. Introducing slack variables s_1, s_2, s_3, the given LPP becomes

$$\text{max. } Z = \frac{3}{4}x_1 - 150x_2 + \frac{1}{50}x_3 - 6x_4 + 0s_1 + 0s_2 + 0s_3$$

$$\text{s.t. } \frac{1}{4}x_1 - 60x_2 - \frac{1}{25}x_3 + 9x_4 + s_1 + 0s_2 + 0s_3 = 0$$

$$\frac{1}{2}x_1 - 90x_2 - \frac{1}{50}x_3 + 3x_4 + 0s_1 + s_2 + 0s_3 = 0$$

$$x_3 + 0s_1 + 0s_2 + s_3 = 1$$

and $$x_1, x_2, x_3, s_1, s_2, s_3 \geq 0$$

Firstly, putting $x_1 = 0, x_2 = 0, x_3 = 0, x_4 = 0$ in the above constraints we get

$$s_1 = 0, s_2 = 0, s_3 = 1$$

So, the starting basic feasible solution is given by

$$x_1 = x_2 = x_3 = x_4 = s_1 = s_2 = 0, s_3 = 1$$

Now, apply the simplex method in a usual manner, we obtained the following simplex table.

Simplex Table-1

		$c_j \rightarrow$	3/4	−150	1/50	−6	0	0	0	
B.V.	C_B	x_B	$x_1(\bar{x}_4)$	$x_2(\bar{s}_1)$	$x_3(\bar{s}_2)$	$x_4(\bar{s}_3)$	$s_1(\bar{x}_1)$	$s_2(\bar{x}_2)$	$s_3(\bar{x}_3)$	Min Ratio
s_1	0	0	$\frac{1}{4}(\bar{x}_{14})$	−60	−1/25	9	$1(\bar{x}_{11})$	0	0	0
s_2	0	0	$\frac{1}{2}(\bar{x}_{24})$	−90	−1/50	3	$0(\bar{x}_{21})$	1	0	$0 \rightarrow$
s_3	0	1	0	0	1	0	0	0	1	—
$Z = C_B x_B = 0$, $\Delta_j \rightarrow$			3/4	−150	1/50	−6	0	0	0	
			$\uparrow$					$\downarrow$		

In the above table, using $\Delta_j = c_j - C_B x_j$, we obtained

$$\Delta_1 = \frac{3}{4}, \Delta_2 = -150, \Delta_3 = \frac{1}{50}, \Delta_4 = -6, \Delta_5 = 0, \Delta_6 = 0, \Delta_7 = 0$$

Clearly, $\Delta_1 = \max(\Delta_j) = \alpha_k \quad \Rightarrow \quad k = 1$

$\Rightarrow \quad x_1$ is the incoming vector.

Now, for outgoing vector, we have

$$\frac{x_{B_r}}{x_{r1}} = \min_i \left\{ \frac{x_{B_i}}{x_{i1}}, x_{i1} > 0 \right\}$$

$$= \min \left\{ \frac{x_{B_1}}{x_{11}}, \frac{x_{B_2}}{x_{21}} \right\}$$

$$= \min \left\{ \frac{0}{1/4}, \frac{0}{1/2} \right\} = \min\{0, 0\}$$

$\Rightarrow$ minimum is not unique as it occurs at $i = 1, 2$.

So, this is a problem of degeneracy, so for finding the outgoing vector, we use the following procedure.

First of all renumber the columns in simplex table such that

$$s_1 = \bar{x}_1, s_2 = \bar{x}_2, s_3 = \bar{x}_3, x_1 = \bar{x}_4, x_2 = \bar{s}_1, x_3 = \bar{s}_2, x_4 = \bar{s}_3$$

Here, $I_1 = \{1, 2\}$ and x_1 is the incoming vector, *i.e.*, $x_1 = \bar{x}_4$ so we take $k = 4$ (new)

Now for $i \in I_1$, we have

$$\min_{i \in I_1}\left\{\frac{\bar{x}_{i1}}{\bar{x}_{i4}}, \bar{x}_{i4} > 0\right\} = \min\left\{\frac{\bar{x}_{11}}{\bar{x}_{14}}, \frac{\bar{x}_{21}}{\bar{x}_{24}}\right\}$$

$$= \min\left\{\frac{1}{1/4}, \frac{0}{1/2}\right\} = \min\{4, 0\} = 0 = \frac{\bar{x}_{21}}{\bar{x}_{24}}$$

This minimum is unique and it occurs at $i = 2$.

$\Rightarrow \quad \bar{x}_2(= s_2)$ is the outgoing vector.

Then by usual procedure, we have the following simplex table.

Simplex Table-2

B.V.	C_B	x_B	$c_j \to$ 3/4 x_1	−150 x_2	1/50 x_3	−6 x_4	0 s_1	0 s_2	0 s_3	Min Ratio
s_1	0	0	0	−15	−3/100	15/2	1	−1/2	0	—
x_1	3/4	0	1	−180	−1/25	6	0	2	0	—
s_3	0	1	0	0	①	0	0	0	1	1 (Min) →
$Z = C_B x_B = 0, \Delta_j \to$			0	−15	1/20 ↑	−21/2	0	−3/2 ↓	0	
s_1	0	3/10	0	−15	0	15/2	1	−1/2	3/100	
x_1	3/4	1/25	1	−180	0	6	0	2	11/25	
x_3	1/50	1	0	0	1	0	0	0	0	
$Z = C_B x_B = 1/20, \Delta_j \to$			0	−15	0	−21/2	0	−3/2	−1/20	

In the last row of the above table all $\Delta_j \leq 0$. Hence, solution is optimal and is given by

$$x_1 = \frac{1}{25}, x_2 = 0, x_3 = 1, x_4 = 0 \text{ and max. } Z = \frac{1}{20}.$$

EXAMPLE 5. *Solve the following LPP*

$$\text{max. } Z = 5x_1 + 3x_2$$

subject to the constraints

$$x_1 + x_2 \leq 2$$
$$5x_1 + 2x_2 \leq 10$$
$$3x_1 + 8x_2 \leq 12$$

and

$$x_1, x_2 \geq 0$$

SOLUTION. Using the slack variables s_1, s_2 and s_3, the given LPP becomes

max. $Z = 5x_1 + 3x_2 + 0s_1 + 0s_2 + 0s_3$

s.t. $x_1 + x_2 + s_1 = 2$

            $5x_1 + 2x_2 + s_2 = 10$

            $3x_1 + 8x_2 + s_3 = 12$

and $x_1, x_2, s_1, s_2, s_3 \geq 0$

Now, apply the simplex method in a usual manner, we have the following simplex table

Simplex Table-1

B.V.	C_B	x_B	x_1 $(\overline{s_2})$	x_2 $(\overline{s_3})$	s_1 $(\overline{x_1})$	s_2 $(\overline{x_2})$	s_3 $(\overline{s_1})$	Min Ratio
		$c_j \rightarrow$	5	3	0	0	0	
s_1	0	2	1	1	$1(\overline{x}_{11})$	0	0	$2/1 = 2$
s_2	0	10	⑤$(\overline{x}_{24})$	2	$0(\overline{x}_{21})$	1	0	$10/5 = 2 \rightarrow$
s_3	0	12	3	8	0	0	1	$12/3 = 4$
$Z = C_B x_B = 0,$		$\Delta_j \rightarrow$	5 $\uparrow$	3	0 $\downarrow$	0	0	

In the above table, by using $\Delta_j = c_j - C_B x_j$ we obtained

$$\Delta_1 = c_1 - C_B x_1 = 5 - (0,0,0) \begin{pmatrix} 1 \\ 5 \\ 3 \end{pmatrix} = 5$$

$$\Delta_2 = c_2 - C_B x_2 = 3 - (0,0,0) \begin{pmatrix} 1 \\ 2 \\ 8 \end{pmatrix} = 3$$

$$\Delta_3 = c_3 - C_B x_3 = 0 - (0,0,0) \begin{pmatrix} 1 \\ 0 \\ 0 \end{pmatrix} = 0$$

$$\Delta_4 = c_4 - C_B x_4 = 0 - (0,0,0) \begin{pmatrix} 0 \\ 1 \\ 0 \end{pmatrix} = 0$$

and $$\Delta_5 = c_5 - C_B x_5 = 0 - (0,0,0) \begin{pmatrix} 0 \\ 0 \\ 1 \end{pmatrix} = 0$$

Clearly, $\Delta_j \geq 0 \ \forall \ j$, so solution is not optimal.

Now, $\alpha_k = \max\{\Delta_j\}$ for all non-basic variables.

            $= \max\{\Delta_1, \Delta_2\} = \max\{5, 3\} = 5$

$\Rightarrow$ $k = 1$

$\Rightarrow$ x_1 is the incoming vector.

Further, by minimum ratio column in starting simplex table, we find that the minimum is not unique and it occurs at $i = 1$ and $i = 2$. So, it is a problem of degeneracy.

To resolve the degeneracy, firstly renumber the columns in the starting table such that

$$s_1 = \bar{x}_1, s_2 = \bar{x}_2, s_3 = \bar{s}_1, x_1 = \bar{s}_2 \text{ and } x_2 = \bar{s}_3$$

Here, $I_1 = \{1, 2\}$ ($\because$ minimum ratio occurs at $i = 1$ and $i = 2$)

Now, for $i = 1, 2 \in I_1$

$$\min_{i \in I_1}\left\{\frac{\bar{x}_{i1}}{\bar{x}_{iK}}, \bar{x}_{iK} > 0\right\} = \min\left\{\frac{\bar{x}_{i1}}{\bar{x}_{i4}}, \bar{x}_{14} > 0\right\}$$

$$= \min\left\{\frac{\bar{x}_{11}}{\bar{x}_{14}}, \frac{\bar{x}_{21}}{\bar{x}_{24}}\right\}$$

$$= \min\left\{\frac{1}{1}, \frac{0}{5}\right\} = 0 = \frac{\bar{x}_{21}}{\bar{x}_{24}}$$

This minimum is unique and occurs at $i = 2$. Therefore, the vector $s_2\ (= \bar{x}_2)$ is the outgoing vector. And key element = 5

Now, apply the usual procedure, we have the following simplex table.

Simplex Table-2

B.V.	C_B	x_B	$c_j \rightarrow$					Min Ratio
			5	3	0	0	0	
			x_1	x_2	s_1	s_2	s_3	
s_1	0	0	0	(3/5)	1	–1/5	0	0 (Min) $\rightarrow$
x_1	5	2	1	2/5	0	1/5	0	5
s_3	0	6	0	34/5	0	–3/5	1	15/17
$Z = C_B x_B = 10$,		$\Delta_j \rightarrow$	0	1	0	–1	0	
				$\uparrow$	$\downarrow$			
x_2	3	0	0	1	5/3	–1/3	0	
x_1	5	2	1	0	–2/3	1/3	0	
s_3	0	6	0	0	–34/3	5/3	1	
$Z = C_B x_B = 10$,		$\Delta_j \rightarrow$	0	0	–5/3	–2/3	0	

We observe that all $\Delta_j \leq 0$. Hence, solution is optimal and is given by

$$x_1 = 2, x_2 = 0 \text{ and max. } Z = 10.$$

 **Exercise-4.1**

Solve the following LPP:

1. max. $Z = x_1 + 2x_2 + x_3$
s.t. $2x_1 + x_2 - x_3 \leq 2$
  $-2x_1 + x_2 - 5x_3 \leq -6$
  $4x_1 + x_2 + x_3 \leq 6$
and $x_1, x_2, x_3 \geq 0$

2. max. $Z = 3x_1 + 9x_2$
s.t. $x_1 + 4x_2 \leq 8$
     $x_1 + 2x_2 \leq 4$
and $x_1, x_2 \geq 0$

3. max. $Z = 3x_1 + 5x_2 + 4x_3$
s.t. $2x_1 + 3x_3 \leq 18$
     $2x_2 + 5x_3 \leq 18$
     $3x_1 + 2x_2 + 4x_3 \leq 25$
and $x_1, x_2, x_3 \geq 0$

4. max. $Z = x_1 - x_2 + 3x_3$
s.t. $x_1 + x_2 + x_3 \leq 10$
     $2x_1 - x_3 \leq 12$
     $2x_1 - 2x_2 + 3x_3 \leq 0$
and $x_1, x_2, x_3 \geq 0$

5. max. $Z = 3x_1 + 5x_2$

s.t. $\quad x_1 + x_3 = 4$

$\quad\quad\quad x_2 + x_4 = 6$

$\quad 3x_1 + 2x_2 + x_5 = 12$

and $x_1, x_2, x_3, x_4, x_5 \geq 0$

6. max. $Z = 3x_1 + 4x_2$

s.t. $\quad 7x_1 + 5x_2 + x_3 = 40$

$\quad\quad 3x_1 + 4x_2 + x_4 = 20$

$\quad\quad\quad x_1 - x_2 + x_5 = 0$

and $\quad x_1, x_2, x_3, x_4, x_5 \geq 0$

ANSWERS

1. $x_1 = 0, x_2 = 4, x_3 = 2$, max. $Z = 10$ **2.** $x_1 = 0, x_2 = 2$, max. $Z = 18$

3. $x_1 = \dfrac{7}{3}, x_2 = 9, x_3 = 0$, max. $Z = 52$ **4.** $x_1 = 0, x_2 = 6, x_3 = 4$, max. $Z = 6$

5. $x_1 = 0, x_2 = 6, x_3 = 4, x_4 = 0, x_5 = 0$, max. $Z = 30$

6. $x_1 = 0, x_2 = 5, x_3 = 15, x_4 = 0, x_5 = 5$, max. $Z = 20$

Glossary

- **Basic and Non-basic variables:** If there are m equations in n variables ($n > m$), a basic solution is one obtained by solving for m variables in terms of the remaining $(n - m)$ variables and setting these $(n - m)$ variables equal to zero. Then remaining $(n - m)$ variables are known as non-basic variables, whereas the m variables are known as basic.

- **Basic Feasible solution:** A basic solution is said to be basic feasible if all m basic variables are non-negative.

- **Degeneracy in LP:** The phenomenon of obtaining a degenerate basic feasible solution in LP is called degeneracy in LP.

- **Degenerate solution:** At any iteration of simplex method, if there is a tie in selecting the departing vector, then the next simplex iteration produces a degenerate solution.

REVIEW QUESTIONS

1. Explain the degeneracy in linear programming problem.

2. Explain the simplex method to resolve degeneracy.

MULTIPLE CHOICE QUESTIONS (CHOOSE THE MOST APPROPRIATE ONE)

1. The degeneracy may appear in a LPP at the very first iteration when some component of **b** is:
   (a) one   (b) zero
   (c) non-zero   (d) None of these

2. If one or more of the basic variables in a BFS is zero then the solution is known as:
   (a) degenerate   (b) non-degenerate
   (c) optimal   (d) None of these

3. A non-degenerate optimal solution may be obtained from a degenerate:
   (a) feasible solution
   (b) basic feasible solution
   (c) can't say
   (d) None of these

4. An optimal degenerate solution may be obtained from a/an:
   (a) optimal solution
   (b) non-degenerate solution
   (c) degenerate solution
   (d) None of these

5. If the choice of outgoing vector at any iteration in simplex method is not unique then the next solution is bound to be:
   (a) degenerate   (b) non-degenerate
   (c) optimal   (d) None of these

6. The procedure which prevent cycling within the simplex routine is called the resolution of:
   (a) non-degeneracy   (b) degeneracy
   (c) can't say   (d) None of these

7. A non-degenerate solution becomes degenerate in the next iteration if the outgoing vector is:
   (a) unique   (b) not unique
   (c) can't say   (d) None of these

8. If some components of vector **b** of LPP is zero, then degeneracy may occur at the first iteration and the next solution:
   (a) may be degenerate
   (b) may not be degenerate
   (c) can't say
   (d) None of these

ANSWERS

1. (b)	**2.** (a)	**3.** (b)	**4.** (c)	**5.** (a)	**6.** (b)	**7.** (b)	**8.** (a)

Revised Simplex Method 5

5.1 INTRODUCTION

Revised simplex method is an another efficient method to solve an LPP. It is also developed by G.B. Dantzig. In simplex method, at each iteration, we had to calculate $c_j - z_j$ corresponding to non-basic variables. Columns to decide whether the current solution is optimal or not. In revised simplex method we need to recompute the values of B^{-1} (**B** is the basis matrix), x_B and Z.

In revised simplex method the word 'revised' refers to the procedure of changing or updating the simplex table. In revised simplex method, we have to calculate B^{-1} at each iteration from its previous value only when y_i is changed at each iteration for which non-basic variable is entered into the basis.

5.2 STANDARD FORM OF REVISED SIMPLEX METHOD

There are following two standard forms:

(1) Standard form I: In this form, only slack and surplus variables are introduced to form basis matrix and there is no need for artificial variables, *i.e.*, an identity matrix is obtained by adding slack variables.

(2) Standard form II: In this form, the artificial variables are used to form basis matrix. In this case two phase method is used in a slightly different way to remove artificial variables.

5.3 REVISED SIMPLEX METHOD FOR STANDARD FORM-I

In the revised simplex method, we treat the objective function of the given LPP as additional constraints along with other constraints. So, the number of constraints in revised simplex method should be $(m + 1)$ while in simple simplex method, they were m in numbers (m is the no. of constraints in the given LPP).

Consider an LPP in the standard form:

$$\max. Z = c_1 x_1 + c_2 x_2 + \ldots + c_n x_n$$

subject to the constraints

$$\left. \begin{array}{l} a_{11}x_1 + a_{12}x_2 + \ldots + a_{1n}x_n \le b_1 \\ a_{21}x_1 + a_{22}x_2 + \ldots + a_{2n}x_n \le b_2 \\ \vdots \\ a_{m1}x_1 + a_{m2}x_2 + \ldots + a_{mn}x_n \le b_m \end{array} \right] \qquad \ldots(1)$$

and $x_1, x_2, \ldots, x_n \ge 0$

To write the above LPP in standard form I, we proceed as follows:

Firstly consider the objective function as an additional constraints and then use slack (or surplus) variables. Then we have the following form,

$$\left.\begin{array}{l} Z - c_1 x_1 - c_2 x_2 - \ldots - c_n x_n - 0 \cdot x_{n+1} - 0 \cdot x_{n+2} - \ldots - 0 \cdot x_{n+m} = 0 \\ a_{11} x_1 + a_{12} x_2 + a_{13} x_3 + \ldots + a_{1n} x_n + x_{n+1} = b_1 \\ a_{21} x_1 + a_{22} x_2 + a_{23} x_3 + \ldots + a_{2n} x_n + x_{n+2} = b_2 \\ \vdots \\ a_{m1} x_1 + a_{m2} x_2 + a_{m3} x_3 + \ldots + x_{m+n} = b_m \end{array}\right] \qquad \ldots(2)$$

Here, $x_{n+1}(= s_1)$, $x_{n+2}(= s_2)$, ..., $x_{n+m}(= s_m)$ are slack variables.

Here the system (2) contains $(m + 1)$ simultaneous equations in $(n + m + 1)$ variables, i.e., $Z, x_1, x_2, \ldots, x_n, x_{n+1}, \ldots, x_{n+m}$.

In matrix form the system (2) can be written as

$$\begin{bmatrix} 1 & -c_1 & -c_2 & \cdots & -c_n & 0 & 0 & \cdots & 0 \\ 0 & a_{11} & a_{12} & \cdots & a_{1n} & 1 & 0 & \cdots & 0 \\ 0 & a_{21} & a_{22} & \cdots & a_{2n} & 0 & 1 & \cdots & 0 \\ \vdots & \vdots & \vdots & & \vdots & \vdots & \vdots & & \vdots \\ 0 & a_{m1} & a_{m2} & \cdots & a_{mn} & 0 & 0 & \cdots & 1 \end{bmatrix} \begin{bmatrix} Z \\ x_1 \\ x_2 \\ \vdots \\ x_{n+m} \end{bmatrix} = \begin{bmatrix} 0 \\ b_1 \\ b_2 \\ \vdots \\ b_m \end{bmatrix}$$

which can also be written as

$$\begin{bmatrix} 1 & -C \\ 0 & A \end{bmatrix} \begin{bmatrix} Z \\ X \end{bmatrix} = \begin{bmatrix} 0 \\ b \end{bmatrix} \qquad \ldots(3)$$

where,
$$0 = \begin{bmatrix} 0 \\ 0 \\ \vdots \\ 0 \end{bmatrix}_{m \times 1}, \quad -C = \begin{bmatrix} -c_1 & -c_2 & -c_3 & -\ldots - c_n & 0 \ldots 0 \end{bmatrix}_{1 \times (m+n)}$$

$$A = \begin{bmatrix} a_{11} & a_{12} & \cdots & a_{1n} & 1 & 0 & \cdots & 0 \\ a_{21} & a_{22} & \cdots & a_{2n} & 0 & 1 & \cdots & 0 \\ \vdots & \vdots & & \vdots & \vdots & \vdots & & \vdots \\ a_{m1} & a_{m2} & \cdots & a_{mn} & 0 & 0 & \cdots & 1 \end{bmatrix}_{(m+n) \times (m+n)}$$

and
$$b = \begin{bmatrix} b_1 \\ b_2 \\ \vdots \\ b_m \end{bmatrix}$$

Now we can consider an identity matrix I_m as the basis in the simplex method.

Further $\qquad A = (\alpha_1, \alpha_2, \ldots, \alpha_n, \alpha_{n+1}, \ldots, \alpha_{n+m})$
where each α_j is m-dimensional column vector.

Now corresponding to each α_j in A, we can define a new $(m + 1)$ component vector by $\alpha_j^{(1)}$ such that $\qquad \alpha_j^{(1)} = [-c_j \alpha_j]'; j = 1, 2 \ldots (n + m)$

$$= \begin{bmatrix} -c_j \\ \alpha_j \end{bmatrix} \qquad \ldots(4)$$

and corresponding to $\boldsymbol{b}$, we can define $(m + 1)$ component vector by

$$\boldsymbol{b}^{(1)} = [0, b_1, b_2, \ldots, b_m]' = [0, \boldsymbol{b}]'$$

and the column corresponding to Z is the $(m + 1)$ component unit vector denoted by e_1 such that

$$e_1 = \begin{bmatrix} 1 \\ 0 \\ 0 \\ \vdots \\ 0 \end{bmatrix}_{m \times 1}$$

Also, in revised simplex method for standard form-I, the basis matrix be denoted by $\boldsymbol{B}_1$ of order $(m + 1) \times (m + 1)$ and e_1 is the first column in it and remaining m columns are any m $\alpha_j^{(1)}$ which are linearly independent and denoted by

$$\beta_i^{(1)}; i = 1, 2, \ldots, m$$

Therefore, we can write

$$\boldsymbol{B}_1 = [e_1, \beta_1^{(1)}, \beta_2^{(1)}, \ldots, \beta_m^{(1)}]$$

$$= \begin{bmatrix} 1 & -C_{B_1} & -C_{B_2} & \cdots & \cdots & -C_{B_m} \\ 0 & \beta_{11} & \beta_{12} & \cdots & \cdots & \beta_{1m} \\ 0 & \beta_{21} & \beta_{22} & \cdots & \cdots & \beta_{2m} \\ \vdots & \vdots & \vdots & & & \vdots \\ 0 & \beta_{m1} & \beta_{m2} & \cdots & \cdots & \beta_{mm} \end{bmatrix} = \begin{bmatrix} 1 & -C_B \\ \boldsymbol{0} & \boldsymbol{B} \end{bmatrix}$$

Here, $C_B = [C_{B_1} \quad C_{B_2} \quad \cdots \quad C_{B_m}], \boldsymbol{0} = \begin{bmatrix} 0 \\ 0 \\ \vdots \\ 0 \end{bmatrix}_{m \times 1}$

and $\boldsymbol{B} = \begin{bmatrix} \beta_{11} & \beta_{12} & \cdots & \beta_{1m} \\ \beta_{21} & \beta_{22} & \cdots & \beta_{2m} \\ \vdots & \vdots & & \vdots \\ \beta_{m1} & \beta_{m2} & \cdots & \beta_{mm} \end{bmatrix} = [\beta_1, \beta_2, \ldots, \beta_m]$

Further we proceed as follows:

(i) To find $\boldsymbol{B}_1^{-1}$: We have

$$\boldsymbol{B}_1 = \begin{bmatrix} 1 & -C_B \\ \boldsymbol{0} & \boldsymbol{B} \end{bmatrix}$$

$$\Rightarrow \qquad \boldsymbol{B}_1^{-1} = \begin{bmatrix} 1 & C_B \boldsymbol{B}^{-1} \\ \boldsymbol{0} & \boldsymbol{B}^{-1} \end{bmatrix} \qquad \ldots(5)$$

But $\boldsymbol{B} = \boldsymbol{I}_m$ then $\boldsymbol{B}^{-1} = \boldsymbol{I}_m^{-1} = \boldsymbol{I}_m$

$$\Rightarrow \qquad \boldsymbol{B}_1^{-1} = \begin{bmatrix} 1 & C_B \boldsymbol{I}_m \\ \boldsymbol{0} & \boldsymbol{I}_m \end{bmatrix} = \begin{bmatrix} 1 & C_B \\ \boldsymbol{0} & \boldsymbol{I}_m \end{bmatrix}$$

If all $b_i \geq 0$, then we use only slack variables to form the basis matrix so we have

$$B = I_m, C_B = 0$$

$$\Rightarrow \qquad B_1^{-1} = \begin{bmatrix} 1 & 0 \\ 0 & I_m \end{bmatrix} = I_{m+1}$$

(ii) To find $\alpha_j^{(1)}$ (not in the basis matrix B_1):

If $\alpha_j^{(1)}$ is not in B_1, then it can be expressed as the linear combination of column vectors of B_1. Therefore,

$$\alpha_j^{(1)} = x_{0j}e_1 + x_{1j}\beta_1^{(1)} + x_{2j}\beta_2^{(1)} + \ldots + x_{mj}\beta_m^{(1)}$$

$$= [e_1, \beta_1^{(1)}, \beta_2^{(1)}, \ldots, \beta_m^{(1)}] \begin{bmatrix} x_{0j} \\ x_{1j} \\ \vdots \\ x_{mj} \end{bmatrix}$$

$$\Rightarrow \qquad \alpha_j^{(1)} = B_1 x_j^{(1)}, x_j^{(1)} = \begin{bmatrix} x_{0j} \\ x_{1j} \\ \vdots \\ x_{mj} \end{bmatrix}$$

$$\Rightarrow \qquad x_j^{(1)} = B_1^{-1}\alpha_j^{(1)} \qquad\qquad \ldots(6)$$

$$= \begin{bmatrix} 1 & C_B B^{-1} \\ 0 & B^{-1} \end{bmatrix} \begin{bmatrix} -c_j \\ \alpha_j \end{bmatrix} \qquad\qquad \text{[By (4) and (5)]}$$

$$= \begin{bmatrix} -c_j + C_B B^{-1}\alpha_j \\ 0 + B^{-1}\alpha_j \end{bmatrix} = \begin{bmatrix} -c_j + C_B x_j \\ 0 + x_j \end{bmatrix} \qquad [\because B^{-1}\alpha_j = X_j]$$

$$= \begin{bmatrix} -c_j + z_j \\ x_j \end{bmatrix} \qquad\qquad [\because z_j = C_B x_j]$$

$$= \begin{bmatrix} -\Delta_j \\ x_j \end{bmatrix} \qquad\qquad [\because \Delta_j = c_j - z_j] \qquad \ldots(7)$$

From (6) and (7), we observed that $-\Delta_j$ is obtained when the first row of B_1^{-1} is multiplied by $\alpha_j^{(1)}$ (not in the basis B_1)

$$\Rightarrow \qquad -\Delta_j = (\text{first row of } B_1^{-1}) \times \alpha_j^{-1} (\text{not in the basis } B_1)$$

The first element in $x_j^{(1)}$ is $-\Delta_j$, decides the optimality of the solution.

(iii) To find the Initial Basic Feasible Solution:

If x_B^{-1} is the basic solution corresponding to the basis matrix B_1, then

$$x_B^{-1} = B_1^{-1}b^{(1)} = \begin{bmatrix} 1 & C_B B^{-1} \\ 0 & B^{-1} \end{bmatrix} \begin{bmatrix} 0 \\ b \end{bmatrix}$$

$$= \begin{bmatrix} C_B B^{-1} b \\ B^{-1} b \end{bmatrix} = \begin{bmatrix} C_B x_B \\ x_B \end{bmatrix} \qquad [\because x_B = B^{-1} b]$$

$$= \begin{bmatrix} Z \\ x_B \end{bmatrix} \qquad [\because Z = C_B x_B] \qquad \dots(8)$$

In equation (8), the first component is x_B^{-1} in Z which is the value of objective function and the second component is x_B which is the basic feasible solution of the original LPP. Now since Z can be negative so that the initial solution x_B^{-1} is not necessarily basic feasible solution.

WORKING PROCEDURE

STEP 1. Write the given LPP in the standard form of maximization.

STEP 2. Write the given LPP in standard form-I of revised simplex method.

STEP 3. Write basis matrix B_1 and B_1^{-1} and then find IBFS.

$$x_B^{-1} = B_1^{-1} b^{(1)}$$

If for all $b_i \geq 0$ then $B_1 = I_{m+1}$, $B_1^{-1} = I_{m+1}$ then

$$x_B^{-1} = I_{m+1} b^{(1)} = b^{(1)} = \begin{bmatrix} 0 \\ b_1 \\ b_2 \\ \vdots \\ b_m \end{bmatrix}$$

STEP 4. Construct the starting revised simplex table as given below:

B.V.	x_B^{-1}	B_1^{-1}					$x_k^{-1} = B^{-1}\alpha_k^{(1)}$	Min Ratio $\dfrac{x_{B_i}}{x_{ik}}, x_{ik} > 0$
		e_1	$\beta_1^{(1)}$	$\beta_2^{(1)}$	...	$\beta_m^{(1)}$		
Z	0	0	0	0	...	0	$z_k - c_k = -\Delta_k$	
s_1	$b_1(x_{B_1})$	1	0	0	...	0	x_{1k}	
s_2	$b_2(x_{B_2})$	0	1	0		0	x_{2k}	
s_3	$b_3(x_{B_3})$	0	0	1		0	x_{3k}	
$\vdots$	$\vdots$	$\vdots$	$\vdots$	$\vdots$		$\vdots$	$\vdots$	
s_m	$b_m(x_{B_m})$	0	0	0		1	x_{mk}	

STEP 5. Now compute Δ_j for all $\alpha_j^{(1)}$ (not in the basis B_1) by using the following formula.

$$\Delta_j = -(\text{first row of } B_1^{-1}) \times \alpha_j^{(1)}$$

and if all $\Delta_j \leq 0$ then the obtained BFS is optimal, but if atleast one of $\Delta_j > 0$, then solution is not optimal and go to next step.

STEP 6. Now we have to find incoming and outgoing vector. For this, compute

$$\max_j(\Delta_j) \, \forall j \text{ for which } \alpha_j^{(1)} \text{ are not in } B_1.$$

If $\max_{j}\{\Delta_j\} = \Delta_k$, then $\alpha_k^{(1)}$ will be incoming vector

To find outgoing vector, use the formula

$$x_k^{(1)} = B_1^{-1}\alpha_k^{(1)} = \begin{bmatrix} -\Delta_k \\ x_{1k} \\ x_{2k} \\ \vdots \\ x_{mk} \end{bmatrix}$$

Now by minimum ratio rule, we may find

$$\max_{j}\left\{\frac{x_{B_i}}{x_{ik}}, x_{ik} > 0\right\}$$

If $\min_{j}\left\{\frac{x_{B_i}}{x_{ik}}, x_{ik} > 0\right\} = \frac{x_{B_r}}{x_{rk}}$, then the vector $\beta_r^{(1)}$ will be the outgoing vector.

STEP 7.   In this step, we have to find the improved solution by finding the key element as in the ordinary simplex method. If x_{rk} is the key element then bring $\alpha_k^{(1)}$ in place for $\beta_r^{(1)}$ and construct the new revised simplex table, from which one can find improved basic feasible solution.

Repeat step (6) and (7) untill we get an optimal solution of the given LPP.

 Solved Examples

EXAMPLE 1.   *Solve the following LPP by revised simplex method.*

$$max. \ Z = 6x_1 - 2x_2 + 3x_3$$

subject to the constraints

$$2x_1 - x_2 + 2x_3 \leq 2$$
$$x_1 + 4x_3 \leq 4$$

and $x_1, x_2, x_3 \geq 0$

SOLUTION.   The given LPP of maximization can be written in the revised simplex form as follows:

$$max. \ Z = 6x_1 - 2x_2 + 3x_3 + 0s_1 + 0s_2$$

$$\left. \begin{array}{r} Z - 6x_1 + 2x_2 - 3x_3 + 0s_1 + 0s_2 = 0 \\ 2x_1 - x_2 + 2x_3 + s_1 = 2 \\ x_1 + 4x_3 + s_2 = 4 \end{array} \right] \qquad \ldots(1)$$

s.t.

In the matrix form, the above system can be written as

$$\begin{array}{cccccc} e_1 & \alpha_1^{(1)} & \alpha_2^{(1)} & \alpha_3^{(1)} & \alpha_4^{(1)} & \alpha_5^{(1)} \end{array}$$

$$\begin{bmatrix} 1 & -6 & 2 & -3 & 0 & 0 \\ 0 & 2 & -1 & 2 & 1 & 0 \\ 0 & 1 & 0 & 4 & 0 & 1 \end{bmatrix} \begin{bmatrix} Z \\ x_1 \\ x_2 \\ x_3 \\ s_1 \\ s_2 \end{bmatrix} = \begin{bmatrix} 0 \\ 2 \\ 4 \end{bmatrix} \qquad \ldots(2)$$

$$\Rightarrow \qquad B_1 = \begin{bmatrix} 1 & 0 & 0 \\ 0 & 1 & 0 \\ 0 & 0 & 1 \end{bmatrix} = [e_1, \alpha_4^{(1)}, \alpha_5^{(1)}]$$

By hypothesis, we take $\beta_1^{(1)} = \alpha_4^{(1)}, \beta_2^{(1)} = \alpha_5^{(1)}$

Therefore, $\qquad B_1 = \begin{bmatrix} e_1 & \beta_1^{(1)} & \beta_2^{(1)} \end{bmatrix} = I_3$

$$\Rightarrow \qquad B_1^{-1} = I_3^{-1} = I_3$$

and the initial basic feasible solution is given by

$$x_B^{(1)} = B_1^{-1} \begin{bmatrix} 0 \\ 2 \\ 4 \end{bmatrix} = I_3 \begin{bmatrix} 0 \\ 2 \\ 4 \end{bmatrix} = \begin{bmatrix} 0 \\ 2 \\ 4 \end{bmatrix}$$

Now, we have the following revised simplex table.

Revised Simplex table-1

Variables in the basis	solution $x_B^{(1)}$	B_1^{-1}			$x_1^{(1)} = B_1^{-1}\alpha_1^{(1)}$	Min Ratio
		e_1	$\beta_1^{(1)}(\alpha_4^{(1)})$	$\beta_2^{(1)}(\alpha_5^{(1)})$		
Z	0	1	0	0	-6	$\min\limits_{i} \left\{ \dfrac{x_{B_i}}{x_{i1}}, x_{i1} > 0 \right\}$
s_1	$2(x_{B_1})$	0	1	0	②(x_{11})	2/2 (min) →
s_2	$4(x_{B_2})$	0	0	1	$1(x_{21})$	4/1 = 4

Since $\alpha_1^{(1)}, \alpha_2^{(1)}$ and $\alpha_3^{(1)}$ are not in the basis B_1 therefore,

$$\Delta_1 = -(\text{first row of } B_1^{-1}) \times \alpha_1^{(1)}$$

$$= -[1,0,0] \begin{bmatrix} -6 \\ 2 \\ 1 \end{bmatrix} = 6$$

$$\Delta_2 = -[1,0,0] \begin{bmatrix} 2 \\ -1 \\ 0 \end{bmatrix} = -2$$

$$\Delta_3 = -[1,0,0] \begin{bmatrix} -3 \\ 2 \\ 4 \end{bmatrix} = 3$$

Clearly, $\Delta_1, \Delta_3 > 0$ therefore, the initial basic feasible solution is not optimal.

To find incoming vector:

Since, $\qquad \Delta_1 = \max\{\Delta_1, \Delta_2, \Delta_3\} = \Delta_k$

$$\Rightarrow \qquad k = 1$$

$\Rightarrow \quad \alpha^{(1)}$ is the incoming vector.

To find outgoing vector:

Using minimum ratio rule

$$\frac{x_{B_r}}{x_{r1}} = \min_i \left\{ \frac{x_{B_i}}{x_{i1}}, x_{i1} > 0 \right\}$$

$$= \min \left\{ \frac{x_{B_1}}{x_{11}}, \frac{x_{B_2}}{x_{21}} \right\} = \min \left\{ \frac{2}{2}, \frac{4}{1} \right\} = \frac{2}{2} = \frac{x_{B_1}}{x_{11}}$$

$\Rightarrow$ $\quad\quad\quad r = 1$

$\Rightarrow$ $\quad \beta_1^{(1)} (= \alpha_4^{(1)})$ is the outgoing vector.

Hence, $x_{11} = 2$ is the required key element.

Now, apply the usual procedure, we have the following revised simplex table.

Revised Simplex table-2

Variables in the basis	solution $x_B^{(1)}$	B_1^{-1}			$x_2^{(1)} = B_1^{-1}\alpha_2^{(1)}$	Minimum Ratio
		e_1	$\beta_1^{(1)}\left(\alpha_1^{(1)}\right)$	$\beta_2^{(1)}\left(\alpha_5^{(1)}\right)$		
Z	6	1	3	0	-1	$\min_i\left\{\dfrac{x_{B_i}}{x_{i2}}, x_{i2} > 0\right\}$
x_1	1	0	1/2	0	$-1/2(x_{12})$	$-$ve
s_2	3	0	$-1/2$	1	①/2 (x_{22})	3/1 (min) $\rightarrow$

Clearly, from the above table

$\quad x_1 = 1, x_2 = 0$ and max $Z = 6$ is the new improved solution.

Now, we will check the optimality of this solution.

Since $\alpha_2^{(1)}, \alpha_3^{(1)}$ and $\alpha_4^{(1)}$ are not in the basis B_1 so

$$\Delta_2 = -(\text{first row of } B_1^{-1} \text{ in second table}) \times \alpha_2^{(1)}$$

$$= -[1,3,0]\begin{bmatrix} 2 \\ -1 \\ 0 \end{bmatrix} = 1$$

$$\Delta_3 = -[1,3,0]\begin{bmatrix} -3 \\ 2 \\ 4 \end{bmatrix} = -3$$

$$\Delta_4 = -[1,3,0]\begin{bmatrix} 0 \\ 1 \\ 0 \end{bmatrix} = 3$$

Clearly, $\Delta_2 > 0 \Rightarrow$ improved solution is not optimal.

To find incoming vector:

$\because$ $\quad\quad\quad \Delta_2 = \max\{\Delta_2, \Delta_3, \Delta_4\} = \Delta_k \quad \Rightarrow \quad k = 2$

$\Rightarrow$ $\quad \alpha_2^{(1)}$ is the incoming vector.

Now, $\quad x_2^{(1)} = B_1^{-1}\alpha_2^{(1)} = \begin{bmatrix} 1 & 3 & 0 \\ 0 & 1/2 & 0 \\ 0 & -1/2 & 1 \end{bmatrix}\begin{bmatrix} 2 \\ -1 \\ 0 \end{bmatrix} = \begin{bmatrix} -\Delta_2 \\ x_{12} \\ x_{22} \end{bmatrix}$

To find the outgoing vector:

Apply minimum ratio rule, we have

$$\frac{x_{B_r}}{x_{r2}} = \min_i \left\{ \frac{x_{B_i}}{x_{i2}}, x_{i2} > 0 \right\} = \min \left\{ \frac{x_{B_2}}{x_{22}} \right\} \qquad [\because x_{12} = -1]$$

$$\Rightarrow \qquad \frac{x_{B_r}}{x_{r2}} = \frac{x_{B_2}}{x_{22}} \qquad \Rightarrow \qquad r = 2$$

Therefore, $\beta_2^{(1)}(= \alpha_5^{(1)})$ is the outgoing vector and hence $x_{22} = \dfrac{1}{2}$ is the key element.

Further to find the improved solution, we have the following simplex table:

Revised Simplex table-3

Variables in the basis	solution $x_B^{(1)}$	B_1^{-1}			$x_k^{(1)}$	Min. Ratio
		e_1	$\beta_1^{(1)}\left(\alpha_1^{(1)}\right)$	$\beta_2^{(1)}\left(\alpha_2^{(1)}\right)$		
Z	12	1	2	2		
x_1	4	0	0	1		
x_2	6	0	−1	2		

From the above table, we obtained

$$x_1 = 4, x_2 = 6, x_3 = 0 \text{ and max. } Z = 12$$

Now we shall check the optimality of this solution.

Clearly, $\alpha_3^{(1)}, \alpha_4^{(1)}$ and $\alpha_5^{(1)}$ are not in the basis $\boldsymbol{B}_1$. Therefore,

$$\Delta_3 = -(\text{first row of } \boldsymbol{B}_1^{-1} \text{ of third table}) \times \alpha_3^{(1)}$$

$$= -[1,2,2] \begin{bmatrix} -3 \\ 2 \\ 4 \end{bmatrix} = -9$$

$$\Delta_4 = -[1,2,2] \begin{bmatrix} 0 \\ 1 \\ 0 \end{bmatrix} = -2$$

$$\Delta_5 = -[1,2,2] \begin{bmatrix} 0 \\ 0 \\ 1 \end{bmatrix} = -2$$

We observe that $\Delta_3, \Delta_4, \Delta_5 \leq 0$. Hence, this improved solution is optimal and is given by

$$x_1 = 4, x_2 = 6, x_3 = 0 \text{ and max. } Z = 12$$

EXAMPLE 2. *Solve the following LPP by revised simplex method:*

$$\text{max. } Z = x_1 + 2x_2$$

subject to the constraints

$$x_1 + x_2 \le 3$$
$$x_1 + 2x_2 \le 5$$
$$3x_1 + x_2 \le 6$$

and $\quad x_1, x_2 \ge 0$

SOLUTION. The given maximization LPP can be written in revised simplex form as follows:

$$\text{max. } Z = x_1 + 2x_2 + 0s_1 + 0s_2 + 0s_3$$

s.t.

$$\left. \begin{array}{l} Z - x_1 - 2x_2 + 0s_1 + 0s_2 + 0s_3 = 0 \\ x_1 + x_2 + s_1 = 3 \\ x_1 + 2x_2 + s_2 = 5 \\ 3x_1 + x_2 + s_3 = 6 \end{array} \right\} \qquad \ldots (1)$$

and $\quad x_1, x_2, s_1, s_2, s_3 \ge 0$

The above system can be written in matrix form as follows:

$$\begin{array}{cccccc} e_1 & \alpha_1^{(1)} & \alpha_2^{(1)} & \alpha_3^{(1)} & \alpha_4^{(1)} & \alpha_5^{(1)} \end{array}$$

$$\begin{bmatrix} 1 & -1 & -2 & 0 & 0 & 0 \\ 0 & 1 & 1 & 1 & 0 & 0 \\ 0 & 1 & 2 & 0 & 1 & 0 \\ 0 & 3 & 1 & 0 & 0 & 1 \end{bmatrix} \begin{bmatrix} Z \\ x_1 \\ x_2 \\ s_1 \\ s_2 \\ s_3 \end{bmatrix} = \begin{bmatrix} 0 \\ 3 \\ 5 \\ 6 \end{bmatrix} \qquad \ldots (2)$$

Here, $\qquad \mathbf{B}_1 = \begin{bmatrix} 1 & 0 & 0 & 0 \\ 0 & 1 & 0 & 0 \\ 0 & 0 & 1 & 0 \\ 0 & 0 & 0 & 1 \end{bmatrix} = \begin{bmatrix} e_1 & \alpha_3^{(1)} & \alpha_4^{(1)} & \alpha_5^{(1)} \end{bmatrix}$

Now, by hypothesis we have

$$\alpha_3^{(1)} = \beta_1^{(1)}, \alpha_4^{(1)} = \beta_2^{(1)}, \alpha_5^{(1)} = \beta_3^{(1)}$$

So, $\qquad \mathbf{B}_1 = \begin{bmatrix} e_1 & \beta_1^{(1)} & \beta_2^{(1)} & \beta_3^{(1)} \end{bmatrix} = I_4$

$$\Rightarrow \quad \mathbf{B}_1^{-1} = I_4^{-1} = I_4$$

Also, the initial solution is given by

$$x_B^{(1)} = \begin{bmatrix} 0 \\ 3 \\ 5 \\ 6 \end{bmatrix}$$

Now we have the following revised simplex table.

Revised Simplex table-1

Variables in the basis	solution $x_B^{(1)}$	e_1	$\beta_1^{(1)}$	$\beta_2^{(1)}$	$\beta_3^{(1)}$	$x_2^{(1)} = B_1^{-1}\alpha_2^{(1)}$	Min Ratio $\dfrac{x_{B_i}}{x_{i2}}, x_{i2} > 0$
				B_1^{-1}			
Z	0	1	0	0	0	-2	$\min\limits_i\left\{\dfrac{x_{B_i}}{x_{i2}}, x_{i2} > 0\right\}$
s_1	$3(x_{B_1})$	0	1	0	0	$1\ (x_{12})$	$3/1$
s_2	$5(x_{B_2})$	0	0	1	0	②$\ (x_{22})$	$5/2$ (min) $\rightarrow$
s_3	$6(x_{B_3})$	0	0	0	1	$1\ (x_{32})$	$6/1$

Now, since $\alpha_1^{(1)}$ and $\alpha_2^{(1)}$ are not in the basis therefore,

$$\Delta_1 = -\begin{bmatrix} 1 & 0 & 0 & 0 \end{bmatrix}\begin{bmatrix} -1 \\ 1 \\ 1 \\ 3 \end{bmatrix} = 1$$

$$\Delta_2 = -\begin{bmatrix} 1 & 0 & 0 & 0 \end{bmatrix}\begin{bmatrix} -2 \\ 1 \\ 2 \\ 1 \end{bmatrix} = 2$$

which implies that $\Delta_1, \Delta_2 > 0$ so solution is not optimal.

To find incoming vector:

Here, $\max(\Delta_1, \Delta_2) = \Delta_2 = \Delta_k \qquad \Rightarrow \qquad k = 2$

$\Rightarrow \quad \alpha_2^{(1)}$ is the incoming vector.

Also, $\quad x_K^{(1)} = x_2^{(1)} = B_1^{-1}\alpha_2^{(1)}$

$$= \begin{bmatrix} 1 & 0 & 0 & 0 \\ 0 & 1 & 0 & 0 \\ 0 & 0 & 1 & 0 \\ 0 & 0 & 0 & 1 \end{bmatrix}\begin{bmatrix} -2 \\ 1 \\ 2 \\ 1 \end{bmatrix} = \begin{bmatrix} -2 \\ 1 \\ 2 \\ 1 \end{bmatrix}$$

To find outgoing vector:

Apply the minimum ratio rule, we have

$$\frac{x_{B_r}}{x_{r2}} = \min_i\left\{\frac{x_{B_i}}{x_{i2}}, x_{i2} > 0\right\} = \min\left\{\frac{3}{1}, \frac{5}{2}, \frac{6}{1}\right\} = \frac{5}{2} = \frac{x_{B_2}}{x_{22}}$$

$\Rightarrow \qquad r = 2$

$\Rightarrow \quad \beta_2^{(1)}(= \alpha_4^{(1)})$ is the outgoing vector.

Hence, $x_{22} = 2$ is the key element.

Now, to find improved solution, apply the usual procedure we have the following revised simplex table:

Revised Simplex table-2

Variables in the basis	solution $x_B^{(1)}$	e_1	$B_1^{(1)}\left(\alpha_3^{(1)}\right)$	$B_2^{(1)}\left(\alpha_2^{(1)}\right)$	$B_3^{(1)}\left(\alpha_5^{(1)}\right)$	$x_k^{-1} = B_1^{-1}\alpha_k^{(1)}$	Min Ratio
Z	5	1	0	1	0		
s_1	1/2	0	1	−1/2	0		
x_2	5/2	0	0	1/2	0		
s_3	7/2	0	0	−1/2	1		

From the above table we observe that

$$x_2 = \frac{5}{2}, s_1 = \frac{1}{2}, s_3 = \frac{7}{2} \text{ and max. } Z = 5$$

Now, we have to check the optimality of this improved solution.

Since $\alpha_1^{(1)}$ and $\alpha_4^{(1)}$ are not in the basis B_1. Then

$$\Delta_1 = -(\text{first row of } B_1^{-1} \text{ in the second table}) \times \alpha_1^{(1)}$$

$$= -\begin{bmatrix} 1 & 0 & 1 & 0 \end{bmatrix} \begin{bmatrix} -1 \\ 1 \\ 1 \\ 3 \end{bmatrix} = 0$$

$$\Delta_4 = -\begin{bmatrix} 1 & 0 & 1 & 0 \end{bmatrix} \begin{bmatrix} 0 \\ 0 \\ 1 \\ 0 \end{bmatrix} = -1$$

We observe that $\Delta_1, \Delta_4 \leq 0 \Rightarrow$ solution is optimal and is given by

$$x_1 = 0, x_2 = \frac{5}{2} \text{ and max. } Z = 5$$

EXAMPLE 3.   *Solve the following LPP by revised simplex method:*

$$max. \ Z = 3x_1 + x_2 + 2x_3 + 7x_4$$

subject to the constraints

$$2x_1 + 3x_2 - x_3 + 4x_4 \leq 40$$
$$- 2x_1 + 2x_2 + 5x_3 - x_4 \leq 35$$
$$x_1 + x_2 - 2x_3 + 3x_4 \leq 100$$
$$and \ x_1 \leq 2, x_2 \geq 1, x_3 \geq 3, x_4 \geq 1$$

SOLUTION.   In the given LPP, the non-negative restrictions are given by

$$x_1 \geq 2, x_2 \geq 1, x_3 \geq 3 \text{ and } x_4 \geq 4$$

$\Rightarrow$ lower bounds of x_1, x_2, x_3 and x_4 are not zero.

Thus, introduce new variables y_1, y_2, y_3 and y_4 such that

$$x_1 = y_1 + 2, x_2 = y_2 + 1, x_3 = y_3 + 3, x_4 = y_4 + 4$$

such that $y_1 \geq 0, y_2 \geq 0, y_3 \geq 0, y_4 \geq 0$

Therefore, the given LPP reduces to
$$\max. Z = 3(y_1 + 2) + (y_2 + 1) + 2(y_3 + 3) + 7(y_4 + 4)$$
$$= 3y_1 + y_2 + 2y_3 + 7y_4 + 41$$
or
$$\max. Z' = 3y_1 + y_2 + 2y_3 + 7y_4 = Z - 41$$
s.t.
$$2y_1 + 3y_2 - y_3 + 4y_4 \le 20$$
$$-2y_1 + 2y_2 + 5y_3 - y_4 \le 26$$
$$y_1 + y_2 - 2y_3 + 3y_4 \le 91$$
and
$$y_1, y_2, y_3, y_4 \ge 0$$

The above LPP is of maximization. This LPP in revised simplex form can be written as
$$\max. Z' = 3y_1 + y_2 + 2y_3 + 7y_4 + 0s_1 + 0s_2 + 0s_3$$
s.t.

$$\left.\begin{array}{l} Z' - 3y_1 - y_2 - 2y_3 - 7y_4 + 0s_1 + 0s_2 + 0s_3 = 0 \\ 2y_1 + 3y_2 - y_3 + 4y_4 + s_1 = 20 \\ -2y_1 + 2y_2 + 5y_3 - y_4 + s_2 = 26 \\ y_1 + y_2 - 2y_3 + 3y_4 + s_3 = 91 \end{array}\right] \qquad \dots(1)$$

and
$$y_i, s_i \ge 0$$

The above system of equations can be written in matrix form as follows:

$$\begin{array}{cccccccc} e_1 & \alpha_1^{(1)} & \alpha_2^{(1)} & \alpha_3^{(1)} & \alpha_4^{(1)} & \alpha_5^{(1)} & \alpha_6^{(1)} & \alpha_7^{(1)} \end{array}$$

$$\begin{bmatrix} 1 & -3 & -1 & -2 & -7 & 0 & 0 & 0 \\ 0 & 2 & 3 & -1 & 4 & 1 & 0 & 0 \\ 0 & -2 & 2 & 5 & -1 & 0 & 1 & 0 \\ 0 & 1 & 1 & -2 & 3 & 0 & 0 & 1 \end{bmatrix} \begin{bmatrix} Z' \\ y_1 \\ y_2 \\ y_3 \\ y_4 \\ s_1 \\ s_2 \\ s_3 \end{bmatrix} = \begin{bmatrix} 0 \\ 20 \\ 26 \\ 91 \end{bmatrix} \qquad \dots(2)$$

$$\Delta_4 = -\begin{bmatrix} 1 & 0 & 0 & 0 \end{bmatrix} \begin{bmatrix} -7 \\ 4 \\ -1 \\ 3 \end{bmatrix} = 7$$

Clearly all $\Delta_j > 0 \Rightarrow$ solution is not optimal.

To find incoming vector:

$\because \; \max\{\Delta_1, \Delta_2, \Delta_3, \Delta_4\} = \max\{3, 1, 2, 7\} = 7 = \Delta_4$

$\Rightarrow \qquad\qquad k = 4$

$\Rightarrow \quad \alpha_4^{(1)}$ is the incoming vector which enter in the basis matrix.

Also,
$$y_4^{(1)} = B_1^{-1}\alpha_4^{(1)} = I_4\alpha_4^{(1)} = \alpha_4^{(1)} = \begin{bmatrix} -7 \\ 4 \\ -1 \\ 3 \end{bmatrix} = \begin{bmatrix} -\Delta_4 \\ y_{14} \\ y_{24} \\ y_{34} \end{bmatrix}$$

To find outgoing vector: Apply minimum ratio, we get

$$\frac{y_{B_r}}{y_{r4}} = \min_i \left\{ \frac{y_{B_i}}{y_{i4}}, y_{i4} > 0 \right\}$$

$$= \min \left\{ \frac{y_{B_1}}{y_{14}}, \frac{y_{B_2}}{y_{34}} \right\} = \min \left\{ \frac{20}{4}, \frac{91}{3} \right\} = \frac{20}{4} = \frac{y_{B_1}}{y_{14}}$$

$$\Rightarrow \qquad r = 1$$

So, $\beta_1^{(1)} (= \alpha_5^{(1)})$ is the outgoing vector.

$$\Rightarrow \quad y_{14} = 4 \text{ is the key element.}$$

Revised Simplex table-1

Variables in the basis	solution $y_B^{(1)}$	B_1^{-1}				$y_3^{(1)} = B_1^{-1} \alpha_3^{(1)}$	Min Ratio
		e_1	$\beta_1^{(1)} \left(\alpha_4^{(1)} \right)$	$\beta_2^{(1)} \left(\alpha_6^{(1)} \right)$	$\beta_3^{(1)} \left(\alpha_7^{(1)} \right)$		
Z'	35	1	7/4	0	0	–15/4	$\min_i \left\{ \dfrac{y_{B_i}}{y_{i3}}, y_{i3} > 0 \right\}$
y_4	$5(y_{B_1})$	0	1/7	0	0	–1/4 (y_{13})	–ve
s_2	$31(y_{B_2})$	0	1/4	1	0	⑲/4 (y_{23})	124/19 (min) →
s_3	$76(y_{B_3})$	0	–3/4	0	1	–5/4 (y_{33})	–ve

$$\downarrow$$

From the above table, we observe that

$$y_4 = 5, \ s_2 = 31, \ s_3 = 76 \text{ and max. } Z' = 35$$

From (2), the basis matrix

$$B_1 = \begin{bmatrix} 1 & 0 & 0 & 0 \\ 0 & 1 & 0 & 0 \\ 0 & 0 & 1 & 0 \\ 0 & 0 & 0 & 1 \end{bmatrix} = \begin{bmatrix} e_1 & \alpha_5^{(1)} & \alpha_6^{(1)} & \alpha_7^{(1)} \end{bmatrix}$$

By hypothesis $\alpha_5^{(1)} = \beta_1^{(1)}, \alpha_6^{(1)} = \beta_2^{(1)}, \alpha_7^{(1)} = \beta_3^{(1)}$

$$\Rightarrow \qquad B_1 = [e_1 \ \ \beta_1^{(1)} \ \ \beta_2^{(1)} \ \ \beta_3^{(1)}] = I_4$$

$$\Rightarrow \qquad B_1^{-1} = I_4^{-1} = I_4$$

Now apply the revised simplex method in a usual manner, we have the following revised simplex table.

Revised Simplex table-2

Variables in the basis	solution $y_B^{(1)}$	B_1^{-1}				$y_4^{(1)} = B_1^{-1}\alpha_4^{(1)}$	Minimum Ratio
		e_1	$\beta_1^{(1)}\left(\alpha_5^{(1)}\right)$	$\beta_2^{(1)}\left(\alpha_6^{(1)}\right)$	$\beta_3^{(1)}\left(\alpha_7^{(1)}\right)$		
Z'	0	1	0	0	0	−7	$\min\limits_{i}\left\{\dfrac{y_{B_i}}{y_{i4}}, y_{i4} > 0\right\}$
s_1	$20(y_{B_1})$	0	1	0	0	④(y_{14})	$20/4 = 5(\min) \rightarrow$
s_2	$26(y_{B_2})$	0	0	1	0	−1 (y_{24})	−ve
s_3	$91(y_{B_3})$	0	0	0	1	3 (y_{34})	91/3

Clearly, $\alpha_1^{(1)}, \alpha_2^{(1)}, \alpha_3^{(1)}, \alpha_4^{(1)}$ are not in the basis matrix B_1 then

$$\Delta_1 = -(\text{first row of } B_1^{-1} \text{ of first table}) \times \alpha_1^{(1)}$$

$$= -[1 \quad 0 \quad 0 \quad 0]\begin{bmatrix} -3 \\ 2 \\ -2 \\ 1 \end{bmatrix} = 3$$

$$\Delta_2 = -[1 \quad 0 \quad 0 \quad 0]\begin{bmatrix} -1 \\ 3 \\ 2 \\ 1 \end{bmatrix} = 1$$

$$\Delta_3 = -[1 \quad 0 \quad 0 \quad 0]\begin{bmatrix} -2 \\ -1 \\ 5 \\ -2 \end{bmatrix} = 2$$

Now we check the optimality of this solution.

$\because \quad \alpha_1^{(1)}, \alpha_2^{(1)}, \alpha_3^{(1)}$ and $\alpha_5^{(1)}$ are not in the basis matrix B_1.

So, now basis matrix is

$$B_1 = [e_1 \quad \alpha_4^{(1)} \quad \alpha_6^{(1)} \quad \alpha_7^{(1)}]$$

Using $\qquad \Delta_j = -(\text{first row of } B_1^{-1} \text{ in second table}) \times \alpha_j^{(1)}$, we get

$$\Delta_1 = -\left[1 \quad \frac{7}{4} \quad 0 \quad 0\right]\begin{bmatrix} -3 \\ 2 \\ -2 \\ 1 \end{bmatrix} = -\frac{1}{2}$$

$$\Delta_2 = -[1 \quad \frac{7}{4} \quad 0 \quad 0] \begin{bmatrix} -1 \\ 3 \\ 2 \\ 1 \end{bmatrix} = -\frac{19}{4}$$

$$\Delta_3 = -[1 \quad \frac{7}{4} \quad 0 \quad 0] \begin{bmatrix} -2 \\ -1 \\ 5 \\ -2 \end{bmatrix} = \frac{15}{4}$$

and

$$\Delta_5 = -[1 \quad \frac{7}{4} \quad 0 \quad 0] \begin{bmatrix} 0 \\ 1 \\ 0 \\ 0 \end{bmatrix} = -\frac{7}{4}$$

Clearly, $\Delta_3 > 0 \Rightarrow$ solution is not optimal.

To find incoming vector:

$$\because \quad \max\{\Delta_1, \Delta_2, \Delta_3, \Delta_4\} = \max\left\{\frac{-1}{2}, \frac{-19}{4}, \frac{15}{4}, \frac{-7}{4}\right\} = \frac{15}{4} = \Delta_3$$

$$\Rightarrow \quad\quad\quad k = 3$$

$\Rightarrow \quad \alpha_3^{(1)}$ is the incoming vector.

and

$$y_3^{(1)} = \boldsymbol{B}_1^{-1}\alpha_3^{(1)} = \begin{bmatrix} 1 & 7/4 & 0 & 0 \\ 0 & 1/4 & 0 & 0 \\ 0 & 1/4 & 1 & 0 \\ 0 & -3/4 & 0 & 1 \end{bmatrix} \begin{bmatrix} -2 \\ -1 \\ 5 \\ -2 \end{bmatrix}$$

$$= \begin{bmatrix} -15/4 \\ -1/4 \\ 19/4 \\ -5/4 \end{bmatrix} = \begin{bmatrix} -\Delta_3 \\ y_{13} \\ y_{23} \\ y_{33} \end{bmatrix}$$

To find outgoing vector: Apply minimum ratio rule, we get

$$\frac{y_{B_r}}{y_{r3}} = \min_i\left\{\frac{y_{B_i}}{y_{i3}}, y_{i3} > 0\right\} = \min\left\{\frac{y_{B_1}}{y_{13}}, \frac{y_{B_2}}{y_{23}}, \frac{y_{B_3}}{y_{33}}\right\}$$

$$= \left\{\frac{31}{19/4}\right\} = \frac{y_{B_2}}{y_{23}}$$

$$\Rightarrow \quad\quad\quad r = 2$$

$\Rightarrow \quad \beta_2^{(1)}(= \alpha_6^{(1)})$ is the outgoing vector.

$$\Rightarrow \quad y_{23} = \frac{19}{4} \text{ is the key element.}$$

Revised Simplex table-3

Variables in the basis	solution $y_B^{(1)}$	B_1^{-1}				$y_1^{(1)} = B_1^{-1}\alpha_1^{(1)}$	Minimum Ratio
		e_1	$\beta_1^{(1)}\left(\alpha_4^{(1)}\right)$	$\beta_2^{(1)}\left(\alpha_3^{(1)}\right)$	$\beta_3^{(1)}\left(\alpha_7^{(1)}\right)$		
Z'	1130/19	1	37/19	15/19	0	–13/19	$\min\limits_{i}\left\{\dfrac{y_{B_i}}{y_{i1}}, y_{i1} > 0\right\}$
y_4	126/19	0	5/19	1/19	0	(8/19) (y_{11})	63/4 (min) →
y_3	124/19	0	1/19	4/19	0	–6/19 (y_{21})	–ve
s_3	1599/19	0	–13/19	5/19	1	–17/19 (y_{31})	–ve

We observe from the above table that

$$y_4 = \frac{126}{19}, y_3 = \frac{124}{19}, s_3 = \frac{1599}{19} \text{ and max } Z = \frac{1130}{19}$$

Now, we check the optimality of this solution

New basis $B_1 = [e_1 \quad \alpha_4^{(1)} \quad \alpha_3^{(1)} \quad \alpha_7^{(1)}]$

$\Rightarrow \quad \alpha_1^{(1)}, \alpha_2^{(1)}, \alpha_5^{(1)}$ and $\alpha_6^{(1)}$ are not in B_1.

Using $\qquad \Delta_j = -(\text{first row of } B_1^{-1} \text{ of third table}) \times \alpha_j^{(1)}, j = 1,2,5,6,$

we get

$$\Delta_1 = -\begin{bmatrix} 1 & \dfrac{37}{19} & \dfrac{15}{19} & 0 \end{bmatrix}\begin{bmatrix} -3 \\ 2 \\ -2 \\ 1 \end{bmatrix} = \frac{13}{19}$$

$$\Delta_2 = -\begin{bmatrix} 1 & \dfrac{37}{19} & \dfrac{15}{19} & 0 \end{bmatrix}\begin{bmatrix} -1 \\ 3 \\ 2 \\ 1 \end{bmatrix} = -\frac{122}{19}$$

$$\Delta_5 = -\begin{bmatrix} 1 & \dfrac{37}{19} & \dfrac{15}{19} & 0 \end{bmatrix}\begin{bmatrix} 0 \\ 1 \\ 0 \\ 0 \end{bmatrix} = -\frac{37}{19}$$

$$\Delta_6 = -\begin{bmatrix} 1 & \dfrac{37}{19} & \dfrac{15}{19} & 0 \end{bmatrix}\begin{bmatrix} 0 \\ 0 \\ 1 \\ 0 \end{bmatrix} = -\frac{15}{19}$$

$\because \quad \Delta_1 > 0 \qquad \Rightarrow \quad$ solution is not optimal.

To find incoming vector:

$\because \quad \max \{\Delta_1, \Delta_2, \Delta_5, \Delta_6\} = \Delta_k$

$$\Rightarrow \quad \max\left\{\frac{13}{19}, \frac{-122}{19}, \frac{-37}{19}, \frac{-15}{19}\right\} = \frac{13}{19} = \Delta_1$$

$$\Rightarrow \qquad k = 1$$

$$\Rightarrow \quad \alpha_1^{(1)} \text{ is incoming vector.}$$

Also,
$$y_1^{(1)} = B_1^{-1}\alpha_1^{(1)} = \begin{bmatrix} 1 & 37/19 & 15/19 & 0 \\ 0 & 5/19 & 1/19 & 0 \\ 0 & 1/19 & 4/19 & 0 \\ 0 & -13/19 & 5/19 & 1 \end{bmatrix}\begin{bmatrix} -3 \\ 2 \\ -2 \\ 1 \end{bmatrix}$$

$$= \begin{bmatrix} -13/19 \\ 8/19 \\ -6/19 \\ -17/19 \end{bmatrix} = \begin{bmatrix} -\Delta_1 \\ y_{11} \\ y_{21} \\ y_{31} \end{bmatrix}$$

To find outgoing vector:

$$\because \qquad \frac{y_{B_r}}{y_{r1}} = \min_i\left\{\frac{y_{B_i}}{y_{i1}}, y_{i1} > 0\right\} = \min\left\{\frac{y_{B_1}}{y_{11}}, \frac{y_{B_2}}{y_{21}}, \frac{y_{B_3}}{y_{31}}\right\}$$

$$= \min\left\{\frac{126/19}{8/19}\right\} \qquad\qquad \{\because\ y_{21}, y_{31} < 0\}$$

$$= \frac{126}{8} = \frac{y_{B_1}}{y_{11}}$$

$$\Rightarrow \qquad r = 1$$

$$\Rightarrow \quad \beta_1^{(1)} \text{ is outgoing vector.}$$

$$\Rightarrow \quad y_{11} \text{ is the key element.}$$

Further, we have to improved this basic feasible solution. Apply the usual procedure, we have the following revised simplex table.

Revised Simplex table-4

Variables in the basis	solution $y_B^{(1)}$	B_1^{-1}				
		e_1	$\beta_1^{(1)}\left(\alpha_1^{(1)}\right)$	$\beta_2^{(1)}\left(\alpha_3^{(1)}\right)$	$\beta_3^{(1)}\left(\alpha_7^{(1)}\right)$	
Z'	281/4	1	19/8	7/8	0	
y_1	63/4	0	5/8	1/8	0	
y_3	23/2	0	1/4	1/4	0	
s_3	393/4	0	−1/8	3/8	1	

From the above table, we observed that the improved BFS is given by

$$y_1 = \frac{63}{4}, y_2 = 0, y_3 = \frac{23}{2}, y_4 = 0 \text{ and } \max Z' = \frac{281}{4}$$

Now, we check the optimality of this solution

$$\because \quad \text{New } B_1 = [e_1 \quad \alpha_1^{(1)} \quad \alpha_3^{(1)} \quad \alpha_7^{(1)}]$$

$\Rightarrow \quad \alpha_2^{(1)}, \alpha_4^{(1)}, \alpha_5^{(1)}$ and $\alpha_6^{(1)}$ are not in the basis $\boldsymbol{B}_1$.

Then by using $\Delta_j = -(\text{first row of } \boldsymbol{B}_1^{-1} \text{ of fourth table}) \times \alpha_j^{(1)}, j = 2,4,5,6$

We get
$$\Delta_2 = -\left[1 \quad \frac{19}{8} \quad \frac{7}{8} \quad 0\right]\begin{bmatrix} -1 \\ 3 \\ 2 \\ 1 \end{bmatrix} = -\frac{63}{8}$$

$$\Delta_4 = -\left[1 \quad \frac{19}{8} \quad \frac{7}{8} \quad 0\right]\begin{bmatrix} -7 \\ 4 \\ -1 \\ 3 \end{bmatrix} = -\frac{13}{8}$$

$$\Delta_5 = -\left[1 \quad \frac{19}{8} \quad \frac{7}{8} \quad 0\right]\begin{bmatrix} 0 \\ 1 \\ 0 \\ 0 \end{bmatrix} = -\frac{19}{8}$$

$$\Delta_6 = -\left[1 \quad \frac{19}{8} \quad \frac{7}{8} \quad 0\right]\begin{bmatrix} 0 \\ 0 \\ 1 \\ 0 \end{bmatrix} = -\frac{7}{8}$$

$\Rightarrow \quad$ all $\Delta_j < 0, j = 2, 4, 5, 6$
$\Rightarrow \quad$ solution is optimal.

Hence, the final optimal solution of the given LPP is

$$x_1 = y_1 + 2 = \frac{63}{4} + 2 = \frac{71}{4}$$
$$x_2 = y_2 + 1 = 0 + 1 = 1$$
$$x_3 = y_3 + 3 = \frac{23}{2} + 3 = \frac{29}{2}$$
$$x_4 = y_4 + 4 = 0 + 4 = 4$$

and max. $Z = \text{max.} Z' + 41 = \frac{281}{4} + 41 = \frac{445}{4}$

5.4 REVISED SIMPLEX METHOD FOR STANDARD FORM-II

We have already discussed that in the given LPP if we used artificial variables to form the basis matrix then it is treated as standard form-II. Such type of problems can be solved by two phase method. Phase-I deals with the removal of artificial variables and an IBFS is obtained while phase-II deals with IBFS (obtained in step 1) to obtain optimal solution.

PHASE-I (FOR REVISED SIMPLEX METHOD)

Let $A_1, A_2, ..., A_m$ be the artificial variables and Z_a be the artificial objective function defined by

$$\text{max. } Z_a = -A_1 - A_2 - ... - A_m$$

If the given LPP is of maximization, cost -1 is assign to each variable, but if the problem is of minimization, then we assign the cost $+1$ to each artificial variable.

Here, we have two objective functions with m constraints, so we have to consider the problem in revised simplex form with $(m + 2)$ constraints.

Thus, we can write

$$\left.\begin{array}{r} Z - c_1 x_1 - c_2 x_2 - \ldots - c_n x_n + A_1 + A_2 + \ldots + A_m = 0 \\ Z_a + 0x_1 + 0x_2 + \ldots + 0x_n + A_1 + A_2 + \ldots + A_m = 0 \\ a_{11} x_1 + a_{12} x_2 + \ldots + a_{1n} x_n + A_1 = b_1 \\ a_{21} x_1 + a_{22} x_2 + \ldots + a_{2n} x_n + A_2 = b_2 \\ \ldots\ldots\ldots\ldots\ldots\ldots\ldots\ldots\ldots\ldots\ldots\ldots\ldots\ldots \\ \ldots\ldots\ldots\ldots\ldots\ldots\ldots\ldots\ldots\ldots\ldots\ldots\ldots\ldots \\ a_{m1} x_1 + a_{m2} x_2 + \ldots + a_{mn} x_n + A_m = b_m \end{array}\right\} \qquad \ldots(1)$$

and $\qquad\qquad x_i, A_j \geq 0$

Further, in phase-I the problem is to maximize Z_a first subject to the constraints (1) with Z_a and Z both unrestricted in sign.

Here, we have the following possibilities:

(i) **If max $Z_a = 0$**, then $A_1 = A_2 = \ldots = A_m = 0$, therefore the values of $x_1, x_2, x_3, \ldots, x_n$ will give the IBFS of phase I.

(ii) **If max. $Z_a < 0$**, then at least one artificial variable has a non-negative value. In this case, the original problem has no feasible solution.

PHASE-II

After removing all the artificial variables we get a basic feasible solution of the problem. In phase-II, we proceed to get the optimal solution by using exactly the same procedure as in revised simplex method in standard form-I.

Here, we use the following steps:

(1) Formation of Basis and its inverse:

The above system of equations (1) can be written as:

$$\begin{bmatrix} 1 & 0 & -c_1 & -c_2 & \cdots & -c_n & 0 & 0 & \cdots & 0 \\ 0 & 1 & 0 & 0 & \cdots & 0 & 1 & 1 & \cdots & 1 \\ 0 & 0 & a_{11} & a_{12} & \cdots & a_{1n} & 1 & 0 & \cdots & 0 \\ 0 & 0 & a_{21} & a_{22} & \cdots & a_{2n} & 0 & 1 & \cdots & 0 \\ \vdots & \vdots & \vdots & \vdots & & \vdots & & & & \vdots \\ 0 & 0 & a_{m1} & a_{m2} & \cdots & a_{mn} & 0 & 0 & \cdots & 1 \end{bmatrix} \begin{bmatrix} Z \\ Z_a \\ x_1 \\ x_2 \\ \vdots \\ x_n \\ A_1 \\ A_2 \\ \vdots \\ A_m \end{bmatrix} = \begin{bmatrix} 0 \\ 0 \\ b_1 \\ b_2 \\ \vdots \\ b_m \end{bmatrix} \qquad \ldots(2)$$

or $\quad [e_1^{(2)} \quad e_2^{(2)} \quad \alpha_1^{(2)} \quad \alpha_2^{(2)} \quad \cdots \quad \alpha_n^{(2)} \quad \alpha_{n+1}^{(2)} \quad \alpha_{n+2}^{(2)} \quad \cdots \quad \alpha_{n+m}^{(2)}] \mathbf{x}_B^{(2)} = \mathbf{b}^{(2)}$

$$\text{where } e_1^{(2)} = \begin{bmatrix} 1 \\ 0 \\ 0 \\ \vdots \\ 0 \end{bmatrix}_{(m+2)\times 1} \quad , e_2^{(2)} = \begin{bmatrix} 0 \\ 1 \\ 0 \\ 0 \\ \vdots \\ 0 \end{bmatrix}_{(m+2)\times 1}$$

$$\text{and } \alpha_j^{(2)} = \begin{bmatrix} -c_j \\ 0 \\ a_{1j} \\ a_{2j} \\ \vdots \\ a_{mj} \end{bmatrix} = \begin{bmatrix} -c_j \\ 0 \\ \alpha_j \end{bmatrix} \text{ for } j = 1, 2, 3, ..., n$$

$$\text{where } \alpha_j = \begin{bmatrix} a_{1j} \\ a_{2j} \\ \vdots \\ a_{mj} \end{bmatrix} \text{ and } \alpha_j^{(2)} = \begin{bmatrix} 0 \\ 1 \\ 1 \\ 0 \\ \vdots \\ 0 \end{bmatrix} = \begin{bmatrix} 0 \\ 1 \\ e_j \end{bmatrix} \text{ for } j = n + 1, n + 2, ..., n + m$$

Here, e_j is the unit vector and $\alpha_j^{(2)}$ denotes the columns of the coefficients of artificial variables $A_1, A_2, ..., A_m$.

$$\text{and } \quad x_B^{(2)} = \begin{bmatrix} Z \\ Z_a \\ x_1 \\ x_2 \\ \vdots \\ x_n \\ A_1 \\ A_2 \\ \vdots \\ A_m \end{bmatrix}, b^{(2)} = \begin{bmatrix} 0 \\ 0 \\ b_1 \\ b_2 \\ \vdots \\ b_m \end{bmatrix}$$

and $e_1^{(2)}, e_2^{(2)}$ are the $(m + 2)$ component unit vector corresponding to the objective function Z and Z_a respectively.

(2) To find basis B_2:

Since, there are $(m + 2)$ constraints, so B_2 must be of order $(m + 2) \times (m + 2)$.

such that

$$B_2 = \begin{bmatrix} 1 & 0 & 0 & 0 & \cdots & 0 \\ 0 & 1 & 1 & 1 & \cdots & 1 \\ 0 & 0 & 1 & 0 & & 0 \\ 0 & 0 & 0 & 1 & & 0 \\ \vdots & \vdots & \vdots & \vdots & & \vdots \\ 0 & 0 & 0 & 0 & & 1 \end{bmatrix}_{(m+2)\times(m+2)}$$

$$= [e_1^{(2)}, e_2^{(2)}, \alpha_{n+1}^{(2)}, \alpha_{n+2}^{(2)}, \ldots \alpha_{n+m}^{(2)}]$$

$$= [e_1^{(2)}, e_2^{(2)}, \beta_1^{(2)}, \beta_2^{(2)} \ldots \beta_m^{(2)}]$$

where $\beta_j^{(2)}$ are any of $\alpha_j^{(2)}$ which are linearly independent.

Also, $B_2 = \begin{bmatrix} 1 & 0 & -C_B \\ 0 & 1 & -C_{B_a} \\ \hdashline O & O & B \end{bmatrix} = \begin{bmatrix} 1 & 0 & -C_B \\ 0 & 1 & -C_{B_a} \\ \hdashline O & & B \end{bmatrix}$

where $-C_B = [0 \quad 0 \quad \cdots \quad 0]_{1\times m}$

$\qquad -C_{B_a} = [1 \quad 1 \quad \cdots \quad 1]_{1\times m}$

$$O = \begin{bmatrix} 0 & 0 \\ 0 & 0 \\ \vdots & \vdots \\ 0 & 0 \end{bmatrix}_{m\times 2} \quad \text{and} \quad B = \begin{bmatrix} 1 & 0 & 0 & \cdots & 0 \\ 0 & 1 & 0 & \cdots & 0 \\ \vdots & & & & \\ 0 & 0 & 0 & \cdots & 1 \end{bmatrix} = I_m$$

Also, $C_B = [C_{B_1} \quad C_{B_2} \quad \cdots \quad C_{B_m}]$ and $C_{B_a} = [C_{B_{a_1}} \quad C_{B_{a_2}} \quad \cdots \quad C_{B_{a_m}}]$

Here, c_j are the coefficient of the basic variables $x_1, x_2, \ldots, x_n$ in the given objective function Z and $C_{B_{a_j}}$ denote the coefficients of the artificial variables in artificial objective function Z_a.

(3) To find B_2^{-1} :

We have $\qquad B_2 = \begin{bmatrix} 1 & 0 & -C_B \\ 0 & 1 & -C_{B_a} \\ \hdashline O & & B \end{bmatrix}$

which can be written as

$$B_2 = \begin{bmatrix} I_2 & -C_B^{(2)} \\ O & B \end{bmatrix}$$

where $\qquad I_2 = \begin{bmatrix} 1 & 0 \\ 0 & 1 \end{bmatrix}, C_B^{(2)} = \begin{bmatrix} C_B \\ C_{B_a} \end{bmatrix}$

Clearly, $\qquad B_2^{-1} = \begin{bmatrix} I_2 & C_B^{(2)}B^{-1} \\ O & B^{-1} \end{bmatrix} = \begin{bmatrix} I_2 & C_B^{(2)} \\ O & I_m \end{bmatrix} \qquad [\because B = I_m^{-1} = I_m]$

(4) To find $x_j^{(2)}$ corresponding $\alpha_j^{(2)}$ not in the basis B_2:

Since $\alpha_j^{(2)}$ is not in the basis so it can be written as the linear combination of the components of B_2, so

$$\alpha_j^{(2)} = x_{0j}e_1^{(2)} + x'_{0j}e_2^{(2)} + x_{1j}\beta_1^{(2)} + x_{2j}\beta_2^{(2)} + \dots + x_{mj}\beta_m^{(2)}$$

$$= [e_1^{(2)} \quad e_2^{(2)} \quad \beta_1^{(2)} \quad \beta_2^{(2)} \quad \cdots \quad \beta_m^{(2)}] \begin{bmatrix} x_{0j} \\ x'_{0j} \\ x_{1j} \\ x_{2j} \\ \vdots \\ x_{mj} \end{bmatrix}$$

$$\Rightarrow \qquad \alpha_j^{(2)} = B_2 x_j^{(2)} \text{ where } x_j^{(2)} = \begin{bmatrix} x_{0j} \\ x'_{0j} \\ x_{1j} \\ x_{2j} \\ \vdots \\ x_{mj} \end{bmatrix}$$

(or) $\qquad x_j^{(2)} = B_2^{-1}\alpha_j^{(2)} = \begin{bmatrix} I_2 & C_B^{(2)}B^{-1} \\ O & B^{-1} \end{bmatrix} \begin{bmatrix} -c_j \\ 0 \\ \alpha_j \end{bmatrix}$ for $j = 1, 2, 3, \dots, n$

$$= \begin{bmatrix} 1 & 0 & C_B B^{-1} \\ 0 & 1 & C_{B_a} \cdot B^{-1} \\ O & O & B^{-1} \end{bmatrix} \begin{bmatrix} -c_j \\ 0 \\ \alpha_j \end{bmatrix} = \begin{bmatrix} -c_j + c_B B^{-1}\alpha_j \\ 0 + C_{B_a} B^{-1}\alpha_j \\ B^{-1}\alpha_j \end{bmatrix} = \begin{bmatrix} -c_j + C_B x_j \\ C_{B_a} x_j \\ B^{-1}\alpha_j \end{bmatrix}$$

$$x_j^{(2)} = \begin{bmatrix} z_j - c_j \\ z_{ja} - c_{ja} \\ x_j \end{bmatrix} = \begin{bmatrix} -\Delta_j \\ -\Delta_{ja} \\ x_j \end{bmatrix}$$

Therefore,

(i) $\Delta_j = -(\text{first row of } B_2^{-1}) \times \alpha_j^{(2)}$

(ii) $\Delta_{j_a} = -(\text{second row of } B_2^{-1}) \times \alpha_j^{(2)}$

(iii) x_j is obtained when the last m rows of B_2^{-1} are multiplied with $\alpha_j^{(2)}$.

(5) To find initial BFS $x_B^{(2)}$:

Here we have

$$x_B^{(2)} = B_2^{-1} b^{(2)} = \begin{bmatrix} 1 & 0 & C_B B^{-1} \\ 0 & 1 & C_{B_a} \cdot B^{-1} \\ O & O & B^{-1} \end{bmatrix} \begin{bmatrix} 0 \\ 0 \\ b \end{bmatrix}$$

$$= \begin{bmatrix} C_B B^{-1} b \\ C_{B_a} B^{-1} b \\ B^{-1} b \end{bmatrix} = \begin{bmatrix} C_B x_B \\ C_{B_a} x_b \\ x_B \end{bmatrix} = \begin{bmatrix} Z \\ Z_a \\ x_B \end{bmatrix} \qquad [\because x_B = B^{-1} b]$$

Therefore,

   (i) $Z = $ (first row of B_2^{-1}) $\times b^{(2)}$

  (ii) $Z_a = $ (second row of B_2^{-1}) $\times b^{(2)}$

 (iii) $x_B = $ (last m rows of B_2^{-1}) $\times b^{(2)}$

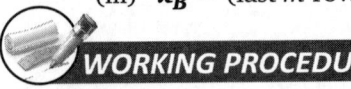

WORKING PROCEDURE

For Phase-1

> **STEP 1.**   Write the given LPP in revised simplex method standard form-II.
>
> **STEP 2.**   Form initial basis matrix B_2 and its inverse B_2^{-1}.
>
> **STEP 3.**   Using $x_B^{(2)} = $ (last m rows of B_2^{-1}) $\times b^{(2)} = \begin{bmatrix} 0 \\ 0 \\ x_B \end{bmatrix}$
>
>   Find the value of $x_B^{(2)}$.
>
>   Here, all $x_i \geq 0$ but no restriction on Z and Z_a.
>
> **STEP 4.**   Construct the initial revised simplex table for standard form-II in the following form.

Variables in the basis	solution $x_B^{(2)}$	B_2^{-1}						$x_k^{(2)} = B_2^{-1} \alpha_k^{(2)}$	Min. Ratio $\min\limits_i \left\{ \dfrac{x_{B_i}}{x_{ik}}, x_{ik} > 0 \right\}$
		$e_1^{(2)}$	$e_2^{(2)}$	$\beta_1^{(2)}$	$\beta_2^{(2)}$	...	$\beta_m^{(2)}$		
Z	0	1	0	...	...	...		$-\Delta_k$	
Z_a	0	0	1	...	...	...		$-\Delta_{k_a}$	
s_1	x_{B_1}	0	0						
s_2	x_{B_2}	0	0						
$\vdots$	$\vdots$	$\vdots$	$\vdots$						
s_m	x_{B_m}	0	0						

STEP 5. Maximize Z_a by using the following results.
  (i) If $Z_a = 0$ then we take $A_1 = 0$, $A_2 = 0$, ..., $A_m = 0$ and use phase II.
  (ii) If $Z_a < 0$ then calculate Δ_{ja} for all $\alpha_j^{(2)}$ not in the basis $\boldsymbol{B}_2$ by using the following formula
$$\Delta_{ja} = -(\text{second row of } \boldsymbol{B}_2^{-1}) \times \alpha_j^{(2)}$$
  Here,
  (i) if all $\Delta_{ja} \leq 0$, then Z_a is maximum and no feasible solution exists.
  (ii) if at least one $\Delta_{ja} > 0$, then go to the next step.

STEP 6. **To find incoming and outgoing vectors**

If $\Delta_{ka} = \max_{j}\{\Delta_{ja}\}$ $\forall j$ such that $\alpha_j^{(2)} \notin \boldsymbol{B}_2$

Then $\alpha_k^{(2)}$ will be the incoming vector.

Now calculate $x_k^{(2)} = \boldsymbol{B}_2^{-1}\alpha_k^{(2)}$

Then find outgoing vector by using the following formula
$$\frac{x_{B_r}}{x_{rk}} = \min_{i}\left\{\frac{x_{B_i}}{x_{ik}}, x_{ik} > 0\right\}$$

If minimum occur at $i = r$ then $\beta_r^{(2)}$ will be the outgoing vector.

STEP 7. Bringing $\alpha_k^{(2)}$ in place of $\beta_r^{(2)}$ to get next revised simplex table. Repeat step (1) to (7) till we get max. $Z_a = 0$ $\forall \Delta_{ja} \leq 0$.

For Phase-2

Here, we have to maximize Z by removing Z_a row and $e_2^{(2)}$ column from the last revised simplex table of phase 1. During this process, the basis matrix $\boldsymbol{B}_2$ of order $(m + 2) \times (m + 2)$ is now reduced to basis matrix $\boldsymbol{B}_1$ of order $(m + 1) \times (m + 1)$.

Then proceed exactly the same way as in revised simplex method of standard form I.

 Solved Examples

EXAMPLE 1. *Solve the following LPP by revised simplex method:*
$$\text{max. } Z = x_1 + 2x_2 + 3x_3 - x_4$$
  subject to the constraints
$$x_1 + 2x_2 + 3x_3 = 15$$
$$2x_1 + x_2 + 5x_3 = 20$$
$$x_1 + 2x_2 + x_3 + x_4 = 10$$
  and $x_1, x_2, x_3, x_4 \geq 0$

SOLUTION. Using artificial variables, the constraints of given LPP can be written as
$$x_1 + 2x_2 + 3x_3 + A_1 = 15$$
$$2x_1 + x_2 + 5x_3 + A_2 = 20$$
$$x_1 + 2x_2 + x_3 + x_4 + A_3 = 10$$

and $\quad x_1, x_2, x_3, x_4, A_1, A_2, A_3 \geq 0$

Now the given LPP can be written in revised simplex standard form II as follows:

$$\left.\begin{array}{l} Z - x_1 - 2x_2 - 3x_3 + x_4 + 0A_1 + 0A_2 + 0A_3 = 0 \\ Z_a + 0x_1 + 0x_2 + 0x_3 + 0x_4 + A_1 + A_2 + A_3 = 0 \\ x_1 + 2x_2 + 3x_3 + A_1 = 15 \\ 2x_1 + x_2 + 5x_3 + A_2 = 20 \\ x_1 + 2x_2 + x_3 + x_4 + A_3 = 10 \end{array}\right] \qquad \ldots(1)$$

Phase 1: We can write the above system (1) in the following form.

$$\begin{array}{ccccccccc} e_1^{(2)} & e_2^{(2)} & \alpha_1^{(2)} & \alpha_2^{(2)} & \alpha_3^{(2)} & \alpha_4^{(2)} & \alpha_5^{(2)} & \alpha_6^{(2)} & \alpha_7^{(2)} \end{array}$$

$$\begin{bmatrix} 1 & 0 & -1 & -2 & -3 & 1 & 0 & 0 & 0 \\ 0 & 1 & 0 & 0 & 0 & 0 & 1 & 1 & 1 \\ 0 & 0 & 1 & 2 & 3 & 0 & 1 & 0 & 0 \\ 0 & 0 & 2 & 1 & 5 & 0 & 0 & 1 & 0 \\ 0 & 0 & 1 & 2 & 1 & 1 & 0 & 0 & 1 \end{bmatrix} \begin{array}{c} Z \\ Z_a \\ x_1 \\ x_2 \\ x_3 \\ x_4 \\ A_1 \\ A_3 \\ A_3 \end{array} = \begin{bmatrix} 0 \\ 0 \\ 15 \\ 20 \\ 10 \end{bmatrix}$$

with $x_B^{(2)}$ and $b^{(2)}$.

Therefore, $\quad B_2 = \begin{bmatrix} 1 & 0 & 0 & 0 & 0 \\ 0 & 1 & 1 & 1 & 1 \\ 0 & 0 & 1 & 0 & 0 \\ 0 & 0 & 0 & 1 & 0 \\ 0 & 0 & 0 & 0 & 1 \end{bmatrix} = \begin{bmatrix} e_1^{(2)} & e_2^{(2)} & \alpha_5^{(2)} & \alpha_6^{(2)} & \alpha_7^{(2)} \end{bmatrix}$

$$= [e_1^{(2)} \quad e_2^{(2)} \quad \beta_1^{(2)} \quad \beta_2^{(2)} \quad \beta_3^{(2)}]$$

where $\alpha_5^{(2)} = \beta_1^{(2)}, \alpha_6^{(2)} = \beta_2^{(2)}, \alpha_7^{(2)} = \beta_3^{(2)}$

$\Rightarrow \qquad B_2 = \begin{bmatrix} 1 & 0 & -C_B \\ 0 & 1 & -C_{B_a} \\ O & O & B \end{bmatrix}$ where $C_B = [0 \quad 0 \quad 0], C_{B_a} = [-1 \quad -1 \quad -1]$

and $\qquad B = \begin{bmatrix} 1 & 0 & 0 \\ 0 & 1 & 0 \\ 0 & 0 & 1 \end{bmatrix} = I_3$

So, $\qquad B^{-1} = I_3^{-1} = I_3 = \begin{bmatrix} 1 & 0 & C_B B^{-1} \\ 0 & 1 & C_{B_a} B^{-1} \\ O & O & B^{-1} \end{bmatrix} = \begin{bmatrix} 1 & 0 & C_B \\ 0 & 1 & C_{B_a} \\ O & O & I_3 \end{bmatrix}$

$$= \begin{bmatrix} 1 & 0 & 0 & 0 & 0 \\ 0 & 1 & -1 & -1 & -1 \\ 0 & 0 & 1 & 0 & 0 \\ 0 & 0 & 0 & 1 & 0 \\ 0 & 0 & 0 & 0 & 1 \end{bmatrix}$$

The initial solution $x_B^{(2)}$ is given by

$$x_B^{(2)} = B_2^{-1} \cdot b^{(2)} = \begin{bmatrix} 1 & 0 & 0 & 0 & 0 \\ 0 & 1 & -1 & -1 & -1 \\ 0 & 0 & 1 & 0 & 0 \\ 0 & 0 & 0 & 1 & 0 \\ 0 & 0 & 0 & 0 & 1 \end{bmatrix} \begin{bmatrix} 0 \\ 0 \\ 15 \\ 20 \\ 10 \end{bmatrix} = \begin{bmatrix} 0 \\ -45 \\ 15 \\ 20 \\ 10 \end{bmatrix}$$

Then we have the following table:

Revised Simplex Table-1

Variables in the basis	solution $x_B^{(2)}$	B_2^{-1}						$x_3^{(2)} = B_2^{-1}\alpha_3^{(2)}$	Min. Ratio $\left\{ \dfrac{x_{B_i}}{x_{i3}}, x_{i3} > 0 \right\}$
		$e_1^{(2)}$	$e_2^{(2)}$	$\beta_1^{(2)}$ $(\alpha_5^{(2)})$	$\beta_2^{(2)}$ $(\alpha_6^{(2)})$	$\beta_3^{(2)}$ $(\alpha_7^{(2)})$			
Z	0	1	0	0	0	0	-3		
Z_a	-45	0	1	-1	-1	-1	-9		
A_1	$15\ (x_{B_1})$	0	0	1	0	0	3 (x_{13})	$15/3 = 5$
A_2	$20\ (x_{B_2})$	0	0	0	1	0	5 (x_{23})	$20/5 = 4$ (min) $\rightarrow$
A_3	$10\ (x_{B_3})$	0	0	0	0	1	1 (x_{33})	$10/1 = 10$
				$\downarrow$				

Clearly, from the above table

max. $Z_a = -45 < 0$, so we have to find $\Delta_{ja}\ \forall\ j$ for which $\alpha_j^{(2)}$ are not in the basis B_2.

Since $\qquad B_2 = [e_1^{(2)}\ \ e_2^{(2)}\ \ \alpha_5^{(2)}\ \ \alpha_6^{(2)}\ \ \alpha_7^{(2)}]$

$\alpha_j^{(2)} \notin B_2, j = 1, 2, 3, 4.$

Now, using $\Delta_{ja} = -(\text{second row of } B_2^{-1}) \times \alpha_j^{(2)}, j = 1, 2, 3, 4$ we get

$$\Delta_{1a} = -[0\ \ 1\ \ -1\ \ -1\ \ -1] \begin{bmatrix} -1 \\ 0 \\ 1 \\ 2 \\ 1 \end{bmatrix} = 4$$

$$\Delta_{2a} = -[0 \quad 1 \quad -1 \quad -1 \quad -1] \begin{bmatrix} -2 \\ 0 \\ 2 \\ 1 \\ 2 \end{bmatrix} = 5$$

$$\Delta_{3a} = -[0 \quad 1 \quad -1 \quad -1 \quad -1] \begin{bmatrix} -3 \\ 0 \\ 3 \\ 5 \\ 1 \end{bmatrix} = 9$$

$$\Delta_{4a} = -[0 \quad 1 \quad -1 \quad -1 \quad -1] \begin{bmatrix} 1 \\ 0 \\ 0 \\ 0 \\ 1 \end{bmatrix} = 1$$

Clearly $\Delta_{1a}, \Delta_{2a}, \Delta_{3a}, \Delta_{4a} > 0$. So, we have to find the improved value of Z_a.

To find the incoming vector

$$\Delta_{ka} = \max_{j}\{\Delta_{ja}\}, j = 1, 2, 3, 4$$

$$= \max\{\Delta_{1a}, \Delta_{2a}, \Delta_{3a}, \Delta_{4a}\} = \max\{4, 5, 9, 1\} = 9 = \Delta_{3a}$$

$\Rightarrow \qquad k = 3$

$\Rightarrow \quad \alpha_3^{(2)}$ is the incoming vector.

Now, $\qquad x_3^{(2)} = B_2^{-1}\alpha_3^{(2)} = \begin{bmatrix} 1 & 0 & 0 & 0 & 0 \\ 0 & 1 & -1 & -1 & -1 \\ 0 & 0 & 1 & 0 & 0 \\ 0 & 0 & 0 & 1 & 0 \\ 0 & 0 & 0 & 0 & 1 \end{bmatrix} \begin{bmatrix} -3 \\ 0 \\ 3 \\ 5 \\ 1 \end{bmatrix} = \begin{bmatrix} -3 \\ -9 \\ 3 \\ 5 \\ 1 \end{bmatrix} = \begin{bmatrix} -\Delta_3 \\ -\Delta_{3a} \\ x_{13} \\ x_{23} \\ x_{33} \end{bmatrix}$

To find outgoing vector:

Using $\dfrac{x_{B_r}}{x_{rk}} = \min_{i}\left\{\dfrac{x_{B_i}}{x_{ik}}, x_{ik} > 0\right\}$ we get

$$\frac{x_{B_r}}{x_{r3}} = \min_{i}\left\{\frac{x_{B_i}}{x_{i3}}, x_{i3} > 0\right\} = \min\left\{\frac{x_{B_1}}{x_{13}}, \frac{x_{B_2}}{x_{23}}, \frac{x_{B_3}}{x_{33}}\right\}$$

$$= \min\left\{\frac{15}{3}, \frac{20}{5}, \frac{10}{1}\right\} = \min\{5, 4, 10\} = 4 = \frac{x_{B_2}}{x_{23}}$$

$\Rightarrow \qquad r = 2$

So, $\beta_2^{(2)}(= \alpha_6^{(2)})$ is the outgoing vector.

Therefore, key element $= x_{23} = 5$

Now, to improved the value of Z_a, following the usual procedure we get the following second simplex table.

Revised Simplex Table-2

Variables in the basis	solution $x_B^{(2)}$	B_2^{-1}						$x_2^{(2)} = B_2^{-1}\alpha_2^{(2)}$	Min. Ratio $\left\{ \dfrac{x_{B_i}}{x_{i2}}, x_{i2} > 0 \right\}$
		$e_1^{(2)}$	$e_2^{(2)}$	$\beta_1^{(2)}$ $\left(\alpha_5^{(2)}\right)$	$\beta_2^{(2)}$ $\left(\alpha_3^{(2)}\right)$	$\beta_3^{(2)}$ $\left(\alpha_7^{(2)}\right)$			
Z	12	1	0	0	3/5	0		−7/5	
Z_a	−9	0	1	−1	4/5	−1		−16/5	
A_1	3 (x_{B_1})	0	0	1	−3/5	0		⑦/5 (x_{12})	15/7 (min) →
A_2	4 (x_{B_2})	0	0	0	1/5	0		1/5 (x_{22})	20
A_3	6 (x_{B_3})	0	0	0	−1/5	1		9/5 (x_{32})	10/3

From above table, we observe that

$$\text{max. } Z_a = -9 < 0$$

So, we find Δ_{ja} for those j for which $\alpha_j^{(2)} \notin B_2$

$\therefore$ New B_2 is given by

$$B_2 = [e_1^{(2)} \quad e_2^{(2)} \quad \alpha_5^{(2)} \quad \alpha_3^{(2)} \quad \alpha_7^{(2)}]$$

$$\Rightarrow \quad \alpha_1^{(2)}, \alpha_2^{(2)}, \alpha_4^{(2)}, \alpha_6^{(2)} \notin B_2$$

Then using $\Delta_{ja} = -$ (second row of B_2^{-1} of second table) $\times \alpha_j^{(2)}, j = 1, 2, 4, 6$

$$\Delta_{1a} = - \begin{bmatrix} 0 & 1 & -1 & \dfrac{4}{5} & -1 \end{bmatrix} \begin{bmatrix} -1 \\ 0 \\ 1 \\ 2 \\ 1 \end{bmatrix} = \frac{2}{5}$$

$$\Delta_{2a} = - \begin{bmatrix} 0 & 1 & -1 & \dfrac{4}{5} & -1 \end{bmatrix} \begin{bmatrix} -2 \\ 0 \\ 2 \\ 1 \\ 2 \end{bmatrix} = \frac{16}{5}$$

$$\Delta_{4a} = - \begin{bmatrix} 0 & 1 & -1 & \dfrac{4}{5} & -1 \end{bmatrix} \begin{bmatrix} 1 \\ 0 \\ 0 \\ 0 \\ 1 \end{bmatrix} = 1$$

$$\Delta_{6a} = - \begin{bmatrix} 0 & 1 & -1 & \dfrac{4}{5} & -1 \end{bmatrix} \begin{bmatrix} 0 \\ 1 \\ 0 \\ 1 \\ 0 \end{bmatrix} = -\frac{9}{5}$$

Clearly, $\Delta_{1a}, \Delta_{2a}, \Delta_{4a} > 0$

So to improve the value of Z_a, we proceed as follows.

To find incoming vector:

$$\Delta_{ka} = \max_j \{\Delta_{ja}\}, j = 1, 2, 4, 6$$

$$= \max\{\Delta_{1a}, \Delta_{2a}, \Delta_{4a}, \Delta_{6a}\} = \max\left\{\frac{2}{5}, \frac{16}{5}, 1, -\frac{9}{5}\right\}$$

$$= \frac{16}{5} = \Delta_{2a}$$

$\Rightarrow \qquad k = 2$

$\Rightarrow \qquad \alpha_2^{(2)}$ is the incoming vector.

Now, $\qquad x_2^{(2)} = B_2^{-1}\alpha_2^{(2)} = \begin{bmatrix} 1 & 0 & 0 & 3/5 & 0 \\ 0 & 1 & -1 & 4/5 & -1 \\ 0 & 0 & 1 & -3/5 & 0 \\ 0 & 0 & 0 & 1/5 & 0 \\ 0 & 0 & 0 & -1/5 & 1 \end{bmatrix} \begin{bmatrix} -2 \\ 0 \\ 2 \\ 1 \\ 2 \end{bmatrix} = \begin{bmatrix} -7/5 \\ -16/5 \\ 7/5 \\ 1/5 \\ 9/5 \end{bmatrix} = \begin{bmatrix} -\Delta_2 \\ -\Delta_{2a} \\ x_{12} \\ x_{22} \\ x_{32} \end{bmatrix}$

To find outgoing vector:

Using $\dfrac{x_{B_r}}{x_{rk}} = \min_i \left\{\dfrac{x_{B_i}}{x_{ik}}, x_{ik} > 0\right\}$ we get

$$\frac{x_{B_r}}{x_{r2}} = \min_i \left\{\frac{x_{B_i}}{x_{i2}}, x_{i2} > 0\right\} = \min\left\{\frac{x_{B_1}}{x_{12}}, \frac{x_{B_2}}{x_{22}}, \frac{x_{B_3}}{x_{32}}\right\}$$

$$= \min\left\{\frac{3}{7/5}, \frac{4}{1/5}, \frac{6}{9/5}\right\} = \min\left\{\frac{15}{7}, 20, \frac{30}{9}\right\} = \frac{15}{7} = \frac{x_{B_1}}{x_{12}}$$

$\Rightarrow \qquad r = 1$

$\Rightarrow \qquad \beta_1^{(1)}(= \alpha_5^{(2)})$ is the outgoing vector.

and hence, key element $= x_{12} = \dfrac{7}{5}$

Now apply the same procedure we get the following simplex table-3.

Revised Simplex Table-3

Variables in the basis	solution $x_B^{(2)}$	B_2^{-1}					$x_4^{(2)} = B_2^{-1}\alpha_2^{(2)}$	Min. Ratio $\min_i\left\{\dfrac{x_{B_i}}{x_{i4}}, x_{i4} > 0\right\}$
		$e_1^{(2)}$	$e_2^{(2)}$	$\beta_1^{(2)}$ $\left(\alpha_2^{(2)}\right)$	$\beta_2^{(2)}$ $\left(\alpha_3^{(2)}\right)$	$\beta_3^{(2)}$ $\left(\alpha_7^{(2)}\right)$		
Z	15	1	0	1	0	1	1	
Z_a	$-15/17$	0	1	9/7	$-4/7$	-1	-1	
x_2	$15/7\ (x_{B_1})$	0	0	5/7	$-3/7$	0	0 (x_{14})	—
x_3	$25/7\ (x_{B_2})$	0	0	$-1/7$	2/7	0	0 (x_{24})	—
A_3	$15/7\ (x_{B_3})$	0	0	$-9/7$	4/7	1	①(x_{34})	15/7 (min) →
						↑		

From the above table, we observe that

$$\max.Z_a = -\frac{15}{7} < 0$$

So, we again find Δ_{ja} for all j for which $\alpha_j^{(2)} \notin \boldsymbol{B}_2$.

Thus, new $\boldsymbol{B}_2$ is given by

$$\boldsymbol{B}_2 = [e_1^{(2)} \quad e_2^{(2)} \quad \alpha_2^{(2)} \quad \alpha_3^{(2)} \quad \alpha_7^{(2)}]$$

$$\Rightarrow \qquad \alpha_1^{(2)}, \alpha_4^{(2)}, \alpha_5^{(2)}, \alpha_6^{(2)} \notin \boldsymbol{B}_2$$

Now, using $\Delta_{ja} = -$ (second row of $\boldsymbol{B}_2^{-1}$ of third table) $\times \alpha_j^{(2)}$, for $j = 1, 4, 5, 6$, we get

$$\Delta_{1a} = -\begin{bmatrix} 0 & 1 & \dfrac{9}{7} & -\dfrac{4}{7} & -1 \end{bmatrix} \begin{bmatrix} -1 \\ 0 \\ 1 \\ 2 \\ 1 \end{bmatrix} = \frac{6}{7}; \quad \Delta_{4a} = -\begin{bmatrix} 0 & 1 & \dfrac{9}{7} & -\dfrac{4}{7} & -1 \end{bmatrix} \begin{bmatrix} 1 \\ 0 \\ 0 \\ 0 \\ 1 \end{bmatrix} = 1$$

$$\Delta_{5a} = -\begin{bmatrix} 0 & 1 & \dfrac{9}{7} & -\dfrac{4}{7} & -1 \end{bmatrix} \begin{bmatrix} 0 \\ 1 \\ 1 \\ 0 \\ 0 \end{bmatrix} = -\frac{16}{7}; \quad \Delta_{6a} = -\begin{bmatrix} 0 & 1 & \dfrac{9}{7} & -\dfrac{4}{7} & -1 \end{bmatrix} \begin{bmatrix} 0 \\ 1 \\ 0 \\ 1 \\ 0 \end{bmatrix} = -\frac{3}{7}$$

Clearly, $\Delta_{1a}, \Delta_{4a} > 0 \Rightarrow$ we can improve the value of Z_a.

To find incoming vector:

$$\Delta_{ka} = \max_j\{\Delta_{ja}\} \quad \text{for } j = 1, 4, 5, 6$$

$$= \max\{\Delta_{1a}, \Delta_{4a}, \Delta_{5a}, \Delta_{6a}\}$$

$$= \max\left\{\frac{6}{7}, 1, \frac{-16}{7}, \frac{-3}{7}\right\} = 1 = \Delta_{4a}$$

$$\Rightarrow \qquad k = 4$$

$$\Rightarrow \qquad \alpha_4^{(2)} \text{ is the incoming vector.}$$

Now $\quad x_4^{(2)} = \boldsymbol{B}_2^{-1}\alpha_4^{(2)} = \begin{bmatrix} 1 & 0 & 1 & 0 & 1 \\ 0 & 1 & 9/7 & 4/7 & -1 \\ 0 & 0 & 5/7 & -3/7 & 0 \\ 0 & 0 & -1/7 & 2/7 & 0 \\ 0 & 0 & -9/7 & 4/7 & 1 \end{bmatrix} \begin{bmatrix} 1 \\ 0 \\ 0 \\ 0 \\ 1 \end{bmatrix} = \begin{bmatrix} 1 \\ -1 \\ 0 \\ 0 \\ 1 \end{bmatrix} = \begin{bmatrix} -\Delta_4 \\ -\Delta_{4a} \\ x_{14} \\ x_{24} \\ x_{34} \end{bmatrix}$

To find outgoing vector:

Using $\quad \dfrac{x_{B_r}}{x_{rk}} = \min_i\left\{\dfrac{x_{B_i}}{x_{ik}}, x_{ik} > 0\right\}$ we get $\quad \dfrac{x_{B_r}}{x_{r4}} = \min_i\left\{\dfrac{x_{B_i}}{x_{i4}}, x_{i4} > 0\right\}$

$$= \min\left\{\frac{x_{B_1}}{x_{14}}, \frac{x_{B_2}}{x_{24}}, \frac{x_{B_3}}{x_{34}}\right\}$$

$$= \min\left\{\frac{x_{B_3}}{x_{34}}\right\} \qquad (\because x_{14} = x_{24} = 0)$$

$$= \frac{x_{B_3}}{x_{34}}$$

$$\Rightarrow \qquad r = 3$$

$\therefore \quad \beta_3^{(2)} (= \alpha_7^{(2)})$ is the outgoing vector.

Hence, key element $= x_{34} = 1$

Further, apply the similar procedure, we have the following table.

Revised simplex table-4

Variables in the basis	solution $x_B^{(2)}$	$e_1^{(2)}$	$e_2^{(2)}$	$\beta_1^{(2)}$ $(\alpha_2^{(2)})$	$\beta_2^{(2)}$ $(\alpha_3^{(2)})$	$\beta_3^{(2)}$ $(\alpha_4^{(2)})$		
Z	90/7	1	0	16/7	−4/7	−1		
Z_a	0	0	1	0	0	0		
x_2	15/7	0	0	5/7	−3/7	0		
x_3	25/7	0	0	−1/7	2/7	0		
x_4	15/7	0	0	−9/7	4/7	1		

From the above table, we observe that max. $Z_a = 0$ and all the artificial variables are removed. Hence, phase I is completed.

Now for phase 2, the subscript 2 is changed to 1 and the next simplex table is constructed by removing Z_a-row and $e_2^{(2)}$-column from the fourth table, we get

Revised simplex table-5

Variables in the basis	solution $x_B^{(1)}$	$e_1^{(1)}$	$\beta_1^{(1)}$ $(\alpha_2^{(1)})$	$\beta_2^{(1)}$ $(\alpha_3^{(1)})$	$\beta_3^{(1)}$ $(\alpha_4^{(1)})$	$x_1^{(1)} = B_1^{-1}\alpha_1^{(1)}$	Min. Ratio $\min_i\left\{\frac{x_{B_i}}{x_{i1}}, x_{i1} > 0\right\}$
Z	90/7	1	16/7	− 4/7	−1	−6/7	
x_2	15/7 (x_{B_1})	0	5/7	−3/7	0	−1/7	—
x_3	25/7 (x_{B_2})	0	−1/7	2/7	0	3/7	25/3
x_4	15/7 (x_{B_3})	0	−9/7	4/7	1	⑥/7	5/2 (min) →
					↓		

From above table, we observe that

$$x_1 = 0, x_2 = \frac{15}{7}, x_3 = \frac{25}{7}, x_4 = \frac{15}{7} \text{ and } \max.Z = \frac{90}{7}$$

is the solution of the given LPP.

Now we have to check the optimality of the solution.

Since, $\alpha_1^{(1)} \notin B_1$

Then

$$\Delta_1 = -(\text{first row of } B_1^{-1}) \times \alpha_1^{(1)}$$

$$= -\begin{bmatrix} 1 & \dfrac{16}{7} & -\dfrac{4}{7} & -1 \end{bmatrix} \begin{bmatrix} -1 \\ 1 \\ 2 \\ 1 \end{bmatrix} = \frac{6}{7}$$

$$\Rightarrow \qquad \Delta_1 = \frac{6}{7} > 0$$

$\Rightarrow$ solution is not optimal

To find incoming vector:

$$\Delta_k = \max_j\{\Delta_j\}, j = 1 = \max\{\Delta_1\} = 1 = \Delta_1$$

$$\Rightarrow \qquad k = 1$$

$\Rightarrow \quad \alpha_1^{(1)}$ is the incoming vector.

Now,

$$x_1^{(1)} = B_1^{-1}\alpha_1^{(1)} = \begin{bmatrix} 1 & 16/7 & -4/7 & -1 \\ 0 & 5/7 & -3/7 & 0 \\ 0 & -1/7 & 2/7 & 0 \\ 0 & -9/7 & 4/7 & 1 \end{bmatrix} \begin{bmatrix} -1 \\ 1 \\ 2 \\ 1 \end{bmatrix} = \begin{bmatrix} -6/7 \\ -1/7 \\ 3/7 \\ 6/7 \end{bmatrix} = \begin{bmatrix} -\Delta_1 \\ x_{11} \\ x_{21} \\ x_{31} \end{bmatrix}$$

To find outgoing vector:

Using $\dfrac{x_{B_r}}{x_{rk}} = \min_i\left\{\dfrac{x_{B_i}}{x_{ik}}, x_{ik} > 0\right\}$ we get

$$\frac{x_{B_r}}{x_{r1}} = \min_i\left\{\frac{x_{B_i}}{x_{i1}}, x_{i1} > 0\right\} = \min\left\{\frac{x_{B_2}}{x_{21}}, \frac{x_{B_3}}{x_{31}}\right\} \qquad \left(\because x_{11} = -\frac{1}{7} < 0\right)$$

$$= \min\left\{\frac{25/7}{3/7}, \frac{15/7}{6/7}\right\} = \min\left\{\frac{25}{2}, \frac{5}{2}\right\} = \frac{5}{2} = \frac{x_{B_3}}{x_{31}}$$

$$\Rightarrow \qquad r = 3$$

$\Rightarrow \quad \beta_3^{(1)}(= \alpha_4^{(1)})$ is the outgoing vector and hence, key element $= x_{31} = \dfrac{6}{7}$

Now, to check the optimality, proceed in a usual manner, we get the following simplex table.

Revised simplex table-6

Variables in the basis	solution $x_B^{(1)}$	B_1^{-1}				$x_k^{(1)}$	Min. Ratio
		$e_1^{(1)}$	$\beta_1^{(1)}$ $(\alpha_2^{(1)})$	$\beta_2^{(1)}$ $(\alpha_3^{(1)})$	$\beta_3^{(1)}$ $(\alpha_1^{(1)})$		
Z	15	1	1	0	0		
x_2	5/2	0	1/2	–1/3	1/6		
x_3	5/2	0	1/2	0	–1/2		
x_1	5/2	0	–3/2	2/3	7/6		

From the above table, we observe that, the solution is

$$x_1 = \frac{5}{2}, x_2 = \frac{5}{2}, x_3 = \frac{5}{2}, \max. Z = 15$$

To check the optimality of this solution, we proceed as follows:

$\because \quad \alpha_4^{(1)} \notin \boldsymbol{B}_1$ so,

$$\Delta_4 = -(\text{first row of } \boldsymbol{B}_1^{-1}) \times \alpha_4^{(1)}$$

$$= -[1 \quad 1 \quad 0 \quad 0]\begin{bmatrix}1\\0\\0\\1\end{bmatrix} = -1 < 0$$

$\Rightarrow$ solution is optimal and is given by

$$x_1 = x_2 = x_3 = \frac{5}{2} \text{ and max. } Z = 15$$

EXAMPLE 2. *Using revised simplex method, solve the following LPP*

$$\min.\ Z = x_1 + 2x_2$$

subject to the constraints

$$2x_1 + 5x_2 \geq 6$$
$$x_1 + x_2 \geq 2$$

and $\quad x_1, x_2 \geq 0$

SOLUTION. The given LPP is of minimization, so first we convert it into maximization problem such that

$$\max.\ Z' = -Z = -x_1 - 2x_2$$

Now proceed as usual, using surplus variables s_1, s_2 and artificial variables A_1 and A_2, we have the following form of phase 1

Phase 1. The given LPP in revised simplex form-II can be written as

$$\left.\begin{aligned} Z' + x_1 + 2x_2 + 0s_1 + 0s_2 + 0A_1 + 0A_2 &= 0 \\ Z_a + 0x_1 + 0x_2 + 0s_1 + 0s_2 + A_1 + A_2 &= 0 \\ 2x_1 + 5x_2 - s_1 + A_1 &= 6 \\ x_1 + x_2 - s_2 + A_2 &= 2 \end{aligned}\right] \quad \ldots(1)$$

and $x_i, s_i \geq 0, A_i \geq 0$

Here, the artificial objective function which to be maximized in phase 1 is given by

$$\text{max. } Z_a = -A_1 - A_2$$

In matrix form, system (1) can be written as

$$
\begin{array}{cccccccc}
e_1^{(2)} & e_2^{(2)} & \alpha_1^{(2)} & \alpha_2^{(2)} & \alpha_3^{(2)} & \alpha_4^{(2)} & \alpha_5^{(2)} & \alpha_6^{(2)}
\end{array}
\quad
\begin{bmatrix} Z' \\ Z_a \end{bmatrix}
$$

$$
\begin{bmatrix}
1 & 0 & 1 & 2 & 0 & 0 & 0 & 0 \\
0 & 1 & 0 & 0 & 0 & 0 & 1 & 1 \\
0 & 0 & 2 & 5 & -1 & 0 & 1 & 0 \\
0 & 0 & 1 & 1 & 0 & -1 & 0 & 1
\end{bmatrix}
\begin{bmatrix} x_1 \\ x_2 \\ s_1 \\ s_2 \\ A_1 \\ A_2 \end{bmatrix}
=
\begin{bmatrix} 0 \\ 0 \\ 6 \\ 2 \end{bmatrix}
\quad b^{(2)}
$$

with $x_B^{(2)}$ as indicated.

The basis matrix B_2 is given by

$$
B_2 =
\begin{bmatrix}
1 & 0 & 0 & 0 \\
0 & 1 & 1 & 1 \\
0 & 0 & 1 & 0 \\
0 & 0 & 0 & 1
\end{bmatrix}
= [e_1^{(2)} \quad e_2^{(2)} \quad \alpha_5^{(2)} \quad \alpha_6^{(2)}]
$$

$$
= [e_1^{(2)} \quad e_2^{(2)} \quad \beta_1^{(2)} \quad \beta_2^{(2)}] \text{ where } \alpha_5^{(2)} = \beta_1^{(2)}, \alpha_6^{(2)} = \beta_2^{(2)}
$$

Here, B_2 can be written as

$$
B_2 =
\begin{bmatrix}
1 & 0 & -C_B \\
0 & 1 & -C_{B_a} \\
O & O & B
\end{bmatrix}
$$

where $C_B = [0 \quad 0], C_{B_a} = [-1 \quad -1], B = \begin{bmatrix} 1 & 0 \\ 0 & 1 \end{bmatrix} = I_2 \Rightarrow B_2^{-1} = I_2$

So, $B_2^{-1} = \begin{bmatrix} 1 & 0 & -C_B B^{-1} \\ 0 & 1 & -C_{B_a} B^{-1} \\ O & O & B \end{bmatrix} = \begin{bmatrix} 1 & 0 & -C_B \\ 0 & 1 & -C_{B_a} \\ O & O & B \end{bmatrix} = \begin{bmatrix} 1 & 0 & 0 & 0 \\ 0 & 1 & -1 & -1 \\ 0 & 0 & 1 & 0 \\ 0 & 0 & 0 & 1 \end{bmatrix}$

and initial solution $x_B^{(2)}$ is given by

$$x_B^{(2)} = B_2^{-1} b^{(2)}$$

$$
= \begin{bmatrix}
1 & 0 & 0 & 0 \\
0 & 1 & -1 & -1 \\
0 & 0 & 1 & 0 \\
0 & 0 & 0 & 1
\end{bmatrix}
\begin{bmatrix} 0 \\ 0 \\ 6 \\ 2 \end{bmatrix}
=
\begin{bmatrix} 0 \\ -8 \\ 6 \\ 2 \end{bmatrix}
=
\begin{bmatrix} Z' \\ Z_a \\ 6 \\ 2 \end{bmatrix}
$$

Now, we have the following simplex table

Revised simplex table-1

Variables in the basis	solution $x_B^{(1)}$	$e_1^{(2)}$	$e_2^{(2)}$	$\beta_1^{(2)}$ $\left(\alpha_5^{(2)}\right)$	$\beta_2^{(2)}$ $\left(\alpha_6^{(2)}\right)$	$x_2^{(2)} = B_2^{-1}\alpha_2^{(2)}$	Min. Ratio $\min\limits_i\left\{\dfrac{x_{B_i}}{x_{ik}}, x_{ik} > 0\right\}$
Z'	0	1	0	0	0	2	
Z_a	-8	0	1	-1	-1	-6	
A_1	$6\ (x_{B_1})$	0	0	1	0	$5\ (x_{12})$	$6/5$ (min) $\rightarrow$
A_2	$2\ (x_{B_2})$	0	0	0	1	①(x_{22})	$2/1$
				$\downarrow$			

$$\Rightarrow \qquad Z_a = -8 < 0$$

So, we have to find $\Delta_{ja}\ \forall\ j$ for which $\alpha_j^{(2)}$ are not in the basis.

By definition, $\alpha_1^{(2)}, \alpha_2^{(2)}, \alpha_3^{(2)}, \alpha_4^{(2)}$ are not in the basis.

Now, using $\Delta_{ja} = -(\text{second row of } B_2^{-1}) \times \alpha_j^{(2)}$ for $j = 1, 2, 3, 4$, we get

$$\Delta_{1a} = -[0 \quad 1 \quad -1 \quad -1]\begin{bmatrix}1\\0\\2\\1\end{bmatrix} = 3\ ; \ \Delta_{2a} = -[0 \quad 1 \quad -1 \quad -1]\begin{bmatrix}2\\0\\5\\1\end{bmatrix} = 6$$

$$\Delta_{3a} = -[0 \quad 1 \quad -1 \quad -1]\begin{bmatrix}0\\0\\-1\\0\end{bmatrix} = -1\ ; \ \Delta_{4a} = -[0 \quad 1 \quad -1 \quad -1]\begin{bmatrix}0\\0\\0\\-1\end{bmatrix} = -1$$

Clearly $\Delta_{1a}, \Delta_{2a} > 0 \Rightarrow$ The value of Z_a can be improved as follows:

To find incoming vector:

We have

$$\Delta_{ka} = \max_j\{\Delta_{ja}\}, j = 1, 2, 3, 4$$

$$= \max\{\Delta_{1a}, \Delta_{2a}, \Delta_{3a}, \Delta_{4a}\} = \max\{3, 6, -1, -1\} = 6 = \Delta_{2a}$$

$$\Rightarrow \qquad k = 2$$

$$\Rightarrow \qquad \alpha_2^{(2)} \text{ is the incoming vector.}$$

$$\text{Also,} \quad x_2^{(2)} = B_2^{-1}\alpha_2^{(2)} = \begin{bmatrix}1 & 0 & 0 & 0\\0 & 1 & -1 & -1\\0 & 0 & 1 & 0\\0 & 0 & 0 & 1\end{bmatrix}\begin{bmatrix}2\\0\\5\\1\end{bmatrix} = \begin{bmatrix}2\\-6\\5\\1\end{bmatrix} = \begin{bmatrix}-\Delta_2\\-\Delta_{2a}\\x_{12}\\x_{22}\end{bmatrix}$$

To find outgoing vector:

Using $\dfrac{x_{B_r}}{x_{rk}} = \min_i \left\{ \dfrac{x_{B_i}}{x_{ik}}, x_{ik} > 0 \right\}$ we get

$$\frac{x_{B_r}}{x_{r2}} = \min\left\{ \frac{x_{B_1}}{x_{12}}, \frac{x_{B2}}{x_{22}} \right\} = \min\left\{ \frac{6}{5}, \frac{2}{1} \right\} = \frac{6}{5} = \frac{x_{B_1}}{x_{12}}$$

$\Rightarrow \qquad r = 1$

$\Rightarrow \quad \beta_1^{(2)}(= \alpha_5^{(2)})$ is the outgoing vector and key element $= x_{12} = 5$

Now, to improve the value of Z_a, proceed as usual, we have the following table.

Revised simplex table-2

Variables in the basis	solution $x_B^{(1)}$	B_2^{-1}				$x_1^{(2)} = B_2^{-1}\alpha_1^{(2)}$	Min. Ratio $\min_i\left\{\dfrac{x_{B_i}}{x_{i1}}, x_{i1} > 0\right\}$
		$e_1^{(2)}$	$e_2^{(2)}$	$\beta_1^{(2)}$ $\left(\alpha_2^{(2)}\right)$	$\beta_2^{(2)}$ $\left(\alpha_6^{(2)}\right)$		
Z'	$-12/5$	1	0	$-2/5$	0	$1/5$	
Z_a	$-4/5$	0	1	$1/5$	-1	$-3/5$	
x_2	$6/5\ (x_{B_1})$	0	0	$1/5$	0	$2/5\ (x_{11})$	3
A_2	$4/5\ (x_{B_2})$	0	0	$-1/5$	1	$③/5\ (x_{22})$	$4/3$ (min) $\rightarrow$

We observe that

$$\max Z_a = -\frac{4}{5} < 0$$

So, we have to find $\Delta_{ja}\ \forall\ j$ for which $\alpha_j^{(2)}$ are not in B_2, which becomes

$$B_2 = [e_1^{(2)} \quad e_2^{(2)} \quad \alpha_2^{(2)} \quad \alpha_6^{(2)}]$$

$$\Delta_{1a} = -(\text{second row of } B_2^{-1} \text{ of second table}) \times \alpha_1^{(2)}$$

$$= -\begin{bmatrix} 0 & 1 & \dfrac{1}{5} & -1 \end{bmatrix} \begin{bmatrix} 1 \\ 0 \\ 2 \\ 1 \end{bmatrix} = \frac{3}{5}$$

$$\Delta_{3a} = -\begin{bmatrix} 0 & 1 & \dfrac{1}{5} & -1 \end{bmatrix} \begin{bmatrix} 0 \\ 0 \\ -1 \\ 0 \end{bmatrix} = \frac{1}{5}$$

$$\Delta_{4a} = -\begin{bmatrix} 0 & 1 & \dfrac{1}{5} & -1 \end{bmatrix} \begin{bmatrix} 0 \\ 0 \\ 0 \\ -1 \end{bmatrix} = -1$$

$$\Delta_{5a} = -\begin{bmatrix} 0 & 1 & \dfrac{1}{5} & -1 \end{bmatrix} \begin{bmatrix} 0 \\ 1 \\ 1 \\ 0 \end{bmatrix} = -\dfrac{6}{5}$$

$\Rightarrow \quad \Delta_{1a}, \Delta_{3a} > 0$, so we can improved the value of Z_a.

To find incoming vector:

$$\Delta_{ka} = \max_{j}\{\Delta_{ja}\}, j = 1, 3, 4, 5$$

$$= \max\{\Delta_{1a}, \Delta_{3a}, \Delta_{4a}, \Delta_{5a}\} = \max\left\{\dfrac{3}{5}, \dfrac{1}{5}, -1, -\dfrac{6}{7}\right\}$$

$$= \dfrac{3}{5}$$

$\Rightarrow \qquad k = 1$

$\Rightarrow \quad \alpha_1^{(2)}$ is the incoming vector.

Now, $\quad x_1^{(1)} = B_2^{-1}\alpha_1^{(2)} = \begin{bmatrix} 1 & 0 & -2/5 & 0 \\ 0 & 1 & 1/5 & -1 \\ 0 & 0 & 1/5 & 0 \\ 0 & 0 & -1/5 & 1 \end{bmatrix} = \begin{bmatrix} 1/5 \\ -3/5 \\ 2/5 \\ 3/5 \end{bmatrix} = \begin{bmatrix} -\Delta_1 \\ -\Delta_{1a} \\ x_{11} \\ x_{21} \end{bmatrix}$

To find outgoing vector:

Using $\quad \dfrac{x_{B_r}}{x_{rk}} = \min_{i}\left\{\dfrac{x_{B_i}}{x_{ik}}, x_{ik} > 0\right\}$ we get

$$\dfrac{x_{B_r}}{x_{r1}} = \min_{i}\left\{\dfrac{x_{B_i}}{x_{i1}}, x_{i1} > 0\right\} = \min\left\{\dfrac{x_{B1}}{x_{11}}, \dfrac{x_{B2}}{x_{21}}\right\}$$

$$= \min\left\{\dfrac{6/5}{2/5}, \dfrac{4/5}{3/5}\right\} = \min\left\{3, \dfrac{4}{3}\right\} = \dfrac{4}{3} = \dfrac{x_{B2}}{x_{21}}$$

$\Rightarrow \qquad r = 2$

$\Rightarrow \quad \beta_2^{(2)}(= \alpha_6^{(2)})$ is the outgoing vector.

$\Rightarrow \quad x_{21} = \dfrac{3}{5}$ is the key element.

Now again apply the usual procedure to improve the value of Z_a, we get the following table.

Revised simplex table-3

Variables in the basis	solution $x_B^{(2)}$	B_2^{-1}				$x_k^{(2)}$	Min. Ratio
		$e_1^{(2)}$	$e_2^{(2)}$	$\beta_1^{(2)}$ $\left(\alpha_2^{(2)}\right)$	$\beta_2^{(2)}$ $\left(\alpha_1^{(2)}\right)$		
Z'	$-8/3$	1	0	$-1/3$	$-1/3$		
Z_a	0	0	1	0	0		
x_2	$2/3$	0	0	$1/3$	$-2/3$		
x_1	$4/3$	0	0	$-1/3$	$5/3$		

Clearly, from the above table, max. $Z_a = 0$
$\Rightarrow$ Process of Phase 1 comes to an end.
Now, we enter phase 2 by making a change in subscript 2 to 1.
Phase 2. Now construct the new table by removing second row and $e_2^{(2)}$ column from third table as follows.

Revised simplex table-4

Variables in the basis	solution $x_B^{(1)}$	B_1^{-1}			$x_k^{(1)}$	Min. Ratio
		$e_1^{(2)}$	$\beta_1^{(2)}\left(\alpha_2^{(2)}\right)$	$\beta_2^{(2)}\left(\alpha_1^{(2)}\right)$		
Z'	$-8/5$	1	$-1/3$	$-1/3$		
x_2	$2/3$	0	$1/3$	$-2/3$		
x_1	$4/3$	0	$-1/3$	$5/3$		

The basis B_1 is given by $B_1 = [e_1^{(1)} \quad \alpha_2^{(1)} \quad \alpha_1^{(1)}] = [e_1^{(1)} \quad \beta_1^{(1)} \quad \beta_2^{(1)}]$

$\Rightarrow \quad \alpha_3^{(1)}, \alpha_4^{(1)} \notin B_1$

Now, $\quad \Delta_3 = -(\text{first row of } B_1^{-1}) \cdot \alpha_3^{(1)} = -\begin{bmatrix} 1 & -\dfrac{1}{3} & -\dfrac{1}{3} \end{bmatrix} = \begin{bmatrix} 0 \\ -1 \\ 0 \end{bmatrix} = -\dfrac{1}{3}$

$$\Delta_4 = -\begin{bmatrix} 1 & -\dfrac{1}{3} & -\dfrac{1}{3} \end{bmatrix}\begin{bmatrix} 0 \\ 0 \\ -1 \end{bmatrix} = -\dfrac{1}{3}$$

Clearly $\Delta_3, \Delta_4 < 0$
$\Rightarrow$ solution obtained in table-4 is optimal and is given by

$$x_1 = \frac{4}{3}, x_2 = \frac{2}{3} \text{ and } \max. Z' = -\frac{8}{5}, i.e., \min Z = \frac{8}{5}$$

5.5 COMPARISON OF SIMPLEX METHOD AND REVISED SIMPLEX METHOD

When we consider an LPP with constraints $Ax = b$ where A is a matrix of order $m \times n$. Then for solving this LP by simplex method, we have to transform $(n + 1)$ columns at each iteration. Further at each iteration one variable is introduced and one is removed from the basis, so we have to compute total $n - m + 1$ columns. Also, for each of these columns we have to transforms $m + 1$ elements and for moving one iteration to another, we also need to calculate minimum ratio. Hence, we have to perform multiplication operation $(m+1)(n-m+1)$ times and addition $m(n - m + 1)$ times.

But in revised simplex method, there are total $(m + 1)$ rows and $(m + 2)$ columns. Therefore, when we move from one iteration to another we have to perform $(m + 1)^2$ multiplication operation in addition to $m(n - m)$ operation for calculating $c_j - z_j$. Also, in revised simplex method, while updating the table to move from one solution to another, an additional table of original non-basic variables, not in the basis is required, which may cause of some error.

Though the revised simplex method has not met with wide acceptance for hand computation even then there are a few advantages of revised simplex method over the simplex method.

(1) The total no. of operations for the revised simplex method is approximately $mn + m^2 + 3m$, whereas in simplex method the total no. of operations is approximately $mn - m^2 + n + m$. Thus, revised simplex method deals with less computations than that of simplex method.

(2) In revised simplex method, the inverse of the current basis is automatically generated and the next BFS is also obtained, whereas it is not so in simplex method.

(3) In revised simplex method, we introduce $(m + 1)(m + 2)$ entries in each table while, in simplex method these are only $(n + 1)(m + 1)$. If $n << m$, a lot of labour and so error can be avoided.

Exercise-5.1

Using revised simplex method, solve the following LPP:

1. max. $Z = x_1 + 2x_2$
   s.t. $x_1 + x_2 \le 3$
   $x_1 + 2x_2 \le 5$
   $3x_1 + x_2 \le 6$
   and $x_1, x_2 \ge 0$

2. max. $Z = 6x_1 - 2x_2 + 3x_3$
   s.t. $2x_1 - x_2 + 2x_3 \le 2$
   $x_1 + 4x_3 \le 4$
   and $x_1, x_2, x_3 \ge 0$

3. max. $Z = 3x_1 + x_2 + 2x_3 + 7x_4$
   s.t. $2x_1 + 3x_2 - x_3 + 4x_4 \le 40$
   $-2x_1 + 2x_2 - 5x_3 - 4x_4 \le 35$
   $x_1 + x_2 - 2x_3 + 3x_4 \le 100$
   and $x_1, x_2, x_3, x_4 \ge 0$

4. max. $Z = 2x_1 + x_2$
   s.t. $3x_1 + 4x_2 \le 6$
   $6x_1 + x_2 \le 3$
   and $x_1, x_2 \ge 0$

5. max. $Z = x_1 + x_2 + 3x_3$
   s.t. $3x_1 + 2x_2 + x_3 \le 3$
   $2x_1 + x_2 + 2x_3 \le 2$
   and $x_1, x_2, x_3 \ge 0$

6. max. $Z = 3x_1 + 6x_2 + 2x_3$
   s.t. $3x_1 + 4x_2 + x_3 \le 2$
   $x_1 + 3x_2 + 2x_3 \le 1$

and $x_1, x_2, x_3 \ge 0$

7. min. $Z = 2x_1 + x_2$
   s.t. $3x_1 + x_2 \le 3$
   $4x_1 + 3x_2 \ge 6$
   $x_1 + 2x_2 \le 3$
   and $x_1, x_2 \ge 0$

8. min. $Z = x_1 + x_2$
   s.t. $x_1 + 2x_2 \ge 7$
   $4x_1 + x_2 \ge 6$
   and $x_1, x_2 \ge 0$

9. max. $Z = 5x_1 + 3x_2$
   s.t. $4x_1 + 5x_2 \ge 10$
   $5x_1 + 2x_2 \le 10$
   $3x_1 + 8x_2 \le 12$
   and $x_1, x_2 \ge 0$

10. max. $Z = x_1 + 2x_2$
    s.t. $3x_1 + 2x_2 \ge 6$
    $x_1 + 6x_2 \ge 3$
    and $x_1, x_2 \ge 0$

11. max. $Z = 2x_1 + 4x_2 + 6x_3 - 2x_4$
    s.t. $x_1 + 2x_2 + 3x_3 = 15$
    $2x_1 + x_2 + 5x_3 = 20$
    $3x_1 + 6x_2 + 3x_3 + 3x_4 = 30$
    and $x_1, x_2, x_3 \ge 0$

12. max. $Z = -5x_2$
    s.t. $x_1 + x_2 \le 2$
    $x_1 + 5x_2 \ge 10$
    and $x_1, x_2 \ge 0$

ANSWERS

1. $x_1 = 0, x_2 = \dfrac{2}{5}, \text{max}.Z = 5$

2. $x_1 = 4, x_2 = 6, x_3 = 0, \text{max}. Z = 12$

3. $x_1 = \dfrac{71}{4}, x_2 = 1, x_3 = \dfrac{29}{2}, x_4 = 4, \text{max}.Z = \dfrac{445}{4}$

4. $x_1 = \dfrac{2}{7}, x_2 = \dfrac{9}{7}, \text{max}.Z = \dfrac{13}{7}$

5. $x_1 = x_2 = 0, x_3 = 1, \text{max}. Z = 3$

6. $x_1 = \dfrac{2}{5}, x_2 = \dfrac{1}{3}, x_3 = 0, \text{max} Z = \dfrac{12}{5}$

7. $x_1 = \dfrac{3}{5}, x_2 = \dfrac{6}{5}, \text{min}.Z = \dfrac{12}{5}$

8. $x_1 = \dfrac{5}{7}, x_2 = \dfrac{22}{7}, \text{min}.Z = \dfrac{2}{7}$

9. $x_1 = \dfrac{28}{17}, x_2 = \dfrac{15}{17}, \text{max}.Z = \dfrac{185}{17}$

10. $x_1 = 0, x_2 = 3, \text{max}.Z = 6$

11. $x_1 = \dfrac{5}{2}, x_2 = \dfrac{5}{2}, x_3 = \dfrac{5}{2}, \text{max}.Z = 30$

12. $x_1 = 0, x_2 = 2, \text{max}.Z = -10$

🖈 Glossary

- **Standard form I:** In this form, only slack and surplus variables are introduced to form basis matrix and there is no need for artificial variables, *i.e.*, an identity matrix is obtained by adding slack variables.

- **Standard form II:** In this form, the artificial variables are used to form basis matrix. In this case two phase method is used in a slightly different way to remove artificial variables.

🔧 REVIEW QUESTIONS

1. Explain the Revised simplex method and compare it with simplex method.

2. Formulate a linear programming problem in the form of revised simplex method.

3. Explain the revised simplex method of standard form I and II.

4. Write the differences of revised simplex method and simplex method.

5. Describe the revised simplex method when artificial vectors are added to obtain an identity matrix for the initial basis matrix.

✎ MULTIPLE CHOICE QUESTIONS (CHOOSE THE MOST APPROPRIATE ONE)

1. In LPP if FS is optimal then basic feasible solution will be:
   - (a) non-zero solution
   - (b) may or may not be optimal
   - (c) optimal
   - (d) None of these

2. In standard form-II of revised simplex method the basis matrix is denoted by:
   - (a) B_1
   - (b) B_0
   - (c) B_2
   - (d) None of these

3. In LPP, in simplex table, $\Delta j \leq 0$, then solution under test:
   - (a) optimal
   - (b) minimum
   - (c) unbounded
   - (d) none of these

4. Extreme points of $\{(x, y)/m \leq 1, \|y\| \leq 1\}$ is:
   - (a) $(1, -1)$
   - (b) $(-1, 1)$
   - (c) $(1, 1)$
   - (d) $(-1, -1)$

5. The number of additional constraints in standard form II of revised simplex method:
   - (a) 0
   - (b) 2
   - (c) 1
   - (d) None of these

6. Simplex method was developed by:
   - (a) Mody
   - (b) Maxwell
   - (c) Dantzig
   - (d) None of these

7. The Union of two convex sets is:
   - (a) not a convex
   - (b) a line segment
   - (c) a convex set
   - (d) may or may not be a convex set

8. In standard form II of revised simplex method we need:
   - (a) slack variables
   - (b) surplus variables
   - (c) artificial variables
   - (d) none of these

9. Matrix form of a LPP is
   Max. $Z = Cx$ subject to $AX = b, X \geq 0$, $A = [a_{ij}]_{m \times n}, N > m + n$. Suppose B be a submatrix of order $m \times n$ which is non-singular selected from A then $B^{-1}b$ is called:
   - (a) Optimal
   - (b) BFS of LPP
   - (c) Non basic feasible solution of LPP
   - (d) None of these

10. If a LPP max $Z = Cx$ s.t. $Ax = b, X \geq 0$, where $A = \{\lambda_1, \lambda_2, ..., \lambda_n\}$ is coefficient matrix of order $m \times N, N = m + n$, has at least one feasible solution then it has:
    - (a) zero solution
    - (b) one feasible solution but not basic
    - (c) one BFS
    - (d) none of these

11. In LPP, max $Z = Cx, AX \leq b$, then all:
    - (a) $b_i \geq 0$
    - (b) $b_i \leq 0$
    - (c) $b_i = 0$
    - (d) None of these

12. If the solution contains one or more artificial variable as basic and variable LPP has:
    - (a) No F.S.
    - (b) F.S.
    - (c) N.S.
    - (d) None of these

13. Simplex method was developed in the year:
    - (a) 1947
    - (b) 1747
    - (c) 1917
    - (d) None of these

14. In Final simplex table if $\Delta_j \leq 0$ then optimal solution is:
    (a) optimal
    (b) unique
    (c) bounded
    (d) none of these

15. The set of all the internal points of a convex set is a:
    (a) line segment
    (b) convex
    (c) concave
    (d) none of these

16. If max position Δj are minimum ratio $\to$ –ve or $\to \infty$ non-solution under test is:
    (a) unbounded
    (b) bounded
    (c) rational
    (d) none of these

17. In standard form-I of revised simplex method the basis matrix is denoted by:
    (a) B_0
    (b) B_1
    (c) B_2
    (d) None of these

18. In standard form-I of revised simplex method we do not need:
    (a) surplus variables
    (b) artificial variables
    (c) slack variables
    (d) none of these

19. The number of additional constraints in standard form I of revised simplex method is:
    (a) 0
    (b) 1
    (c) 2
    (d) None of these

ANSWERS

1. (c)	**2.** (c)	**3.** (a)	**4.** (c)	**5.** (b)	**6.** (c)	**7.** (c)	**8.** (c)	**9.** (b)
10. (d)	**11.** (a)	**12.** (a)	**13.** (a)	**14.** (b)	**15.** (b)	**16.** (a)	**17.** (b)	**18.** (b)
19. (b)								

□□□□

Duality in Linear Programming 6

6.1 INTRODUCTION

The concept of duality is most important and useful tools in mathematics and many branches of engineering. It explains that for every LPP there is a related unique LPP involving the same data that also describes the original LPP, *i.e.*, each linear programming can be analysed in two different ways but having equivalent solutions. The original LPP is called 'primal' and other related problem is called 'dual'. In this chapter we shall discuss the duality in linear programming along with some related theorems.

6.2 RELATIONSHIP BETWEEN PRIME AND DUAL

The relationship between primal and dual of an LPP contains the following points:

(i) The maximization (minimization) in primal becomes the minimization (maximization) in its dual and conversely.

(ii) If the primal contains n variables and m constraints, then the dual will contains m variables and n constraints.

(iii) The coefficients $c_1, c_2, ..., c_n$ in the objective function of the primal appear in the constraints in the dual.

(iv) The constants $b_1, b_2, ..., b_m$ in the constraints of the primal appear in the objective function of the dual.

(v) If any of the two (either primal or dual) has an infeasible solution, then the value of objective function of other is unbounded.

(vi) If either the primal or dual has an unbounded solution, then the solution to the other problem is infeasible.

(vii) If some primal variables are unrestricted in sign, then these dual constraints will be equations that correspond to the said primal variables.

(viii) If a primal contains a constraint as an equation, then the variable in its dual corresponding to that equation will be unrestricted in sign.

This relationship between prime and dual can be summarized in the following table.

S.No.	Primal	Dual
1.	Objective function is of maximization	Objective function is of minimization
2.	i^{th} primal variable, x_i	i^{th} dual constraints
3.	i^{th} primal constraints	i^{th} dual variable, y_i
4.	Primal constraints $\leq$ type	Dual constraints $\geq$ type
5.	Primal variable x_i unrestricted in sign	Dual constraints i is $=$ type
6.	Primal constraints i is $=$ type	Dual variable y_i is unrestricted in sign

☛ REMARKS
- The variables in both primal and dual are non-negative.
- The value of the objective function for any feasible solution of the primal is less than the value of the objective function for any feasible solution of the dual.

6.3 SYMMETRIC PRIMAL-DUAL PROBLEMS

There are two important forms of primal and dual problems namely the standard form and symmetrical (or canonical) form.

6.3.1 STANDARD FORM OF A PRIMAL PROBLEM
[MEERUT–2007, 09, 12]

The given LPP is said to be in standard form if:
   (i) for maximization problem, all the constraints have $\leq$ sign.
  (ii) for minimization problem, all the constraints have $\geq$ sign.

If the primal is in standard form, then we call it symmetric primal and its dual is symmetric dual.

6.3.2 FORMULATION OF DUAL LINEAR PROGRAMMING PROBLEM
[GORAKHPUR–2011]

Let us suppose the primal LPP is given in the form of
$$\text{max. } Z_x = c_1 x_1 + c_2 x_2 + \ldots + c_n x_n$$
subject to the constraints
$$a_{11} x_1 + a_{12} x_2 + \ldots + a_{1n} x_n \leq b_1$$
$$a_{21} x_1 + a_{22} x_2 + \ldots + a_{2n} x_n \leq b_2$$
$$\vdots$$
$$a_{m1} x_1 + a_{m2} x_2 + \ldots + a_{mn} x_n \leq b_m$$
and $\qquad x_1, x_2, \ldots, x_n \geq 0$

Then to obtain the dual of the above primal, we use the following steps:
(1) minimize the objective function instead of maximization
(2) interchange $c_1, c_2, \ldots, c_n$ and $b_1, b_2, \ldots, b_n$, i.e., interchange the role of constant terms and the coefficients of the objective function.
(3) Replace A (the coefficient matrix) by A' (transpose of A).
(4) Change $\leq$ by $\geq$ sign.

Then the corresponding dual LPP is defined as
$$\text{min. } Z_y = b_1 y_1 + b_2 y_2 + \ldots + b_m y_m$$

subject to the constraints

$$a_{11}y_1 + a_{21}y_2 + \dots + a_{m1}y_m \geq c_1$$
$$a_{12}y_1 + a_{22}y_2 + \dots + a_{m2}y_m \geq c_2$$
$$\vdots$$
$$a_{1n}y_1 + a_{2n}y_2 + \dots + a_{mn}y_m \geq c_n$$

and

$$y_1, y_2, \dots, y_m \geq 0$$

The above conversion (primal to dual) can be expressed in the following figure.

Primal	Dual
$\max . Z_x = \sum\limits_{j=1}^{n} c_j x_j$	$\min . Z_y = \sum\limits_{i=1}^{m} b_i y_i$
subject to the constraints	subject to the constraints
$\sum\limits_{j=1}^{n} a_{ij} x_j \leq b_i$	$\sum\limits_{i=1}^{m} a_{ji} y_i \geq c_j \quad j = 1, 2, \dots, n$
$i = 1, 2, \dots, m$	and $\quad y_i \geq 0, i = 1, 2, \dots, m$
$a_{ij} = a_{ji}$	
and $\quad x_j \geq 0, j = 1, 2, \dots, n$	

6.3.3 MATRIX FORM OF PRIMAL DUAL PROBLEM [MEERUT–2004]

Consider the above primal dual problem in matrix form

$$\max . Z_x = \boldsymbol{CX}$$

subject to the constraints

$$\boldsymbol{AX} \leq \boldsymbol{b}$$
$$\text{and} \quad \boldsymbol{X} \geq 0$$

$$\quad \dots (1)$$

where,

$$\boldsymbol{C} = [c_1 \quad c_2 \quad \cdots \quad c_n]_{1 \times n}$$

$$\boldsymbol{X} = \begin{bmatrix} x_1 \\ x_2 \\ \vdots \\ x_n \end{bmatrix}_{n \times 1} = [x_1 \quad x_2 \quad \cdots \quad x_n]'_{1 \times n}$$

$$\boldsymbol{A} = \begin{bmatrix} a_{11} & a_{12} & \cdots & a_{1n} \\ a_{21} & a_{22} & \cdots & a_{2n} \\ \vdots & & & \\ a_{m1} & a_{m2} & \cdots & a_{mn} \end{bmatrix}_{m \times n}, \boldsymbol{b} = \begin{bmatrix} b_1 \\ b_2 \\ \vdots \\ b_m \end{bmatrix}_{m \times 1}$$

Let $\boldsymbol{O}$ be a zero matrix of order $m \times 1$. Then dual of the above problem (1) be given by

$$\min . Z_y = \boldsymbol{b}'\boldsymbol{Y}$$

subject to the constraints

$$\boldsymbol{A}'\boldsymbol{Y} \geq \boldsymbol{C}'$$
$$\text{and} \quad \boldsymbol{Y} \geq 0$$

Here, $\boldsymbol{A}'$, $\boldsymbol{b}'$ and $\boldsymbol{C}'$ are the transpose of $\boldsymbol{A}$, $\boldsymbol{b}$ and $\boldsymbol{C}$ respectively.

6.3.4 Unsymmetric Primal-Dual Form

If in the given LPP, all the constraints are equations, then it is called unsymmetric primal problem.

Consider the unsymmetric primal

$$Z_x = \mathbf{CX}$$

subject to the constraints

$$\mathbf{AX} = \mathbf{b}$$

and

$$\mathbf{X} \geq 0$$

Then dual of the above primal is given by

$$Z_y = \mathbf{b'Y}$$

subject to the constraints

$$\mathbf{A'Y} \geq \mathbf{C'}$$

and

$$\mathbf{Y} \geq 0$$

Here, ′ (dash) denote the transpose.

☞ **Remark**
- In the above case, the dual variables are unrestricted in sign.

6.4 DUAL OF AN LPP WITH MIXED RESTRICTIONS

There are some situations when LPP contains a mixture of inequalities (*i.e.*, $\leq$ and $\geq$) non-negative variables and unrestricted variables. To find the dual of such case, we use the following steps:

STEP 1. If the given LP is a problem of maximization and contains some constraints of the type $\geq$, then to make all constraints $\leq$ type, multiply both sides of the such constraints by –1 and make the sign $\leq$.

STEP 2. If the given LP is a problem of minimization and contains some constraints $\leq$ type then to make all constraints $\geq$ type, multiply both sides of such constraints by –1 and make the sign $\geq$.

STEP 3. If a constraint is an equation (has = sign), then replace it by two constraints involving both sign $\leq$ and $\geq$ by using the fact that $a = b \Leftrightarrow a \leq b$ or $a \geq b$.

STEP 4. If the given LPP has some unrestricted variables, then replace it by the difference of two non-negative variables.

Solved Examples

EXAMPLE 1. *Write the dual of the following LPP:*

$$max.\ Z = x_1 - x_2 + 3x_3$$

subject to the constraints

$$x_1 + x_2 + x_3 \leq 10$$
$$2x_1 - x_2 - x_3 \leq 2$$
$$2x_1 - 2x_2 - 3x_3 \leq 6$$

and $\qquad x_1, x_2, x_3 \geq 0$

SOLUTION. We observe that in the given LPP there are $m = 3$ constraints and $n = 3$ variables which shows that there should be $m = 3$ dual variables and $n = 3$ constraints.

Further, the coefficient of the primal variables $c_1 = 1$, $c_2 = -1$, $c_3 = 3$ become right hand side constraints of the dual and right hand constraints $b_1 = 0$, $b_2 = 2$, $b_3 = 6$ becomes the coefficients of the dual objective function. Also, the required dual must have a minimizing objective function with all $\geq$ constraints in another variables say y_1, y_2, y_3.

Hence, the resultant dual is given by

$$\min Z = 10y_1 + 2y_2 + 6y_3$$

subject to the constraints

$$y_1 + 2y_2 + 2y_3 \geq 1$$
$$y_1 - y_2 - 2y_3 \geq -1$$
$$y_1 - y_2 - 3y_3 \geq 3$$

and $y_1, y_2, y_3 \geq 0$

EXAMPLE 2.   *Find the dual of the following LPP:*

$$max. \ Z = 3x_1 + 5x_2 + 4x_3$$

subject to the constraints

$$2x_1 + 3x_2 \leq 8$$
$$2x_2 + 5x_3 \leq 10$$
$$3x_1 + 2x_2 + 4x_3 \leq 15$$

and $x_1, x_2, x_3 \geq 0$ [MEERUT–2012]

SOLUTION.   We shall find the dual of this LPP by alternative matrix method.

Clearly, the given LPP is written in the standard primal form. In matrix form it can be written as

$$max. \ Z = [3 \quad 5 \quad 4][x_1 \quad x_2 \quad x_3] = \boldsymbol{C} \cdot \boldsymbol{x}$$

s.t.

$$\begin{bmatrix} 2 & 3 & 0 \\ 0 & 2 & 5 \\ 3 & 2 & 4 \end{bmatrix} \begin{bmatrix} x_1 \\ x_2 \\ x_3 \end{bmatrix} \leq \begin{bmatrix} 8 \\ 10 \\ 15 \end{bmatrix}$$

or $\boldsymbol{Ax} \leq \boldsymbol{b}$

and $x_1, x_2, x_3 \geq 0$

Now, dual of this LPP is given by

$$\min Z_y = b' \cdot y = [8 \quad 10 \quad 15][y_1 \quad y_2 \quad y_3]$$

$$= 8y_1 + 10y_2 + 15y_3$$

s.t. $\boldsymbol{A'y} \geq \boldsymbol{C'}$

or $\begin{bmatrix} 2 & 0 & 3 \\ 3 & 2 & 2 \\ 0 & 5 & 4 \end{bmatrix} \begin{bmatrix} y_1 \\ y_2 \\ y_3 \end{bmatrix} \geq \begin{bmatrix} 3 \\ 5 \\ 4 \end{bmatrix}$ $\Rightarrow$ $\begin{bmatrix} 2y_1 + 0y_2 + 3y_3 \\ 3y_1 + 2y_2 + 2y_3 \\ 0y_1 + 5y_2 + 4y_3 \end{bmatrix} \geq \begin{bmatrix} 3 \\ 5 \\ 4 \end{bmatrix}$

Finally the dual problem of the given LPP is given by

$$min. \ Z_y = 8y_1 + 10y_2 + 15y_3$$

subject to the constraints

$$2y_1 + 3y_3 \geq 3$$
$$3y_1 + 2y_2 + 2y_3 \geq 5$$
$$5y_2 + 4y_3 \geq 4$$

and $y_1, y_2, y_3 \geq 0$

EXAMPLE 3. *Find the dual of the following LPP:*

$$\text{min. } Z = 3x_1 + x_2$$

subject to the constraints

$$2x_1 + 3x_2 \geq 2$$
$$x_1 + x_2 \geq 1$$

and $\qquad x_1, x_2 \geq 0$ $\hfill$ [KANPUR–2012]

SOLUTION. The given LPP of minimization is in its standard form. In the matrix form it can be written as

$$\text{min. } Z = 3x_1 + x_2 = [3 \quad 1][x_1 \quad x_2] = \boldsymbol{C} \cdot \boldsymbol{x}$$

s.t.
$$\begin{bmatrix} 2 & 3 \\ 1 & 2 \end{bmatrix}\begin{bmatrix} x_1 \\ x_2 \end{bmatrix} \geq \begin{bmatrix} 2 \\ 1 \end{bmatrix}$$

or $\qquad\qquad\qquad \boldsymbol{Ax} \geq \boldsymbol{b}$

and $\qquad\qquad\qquad x_1, x_2 \geq 0$

Now proceed same as in example 2, the dual of the given LPP is

$$\text{max. } Z_y = \boldsymbol{b}' \cdot \boldsymbol{y} = [3 \quad 1][y_1 \quad y_2] = 3y_1 + y_2$$

such that $\qquad\qquad A'\boldsymbol{y} \leq \boldsymbol{C}'$

or
$$\begin{bmatrix} 2 & 1 \\ 3 & 2 \end{bmatrix}\begin{bmatrix} y_1 \\ y_2 \end{bmatrix} \leq \begin{bmatrix} 3 \\ 1 \end{bmatrix}$$

$$\Rightarrow \qquad \begin{bmatrix} 2y_1 + y_2 \\ 3y_1 + 2y_2 \end{bmatrix} \leq \begin{bmatrix} 3 \\ 1 \end{bmatrix}$$

or $\qquad\qquad 2y_1 + y_2 \leq 3, \ 3y_1 + 2y_2 \leq 1$

and $\qquad\qquad y_1, y_2 \geq 0$

Hence, the dual of the given LPP is given by

$$\text{max. } Z_y = 3y_1 + y_2$$

s.t. $\qquad\qquad 2y_1 + y_2 \leq 3$
$$3y_1 + 2y_2 \leq 1$$

and $\qquad\qquad y_1, y_2 \geq 0$

EXAMPLE 4. *Find the dual of the following LPP:*

$$\text{max. } Z = 3x_1 + 4x_2$$

subject to the constraints

$$2x_1 + 6x_2 \leq 16$$
$$5x_1 + 2x_2 \geq 20$$

and $\qquad\qquad x_1, x_2 \geq 0$ $\hfill$ [KANPUR–2007]

SOLUTION. The given LPP of maximization is not in standard form. Firstly we shall write the given LPP in the standard primal form as follows.

$$\text{max. } Z = 3x_1 + 4x_2$$

s.t. $\qquad 2x_1 + 6x_2 \leq 16$
$$- 5x_1 - 2x_2 \leq - 20$$

and $\qquad\qquad x_1, x_2 \geq 0$

In matrix form it can be written as

max. $Z = 3x_1 + 4x_2 = [3 \quad 4][x_1 \quad x_2] = \boldsymbol{C} \cdot \boldsymbol{x}$

such that

$\begin{bmatrix} 2 & 6 \\ -5 & -2 \end{bmatrix}\begin{bmatrix} x_1 \\ x_2 \end{bmatrix} \leq \begin{bmatrix} 16 \\ -20 \end{bmatrix}$, i.e., $\boldsymbol{Ax} \leq \boldsymbol{b}$, $x_1, x_2 \geq 0$

Now, the dual of the above problem is given by

$\min Z_y = \boldsymbol{b}' \cdot \boldsymbol{y} = [16 \quad -20][y_1 \quad y_2] = 16y_1 - 20y_2$

s.t. $\qquad \boldsymbol{A}' \cdot \boldsymbol{y} \geq \boldsymbol{C}'$

$\Rightarrow \begin{bmatrix} 2 & -5 \\ 6 & -2 \end{bmatrix}\begin{bmatrix} y_1 \\ y_2 \end{bmatrix} \geq \begin{bmatrix} 3 \\ 4 \end{bmatrix}$

or $\begin{bmatrix} 2y_1 - 5y_2 \\ 6y_1 - 2y_2 \end{bmatrix} \geq \begin{bmatrix} 3 \\ 4 \end{bmatrix}$

$\Rightarrow 2y_1 - 5y_2 \geq 3$ and $6y_1 - 2y_2 \geq 4$

Hence, the dual of the given LPP is

$\min Z_y = 16y_1 - 20y_2$

s.t. $\quad 2y_1 - 5y_2 \geq 3$

$\qquad 6y_1 - 2y_2 \geq 4$

and $\qquad y_1, y_2 \geq 0$

EXAMPLE 5. *Find the dual of the following LPP:*

$$\min Z = 2x_1 + 2x_2 + 4x_3$$

subject to the constraints

$$2x_1 + 3x_2 + 5x_3 \geq 2$$

$$3x_1 + x_2 + 7x_3 \leq 3$$

$$x_1 + 4x_2 + 6x_3 \leq 5$$

and $\quad x_1, x_2, x_3 \geq 0$ [MEERUT–2011]

SOLUTION. Firstly, we shall write the given LPP in the standard primal form as follows:

$\min. Z = 2x_1 + 2x_2 + 4x_3$

s.t $\quad 2x_1 + 3x_2 + 5x_3 \geq 2$

$\qquad - 3x_1 - x_2 - 7x_3 \geq -3$

$\qquad - x_1 - 4x_2 - 6x_3 \geq -5$

and $\qquad x_1, x_2, x_3 \geq 0$

which can be written in matrix form as follows:

$\min Z = 2x_1 + 2x_2 + 4x_3 = [2 \quad 2 \quad 4][x_1 \quad x_2 \quad x_3] = \boldsymbol{C} \cdot \boldsymbol{x}$

s.t. $\begin{bmatrix} 2 & 3 & 5 \\ -3 & -1 & -7 \\ -1 & -4 & -6 \end{bmatrix}\begin{bmatrix} x_1 \\ x_2 \\ x_3 \end{bmatrix} \geq \begin{bmatrix} 2 \\ -3 \\ -5 \end{bmatrix}$

i.e., $\boldsymbol{Ax} \geq \boldsymbol{b}$ and $x_1, x_2, x_3 \geq 0$

Now, dual of the given LPP is

$\max Z_y = \boldsymbol{b}' \cdot \boldsymbol{y} = [2 \quad -3 \quad 5][y_1 \quad y_2 \quad y_3] = 2y_1 - 3y_2 - 5y_3$

such that $\qquad A'y \leq C'$

or $\begin{bmatrix} 2 & -3 & -1 \\ 3 & -1 & -4 \\ 5 & -7 & -6 \end{bmatrix} \begin{bmatrix} y_1 \\ y_2 \\ y_3 \end{bmatrix} \leq \begin{bmatrix} 2 \\ 2 \\ 2 \end{bmatrix} \Rightarrow \begin{bmatrix} 2y_1 - 3y_2 - y_3 \\ 3y_1 - y_2 - 4y_3 \\ 5y_1 - 7y_2 - 6y_3 \end{bmatrix} \leq \begin{bmatrix} 2 \\ 2 \\ 4 \end{bmatrix}$

$\Rightarrow \quad 2y_1 - 3y_2 - y_3 \leq 2; 3y_1 - y_2 - 4y_3 \leq 2; 5y_1 - 7y_2 - 6y_3 \leq 4$

Hence, the required dual of the given LPP is

$$\text{max. } Z_y = 2y_1 - 3y_2 + 5y_3$$
$$\text{s.t.} \qquad 2y_1 - 3y_2 - y_3 \leq 2$$
$$3y_1 - y_2 - 4y_3 \leq 2$$
$$5y_1 - 7y_2 - 6y_3 \leq 4$$
$$\text{and} \qquad y_1, y_2, y_3 \geq 0$$

EXAMPLE 6. *Find the dual of the following LPP:*

$$\text{min. } Z = 7x_1 + 3x_2 + 8x_3$$

subject to the constraints

$$8x_1 + 2x_2 + x_3 \geq 3$$
$$3x_1 + 6x_2 + 4x_3 \geq 4$$
$$4x_1 + x_2 + 5x_3 \geq 1$$
$$x_1 + 5x_2 + 2x_3 \geq 7$$

and $\qquad x_1, x_2, x_3 \geq 0$

SOLUTION. The given LPP is in standard form. Since there are four constraints in the problem, so in the dual problem there will be four variables.

The given LPP can be written in matrix form as follows:

$$\text{min. } Z = 7x_1 + 3x_2 + 8x_3 = \begin{bmatrix} 7 & 3 & 8 \end{bmatrix} \begin{bmatrix} x_1 & x_2 & x_3 \end{bmatrix} = C \cdot x$$

such that

$$\begin{bmatrix} 8 & 2 & 1 \\ 3 & 6 & 4 \\ 4 & 1 & 5 \\ 1 & 5 & 2 \end{bmatrix} \begin{bmatrix} x_1 \\ x_2 \\ x_3 \end{bmatrix} \leq \begin{bmatrix} 3 \\ 4 \\ 1 \\ 7 \end{bmatrix}$$

or $\qquad Ax \geq b, x_1, x_2, x_3 \geq 0$

Now, the dual of this LPP is given by

$$\text{max } Z_y = b' \cdot y = \begin{bmatrix} 3 & 4 & 1 & 7 \end{bmatrix} \begin{bmatrix} y_1 & y_2 & y_3 & y_4 \end{bmatrix}$$
$$= 3y_1 + 4y_2 + y_3 + 7y_4$$

s.t. $\qquad A'y \leq C'$

$$\Rightarrow \begin{bmatrix} 8 & 3 & 4 & 1 \\ 2 & 6 & 1 & 5 \\ 1 & 4 & 5 & 2 \end{bmatrix} \begin{bmatrix} y_1 \\ y_2 \\ y_3 \\ y_4 \end{bmatrix} \leq \begin{bmatrix} 7 \\ 3 \\ 8 \end{bmatrix}$$

$\Rightarrow \quad 8y_1 + 3y_2 + 4y_3 + y_4 \leq 7;$

$$2y_1 + 6y_2 + y_3 + 5y_4 \le 3$$
$$y_1 + 4y_2 + 5y_3 + 2y_4 \le 8$$

and $\qquad y_1, y_2, y_3, y_4 \ge 0$

Hence, the dual problem of the given LPP is

max. $Z_y = 3y_1 + 4y_2 + y_3 + 7y_4$

s.t. $8y_1 + 3y_2 + 4y_3 + y_4 \le 7$

$\qquad 2y_1 + 6y_2 + y_3 + 5y_4 \le 3$

$\qquad y_1 + 4y_2 + 5y_3 + 2y_4 \le 8$

and $\qquad y_1, y_2, y_3, y_4 \ge 0$

EXAMPLE 7. *Write the dual of the following LPP:*

$$\text{min. } Z = 2x_2 + 5x_3$$

subject to the constraints

$$x_1 + x_2 \ge 2$$
$$2x_1 + x_2 + 6x_3 \le 6$$
$$x_1 - x_2 + 3x_3 = 4$$

and $\qquad x_1, x_2, x_3 \ge 0$ $\qquad$ [MEERUT–2005]

SOLUTION. To write the given LPP in standard form, we proceed as follows:

(i) Multiply the second constraint by –1, then it becomes
$$-2x_1 - x_2 - 6x_3 \ge -6$$

(ii) The third constraint is an equality. Replace it by two constraints such that
$$x_1 - x_2 + 3x_3 \ge 4$$
and $\qquad x_1 - x_2 + 3x_3 \le 4 \qquad \Rightarrow \qquad -x_1 + x_2 - 3x_3 \ge -4$

Thus, the given LPP can be written as (in standard form)

min. $Z = 0x_1 + 2x_2 + 5x_3$

s.t. $\qquad x_1 + x_2 \ge 2$

$\qquad -2x_1 - x_2 - 6x_3 \ge -6$

$\qquad x_1 - x_2 + 3x_3 \ge 4$

$\qquad -x_1 + x_2 - 3x_3 \ge -4$

and $\qquad x_1, x_2, x_3 \ge 0$

The above LPP can be written in matrix form as

min. $Z = [0 \quad 2 \quad 5][x_1 \quad x_2 \quad x_3] = \mathbf{C} \cdot \mathbf{x}$

such that $\begin{bmatrix} 1 & 1 & 0 \\ -2 & -1 & -6 \\ 1 & -1 & 3 \\ -1 & 1 & -3 \end{bmatrix} \begin{bmatrix} x_1 \\ x_2 \\ x_3 \end{bmatrix} \ge \begin{bmatrix} 2 \\ -6 \\ 4 \\ -4 \end{bmatrix}$

$\Rightarrow \qquad A\mathbf{x} \ge \mathbf{b}$ and $x_1, x_2, x_3 \ge 0$

∴ Dual is

max. $Z_y = \mathbf{b}' \cdot \mathbf{y} = [2 \quad -6 \quad 4 \quad -4][y_1 \quad y_2 \quad y_3' \quad y_3'']$

$= 2y_1 - 6y_2 + 4(y_3' - y_3'')$

(y_3', y_3'' are taken because the third constraint of the primal have = sign) such that $A'y \leq C'$

$$\Rightarrow \quad \begin{bmatrix} 1 & -2 & 1 & 1 \\ 1 & -1 & -1 & -1 \\ 0 & -6 & 3 & 3 \end{bmatrix} \begin{bmatrix} y_1 \\ y_2 \\ y_3' \\ y_3'' \end{bmatrix} \leq \begin{bmatrix} 0 \\ 2 \\ 5 \end{bmatrix}$$

or $$\begin{bmatrix} y_1 - 2y_2 + y_3' - y_3'' \\ y_1 - y_2 - y_3' + y_3'' \\ 0y_1 - 6y_2 + 3y_3' - 3y_3'' \end{bmatrix} \leq \begin{bmatrix} 0 \\ 2 \\ 5 \end{bmatrix}$$

$$y_1, y_2, y_3', y_3'' \geq 0$$

or max. $Z_y = 2y_1 - 6y_2 - 4(y_3' - y_3'')$

subject to the constraints

$$y_1 - 2y_2 + (y_3' - y_3'') \leq 0$$
$$y_1 - y_2 - (y_3' - y_3'') \leq 2$$
$$-6y_1 + 3(y_3' - y_3'') \leq 5$$

and $y_1, y_2, y_3', y_3'' \geq 0$

Finally, put $y_3 = y_3' - y_3''$, the required dual be given by

max. $Z_y = 2y_1 - 6y_2 + 4y_3$
subject to

$$y_1 - 2y_2 + y_3 \leq 0$$
$$y_1 - y_2 - y_3 \leq 2$$
$$-6y_1 + 3y_3 \leq 5$$

and $y_1, y_2 \geq 0, y_3$ is unrestricted.

EXAMPLE 8. *Write the dual of the following LPP:*

$$max. \ Z = 2x_1 + 3x_2 + x_3$$

subject to the constraints

$$4x_1 + 3x_2 + x_3 = 6$$
$$x_1 + 2x_2 + 5x_3 = 4$$

and $x_1, x_2, x_3 \geq 0$

SOLUTION. The given primal problem is not in standard form. Following the usual procedure, the standard form of the given LPP is

max. $Z = 2x_1 + 3x_2 + x_3$
s.t. $\quad 4x_1 + 3x_2 + x_3 \leq 6$
$\quad -4x_1 - 3x_2 - x_3 \leq -6$
$\quad x_1 + 2x_2 + 5x_3 \leq 4$
$\quad -x_1 - 2x_2 - 5x_3 \leq -4$

and $\quad x_1, x_2, x_3 \geq 0$

The above problem can be written in matrix form as

max. $Z = 2x_1 + 3x_2 + x_3 = \begin{bmatrix} 2 & 3 & 1 \end{bmatrix} \begin{bmatrix} x_1 & x_2 & x_3 \end{bmatrix} = C \cdot x$

subject to

$$\begin{bmatrix} 4 & 3 & 1 \\ -4 & -3 & -1 \\ 1 & 2 & 5 \\ -1 & -2 & -5 \end{bmatrix} \begin{bmatrix} x_1 \\ x_2 \\ x_3 \end{bmatrix} \leq \begin{bmatrix} 6 \\ -6 \\ 4 \\ -4 \end{bmatrix}$$

or $\quad Ax \leq b$

and $x_1, x_2, x_3 \geq 0$

Now, the required dual is

$$\min Z_y = b'y = [6 \quad -6 \quad 4 \quad -4] = [y_1' \quad y_1'' \quad y_2' \quad y_2'']$$

$$= 6(y_1' - y_1'') + 4(y_2' - y_2'')$$

such that $A'y \geq C'$

$$\Rightarrow \qquad \begin{bmatrix} 4 & -4 & 1 & -1 \\ 3 & -3 & 2 & -2 \\ 1 & -1 & 5 & -5 \end{bmatrix} \begin{bmatrix} y_1' \\ y_1'' \\ y_2' \\ y_2'' \end{bmatrix} \geq \begin{bmatrix} 2 \\ 3 \\ 1 \end{bmatrix}$$

$$\Rightarrow \qquad \begin{bmatrix} 4y_1' - 4y_1'' + y_2' - y_2'' \\ 3y_1' - 3y_1'' + 2y_2' - 2y_2'' \\ y_1' - y_1'' + 5y_2' - 5y_2'' \end{bmatrix} \geq \begin{bmatrix} 2 \\ 3 \\ 1 \end{bmatrix} \quad y_1', y_1'', y_2', y_2'' \geq 0$$

$$\Rightarrow \qquad 4(y_1' - y_1'') + y_2' - y_2'' \geq 2$$

$$3(y_1' - y_1'') + 2(y_2' - y_2'') \geq 3$$

$$(y_1' - y_1'') + 5(y_2' - y_2'') \geq 1$$

Putting $y_1 = y_1' - y_1''$ and $y_2 = y_2' - y_2''$, we get the required dual is

min. $Z_y = 6y_1 + 4y_2$

s.t. $\quad 4y_1 + y_2 \geq 2$

$\qquad 3y_1 + 2y_3 \geq 3$

$\qquad y_1 + 5y_2 \geq 0$

and y_1, y_2 are unrestricted.

EXAMPLE 9. *Find the dual of the following LPP:*

$$min. \ Z = x_1 - 3x_2 - 2x_3$$

subject to the constraints

$$3x_1 - x_2 + 2x_3 \leq 7$$

$$2x_1 - 4x_2 \geq 12$$

$$-4x_1 + 3x_2 + 8x_3 = 10$$

and $x_1, x_2 \geq 0$, x_3 is unrestricted.

SOLUTION. Following the usual procedure, the standard form of the given LPP is

$$min. Z = x_1 - 3x_2 - 2(x_3' - x_3'') = x_1 - 3x_2 - 2x_3' + 2x_3''$$

s.t. $\quad -3x_1 + x_2 - 2x_3' + 2x_3'' \geq -7$

$$2x_1 - 4x_2 \geq 12$$
$$-4x_1 + 3x_2 + 8x_3' - 8x_3'' \geq 10$$
$$4x_1 - 3x_2 - 8x_3' + 8x_3'' \geq -10$$

and $x_1, x_2, x_3', x_3'' \geq 0$

The above LPP can be written in matrix from as follows:

$$\text{min. } Z = [1 \quad -3 \quad -2 \quad 2][x_1 \quad x_2 \quad x_3' \quad x_3''] = C \cdot x$$

such that

$$\begin{bmatrix} -3 & 1 & -2 & 2 \\ 2 & -4 & 0 & 0 \\ -4 & 3 & 8 & -8 \\ 4 & -3 & -8 & 8 \end{bmatrix} \begin{bmatrix} x_1 \\ x_2 \\ x_3' \\ x_3'' \end{bmatrix} \geq \begin{bmatrix} -7 \\ 12 \\ 10 \\ -10 \end{bmatrix} \Rightarrow Ax \geq b$$

and $x_1, x_2, x_3', x_3'' \geq 0$

Now the dual of the above problem is given by

$$\text{max. } Z_y = b' \cdot y = [-7 \quad 12 \quad 10 \quad -10][y_1 \quad y_2 \quad y_3' \quad y_3'']$$
$$= -7y_1 + 12y_2 + 10(y_3' - y_3'')$$

such that $A'y \leq C'$

$$\Rightarrow \begin{bmatrix} -3 & 2 & -4 & 4 \\ 1 & -4 & 3 & -3 \\ -2 & 0 & 8 & -8 \\ 2 & 0 & -8 & 8 \end{bmatrix} \begin{bmatrix} y_1 \\ y_2 \\ y_3' \\ y_3'' \end{bmatrix} \leq \begin{bmatrix} 1 \\ -3 \\ -2 \\ 2 \end{bmatrix}$$

$$\Rightarrow \quad -3y_1 + 2y_2 - 4(y_3' - y_3'') \leq 1$$
$$y_1 - 4y_2 + 3(y_3' - y_3'') \leq -3$$
$$-2y_1 - 4y_2 + 8(y_3' - y_3'') \leq -2$$
$$2y_1 - 8(y_3' - y_3'') \leq 2$$

and $y_1, y_2, y_3', y_3'' \geq 0$

Now using $y_3 = y_3' - y_3''$, we get the required dual as

$$\text{max } Z_y = -7y_1 + 12y_2 + 10y_3$$

s.t. $-3y_1 + 2y_2 - 4y_3 \leq 1$
$$y_1 - 4y_2 + 3y_3 \leq -3$$
$$-2y_1 + 8y_3 \leq -2 \Rightarrow 2y_1 - 8y_3 \geq 2$$
$$2y_1 - 8y_3 \leq 2$$

and $y_1, y_2 \geq 0, y_3$ is unrestricted.

Hence, the required dual of the given LPP is

$$\text{max. } Z_y = -7y_1 + 12y_2 + 10y_3$$

s.t. $-3y_1 + 2y_2 - 4y_3 \leq 1$
$$-y_1 + 4y_2 - 3y_3 \geq 3$$
$$2y_1 - 8y_3 = 2$$

and $y_1, y_2 \geq 0, y_3$ is unrestricted.

EXAMPLE 10. *Write the dual of the following problem*

$$\min Z = x_1 + x_2 + x_3$$
$$s.t. \quad x_1 - 3x_2 + 4x_3 = 5$$
$$x_1 - 2x_2 \leq 3$$
$$2x_2 - x_3 \geq 4$$

and $x_1, x_2 \geq 0$; x_3 is unrestricted.

SOLUTION. Using $x_3 = x_3' - x_3''$, the given LPP can be written in standard form as follows.

$$\min. \ Z = x_1 + x_2 + x_3' - x_3''$$

subject to the constraints

$$-x_1 + 3x_2 - 4(x_3' - x_3'') \geq -5$$
$$x_1 - 3x_2 + 4(x_3' - x_3'') \geq 5$$
$$-x_1 + 2x_2 \geq -3$$
$$2x_2 - (x_3' - x_3'') \geq 4$$

and $\quad x_1, x_2, x_3', x_3'' \geq 0$

The above problem can be written in matrix form as follows:

$$\min. \ Z = [1 \ \ 1 \ \ 1 \ \ -1][x_1 \ \ x_2 \ \ x_3' \ \ x_3''] = \mathbf{C} \cdot \mathbf{x}$$

$$\Rightarrow \quad A\mathbf{x} \geq \mathbf{b}, \ x_1, x_2, x_3', x_3'' \geq 0$$

Now, the dual of the given primal is

$$\max. \ Z_y = [-5 \ \ 5 \ \ -3 \ \ 4][y_1' \ \ y_1'' \ \ y_2 \ \ y_3]$$

$$= -5(y_1' - y_1'') - 3y_2 + 4y_3$$

subject to the constraints $A'\mathbf{y} \leq \mathbf{C}'$

$$\Rightarrow \quad \begin{bmatrix} -1 & 1 & -1 & 0 \\ 3 & -3 & 2 & 2 \\ -4 & 4 & 0 & -1 \\ 4 & -4 & 0 & 1 \end{bmatrix} \begin{bmatrix} y_1' \\ y_1'' \\ y_2 \\ y_3 \end{bmatrix} \leq \begin{bmatrix} 1 \\ 1 \\ 1 \\ -1 \end{bmatrix}$$

$$\Rightarrow \quad -(y_1' - y_1'') - y_2 \leq 1$$
$$3(y_1' - y_1'') + 2y_2 + 2y_3 \leq 1$$
$$-4(y_1' - y_1'') + y_3 \leq -1$$

and $\quad y_1', y_1'', y_2, y_3 \geq 0$

Using $y_1 = y_1' - y_1''$, we get the dual is

$$\max. \ Z_y = -5y_1 - 3y_2 + 4y_3$$

subject to

$$-y_1 - y_2 \leq 1; \ 3y_1 + 2y_2 + 2y_3 \leq 1$$
$$-4y_1 - y_3 \leq 1, \ 4y_1 + y_3 \leq -1 \Rightarrow -4y_1 - y_3 \geq 1$$
$$y_2, y_3 \geq 0 \text{ and } y_1 \text{ is unrestricted.}$$

Finally, the required dual is

$$\max. \ Z_y = -5y_1 - 3y_2 + 4y_3$$

s.t. $\qquad -y_1 - y_2 \le 1$
$$3y_1 + 2y_2 + 2y_3 \le 1$$
$$-4y_1 - y_3 = 1$$
and $y_2, y_3 \ge 0$, y_1 is unrestricted in sign.

 ## Exercise-6.1

Find the dual of the following LPP:

1. max. $Z = x_1 - x_2 + 3x_3$
subject to the constraints
$$x_1 + x_2 + x_3 \le 10$$
$$2x_1 - x_3 \le 2$$
$$2x_1 - 2x_2 + 3x_3 \le 3$$
and $\qquad x_1, x_2, x_3 \ge 0$

2. min $Z = 4x_1 + 6x_2 + 18x_3$
subject to the constraints
$$x_1 + 3x_3 \ge 3$$
$$x_2 + 2x_3 \ge 5$$
and $\qquad x_1, x_2, x_3 \ge 0$

3. min. $Z = 10x_1 + 20x_2$
subject to the constraints
$$3x_1 + 2x_2 \ge 18$$
$$2x_1 - x_2 \le 6$$
and $\qquad x_1, x_2 \ge 0$

4. max. $Z = 3x_1 + x_2 + 4x_3 + x_4 + 9x_5$
subject to the constraints
$$4x_1 - 5x_2 - 9x_3 + x_4 - 2x_5 \le 6$$
$$2x_1 + 3x_2 + 4x_3 - 5x_4 + x_5 \le 9$$
$$x_1 + x_2 - 5x_3 - 7x_4 + 11x_5 \le 10$$
and $\qquad x_1, x_2, x_3, x_4, x_5 \ge 0$

5. min. $Z = 3x_1 - 2x_2 + 4x_3$
subject to the constraints
$$3x_1 + 5x_2 + 4x_3 \ge 7$$
$$6x_1 + x_2 + 3x_3 \ge 4$$
$$7x_1 - 2x_2 - x_3 \le 10$$
$$x_1 - 2x_2 + 5x_3 \ge 3$$
$$4x_1 + 7x_2 - 2x_3 \ge 2$$
and $\qquad\qquad\qquad x_1, x_2, x_3 \ge 0$
[GORAKHPUR–2009, 11]

6. max. $Z = x_1 + 3x_2$
subject to the constraints
$$3x_1 + 2x_2 \le 6$$
$$3x_1 + x_2 = 4$$
and $\qquad\qquad x_1, x_2 \ge 0$

7. min. $Z = x_1 + x_2 + x_3$
subject to the constraints
$$x_1 - 3x_2 + 4x_3 = 5$$
$$x_1 - 2x_2 \le 3$$
$$2x_2 - x_3 \ge 4$$
and $x_1, x_3 \ge 0$, x_2 is unrestricted
[MEERUT–2006]

8. max. $Z = 3x_1 + x_2 + x_3 - x_4$
subject to the constraints
$$x_1 + 5x_2 + 3x_3 + 4x_4 \le 5$$
$$x_1 + x_2 = -1$$
$$x_3 - x_4 \ge -5$$
and $\qquad x_1, x_2, x_3, x_4 \ge 0$

9. max. $Z = 3x_1 + 5x_2 + 7x_3$
subject to the constraints
$$x_1 + x_2 + 3x_3 \le 10$$
$$4x_1 - x_2 + 2x_3 \ge 15$$
and $x_1, x_2 \ge 0$; x_3 is unrestricted

10. max. $Z = 3x_1 + x_2 + 2x_3 - x_4$
subject to the constraints
$$2x_1 - x_2 + 3x_3 + x_4 = 1$$
$$x_1 + x_2 - x_3 + x_4 = 3$$
and $x_1, x_2, x_3 \ge 0$; x_4 is unrestricted.

Answers

1. max. $Z_y = 10y_1 + 2y_2 + 3y_3$
s.t.

$y_1 + 2y_2 + 2y_3 \ge 1$
$y_1 - 2y_3 \ge -1$
$y_1 - y_2 + 3y_3 \ge 3$
and $y_1, y_2, y_3 \ge 0$

2. max. $Z_y = 3y_1 + 5y_2$
s.t.

$y_1 \le 4$
$y_2 \le 6$
$3y_1 + 2y_2 \le 18$
and $y_1, y_2 \ge 0$

3. max. $Z_y = 18y_1 + 8y_2 - 6y_3$
s.t.
$3y_1 + y_2 - 2y_3 \leq 10$
$2y_1 + 3y_2 + y_3 \leq 20$
and $y_1, y_2, y_3 \geq 0$

4. min. $Z_y = 6y_1 + 9y_2 + 10y_3$
s.t.
$4y_1 + 2y_2 + y_3 \geq 3$
$-5y_1 + 3y_2 + y_3 \geq 1$
$-9y_1 + 4y_2 - 5y_3 \geq 4$
$y_1 - 5y_2 - 7y_3 \geq 1$
$-2y_1 + y_2 + 11y_3 \geq 9$
and $y_1, y_2, y_3 \geq 0$

5. max. $Z_y = 7y_1 + 4y_2 - 10y_3 + 3y_4 + 2y_5$
s.t.
$3y_1 + 6y_2 - 7y_3 + y_4 + y_5 \leq 3$
$5y_1 + y_2 + 2y_3 - 2y_4 + 7y_5 \leq -2$
$4y_1 + 3y_2 + y_3 + 5y_4 - 2y_5 \leq 4$
and $y_1, y_2, y_3, y_4, y_5 \geq 0$

6. min. $Z_y = 6y_1 + 4y_2$
s.t.
$3y_1 + 3y_2 \geq 1$
$2y_1 + y_2 \geq 3$
and $y_1 \geq 0, y_2$ is unrestricted in sign

7. max. $Z_y = 5y_1 - 3y_2 + 4y_3$
s.t.
$y_1 - y_2 \leq 1$
$-3y_1 + 2y_2 + 2y_3 = 1$
$4y_1 - y_3 \leq 1$
and $y_2, y_3 \geq 0$; y_1 is unrestricted

8. min. $Z_y = 5y_1 - y_2 + 5y_3$
s.t.
$y_1 + y_2 \geq 3$
$5y_1 + y_2 \geq 1$
$3y_1 - y_3 \geq 1$
$4y_1 + y_3 \geq -1$
and $y_1, y_3 \geq 0$; y_2 is unrestricted

9. min. $Z_y = 10y_1 - 15y_2$
s.t.
$y_1 - 4y_2 \geq 3$
$y_1 + y_2 \geq 5$
$3y_1 - 2y_2 = 7$
and $y_1, y_2 \geq 0$

10. min. $Z_y = y_1 + 3y_2$
s.t.
$2y_1 + y_2 \geq 3$
$-y_1 + y_2 \geq 1$
$3y_1 - y_2 \geq 2$
$y_1 + y_2 = -1$
and y_1, y_2 are unrestricted.

6.5 SOME RESULTS ON DUALITY

THEOREM I.   *The dual of a dual of the given primal is the primal itself.*

[KANPUR–2011; MEERUT–2010, 14]

PROOF. Consider the primal problem

$$\text{max.} Z_x = c_1 x_1 + c_2 x_2 + \ldots + c_n x_n$$

subject to the constraints

$$a_{11}x_1 + a_{12}x_2 + \ldots + a_{1n}x_n \leq b_1$$
$$a_{21}x_1 + a_{22}x_2 + \ldots + a_{2n}x_n \leq b_2$$
$$\vdots$$
$$a_{m1}x_1 + a_{m2}x_2 + \ldots + a_{mn}x_n \leq b_m$$

$$\ldots(1)$$

and
$$x_1, x_2, \ldots, x_n \geq 0$$

Following the usual procedure, the dual of the above primal is

$$\text{min.} Z_w = b_1 w_1 + b_2 w_2 + \ldots + b_m w_m$$

subject to the constraints

$$a_{11}w_1 + a_{21}w_2 + \ldots + a_{m1}w_m \geq c_1$$
$$a_{12}w_1 + a_{22}w_2 + \ldots + a_{m2}w_m \geq c_2$$
$$\vdots$$
$$a_{1n}w_1 + a_{2n}w_2 + \ldots + a_{mn}w_m \geq c_n$$

$$\ldots(2)$$

and
$$w_1, w_2, \ldots, w_m \geq 0$$

Next to find the dual of the above dual, firstly we shall write this into standard form as follows:

The standard maximization form is

$$\max.(-Z_w) = -b_1 w_1 - b_2 w_2 - \ldots - b_m w_m$$

subject to the constraints

$$-a_{11}w_1 - a_{21}w_2 - \ldots - a_{m1}w_m \leq -c_1$$
$$-a_{12}w_1 - a_{22}w_2 - \ldots - a_{m2}w_m \leq -c_2$$
$$\vdots$$
$$-a_{1n}w_1 - a_{2n}w_2 - \ldots - a_{mn}w_m \leq -c_n$$

...(3)

and
$$w_1, w_2, \ldots, w_m \geq 0$$

Following the usual procedure, the dual of (3) is given by

$$\min.Z_y = -c_1 y_1 - c_2 y_2 - \ldots - c_n y_n$$

subject to the constraints

$$-a_{11}y_1 - a_{12}y_2 - \ldots - a_{1n}y_n \geq -b_1$$
$$-a_{21}y_1 - a_{22}y_2 - \ldots - a_{2n}y_n \geq -b_2$$
$$\vdots$$
$$-a_{m1}y_1 - a_{m2}y_2 - \ldots - a_{mn}y_n \geq -b_m$$

...(4)

and
$$y_1, y_2, \ldots, y_n \geq 0$$

Finally system (4) can be written in standard maximization form as given below:

$$\max.Z_y' = c_1 y_1 + c_2 y_2 + \ldots + c_n y_n \quad (\text{Here } Z_y' = -Z_y)$$

subject to the constraints

$$a_{11}y_1 + a_{12}y_2 + \ldots + a_{1n}y_n \leq b_1$$
$$a_{21}y_1 + a_{22}y_2 + \ldots + a_{2n}y_n \leq b_2$$
$$\vdots$$
$$a_{m1}y_1 + a_{m2}y_2 + \ldots + a_{mn}y_n \leq b_m$$

...(5)

and
$$y_1, y_2, \ldots, y_n \geq 0$$

We observe that given primal system (1) and dual of dual (5) are identical.

Hence, we conclude that dual of a dual is the primal.

THEOREM 2. (Weak LP duality Theorem)

If x is any feasible solution to the primal given by $Z_p = Cx$ subject to $Ax \leq b$, $x \geq 0$ and w is any feasible solution to the dual problem.

$$\min. Z_D = b' \cdot w \text{ subject to } A'w \geq C', w \geq 0$$

Then $C \cdot x \leq b'w$, i.e., $Z_p \leq Z_D$

PROOF. Consider the given primal

$$\max.Z_p = C \cdot x$$

subject to the constraints
$$Ax \leq b$$

...(1)

and
$$x \geq 0$$

Following the usual procedure, the dual of primal (1) is given by

$$\min . Z_D = \boldsymbol{b}'\boldsymbol{w}$$

subject to the constraints

$$A'\boldsymbol{w} \geq \boldsymbol{C}'$$

and

$$\boldsymbol{w} \geq 0 \qquad \qquad \qquad \dots (2)$$

Further, suppose that $\boldsymbol{w} = (w_1, w_2, \dots, w_m)$ be any feasible solution of the dual (2). Since, $A\boldsymbol{x} \leq \boldsymbol{b}$ are the constraints of the primal (1), multiply both sides of $A\boldsymbol{x} \leq \boldsymbol{b}$ with $\boldsymbol{w}'$, we get

$$\boldsymbol{w}'(A\boldsymbol{x}) \leq \boldsymbol{w}' \cdot \boldsymbol{b}$$

$$\Rightarrow \qquad (A'\boldsymbol{w})'\boldsymbol{x} \leq (\boldsymbol{b}'\boldsymbol{w})' \qquad \qquad \dots (3)$$

Further, $\boldsymbol{x}$ is the feasible solution of the primal (1) and $A'\boldsymbol{w} \geq \boldsymbol{C}'$ denote the constraints of the dual (2). Thus,

$$\boldsymbol{x}'(A'\boldsymbol{w}) \geq \boldsymbol{x}' \cdot \boldsymbol{C}'$$

$$\Rightarrow \qquad \boldsymbol{x}'(\boldsymbol{w}'A)' \geq (\boldsymbol{C} \cdot \boldsymbol{x})' \qquad \qquad [\because \ (ab)' = b'a']$$

$$\Rightarrow \qquad [(\boldsymbol{w}'A)\boldsymbol{x}]' \geq (\boldsymbol{C}\boldsymbol{x})'$$

$$\Rightarrow \qquad (\boldsymbol{w}'A)\boldsymbol{x} \geq \boldsymbol{C} \cdot \boldsymbol{x}$$

$$\Rightarrow \qquad (A'\boldsymbol{w})' \geq \boldsymbol{C} \cdot \boldsymbol{x} \qquad \qquad \dots (4)$$

On combining (3) and (4), we get

$$\boldsymbol{C} \cdot \boldsymbol{x} \leq (A'\boldsymbol{w})'\boldsymbol{x} \leq (\boldsymbol{b}'\boldsymbol{w})'$$

$$\Rightarrow \qquad \boldsymbol{C} \cdot \boldsymbol{x} \leq (\boldsymbol{b}'\boldsymbol{w})'$$

$$\Rightarrow \qquad \boldsymbol{C}\boldsymbol{x} \leq \boldsymbol{b}'\boldsymbol{w}$$

Hence, $Z_p \leq Z_D$.

<u>THEOREM 3.</u> **(Basic Duality Theorem)** *If $\boldsymbol{x_0}$ is an optimal solution to the primal then there exists a feasible solution $\boldsymbol{w_0}$ to the dual such that $\boldsymbol{C}\boldsymbol{x_0} = \boldsymbol{b}'\boldsymbol{w_0}$, $\boldsymbol{b}'$ is the transpose of b.*

<u>PROOF.</u> Consider the primal problem

$$\max Z_p = \boldsymbol{C}\boldsymbol{x}$$

subject to the constraints

$$A\boldsymbol{x} \leq \boldsymbol{b}$$

and

$$\boldsymbol{x} \geq 0 \qquad \qquad \qquad \dots (1)$$

Let the dual of (1) be given by

$$\min Z_d = \boldsymbol{b}'\boldsymbol{w}$$

subject to the constraints

$$A'\boldsymbol{w} \geq \boldsymbol{C}'$$

and

$$\boldsymbol{w} \geq 0 \qquad \qquad \qquad \dots (2)$$

As per given, $\boldsymbol{x_0}$ is an optimal solution to the primal (1), then as in simplex method system (1), can be written as

$$\max . Z_p = \boldsymbol{C}\boldsymbol{x}$$

subject to the constraints

$$Ax + IS = b$$

where S is the vector of slack variables and I is the associated $m \times n$ identity matrix. If $x_0 \; (=(x_B, 0))$ is an optimal solution to the primal (1) where x_B denote the optimal basic feasible solution such that

$$x_B = B^{-1}b, \; B \text{ is the optimal basis of } A$$

then the optimal primal objective function is

$$Z = Cx_0 = C_B x_B, \; C_B \text{ is the vector containing the prices of the basic variables.}$$

Now, consider

$$z_j - c_j = C_B y_j - c_j = \begin{cases} C_B \cdot B^{-1}\alpha_j - c_j \; \forall \; \alpha_j \in A \\ C_B \cdot B^{-1}e_j - 0 \; \forall \; e_j \in I \end{cases}$$

Now, since x_0 is the optimal solution, so we must have

$$z_j - c_j \geq 0 \; \forall \; j$$

which implies $C_B B^{-1}\alpha_j \geq c_j, \; C_B B^{-1}e_j > 0 \; \forall \; j$

In matrix form, it can be written as

$$C_B B^{-1}A \geq C, \; C_B B^{-1} \geq 0$$
$$\Rightarrow \qquad A'B^{-1}C_B{}' \geq C'B^{-1}C_B \geq 0$$
$$\Rightarrow \qquad A'w_0 \geq C'; \; w_0 \geq 0 \qquad\qquad\qquad (B^{-1}C_B = w_0)$$

which shows that w_0 is a feasible solution of the dual problem and corresponding dual objective function is

$$b'w_0 = w_0' \cdot b = C_B \cdot B^{-1}b = C_B x_B = C \cdot x_0$$

Hence, we conclude that corresponding to a given optimal solution x_0 of the primal, there exists a feasible solution w_0 of dual such that

$$Cx_0 = b'w_0$$

☛ **REMARK**

• In a similar manner, we may prove the following result:

"If w_0 is an optimal solution to the dual, then there exists a feasible solution x_0 to the primal such that $Cx_0 = b'w_0$."

THEOREM 4. (The necessary and sufficient condition for an LPP and its dual to have optimal solution)

The necessary and sufficient condition for any linear programming problem and its dual to have optimal solution is that both have feasible solution.

PROOF. Consider the primal problem

$$\left. \begin{aligned} \max. Z_p &= C \cdot x \\ \text{subject to the constraints} \quad & \\ Ax &\leq b \\ \text{and} \qquad x &\geq 0 \end{aligned} \right] \qquad \ldots(1)$$

Let x be any feasible solution of (1)

The dual of (1) is given by

$$\left.\begin{array}{c} \min. Z_D = \boldsymbol{b}' \cdot \boldsymbol{w} \\ \text{subject to the constraints} \\ A'\boldsymbol{w} \geq \boldsymbol{C}' \\ \text{and} \qquad \boldsymbol{w} \geq 0 \end{array}\right] \qquad \dots(2)$$

Let $\hat{\boldsymbol{w}}$ be any feasible solution of (2) such that $\boldsymbol{C} \cdot \hat{\boldsymbol{x}} = \boldsymbol{b} \cdot \hat{\boldsymbol{w}}$. We have to prove that $\hat{\boldsymbol{x}}$ is the optimal solution of the primal and $\hat{\boldsymbol{w}}$ is the optimal solution of the dual.

Using weak LP duality theorem, we can write

$$Z_p \leq Z_D$$

$$\Rightarrow \qquad \boldsymbol{Cx} \leq \boldsymbol{b}' \cdot \boldsymbol{w}'$$

$$\Rightarrow \qquad \boldsymbol{Cx} \leq C\hat{x} \qquad\qquad [\because \ \boldsymbol{C}\hat{x} = \boldsymbol{b}'\hat{\boldsymbol{w}} \ \text{(given)}]$$

which shows that the value of the objective function of the primal at the feasible solution $\hat{x}$ is greater than its value at any other feasible solution $\boldsymbol{x}$.

$$\Rightarrow \qquad\qquad \hat{\boldsymbol{x}} \ \text{is the optimal solution of the primal maximization problem.}$$

Further, suppose that $\boldsymbol{w}$ is any feasible solution of dual (2), $\hat{\boldsymbol{x}}$ is the given feasible solution of the primal (1). Then again by weak LP duality theorem.

$$\boldsymbol{C}\hat{\boldsymbol{x}} \leq \boldsymbol{b}'\boldsymbol{w}$$

$$\Rightarrow \qquad \boldsymbol{b}'\hat{\boldsymbol{w}} \leq \boldsymbol{b}'\boldsymbol{w} \qquad (\because \ \boldsymbol{C}\hat{\boldsymbol{x}} = \boldsymbol{b}'\hat{\boldsymbol{w}}) \qquad\qquad \text{(By basic duality theorem)}$$

which shows that the value of the objective function of the dual problem at the given feasible solution $\hat{\boldsymbol{w}}$ is less than its value at any other feasible solution $\boldsymbol{w}$.

$$\Rightarrow \qquad\qquad \hat{\boldsymbol{w}} \ \text{is the optimal solution of the dual minimization problem.}$$

☛ **REMARK**

• The above theorem can be restated as:

"If $\hat{\boldsymbol{x}}$ is a feasible solution to the primal problem given by max. $Z_p = \boldsymbol{C}\cdot\boldsymbol{x}$ subject to $A\boldsymbol{x} \leq \boldsymbol{b}$, $\boldsymbol{x} \geq 0$ and $\hat{\boldsymbol{w}}$ is a feasible solution to its dual min. $Z_D = \boldsymbol{b}'\boldsymbol{w}$ subject to $A'\boldsymbol{w} \geq \boldsymbol{C}'$, $\boldsymbol{w} \geq 0$ such that $\boldsymbol{C}\hat{\boldsymbol{x}} = \boldsymbol{b}' \cdot \hat{\boldsymbol{w}}$, then $\hat{x}$ is the optimal solution of the primal and $\hat{\boldsymbol{w}}$ is the optimal solution of the dual problem."

THEOREM 5. **(Fundamental Theorem of Duality)** *If either the primal or the dual problem has a finite optimal solution, then the other problem has a finite optimal solution and the optimal values of the objective function in both the problems are the same.* [DELHI–2004; PATNA–2007]

PROOF. Consider the primal problem

$$\left.\begin{array}{c} \max Z_p = \boldsymbol{Cx} \\ \text{subject to the constraints} \\ A\boldsymbol{x} \leq \boldsymbol{b} \\ \text{and} \qquad \boldsymbol{x} \geq 0 \end{array}\right] \qquad \dots(1)$$

The dual of (1) is given by

$$\left.\begin{array}{c} \min Z_d = b'w \\ \text{subject to the constraints} \\ A'w \geq C' \\ \text{and} \qquad w \geq 0 \end{array}\right] \qquad \dots(2)$$

Let us assume that the primal has a finite optimal feasible solution x_B. Now as in simplex method, introduce slack variables to each of the constraints (1) then primal (1) becomes

$$\left.\begin{array}{c} \max Z_p = Cx \\ \text{subject to the constraints} \\ Ax + IS = b \\ \text{and} \qquad x \geq 0, S \geq 0 \end{array}\right] \qquad \dots(3)$$

Here, S is the vector of slack variables and I is the associated $m \times n$ identity matrix. B is the basis matrix and C_B is the m-component row vector containing the prices of the basis variables.

Now, since x_B is the optimal solution of (1) therefore, we must have

$$c_j - z_j \leq 0 \ \forall \ j$$

Here, $Z_j = C_B y_i = C_B B^{-1} \alpha_j$

Therefore, $c_j - C_B B^{-1} \alpha_j \leq 0 \ \forall \ \alpha_j$

or $\qquad C_B B^{-1} \alpha_j \geq c_j \qquad \qquad \dots(4)$

$\Rightarrow \qquad C_B B^{-1}(\alpha_1, \alpha_2, \dots, \alpha_n) \geq (c_1, c_2, \dots, c_n)$

which implies that

$$C_B \cdot B^{-1} \cdot A \geq C \qquad \dots(5)$$

Let us take $(\hat{w})' = C_B \cdot B^{-1}$ where $\hat{w} = [w_1, w_2, \dots, w_m]$

Then (5) reduces to

$$(\hat{w})A \geq C \qquad \Rightarrow \qquad [(\hat{w})' \cdot A]' \geq C'$$

$\Rightarrow \qquad A'(\hat{w}) \geq C' \qquad \Rightarrow \qquad \hat{w} \text{ satisfy (2)}$

$\Rightarrow \qquad \hat{w} \text{ is the solution of (2).}$

Next, we show that $\hat{w}$ is the optimal solution to the dual (2).

Consider, $\min Z_d = b'\hat{w} = [(\hat{w})'b]' = (\hat{w})'b$

$$= [C_B B^{-1}]b = C_B(B^{-1} \cdot b) = C_B \cdot x_B = \max Z_p$$

Here, $\hat{w}$ and x_B are the feasible solution of the dual (2) and the primal (1) respectively and

$$\min Z_d = \max. Z_p$$

THEOREM 6. *If the primal problem has an unbounded solution then its dual is infeasible.*

PROOF. Let if possible, the dual problem has feasible solution, then by weak duality theorem, any feasible solution to the dual would provide an upper bound on the primal objective function, which is a contradiction, because primal problem is unbounded. Hence, dual has infeasible solution.

In a similar manner, we can prove that if the primal problem has infeasible solution, then its dual has unbounded solution.

☛ **REMARK**
  • If primal (dual) problem has unbounded optimum solution, the other problem has either no solution at all or an unbounded solution.

THEOREM 7. *If any of the constraints in the primal is a perfect equality, the corresponding dual variable is unrestricted in sign.*

PROOF. Let us suppose in the given primal, the kth constraint is an equality. Then we can write its primal in standard form as follows.

$$\max. \ Z = c_1 x_1 + c_2 x_2 + \ldots + c_n x_n$$

subject to the constraints

$$a_{11} x_1 + a_{12} x_2 + \ldots + a_{1n} x_n \leq b_1$$
$$a_{21} x_1 + a_{22} x_2 + \ldots + a_{2n} x_n \leq b_2$$
$$\vdots$$
$$a_{k1} x_1 + a_{k2} x_2 + \ldots + a_{kn} x_n \leq b_k$$
$$-a_{k1} x_1 - a_{k2} x_2 - \ldots - a_{kn} x_n \leq -b_k$$
$$\vdots$$
$$a_{m1} x_1 + a_{m2} x_2 + \ldots + a_{mn} x_n \leq b_m$$

and $\quad x_1, x_2, \ldots, x_n \geq 0$

We can write the dual of the above primal as follows:

$$\min. \ Z_d = b_1 y_1 + b_2 y_2 + \ldots + b_k (y'_k - y''_k) + \ldots + b_m y_m$$

subject to the constraints

$$a_{11} y_1 + a_{21} y_2 + \ldots + a_{k1}(y'_k - y''_k) + \ldots + a_{m1} y_m \geq c_1$$
$$a_{12} y_1 + a_{22} y_2 + \ldots + a_{k2}(y'_k - y''_k) + \ldots + a_{m2} y_m \geq c_2$$
$$\vdots$$
$$a_{1n} y_1 + a_{2n} y_2 + \ldots + a_{kn}(y'_k - y''_k) + \ldots + a_{mn} y_m \geq c_n$$

and $\quad y_1, y_2, \ldots, y'_k, y''_k, \ldots, y_m \geq 0$

Using $y_k = y'_k - y''_k$ in the above dual we get

$$\min. \ Z_d = b_1 y_1 + b_2 y_2 + \ldots + b_k y_k + \ldots + b_m y_m$$

subject to the constraints

$$a_{11} y_1 + a_{21} y_2 + \ldots + a_{k1} y_k + \ldots + a_{m1} y_m \geq c_1$$
$$a_{12} y_1 + a_{22} y_2 + \ldots + a_{k2} y_k + \ldots + a_{m2} y_m \geq c_2$$
$$\vdots$$
$$a_{1n} y_1 + a_{2n} y_2 + \ldots + a_{kn} y_k + \ldots + a_{mn} y_m \geq c_n$$

and $\quad y_1, y_2, \ldots, y_{k-1}, y_{k+1}, \ldots, y_m \geq 0$ and y_k is unrestricted in sign because

$$y_k > 0 \text{ if } y'_k > y''_k \text{ and } y_k < 0 \text{ if } y'_k < y''_k.$$

THEOREM 8. *If any variable of the primal is unrestricted in sign, the corresponding constraints in the dual will be a strict equality.*

PROOF. Let us consider the primal problem (with k^{th} variable unrestricted) as

$$\max. \ Z = c_1 x_1 + c_2 x_2 + \ldots + c_k x_k + \ldots + c_n x_n$$

subject to the constraints

$$a_{11}x_1 + a_{12}x_2 + \ldots + a_{1k}x_k + \ldots + a_{1n}x_n \leq b_1$$
$$a_{21}x_1 + a_{22}x_2 + \ldots + a_{2k}x_k + \ldots + a_{2n}x_n \leq b_2$$
$$\vdots$$
$$a_{m1}x_1 + a_{m2}x_2 + \ldots + a_{mk}x_k + \ldots + a_{mn}x_n \leq b_m$$

and $x_1, x_2, \ldots, x_{k-1}, x_{k+1}, \ldots, x_m \geq 0$ and x_k is unrestricted.

Let us put $x_k = x_k' - x_k'', x_k', x_k'' \geq 0$ in the above primal, then we get

$$\text{max. } Z = c_1 x_1 + c_2 x_2 + \ldots + c_k(x_k' - x_k'') + \ldots + c_n x_n$$

subject to the constraints

$$a_{11}x_1 + a_{12}x_2 + \ldots + a_{1k}(x_k' - x_k'') + \ldots + a_{1n}x_n \leq b_1$$
$$a_{21}x_1 + a_{22}x_2 + \ldots + a_{2k}(x_k' - x_k'') + \ldots + a_{2n}x_n \leq b_2$$
$$\vdots$$
$$a_{m1}x_1 + a_{m2}x_2 + \ldots + a_{mk}(x_k' - x_k'') + \ldots + a_{mn}x_n \leq b_m$$

and $x_1, x_2, \ldots, x_{k-1}, x_k', x_k'', x_{k+1}, \ldots, x_m \geq 0$

Now, we can write the dual of the above LPP in a usual manner as follows.

$$\text{min. } Z_d = b_1 y_1 + b_2 y_2 + \ldots + b_m y_m$$

subject to the constraints

$$a_{11}y_1 + a_{21}y_2 + \ldots + a_{m1}y_m \geq c_1$$
$$a_{12}y_1 + a_{22}y_2 + \ldots + a_{m2}y_m \geq c_2$$
$$\vdots$$
$$a_{1k}y_1 + a_{2k}y_2 + \ldots + a_{mk}y_m \geq c_k$$
$$-a_{1k}y_1 - a_{2k}y_2 - \ldots - a_{mk}y_m \geq -c_k$$
$$\vdots$$
$$a_{1n}y_1 + a_{2n}y_2 + \ldots + a_{mn}y_m \geq c_n$$

and $y_1, y_2, \ldots, y_m \geq 0$

Clearly, the two constraints

$$a_{1k}y_1 + a_{2k}y_2 + \ldots + a_{mk}y_k \geq c_k$$

and $\quad -a_{1k}y_1 - a_{2k}y_2 - \ldots - a_{mk}y_K \geq -c_k$

will be equivalent to the single equation

$$a_{1k}y_1 + a_{2k}y_2 + \ldots + a_{mk}y_k = c_k$$

Hence, we conclude that, if the kth-variable of the primal is unrestricted, then the kth-constraints in the dual will be an equality.

<u>THEOREM 9.</u> **(Existence Theorem)** *There exists a bounded (finite) optimum solution to a linear programming problem if and only if there exists a feasible solution to both primal and its dual.*

<u>PROOF.</u> Consider the primal problem

$$\left. \begin{aligned} \text{max.} Z_p &= \boldsymbol{Cx} \\ \text{subject to the constraints} \\ \boldsymbol{Ax} &\leq \boldsymbol{b} \\ \text{and} \qquad \boldsymbol{x} &\geq 0 \end{aligned} \right\} \qquad \ldots(1)$$

The dual of (1) can be written as

$$\left.\begin{array}{c} \min. Z_d = \boldsymbol{b'w} \\ \text{subject to the constraints} \\ A'\boldsymbol{w} \geq \boldsymbol{C'} \\ \text{and} \qquad \boldsymbol{w} \geq 0 \end{array}\right] \qquad \ldots(2)$$

Now, suppose that there exists an optimum feasible solution to the primal problem, then by fundamental theorem of duality, we can say that the dual problem has at least one feasible solution.

Conversely, let us suppose that both the primal and the dual possess feasible solution $\hat{\boldsymbol{x}}$ and $\hat{\boldsymbol{w}}$ respectively.

Then clearly $\boldsymbol{C\hat{x}}$ and $\boldsymbol{b'\hat{w}}$ both are finite and $\boldsymbol{C\hat{x}} \leq \boldsymbol{b'\hat{w}}$

$\Rightarrow$ $\boldsymbol{b'\hat{w}}$ acts as an upper bound on $\boldsymbol{Cx}$ although not necessarily least upper bound (supremum).

Hence, the primal must have finite optimum solution.

☛ REMARK
- In the above manner, we may prove the following results:
  (i) If there does not exist any feasible solution to the dual (primal) but there exists at least one of the primal (dual) then there does not exist any finite optimum solution to the primal (dual).
  (ii) If there does not exist any finite optimum solution to the primal (dual) then there does not exists any feasible solution to the dual (primal).

THEOREM 10. **(Complimentry Slackness Theorem)** *In an optimal solution of a linear programming problem*
  (i) if i^{th} dual variable is positive, then i^{th} constraint must be an equation.
  (ii) if the ith constraint in the given primal is a strict inequality, then i^{th} dual variable must be zero.

PROOF. Consider a primal problem

$$\left.\begin{array}{c} \max. Z_p = \sum\limits_{j=1}^{n} c_j x_j \\ \text{subject to the constraints} \\ \sum\limits_{j=1}^{n} a_{ij} x_j \leq b_i \quad (i=1,2,\ldots,m) \end{array}\right] \qquad \ldots(1)$$

and $x_j \geq 0 \; \forall \, j$
Following the usual procedure, the dual of (1) is given by

$$\left.\begin{array}{c} \min. Z_d = \sum\limits_{i=1}^{n} b_i y_i \\ \text{subject to the constraints} \\ \sum\limits_{i=1}^{m} a_{ij} y_i \geq c_j \quad (j=1,2,\ldots,n) \end{array}\right] \qquad \ldots(2)$$

and $y_i \geq 0 \; \forall \, i$

Using weak duality theorem, we have

$$\sum_{j=1}^{n} c_j x_j \le \sum_{i=1}^{m} \sum_{j=1}^{n} a_{ij} x_j y_i \le \sum_{i=1}^{m} b_i y_i \qquad \ldots(3)$$

for any x_j and y_i feasible to primal and dual problem respectively.

Since, these solutions are not only feasible but also optimal to these problem, so by fundamental theorem of duality, we must have

$$\sum_{i=1}^{m} \sum_{j=1}^{n} a_{ij} x_j y_i = \sum_{i=1}^{m} b_i y_i$$

which implies

$$\sum_{i=1}^{m} \left[\sum_{j=1}^{n} a_{ij} x_j - b_i \right] y_i = 0 \qquad \ldots(4)$$

Equation (4) holds only if each of its term is equal to zero, *i.e.*,

$$\left(\sum_{j=1}^{n} a_{ij} x_j - b_i \right) y_i = 0 \qquad \ldots(5)$$

(i) Here, since the ith dual variable y_i is positive then from (5), we get

$$\sum_{j=1}^{n} a_{ij} x_j = b_i, \text{ which is the } i^{\text{th}} \text{ constraint of the given primal problem}$$

(ii) If the ith constraint of the primal problem is a strict inequality, *i.e.*,

$$\sum_{j=1}^{n} a_{ij} x_j < b_i$$

Then using (5) we get $y_i = 0$.

Converse of (i)

If the jth primal variable is positive, then the jth constraint of the dual must be an equation. For the proof of it, we proceed as follows.

Clearly, at the optimum solution, (3) reduces to

$$\sum_{j=1}^{n} c_j x_j = \sum_{j=1}^{n} \sum_{i=1}^{m} a_{ij} y_i x_j$$

$$\Rightarrow \qquad \sum_{j=1}^{n} \left[c_j - \sum_{i=1}^{m} a_{ij} y_i \right] \cdot x_j = 0$$

$$\Rightarrow \qquad \left[c_j - \sum_{i=1}^{m} a_{ij} y_i \right] x_j = 0 \qquad \ldots(6)$$

Now since $x_j \ge 0$, therefore, for the consistency of (6) we must have

$$c_j - \sum_{i=1}^{m} a_{ij} y_i = 0$$

$$\Rightarrow \qquad \sum_{i=1}^{m} a_{ij} y_i = c_j, \text{ which is the } j^{\text{th}} \text{ constraint of the primal problem.}$$

Converse of (ii)

If the jth constraint in dual problem is a strict inequality, then j^{th}-primal variable must be zero.

To prove it, we proceed as follows:

∵ the jth constraint in dual problem is a strict inequality, *i.e.*,

$$\sum_{i=1}^{m} a_{ij} y_i > c_j$$

Then, for the consistency of (6) we must have

$$x_j = 0$$

☞ **REMARK**

• The above theorem can be restated as follows:

"A necessary and sufficient condition for any pair of feasible solutions to the primal and dual to the optimal is that $y_i x_{n+i} = 0, j = 1, 2, ..., m$ where x_{n+i} is the slack variable in the primal and $x_j y_{m+j} = 0, j = 1, 2, ..., n$ where y_{m+j} is the surplus variable for the dual."

6.6 METHOD FOR OBTAINING THE SOLUTION TO THE DUAL FROM THE FINAL SIMPLEX TABLE OF THE PRIMAL AND VICE-VERSA

To obtain the solution to the dual problem from final simplex table of the primal problem we proceed as follows:

(i) Firstly we write optimal solution of the primal problem from final simplex table which gives the optimal solution to its dual problem by using duality theorem in which we have proved that

$$\min Z_p = \max Z_d$$

(ii) Next, we write the value of $\Delta_j \; (= c_j - z_j)$ corresponding to the slack (or surplus) variables from final simplex table, these value of Δ_j with different sign give the values of dual variables.

(iii) If either problem has unbounded solution, then the other will have no feasible solution.

WORKING PROCEDURE

STEP 1.	Apply the simplex method to the problem (primal or dual) with lessor number of constraints.
STEP 2.	Read the solution of the other from the final simplex table according to the rule discussed above.

Solved Examples

EXAMPLE 1. *Find the solution of the following problem and its dual by simplex method. Also, read the solution of each problem from the final simplex table of the other*

$$max. \; Z = 40x_1 + 50x_2$$

subject to the constraints

$$2x_1 + 3x_2 \le 3$$
$$8x_1 + 4x_2 \le 5$$

and $\; x_1, x_2 \ge 0$

SOLUTION. Clearly, the given LPP of maximization is in standard form. Using slack variables s_1, s_2 the problem reduces to max. $Z = 40x_1 + 50x_2 + 0s_1 + 0s_2$

s.t. $\quad 2x_1 + 3x_2 + s_1 = 3$

$\qquad 8x_1 + 4x_2 + s_2 = 5$

and $\qquad x_1, x_2, s_1, s_2 \geq 0$

Apply the simplex method in a usual manner, we have the following simplex table.

Simplex table-1

B.V.	C_B	x_B	x_1	x_2	s_1	s_2	Min Ratio
		$c_j \rightarrow$	40	50	0	0	
s_1	0	3	2	③	1	0	$3/3 = 1$ (min)$\rightarrow$
s_2	0	5	8	4	0	1	$5/4$
$Z = C_B x_B = 0, \Delta_j \rightarrow$			40	50	0	0	
				$\uparrow$	$\downarrow$		
x_2	50	1	2/3	1	1/3	0	$3/2$
s_2	0	1	⑯/3	0	$-4/3$	1	$3/16$ (min)$\rightarrow$
$Z = C_B x_B = 50, \Delta_j \rightarrow$			20/3	0	$-50/3$	0	
			$\uparrow$			$\downarrow$	
x_2	50	7/8	0	1	1/2	$-1/8$	
x_1	40	3/16	1	0	$-1/4$	3/16	
$Z = C_B x_B = \dfrac{205}{4}, \Delta_j \rightarrow$			0	0	-15	$-5/4$	

We observe that in the last row of last table, $\Delta_j \leq 0$ for all j.

$\Rightarrow \quad$ solution is optimal and is given by

$$x_1 = \frac{3}{16}, x_2 = \frac{7}{8} \text{ and max. } Z = \frac{205}{4}$$

Dual Problem. The dual of the given primal problem is

$$\text{max. } Z_y = b' \cdot y = [3 \quad 5][y_1 \quad y_2] = 3y_1 + 5y_2$$

such that $\quad A'y \geq C' \quad$ or $\quad \begin{bmatrix} 2 & 8 \\ 3 & 4 \end{bmatrix}\begin{bmatrix} y_1 \\ y_2 \end{bmatrix} \geq \begin{bmatrix} 40 \\ 50 \end{bmatrix}$

$\Rightarrow \qquad 2y_1 + 8y_2 \geq 40$

$\qquad\qquad 3y_1 + 4y_2 \geq 50$

and $\qquad y_1, y_2 \geq 0$

Now, converting this dual problem to maximization problem and introducing surplus variables s_1, s_2 and artificial variables A_1, A_2, we get

$$\text{max.} Z_y' = -Z_y = -3y_1 - 5y_2 + 0s_1 + 0s_2 - MA_1 - MA_2$$

s.t. $\quad 2y_1 + 8y_2 - s_1 + A_1 = 40$

$\qquad 3y_1 + 4y_2 - s_2 + A_2 = 50$

and $\quad y_1, y_2, s_1, s_2, A_1, A_2 \geq 0$

Now, apply the simplex method in a usual manner, we have the following simplex table.

Simplex table-2

B.V.	C_B	x_B	$c_j \rightarrow$ y_1	y_2	s_1	s_2	A_1	A_2	Min Ratio
			-3	-5	0	0	$-M$	$-M$	
A_1	$-M$	40	2	⑧	-1	0	1	0	$40/8 = 5$ (Min) $\rightarrow$
A_2	$-M$	50	3	4	0	-1	0	1	$50/4$
$Z'_y = C_B x_B,$ $= -90M$		$\Delta_j \rightarrow$	$(5M - 3)$	$(12M-5)$ $\uparrow$	$-M$	$-M$ $\downarrow$	0	0	
y_2	-5	5	1/4	1	$-1/8$	0	—	0	20
A_2	$-M$	30	②	0	1/2	-1	—	1	$30/2 = 15$ (Min) $\rightarrow$
$Z'_y = -30M - 25\Delta_j \rightarrow$			$(8M - 7)/4$ $\uparrow$	0	$(4M-5)/8$	$-M$	—	0 $\downarrow$	
y_2	-5	5/4	0	1	$-3/16$	1/8	—	—	
y_1	-3	15	1	0	1/4	$-1/2$	—	—	
$Z'_y = -205/4, \Delta_j \rightarrow$			0	0	$-3/16$	$-7/8$			

Clearly, all $\Delta_j \leq 0 \Rightarrow$ solution is optimal and is given by

$$y_1 = 15, y_2 = \frac{5}{4} \text{ and min. } Z_y = -\max . Z'_y = \frac{205}{4}$$

How to Read the solution? In the final simplex table of the primal, negative of the values of $\Delta_j's$ for slack variables s_1 and s_2 are the values of y_1, y_2 of the dual variables.

i.e., $y_1 = -\Delta_3 = 15, y_2 = -\Delta_4 = \frac{5}{4}$

and in the final simplex table of the dual, negative of the values of $\Delta_j's$ for surplus variables are the values of x_1, x_2 of the primal, i.e.,

$$x_1 = -\Delta_3 = \frac{3}{16}, x_2 = -\Delta_4 = \frac{7}{8}, \max . Z = \frac{205}{4} = \min . Z_y$$

EXAMPLE 2. Write the dual of the following LPP and hence solve it.

max. $Z = 3x_1 - 2x_2$

subject to the constraints

$$x_1 \leq 4$$
$$x_2 \leq 6$$
$$x_1 + x_2 \leq 5$$
$$-x_2 \leq -1$$

and $x_1, x_2 \geq 0$ [MEERUT–2007, 08; DELHI–2011]

SOLUTION. The given LPP of maximization is in standard form.

Following the usual procedure, the dual of the given LPP is given by

$$\text{min. } Z_d = 4y_1 + 6y_2 + 5y_3 - y_4$$
$$\text{s.t.} \qquad y_1 + y_3 \geq 3$$
$$y_2 + y_3 - y_4 \geq -2$$
$$\text{and } y_1, y_2, y_3, y_4 \geq 0$$

Changing this dual problem to maximization, introduce surplus variable s_1 and slack variable s_2, the dual problem reduces to

$$\text{max.} Z'_d = -4y_1 - 6y_2 - 5y_3 + y_4 + 0s_1 + 0s_2$$
$$\text{s.t.} \qquad y_1 + y_3 - s_1 = 3$$
$$-y_2 - y_3 + y_4 + s_2 = 2$$
$$\text{and} \qquad y_i, s_i \geq 0$$

(Here, artificial variable is not used because in the first constraint y_1 will serve the purpose of unit matrix)

Now, apply the simplex method in a usual manner, we have the following simplex table.

Simplex table

		$c_j \rightarrow$	-4	-6	-5	1	0	0	
B.V.	C_B	x_B	y_1	y_2	y_3	y_4	s_1	s_2	Min Ratio
y_1	-4	3	1	0	1	0	-1	0	—
s_2	0	2	0	-1	-1	①	0	1	2 (Min) →
$Z'_d = C_B x_B,$ $= -12$		$\Delta_j \rightarrow$	0	-6	-1	1	-4	0	
						↑	↓		
y_1	-4	3	1	0	1	0	-1	0	
y_4	1	2	0	-1	-1	1	0	1	
$Z'_d = -10,$		$\Delta_j \rightarrow$	0	-5	0	0	-4	-1	

$\because$ in the last row of the above table, all $\Delta_j \leq 0$

$\Rightarrow$ solution is optimal for dual problem and is given by

$$y_1 = 3, y_2 = 0, y_3 = 0, y_4 = 2, \text{ min.} Z_d = -\text{max} Z'_d = 10$$

Solution of the primal: From the final simplex table of the dual the required solution of the primal problem is

$$x_1 = -\Delta_5 = 4; x_2 = -\Delta_6 = 1 \text{ and max. } Z = -\text{ min. } Z_d = 10$$

EXAMPLE 3. *Using the principle of duality, solve the following LPP*

$$\text{max. } Z = 2x_1 + x_2$$
$$\text{subject to the constraints}$$
$$x_1 + 2x_2 \leq 10$$
$$x_1 + x_2 \leq 6$$
$$x_1 - x_2 \leq 2$$
$$x_1 - 2x_2 \leq 1$$
$$\text{and} \qquad x_1, x_2 \geq 0 \qquad \text{[GORAKHPUR–2007; MEERUT–2009]}$$

SOLUTION. Since, the given LPP of maximization is in standard form, so following the usual procedure, the dual of the given primal problem is

$$\text{min } Z_d = 10y_1 + 6y_2 + 2y_3 + y_4$$

s.t.
$$y_1 + y_2 + y_3 + y_4 \geq 2$$
$$2y_1 + y_2 - y_3 - 2y_4 \geq 1$$
and
$$y_1, y_2, y_3, y_4 \geq 0$$

Now, changing this minimization problem into standard maximization and then introduce surplus variables s_1, s_2 and artificial variables A_1 and A_2, the dual problem becomes

$$\max. Z_d' = -Z_d' = -10y_1 - 6y_2 - 2y_3 - y_4 + 0s_1 + 0s_2 - MA_1 - MA_2$$

s.t.
$$y_1 + y_2 + y_3 + y_4 - s_1 + A_1 = 2$$
$$2y_1 + y_2 - y_3 - 2y_4 - s_2 + A_2 = 1$$

and
$$y_1, y_2, y_3, y_4, s_1, s_2, A_1, A_2 \geq 0$$

Now apply the usual procedure, we have the following simplex table.

Simplex table

B.V.	C_B	x_B	$c_j \rightarrow$ -10 y_1	-6 y_2	-2 y_3	-1 y_4	0 s_1	0 s_2	$-M$ A_1	$-M$ A_2	Min Ratio
A_1	$-M$	2	1	1	1	1	-1	0	1	0	2/1=2
A_2	$-M$	1	②	1	-1	-2	0	-1	0	1	1/2 (min)→
$Z_d' = -3M$,	$\Delta_j \rightarrow$		(3M−10) ↑	(2M−6)	-2	$(-M-1)$	$-M$	$-M$	0	0 ↓	
A_1	$-M$	3/2	0	1/2	3/2	②	-1	1/2	1	—	3/4 (min)→
y_1	-10	1/2	1	1/2	$-1/2$	-1	0	$-1/2$	0	—	−ve
$Z_d' = -\frac{1}{2}(3M+10), \Delta_j \rightarrow$			0	$\frac{-1+M}{2}$	$\frac{-7+3M}{2}$	$(-11+2M)$ ↑	$-M$	$\left(\frac{M}{2}-5\right)$ ↓	0	—	
y_4	-1	3/4	0	1/4	3/4	1	$-1/2$	1/4	—	—	3
y_1	-10	5/4	1	③/4	1/4	0	$-1/2$	$-1/4$	—	—	5/3 (min)→
$Z_d' = \frac{-53}{4}$,	$\Delta_j \rightarrow$		0	7/4 ↓	5/4 ↑	0	$\frac{-11}{2}$	$-9/4$	—	—	
y_4	-1	1/3	$-1/3$	0	②/3	1	$-1/3$	1/3	—	—	1/2(min)→
y_2	-6	5/3	4/3	1	1/3	0	$-2/3$	$-1/3$	—	—	5
$Z_d' = \frac{-31}{3}$	$\Delta_j \rightarrow$		$-7/3$	0	2/3 ↑	0 ↓	$-13/3$	$-5/3$			
y_3	-2	1/2	$-1/2$	0	1	3/2	$-1/2$	1/2			
y_2	-6	3/2	3/2	1	0	$-1/2$	$-1/2$	$-1/2$			
$Z_d' = -10$,	$\Delta_j \rightarrow$		-2	0	0	-1	-4	-2			

In the last row of the above table all $\Delta_j \leq 0$

$\Rightarrow$ solution of the dual is optimal and is given by

$$y_1 = 0, y_2 = \frac{3}{2}, y_3 = \frac{1}{2}, y_4 = 0$$

$$\min.Z_d = -\max.Z_d' = 10$$

Further, from the last table the solution of the primal problem is given by

$$x_1 = -\Delta_5 = 4, x_2 = -\Delta_6 = 2 \text{ and } \max.Z = \min Z_d = 10$$

EXAMPLE 4.   *Using the principle of duality, solve the following LPP*

$$max. \ Z = 3x_1 + 2x_2$$

subject to the constraints

$$x_1 + x_2 \geq 1$$
$$x_1 + x_2 \leq 7$$
$$x_1 + 2x_2 \leq 10$$
$$x_2 \leq 3$$
$$and \quad x_1, x_2 \geq 0$$

SOLUTION.   The given problem of maximization can be written in standard form as follows.

max. $Z = 3x_1 + 2x_2$

s.t. $-x_1 - x_2 \leq -1$

$\qquad\qquad x_1 + x_2 \leq 7$

$\qquad\qquad x_1 + 2x_2 \leq 10$

$\qquad\qquad\qquad x_2 \leq 3$

and $\qquad\qquad x_1, x_2 \geq 0$

The dual of the above primal problem is given by

$\qquad$ min. $Z_d = -y_1 + 7y_2 + 10y_3 + 3y_4$

$\qquad$ s.t. $\qquad\qquad -y_1 + y_2 + y_3 \geq 3$

$\qquad\qquad -y_1 + y_2 + 2y_3 + y_4 \geq 2$

$\qquad$ and $\qquad\qquad y_1, y_2, y_3, y_4 \geq 0$

To solve the above dual, first we change it into maximization problem and then introduce the surplus variables s_1, s_2 and artificial variable A_1, then it becomes

$\qquad$ max.$Z_d' = -Z_d = y_1 - 7y_2 - 10y_3 - 3y_4 + 0s_1 + 0s_2 - MA_1$

s.t. $\qquad\qquad -y_1 + y_2 + y_3 - s_1 + A_1 = 3$

$\qquad\qquad -y_1 + y_2 + 2y_3 + y_4 - s_2 = 2$

and $\qquad\qquad y_i, s_i \geq 0$

Now apply the simplex method in a usual manner, we have the following simplex table.

Simplex table

B.V.	C_B	x_B	$c_j \rightarrow$ 1 y_1	−7 y_2	−10 y_3	−3 y_4	0 s_1	0 s_2	−M A_1	Min Ratio
A_1	−M	3	−1	1	1	0	−1	0	1	3/1 =3
y_4	−3	2	−1	①	2	1	0	−1	0	2/1 =2 (min) →
$Z_d' = -3M - 6, \Delta_j \rightarrow$			(−M−2)	(M − 4) ↑	(M − 4)	0 ↓	− M	− 3	0	
A_1	−M	1	0	0	−1	−1	−1	①	1	1 (min)
y_2	−7	2	−1	1	2	1	0	−1	0	−ve
$Z_d' = -M - 14, \Delta_j \rightarrow$			−6	0	−M + 4	−M + 4	−M + 7 ↑	M − 7 ↓	0	
s_2	0	1	0	0	−1	−1	−1	1	—	
y_2	−7	3	−1	1	1	0	−1	0	—	
$Z_d' = -21, \quad \Delta_j \rightarrow$			−6	0	−3	−3	− 7	0	—	

In the last row of last table all $\Delta_j \leq 0 \Rightarrow$ solution of the dual is optimal and is given by

$y_1 = 0, y_2 = 3, y_3 = 0, y_4 = 0$ and $\min Z_d = -\max Z_d' = 21$

Also, from the last table, the optimal solution of the given primal problem is

$x_1 = -\Delta_5 = 7, x_2 = -\Delta_6 = 0$ and max. $Z = \min Z_d = 21$

EXAMPLE 5. *Write the dual of the following LPP:*

$$max. \ Z = 2x_1 + 3x_2$$

subject to the constraints

$$2x_1 + 2x_2 \leq 10$$
$$2x_1 + x_2 \leq 6$$
$$x_1 + 2x_2 \leq 6$$

$$and \qquad x_1, x_2 \geq 0$$

Solve the above primal and then find the solution of the dual.

SOUTION. Since the given primal is written in standard form so, the required dual is given by

min. $Z_d = 10y_1 + 6y_2 + 6y_3$

s.t. $2y_1 + 2y_2 + y_3 \geq 2$

$2y_1 + y_2 + 2y_3 \geq 3$

and $y_1, y_2, y_3 \geq 0$

Now, we have to solve the given primal problem.

Introducing the slack variables s_1, s_2, s_3, we get

max. $Z = 2x_1 + 3x_2 + 0s_1 + 0s_2 + 0s_3$

s.t. $2x_1 + 2x_2 + s_1 = 10$

$2x_1 + x_2 + s_2 = 6$

$$x_1 + 2x_2 + s_3 = 6$$

and $\qquad x_1, x_2, s_1, s_2, s_3 \geq 0$

Now apply the simplex method in a usual manner, we have the following simplex table.

Simplex table

B.V.	C_B	x_B	x_1	x_2	s_1	s_2	s_3	Min Ratio
		$c_j \rightarrow$	2	3	0	0	0	
s_1	0	10	2	2	1	0	0	5
s_2	0	6	2	1	0	1	0	6
s_3	0	6	1	②	0	0	1	3 (min) $\rightarrow$
$Z = C_B x_B = 0,\ \Delta_j \rightarrow$			2	3 $\uparrow$	0	0	0 $\downarrow$	
s_1	0	4	1	0	1	0	–1	4
s_2	0	3	③/2	0	0	1	–1/2	2 (min) $\rightarrow$
x_2	3	3	1/2	1	0	0	1/2	6
$Z = C_B x_B = 9,\ \Delta_j \rightarrow$			1/2 $\uparrow$	0	0	0	–3/2 $\downarrow$	
s_1	0	2	0	0	1	–2/3	–2/3	
x_1	2	2	1	0	0	2/3	–1/3	
x_2	3	2	0	1	0	–1/3	2/3	
$Z = C_B x_B = 10,\ \Delta_j \rightarrow$			0	0	0	–1/3	– 4/3	

We observe that in the last row of last table all $\Delta_j \leq 0$

$\Rightarrow$ solution of the primal is optimal and is given by $x_1 = 2$, $x_2 = 2$ and max. $Z = 10$

Solution of the dual: The solution of the dual is given by the negative Δ_3, Δ_4, Δ_5 which corresponds to s_1, s_2 and s_3 in the last simplex table.

$\therefore \qquad y_1 = -\Delta_3 = 0, y_2 = -\Delta_4 = \dfrac{1}{3}, y_3 = -\Delta_5 = \dfrac{4}{3}$ and min. Z_d = max. $Z = 10$

Summary of the Correspondance between the Primal and its Dual

The following table shows the various correspondance between the primal and dual problems.

S.No.	Primal (maximize)	Dual (minimize)
1.	Objective function max. Z_p	Objective function min Z_d
2.	Coefficient matrix $A = [a_{ij}]$	Transpose A' of A
3.	Requirement vector b	Price vector C
4.	i^{th} constraints $\leq$	i^{th} variable ≥ 0
5.	i^{th} constraints $\geq$	i^{th} variable ≥ 0
6.	i^{th} constraints $=$	i^{th} variable unrestricted
7.	i^{th} variable ≥ 0	i^{th} constraints $\geq$
8.	i^{th} variable ≤ 0	i^{th} constraints $\leq$
9.	i^{th} variable unrestricted	i^{th} constraints $=$
10.	Finite optimal solution	Equal finite optimal solution
11.	Feasible and unbounded solution	Infeasible solution
12.	Infeasible solution	Feasible and unbounded solution

Exercise-6.2

1. Find the solution of the following LPP and its dual by simplex method

min. $Z = x_1 - x_2$

s.t. $2x_1 + x_2 \geq 2$; $-x_1 - x_2 \geq 1$

and $x_1, x_2 \geq 0$

2. Write the dual of the following LPP and solve it

max. $Z = 4x_1 + 2x_2$

s.t. $-x_1 - x_2 \leq -3$,

$-x_1 + x_2 \leq -2$, $x_1, x_2 \geq 0$

3. Use principle of duality, solve the following LPP

max. $Z = 3x_1 + 2x_2$

s.t. $2x_1 + x_2 \leq 5$; $x_1 + x_2 \leq 3$

and $x_1, x_2 \geq 0$

Use principle of duality, solve the following LPP (Qus 4-7)

4. min $Z = 2x_1 + 2x_2$

s.t. $2x_1 + 4x_2 \geq 1$; $x_1 + 2x_2 \geq 1$

$2x_1 + x_2 \geq 1$

and $x_1, x_2 \geq 0$

5. min. $Z = 3x_1 + x_2$

s.t. $x_1 + x_2 \geq 1$

$2x_1 + 3x_2 \geq 2$

and $x_1, x_2 \geq 0$

6. max. $Z = 3x_1 - 2x_2$

s.t. $x_1 + x_2 \leq 5$, $x_1 \leq 4$, $1 \leq x_2 \leq 6$

and $x_1, x_2 \geq 0$

7. max. $Z = 4x_1 + 3x_2$

s.t. $x_1 \leq 6$; $x_2 \leq 8$; $x_1 + x_2 \leq 7$, $3x_1 + x_2 \leq 15$;

$-x_2 \leq 1$

and $x_1, x_2 \geq 0$

8. Solve the following primal and its dual by simplex method

max. $Z = 40x_1 + 35x_2$

s.t. $2x_1 + 3x_2 \leq 60$; $4x_1 + 3x_2 \leq 96$

and $x_1, x_2 \geq 0$

9. Solve the following primal problem by simplex method and deduce from it the solution to the dual problem

max. $Z = 2x_1 - x_2$

s.t. $x_1 + x_2 \leq 10$; $-2x_1 + x_2 \leq 2$

$4x_1 + 3x_2 \geq 12$

and $x_1, x_2 \geq 0$

10. Solve the following primal problem and its dual by simplex method

max. $Z = 5x_1 + 12x_2 + 4x_3$

subject to $x_1 + 2x_2 + x_3 \leq 5$;

$2x_1 - x_2 + 3x_3 = 2$

and $x_1, x_2 \geq 0$

11. Apply simplex method to solve the following LPP

max. $Z = 30x_1 + 23x_2 + 29x_3$

s.t. $6x_1 + 5x_2 + 3x_3 \leq 26$

$4x_1 + 2x_2 + 5x_3 \leq 7$

and $x_1, x_2, x_3 \geq 0$

Hence, or otherwise find the solution to the dual of the above problem.

12. Using the principle of duality, solve the following LPP

max. $Z = 5x_1 - 2x_2 + 3x_3$

s.t. $2x_1 + 2x_2 - x_3 \geq 2$

$3x_1 - 4x_2 \leq 3$

$x_1 + 2x_3 \leq 5$

and $x_1, x_2, x_3 \geq 0$

13. Using duality, solve the following LPP

min. $Z = 3x_1 - 2x_2 + 4x_3$

s.t. $3x_1 + 5x_2 + 4x_3 \geq 7$; $6x_1 + x_2 + 3x_3 \geq 4$;

$7x_1 - 2x_2 - x_3 \leq 10$; $x_1 - 2x_2 + 5x_3 \geq 3$;

$4x_1 + 7x_2 - 2x_3 \geq 2$

and $x_1, x_2, x_3 \geq 0$

14. Using duality, solve the following primal problem

min. $Z = 6x + 5y + 2z$

s.t. $x + 3y + 2z \geq 5$; $2x + 2y + z \geq 2$

$4x - 2y + 3z \geq -1$

and $x, y, z \geq 0$

15. Use duality theory, solve the following primal problem

min $Z = 10x_1 + 6x_2 + 2x_3$

s.t. $-x_1 + x_2 + x_3 \geq 1$; $3x_1 + x_2 - x_3 \geq 2$

and $x_1, x_2, x_3 \geq 0$

ANSWERS

1. No feasible solution for primal, unbounded solution for dual.

2. No finite optimal solution

3. $x_1 = 2, x_2 = 1$, max $Z = 8$

4. $x_1 = \dfrac{1}{3}, x_2 = \dfrac{1}{3}, \min Z = \dfrac{4}{3}$

5. $x_1 = 0, x_2 = 1, \min. Z = 1$

6. $x_1 = 4, x_2 = 1$ and max. $Z = 10$ **7.** $x_1 = 4, x_2 = 3$, max. $Z = 25$

8. For primal: $x_1 = 18, x_2 = 8$ and max. $Z = 1000$; **For dual:** $y_1 = \dfrac{10}{3}, y_2 = \dfrac{25}{3}$ and min. $Z = 1000$

9. For primal: $x_1 = 10, x_2 = 0$, max. $Z = 20$; **For dual:** $y_1 = 2, y_2 = 0, y_3 = 0$, min. $Z_d = 20$

10. For primal: $x_1 = \dfrac{9}{5}, x_2 = \dfrac{8}{5}, x_3 = 0$ and max. $Z = \dfrac{141}{5}$; **For dual:** $y_1 = \dfrac{29}{5}, y_2 = \dfrac{-2}{5}$, min. $Z_d = \dfrac{141}{5}$

11. For Primal: $x_1 = 0, x_2 = \dfrac{7}{2}, x_3 = 0$ and max. $Z = \dfrac{161}{2}$; **For dual:** $y_1 = 0, y_2 = \dfrac{23}{2}$, min. $Z_d = \dfrac{161}{2}$

12. $x_1 = \dfrac{23}{3}, x_2 = 5, x_3 = 0, \max Z = \dfrac{85}{3}$ **13.** No feasible solution

14. $x_1 = 0, x_2 = 0, x_3 = \dfrac{5}{2}$ and min. $Z = 5$ **15.** $x_1 = \dfrac{1}{4}, x_2 = \dfrac{5}{4}, x_3 = 0$ and min $Z = 10$

 **Glossary**

- **Unsymmetric Primal Problem:** If in the given LPP, all the constraints are equations, then it is called unsymmetric primal problem.

REVIEW QUESTIONS

1. Write a short note on Primal-dual relationship.
2. Define the dual of a linear programming problem.
3. Discuss the principle of duality in linear programming.
4. Write the general rule to find the solution of a primal and its dual by simplex method.
5. Write the significance of dual variable in a LP model.

MULTIPLE CHOICE QUESTIONS (CHOOSE THE MOST APPROPRIATE ONE)

1. The relation between the objective function of primal and dual problem of LPP is:
   [MEERUT–2013]
   (a) max. $Z_p = -\min Z_d$
   (b) max. $Z_p = \min Z_d$
   (c) max $Z_p = \min Z_d$
   (d) None of these

2. Dual Simplex method is applied to solve LP problems that start with: [MEERUT–2013]
   (a) Feasible solution
   (b) Both feasible and optimal
   (c) Infeasible solution
   (d) Infeasible solution and Optimal solution

3. In Dual Simplex method the key column is selected by calculating: [MEERUT–2013]
   (a) $\min\{(z_j - c_j)/x_{rj} : x_{rj} > 0\}$
   (b) $\max\{(z_j - c_j)/x_{rj} : x_{rj} < 0\}$
   (c) $\min\{(z_j - c_j)/x_{rj} : x_{rj} > 0\}$
   (d) $\max\{(z_j - c_j)/x_{rj} : x_{rj} \leq 0\}$

4. Dual of the dual is: [MEERUT–2013]
   (a) Dual
   (b) Primal
   (c) Either primal or dual
   (d) None of the above

5. For the dual problem and primal problem, the optimum value of the objective function is:
   [MEERUT–2013]
   (a) Zero
   (b) Different
   (c) Same
   (d) Both (a) and (b) are true

6. If dual has an Unbounded solution, then primal has: [MEERUT–2013]
   (a) An infeasible solution
   (b) A feasible solution
   (c) An unbounded solution
   (d) None of the above

7. If the given primal is a minimization problem then its dual is:

8. (a) minimize problem
   (b) maximize problem
   (c) (a) and (b) both
   (d) None of these

8. If the primal ith inequality, then in dual ith variable:
   (a) $w_i \leq 0$ (b) $w_i = 0$
   (c) $w_i \geq 0$ (d) None of these

9. If a primal of LPP has finite optimal then values of objective function for primal and dual:
   (a) unequal   (b) odd solution
   (c) equal   (d) none of these

10. In LPP if both the primal and dual problem have finite optimal solution and problem object function C prime and dual then:
    (a) $Z_p \cdot 2d$ (b) $Z_p - 2d$
    (c) $Z_p = 2d$ (d) None of these

11. If the problem is in standard primal form of minimization, all the constraints involve the sign:
    (a) $\geq$ (b) $\leq$
    (c) $=$ (d) None of these

12. The requirement vector of the primal is the price vector of the:
    (a) primal   (b) dual
    (c) non-dual   (d) none of these

13. If the primal has unbounded solution then its dual has:
    (a) either no solution or an unbounded solution
    (b) no solution
    (c) an unbounded solution
    (d) (b) and (c) both

14. Standard primal problem is of maximization sign of constraints:
    (a) $\geq$ (b) $=$
    (c) $\leq$ (d) none of these

15. The coefficient matrix of the dual is obtained

by transposing the coefficient matrix of the:
(a) dual   (b) primal
(c) both (a) and (b)   (d) none of these

16. If the primal problem has a finite feasible optimal solution then the dual has:
(a) finite optimal F.S.
(b) no solution
(c) either no solution or an Unbounded solution
(d) none of these

17. In dual simplex method, let the variable x_i leave and the variable x_j enter. Let $x_j > 0$. Then later on:
(a) x_j can become negative
(b) x_j will remain positive
(c) x_j may be negative or positive
(d) None of these

18. Requirement vector of the primal is the _____ of the dual.
(a) row vector   (b) price vector
(c) column vector   (d) none of these

19. If both the primal and dual problems have finite optimal solution and Zp, Zd are the optimal values of the objective functions of both problems respectively then we have:
(a) $Z_p > Z_d$ (b) $Z_p < Z_d$
(c) $Z_p = Z_d$ (d) None of these

20. If the primal problem has a unbounded feasible solution, then the dual problem has:
(a) a finite optimal feasible solution
(b) either no solution or an unbounded solution
(c) no solution
(d) none of these

21. If the primal has finite optimal solution the dual has finite optimal solution with:
(a) equal optimal value of objective function
(b) unbounded solution
(c) no equal optimal value of objective function
(d) none of these

22. The Necessary and Sufficient conditions for any LPP and its dual to have optimal solution is that both have:
(a) B.F.S.
(b) F.S.
(c) optimal Solution

(d) degenerate Solution

23. If the problem in standard primal form is of maximization, all the constraints of its dual involve the sign:
(a) ≥   (b) ≤
(c) =   (d) None of these

24. In standard primal problem of min., sign of constraints:
(a) ≥   (b) ≤
(c) =   (d) None of these

25. If the primal has Unbounded solution, then the dual has either No solution or:
(a) an unbounded solution
(b) bounded solution
(c) feasible solution
(d) basic feasible solution

26. If a primal has optimal feasible solution, then its dual has:
(a) finite optimal solution
(b) optimal
(c) (a) and (b) both
(d) none of these

27. If the ith variable of the primal is positive then ith variable of the dual is:
(a) +ve   (b) −ve
(c) zero   (d) unrestricted

28. If primal coefficient matrix A, then dual transpose of the coefficient matrix:
(a) A (b) A' or A^T
(c) A^{-1} (d) none of these

29. If the problem is standard primal form is of maximization, all the constraints involve the sign:
(a) ≥   (b) ≤
(c) =   (d) none of these

30. The optimal values of the objective function of primal and dual problems if exists are:
(a) one   (b) two
(c) zero   (d) equal

31. If primal/dual problem has a Unbounded optimal solution, the other problem has either _____ or _____.
(a) solution, bounded solution
(b) no solution, bounded solution
(c) no solution, unbounded solution
(d) none of these

ANSWERS

1. (b)	2. (b)	3. (b)	4. (b)	5. (d)	6. (c)	7. (b)	8. (c)	9. (c)
10. (c)	11. (a)	12. (b)	13. (a)	14. (c)	15. (b)	16. (a)	17. (a)	18. (b)
19. (c)	20. (b)	21. (a)	22. (b)	23. (a)	24. (a)	25. (a)	26. (a)	27. (c)
28. (b)	29. (b)	30. (d)	31. (c)					

Dual Simplex Method and Primal-Dual Algorithm

7

7.1 INTRODUCTION

It is a well known fact that simplex method deals with a basic feasible solution and simplex algorithm is terminated after achieving the optimal solution. In this algorithm, the procedure will be stopped when all $\Delta_j \leq 0$ for maximization and $\Delta_j \geq 0$ for minimization problem. But there are some situations when one or more solution variable x_B are negative and optimality condition is satisfied, then current optimal solution may not be feasible. In such type of cases it is possible to find a starting basic but not feasible solution that is dual feasible, *i.e.*, all Δ_j $(= c_j - z_j) \leq 0$ for a maximization problem. In such cases, a type of simplex method is called dual-simplex method.

☞ REMARK

- In the dual simplex method, we always attempt to retain optimality while bringing the primal back to feasibility (*i.e.*, $x_{B_i} \geq 0$ for all i)

7.2 DUAL-SIMPLEX ALGORITHM

Dual simplex method is the technique which deals with only slack variables, while in simplex method we use surplus as well as artificial variables in some cases. Dual simplex method starts with a basic optimal solution not necessarily feasible of the primal and decrease the number of negative variable iteratively step by step maintaining the optimality criterion at each iteration. This method was developed by **C.E. Lemke**.

The steps of dual-simplex algorithm may be given as follows:

WORKING PROCEDURE

STEP 1. Convert the given LPP into standard form (maximization with $\leq$ constraints)

STEP 2. Introduce slack variables whenever needed.

STEP 3. Obtain the initial basic feasible solution by putting all given variables equal to zero.

Let $x_B = (x_{B_1}, x_{B_2}, \ldots, x_{B_n})$ be the initial basic feasible solution.

We observe that if all $x_{B_i} \geq 0$ then this feasible solution is optimal but if at least one of x_{B_i} is negative, then go to the next step.

STEP 4. Construct starting simplex table as usual done in simplex table.

STEP 5. If $x_{B_r} = \min\{x_{B_i} : x_{B_i} < 0\}$ then β_r is the outgoing vector.

STEP 6. If $\dfrac{\Delta_k}{a_{rk}} = \min_j \left\{ \dfrac{\Delta_j}{a_{rj}}, a_{rj} < 0 \right\}$ then α_k is the incoming vector, where $\{\alpha_1, \alpha_2, \ldots, \alpha_{n-m}\}$ is the coefficient matrix.

If all $a_{rj} \geq 0$, $j = 1, 2, \ldots, n$ then primal solution is dual bounded (*i.e.*, infeasible).

STEP 7. To check the optimality, replace β_r by α_k by applying row operation and all basic variables reduce to non-negative values. This solution is the optimal feasible solution. If atleast one basic variable is negative, then repeat the step 5, 6 and 7 iteratively until an optimal feasible solution is achieved.

 Solved Examples

EXAMPLE 1. *Apply dual simplex method, solve the following LPP*

$$\min. Z = 5x_1 + 6x_2$$
$$\text{subject to the constraints}$$
$$x_1 + x_2 \geq 2$$
$$4x_1 + x_2 \geq 4$$
$$and \qquad x_1, x_2 \geq 0$$

SOLUTION. Write the given LPP into standard form as follows

$$\left.\begin{array}{l} \max Z' = -5x_1 - 6x_2 \\ \text{s.t.} \quad -x_1 - x_2 \leq -2 \\ \qquad -4x_1 - x_2 \leq -4 \\ \text{and} \qquad x_1, x_2 \geq 0 \end{array}\right] \qquad ...(1)$$

We observe that all $c_j \leq 0$ and $b_i \leq 0$, so it is possible to obtain the solution by dual simplex method.

Introducing slack variables s_1 and s_2, (1) becomes

$$\max Z' = -5x_1 - 6x_2 + 0s_1 + 0s_2$$
$$\text{s.t.} \qquad -x_1 - x_2 + s_1 = -2$$
$$\qquad -4x_1 - x_2 + s_2 = -4$$
$$\text{and} \qquad x_1, x_2, s_1, s_2 \geq 0$$

Now, we have the following initial simplex table

Simplex table-1

B.V.	C_B	x_B	$c_j \rightarrow$ $x_1(\alpha_1)$	-6 $x_2(\alpha_2)$	0 $s_1(\beta_1)$	0 $s_2(\beta_2)$
			-5			
s_1	0	-2	-1	-1	1	0
s_2	0	-4	$\boxed{-4}$	-1	0	1

To find outgoing vector: We have

$$x_{B_1} = -2, x_{B_2} = -4$$
$$\Rightarrow \qquad \min\{x_{B_1}, x_{B_2}\} = \min\{-2, -4\} = -4 = x_{B_2}$$
$$\Rightarrow \qquad\qquad r = 2$$
$$\Rightarrow \qquad \beta_2 \text{ is the outgoing vector.}$$

To find incoming vector: We have

$$\frac{\Delta_k}{a_{rk}} = \min_{j} \left\{ \frac{\Delta_j}{a_{rj}}, a_{rj} < 0 \right\}$$

$$\Rightarrow \quad \frac{\Delta_k}{a_{2k}} = \min_j \left\{ \frac{\Delta_1}{a_{21}}, \frac{\Delta_2}{a_{22}} \right\} = \min\left\{ \frac{-5}{-4}, \frac{-6}{-1} \right\}$$

$$= \frac{5}{4} = \frac{\Delta_1}{a_{21}}$$

$$\Rightarrow \quad k = 1$$

$\Rightarrow$ α_1 is the incoming vector.
which gives the key element $= a_{22} = -4$
Now, we have the next simplex table.

Simplex table-2

		$c_j \rightarrow$	-5	-6	0	0
B.V.	C_B	x_B	$x_1(\beta_2)$	$x_2(\alpha_2)$	$s_1(\beta_1)$	$s_2(\alpha_4)$
s_1	0	-1	0	$-3/4$	1	($-1/4$)
x_1	-5	1	1	$1/4$	0	$-1/4$
$Z' = C_B x_B = -5,\ \Delta_j \rightarrow$			0	$-19/4$	0	$-5/4$
					$\downarrow$	$\uparrow$

Here, we observe that $x_1 = 1, x_2 = 0, s_1 = -1, s_2 = 0$ which is not feasible and all $\Delta_j \le 0$. Therefore, this solution is optimal but not feasible solution. Hence, this solution can be improved further as follows.

To find outgoing vector: Here,

$$x_{B_1} = -1, x_{B_2} = +1$$

Clearly, $x_{B_1} < 0$, therefore,

$$\min\{x_{B_1}, x_{B_2}\} = \min\{-1, 1\} = -1 = x_{B_1}$$

$\Rightarrow$ β_1 is the outgoing vector.

To find incoming vector:

Since, $$\frac{\Delta_k}{a_{rk}} = \min_j \left\{ \frac{\Delta_j}{a_{rj}}, a_{rj} < 0 \right\} = \min\left\{ \frac{\Delta_2}{a_{12}}, \frac{\Delta_4}{a_{14}} \right\}$$

$$= \min\left\{ \frac{-19/4}{-3/4}, \frac{-5/4}{-1/4} \right\} = \min\left\{ \frac{19}{3}, 5 \right\} = 5 = \frac{\Delta_4}{a_{14}}$$

$\Rightarrow$ $\alpha_4 (= s_2)$ is the incoming vector.
So, we have the following simpex table.

Simplex table-3

		$c_j \rightarrow$	-5	-6	0	0
B.V.	C_B	x_B	x_1	x_2	s_1	s_2
s_2	0	4	0	3	-4	1
x_1	-5	2	1	1	-1	0
$Z' = C_B x_B = -10,\ \Delta_j \rightarrow$			0	-1	-5	0

From the above table we observe that all $\Delta_j \le 0$, $x_1 = 2, x_2 = 0, s_1 = 0, s_2 = 4$, therefore, the obtained solution is feasible and optimal and is given by

$$x_1 = 2, x_2 = 0, \min Z = -\max. Z' = 10$$

EXAMPLE 2. *Use dual simplex method, solve the following LPP*

$$max.\ Z = -3x_1 - x_2$$

subject to the constraints

$$x_1 + x_2 \geq 1$$
$$2x_1 + 3x_2 \geq 2$$

and $\qquad x_1, x_2 \geq 0$ [MEERUT–2006; GORAKHPUR–2008, 11]

SOLUTION. The given LPP can be written in standard form as follows:

$$max.\ Z = -3x_1 - x_2$$
$$s.t. \qquad -x_1 - x_2 \leq -1$$
$$-2x_1 - 3x_2 \leq -2$$
$$and \qquad x_1, x_2 \geq 0$$

Clearly, the above problem is of maximization and all $c_j \leq 0$, so, dual simplex method is applicable.

Introducing slack variables s_1, s_2, the above LPP becomes

$$max.\ Z = -3x_1 - x_2 + 0s_1 + 0s_2$$
$$s.t. \qquad -x_1 - x_2 + s_1 = -1$$
$$-2x_1 - 3x_2 + s_2 = -2$$
$$and \qquad x_1, x_2, s_1, s_2 \geq 0$$

The initial basic solution of this LPP is given by

$$x_1 = x_2 = 0, s_1 = -1, s_2 = -2$$

Simplex table-1

B.V.	C_B	x_B	$x_1(\alpha_1)$	$x_2(\alpha_2)$	$s_1(\beta_1)$	$s_2(\beta_2)$
	$c_j \rightarrow$		-3	-1	0	0
s_1	0	-1	-1	-1	1	0
s_2	0	-2	-2	-3	0	1
$Z = C_B x_B = 0,$		$\Delta_j \rightarrow$	-3	-1	0	0
				$\uparrow$		$\downarrow$

To find outgoing vector:

$\because \quad x_{B_1} = -1, x_{B_2} = -2 \quad \Rightarrow \quad min\{x_{B_1}, x_{B_2}\} = min\{-1, -2\} = -2 = x_{B_2}$

$\Rightarrow \quad \beta_2$ is the outgoing vector.

To find incoming vector: We have

$$\frac{\Delta_k}{a_{2k}} = \min_j \left\{ \frac{\Delta_1}{a_{2j}}, a_{2j} < 0 \right\} = min \left\{ \frac{\Delta_1}{a_{21}}, \frac{\Delta_2}{a_{22}} \right\}$$

$$= min \left\{ \frac{-3}{-2}, \frac{-1}{-3} \right\} = \frac{1}{3} = \frac{\Delta_2}{a_{22}}$$

$\Rightarrow \quad k = 2$, *i.e.*, α_2 is the incoming vector.

So, key element $= a_{22} = -3$

Further, following the usual procedure, we have the following simplex table.

Simplex table-2

	$c_j \rightarrow$		-3	-1	0	0
B.V.	C_B	x_B	$x_1(\alpha_1)$	$x_2(\beta_2)$	$s_1(\beta_1)$	$s_2(\alpha_4)$
s_1	0	$-1/3$	$-1/3$	0	1	$\boxed{-1/3}$
x_2	-1	$2/3$	$2/3$	1	0	$-1/3$
$Z' = -2/3,$		$\Delta_j \rightarrow$	$-7/3$	0	0	$-1/3$
					$\downarrow$	$\uparrow$

Here, we observe that, the solution given by

$$x_1 = 0, s_1 = -\frac{1}{3} \text{ is infeasible but } \Delta_j \le 0.$$

$\Rightarrow$ it can be improved further.

To find outgoing vector:

$$x_{B_1} = -\frac{1}{3}, x_{B_2} = \frac{2}{3}$$

$\Rightarrow$ $\min\{x_{B_1}, x_{B_2}\} = \min\left\{-\frac{1}{3}, \frac{2}{3}\right\} = -\frac{1}{3} = x_{B_1}$

$\Rightarrow$ β_1 is the outgoing vector.

To find incoming vector:

Since, $\dfrac{\Delta_k}{a_{rk}} = \min\limits_j \left\{\dfrac{\Delta_j}{a_{rj}}, a_{rj} < 0\right\}$

$\therefore$ $\dfrac{\Delta_k}{a_{1k}} = \min\limits_j \left\{\dfrac{\Delta_j}{a_{1j}}, a_{1j} < 0\right\} = \min\left\{\dfrac{\Delta_1}{a_{11}}, \dfrac{\Delta_4}{a_{14}}\right\} = \min\left\{\dfrac{-7/3}{-1/3}, \dfrac{-1/3}{-1/3}\right\} = 1 = \dfrac{\Delta_4}{a_{14}}$

$\Rightarrow$ $\alpha_4(= s_2)$ is the incoming vector

And key element $= -\dfrac{1}{3} = a_{14}$

Simplex table-3

	$c_j \rightarrow$		-3	-1	0	0
B.V.	C_B	x_B	$x_1(\alpha_1)$	$x_2(\beta_2)$	$s_1(\alpha_3)$	$s_2(\beta_1)$
s_2	0	1	1	0	-3	1
x_2	-1	1	1	1	-1	0
$Z' = C_B x_B = -1,$		$\Delta_j \rightarrow$	-2	0	-1	0

Here, we observe that all $\Delta_j \le 0$ and $x_1 = 0, x_2 = 1, s_1 = 0, s_2 = 1$.

$\Rightarrow$ Feasible solution is optimal and is given by

$$x_1 = 0, x_2 = 1 \text{ and max. } Z = -1$$

EXAMPLE 3.   *Using dual simplex method, solve the following LPP*

$$\min. Z = 6x_1 + 7x_2 + 3x_3 + 5x_4$$

subject to the constraints

$$5x_1 + 6x_2 - 3x_3 + 4x_4 \ge 12$$
$$x_2 + 5x_3 - 6x_4 \ge 10$$
$$2x_1 + 5x_2 + x_3 + x_4 \ge 8$$

and $$x_1, x_2, x_3, x_4 \ge 0$$

[MEERUT–2006]

SOLUTION. First of all convert the given LPP into standard maximization form as follows.

$$\text{max. } Z' = -6x_1 - 7x_2 - 3x_3 - 5x_4$$

s.t.
$$-5x_1 - 6x_2 + 3x_3 - 4x_4 \leq -12$$
$$-x_2 - 5x_3 + 6x_4 \leq -10$$
$$-2x_1 - 5x_2 - x_3 - x_4 \leq -8$$

and
$$x_1, x_2, x_3, x_4 \geq 0$$

Proceeding as usual, by introducing slack variables, we have

$$\text{max. } Z' = -6x_1 - 7x_2 - 3x_3 - 5x_4 + 0s_1 + 0s_2 + 0s_3$$

s.t.
$$-5x_1 - 6x_2 + 3x_3 - 4x_4 + s_1 = -12$$
$$-x_2 - 5x_3 + 6x_4 + s_2 = -10$$
$$-2x_1 - 5x_2 - x_3 - x_4 + s_3 = -8$$

and
$$x_1, x_2, x_3, x_4, s_1, s_2, s_3 \geq 0$$

The initial basic solution is
$$x_1 = 0, x_2 = 0, x_3 = 0, x_4 = 0, s_1 = -12, s_2 = -10, s_3 = -8$$

which is not feasible.

Proceeding in a usual manner, we have the following simplex table.

		$c_j \rightarrow$	-6	-7	-3	-5	0	0	0
B.V.	C_B	x_B	$x_1(\alpha_1)$	$x_2(\alpha_2)$	$x_3(\alpha_3)$	$x_4(\alpha_4)$	$s_1(\beta_1)$	$s_2(\beta_2)$	$s_3(\beta_3)$
s_1	0	-12	-5	$\boxed{-6}$	3	-4	1	0	0
s_2	0	-10	0	-1	-5	6	0	1	0
s_3	0	-8	-2	-5	-1	-1	0	0	1
$Z' = C_B x_B = 0$,		$\Delta_j \rightarrow$	-6	-7	-3	-5	0	0	0
				$\uparrow$				$\downarrow$	

We observe that, all $\Delta_j \leq 0$ in the above table and
$$x_{B_1} = -12, x_{B_2} = -10, x_{B_3} = -8$$

To find outgoing vector:

Since,
$$x_{B_r} = \min_i \left\{ x_{B_i}, x_{B_i} < 0 \right\}$$

Therefore,
$$x_{B_r} = \min\{ x_{B_1}, x_{B_2}, x_{B_3} \}$$
$$= \min\{-12, -10, -8\} = x_{B_1} = -12$$

$\Rightarrow \quad \beta_1$ is the outgoing vector.

To find the incoming vector:

Since,
$$\frac{\Delta_k}{\alpha_{rk}} = \min_j \left\{ \frac{\Delta_j}{a_{rj}}, a_{rj} < 0 \right\}$$

Therefore,
$$\frac{\Delta_k}{a_{1k}} = \min \left\{ \frac{\Delta_1}{a_{11}}, \frac{\Delta_2}{a_{12}}, \frac{\Delta_4}{a_{14}} \right\} \qquad \{\because a_{13} > 0\}$$

$$= \min \left\{ \frac{-6}{-5}, \frac{-7}{-6}, \frac{-5}{-4} \right\} = \min \left\{ \frac{6}{5}, \frac{7}{6}, \frac{5}{4} \right\}$$

$$= \frac{7}{6} = \frac{\Delta_2}{a_{12}}$$

$\Rightarrow \quad \alpha_2$ is the incoming vector.

$\therefore \quad$ Key element $= a_{12} = -6$

Simplex table-2

B.V.	C_B	x_B	$x_1(\alpha_1)$	$x_2(\beta_1)$	$x_3(\alpha_3)$	$x_4(\alpha_4)$	s_1	$s_2(\beta_2)$	$s_3(\beta_3)$
$c_j \rightarrow$			-6	-7	-3	-5	0	0	0
x_2	-7	2	$5/6$	1	$-1/2$	$2/3$	$-1/6$	0	0
s_2	0	-8	$5/6$	0	$\boxed{-11/2}$	$20/3$	$-1/6$	1	0
s_3	0	2	$13/6$	0	$-7/2$	$7/3$	$-5/6$	0	1
$Z' = C_B x_B = -14, \Delta_j \rightarrow$			$-1/6$	0	$-13/2$	$-1/3$	$-7/6$	0	0
					$\uparrow$			$\downarrow$	

Here, we observe that all $\Delta_j \leq 0$ but

$x_1 = 0, x_2 = 2, x_3 = 0, x_4 = 0, s_1 = 0, s_2 = -8, s_3 = 2$

$\Rightarrow$ solution is optimal but not feasible.

Further, $x_{B_1} = 2, x_{B_2} = -8, x_{B_3} = 2$

$\Rightarrow \qquad\qquad x_{B_2} < 0$

To find outgoing vector:

$$x_{B_r} = \min\{x_{B_1}, x_{B_2}, x_{B_3}\} = x_{B_2}$$

$\Rightarrow \qquad\qquad r = 2$

$\Rightarrow \quad \beta_2$ is the outgoing vector.

To find incoming vector:

Since, $\qquad\qquad \dfrac{\Delta_k}{a_{rk}} = \min_j\left\{\dfrac{\Delta_j}{a_{rj}}, a_{rj} < 0\right\}$

Therefore, $\qquad \dfrac{\Delta_k}{a_{2k}} = \min_j\left\{\dfrac{\Delta_j}{a_{2j}}, a_{2j} < 0\right\} = \min\left\{\dfrac{\Delta_3}{a_{23}}, \dfrac{\Delta_5}{a_{25}}\right\} \quad (\because a_{23} < 0, a_{25} < 0)$

$$= \min\left\{\dfrac{-15/2}{-11/2}, \dfrac{-7/6}{-1/6}\right\} = \min\left\{\dfrac{13}{11}, 7\right\} = \dfrac{13}{11} = \dfrac{\Delta_3}{a_{23}}$$

$\Rightarrow \quad k = 3$, i.e., α_3 is the incoming vector.

and $\qquad$ key element $= a_{23} = \dfrac{-11}{2}$

Simplex table-3

B.V.	C_B	x_B	$x_1(\alpha_1)$	$x_2(\alpha_2)(\beta_1)$	$x_3(\alpha_3)(\beta_2)$	$x_4(\alpha_4)$	s_1	s_2	$s_3(\beta_3)$
$c_j \rightarrow$			-6	-7	-3	-5	0	0	0
x_2	-7	$30/11$	$25/33$	1	0	$2/33$	$-5/33$	$-1/11$	0
x_3	-3	$16/11$	$-5/33$	0	1	$-40/33$	$1/33$	$-2/11$	0
s_3	0	$8/11$	$18/33$	0	0	$-21/33$	$-8/11$	$-7/11$	1
$Z' = -258/11, \Delta_j \rightarrow$			$-38/33$	0	0	$-271/33$	$-32/33$	$-13/11$	0

Here, we observe that all $\Delta_j \leq 0$. So, solution is optimal and feasible and is given by

$$x_1 = 0, x_2 = \frac{30}{11}, x_3 = \frac{16}{11}, x_4 = 0$$

and $\min. Z = -\max Z' = \frac{258}{11}$

EXAMPLE 4. *Using dual simplex method, solve the following LPP*

$$\max. Z = -2x_1 - x_3$$
$$subject\ to\ the\ constraints$$
$$x_1 + x_2 - x_3 \geq 5$$
$$x_1 - 2x_2 + 4x_3 \geq 8$$
$$and \qquad x_1, x_2, x_3 \geq 0$$

SOLUTION. Firstly, we shall write the given LPP into standard form as follows:

$$\max. Z = -2x_1 + 0x_2 - x_3$$
$$\text{s.t.} \qquad -x_1 - x_2 + x_3 \leq -5$$
$$-x_1 + 2x_2 - 4x_3 \leq -8$$
$$and \qquad x_1, x_2, x_3 \geq 0$$

Now, introducing slack variables s_1, s_2 in the above problem, we get

$$\max. Z = -2x_1 + 0x_2 - x_3 + 0s_1 + 0s_2$$
$$\text{s.t.} \qquad -x_1 - x_2 + x_3 + s_1 = -5$$
$$-x_1 + 2x_2 - 4x_3 + s_2 = -8$$
$$and \qquad x_1, x_2, x_3, s_1, s_2 \geq 0$$

Clearly, the initial basic solution is

$$x_1 = 0, x_2 = 0, x_3 = 0, s_1 = -5, s_2 = -8$$

Simplex table-1

		$c_j \rightarrow$	-2	0	-1	0	0
B.V.	C_B	x_B	$x_1(\alpha_1)$	$x_2(\alpha_2)$	$x_3(\alpha_3)$	$s_1(\beta_1)$	$s_2(\beta_2)$
s_1	0	-5	-1	-1	1	1	0
s_2	0	-8	-1	2	-4	0	1
$Z' = C_B x_B = 0$,		$\Delta_j \rightarrow$	-2	0	-1	0	0
					$\uparrow$		$\downarrow$

To find outgoing vector: We have

$$x_{B_r} = \min\{x_{B_1}, x_{B_2}\} = \min\{-5, -8\} = -8 = x_{B_2}$$

$$\Rightarrow \qquad r = 2$$

$\Rightarrow \quad \beta_2$ is the outgoing vector.

To find incoming vector:

Using

$$\frac{\Delta_k}{a_{rk}} = \min_j \left\{ \frac{\Delta_j}{a_{rj}}, a_{rj} < 0 \right\}$$

We get

$$\frac{\Delta_k}{a_{2k}} = \min_j \left\{ \frac{\Delta_j}{a_{rj}}, a_{rj} < 0 \right\}$$

$$= \min\left\{\frac{\Delta_1}{a_{21}},\frac{\Delta_3}{a_{23}}\right\} = \min\left\{\frac{-2}{-1},\frac{-1}{-4}\right\} = \min\left\{2,\frac{1}{4}\right\} = \frac{1}{4} = \frac{\Delta_3}{a_{23}}$$

$\Rightarrow \qquad k = 3$

$\Rightarrow \quad \alpha_3$ is the incoming vector.

and $\qquad$ key element $= a_{23} = -4$

Simplex table-2

B.V.	C_B	x_B	$x_1(\alpha_1)$	$x_2(\alpha_2)$	$x_3(\beta_2)$	$s_1(\beta_1)$	s_2
		$c_j \rightarrow$	-2	0	-1	0	0
s_1	0	-7	$-5/4$	⊝$-1/2$	0	1	$1/4$
x_3	-1	2	$1/4$	$-1/2$	1	0	$-1/4$
$Z' = C_B x_B = -2,$		$\Delta_j \rightarrow$	$-7/4$	$-1/2$	0	0	$-1/4$
				$\uparrow$		$\downarrow$	

Here, we observe that all $\Delta_j \leq 0$ and $x_1 = x_2 = 0, x_3 = 2, s_1 = -7, s_2 = 0$ which is optimal but not feasible.

To find outgoing vector:

$$x_{B_r} = \min\{x_{B_1}, x_{B_2}\} = \min\{-7, 2\} = -7 = x_{B_1}$$

$\Rightarrow \qquad r = 1$

$\Rightarrow \quad \beta_1$ is the outgoing vector.

To find incoming vector:

Using $\qquad \dfrac{\Delta_k}{a_{rk}} = \min_j\left\{\dfrac{\Delta_j}{a_{rj}}, a_{rj} < 0\right\}$

We get $\qquad \dfrac{\Delta_k}{a_{1k}} = \min_j\left\{\dfrac{\Delta_j}{a_{1j}}, a_{1j} < 0\right\} = \min\left\{\dfrac{\Delta_1}{a_{11}},\dfrac{\Delta_2}{a_{12}}\right\}$

$$= \min\left\{\frac{-7/4}{-5/4},\frac{-1/2}{-1/2}\right\} = \min\left\{\frac{7}{5},1\right\} = 1 = \frac{\Delta_2}{a_{12}}$$

$\Rightarrow \qquad k = 2$

$\Rightarrow \quad \alpha_2$ is the incoming vector and key element $= a_{12} = \dfrac{1}{2}$

Simplex table-3

B.V.	C_B	x_B	$x_1(\alpha_1)$	$x_2(\alpha_2)$	$x_3(\beta_2)$	s_1	s_2
		$c_j \rightarrow$	-2	0	-1	0	0
x_2	0	14	$5/2$	1	0	-2	$-1/2$
x_3	-1	9	$3/2$	0	1	-1	$-1/2$
$Z' = C_B x_B = -9,$		$\Delta_j \rightarrow$	$-1/3$	0	0	-1	$-1/2$

Finally, we observe that all $\Delta_j \leq 0$ and $x_1 = 0, x_2 = 14, x_3 = 9, s_1 = s_2 = 0$.

$\Rightarrow \quad$ solution is optimal and feasible and is given by

$\qquad x_1 = 0, x_2 = 14, x_3 = 9$ and max. $Z = -9$

Exercise-7.1

Solve the following LPP by dual simplex method.

1. min. $Z = 2x_1 + x_2$

s.t.

$3x_1 + 2x_2 \geq 3$
$4x_1 + 3x_2 \geq 6$
$x_1 + 2x_2 \geq 3$
and $x_1, x_2 \geq 0$

2. max. $Z = 2x_1 + 3x_2$

s.t.

$2x_1 - x_2 - x_3 \geq 3$
$x_1 - x_2 + x_3 \geq 2$
and $x_1, x_2, x_3 \geq 0$

3. max. $Z = x_1 + 2x_2$

s.t.

$3x_1 + 2x_2 \geq 6$
$x_1 + 6x_2 \geq 3$
and $x_1, x_2 \geq 0$

4. max. $Z = -3x_1 - 2x_2$

s.t.

$x_1 + x_2 \geq 1$
$x_1 + x_2 \leq 7$
$x_1 + 2x_2 \geq 10$
$x_2 \leq 3$

and $x_1, x_2 \geq 0$

5. min. $Z = x_1 + 2x_2$

s.t.

$2x_1 + x_2 \geq 4$
$x_1 + 7x_2 \geq 7$
and $x_1, x_2 \geq 0$

6. min. $Z = x_1 + 2x_2 + 3x_3$

s.t.

$2x_1 - x_2 + x_3 \geq 4$
$x_1 + x_2 + 2x_3 \leq 8$
$x_2 - x_3 \geq 2$
and $x_1, x_2, x_3 \geq 0$

7. min. $Z = 2x_1 + 2x_2 + 4x_3$

s.t.

$2x_1 + 3x_2 + 5x_3 \geq 2$
$3x_1 + x_2 + 7x_3 \leq 3$
$x_1 + 4x_2 + 6x_3 \leq 5$
and $x_1, x_2, x_3 \geq 0$

8. min. $Z = 3x_1 + 2x_2 + x_3 + 4x_4$

s.t.

$2x_1 + 4x_2 + 5x_3 + x_4 \geq 10$
$3x_1 - x_2 + 7x_3 - 2x_4 \geq 2$
$5x_1 + 2x_2 + x_3 + 6x_4 \geq 15$
and $x_1, x_2, x_3, x_4 \geq 0$

ANSWERS

1. $x_1 = \dfrac{3}{5}, x_2 = \dfrac{6}{5}, \min. Z = \dfrac{12}{5}$

2. $x_1 = \dfrac{5}{3}, x_2 = 0, x_3 = \dfrac{1}{3}, \max. Z = \dfrac{13}{3}$

3. $x_1 = \dfrac{15}{8}, x_2 = \dfrac{3}{16}, \max Z = \dfrac{9}{4}$

4. $x_1 = 4, x_2 = 3$ and $\max. Z = -18$

5. $x_1 = \dfrac{21}{13}, x_2 = \dfrac{10}{13}, \max. Z = \dfrac{41}{13}$

6. $x_1 = 3, x_2 = 2, x_3 = \dfrac{1}{3}, \max. Z = 7$

7. $x_1 = 0, x_2 = \dfrac{2}{3}, x_3 = 0, \min. Z = \dfrac{4}{3}$

8. $x_1 = \dfrac{65}{23}, x_2 = 0, x_3 = \dfrac{20}{23}, x_4 = 0, \min. Z = \dfrac{215}{23}$

7.3 PRIMAL-DUAL ALGORITHM

In dual simplex method, we observe that in some situations the necessity of introducing artificial variables removed. But in general, it is not always easy to find a basic solution with all $c_j - z_j \leq 0$ by dual simplex method. The removal of artificial variables are not concerned with the optimality of the given problem. At the end of phase-1 (in two phase method) or when all the artificial variables are removed in Big-M method the feasible solution may not anywhere near optimal. So, to obtain optimality we requires a large number of iterations in phase-II. To overcome such type of difficulties we use primal-dual algorithm, which deals directly with both the primal and dual in the following manner:

STEP 1. Start with a feasible dual solution.

STEP 2. Find infeasible solution to the primal such that following results must satisfied.

(i) Obtain restricted primal (RP) from given primal problem by taking some variables equal to zero so that complimentary slackness theorem is satisfied and the objective function is taken as the sum of artificial variables.

(ii) Find optimal solution of restricted primal problem by using simplex method.

(iii) If the optimal solution of restricted primal is not optimal solution of the given primal, then again find the dual of restricted primal problem and use steps (i) and (ii).

Above process is repeated successively until the optimal solution of the given primal is obtained.

Here, it must be noted that at any stage when a feasible solution to the primal is obtained then it is also optimal to the given primal.

7.3.1 THEORY OF PRIMAL-DUAL ALGORITHM

(1) To find the initial dual solution:

Consider the primal

$$\left. \begin{aligned} \max. Z_p = \mathbf{Cx} \\ \text{subject to the constraints} \\ \mathbf{Ax} = \mathbf{b} \\ \text{and} \qquad \mathbf{x} \geq 0 \end{aligned} \right] \qquad \ldots(1)$$

where $\mathbf{x} = [x_1, x_2, \ldots, x_n]', \mathbf{C} = [c_1, c_2, \ldots, c_n]$

$\mathbf{A} = [a_{ij}]_{m \times n}, \mathbf{b} = [b_1 \quad b_2 \quad \cdots \quad b_m]'$

Following the usual procedure, the dual of (1) is given by

$$\left. \begin{aligned} \min. Z_d = \mathbf{b'y} \\ \text{subject to the constraints} \\ \mathbf{A'y} \geq \mathbf{C'} \end{aligned} \right] \qquad \ldots(2)$$

Here, $\mathbf{y} = [y_1, y_2, \ldots, y_m]'$ is the dual variable vector which is unrestricted in sign.

Now, to find the initial dual solution, we introduce an additional constraints to (1) such that

$x_0 + x_1 + \ldots + x_n = b_0$; where b_0 is arbitrary large.

Now (1) reduces to

$$\left. \begin{aligned} \max. Z_p = 0x_0 + \mathbf{c} \cdot \mathbf{x} = (0, \mathbf{c}) \begin{bmatrix} x_0 \\ \mathbf{x} \end{bmatrix} \\ \text{s.t.} \\ x_0 + x_1 + \ldots + x_n = b_0 \\ \mathbf{Ax} = \mathbf{b} \\ \text{and} \qquad x_0, x_1, \ldots, x_n \geq 0 \end{aligned} \right] \qquad \ldots(3)$$

The primal (3) can be written in vector form as follows

$$\max. Z_p = [0, C]\begin{bmatrix} x_0 \\ x \end{bmatrix}$$

s.t.

$$\begin{bmatrix} 1 & 1 \\ O & A \end{bmatrix}\begin{bmatrix} x_0 \\ x \end{bmatrix} = \begin{bmatrix} b_0 \\ b \end{bmatrix}$$
...(4)

and $(x_0, x) \geq 0$

Now, modified dual of dual (4) is given by

$$\min. Z_d = (b_0, b')\begin{bmatrix} y_0 \\ y \end{bmatrix}$$

s.t. $$\begin{bmatrix} 1 & 0 \\ 1 & A' \end{bmatrix}\begin{bmatrix} y_0 \\ y \end{bmatrix} \geq \begin{bmatrix} 0 \\ C' \end{bmatrix}$$
...(5)

Now, (5) $\Rightarrow$ $y_0 \geq 0$ and $y_0 + A'y \geq C'$
...(6)

We observe that y_0 is non-negative and $y_1, y_2, ..., y_m$ are unrestricted in sign.

If $y_0 = 0$, then (5) reduced to the dual (2) of given primal (1).

In this case, from (6) we can easily find a solution of dual (5) which is given by

$$y = 0$$

$\therefore$ From (6) $y \geq 0$ and $y_0 \geq c_j, j = 1, 2, ..., n$

$$y_0 = \max\{0, c_j\}, j = 1, 2, ..., n$$

and $y_1 = 0 = y_2 = ... = y_m$
...(7)

Further, introducing surplus variables $s_0, s_1, ..., s_n$, then (5) becomes

$$\min Z_d = b_0 y_0 + b_1 y_1 + b_2 y_2 + ... + b_m y_m + 0 s_0 + 0 s_1 + ... + 0 s_n$$

s.t.

$$1 \cdot y_0 + 0 y_1 + 0 y_2 + ... + 0 y_m - s_0 + 0 s_1 + ... + 0 s_n = 0$$
$$1 \cdot y_0 + a_{11} y_1 + a_{21} y_2 + ... + a_{m1} y_m + 0 s_0 - y s_1 + 0 s_2 + ... + 0 s_n = c_1$$
$$\vdots$$
$$1 \cdot y_0 + a_{1n} y_1 + a_{2n} y_2 + ... + a_{mn} y_m + 0 s_0 + 0 s_1 + ... - s_n = c_n$$
...(8)

and $y_0 \geq 0$, $s_0, s_j \geq 0, j = 1, 2, ..., n$

and $y_1, y_2, ..., y_m$ are unrestricted in sign.

Clearly, the initial solution of (8) is given by

$$s_0 = y_0, s_1 = y_0 - c_1, s_2 = y_0 - c_2, ..., s_n = y_0 - c_n$$

$\because$ $y_0 = \max. \{0, c_j\}, j = 1, 2, ..., n$. Then we get at least one $s_0, s_1, ..., s_n$ is zero.

Therefore, solution (7) is the initial dual solution with which we start primal-dual algorithm.

And if $x_0 s_0 + x_1 s_1 + ... + x_n s_n = 0$
...(9)

Then $Z_p = Z_d$

If along (9), $\begin{bmatrix} x_0 \\ x \end{bmatrix}$ is a feasible solution to modified primal (3) or (4), then $\begin{bmatrix} x_0 \\ x \end{bmatrix}$ will

be an optimal solution to modified form (3) and $\begin{bmatrix} y_0 \\ y \end{bmatrix}$ is an optimal solution to the modified dual (5). But if $y_0 = y$ then $Z_p = Z_d$ with condition (9).

Hence x is an optimal solution to the given primal (1) and y is an optimal solution to its dual (2). Further, if we obtain a solution to the modified dual with $y_0 \neq 0$ and a feasible solution to the modified primal such that (9) is satisfied then $x_0 = 0$ and then

$$Z_p = Z_d = b_0 y_0 + b'y$$

(2) To find Restricted Primal (RP):

Here, we add artificial variables to each of its constraints and assign the cost -1 to each artificial variable and zero price corresponding to every legistimate variable. Then the modified primal (5) so obtained is called the restricted primal and is given by

$$\left. \begin{aligned} \max. Z^* &= A_0 - A_1 - A_2 - \ldots - A_m \\ &= 1 \cdot A_0 - 1 \cdot x_a \\ \text{subject to the constraints} \\ \begin{bmatrix} 1 & 1 \\ 0 & A \end{bmatrix} \begin{bmatrix} x_0 \\ x \end{bmatrix} + \begin{bmatrix} 1 & 0 \\ 0 & I_m \end{bmatrix} \begin{bmatrix} x_{a_0} \\ x_a \end{bmatrix} = \begin{bmatrix} b_0 \\ b \end{bmatrix} \end{aligned} \right\} \quad \ldots(10)$$

and $[x_0, x] \geq 0$ and $[x_{a_0}, x_a] \geq 0$

(3) To find the solution of restricted Primal:

For this, apply simplex method successively till we get $Z^* = 0$, i.e., $A_0 = A_1 = A_2 = \ldots = A_m = 0$. Here, we choose the outgoing vector and incoming vector in the following manner:

(i) The legismate activity vectors are taken as $\alpha_j^{(1)}$ for $j = 1, 2, \ldots, n$.

(ii) A basis matrix for the constraints is taken as B_1.

(iii) The price vector corresponding to the variables as $C_{B_1}^*$.

(iv) Calculate $\Delta_j^* = c_j^* - z_j^* = c_j^* - C_{B_1}^* B_1^{-1} \alpha_j^{(1)}$

$$= c_j^* - C_{B_1}^* x_j^{(1)} = B_1^{-1} \alpha_j^{(1)}$$

Rule to find incoming vector: The vector $\alpha_k^{(1)}$ is the incoming vector for which $s_k = 0$. So, all x_j for which s_j are strictly positive, remains zero.

Rule to find outgoing vector: Corresponding to the incoming vector $\alpha_k^{(1)}$, the outgoing vector from the basis is obtained one by one in the same manner as in phase 1.

(4) To obtain new Dual solution $\hat{s}_j$ and the improved value $\hat{Z}_d$ and $\hat{Z}^*$ of Z_d and Z^* respectively:

(i) $\hat{s}_j = s_j - \theta \Delta_j^*$ where $\theta = \min \left\{ \dfrac{s_j}{\Delta_j^*}, \Delta_j^* > 0 \right\}$

If at least one $\hat{s}_j$ is zero then again find $\alpha_k^{(1)}$. The vector entering the primal basis at the next iteration.

(ii) The improved value of Z^* can be calculated by using the formula

$$\hat{Z}^* = C_{B_1} x_B^{(1)}$$

(iii) The improved value of Z_d can be calculated by using the formula

$$\hat{Z}_d = Z_d + \theta \hat{Z}^*$$

(5) **Optimality Test:** The primal-dual algorithm will be stopped if any one of the following is satisfied:

(i) If we obtain $Z^* = 0$ then an optimal solution to the given primal is achieved and is given by $Z_p = Z_d$ provided $s_0 = 0 = y_0$.

(ii) If $Z^* = 0$ is obtained for $s_0 \neq 0$ then the given primal will have unbounded solution.

(iii) If $Z^* < 0$, $\Delta_j^* \leq 0 \ \forall j$ at any iteration, then primal has no solution.

7.4 STEPS OF PRIMAL-DUAL ALGORITHM

STEP 1.   Find restricted primal (RP) from the given primal.

STEP 2.   Find initial dual solution of the dual of restricted primal as given below

$$y_0 = \max\{0, c_j\}, c_j > 0; j = 1, 2, 3, \ldots, n$$

and $$y_1 = 0 = y_2 = \ldots = y_m$$

$$s_0 = y_0, s_j = y_0 - c_j : j = 1, 2, 3, \ldots, n$$

STEP 3.   Find incoming vector $\alpha_k^{(1)}$ corresponding to k, for which $s_k = 0$.

STEP 4.   Find outgoing vector by proceeding same as in phase I of simplex method.

STEP 5.   Compute Δ_j^* using the formula

$$\Delta_j^* = c_j^* - C_{B_1} x_j^{*(1)}, x_j^{*(1)} = B_1^{-1}\alpha_j^{(1)}$$

and obtain Z^* by $Z^* = C_{B_1}^* x_j^{(1)}$

and $Z_d = b_0 y_0 + b_1 y_1 + \ldots + b_m y_m$

STEP 6.   Construct the simplex table as given below

B_1	C_B	c_j^* / $x_B^{(1)}$	c_0^* / $x_0^{(1)}$	c_1^* / $x_1^{(1)}$	c_2^* / $x_2^{(1)}$	$\cdots$	c_n^* / $x_n^{(1)}$	$\cdots$	c_{n+m}^* / $x_{n+m}^{(1)}$	Min.Ratio $x_B^{(1)} / x_k^{(1)}$
$x_{n+1}^{(1)}$	-1	b_0								
$x_{n+2}^{(1)}$	-1									
$\vdots$	$\vdots$									
$x_{n+m}^{(1)}$	-1									
$Z^* = C_{B_1} x_B^{(1)}$		x_j								
		Δ_j^*								
$Z^* = \sum\limits_{j=0}^{m} b_j y_j$		s_j								

STEP 7.   If $Z^* = 0$ and $\Delta_j^* \leq 0 \ \forall j$, then solution is optimal, if not so, go to next step.

STEP 8.   Find new dual solution by using the formula

$$\hat{s}_j = s_j - \theta \Delta_j^*$$

where $\theta = \min\left\{\dfrac{S_j}{\Delta_j^*}, \Delta_j^* > 0\right\}$

STEP 9. Repeat step (5) to (8) till we obtain the optimal solution.

Solved Examples

EXAMPLE 1. *Using primal-dual algorithm, solve the following LPP*

$$\text{max. } Z = x_1 + 6x_2$$
$$\text{subject to the constraints}$$
$$x_1 + x_2 \geq 2$$
$$x_1 + 3x_2 \leq 3$$
$$\text{and} \qquad x_1, x_2 \geq 0$$

SOLUTION. Introducing surplus variable s_1 and slack variable s_2, we can write the given primal as

$$\text{max. } Z = x_1 + 6x_2 + 0s_1 + 0s_2$$
$$\text{s.t.}$$

$$\left. \begin{array}{c} x_1 + x_2 - s_1 = 2 \\ x_1 + 3x_2 + s_2 = 3 \\ \text{and} \qquad x_1, x_2, s_1, s_2 \geq 0 \end{array} \right] \qquad \ldots(1)$$

Now, the modified form of primal (1) is

$$\text{max.} Z = 0 \cdot x_0 + x_1 + 6x_2 + 0s_1 + 0s_2$$
$$\text{s.t.}$$

$$\left. \begin{array}{c} x_0 + x_1 + x_2 + s_1 + s_2 = b_0 \\ 0x_0 + x_1 + x_2 - s_1 + 0s_2 = 2 \\ 0x_0 + x_1 + 3x_2 + 0s_1 + s_2 = 3 \\ \text{and} \qquad x_0, x_1, x_2, s_1, s_2 \geq 0 \end{array} \right] \qquad \ldots(2)$$

Now, the restricted primal of (1) is given by

$$\text{max.} Z^* = -A_0 - A_1 - A_2$$
$$\text{s.t.}$$

$$\left. \begin{array}{c} x_0 + x_1 + x_2 + s_1 + s_2 = b_0 \\ 0x_0 + x_1 + x_2 - s_1 + 0s_2 + A_0 = 2 \\ 0x_0 + x_1 + 3x_2 + 0s_1 + s_2 + A_2 = 3 \\ \text{and} \quad x_0, x_1, x_2, s_1, s_2, A_0, A_1, A_2 \geq 0 \end{array} \right] \qquad \ldots(3)$$

The initial dual solution of (2) is given by

$$y_0 = \max\{0, c_j\}, c_j > 0$$
$$= \max\{0, 1, 6\} = 6$$

and
$$y_1 = y_2 = y_3 = y_4 = 0$$

So,
$$s_0 = y_0 = 6, s_1 = y_0 - c_1 = 6 - 1 = 5$$
$$s_2 = y_0 - c_2 = 6 - 6 = 0, s_3 = y_0 - c_3 = 6 - 0 = 6$$
$$s_4 = y_0 - c_4 = 6 - 0 = 6$$

and
$$Z_d = b_0 y_0 + b_1 y_1 + b_2 y_2 + b_3 y_3 + b_4 y_4 = 6b_0$$

Simplex table-1

B_1	$C_{B_1}^*$	$x_B^{(1)}$	$x_0^{(1)}$ $(\alpha_0^{(1)})$	$x_1^{(1)}$ $(\alpha_1^{(1)})$	$x_2^{(1)}$ $(\alpha_2^{(1)})$	$s_1^{(1)}$ $(\alpha_3^{(1)})$	$s_2^{(1)}$ $(\alpha_4^{(1)})$	$A_0^{(1)}$ (β_1)	$A_1^{(1)}$ (β_2)	$A_2^{(1)}$ (β_3)	Min Ratio $x_B^{(1)}/x_2^{(1)}$
c_j^*			0	0	0	0	0	-1	-1	-1	
$A_0^{(1)}$	-1	b_0	1	1	1	1	1	1	0	0	$b_0/1$
$A_1^{(1)}$	-1	2	0	1	1	-1	0	0	1	0	2/1
$A_2^{(1)}$	-1	3	0	1	3	0	1	0	0	1	3/3 (min)
$Z^* = C_{B_1}^* x_B^{(1)}$		x_j	0	0	0	0	0	b_0	2	3	
$= -b_0 - 5$		Δ_j^*	1	3	5	0	2	0	0	0	
$Z_d^* = 6b_0$		s_j	6	5	0	6	6	0	0	0	
				↑						↓	

In the above table, Δ_j^* is calculated by the following formula

$$\Delta_j^* = c_j^* - C_{B_1}^* x_j^{(1)}$$

and $\Delta_j^* \geq 0 \Rightarrow$ The obtained solution is not optimal.

Now, we shall improve this solution.

We have $\Delta_j = 0$ for $j = 2$ (not in basis)

$\Rightarrow s_2 = 0 \Rightarrow \alpha_2^{(1)}$ is the incoming vector.

Also, by minimum ratio rule, β_3 is the outgoing vector and key element $a_{23} = 3$

Simplex table-2

B_1	$C_{B_1}^*$	$x_B^{(1)}$	$x_0^{(1)}$	$x_1^{(1)}$	$x_2^{(1)}$	$s_1^{(1)}$	$s_2^{(1)}$	$A_0^{(1)}$ (β_1)	$A_1^{(1)}$ (β_2)	Min Ratio $x_B^{(1)}/x_1^{(1)}$
c_j^*			0	0	0	0	0	-1	-1	
$A_0^{(1)}$	-1	b_0-1	1	2/3	0	1	2/3	1	0	$(b_0-1)/(2/3)$
$A_1^{(1)}$	-1	1	0	2/3	0	-1	$-1/3$	0	1	3/2(min)→
$x_2^{(1)}$	0	1	0	1/3	1	0	1/3	0	0	3
$Z^* = C_{B_1}^* x_B^{(1)} = -b_0$		x_j	0	0	1	0	0	b_0-1	1	
		Δ_j^*	1	4/3	0	0	1/3	0	0	
$Z_d^* = Z_d + \theta Z^*$		s_j	6	5	0	6	6	0	0	
$= 9b_0/2$		$\hat{s}_j$	9/4	0	0	6	19/4	0	0	
				↑					↓	

In the above table Δ_j^* is calculated by the following formula

$$\Delta_j^* = c_j^* - C_{B_1}^* x_j^* \text{ for all } j \text{ not in the basis.}$$

So, $\quad \Delta_0^* = 1, \Delta_1^* = \dfrac{4}{3}, \Delta_3^* = 0, \Delta_4^* = \dfrac{1}{3}$

Clearly, $\Delta_j^* \ge 0 \Rightarrow$ solution is not optimal.

We have to again improve this solution.

The new dual solution $\hat{s}_j$ be given by

$$\hat{s}_j = s_j - \theta\Delta_j^*$$

where, $\quad \theta = \min\left\{\dfrac{s_j}{\Delta_j^*}, \Delta_j^* > 0\right\}$

$$= \min\left\{\dfrac{s_0}{\Delta_0^*}, \dfrac{s_1}{\Delta_1^*}, \dfrac{s_4}{\Delta_4^*}\right\}$$

$$= \min\left\{\dfrac{6}{1}, \dfrac{5}{4/3}, \dfrac{6}{1/3}\right\}$$

$$= \min\left\{6, \dfrac{15}{4}, 18\right\} = \dfrac{15}{4}$$

So, $\quad \hat{Z}_d = Z_d + \theta Z^* = 6b_0 + \dfrac{15}{4}(-b_0) = \dfrac{9b_0}{4}$

and $\quad \hat{s}_0 = s_0 - \theta\Delta_0^* = 6 - \dfrac{15}{4} \times 1 = \dfrac{9}{4}$

$$\hat{s}_1 = s_1 - \theta\Delta_1^* = 5 - \dfrac{15}{4} \times \dfrac{4}{3} = 0$$

$$\hat{s}_2 = s_2 - \theta\Delta_2^* = 0 - \dfrac{15}{4} \times 0 = 0$$

$$\hat{s}_3 = s_3 - \theta\Delta_3^* = 6 - \dfrac{15}{4} \times 0 = 6$$

$$\hat{s}_4 = s_4 - \theta\Delta_4^* = 6 - \dfrac{15}{4} \times \dfrac{1}{3} = \dfrac{19}{4}$$

Clearly, $\hat{s}_2 = 0$ as $j = 2$ is not in the basis.

$\Rightarrow \quad \alpha_1^{(1)}(= x_1^{(1)})$ is the incoming vector.

Further, by minimum ratio rule, the outgoing vector is $\beta_2(= A_1^{(1)})$

$\Rightarrow \quad$ Key element $= \dfrac{2}{3} = a_{22}$

Simplex table-3

B_1	$C^*_{B_1}$	$x_B^{(1)}$	$x_0^{(1)}$	$x_1^{(1)}$ (β_2)	$x_2^{(1)}$ (β_3)	$s_1^{(1)}$	$s_2^{(1)}$	$A_0^{(1)}$ (β_1)	Min Ratio $x_B^{(1)}/x_0^{(1)}$
	c^*_j		0	0	0	0	0	−1	
$A_0^{(1)}$	−1	b_0-2	1	0	0	2	1	1	$(b_0-2)/1 \rightarrow$
$A_1^{(1)}$	0	3/2	0	1	0	−3/2	−1/2	0	—
$x_2^{(1)}$	0	1/2	0	0	1	1/2	1/2	0	—
$Z^* = C^*_{B_1} x_B^{(1)}$		x_j	0	3/2	1/2	0	0	b_0-1	
$= 2 - b_0$		Δ^*_j	1	0	0	2	1	0	
$\hat{Z}'_d = \hat{Z}_d + \theta Z^*$		s_j	9/4	0	0	6	19/4	0	
$= 9/2$		$\hat{s}_j$	0	0	0	3/2	5/2	0	
		↑							

Here, Δ^*_j is calculated by the formula given by

$$\Delta^*_j = c^*_j - C^*_{B_1} x_B^{(1)}, \text{ for } j \text{ not in the basis.}$$

Therefore, $\Delta^*_0 = 1, \Delta^*_3 = 2, \Delta^*_4 = 1$

Clearly, $\Delta^*_j > 0 \Rightarrow$ solution is not optimal. To improve it, proceed as follows.

$$\hat{s}_j = s_j - \theta \Delta^*_j$$

where

$$\theta = \min\left\{\frac{\hat{s}_j}{\Delta^*_j}, \Delta^*_j > 0\right\} = \min\left\{\frac{\hat{s}_0}{\Delta^*_0}, \frac{\hat{s}_3}{\Delta^*_3}, \frac{\hat{s}_4}{\Delta^*_4}\right\}$$

$$= \min\left\{\frac{9/4}{1}, \frac{6}{2}, \frac{19/4}{1}\right\} = \frac{9}{4} = \frac{\hat{s}_0}{\Delta^*_0}$$

So,

$$\hat{s}'_0 = s_0 - \theta\Delta^*_0 = \frac{9}{4} - \frac{9}{4} \times 1 = 0$$

$$\hat{s}'_1 = s_1 - \theta\Delta^*_1 = 0 - \frac{9}{4} \times 0 = 0$$

$$\hat{s}'_2 = s_2 - \theta\Delta^*_2 = 0 - \frac{9}{4} \times 0 = 0$$

$$\hat{s}'_3 = s_3 - \theta\Delta^*_3 = 6 - \frac{9}{4} \times 2 = \frac{3}{2}$$

$$\hat{s}'_4 = s_4 - \theta\Delta^*_4 = \frac{19}{4} - \frac{9}{4} \times 1 = \frac{5}{2}$$

which implies $\hat{s}_0' = 0$ which is corresponding to $j = 0$ not in the basis. Therefore,

$$\hat{Z}_d = Z_d + \theta \cdot Z^* = \frac{9}{4}b_0 + \frac{9}{4}(-b_0 + 2) = \frac{9}{2}$$

Also, new $\hat{s}_0 = 0 \Rightarrow \alpha_0^{(1)}(= x_0^{(1)})$ is the incoming vector and by minimum ratio rule, $\beta(= A_0^{(1)})$ is the outgoing vector.

$\Rightarrow$ Key element, $a_{11} = 1$

Now, we have the following simplex table

Simplex table-4

B_1	$C_{B_1}^*$	$x_B^{(1)}$	$x_0^{(1)}$ (β_1)	$x_1^{(1)}$ (β_2)	$x_2^{(1)}$ (β_3)	$s_1^{(1)}$	$s_2^{(1)}$
	c_j^*		0	0	0	0	0
$x_0^{(1)}$	0	b_0-2	1	0	0	2	1
$x_1^{(1)}$	0	3/2	0	1	0	–3/2	–1/2
$x_2^{(1)}$	0	1/2	0	0	1	1/2	1/2
$Z^* = C_{B_1}^* x_B^{(1)} = 0$		x_j	b_0-2	3/2	1/2	0	0
		Δ_j^*	0	0	0	0	0
$\hat{Z}_d' = \hat{Z}_d + \theta Z^*$ $= 9/2$		$\hat{s}''$	0	0	0	3/2	5/2

Here, we observe that all $\Delta_j^* \leq 0 \Rightarrow$ solution is optimal.

Also, $Z^* = 0$ and $s_0'' = 0 = s_0$ therefore, the solution of the given primal is also optimal with $Z_p = Z_d''$

Hence, the optimal solution of the given primal is

$$x_1 = \frac{3}{2}, x_2 = \frac{1}{2} \text{ and } \max. Z_p = Z_d'' = \frac{9}{2}$$

 Exercise-7.2

Solve the following LPP by primal dual algorithm:

1. min. $Z = 3x_1 + 4x_2$
   subject to the constraints
   $2x_1 + 3x_2 \geq 8$
   $5x_1 + 2x_2 \geq 12$

and $x_1, x_2 \geq 0$

2. min. $Z = 2x_1 + 2x_2 + 3x_3$
   subject to the constraints
   $x_1 + x_3 \geq 1$
   $x_2 + x_3 \geq 2$
   and $x_1, x_2, x_3 \geq 0$

ANSWERS

1. $x_1 = \frac{20}{11}, x_2 = \frac{16}{11}$ and min. $Z = \frac{104}{11}$ $\qquad$ 2. $x_1 = 0, x_2 = 1, x_3 = 1$ and min. $Z = 5$

REVIEW QUESTIONS

1. Write the difference between regular simplex and dual simplex method.
2. Write the dual simplex algorithm.
3. Write the primal-dual simplex algorithm.
4. Write the method to choose the incoming and outgoing vector in dual simplex method.

MULTIPLE CHOICE QUESTIONS (CHOOSE THE MOST APPROPRIATE ONE)

1. Which of the following is true?
   (a) If primal has a feasible solution, then its dual will also have a feasible solution
   (b) If both primal and dual have feasible solution then both will have bounded optimal solution
   (c) If primal has no solution, then its dual will have an unbounded solution
   (d) None of these

2. In application of dual simplex method, the availability vector (RHS vector) must be:
   (a) ≥ 0
   (b) ≤ 0
   (c) no such type of restriction is required
   (d) None of these

3. If in the primal the number of constraints are m and in dual the variables are n, then:
   (a) $m \geq n$ (b) $m \leq n$
   (c) $m = n$ (d) None of these

4. Consider max. $Z = x_1$ subject to $x_1 + x_2 \leq 1$, $x_1 - x_2 \geq 1$, $x_1 \geq 0$, $x_2 \geq 0$. The solution of its dual is:
   (a) bounded   (b) unbounded
   (c) degenerate   (d) None of these

5. Let x be a non-optimal feasible solution of a LPP of maximization and y is a dual feasible solution then:
   (a) the primal objective value at x is greater than the dual objective value at y
   (b) the primal objective value at x is less than the dual objective value at y
   (c) the dual can be unbounded
   (d) None of these

6. If a slack or surplus variable s_1 is in optimal primal basis then in the optimal dual solution:
   (a) the dual variable corresponding to i^{th} primal constraints is zero
   (b) the dual variable corresponding to i^{th} primal constraints is unrestricted
   (c) both (a) and (b) are true
   (d) None of these

7. If there exist feasible solutions to both the primal and its dual then:
   (a) the primal may have an unbounded solution while the dual a bounded solution

(b) the primal may have a bounded solution while the dual an unbounded solution
   (c) both (a) and (b) are true
   (d) None of these

8. If primal: max. $Z_x = x_1 + x_2 + x_3, x_1, x_2, x_3 \geq 0$ then the solution of primal and its dual are respectively:
   (a) infeasible, feasible
   (b) feasible, infeasible
   (c) feasible, feasible
   (d) None of these

9. If the dual of the problem has infeasible solution then the value of objective function is:
   (a) bounded   (b) unbounded
   (c) no solution   (d) None of these

10. In LPP, if primal objective function is of maximization then for its dual the objective function is of:
    (a) maximization   (b) minimization
    (c) optimization   (d) None of these

11. If a slack or surplus variable s_j is in optimal primal basis, then the optimal dual solution:
    (a) the dual variable corresponding to j^{th} primal constraint is zero
    (b) the slack or surplus variable attached to j^{th} dual constraint is zero
    (c) both (a) and (b) are true
    (d) None of these

12. If the primal of LPP has bounded solution, then dual of the problem has:
    (a) optimal solution
    (b) no solution
    (c) unbounded solution
    (d) None of these

13. If the primal of LPP has no solution, then dual of the problem has:
    (a) unbounded solution
    (b) bounded solution
    (c) either no solution or is unbounded
    (d) None of these

14. If the primal has an unbounded solution then the dual has:
    (a) optimal solution   (b) no solution
    (c) bounded solution   (d) None of these

ANSWERS

| 1. (b) | 2. (c) | 3. (c) | 4. (a) | 5. (b) | 6. (a) | 7. (b) | 8. (b) | 9. (b) |
| 10. (b) | 11. (a) | 12. (b) | 13. (c) | 14. (b) | | | | |

Sensitivity Analysis

8.1 INTRODUCTION

Consider the linear programming problem
$$\max. Z = Cx$$
subject to the constraints
$$Ax = b$$
and
$$x \geq 0$$

The optimal solution of the above LPP depends upon the three parameters a_{ij}, b_i and c_j. In most of the cases, we have assumed that these parameters are constant, but there are some situations for which the values of these parameters vary time to time. Due to the variation in these parameters, optimal solution will affect. We shall discuss these changes in sensitivity analysis.

Definition. *The investigation that deals with changes in the optimal solutions due to discrete variations in the parameters a_{ij}, b_i and c_j are called sensitivity analysis.*

In this chapter, we shall discuss the effect on optimal solution due to the following changes in the given LPP.

(1) Variation in the coefficient (price vector) in the objective function, c_j.
(2) Variation in the right hand side constants (requirement vectors), b_j.
(3) Variation in the coefficient of decision variables on the left hand side of constraints, a_{ij}.
(4) Addition of a new variable to the existing list of variables in LPP.
(5) Addition of a new constraints to the original LP constraints.

☛ Remarks

- The sensitivity analysis is also known as **post-optimality analysis**, because it does not begin until the optimal solution to the given LPP is obtained.
- The sensitivity analysis is the study of knowing the effect on optimal solution of the LP model due to variations in the input coefficients or parameters.
- The sensitivity analysis provide the sensitivity range within which the LP model parameters can vary without changing the optimality of the current optimal solution.

8.2 CHANGE IN THE OBJECTIVE FUNCTION COEFFICIENTS (PRICE VECTORS), c_j

Change in the profit or cost coefficients in the objective function can occur for any basic variable or any non-basic variables. The sensitivity range for these variables is determined differently. So, there are two cases will be discussed seperately as given below:

Let x_B be the optimal basic feasible solution of an LPP, then we have

$$x_B = B^{-1} \cdot b \qquad \qquad ...(1)$$

where B is the optimal basis matrix and b is the requirement vector.

Obviously, the change in c_j does not affect x_B because (1) is independent of c_j but $\Delta_j = c_j - z_j$ depends on c_j.

Therefore, the change in c_j will affect the Δ_j, *i.e.*, the optimality condition.

While computing Δ_j $(= c_j - z_j)$ values, following two situations may arises:

 (i) The new Δ_j values satisfy, optimality condition and the solution remains unchanged, however, optimal value of the objective function may change.

 (ii) The optimality condition is not satisfied. In such a case we will use simplex method to achieve optimality.

Case I. Change in the coefficient of a non-basic variable, *i.e.*, change in c_j when it is not in C_B

If $\Delta_j \leq 0$ for all non-basic variables in a maximization LPP, then the current optimal solution remains unchanged. Let $c_j \notin C_B$, then $\Delta_j = c_j - z_j \leq 0$ for all j not in the basis, so x_B is an optimal solution.

Suppose that Δc_j is a change in c_j which is not in the basis, then there is no change in C_B and hence there is no change in B. Therefore,

$$z_j = C_B \cdot B^{-1} \alpha_j$$

will remains unchanged.

$\Rightarrow$ x_B remains optimal.

Now, if c_j changes to $c_j' = c_j + \Delta c_j$, then by optimality, we must have

$$(c_j + \Delta c_j) - z_j \leq 0 \qquad \qquad (\because \Delta_j \leq 0 \; \forall \, j \text{ not in the basis})$$

$\Rightarrow$ $\qquad \qquad \qquad \Delta c_j \leq -(c_j - z_j)$

$\Rightarrow$ $\qquad \qquad \qquad \Delta c_j \leq -\Delta_j$

Thus, we conclude that

> To retain the optimality of the current optimal solution for a change in Δc_j in c_j, we must have
> $$(c_j + \Delta c_j) - z_k \leq 0$$
> or $\qquad (c_j + \Delta c_j) \leq z_k$

☛ **Remarks**

 • For any LPP of maximization, the value of c_j may be increased upto the value of z_k and decrease to $-\infty$ without affecting the optimal solution.

 • In this case, there is no lower bound of Δc_j.

Case II. Change in the coefficient of a basic variable, *i.e.*, change in c_j when it is in C_B

We know that, in a maximization LPP, the change in the coefficient say, c_j of a basic variable x_j affect the Δ_j $(= c_j - z_j)$ values corresponding to all non-basic variably in the optimal simplex table.

If $c_j \in C_B$ and we make a change in c_j, then z_j will be changed because

$$z_j = C_B B^{-1} \alpha_j$$

Since, $$z_j = C_B B^{-1} \alpha_j = C_B \alpha_j = \sum_{i=1}^{m} C_{B_i} x_{ij}$$

Further, suppose C_{B_k} changes to $C'_{B_k} = C_{B_k} + \Delta C_{B_k}$

Then new value of z_j, i.e., z_j^* is given by

$$z_j^* = \sum_{i=1}^{m} C_{B_i} x_{ij} + C'_{B_k} x_{kj} \qquad (i \neq k)$$

$$= \sum_{\substack{i=1 \\ (i \neq k)}}^{m} C_{B_i} x_{ij} + (C_{B_k} + \Delta C_{B_k}) x_{kj}$$

$$= \sum_{i=1}^{m} C_{B_i} x_{ij} + \Delta C_{B_k} x_{kj}$$

$$\Rightarrow \qquad z_j^* = z_j + \Delta C_{B_k} x_{kj}$$

$$\Rightarrow \qquad c_j - z_j^* = (c_j - z_j) - \Delta C_{B_k} x_{kj}$$

Now for optimal solution, $c_j - z_j^* \leq 0 \; \forall j$ not in the basis.

$$\Rightarrow \qquad (c_j - z_j) - \Delta C_{B_k} x_{kj} \leq 0$$

$$\Rightarrow \qquad \Delta C_{B_k} x_{kj} \leq c_j - z_j$$

So, $\Delta C_{B_k} \geq \dfrac{c_j - z_j}{x_{kj}}$, when $x_{kj} > 0$

and $\Delta C_{B_k} \leq \dfrac{c_j - z_j}{x_{kj}}$, when $x_{kj} < 0$

On combining the above two inequalities, we have

$$\max_{x_{kj} > 0} \left\{ \frac{c_j - z_j}{x_{kj}} \right\} \leq C_{B_k} \leq \min_{x_{kj} < 0} \left\{ \frac{c_j - z_j}{x_{kj}} \right\} \qquad \ldots(2)$$

for all j corresponding to which α_j is not in the optimal basis. Thus we conclude that

If we make a change in $C_{B_k} \in C_B$, then the solution will remain optimal when the change ΔC_{B_k} in C_{B_k} satisfy the inequality (2).

New value of Objective Function

The new value of the objective function Z^* is given by

$$Z^* = C_B x_B = \sum_{\substack{i=1 \\ (i \neq k)}}^{m} C_{B_i} \cdot x_{B_i} + (C_{B_k} + \Delta C_{B_k}) x_{B_k}$$

$$= \sum_{\substack{i=1 \\ (i \neq k)}}^{m} C_{B_i} x_{B_i} + C_{B_k} \cdot x_{B_k} + \Delta C_{B_k} \cdot x_{B_k}$$

$$= \sum_{i=1}^{m} C_{B_i} x_{B_i} + \Delta C_{B_k} x_{B_k}$$

$$\Rightarrow \qquad Z^* = Z + \Delta C_{B_k} x_{B_k}$$

which conclude that the new value of the objective function is improved by $\Delta C_{B_k} x_{B_k}$, when $C_{B_k} \in C_B$ changes to $C_{B_k} + \Delta C_{B_k}$, where x_{B_k} is the basic variable corresponding to C_{B_k}.

Aliter: The sensitivity limits for the contribution per unit a basic variable can be calculated as follows:

lower limit = original value c_k – {lowest absolute value of improvement ratio or
$$- \infty \text{ (if no ratio is negative)}\}$$

upper limit = original value c_k + {lowest positive value of improvement ratio or
$$\infty \text{ (if no ratio is positive)}\}$$

$$\text{where improvement ratio} = \frac{\text{per unit improvement value}}{\text{input output coefficients in the variable row}}$$

$$= \frac{c_j - z_j}{a_{kj}}$$

☞ **Remarks**

- If the range of c_k is $p \le c_k \le q$ then we check the optimality at $c_k = p$ and $c_k = q$. If solution is not optimal at $c_k = p$, then range of c_k will be $p < c_k \le q$ and if solution is not optimal at $c_k = q$, then range of c_k will be $p \le c_k < q$.
- In the calculation of sensitivity analysis, the artificial variables column in the simplex table are ignored.

 Solved Examples

EXAMPLE 1. *Find the optimal solution of the LPP given by*
$$max. Z = 3x_1 + 5x_2$$
subject to the constraints
$$x_1 + x_2 \le 1$$
$$2x_1 + 3x_2 \le 1$$
and $\qquad x_1, x_2 \ge 0$

Obtain the variation in $c_1(= 3)$ and $c_2(= 5)$ without affecting the above optimal solution.

SOLUTION. Apply the simplex method in a usual manner, we have the following simplex table.

B.V.	C_B	x_B	$c_j \rightarrow$ 3 x_1	5 x_2	0 s_1	0 s_2	Min Ratio
s_1	0	1	1	1	1	0	1/1
s_2	0	1	2	③	0	1	1/3 (min) →
$Z = C_B x_B = 0$		$\Delta_j \rightarrow$	3	5	0	0	
				↑		↓	
s_1	0	2/3	1/3	0	1	–1/3	
x_2	5	1/3	1/3	1	0	1/3	
$Z = C_B x_B = 5/3$		$\Delta_j \rightarrow$	–1/3	0	0	–5/3	

From the above table we observe that all $\Delta_j \leq 0$

$\Rightarrow$ solution is optimal.

The optimal solution is given by

$$x_1 = 0, x_2 = \frac{1}{3} \text{ and } \max Z = \frac{5}{3}$$

To find variation in $c_1 = 3$

From the final simplex table, we have that c_1 is not in the basis B. So,

$$\Delta c_1 \leq -\Delta_1$$

$\Rightarrow$ $\qquad \Delta c_1 \leq -\left(-\frac{1}{3}\right), i.e., \Delta c_1 \leq \frac{1}{3}$

So, without affecting the optimality, the range of c_1 is given by

$$-\infty < c_1 < c_1 + \Delta c_1$$

$\Rightarrow$ $\qquad -\infty < c_1 \leq 3 + \frac{1}{3} = \frac{10}{3}$

Hence, c_1 can vary between $-\infty$ and $\frac{10}{3}$ without affecting optimal solution.

To find the variation in $c_2 = 5$

Clearly, $c_2 \in \boldsymbol{B}$

Now, since $\qquad \boldsymbol{C_B} = (0,5) = (C_{B_1}, C_{B_2})$

$\Rightarrow$ $\qquad C_{B_2} = 5(= c_2)$

$\Rightarrow$ $\qquad k = 2$

We know that the range of ΔC_{B_2} is given by

$$\max_{x_{2j}>0}\left\{\frac{c_j - z_j}{x_{2j}}\right\} \leq \Delta C_{B_2} \leq \min_{x_{2j}<0}\left\{\frac{c_j - z_j}{x_{2j}}\right\}$$

Here, $x_{21} = \frac{2}{3}, x_{24} = \frac{1}{3}$. we can take x_{22} and x_{23} because they are corresponding to the basis. Clearly, $x_{2j} > 0$ for $j = 1, 4$ but there is no $x_{2j} < 0$ which implies that ΔC_{B_2} has no upper bound.

$\therefore$ Range of C_{B_2} is given by

$$\max\left\{\frac{c_1 - z_1}{x_{21}}, \frac{c_4 - z_4}{x_{24}}\right\} < \Delta C_{B_2} \leq \infty \qquad \Rightarrow \max\left\{\frac{\Delta_1}{x_{21}}, \frac{\Delta_4}{x_{24}}\right\} \leq \Delta C_{B_2} \leq \infty$$

$\Rightarrow \max\left\{\frac{-1/3}{2/3}, \frac{-5/3}{1/3}\right\} \leq \Delta C_{B_2} \leq \infty \qquad \Rightarrow \max\left\{-\frac{1}{2}, -5\right\} \leq \Delta C_{B_2} < \infty$

$\Rightarrow \qquad -\frac{1}{2} \leq \Delta C_{B_2} < \infty$

$\Rightarrow \qquad C_{B_2} - \frac{1}{2} \leq C_{B_2} + \Delta C_{B_2} < C_{B_2} + \infty$

$\Rightarrow \qquad 5 - \frac{1}{2} \leq C_{B_2} + \Delta C_{B_2} < \infty \text{ or } \frac{9}{2} \leq C_{B_2} + \Delta C_{B_2} < \infty$

Hence, we conclude that c_2 can vary between $\frac{9}{2}$ and ∞ without affecting the optimum solution.

EXAMPLE 2. *Find an optimal solution to the following LPP*

$$max. \ Z = 3x_1 + 5x_2$$

subject to the constraints

$$x_1 \le 4$$
$$x_2 \le 6$$
$$3x_1 + 2x_2 \le 18$$
and $x_1, x_2 \ge 0$

What happens to this solution if the objective function is changed to $Z^* = 3x_1 + x_2$ *and* $Z^* = 3x_1 + 4x_2$.

SOLUTION. Using slack variables s_1, s_2, s_3 the given LPP becomes

max. $Z = 3x_1 + 5x_2 + 0s_1 + 0s_2 + 0s_3$

s.t. $\quad x_1 + x_2 + s_1 = 4$
$$x_2 + s_2 = 6$$
$$3x_1 + 2x_2 + s_3 = 18$$

and $\quad x_1, x_2, s_1, s_2, s_3 \ge 0$

The initial basic feasible solution is given by

$$x_1 = 0, x_2 = 0, s_1 = 4, s_2 = 6, s_3 = 18$$

Now, apply the simplex method in a usual manner, we have the following simplex table.

Simplex table-1

		$c_j \rightarrow$	3	5	0	0	0	
B.V.	C_B	x_B	x_1	x_2	s_1	s_2	s_3	Min Ratio x_B/x_2
s_1	0	4	1	0	1	0	0	—
s_2	0	6	0	①	0	1	0	6/11(min)→
s_3	0	18	3	2	0	0	1	18/2
$Z = C_B x_B = 0$		$\Delta_j \rightarrow$	3	5 ↑	0	0 ↓	0	
s_1	0	4	1	0	1	0	0	4/1
x_2	5	6	0	1	0	1	0	—
s_3	0	6	③	0	0	-2	1	6/3(min) →
$Z = C_B x_B = 30$		$\Delta_j \rightarrow$	3 ↑	0	0	-5	0 ↓	
s_1	0	2	0	0	1	2/3	-1/3	
x_2	5	6	0	1	0	1	0	
x_1	3	2	1	0	0	-2/3	1/3	
$Z = C_B x_B = 36$		$\Delta_j \rightarrow$	0	0	0	-3	-1	

In the last row of the above table all $\Delta_j \le 0$

$\Rightarrow$ solution is optimal and is given by

$$x_1 = 2, x_2 = 6 \text{ and max. } Z = 36$$

Variation in c_2

Using final table

$$C_B = (0,5,3) = (C_{B_1}, C_{B_2}, C_{B_3})$$

$$\Rightarrow \quad C_{B_2} = 5 = C_2$$

$$\Rightarrow \quad k = 2$$

The range of ΔC_{B_2} is given by

$$\max_{x_{kj} \geq 0} \left\{ \frac{c_j - z_j}{x_{kj}} \right\} \leq \Delta C_{B_2} \leq \max_{x_{kj} < 0} \left\{ \frac{c_j - z_j}{x_{kj}} \right\}$$

$$\Rightarrow \quad \max_{x_{2j} > 0} \left\{ \frac{\Delta_j}{x_{kj}} \right\} \leq \Delta C_{B_2} \leq \min_{x_{2j} < 0} \left\{ \frac{\Delta_j}{x_{kj}} \right\}$$

Here, $j = 4, 5$ (not in the basis)

$$x_{24} = 1 \text{ and } x_{25} = 0$$

We observe that $x_{24} > 0$ and there is no $x_{2j} < 0$, so that ΔC_{B_2} has no upper bound.

$\therefore$ The range of ΔC_{B_2} is given by

$$\frac{\Delta_4}{x_{24}} \leq \Delta C_{B_2} < \infty$$

$$\Rightarrow \quad -\frac{3}{1} \leq \Delta C_{B_2} < \infty \quad \Rightarrow \quad -3 \leq \Delta C_{B_2} < \infty$$

$$\Rightarrow \quad -3 \leq \Delta c_2 < \infty \qquad (\because \; C_{B_2} = c_2)$$

$$\Rightarrow \quad 5 - 3 \leq \Delta c_2 < 5 + \infty \quad \Rightarrow \quad 2 \leq c_2 < \infty$$

$\Rightarrow$ c_2 lies between 2 and ∞ without affecting the optimal solution.

Now, if c_2 changes to 1, *i.e.*, if the function is $Z^* = 3x_1 + x_2$ then optimal solution will be changed.

To find new optimal solution

If the objective function is changed from Z to $Z^* = 3x_1 + x_2$ then C_B in the final simplex table will be $C_B = (0, 1, 3)$. So, modified final simplex table is given by:

Simplex table-2

		$c_j \rightarrow$	3	1	0	0	0	
B.V.	C_B	x_B	x_1	x_2	s_1	s_2	s_3	Min Ratio x_B/s_2
s_1	0	2	0	0	1	(2/3)	–1/3	2/(2/3)=3 (min)→
x_2	1	6	0	1	0	1	0	6/1 = 6
x_1	3	2	1	0	0	–2/3	1/3	—
$Z^* = C_B x_B = 12$		$\Delta_j \rightarrow$	0	0	0 ↑	1	–1 ↓	
s_2	0	3	0	0	3/2	1	–1/2	
x_2	1	3	0	1	–3/2	0	1/2	
x_1	3	4	1	0	1	0	0	
$Z^* = C_B x_B = 15$		$\Delta_j \rightarrow$	0	0	–3/2	0	–1/2	

From the last row of the above table, we observe that all $\Delta_j \leq 0$. Hence, new optimal solution is given by

$$x_1 = 4, x_2 = 3 \text{ and max } Z^* = 15$$

Also, if the objective function changed to $Z^* = 3x_1 + 4x_2$ then max $Z = $ max Z^*, because c_2 lies between 2 and ∞ ($c_2 = 4$).

EXAMPLE 3. *Find the limits of the variations of c_1, c_2, c_3, c_4, c_5 and c_6 respectively of the following LPP for which the optimal solution remains optimal.*

$$\text{max. } Z = -x_2 + 3x_2 - 2x_5$$

subject to the constraints

$$x_1 + 3x_2 - x_3 + 2x_5 = 7$$
$$- 2x_2 + 4x_3 + x_4 = 12$$
$$- 4x_2 + 3x_3 + 8x_5 + x_6 = 10$$
$$\text{and } x_j \geq 0 \ \forall \ j = 1, 2, ..., 6$$

SOLUTION. Apply the simplex method, we have the following simplex table

Simplex table

	$c_j \rightarrow$		0	−1	3	0	−2	0	
B.V.	C_B	x_B	x_1	x_2	x_3	x_4	x_5	x_6	Min Ratio
x_1	0	7	1	3	−1	0	2	0	—
x_4	0	12	0	−2	4	1	0	0	12/4 = 3(min)→
x_6	0	10	0	− 4	3	0	8	1	10/3
$Z = C_B x_B = 0$		$\Delta_j \rightarrow$	0	−1	3	0	−2	0	
					↑	↓			
x_1	0	10	1	⑤/2	0	1/4	2	0	4 (min) →
x_3	3	3	0	−1/2	1	1/4	0	0	—
x_6	0	1	0	−5/2	0	−3/4	8	1	—
$Z = C_B x_B = 9$		$\Delta_j \rightarrow$	0	1/2	0	−3/4	−2	0	
				↓		↑			
x_2	−1	4	2/5	1	0	1/10	4/5	0	
x_3	3	5	1/5	0	1	3/10	2/5	0	
x_6	0	11	1	0	0	−1/2	10	1	
$Z = C_B x_B = 11$		$\Delta_j \rightarrow$	−1/5	0	0	− 4/5	−12/5	0	

Clearly all $\Delta_j \leq 0$ ⇒ solution is optimal and is given by

$$x_1 = 0, x_2 = 4, x_3 = 5, x_4 = 0, x_5 = 0, x_6 = 11, \text{ max. } Z = 11$$

Now, from final table, we observe that

$$C_B = (-1, 3, 0) = (c_2, c_3, c_6)$$

Clearly, c_1, c_4 and c_5 are not in C_B. So, limits of c_1, c_4 and c_5 are given by

$$\Delta c_1 \leq -\Delta_1 \Rightarrow \Delta c_1 \leq \frac{1}{5}$$

$$\Delta c_4 \leq -\Delta_4 \Rightarrow \Delta c_4 \leq \frac{4}{5}$$

$$\Delta c_5 \leq -\Delta_5 \Rightarrow \Delta c_5 \leq \frac{12}{5}$$

for which the optimal solution remains the same.

Further, c_2, c_3 and c_6 are in C_B

Now $\qquad C_B = (-1, 3, 0) = (c_2, c_3, c_6) = (C_{B_1}, C_{B_2}, C_{B_3})$

i.e., $\qquad C_{B_1} = c_2, C_{B_2} = c_3, C_{B_3} = c_6$

The limit in $C_{B_1} = c_2$ is given by

$$\max_{x_{ij} > 0} \left\{ \frac{\Delta_j}{x_{ij}} \right\} \le \Delta C_{B_1} \le \min_{x_{ij} < 0} \left\{ \frac{\Delta_j}{x_{ij}} \right\} \text{ for } j = 1, 4, 5 \text{ (which are not in the basis)}$$

Now, $x_{11} = \dfrac{2}{5}, x_{14} = \dfrac{1}{10}, x_{15} = \dfrac{4}{5}$ and there is no $x_{ij} < 0$.

$\therefore$ limit in $C_{B_1} = c_2$ is given by

$$\max \left\{ \frac{\Delta_1}{x_{11}}, \frac{\Delta_4}{x_{14}}, \frac{\Delta_5}{x_{15}} \right\} \le \Delta c_2 < \infty \Rightarrow \max \left\{ \frac{-1/5}{2/5}, \frac{-4/5}{1/10}, \frac{-12/5}{4/5} \right\} \le \Delta c_2 < \infty$$

$$\Rightarrow \quad \max \left\{ -\frac{1}{2}, -8, -3 \right\} \le \Delta c_2 < \infty \qquad \Rightarrow \qquad -\frac{1}{2} \le \Delta c_2 < \infty$$

and limit in $C_{B_2} = c_3$ is given by

$$\max_{x_{2j} > 0} \left\{ \frac{\Delta_j}{x_{2j}} \right\} \le \Delta C_{B_3} \le \min \left\{ \frac{\Delta_j}{x_{2j}} \right\} \text{ for } j = 1, 4, 5 \text{ not in the basis.}$$

Now, $x_{21} = \dfrac{1}{5}, x_{24} = \dfrac{3}{10}, x_{25} = \dfrac{2}{5}$

and there is no $x_{2j} < 0$, therefore the limit in $C_{B_2} = c_3$ is given by

$$\max \left\{ \frac{\Delta_1}{x_{21}}, \frac{\Delta_4}{x_{24}}, \frac{\Delta_5}{x_{25}} \right\} \le \Delta c_3 < \infty$$

$$\Rightarrow \quad \max \left\{ \frac{-1/5}{1/5}, \frac{-4/5}{3/10}, \frac{-12/5}{2/5} \right\} \le \Delta c_3 < \infty$$

$$\Rightarrow \quad \max \left\{ -1, -\frac{8}{3}, -6 \right\} \le \Delta c_3 < \infty \quad \Rightarrow \quad -1 \le \Delta c_3 < \infty$$

Also, the limit in $C_{B_2} = c_6$ is given by

$$\max_{x_{3j} > 0} \left\{ \frac{\Delta_j}{x_{3j}} \right\} \le \Delta C_{B_3} \le \min_{x_{3j} < 0} \left\{ \frac{\Delta_j}{x_{3j}} \right\}, \text{ for } j = 1, 4, 5$$

Now, $x_{31} = 1, x_{34} = -\dfrac{1}{2}, x_{35} = 10$, so limit in $C_{B_3} = c_6$ is given by

$$\max \left\{ \frac{\Delta_1}{x_{31}}, \frac{\Delta_5}{x_{35}} \right\} \le \Delta c_6 \le \min \left\{ \frac{\Delta_4}{x_{34}} \right\}$$

$$\Rightarrow \quad \max \left\{ \frac{-1/5}{1}, \frac{-12/5}{10} \right\} \le \Delta c_6 \le \min \left\{ \frac{-4/5}{-1/2} \right\}$$

$$\Rightarrow \quad -\frac{1}{5} \le \Delta c_6 \le \frac{8}{5}$$

8.3 VARIATION IN THE REQUIREMENT VECTOR, b_i

Consider an LPP of maximization

$$\left.\begin{array}{ll} & \max .Z = \boldsymbol{Cx} \\ \text{subject to the constraints} & \\ & \boldsymbol{Ax} \geq \boldsymbol{b} \\ \text{and} & \boldsymbol{x} \geq 0 \end{array}\right] \qquad \ldots(1)$$

Clearly, the optimality condition of (1) is given by

$$\Delta_j = c_j - z_j = c_j - \boldsymbol{C_B} \cdot \boldsymbol{B}^{-1} \cdot \alpha_j \leq 0 \quad \forall\, j \text{ not appearing in the basis and its solution is}$$

given by $\boldsymbol{x_B} = \boldsymbol{B}^{-1}\boldsymbol{b}$. Since, b_i values are not associated with Δ_j, therefore any change in b_i does not affect the optimality condition. However, it affects the values of the basic variables and the value of the objective function because the determination of solution values $(\boldsymbol{x_B} = \boldsymbol{B}^{-1} \cdot \boldsymbol{b})$, values of b is involved.

Let b_k is changed to $b_k + \Delta b_k$

Then new value of $\boldsymbol{b}$ is given by

$$\begin{aligned} \boldsymbol{b}^* &= [b_1, b_2, \ldots, b_k + \Delta b_k, \ldots, b_m]' \\ &= [b_1 + 0, b_2 + 0, \ldots, b_k + \Delta b_k, \ldots, 0 + b_m]' \\ &= [b_1, b_2, \ldots, b_m]' + [0, 0, \ldots, 0 b_k, \ldots, b_m]' \\ &= \boldsymbol{b} + [0, 0, \ldots, \Delta b_k, \ldots, b_m]' = \boldsymbol{b} + \Delta b_k \end{aligned}$$

Further if $\boldsymbol{x_B^*}$ is the new solution corresponding to $\boldsymbol{b}^*$, then we have

$$\boldsymbol{x_B^*} = \boldsymbol{B}^{-1} \cdot \boldsymbol{b}^* \text{ where } \boldsymbol{B} \text{ is the optimal basis.}$$

Let $\boldsymbol{B}^{-1} = \{\beta_1, \beta_2, \ldots, \beta_k, \ldots \beta_m\}$

where $\beta_i = \begin{bmatrix} \beta_{1i} \\ \beta_{2i} \\ \vdots \\ \beta_{mi} \end{bmatrix}$ for i = 1, 2, ..., m

Then
$$\begin{aligned} x_B^* &= [\beta_1, \beta_2, \ldots, \beta_k, \ldots, \beta_m] \cdot \{\boldsymbol{b} + [0, 0, \ldots, \Delta b_k, 0, \ldots, 0]'\} \\ &= [\beta_1, \beta_2, \ldots, \beta_k, \ldots, \beta_m] \cdot \boldsymbol{b} + [\beta_1, \beta_2, \ldots, \beta_k, \ldots, \beta_m] \cdot [0, 0, \ldots, \Delta b_k, 0, \ldots, 0]' \\ &= \boldsymbol{B}^{-1} \cdot \boldsymbol{b} + \beta_k \Delta b_k \\ &= [x_{B_1}, x_{B_2}, \ldots, x_{B_k}, \ldots, x_{B_m}]' + [\beta_{1k}, \beta_{2k}, \ldots, \beta_{kk}, \ldots, \beta_{mk}]' \cdot \Delta b \\ &= [x_{B_1} + \beta_{1k} \Delta b_k, x_{B_2} + \beta_{2k} \Delta b_k, \ldots, x_{B_k} + \beta_{kk} \Delta b_k, \ldots, x_{B_m} + \beta_{mk} \Delta b_k]' \end{aligned}$$

$$\Rightarrow \quad \boldsymbol{x_{B_i}^*} = x_{B_i} + \beta_{ik} \Delta b_k \text{ for i = 1, 2, ..., m}$$

Now, if x_B^* is feasible then $x_B^* \geq 0$, which implies that

$$x_{B_i} + \beta_{ik} \Delta b_k \geq 0 \quad \forall i = 1, 2, \ldots, m$$

$$\Rightarrow \qquad \Delta b_k \geq -\frac{x_{B_i}}{\beta_{ik}}, \text{ when } \beta_{ik} > 0$$

and $$\Delta b_k \leq -\frac{x_{B_i}}{\beta_{ik}}, \text{ when } \beta_{ik} < 0$$

On combining both the above inequalities, we get

$$\max_{\beta_{ik}>0}\left\{-\frac{x_{B_i}}{\beta_{ik}}\right\} \leq \Delta b_k \leq \min_{\beta_{ik}<0}\left\{-\frac{x_{B_i}}{\beta_{ik}}\right\} \qquad \qquad ...(1)$$

which shows that the new solution x_B^* remains feasible when the range of Δb_k is given by (1)

☛ **Remark**
- If one or more values in the x_B column of the simplex table are negative, the dual simplex method can be used to get an optimal solution to the new problem by maintaining feasibility.

New Value of the Objective Function

Let $$Z = C_B x_B = \sum_{i=1}^{m} C_{B_i} \cdot x_{B_i} \qquad \qquad ...(2)$$

If b_k is changed to $b_k + \Delta b_k$ then new value of the objective function is given by

$$Z^* = C_B x_B^* = \sum_{i=1}^{m} C_{B_i} \cdot x_{B_i}^*$$

$$= \sum_{i=1}^{m} C_{B_i}(x_{B_i} + \beta_{ik}\Delta b_k)$$

$$= \sum_{i=1}^{m} C_{B_i} \cdot x_{B_i} + \sum_{i=1}^{m} \beta_{ik}\Delta b_k = Z + \sum_{i=1}^{m} \beta_{ik}\Delta b_k$$

Hence, we conclude that the new solution x_B^* remains optimal and feasible when the range of Δb_k is given by (1) and the objective function is increased by $\sum_{i=1}^{m} \beta_{ik}\Delta b_k$

☛ **Remark**
- The range of variation in the availability of b_i can also be obtained by using condition of feasibility of the current optimal solution, i.e., $x_B = B^{-1}b \geq 0$

Solved Examples

EXAMPLE 1.   *Solve the following LPP:*

$$max. \ Z = 3x_1 + 5x_2$$

subject to the constraints

$$x_1 \leq 4$$
$$3x_1 + 2x_2 \leq 18$$
and $\quad x_1, x_2 \geq 0$

Find the range of b_1 and b_2 so that the solution remains optimal feasible.

SOLUTION.   The given LPP of maximization is written in standard form. So, using slack variables s_1, s_2 the given problem becomes

$$max. \ Z = 3x_1 + 5x_2 + 0s_1 + 0s_2$$
s.t. $\quad x_1 + 0x_2 + s_1 = 4$
$$3x_1 + 2x_2 + s_2 = 18$$
and $\quad x_1, x_2, s_1, s_2 \geq 0$

Now, apply the simplex method in a usual manner, we have the following simplex table.

Simplex table

B.V.	C_B	x_B	x_1 (α_1)	x_2 (α_2)	s_1 (β_1)	s_2 (β_2)	Min Ratio x_B/x_2
		$c_j \rightarrow$	3	5	0	0	
s_1	0	4	1	0	1	0	—
s_2	0	18	3	②	0	1	18/2 (min) $\rightarrow$
$Z = C_B x_B = 0$		$\Delta_j \rightarrow$	3	5 $\uparrow$	0	0 $\downarrow$	
s_1	0	$4 (x_{B_1})$	1	0	1	0	
x_1	5	$9 (x_{B_2})$	3/2	1	0	1/2	
$Z = C_B x_B = 45$		$\Delta_j \rightarrow$	−9/2	0	0	−5/2	

Clearly, all $\Delta_j \leq 0 \Rightarrow$ solution is optimal and is given by
$$x_1 = 0, x_2 = 9 \text{ and max. } Z = 45$$
Here, we observe that
$$\boldsymbol{B} = (\alpha_3, \alpha_2) \text{ and } \boldsymbol{b} = [4 \quad 18]' = [b_1 \quad b_2]'$$

So, $\boldsymbol{B}^{-1} = (\beta_1, \beta_2) = \begin{bmatrix} 1 & 0 \\ 0 & 1/2 \end{bmatrix} = \begin{bmatrix} \beta_{11} & \beta_{12} \\ \beta_{21} & \beta_{22} \end{bmatrix}$

To find the variation in b_1:

We have $\max\limits_{\beta_{i1}>0}\left\{\dfrac{-x_{B_i}}{\beta_{i1}}\right\} \leq \Delta b_1 \leq \min\limits_{\beta_{i1}<0}\left\{\dfrac{-x_{Bi}}{\beta_{i1}}\right\}$

Here, $\beta_{11} = 1 > 0$ and no $\beta_{i1} < 0$ so that Δb_1 has no upper bound, so range of Δb_1 is given by
$$-\frac{x_{B1}}{\beta_{11}} \leq \Delta b_1 < \infty$$
$$\Rightarrow \quad -\frac{4}{1} \leq \Delta b_1 < \infty \qquad (\because x_{B_1} = 4)$$
$$\Rightarrow \quad -4 \leq \Delta b_1 < \infty$$
Thus, range in b_1 is given by
$$4 - 4 \leq b_1 < 4 + \infty$$
$$\Rightarrow \quad 0 \leq b_1 < \infty$$

To find the variation in b_2:

We have the range of Δb_2 is
$$\max\limits_{\beta_{i2}>0}\left\{\dfrac{-x_{B_i}}{\beta_{i2}}\right\} \leq \Delta b_2 \leq \min\limits_{\beta_{i2}<0}\left\{\dfrac{-x_{B_i}}{\beta_{i2}}\right\}$$

Here, we have
$$\beta_{12} = 0, \beta_{22} = \frac{1}{2} > 0 \text{ and no } \beta_{i2} < 0$$
Therefore, Δb_2 has no upper bound.

So range of Δb_2 is given by

$$\frac{-x_{B_2}}{\beta_{22}} \leq \Delta b_2 < \infty$$

$$\Rightarrow \quad \frac{-9}{1/2} \leq \Delta b_2 < \infty \quad \Rightarrow \quad -18 \leq \Delta b_2 < \infty$$

Hence, the range of b_2 is given by

$$18 - 18 \leq b_2 < 18 + \infty, \text{ i.e., } 0 \leq b_2 < \infty$$

EXAMPLE 2. *Find the optimal solution of the problem*

$$\text{max. } Z = 6x_1 + 8x_2$$

subject to the constraints

$$5x_1 + 10x_2 \leq 60$$
$$4x_1 + 4x_2 \leq 40$$

and $\quad x_1, x_2 \geq 0$

Also, apply sensitivity analysis to find the solution of the given LPP if

(i) *RHS vector* $\begin{bmatrix} 60 \\ 40 \end{bmatrix}$ *of the constraints of the LPP is changed to* $\begin{bmatrix} 40 \\ 20 \end{bmatrix}$.

(ii) *the RHS vector* $\begin{bmatrix} 60 \\ 40 \end{bmatrix}$ *of the constraints is changed to* $\begin{bmatrix} 20 \\ 40 \end{bmatrix}$. [MEERUT–2007]

SOLUTION. Using slack variables s_1, s_2 the given LPP becomes

$$\text{max. } Z = 6x_1 + 8x_2 + 0s_1 + 0s_2$$

s.t.

$$5x_1 + 10x_2 + s_1 = 60$$
$$4x_1 + 4x_2 + s_2 = 40$$

and $\quad x_1, x_2, s_1, s_2 \geq 0$

Apply the simplex method in a usual manner, we have the following simplex table

Simplex table-1

B.V.	$c_j \rightarrow$		6	8	0	0	Min Ratio
	C_B	x_B	x_1	x_2	s_1	s_2	
s_1	0	60	5	⑩	1	0	$60/10 = 6(\text{min}) \rightarrow$
s_2	0	40	4	4	0	1	$40/4 = 10$
$Z = 0$		$\Delta_j \rightarrow$	6 $\uparrow$	8 $\downarrow$	0	0	
x_2	8	6	1/2	1	1/10	0	12
s_2	0	16	②	0	$-2/5$	1	$16/2 (\text{min}) \rightarrow$
$Z = 48$		$\Delta_j \rightarrow$	2 $\uparrow$	0	$-4/5$	0	
x_2	8	2	0	1	1/5	$-1/4$	
x_1	6	8	1	0	$-1/5$	1/2	
$Z = C_B x_B = 64$		$\Delta_j \rightarrow$	0	0	$-2/5$	-1	

Clearly, all $\Delta_j \leq 0 \Rightarrow$ solution is optimal and is given by
$$x_1 = 8, x_2 = 2 \text{ and max. } Z = 64$$

and
$$B^{-1} = \begin{bmatrix} -1/5 & 1/2 \\ 1/5 & -1/4 \end{bmatrix} = \frac{1}{20}\begin{bmatrix} -4 & 10 \\ 4 & -5 \end{bmatrix}$$

(i) If $b = \begin{bmatrix} 60 \\ 40 \end{bmatrix}$ is changed to $b' = \begin{bmatrix} 40 \\ 20 \end{bmatrix}$, the new value of the basic variables

will become
$$x_B = B^{-1}b'$$

i.e.,
$$x_B = \begin{bmatrix} x_1 \\ x_2 \end{bmatrix} = \frac{1}{20}\begin{bmatrix} -4 & 10 \\ 4 & -5 \end{bmatrix}\begin{bmatrix} 40 \\ 20 \end{bmatrix} = \begin{bmatrix} 2 \\ 3 \end{bmatrix}$$

$\Rightarrow \qquad x_1 = 2, x_2 = 3$

$\because$ x_1 and x_2 both are non-negative therefore, this solution is basic feasible solution. Hence, new optimal value of Z is given by
$$Z = 6 \times 2 + 8 \times 3 = 36$$

(ii) If $b = \begin{bmatrix} 60 \\ 40 \end{bmatrix}$ is changed to $b'' = \begin{bmatrix} 20 \\ 40 \end{bmatrix}$, the new value of the basic variables in

final iteration in the above simplex table is
$$x_B = B^{-1} \cdot b'', i.e., \begin{bmatrix} x_1 \\ x_2 \end{bmatrix} = \frac{1}{20}\begin{bmatrix} -4 & 10 \\ 4 & -5 \end{bmatrix}\begin{bmatrix} 20 \\ 40 \end{bmatrix} = \begin{bmatrix} 16 \\ -6 \end{bmatrix}$$

$\Rightarrow \qquad x_1 = 16, x_2 = -6$

$\Rightarrow \qquad x_2 < 0 \Rightarrow$ solution is infeasible.

Hence, by dual simplex method, the modified simplex table (from the last table) can be written as

Simplex table-2

B.V.	C_B	x_B	$c_j \rightarrow$ 6 x_1	8 x_2	0 s_1	0 s_2
x_1	6	16	1	0	-1/5	1/2
x_2	8	-6	0	1	1/5	-1/4
$Z = C_B x_B = 48$		$\Delta_j \rightarrow$	0	0	-2/5 $\downarrow$	-1 $\uparrow$
x_1	6	4	1	2	1/5	0
s_2	0	24	0	-4	-4/5	1
$Z = C_B x_B = 24$		$\Delta_j \rightarrow$	0	-4	-6/5	0

Clearly in the last row of the above table all $\Delta_j \leq 0$.

$\Rightarrow$ solution is optimal and feasible and is given by
$$x_1 = 4, x_2 = 0 \text{ and max. } Z = 24$$

EXAMPLE 3. *For the following LPP*

$$max. \; Z = -x_1 + 2x_2 - x_3$$

subject to the constraints

$$3x_1 + x_2 - x_3 \le 10$$
$$-x_1 + 4x_2 + x_3 \ge 6$$
$$x_2 + x_3 \le 4$$

and $\qquad x_1, x_2, x_3 \ge 0$

Find the seperate range of b_1, b_2, b_3 consistent with the optimal solution.

SOLUTION. Using slack variables s_1, s_3 and surplus variable s_2 and artificial variable A in the given LPP. Then we have

$$max. \; Z = -x_1 + 2x_2 - x_3 + 0s_1 + 0s_2 + 0s_3 - MA$$

subject to

$$3x_1 + x_2 - x_3 + s_1 = 10$$
$$-x_1 + 4x_2 + x_3 - s_2 + A = 6$$
$$x_2 + x_3 + s_3 = 4$$

and $\qquad x_1, x_2, x_3, s_1, s_2, s_3, A \ge 0$

Now apply the simplex method in a usual manner, we have the following simplex table.

Simplex table

		$c_j \rightarrow$	-1	$+2$	-1	0	0	0	$-M$	
B.V.	C_B	x_B	x_1	x_2	x_3	s_1	s_2	s_3	A_1	Min Ratio
s_1	0	10	3	1	-1	1	0	0	0	10/1
A	$-M$	6	-1	4	1	0	-1	0	1	6/4 (Min) $\rightarrow$
s_3	0	4	0	①	1	0	0	1	0	4/1
$Z = C_B x_B,$ $= -6M$		$\Delta_j \rightarrow$	$-1 - M$	$4M + 2$	$M - 1$	0	$-M$	0	0	
				$\uparrow$					$\downarrow$	
s_1	0	$17/2$	$13/4$	0	$-5/4$	1	$1/4$	0	—	(17/2)/(1/4)
x_2	2	$3/2$	$-1/4$	1	$1/4$	0	$-1/4$	0	—	—
s_3	0	$5/2$	$1/4$	0	$3/4$	0	⟨1/4⟩	1	—	(5/2)/(1/4)
$Z = 3$		$\Delta_j \rightarrow$	$-1/2$	0	$-3/2$	0	$1/2$	0		
							$\uparrow$	$\downarrow$		
s_1	0	6	3	0	-2	1	0	-1		
x_2	2	4	0	1	1	0	0	1		
s_2	0	10	1	0	3	0	1	4		
$Z = C_B x_B = 8,$		$\Delta_j \rightarrow$	-1	0	-3	0	0	-2		

We observe that all $\Delta_j \le 0 \Rightarrow$ solution is optimal, feasible and is given by

$$x_1 = 0, x_2 = 4, x_3 = 0 \text{ and max. } Z = 8$$

Further, $\quad C_B = (0, 2, 0), B = [s_1, x_2, s_2]$

$$x_B = (6, 4, 10) = (x_{B_1}, x_{B_2}, x_{B_3})$$

$$b = (10, 6, 4) = (b_1, b_2, b_3)$$

and $\quad B^{-1} = \begin{bmatrix} 1 & 0 & -1 \\ 0 & 1 & 1 \\ 0 & 0 & 4 \end{bmatrix} = (\beta_1, \beta_2, \beta_3)$

To find the variation in b_1:

$\because \quad b_1 = 10$, then the range of Δb_1 is given by

$$\max_{\beta_{i1} > 0} \left\{ \frac{-x_{B_i}}{\beta_{i1}} \right\} \leq \Delta b_1 \leq \min_{\beta_{i1} < 0} \left\{ \frac{-x_{B_i}}{\beta_{i1}} \right\}$$

Also $\beta_{11} = 1 > 0$ and there is no $\beta_{i1} < 0$. Therefore, range of Δb_1 is given by

$$\frac{-x_{B_1}}{\beta_{11}} \leq \Delta b_1 < \infty$$

$\Rightarrow \qquad \dfrac{-6}{1} \leq \Delta b_1 < \infty$

$\Rightarrow \qquad -6 \leq \Delta b_1 < \infty$

So, variation in b_1 is given by

$$10 - 6 \leq b_1 < 10 + \infty$$

$\Rightarrow \qquad 4 \leq b_1 < \infty$

To find the variation in b_2:

Clearly, $b_2 = 6$. Then range of Δb_2 is given by

$$\max_{\beta_{i2} > 0} \left\{ \frac{-x_{B_i}}{\beta_{i2}} \right\} \leq \Delta b_2 \leq \min_{\beta_{i2} < 0} \left\{ \frac{-x_{B_i}}{\beta_{i2}} \right\}$$

Here, we have $\beta_{11} = 0$, $\beta_{22} = 1$ and $\beta_{32} = 0$ and there is no $\beta_{i2} < 0 \Rightarrow$ There is no upper bound of Δb_2. Hence, the range of Δb_2 is given by

$$\frac{-x_{B_2}}{\beta_{22}} \leq \Delta b_2 < \infty \qquad \Rightarrow \qquad \frac{-4}{1} \leq \Delta b_2 < \infty$$

$\Rightarrow \qquad -4 \leq \Delta b_2 < \infty$

Thus, the variation in b_2 is given by

$$6 - 4 \leq b_2 < 6 + \infty$$

$\Rightarrow \qquad 2 \leq b_2 < \infty$

To find the variation in b_3:

Clearly, $b_3 = 4$. Then range of Δb_3 is given by

$$\max_{\beta_{i3} > 0} \left\{ -\frac{x_{B_i}}{\beta_{i3}} \right\} \leq \Delta b_3 \leq \min_{\beta_{i3} < 0} \left\{ \frac{-x_{B_i}}{\beta_{i3}} \right\}$$

Here, we have $\beta_{13} = -1$, $\beta_{23} = 1$, $\beta_{33} = 4$

Then range of Δb_3 is given by

$$\max_{\beta_{i3}>0}\left\{\frac{-x_{B_2}}{\beta_{23}},\frac{-x_{B_3}}{\beta_{33}}\right\}\leq \Delta b_3 \leq \frac{-x_{B_1}}{\beta_{13}}$$

$$\Rightarrow \quad \max\left\{\frac{-4}{1},\frac{-10}{4}\right\}\leq \Delta b_3 \leq \frac{-6}{-1}$$

$$\Rightarrow \qquad\qquad -\frac{5}{2}\leq \Delta b_3 \leq 6$$

Hence, the variation in b_3 is given by

$$4-\frac{5}{2}\leq b_3 \leq 4+6$$

$$\Rightarrow \qquad\qquad \frac{3}{2}\leq b_3 \leq 10$$

8.4 VARIATION IN THE ELEMENTS a_{ij} OF THE COEFFICIENT MATRIX A

Let us suppose the elements of coefficient matrix A is changed. Then we have the following two cases:

(i) Change in a coefficient when variable is a non-basic variable.

(ii) Change in a coefficient when variable is a basic variable.

CASE I. When a non-basic column $a_k \in \boldsymbol{B}$ changed to a_k^*. Then solution will remain optimal if following condition is satisfied

$$c_k - z_k^* = c_k - \boldsymbol{C_B}\cdot \boldsymbol{B}^{-1}a_k^* \leq 0$$

Otherwise the simplex method is continued, after column k of the simplex table is updated, by introducing the non-basic variable x_k into the basis.

The range for the discrete change Δa_{ij} in the coefficient of non-basic variable x_j in the constraint, i can be obtained by solving following linear inequalities

$$\max_{C_B\cdot\beta_i>0}\left\{\frac{\Delta_j}{C_B\cdot\beta_i}\right\}\leq \Delta a_{ij} \leq \min_{C_B\cdot\beta_i<0}\left\{\frac{\Delta_j}{C_B\cdot\beta_i}\right\}$$

Here, β_i is i^{th} column of B^{-1}.

☞ **Remark**

• If a $C_B\cdot\beta_i = 0$ then, Δa_{ij} is unrestricted in sign.

CASE II. If a basic variable $a_k \in \boldsymbol{B}$ is changed to a_k^*. Then to maintain both feasibility and optimality of the current solution, the following conditions are satisfied.

$$(i)\quad \max_{k\neq p}\left\{\frac{-x_{B_k}}{x_{B_k}\beta_{pi}-x_{Bp}\beta_{ki}>0}\right\}\leq \Delta a_{ij}\leq \min_{k\neq p}\left\{\frac{-x_{B_k}}{x_{B_k}\beta_{pi}-x_{Bp}\beta_{ki}<0}\right\}$$

$$(ii)\quad \max\left\{\frac{\Delta_j}{\Delta_j\beta_{pi}-y_{pj}C_B\cdot\beta_i>0}\right\}\leq \Delta a_{ij}\leq \min\left\{\frac{\Delta_j}{\Delta_j\beta_{pi}-y_{pj}C_B\cdot\beta_i<0}\right\}$$

 **Solved Examples**

EXAMPLE 1. Solve the following LPP

$$\text{max. } Z = -x_1 + 3x_2 - 2x_3$$

subject to the constraints

$$3x_1 - x_2 + 2x_3 \le 7$$
$$-2x_1 + 4x_2 \le 12$$
$$-4x_1 + 3x_2 + 8x_3 \le 10$$

and $\quad\quad x_1, x_2, x_3 \ge 0$

Hence, discuss the effect of the following changes in the optimal solution. Also,

(i) Find the range for discrete changes in the coefficients a_{13} and a_{23} consistent with the optimal solution of the given LPP

(ii) x_1-column in the problem is changed from $[3, -2, -4]'$ to $[3, 2, -4]'$.

(iii) x_3-column in the problem is changed from $[2, 0, 8]'$ to $[3, 1, 6]'$.

SOLUTION. Using slack variables s_1, s_2, s_3 we can write the given LPP as follows.

$$\text{max. } Z = -x_1 + 3x_2 - 2x_3 + 0s_1 + 0s_2 + 0s_3$$

subject to the constraints

$$3x_1 - x_2 + 2x_3 + s_1 = 7$$
$$-2x_1 + 4x_2 + s_2 = 12$$
$$-4x_1 + 3x_2 + 8x_3 + s_3 = 10$$

and $\quad\quad x_1, x_2, x_3, s_1, s_2, s_3 \ge 0$

Now, apply the Big-M method in a usual manner, we have the following final simplex table (of optimal solution)

Optimal simplex table

B.V.	$c_j \to$		-1	3	-2	0	0	0
	C_B	x_B	x_1	x_2	x_3	s_1	s_2	s_3
x_1	-1	4	1	0	$4/5$	$2/5$	$1/10$	0
x_2	3	5	0	1	$2/4$	$1/5$	$3/10$	0
s_3	0	11	0	0	10	1	$-1/2$	1
$Z = 11$		$\Delta_j \to$	0	0	$-12/5$	$-1/5$	$-4/5$	0

From the above table, the optimal feasible solution is given by

$$x_1 = 4, x_2 = 5, x_3 = 0 \text{ and max. } Z = 11$$

Here, we observe that

$$B^{-1} = \begin{bmatrix} 2/5 & 1/10 & 0 \\ 1/5 & 3/10 & 0 \\ 1 & -1/2 & 1 \end{bmatrix} = [\beta_1 \quad \beta_2 \quad \beta_3]$$

Then $\quad C_B \cdot \beta_1 = -1\left(\dfrac{2}{5}\right) + 3\left(\dfrac{1}{5}\right) + 0(1) = \dfrac{1}{5}$

$$C_B \cdot \beta_2 = -1\left(\dfrac{1}{10}\right) + 3\left(\dfrac{3}{10}\right) + 0\left(\dfrac{-1}{2}\right) = \dfrac{8}{10}$$

$$C_B \cdot \beta_3 = -1(0) + 3(0) + 0(1) = 0$$

Clearly, x_1, x_2 and s_3 belong to the basis, so any discrete change in coefficients belonging to any of these column vector may affect both feasibility and optimality, while any change in the non-basic variables (x_3, s_1 and s_2) column vector may affect only optimality.

(i) Range for change in a_{13} and a_{23} in the x_3-column is given by

$$\max\left(\frac{c_3 - z_3}{C_B \cdot \beta_1}\right) = \max\left\{\frac{-12/5}{1/5}\right\} \le \Delta a_{13} \implies \Delta a_{13} \ge -12$$

and $\quad \max\left(\frac{c_3 - z_3}{C_B \cdot \beta_2}\right) = \max\left\{\frac{-12/5}{8/10}\right\} \le \Delta a_{23} \implies \Delta a_{23} \ge -3$

(ii) **(a) Feasibility condition:** For $i = 2$ (constraints), $p = 1$ (column) and $k = 2, 3$ (column of B^{-1}) we have

For $k = 2$

$$x_{B_k}\beta_{pi} - x_{B_p}\beta_{ki} = x_{B_2}\beta_{12} - x_{B_1}\beta_{22}$$

$$= 5\left(\frac{1}{10}\right) - 4\left(\frac{3}{10}\right) = -\frac{7}{10}$$

For $k = 3$, $x_{B_3}\beta_{12} - x_{B_1}\beta_{32} = 11\left(\frac{1}{10}\right) - 4\left(-\frac{1}{2}\right) = \frac{31}{10}$

Therefore, the range to maintain feasibility of the existing optimal solution is

$$-\frac{5}{31/10} \le \Delta a_{21} \le \frac{-5}{-7/10}$$

$$2 - \left(\frac{50}{31}\right) \le a_{21} < 2 + \left(\frac{50}{7}\right) \quad \implies \quad \frac{-12}{31} \le a_{21} \le \frac{64}{7}$$

(b) Optimality condition:

$$\Delta_3\beta_{12} - y_{13}C_B \cdot \beta_2 = -\frac{12}{5}\left(\frac{1}{10}\right) - \frac{4}{5}\left(\frac{8}{10}\right) = -\frac{44}{50}$$

$$\Delta_4\beta_{12} - y_{14}C_B \cdot \beta_2 = -\frac{1}{5}\left(\frac{1}{10}\right) - \frac{2}{5}\left(\frac{8}{10}\right) = -\frac{17}{50}$$

$$\Delta_5\beta_{12} - y_{15}C_B \cdot \beta_2 = -\frac{4}{5}\left(\frac{1}{10}\right) - \frac{1}{10}\left(\frac{8}{10}\right) = -\frac{16}{100}$$

$\therefore$ Required range is given by

$$-\infty \le \Delta a_{21} \le \min\left\{\frac{-12/5}{-44/50}, \frac{-1/5}{-17/50}, \frac{-4/5}{-16/100}\right\}$$

$$\implies \quad -\infty \le \Delta a_{21} \le \frac{10}{17}$$

$$\implies \quad -\infty \le a_{21} \le \frac{44}{17}$$

(iii) Let us suppose column vector a_3 of original LPP is changed from $[2, 0, 8]'$ to $[3, 1, 6]'$. Then modified value of $c_3 - z_3^*$ for this column is

$$B^{-1} \cdot a_3^* = \begin{bmatrix} 2/5 & 1/10 & 0 \\ 1/5 & 3/10 & 0 \\ 1 & -1/2 & 1 \end{bmatrix}\begin{bmatrix} 3 \\ 1 \\ 0 \end{bmatrix} = \begin{bmatrix} 13/10 \\ 9/10 \\ 17/2 \end{bmatrix}$$

$$c_3 - z_3^* = c_3 - C_B \cdot B^{-1} \cdot a_3^* = -2 - [-1, 3, 0] \begin{bmatrix} 13/10 \\ 9/10 \\ 17/2 \end{bmatrix} = -\frac{34}{10}$$

8.5 ADDITION OF A NEW VARIABLE

Let us suppose an extra variable x_{n+1} be added. Then solution will remain feasible but it may no longer feasible. If a new variable x_{n+1} is added to the problem, then it will introduce an additional column say α_{n+1} to the coefficient matrix A and an extra cost c_{n+1} will introduced in C. Due to this, the optimality of the solution will be affected. To see the impact of this addition on the current optimal solution, we compute the following

$$y_{n+1} = B^{-1} a_{n+1}$$

and $c_{n+1} - z_{n+1} = c_{n+1} - C_B y_{n+1}$

Then there are following two cases:

  (i)   If $c_{n+1} - z_{n+1} \le 0$ then $x_B = 0$, hence current solution is optimal.

  (ii)   If $c_{n+1} - z_{n+1} > 0$ then current optimal solution can be improved by introducing a new column α_{n+1} in the basis to find the new optimal solution. To improve the solution, we start with the last simplex table of original problem by introducing one extra column α_{n+1} to new variable x_{n+1}.

8.6 ADDITION OF A NEW CONSTRAINT

Let Z be the optimal value of the objective function of given LPP and Z^*, the optimal value of the objective fucntion of new problem which is obtained by adding an extra constraint to the original LPP.

If $Z^* > Z$. Then since, new optimal solution satisfies all the constraints (including additional constraints) so that it is also the optimal solution of the original problem, so, $Z^* > Z$ gives a contradiction.

Hence $Z^* \le Z$ (in maximization case).

Here, we have the following two possibilities,

  (i)   If the optimal solution of the original LPP satisfies the additional constraints, then it is also the optimal solution of new problem and additional new constraint is called redundant constraint.

  (ii)   If the optimal solution of the original problem does not satisfy the additional constraint, then we find the optimal solution of new problem in the following manner.

New optimal solution: Let us suppose B is the optimal basis for the original LPP and B_1, the optimal basis for new problem. Clearly, B_1 will be a square matrix of order $(m+1)$ ($\because$ an extra constraint is added to the original problem) and B_1 is a square matrix of order m. So,

$$B_1 = \begin{bmatrix} B & 0 \\ \alpha & \pm 1 \end{bmatrix} \qquad \qquad \dots (1)$$

Here the last column of B_1 is correspond to slack, surplus and artificial vector associated with the extra constraints and r is a row vector of the coefficients in extra constraints of the variable corresponding to the vector in the optimal basis B.

Now, we can easily obtained B_1^{-1} (by using partition method) such that

$$B_1^{-1} = \begin{bmatrix} \boldsymbol{B}^{-1} & 0 \\ \mp\alpha\boldsymbol{B}^{-1} & \pm 1 \end{bmatrix} \qquad \ldots(2)$$

Now, let $a_{m+1,j}$ be the coefficient of x_j in new $(m+1)^{th}$ constraints and α_j^*, the column vector of the coefficient x_j in the new problem.

Then $x_j^* = B_1^{-1}\alpha_j^* = \begin{bmatrix} \boldsymbol{B}^{-1} & 0 \\ \mp\boldsymbol{B}^{-1} & \pm 1 \end{bmatrix}\alpha_j^*$, where x_j^* is correspond to x_j for new problem.

Further, since $\alpha_j^* = \begin{bmatrix} \alpha_j \\ a_{m+1,j} \end{bmatrix}$ then we have

$$x_j^* = \begin{bmatrix} \boldsymbol{B}^{-1} & 0 \\ \mp\alpha\boldsymbol{B}^{-1} & \pm 1 \end{bmatrix}\begin{bmatrix} \alpha_j \\ a_{m+1,j} \end{bmatrix} = \begin{bmatrix} \boldsymbol{B}^{-1}\alpha_j \\ \mp\alpha\boldsymbol{B}^{-1}\alpha_j \pm a_{m+1,j} \end{bmatrix} \qquad (\because\ x_j = \boldsymbol{B}^{-1}\alpha_j)$$

and $\quad z_j^* = C_{B_1}x_j^*$

$$= (C_B, C_{B_{m+1}})\begin{bmatrix} x_j \\ \mp\alpha x_j \pm a_{m+1,j} \end{bmatrix}$$

$$= C_B x_j + C_{B_{m+1}}(\mp\alpha x_j \pm a_{m+1,j})$$

$$\Rightarrow \qquad z_j^* = z_j + C_{B_{m+1}}(\mp\alpha x_j \pm a_{m+1,j}) \qquad \ldots(3)$$

Now we have the following cases:

(i) If slack or surplus variables is introduced in the additional new constraints, then
$$C_{B_{m+1}} = 0$$
Then from (3)

$$z_j^* = z_j, \text{ i.e., } c_j - z_j^* = c_j - z_j$$

$\Rightarrow \quad c_j - z_j^*$ remains the same (unchanged)

Finally, since the optimal solution of the original problem does not satisfy the new constraints, so that slack or surplus variables, introduced in the new constraints is negative. Therefore, in this case we can apply the dual simplex method to find an optimum solution of the new problem.

(ii) If the new constraints is a perfect inequality, i.e., an artificial variable is introduced, then an additional vector is an artificial vector. Then we have the following two possibilities:

(a) If the artificial variable in the basis solution is negative, we can assign a price zero to the artificial variable and the dual simplex method is used to remove the artificial variable from the basis.

(b) If the artificial variable in the basis solution is positive, we can assign a cost of $-M$ to it and use simplex (standard) method for removal of the artificial variable from the basis. In this case the value of $c_j - z_j$ will be changed.

Solved Examples

EXAMPLE 1. *Solve the following LPP*

$$max. \ Z = 3x_1 + 5x_2$$

subject to the constraints

$$x_1 + x_3 = 4$$
$$3x_1 + 2x_2 + x_4 = 18$$

and $x_1, x_2, x_3, x_4 \geq 0$

If a new variable x_5 is added to the above LPP with price 7 then we have the following problem

$$max \ Z' = 3x_1 + 5x_2 + 7x_5$$

s.t. $x_1 + x_3 + x_5 = 4$
$$3x_1 + 2x_2 + x_4 + 2x_5 = 18$$
and $x_1, x_2, x_3, x_4, x_5 \geq 0$

Find the solution of the new problem.

SOLUTION. **Solution of the original problem:**
The given problem can be written as

$$max. \ Z = 3x_1 + 5x_2 + 0x_3 + 0x_4$$
s.t.

$$x_1 + 0x_2 + x_3 + 0x_4 = 4$$
$$3x_1 + 2x_2 + 0x_3 + x_4 = 18$$
and $x_1, x_2, x_3, x_4 \geq 0$

Now, apply the simplex method in a usual manner, we have the following simplex table.

Simplex table-1

B.V.	C_B	x_B	$c_j \rightarrow$	3	5	0	0	Min Ratio
				x_1 (α_1)	x_2 (α_2)	x_3 (α_3)	x_4 (α_4)	
x_3	0	4		1	0	1	0	—
x_4	0	18		3	②	0	1	18/2 (min) →
$Z = C_B x_B = 0$			$\Delta_j \rightarrow$	3	5 ↑	0	0 ↓	
x_3	0	4		1	0	1	0	
x_2	5	9		3/2	1	0	1/2	
$Z = C_B x_B = 45$			$\Delta_j \rightarrow$	−9/2	0	0	−5/2	

Clearly, all $\Delta_j \leq 0 \Rightarrow$ solution is optimal and is given by
$x_1 = 0, x_2 = 9, x_3 = 4, x_4 = 0$ and max. $Z = 45$
Now, the initial basis of the original LPP is

$$B = (\alpha_3, \alpha_4) \ and \ B^{-1} = \begin{bmatrix} 1 & 0 \\ 0 & 1/2 \end{bmatrix}$$

If one extra variable x_5 with price 7 is added to the original LPP, then new LPP is

max. $Z' = 3x_1 + 5x_2 + 7x_5$

s.t.

$$x_1 + x_2 + x_5 = 4$$
$$3x_1 + 2x_2 + x_4 + 2x_5 = 18$$

and $x_1, x_2, x_3, x_4, x_5 \geq 0$

Here, we have

$$\alpha_5 = \begin{bmatrix} 1 \\ 2 \end{bmatrix}, \text{ then } x_5 = B^{-1}\alpha_5 = \begin{bmatrix} 1 & 0 \\ 0 & 1/2 \end{bmatrix}\begin{bmatrix} 1 \\ 2 \end{bmatrix} = \begin{bmatrix} 1 \\ 1 \end{bmatrix}$$

Now, $\Delta_5 = c_5 - z_5 = c_5 - C_B B^{-1}\alpha_5$

$$= 7 - (0,5)\begin{bmatrix} 1 & 0 \\ 0 & 1/2 \end{bmatrix}\begin{bmatrix} 1 \\ 2 \end{bmatrix}$$

$$= 7 - (0,5)\begin{bmatrix} 1 \\ 1 \end{bmatrix} = 7 - 5 = 2 > 0$$

$\Rightarrow$ solution is not optimal.

Thus, it can be improved by introducing α_5 the column corresponding to new variable in the last simplex table. So, we start with the last simplex table as follows:

Simplex table-2

B.V.	C_B	x_B	x_1	x_2	x_3	x_4	x_5	Min Ratio
		$c_j \rightarrow$	3	5	0	0	7	
x_3	0	4	1	0	1	0	①	4/1 (min) →
x_2	5	9	3/2	1	0	1/2	1	9/2
$Z = C_B x_B = 45$			$\Delta_j \rightarrow$ −9/2	0	0	−5/2	2	
			↓				↑	
x_5	7	4	1	0	1	0	1	
x_2	5	5	1/2	1	−1	1/2	0	
$Z = C_B x_B = 53$			$\Delta_j \rightarrow$ −13/2	0	−2	−5/2	0	

Clearly, in the last row of the above table all $\Delta_j \leq 0$

$\Rightarrow$ solution is optimal and is given by

$x_1 = 0, x_2 = 5, x_3 = 0, x_4 = 0, x_5 = 4$ and max $Z' = 53$

EXAMPLE 2. *Consider the following LPP*

max. $Z = 3x_1 + 5x_2$

subject to the constraints

$$3x_1 + 2x_2 \leq 18$$
$$x_1 + 2x_2 \leq 4$$
$$x_2 \leq 6$$

and $x_1, x_2 \geq 0$

Find the optimal solution of the given LPP. Also

(i) *if the variable x_6 is added to the given LPP, then find an optimal solution to the new LPP. It is given that the coefficient of x_6 in the constraints of the problem are 1, 1 and 1 and its coefficients in the objective function is 2.*

(ii) *Discuss the effect on the optimal basic feasible solution by adding a new constraints $2x_1 + x_2 \leq 8$ to the given set of constraints.*

SOLUTION. Apply the simplex method in a usual manner, the final simplex table is given as under

Simplex table-1

		$c_j \rightarrow$	3	5	0	0	0
B.V.	C_B	x_B	x_1	x_2	s_1	s_2	s_3
x_1	3	2	1	0	1/3	0	−2/3
s_2	0	0	0	0	−2/3	1	4/3
x_2	5	6	0	1	0	0	1
$Z = C_B x_B = 36$		$\Delta_j \rightarrow$	0	0	−1	0	−3

Clearly, all $\Delta_j \leq 0 \Rightarrow$ solution is optimal and is given by

$$x_1 = 2, x_2 = 6 \text{ and max. } Z = 36$$

(i) After adding a new variable x_6 as given, the new LPP is

$$\text{max. } Z = 3x_1 + 5x_2 + 2x_6$$
$$\text{s.t.} \quad 3x_1 + 2x_2 + x_6 \leq 18$$
$$x_1 + 2x_2 + x_6 \leq 4$$
$$x_1 + x_6 \leq 6$$
$$\text{and} \quad x_1, x_2, x_6 \geq 0$$

Clearly, the column vector associated with variable x_6 is $a_6 = (1, 1, 1)$. Then using above table, we have

$$y_6 = \mathbf{B}^{-1} a_6 = \begin{bmatrix} 1/3 & 0 & -2/3 \\ -2/3 & 1 & 4/3 \\ 0 & 0 & 1 \end{bmatrix} \begin{bmatrix} 1 \\ 1 \\ 1 \end{bmatrix} = \begin{bmatrix} -1/3 \\ 5/3 \\ 1 \end{bmatrix}$$

$\because \quad C_B = (3, 0, 5)$. Therefore,

$$c_6 - z_6 = c_6 - C_B y_6$$

$$= 2 - (3, 0, 5) \begin{bmatrix} -1/3 \\ 5/3 \\ 1 \end{bmatrix} = -2 \leq 0$$

$\Rightarrow$ Optimality of the current solution remains unaffected with the addition of x_6.

(ii) Since, the optimal basic feasible solution given in the above table does not satisfy the additional constraint $2x_1 + x_2 \leq 8$, so use additional slack variable for this constraint, the above table becomes

Simplex table-2

		$c_j \rightarrow$	3	5	0	0	0	0
B.V.	C_B	x_B	x_1	x_2	s_1	s_2	s_3	s_4
x_1	3	2	1	0	1/3	0	−2/3	0
s_2	0	0	0	0	−2/3	1	4/3	0
x_2	5	6	0	1	0	0	1	0
s_4	0	8	0	1	0	0	0	1
$Z = C_B x_B = 36$		$\Delta_j \rightarrow$	0	0	−1	0	−3	0

We observe that the matrix $\mathbf{B}$ has been changed due to row 4, so coefficient

in row 4 must be zero, which can be done by using the following operations.
$$R_4(\text{new}) \rightarrow R_4(\text{old}) - 2R_1 - R_3$$
Then we have the following table.

Simplex table-3

B.V.	C_B	x_B	$c_j \rightarrow$ 3 x_1	5 x_2	0 s_1	0 s_2	0 s_3	0 s_4
x_1	3	2	1	0	1/3	0	–2/3	0
s_2	0	0	0	0	–2/3	1	4/3	0
x_2	5	6	0	1	0	0	1	0
s_4	0	–2	0	0	(–2/3)	0	1/3	1
$Z = 36$		$\Delta_j \rightarrow$	0	0	–1 ↑	0	1/3	1 ↓

The solution in the above table is optimal but not feasible. So, we apply the dual simplex method in a usual manner, the new obtained solution is given in the following table.

Simplex table-4

B.V.	C_B	x_B	$c_j \rightarrow$ 3 x_1	5 x_2	0 s_1	0 s_2	0 s_3	0 s_4
x_1	3	1	1	0	0	0	–1/2	1/2
s_2	0	2	0	0	0	1	1	–1
x_2	5	6	0	1	0	0	1	0
s_1	0	3	0	0	1	0	–1/2	–3/2
$Z = C_B x_B = 33$		$\Delta_j \rightarrow$	0	0	0	0	–7/2	–3/2

We observe that all $\Delta_j \le 0 \Rightarrow$ solution is optimal feasible and is given by
$$x_1 = 1, x_2 = 6 \text{ and max. } Z = 33$$

EXAMPLE 3. *Consider the following table which presents an optimal solution to some LPP*

B.V.	C_B	x_B	$c_j \rightarrow$ 2 x_1 (α_1)	4 x_2 (α_2)	1 x_3 (α_3)	3 x_4 (α_4)	2 x_5 (α_5)	0 s_1 (α_6)	0 s_2 (α_7)	0 s_3 (α_8)
x_1	2	3	1	0	0	–1	0	1/2	1/5	–1
x_2	4	1	0	1	0	2	1	–1	0	1/2
x_3	1	7	0	0	1	–1	–2	5	–3/10	2
$Z = C_B x_B = 17$		$\Delta_j \rightarrow$	0	0	0	–2	0	–2	–1/10	–2

(i) *If the additional constraints $2x_1 + 3x_2 - x_3 + 2x_4 - 4x_5 \le 5$ were annexed to the system, would there be any change in the optimal solution.*

(ii) *If the additional constraints $3x_1 + x_2 + 2x_3 + x_4 + 9x_5 \le 19$ were annexed to the system, would there be any change in the optimal solution? If yes, find the new optimal solution.*

SOLUTION.
(i) From the given table, we observe that all $\Delta_j \le 0$ and $x_1 = 3, x_2 = 1, x_3 = 7,$
$x_4 = x_5 = 0$

$\Rightarrow$ solution is optimal feasible.

$\because$ New constraint is $2x_1 + 3x_2 - x_3 + 2x_4 - 4x_5 \leq 5$
clearly satisfies the obtained solution
$(\because 2x_1 + 3x_2 - x_3 + 2x_4 - 4x_5 = 2(3) + 3(1) - 7 + 2(0) - 4(0) = 2 < 5)$
Hence, this is also the optimal solution of the new problem and new constraint is redundant one.

(ii) Clearly, the obtained solution $x_1 = 3, x_2 = 1, x_3 = 7, x_4 = x_5 = 0$ does not satisfy the new constraint $3x_1 + x_2 + 2x_3 + x_4 + 9x_5 \leq 19$ so solution is not optimal for the new problem.

Now, we have to find the optimal solution of the new problem.
Introducing slack variable s_4 to the new constraint, we get
$$3x_1 + x_2 + 2x_3 + x_4 + 9x_5 + 0s_1 + 0s_2 + 0s_3 + s_4 = 19$$
The optimal basis of the original problem is $B = (\alpha_1, \alpha_2, \alpha_3)$
If α_9 is the column vector corresponding to slack variable s_4, then

$$\alpha_9 = \begin{bmatrix} 0 \\ 0 \\ 0 \\ 1 \end{bmatrix}$$

We assume that the cost price $c_9 = 0$ in the new objective function. Now introducing this constraint in optimal table of the original problem, we get the following simplex table.

Simplex table-1

B.V.	C_B	x_B	$c_j \rightarrow$								
			2	4	1	3	2	0	0	0	0
			x_1 (α_1)	x_2 (α_2)	x_3 (α_3)	x_4 (α_4)	x_5 (α_5)	s_1 (α_6)	s_2 (α_7)	s_3 (α_8)	s_4 (α_9)
x_1	2	3	1	0	0	-1	0	1/2	1/5	-1	0
x_2	4	1	0	1	0	2	1	-1	0	1/2	0
x_3	1	7	0	0	1	-1	-2	5	-3/10	2	0
s_4	0	19	3	1	2	1	9	0	0	0	1

Here, we observed that identity matrix is disturbed (due to the inclusion of fourth row). Now identity matrix can be obtained by performing $R_4 - 3R_1$, $R_4 - R_2$ and $R_4 - 2R_3$.
Then we get the following simplex table.

Simplex table-2

B.V.	C_B	x_B	$c_j \rightarrow$								
			2	4	1	3	2	0	0	0	0
			x_1 (β_1)	x_2 (β_2)	x_3 (β_3)	x_4	x_5	s_1	s_2	s_3	s_4 (β_4)
x_1	2	3	1	0	0	-1	0	1/2	1/5	-1	0
x_2	4	1	0	1	0	2	1	-1	0	5	0
x_3	1	-7	0	0	1	-1	-2	5	-3/10	2	0
x_4	0	-5	0	0	0	4	12	-21/2	0	-3/2	1
$Z = C_B x_B = 17$			$\Delta_j \rightarrow$ 0	0	0	-2	0	-2	-1/10	-2	0

Here, we observe that all $\Delta_j \leq 0$.

$\Rightarrow$ Solution is optimal but not feasible (as $s_4 = -5$). Therefore, we apply the dual simplex method to improve the present solution.

To find the outgoing vector:

$$\because \quad x_{B_r} = \min\{x_{B_i} : x_{B_i} < 0\}$$

$$= \min\{x_{B_1}, x_{B_2}, x_{B_3}, x_{B_4}\}$$

$$= x_{B_4} \qquad\qquad (\because x_{B_1}, x_{B_2}, x_{B_3} > 0)$$

$$\Rightarrow \quad r = 4$$

$$\Rightarrow \quad \beta_4 \text{ is the outgoing vector.}$$

To find incoming vector:

$$\because \quad \frac{\Delta_k}{a_{rk}} = \min_j\left\{\frac{\Delta_j}{a_{rj}}, a_{rj} < 0\right\} = \min_j\left\{\frac{\Delta_j}{a_{4j}}, a_{4j} < 0\right\} \text{ for } j = 4,5,6,7,8$$

$$= \min\left\{\frac{\Delta_6}{a_{46}}, \frac{\Delta_7}{a_{47}}\right\} \qquad (\because \text{ only } a_{46}, a_{47} < 0)$$

$$= \min\left\{\frac{-2}{-21/2}, \frac{-2}{-3/2}\right\} = \min\left\{\frac{4}{21}, \frac{4}{3}\right\} = \frac{4}{21} = \frac{\Delta_6}{a_{46}}$$

$$\Rightarrow \quad k = 6$$

$$\Rightarrow \quad \alpha_6(= s_1) \text{ is the incoming vector}$$

and key element $= a_{46} = \dfrac{-21}{2}$

Now, we have the following simplex table

Simplex table-3

B.V.	$c_j \rightarrow$		2	4	1	3	2	0	0	0	0
	C_B	x_B	x_1 (β_1)	x_2 (β_2)	x_3 (β_3)	x_4	x_5	s_1 (β_4)	s_2	s_3	s_4
x_1	2	58/21	1	0	0	−17/21	4/5	0	1/5	−15/14	1/21
x_2	4	31/21	0	1	0	34/21	−1/7	0	0	9/14	−2/21
x_3	1	97/21	0	0	1	19/21	26/7	0	−3/10	9/7	10/21
s_1	0	10/21	0	0	0	−8/21	−8/7	1	0	−3/2	−2/21
$Z = C_B x_B$ $= 337/21$	$\Delta_j \rightarrow$		0	0	0	−58/21	−16/7	0	−1/10	−12/7	−4/21

From the above table, we observe that all $\Delta_j \leq 0$ and

$$x_1 = \frac{58}{21}, x_2 = \frac{31}{21}, x_3 = \frac{97}{21}, x_4 = 0, s_1 = \frac{10}{21}, s_2 = 0, s_3 = 0 \text{ and } s_4 = 0$$

$\Rightarrow$ solution is optimal and feasible

Hence, optimal solution is given by

$$x_1 = \frac{58}{21}, x_2 = \frac{31}{21}, x_3 = \frac{97}{21}, x_4 = 0$$

and $\max Z = \dfrac{337}{21}$, which is less than the original maximum value of Z.

Exercise-8.1

1. Solve the following LPP:

   max. $Z = 15x_1 + 45x_2$

   subject to the constraints

   $$x_1 + 16x_2 \leq 240$$
   $$0.5x_1 + 2x_2 + x_3 = 162$$
   $$x_2 + x_4 = 50$$

   and $x_1, x_2, x_3, x_4 \geq 0$

   Find how much can c_1 be changed without affecting the optimality of the solution.

2. Find an optimal solution of the following LPP

   max. $Z = 5x_1 + 3x_2$

   subject to the constraints

   $$3x_1 + 5x_2 \leq 15$$
   $$5x_1 + 2x_2 \leq 10$$

   and $x_1, x_2 \geq 0$

   Also, find the range of c_1 without affecting the optimality of the solution.

3. Solve the following LPP

   max. $Z = 3x_1 + 4x_2 + x_3 + 7x_4$

   subject to the constraints

   $$8x_1 + 3x_2 + 4x_3 + x_4 \leq 7$$
   $$2x_1 + 6x_2 + x_3 + 5x_4 \leq 3$$
   $$x_1 + 4x_2 + 5x_3 + 2x_4 \leq 8$$

   and $x_1, x_2, x_3, x_4 \geq 0$

   Discuss the effect of discrete changes in the requirements, i.e., b_1, b_2 and b_3 so that the solution remains optimal feasible.

4. Solve the following LPP

   max. $Z = 3x_1 + 5x_2$

   subject to the constraints

   $$x_1 + x_3 = 4$$
   $$3x_1 + 2x_2 + x_4 = 18$$

   and $x_1, x_2, x_3, x_4 \geq 0$

   Would there is any change in the optimal solution if the additional constraints

   (i) $x_2 \leq 10$ or (ii) $x_2 \leq 6$

   is added to the above LPP. In case the optimal solution changes, find the new optimal feasible solution.

5. Solve the following LPP

   max. $Z = 2x_1 + x_2 + 3x_3$

   subject to the constraints

   $$x_1 + x_2 + 2x_3 \leq 5$$
   $$2x_1 + 3x_2 + 4x_3 \leq 12$$

   and $x_1, x_2, x_3 \geq 0$

   What will happen if a new constraint $2x_1 + 2x_2 + 4x_3 \geq 14$ is added?

6. Consider the following table which presents an optimal solution to some LPP

B.V.	C_B	x_B	x_1	x_2	x_3	s_1	s_2
		c_j					
x_1	2	1	1	0	1/2	4	−1/2
x_2	3	2	0	1	1	−1	2
$Z = C_B x_B$ = 8		x_j	1	2	0	0	0
		Δ_j	0	0	−3	−5	−5

For the above problem assuming that s_1 and s_2 were in that order in the initial identity matrix. Calculate the following:

(i) How much can be b_1 and b_2 be increased without effecting the optimality and feasibility of the solution.

(ii) How much c_3 can be increased before the present basic solution will no longer to be optimal.

Answers

1. $x_1 = 184, x_2 = 35, x_3 = 0, x_4 = 15, \max Z = 4335$ and $\dfrac{45}{4} < c_1 \leq \dfrac{225}{8}$

2. $x_1 = \dfrac{20}{19}, x_2 = \dfrac{45}{19}, \max. Z = \dfrac{235}{19}, \dfrac{9}{5} \leq c_1 \leq \dfrac{15}{2}$

3. $x_1 = \dfrac{16}{19}, x_2 = 0, x_3 = 0, x_4 = \dfrac{5}{19}, \max. Z = \dfrac{83}{19}; \dfrac{3}{5} \leq b_1 \leq 12; \dfrac{7}{4} \leq b_2 \leq \dfrac{323}{25}; \dfrac{28}{19} \leq b_3 < \infty$

4. $x_1 = 0, x_2 = 9, x_3 = 4, x_4 = 0, \max. Z = 45$

   (i) No change (ii) $x_1 = 2, x_2 = 6, x_3 = 2, x_4 = 0, \max. Z = 36$

5. $x_1 = 3, x_2 = 2, x_3 = 0, \max. Z = 8$. When $2x_1 + 2x_2 + 4x_3 \geq 14$ is added there is no feasible solution

6. (i) $-\dfrac{1}{4} \leq \Delta b_1 \leq 2; -1 \leq \Delta b_2 \leq 2$ (ii) $-\infty < c_3 \leq 4$

⚙ REVIEW QUESTIONS

1. Write a short note on sensitivity analysis.
2. Discuss the effect of addition of a constraint in sensitivity analysis.
3. Find the limit of variation of elements a_{ik} so that optimal feasible solution of $Ax = b$, $x \geq 0$, max. $Z = Cx$ remains optimal feasible solution when:
   (i) $a_k \in B$ (ii) $a_k \notin B$
4. Discuss the effect of discrete changes in the requirement (on RHS of the inequality) for the LPP of maximization.

🖎 MULTIPLE CHOICE QUESTIONS (CHOOSE THE MOST APPROPRIATE ONE)

1. Sensitivity analysis:
   (i) is also called post optimality analysis as it is carried out after the optimal solution is obtained
   (ii) allows the decision-maker more meaningful information about changes in the LP parameters
   (iii) provides the range within which a parameter may change without affecting optimality
   (a) (i)   (b) (ii)
   (c) (ii), (iii)   (d) All of the above

2. The addition of a new variable to give LPP then problem is called:
   (a) analysis
   (b) sensitivity analysis
   (c) mono analysis
   (d) none of these

3. If $c_k \notin C_B$ changes to $c_k + \Delta c_k$ such that $\Delta c_k = -\Delta k$, there is no lower bound to:
   (a) Δc_j (b) Δc_k
   (c) Δb_j (d) None of these

4. If in a LPP, the changes in the requirement vector b_j is called:
   (a) transportation analysis
   (b) dual analysis
   (c) sensitivity analysis
   (d) none of these

5. In a LPP, the changes in the elements a_{ij} of the coefficient matrix is called:
   (a) sensitivity analysis
   (b) post analysis
   (c) both (a) and (b)
   (d) none of these

6. If $c_k \notin C_B$ changes to $c_k + \Delta c_k$ such that $\Delta c_k \leq z_k - c_k (= -\Delta k)$ the value of the objective function and the optimal solution of the problem remains:
   (a) changed   (b) unchanged

   (c) unbounded   (d) none of these

7. The changes in the optimal solutions due to discrete variation in the parameters a_{ij}, b_j and c_j are called:
   (a) sensitivity analysis
   (b) dual analysis
   (c) simplex analysis
   (d) none of these

8. The addition of a new constraint to given LPP. Then the problem is called:
   (a) post analysis   (b) mono analysis
   (c) analysis   (d) none of these

9. If in a LPP the variations in the price vector C is called:
   (a) simplex analysis
   (b) sensitivity analysis
   (c) transportation analysis
   (d) none of these

10. In a LPP the variation in $c_j \notin C_B$ then its changes to:
    (a) $c_k + \Delta c_k$ (b) $c_j + \Delta c_j$
    (c) c_j (d) None of these

11. If in a LPP the variations in the price vector 'C' and requirement vector 'b' is called:
    (a) post analysis
    (b) sensitivity analysis
    (c) both (a) and (b)
    (d) none of these

12. Which of the following is not correct?
    (a) After the attainment of an optimum solution of an LPP, it is desired to study the effect of changes in the different parameters of the problem on the Current Optimum Solution
    (b) An analysis of post optimal solutions is known as Post-Optimality analysis or Sensitivity analysis
    (c) Post optimality analysis study only the continuous changes in the parameters of LPP

(d) Post optimality analysis form an integral part of formulating an LPP

13. LP context, post-optimal analysis is a technique to:

 (a) determine how optimum solution to an LPP changes is response to problem inputs

 (b) allocate resources optimality

 (c) minimize cost of operations

 (d) spellout the relation between dual and its primal

14. Which of the following is not correct?

 (a) For any changes in the objective function coefficients, the optimal function values of the decision variables would change

 (b) The optimality of the current solution may be affected if right hand side of the constraints is changed

 (c) The feasibility of the current optimum solution may be affected if right hand side of the constraint is changed

 (d) When a new constraint is introduced or one of the current constraint is deleted from an LPP, the post-optimal analysis is due to structural changes

15. Which of the following is not correct?

 (a) Post-optimal analysis is normally carried out after the optimum solution is reached

 (b) Addition of a constraint may affect the current optimum solution

 (c) Addition of a new variable may disturb the feasibility of the current optimum solution

 (d) Addition of new constraints in an LPP can never improve the optimal value of the objective function

16. Which of the following is not correct?

 (a) Changes in the right hand side values of the constraints within the allowable limits would neither change the basis nor the objective function value of an LPP

 (b) Deletion of an existing variable may affect the feasibility of the current optimum Solution

 (c) When multiple changes take place in the objective function or in the RHS values of the constraints, then the cent-percent Rule may be used to determine whether they would affect the current solution

 (d) Changes in the coefficient matrix of the constraints can be analysed to determine their effect on the optimum solution

17. If the optimal solution of the original LPP satisfies the new constraint, it is also an optimal solution of new LPP. In this case the _____ constraint is redundant.

 (a) additional   (b) subtraction

 (c) multiplying   (d) none of these

18. If the solution of LPP is optimal for $c_j = a$ and $c_j = b$ then the variation in c_j will be:

 (a) $a < c_j < b$ (b) $a < c_j \leq b$

 (c) $a \leq c_j < b$ (d) $a \leq c_j \leq b$

19. In a LPP, the variation in $c_j \in C_B$, the range of ΔC_{B_k} such that the solution remains optimal is given by $\max\limits_{y_{kj}>0}\left[\dfrac{c_j - z_j}{y_{kj}}\right] \leq \underline{\quad} \geq \min\limits_{y_{kj}<0}\left[\dfrac{c_j - z_j}{y_{kj}}\right]$:

 (a) C_{B_k} (b) ΔC_{B_k}

 (c) Δc_j (d) None of these

20. If the addition of new constraint alters the nature of the problem, then the new problem must be solved as a:

 (a) fresh problem   (b) above Problem

 (c) no problem   (d) none of these

21. If the solution of LPP is not optimal for $c_j = a$ and $c_j = b$, then the variation in c_j will be:

 (a) $a < c_j < b$ (b) $a < c_j < b$

 (c) $a \leq c_j < b$ (d) $a \leq c_j \leq b$

22. If no $y_{kj} > 0$ there is no _____ to ΔC_{B_k}:

 (a) change   (b) lower bound

 (c) upper bound   (d) None of these

23. If the solution of LPP is not optimal for $c_j = a$, then the variation in c_j will be:

 (a) $a < c_j \leq b$ (b) $a \leq c_j < b$

 (c) $a \leq c_j \leq b$ (d) $a < c_j < b$

24. If the optimal solution of the original LPP does not satisfies the new constraint, it is also an optimal solution of the new constraint, it is also an optimal solution of the new LPP. In this case the _____ constraint is not redundant:

 (a) subtracting   (b) additional

 (c) multiplying   (d) none of these

25. Addition of a new constraint and deletion of an existing constraint simultaneously to a LPP:

 (a) disturbs feasibility only

 (b) disturbs optimality only

 (c) may disturbs both feasibility and optimality

 (d) None of these

26. Change in availability vector and addition of a new constraint simultaneously to a LPP:

 (a) may disturb feasibility

(b) may disturb optimality

(c) may disturb both feasibility and optimality

(d) None of these

27. Change in availabilities and costs of a LPP simultaneouly:

(a) disturb feasibility only

(b) disturb optimality only

(c) may disturb both feasibility and optimality

(d) None of these

28. Addition of a new variable and deletion of an existing variable to LPP simultaneously:

(a) disturb feasibility only

(b) disturb optimality only

(c) may disturb both feasibility and optimality

(d) None of these

29. Let the S_F of LPP be non empty and bounded (that is both from above and below) and let an additional constraint be added to it. Then the new S_F:

(a) may become empty

(b) may become unbounded and consequently the solution may become unbounded

(c) may become empty or may become unbounded

(d) None of these

30. Let the S_F of a LPP be a non-empty and bounded (that is both from above and below) and let a constraint be deleted from it. Then the new S_F:

(a) may become empty

(b) may become unbounded and consequently the solution may become unbounded

(c) may become empty or may become unbounded

(d) None of these

31. The use of cutting plane method:

(a) yields better value of objective function

(b) reduces the Number of constraints in the given problem

(c) require use of standard LP approach between each cutting plane application

(d) all of the above

32. While solving LP problem any Non-integer variable in the solution is picked up to:

(a) enter the solution

(b) obtain the cut constraint

(c) leave the solution

(d) None of these

33. Which of the following is the consequence of adding a new cut constraint to an Optimal Simplex table:

(a) addition of a new variable to the table

(b) makes the previous Optimal Solution infeasible

(c) eliminates non-integer solution from the solution space

(d) all of the above

34. To ensure best marginal increase in the objective function value, a resource value may be increased whose shadow price is comparatively:

(a) larger

(b) smaller

(c) neither (a) nor (b)

(d) both (a) and (b)

35. In sensitivity analysis of the coefficient of the non-basic variable in cost minimization LP problem, the upper sensitivity limit is:

(a) original value + lowest positive value of improvement ratio

(b) original value – lowest absolute value of improvement ratio

(c) positive infinity

(d) negative infinity

36. In a mixed integer programming problem:

(a) all of the decision variables require integer solution

(b) few of the decision variables require integer solution

(c) different objective functions are mixed together

(d) None of these

37. In a Branch and bound minimization tree, the lower bounds on objective function value:

(a) do not decrease in value

(b) do not increase in value

(c) remain constant

(d) None of these

38. The 0-1 integer programming problem:

(a) requires the decision variables to have values between zero and one

(b) requires that the constraints all have coefficients between zero and one

(c) requires that the decision variables have coefficients between zero and one

(d) all of the above

39. Addition of an additional constraint in the existing constraints will cause a:

(a) change in objective function coefficients (c_j)

(b) change in coefficients a_{ij}

(c) both (a) and (b)

(d) none of the above

40. In the Branch and bound approach to a max. problem, a node is terminated if:

(a) a node has an infeasible solution

(b) a node yields a solution that is feasible but not an integer

(c) upper bound is less than the current sub problem's lower bound

(d) all of the above

41. The entering variable in the sensitivity analysis of objective function coefficients is always a:

(a) decision variable (b) non-basic variable

(c) basic variable (d) slack variable

42. The part of the feasible solution space eliminated by plotting a cut contains:

(a) only non-integer solutions

(b) only integer solutions

(c) both (a) and (b)

(d) none of these

43. Branch and bound method divides the feasible solution space into smaller parts by:

(a) branching (b) bounding

(c) enumerating (d) all of the above

44. Sensitivity analysis:

(a) is also called post-optimality analysis as it is carried out after the optimal solution is obtained

(b) allows the decision maker more meaningful information about the changes in the LP model parameter

(c) provides the range within which a parameter may change without affecting optimality

(d) all of the above

45. To obtain Optimality of Current Optimal Solution for a change Δc_k in the coefficient c_k of non-basic variable X_k, we must have:

(a) $\Delta c_k = z_k - c_k$ (b) $\Delta c_k = z_k$

(c) $c_k + \Delta c_k = z_k$ (d) $\Delta c_k \geq z_k$

46. When an additional variable is added in LP model, the existing Optimal Solution can further be improved if:

(a) $z_j - c_j \leq 0$ (b) $z_j - c_j \geq 0$

(c) both (a) and (b) (d) none of the above

47. While performing sensitivity analysis, the Upper bound infinity of the value of the right hand side of a constraint means that:

(a) the Constraint is redundant

(b) the Shadow price for the constraint is zero

(c) there is Slack in the constraint

(d) None of the above

48. Rounding off solution values of decision variables in a LP Problem may not be acceptable because:

(a) it does not satisfy constraints

(b) it violates non-negativity conditions

(c) objective function value is less than the objective function value of LP

(d) None of the above

49. If the additional constraint is added in an equation and an artificial variable appear in the basis of the new problem, the new optimal solution is obtained by:

(a) assigning Zero cost coefficient to the artificial variable if it appears in the basis at negative value

(b) assigning $-M$ cost coefficient to the artificial variable if it appears in the basis at positive value

(c) either (a) or (b)

(d) none of the above

50. A Non-basic variable should be brought into the new solution mix provided its contribution rate (c_j) is:

(a) $c_j^* = c_j + (z_j - c_j)$ (b) $c_j^* > c_j + (z_j - c_j)$

(c) $c_j^* < c_j + (z_j - c_j)$ (d) none of these

Answers

1. (a)	2. (b)	3. (b)	4. (c)	5. (a)	6. (b)	7. (a)	8. (a)	9. (b)
10. (a)	11. (c)	12. (c)	13. (a)	14. (b)	15. (d)	16. (b)	17. (a)	18. (d)
19. (b)	20. (b)	21. (c)	22. (b)	23. (a)	24. (b)	25. (c)	26. (a)	27. (c)
28. (c)	29. (a)	30. (b)	31. (c)	32. (b)	33. (d)	34. (a)	35. (c)	36. (b)
37. (b)	38. (a)	39. (c)	40. (d)	41. (d)	42. (a)	43. (a)	44. (b)	45. (c)
46. (a)	47. (b)	48. (d)	49. (c)	50. (c)				

□□□□

Parametric Linear Programming

9.1 INTRODUCTION

In sensitivity analysis, we study the changes in the problem data that can be made without changing the optimal solution, *i.e.*, we did not concern about the variable that would enter the basis and the variable that would leave the basis, *i.e.*, in sensitivity analysis, we have consider the impact on optimal solution of LP model due to discrete changes in parameters. In this chapter we will discuss another parameter variation analysis, known as parametric analysis to obtain various feasible solution of LP model which become optimal one after the other due to continuous variations in the parameters.

9.2 PARAMETRIC PROGRAMMING

When LP model parameters changes as a linear function of a single parameter then this technique is called linear parametric programming. In this analysis we have to keep a minimum additional efforts required to take care of changes in the optimal solution due to variation in LP model parameters over a range of variation.

Definition. *The investigation which deals with the effect of simultaneous changes of all components of $\boldsymbol{C}$ or $\boldsymbol{b}$ in the optimal solution of the problem is called parametric linear programming.*

In this chapter we shall discuss the parametric analysis only for following two parameters:

(1) systematic variation in the objective function coefficients, c_j
(2) systematic variation in resource availability, b_i

9.3 SYSTEMATIC VARIATION IN THE OBJECTIVE FUNCTION COEFFICIENTS, c_j

Let us define the parametric linear programming as follows:

$$\left.\begin{array}{l} \max.Z = (\boldsymbol{C} + \boldsymbol{C}'\lambda)\boldsymbol{x} = \boldsymbol{C}^* \cdot \boldsymbol{x} \\ \text{subject to the constraints} \\ \qquad \boldsymbol{Ax} = \boldsymbol{b} \\ \text{and } x \geq 0 \end{array}\right] \qquad \ldots(1)$$

We have to find the solution of (1) for each λ such that $\delta \leq \lambda < \phi$ where δ is very small and finite, ϕ finite and large and $\boldsymbol{C}, \boldsymbol{C}', \boldsymbol{b}$ are finite known vectors.

Let us assume that the problem is non-degenerate and has a basic feasible solution, therefore it can be solved for $\lambda = \delta$ by simplex method. Here, we may have following two cases:

(i) the problem has a finite optimal solution for $\lambda = \delta$.

(ii) the problem has no finite optimal solution for $\lambda = \delta$.

CASE (1) WHEN THE PROBLEM HAS A FINITE OPTIMAL SOLUTION FOR $\lambda = \delta$

Let us suppose that B be the optimal basis, x_B the optimal solution, C_B^* the corresponding price vector and z_j^*, the value of z_j at $\lambda = \delta$.

Clearly, $x_B = B^{-1}b$ implies it is independent of $C^* = C + C'\lambda$

$\Rightarrow$ feasibility of the solution remains unaffected when C^* is changed due to the change in λ.

Also, for the optimality, we must have $c_j - z_j^* \leq 0$ for all j not in the basis B.

We observed that, the condition of optimality will be affected due to the variation in C^*, because $c_j^* - z_j^*$ depends upon C^*. Therefore, the change in λ, cause the disturbances in optimality. Thus, in case of λ increases through δ, the solution x_B will remain optimal if $c_j^* - z_j^* \leq 0 \ \forall j$.

Now, consider

$$c_j^* - z_j^* = c_j^* - C_B \cdot B^{-1} \cdot \alpha_j$$

$$= (c_j + c_j'\lambda) - (C_B + C_B'\lambda)B^{-1}\alpha_j$$

$$= c_j - C_B \cdot B^{-1} \cdot \alpha_j + \lambda(c_j - C_B' \cdot B^{-1}\alpha_j)$$

$$= c_j - z_j + \lambda(c_j' - z_j') \qquad\qquad \text{...(2)}$$

Clearly, x_B will remain optimal for those λ for which

$$c_j - z_j + \lambda(c_j' - z_j') \leq 0 \text{, for } j \text{ not in the basis.} \qquad \text{...(3)}$$

But if we consider $\lambda = \delta = 0$ (by shifting the origin at δ). Then from (2)

$$c_j^* - z_j^* = c_j - z_j \qquad\qquad\qquad \text{(for } j \text{ not in the basis } B)$$

$$\leq 0 \qquad\qquad\qquad (\because x_B \text{ is optimal and feasible solution)}$$

Now, we have following two cases:

(i) If $c_j' - z_j' \leq 0 \ \forall j$ not in B, then from (3) $c_j^* - z_j^* \leq 0$. In this case the solution x_B will be optimal and feasible for all values of $\lambda \geq \delta$.

(ii) If $c_j' - z_j' < 0$ for at least one j, then x_B will be optimal for those λ for which $c_j^* - z_j^* < 0$.

$\Rightarrow \ (c_j - z_j) + \lambda(c_j' - z_j') \leq 0$

which can be written as

$$\lambda \geq -\frac{(c_j - z_j)}{(c_j' - z_j')} \text{ for } c_j - z_j > 0$$

and

$$\lambda \leq -\frac{(c_j - z_j)}{(c_j' - z_j')} \text{ for } c_j' - z_j' < 0$$

On combining both the above inequalities, we have

$$\max_{(c_j' - z_j') < 0} \left[-\frac{c_j - z_j}{c_j' - z_j'} \right] \leq \lambda \leq \min_{c_j' - z_j' > 0} \left[-\frac{c_j - z_j}{c_j' - z_j'} \right] \qquad \text{...(4)}$$

Here, we observe that

(i) If there is no $c'_j - z'_j < 0$, then λ has no lower bound.

(ii) If there is no $c'_j - z'_j > 0$, then λ has no upper bound.

Now, let $\overline{\lambda} = \min\limits_{c'_j - z'_j > 0} \left[-\dfrac{c_j - z_j}{c'_j - z'_j} \right]$ and $\underline{\lambda} = \max\limits_{c'_j - z'_j < 0} \left[-\dfrac{c_j - z_j}{c'_j - z'_j} \right]$

Obviously, the solution x_B will remain optimal and feasible when $\underline{\lambda} \leq \lambda \leq \overline{\lambda}$ ...(5)

Further if $\overline{\lambda} = \infty$ then given problem has optimal solution for all $\lambda \geq \underline{\lambda}$ which implies that the problem has optimal and feasible solution at $\lambda = \phi$.

Further, if $\overline{\lambda}$ is finite, then we have to improve the range of λ beyond $\overline{\lambda}$ as given below

If $\overline{\lambda}$ is finite, then suppose that

$$\overline{\lambda} = \min\limits_{c'_j - z'_j > 0} \left\{ -\dfrac{c_j - z_j}{c'_j - z'_j} \right\} = -\dfrac{c_k - z_k}{c'_k - z'_k} \qquad \text{...(6)}$$

Now, if at least one $a_{ik} > 0$, then to improve the range of λ, we assume that

$$\lambda = \overline{\lambda} = -\dfrac{c_k - z_k}{c'_k - z'_k}$$

$\Rightarrow \qquad (c_k - z_k) + \lambda(c'_k - z'_k) = 0$

$\Rightarrow \qquad\qquad c'_k - z'_k = 0 \qquad\qquad\qquad\qquad\qquad$ (Using (2))

Also, for $\lambda > \overline{\lambda}$, we have

$$\lambda > -\dfrac{c_k - z_k}{c'_k - z'_k}$$

$\Rightarrow \qquad (c_k - z_k) + \lambda(c'_k - z'_k) > 0$

$\Rightarrow \qquad\qquad c'_k - z^*_k > 0$

$\Rightarrow \qquad\qquad \Delta^*_k > 0$

$\Rightarrow \;\; \alpha_k$ can be introduced in the basis for the better solution of the problem.

Now, if all corresponding $a_{ik} \leq 0$ then for $\lambda > \overline{\lambda}$, the given problem has no optimal solution

$\Rightarrow$ We cannot improve the range of λ beyond $\overline{\lambda}$.

Therefore, in case of at least one $a_{ik} > 0$, we can improve the range of λ beyond $\overline{\lambda}$ and α_k is introduced in basis B.

If B_1 be the basis obtained by introducing α_k for outgoing vector α_l (selected by minimum ratio rule) and let x_{B_1} be the corresponding solution and let $(c^*_j - z^*_j)^{(1)}$ be the value of $c^*_j - z^*_j$ for new basis B_1. Then

$$\left. \begin{aligned} (c^*_j - z^*_j)^{(1)} = (c^*_j - z^*_j) - (c^*_k - z^*_k)\dfrac{a_{ij}}{a_{jk}} \text{ for } j \neq k \\[2mm] \text{and} \;\; (c^*_k - z^*_k)^{(1)} = 0 \end{aligned} \right] \qquad \text{...(7)}$$

Now, for $\lambda = \overline{\lambda}$, we have

$$(c^*_j - z^*_j)^{(1)} = (c^*_j - z^*_j) - (c^*_k - z^*_k)\dfrac{a_{ij}}{a_{jk}}$$

$$= c_j^* - z_j^*$$ $(\because \ c_k^* - z_k^* = 0 \text{ at } \lambda = \bar{\lambda})$

$$\Rightarrow \qquad (c_j^* - z_j^*)^{(1)} \leq 0 \qquad\qquad\qquad (\because c_j^* - z_j^* \leq 0)$$

Thus, the new solution x_{B_1} is optimal if $\lambda = \bar{\lambda}$. Further, in the next iteration the vector, α_l can not be further introduced if

$$(c_l^* - z_l^*)^{(1)} \leq 0$$

Now, $$(c_l^* - z_l^*)^{(1)} = (c_l^* - z_l^*) - (c_k^* - z_k^*)\frac{a_{il}}{a_{lk}}$$

$$= -\frac{c_k^* - z_k^*}{a_{lk}} \qquad\qquad (\because \text{ when } a_{il} = 1, \ c_l^* - z_l^* = 0)$$

$$= -\frac{1}{a_{lk}}[(c_k - z_k) + \lambda(c_k' - z_k')] \qquad\qquad \text{(By (2))}$$

$$\leq 0$$

$$\Rightarrow \qquad -(c_k - z_k) + \lambda(c_k' - z_k') \leq 0 \qquad\qquad\qquad (\because a_{lk} > 0)$$

$$\Rightarrow \qquad (c_k - z_k) + (c_k' - z_k') \geq 0$$

$$\Rightarrow \qquad \lambda \geq -\left[\frac{c_k - z_k}{c_k' - z_k'}\right]$$

$$\Rightarrow \qquad \lambda \geq \bar{\lambda} \qquad\qquad\qquad \left(\because \ \bar{\lambda} = -\left[\frac{c_k - z_k}{c_k' - z_k'}\right]\right)$$

which is true.

Hence, we proceed in this manner, from one range to another range of λ beyond $\bar{\lambda}$ until we arrive at $\lambda = \phi$. Clearly, above process is valid because no basis is repeated for this.

Now, suppose that $\bar{\lambda}$ is replaced by $\bar{\lambda} + \varepsilon$ where $\varepsilon > 0$ is arbitrary small. Since the problem is non-degenerate so we either solve the problem for $\lambda = \bar{\lambda} + \varepsilon$ or the problem has no finite optimal solution for $\lambda = \bar{\lambda} + \varepsilon$. Therefore, we can not remain indefinitely at a value of λ such that $\bar{\lambda} = \underline{\lambda} = \lambda$ and after getting a basis, we can not we can not return to it to any basis corresponding to lower value of λ as for $\lambda > \bar{\lambda}$, we have new solution.

☞ REMARK

- Here $\bar{\lambda}$ and $\underline{\lambda}$ are known as characteristic value of λ and the optimal solution corresponding to $\underline{\lambda}$ and $\bar{\lambda}$ are called characteristic solution.

CASE (2) IF THE PROBLEM HAS NO FINITE OPTIMAL SOLUTION FOR $\lambda = \delta$

For a basic feasible solution, in simplex method, we get some $c_j^* - z_j^* > 0$ for which all $a_{ik} \leq 0$ at any stage. Suppose $c_k^* - z_k^*$ be the corresponding value of $c_j - z_j$. Therefore, for $\lambda = \delta$, a vector a_k chosen to enter the basis cannot go to the basis as all $a_{ik} \leq 0$.

If a_k is the vector chosen to the basis, then we have

$$c_k^* - z_k^* = (c_k - z_k) + \lambda(c_k^* - z_k^*)$$

Now we have the following possibilities:

(i) If $c_k' - z_k' \geq 0$. Then clearly we have

$$(c_k - z_k) + \lambda(c'_k - z'_k) > 0 \text{ for all } \lambda > \delta$$

which imples that

$$(c^*_k - z^*_k) > 0 \text{ for all } \lambda > \delta$$

$\Rightarrow \quad \Delta^*_k > 0$ for all $\lambda > \delta$

$\Rightarrow$ Problem has no finite optimal solution for any $\lambda \geq \delta$.

(ii) If $c'_k - z'_k < 0$, then

$$(c_k - z_k) + \lambda(c'_k - z'_k) > 0$$

$$\Rightarrow \qquad\qquad (c^*_k - z^*_k) > 0 \text{ for all } \lambda$$

$$\Rightarrow \qquad\qquad \lambda < -\left\{ \frac{c_k - z_k}{c'_k - z'_k} \right\} = \lambda' \text{ (say)}$$

∴ Problem will have no finite optimal solution for all λ such that $\delta \leq \lambda \leq \lambda_1$

For $\lambda = \lambda'$ if $(c_j - z_j) + \lambda'_1(c'_j - z'_j) \leq 0 \ \forall j$ then the problem will have a finite optimal solution. Therefore, upper bound of λ can be taken as

$$\lambda_1 = \min_{(c'_j - z'_j) > 0} -\left\{ \frac{c_j - z_j}{c'_j - z'_j} \right\}$$

So, when $\lambda'_1 \leq \lambda \leq \lambda_1$, the problem has a finite optimal solution and we can proceed as in case 1.

Further, if not all $(c_j - z_j) + \lambda'_1(c'_j - z'_j) \leq 0$, then for some value of j, a vector α_j with $(c_j - z_j) + \lambda'_1(c'_j - z'_j) > 0$ can be introduced to the basis for the improvement of the solution and the criterion for the improvement will have such iteration untill we obtain

$$(c_j - z_j) + \lambda'_1(c'_j - z'_j) \leq 0 \ \forall j$$

or to that iteration until a vector α_p with $(c_p - z_p) + \lambda'_1(c'_p - z'_p) > 0$ and all $\alpha_{ip} \leq 0$ is obtained. Here, in the former case, we shall use the method of case 1 and in the later case if $c'_p - z'_p \geq 0$, then there will be no finite optimal solution for $\lambda \geq \lambda'_1$ and if $c'_p - z'_p < 0$ then no finite optimal solution exists for $\lambda < \lambda'_2$ where

$$\lambda'_2 = -\frac{c_p - z_p}{c'_p - z'_p} \text{ and } \lambda'_1 < \lambda'_2$$

and then proceed to find whether a finite solution exists for $\lambda = \lambda'_2$.

Repeat the above procedure, until a finite optimal solution is obtained for some value of λ in which we can use the case-1 or obtain the information for no finite optimal solution for λ which is case-2.

WORKING PROCEDURE

The above discussion can be summarised as follows:

STEP 1. Start with a finite optimal solution of the problem associated with the basis **B** for $\lambda = \delta$.

STEP 2. Find a set of critical values $\lambda_1, \lambda_2, ..., \lambda_p$ of λ and the corresponding characteristic solutions and a series of basis $B_1, B_2, ..., B_p$ (B_p may not exist) with the following properties:

    (i) Each basis B_i differ from B_{i-1} by a single vector of λ_1 is determined by a unique minimum.

    (ii) Basis B_i is a optimal basis for all values of λ satisfy

$$\lambda_i \leq \lambda \leq \lambda_{i+1} \text{ for } i = 1, 2, ..., p-1$$

    (iii) B_p is an optimal basis for every value of λ such that $\lambda \geq \lambda_p$, if B_p exists (case 1) or there exists no finite optimal solution for $\lambda \geq \lambda_p$.

Solved Examples

EXAMPLE I. *Consider the following LPP:*
$$max. \ Z = (4 + 2\lambda)x_1 + (6 - 2\lambda)x_2 + (2 + 2\lambda)x_3$$
subject to the constraints
$$x_1 + x_2 + x_3 \le 3$$
$$x_1 + 4x_2 + 7x_3 \le 9$$
and $\quad x_1, x_2, x_3 \ge 0$

Find the range of λ *over which the solution remains basic feasible and optimal.*

SOLUTION. Introducing slack variables s_1 and s_2, the given LPP can be written as
$$max. \ Z = (4 + 2\lambda)x_1 + (6 - 2\lambda)x_2 + (2 + 2\lambda)x_3 + 0s_1 + 0s_2$$
s.t. $\quad x_1 + x_2 + x_3 + s_1 = 3$
$$x_1 + 4x_2 + 7x_3 + s_2 = 9$$
and $\quad x_1, x_2, x_3, s_1, s_2 \ge 0$

Now, $C^* = (4 + 2\lambda, 6 - 2\lambda, 2 + 2\lambda, 0, 0)$

The initial basic feasible solution of the given LPP is
$$x_1 = 0, x_2 = 0, x_3 = 0, s_1 = 3 \text{ and } s_2 = 9$$

Simplex table-1

B.V.	C_B^*	x_B	x_1	x_2	x_3	s_1	s_2	Min. Ratio x_B/x_2
	$c_j \to$		4	6	2	0	0	
s_1	0	3	1	1	1	1	0	3/1
s_2	0	9	1	④	7	0	1	9/4 (min) →
$Z = C_B^* X_B$ $= 0$	$c_j^* \to$		$4 + 2\lambda$	$6 - 2\lambda$	$2 + 2\lambda$	0	0	
$c_j^* - z_j^*$	$c_j - z_j$		4	6	2	0	0	
	$c_j' - z_j'$		2	–2	2	0	0	

In the above table
$$c_1^* - z_1^* = c_1^* - C_B^* x_1 = 4 + 2\lambda = (c_1 - z_1) + \lambda(c_1' - z_1')$$
$$\Rightarrow \quad c_1 - z_1 = 4, c_1' - z_1' = 2$$
Also, $c_2^* - z_2^* = c_2^* - C_B^* x_2 = 6 - 2\lambda \quad \Rightarrow \quad c_2 - z_2 = 6, c_2' - z_2' = -2$
$$c_3^* - z_3^* = c_3^* - C_B^* x_3 = 2 + 2\lambda \quad \Rightarrow \quad c_3 - z_3 = 2, c_3' - z_3' = 2$$
$$c_4^* - z_4^* = c_4^* - C_B^* s_1 = 0 \quad \Rightarrow \quad c_4 - z_4 = 0, c_4' - z_4' = 0$$
$$c_5^* - z_5^* = c_5^* - C_B^* s_2 = 0 \quad \Rightarrow \quad c_5 - z_5 = 0, c_5' - z_5' = 0$$

Here, we have that x_2 is the incoming vector and s_2 is the outgoing vector (by minimum ratio rule)

Now, apply simplex method in a usual manner, we have the following simplex table.

Simplex table-2

B.V.	C_B^*	x_B	$c_j \to$ 4 x_1	6 x_2	2 x_3	0 s_1	0 s_2	Min. Ratio x_B/x_1
s_1	0	3/4	3/4	0	–3/4	1	–1/4	$(3/4)/(3/4) = 1$ (min) $\to$
x_2	$6 - 2\lambda$	9/4	1/4	1	7/4	0	1/4	$(9/4)/(1/4)$
$Z = C_B^* X_B$ $= 27/2$		$c_j^* \to$	$4 + 2\lambda$	$6 - 2\lambda$	$2 + 2\lambda$	0	0	
$C_j^* - Z_j^*$	$c_j - z_j$		5/2	0	–17/2	0	–3/2	
	$c_j' - z_j'$		5/2	0	11/2	0	1/2	
			$\uparrow$			$\downarrow$		

In the above table, we have

$$c_1^* - z_1^* = \frac{5}{2} + \lambda\frac{5}{2}, c_2^* - z_2^* = 0$$

$$c_3^* - z_3^* = -\frac{17}{2} + \frac{11\lambda}{2}, c_4^* - z_4^* = 0$$

$$c_5^* - z_5^* = -\frac{3}{2} + \frac{\lambda}{2}$$

Obviously, x_1 is the incoming vector and s_1 is the outgoing vector.

Simplex table-3

B.V.	C_B^*	x_B	$c_j \to$ 4 x_1	6 x_2	2 x_3	0 s_1	0 s_2	Min. Ratio
x_1	$4 + 2\lambda$	1	1	0	–1	4/3	–1/3	
x_2	$6 - 2\lambda$	2	0	1	2	–1/3	1/3	
$Z = C_B^* X_B$ $= 16 - 2\lambda$		$c_j^* \to$	$4 + 2\lambda$	$6 - 2\lambda$	$2 + 2\lambda$	0	0	
$C_j^* - Z_j^*$	$c_j - z_j$		0	0	–6	–10/3	–2/3	
	$c_j' - z_j'$		0	0	8	–10/3	4/3	

Further, we have to calculate $c_j^* - z_j^*$ for $j = 3, 4, 5$ (not in the basis) in the following manner

$$c_3^* - z_3^* = (2 + \lambda) - C_B^* x_3 = (2 + 2\lambda) - (4 + 2\lambda, 6 - 2\lambda)\begin{bmatrix} -1 \\ 2 \end{bmatrix}$$

$$= (2 + 2\lambda) - \{-(4 + 2\lambda) + 2(6 - 2\lambda)\}$$

$$= (2 + 2\lambda) - (8 - 6\lambda) = -6 + 8\lambda$$

$$= (c_3 - z_3) + \lambda(c_3' - z_3')$$

Thus, $c_3 - z_3 = -6, c_3' - z_3' = 8$

$$c_4^* - z_4^* = 0 - C_B^* x_4 = 0 - (4 + 2\lambda, 6 - 2\lambda)\begin{bmatrix} 4/3 \\ -1/3 \end{bmatrix} \qquad (\because x_4 = s_1)$$

$$= -\frac{4}{3}(4 + 2\lambda) + \frac{1}{3}(6 - 2\lambda) = \frac{16}{3} - \frac{8}{3}\lambda + \frac{6}{3} - \frac{2}{3}\lambda$$

$$= \frac{-10}{3} - \frac{10}{3}\lambda = (c_4 - z_4) + \lambda(c_4' - z_4')$$

$\Rightarrow \qquad c_4 - z_4 = -\frac{10}{3}, c_4' - z_4' = -\frac{10}{3}$

Also, $c_5^* - z_5^* = c_5^* - C_B^* x_5 = c_5^* - C_B^* s_2 \qquad (\because x_5 = s_2)$

$$= 0 - (4 + 2\lambda, 6 - 2\lambda)\begin{bmatrix} -1/3 \\ 1/3 \end{bmatrix}$$

$$= \frac{1}{3}(4 + 2\lambda) - \frac{1}{3}(6 - 2\lambda) = \frac{1}{3}(-2 + 4\lambda)$$

$$= -\frac{2}{3} + \frac{4}{3}\lambda = (c_5 - z_5) + \lambda(c_5' - z_5')$$

$\Rightarrow \qquad c_5 - z_5 = -\frac{2}{3}, c_5' - z_5' = \frac{4}{3}$

Now, the third simplex table shows that all $c_j - z_j \le 0$.

$\Rightarrow$ solution is optimal for $\lambda = 0$ and is given by

$$x_1 = 1, x_2 = 2, x_3 = 0 \text{ and max } Z = 16$$

We know that the solution will remain optimal for those value of λ for which

$$\underline{\lambda} \le \lambda \le \overline{\lambda}$$

where $\qquad \underline{\lambda} = \max_{(c_j - z_j) < 0}\left\{ -\frac{c_j - z_j}{c_j' - z_j'} \right\} = \left\{ \frac{-10/3}{-10/3} \right\} = 1$

and $\qquad \overline{\lambda} = \min_{(c_j' - z_j') > 0}\left\{ -\frac{c_j - z_j}{c_j' - z_j'} \right\} = \min\left\{ -\frac{c_3 - z_3}{c_3' - z_3'}, -\frac{c_5 - z_5}{c_5' - z_5'} \right\}$

$$= \min\left\{ -\frac{-6}{7}, -\frac{-2/3}{4/3} \right\} = \min\left\{ \frac{6}{7}, \frac{1}{2} \right\} = \frac{1}{2} = \frac{c_k - z_k}{c_k' - z_k'}$$

$\Rightarrow \qquad k = 5$

Thus, $x_1 = 1, x_2 = 2$ and $x_3 = 0$ is an optimal solution for the given problem for all values of λ such that

$$-1 \le \lambda \le \frac{1}{2}$$

and max. $Z = 16 - 2\lambda$

Now, since $\overline{\lambda} = \frac{1}{2}$, which is finite and $a_{25} = \frac{1}{3} > 0$. Thus, we will try to improve the range of λ beyond $\frac{1}{2}$. When $\lambda > \frac{1}{2}$ then x_5 (s_2) can be introduced in the basis and by minimum ratio rule x_2 is the outgoing vector.

Now, introducing s_2 in place of x_2 in the basis, the next optimal simplex table is as follows.

Simplex table-4

B.V.	C_B^*	x_B	$c_j \to$ 4 x_1	6 x_2	2 x_3	0 s_1	0 s_2
x_1	$4 + 2\lambda$	3	1	1	1	1	0
s_2	0	6	0	3	6	-1	1
$Z = C_B^* X_B$ $= 12 + 6\lambda$	$c_j^* \to$		$4 + 2\lambda$	$6 - 2\lambda$	$2 + 2\lambda$	0	0
$c_j^* - z_j^*$	$c_j - z_j$		0	2	-2	-4	0
	$c_j' - z_j'$		0	-4	0	-2	0

In the above table

$$c_2^* - z_2^* = c_2^* - C_B^* x_2 = (6 - 2\lambda) - (4 + 2\lambda, 0)\begin{bmatrix}1\\3\end{bmatrix}$$

$$= 6 - 2\lambda - 4 - 2\lambda = 2 - 4\lambda = (c_2 - z_2) + \lambda(c_2' - z_2')$$

$\Rightarrow \quad c_2 - z_2 = 0$ and $c_2' - z_2' = -4$

Now, $\quad c_3^* - z_3^* = c_3^* - C_B^* x_3 = (2 + 2\lambda) - (4 + 2\lambda, 0)\begin{bmatrix}1\\6\end{bmatrix}$

$$= (2 + 2\lambda) - (4 + 2\lambda) = -2$$

$$= (c_3 - z_3) + \lambda(c_3' - z_3')$$

$\Rightarrow \quad c_3 - z_3 = -2, c_3' - z_3' = 0$

and $\quad c_4^* - z_4^* = c_4^* - C_B^* x_4 = c_4^* - C_B^* s_1$

$$= 0 - (4 + 2\lambda, 0)\begin{bmatrix}1\\-1\end{bmatrix} = -4 - 2\lambda$$

$$= (c_4 - z_4) + \lambda(c_4' - z_4')$$

$\Rightarrow \quad c_4 - z_4 = -4, c_4' - z_4' = -2$

For $\lambda \geq \dfrac{1}{2}, c_2^* - z_2^* \leq 0, c_3^* - z_3^* < 0, c_4^* - z_4^* \leq 0$

Therefore, the solution for $\lambda \geq \dfrac{1}{2}$ is optimal and this optimal solution is given by

$$x_1 = 3, x_2 = 0, x_3 = 0 \text{ and max. } Z = 12 + 6\lambda$$

Now, for $\lambda < -1$, then from table-3

$$c_4^* - z_4^* = \frac{10}{3}(1 + \lambda) > 0 \text{ and } a_{14} = \frac{4}{3} > 0$$

So, for $\lambda < -1$, the solution in simplex table-3 is not optimal. In this case, the corresponding vector x_4 (s_1) is the incoming vector and by minimum ratio rule x_1 will be the outgoing vector. Introducing s_1 in place of x_1, the next optimal table from table 3 is given below.

	$c_j \rightarrow$		4	6	2	0	0
B.V.	C_B^*	x_B	x_1	x_2	x_3	s_1	s_2
s_1	0	3/4	3/4	0	–3/4	1	–1/4
x_2	$6 - 2\lambda$	9/4	1/4	1	7/4	0	1/4
$Z = C_B^* x_B = \dfrac{27}{2} - \dfrac{9}{2}\lambda$	$c_j^* \rightarrow$		$4 + 2\lambda$	$6 - 2\lambda$	$2 + 2\lambda$	0	0
$c_j^* - z_j^*$	$c_j - z_j$		5/2	0	–17/2	0	–3/2
	$c_j' - z_j'$		5/2	0	11/2	0	1/2

From the above table, we observe that

$$c_1^* - z_1^* = \frac{5}{2}(1 + \lambda), c_3^* - z_3^* = -\frac{17}{3} + \frac{11}{2}\lambda$$

$$c_5^* - z_5^* = -\frac{3}{2} + \frac{1}{2}\lambda$$

Clearly, for $\lambda < -1$, $c_j^* - z_j^* \leq 0 \ \forall j = 1, 3, 5$ (not in the basis). Hence, the solution is optimal and is given by

$$x_1 = 0, x_2 = \frac{9}{4}, x_3 = 0 \text{ and } \max. Z = \frac{27}{2} - \frac{9}{2}\lambda$$

EXAMPLE 2. *For the following LPP:*

$$\max. Z = (3 - 6\lambda)x_1 + (2 - 2\lambda)x_2 + (5 + 5\lambda)x_3$$

subject to the constraints

$$x_1 + 2x_2 + x_3 \leq 430$$
$$3x_1 + 2x_3 \leq 460$$
$$x_1 + 4x_2 \leq 420$$

and $x_1, x_2, x_3 \geq 0$

Find the range of λ for which the solution remain basic feasible and optimal.

SOLUTION. The given LPP can be written as follows:

$$\max. Z = (3 - 6\lambda)x_1 + (2 - 2\lambda)x_2 + (5 + 5\lambda)x_3 + 0s_1 + 0s_2 + 0s_3$$

s.t.
$$x_1 + 2x_2 + x_3 + s_1 = 430$$
$$3x_1 + 2x_3 + s_2 = 460$$
$$x_1 + 4x_2 + s_3 = 420$$

and
$$x_i, s_i \geq 0$$

Simplex table-1

B.V.	C_B^*	x_B	x_1	x_2	x_3	s_1	s_2	s_3	Min. Ratio x_B/x_3
	$c_j \rightarrow$		3	2	5	0	0	0	
s_1	0	430	1	2	1	1	0	0	430/1
s_2	0	460	3	0	②	0	1	0	460/2 (min) →
s_3	0	420	1	4	0	0	0	1	—
$Z = C_B^* X_B$ $= 0$	$c_j^* \rightarrow$		$3-6\lambda$	$2-2\lambda$	$5+5\lambda$	0	0	0	
$c_j^* - z_j^*$	$c_j - z_j$		3	2	5	0	0	0	
	$c_j' - z_j'$		-6	-2	5 ↑	0	0 ↓	0	

In the above table, we have

$$c_1^* - z_1^* = c_1^* - C_B^* x_1 = 3 - 6\lambda - (0,0,0)\begin{bmatrix}1\\3\\1\end{bmatrix} = 3 - 6\lambda = (c_1 - z_1) + \lambda(c_1' - z_1')$$

$$\Rightarrow \quad c_1 - z_1 = 3, c_1' - z_1' = -6$$

Now, $\quad c_2^* - z_2^* = c_2^* - C_B^* x_2 = 2 - 2\lambda - (0,0,0)\begin{bmatrix}2\\0\\4\end{bmatrix} = 2 - 2\lambda$

$$\Rightarrow \quad c_2 - z_2 = 2, c_2' - z_2' = -2$$

Also, $\quad c_3^* - z_3^* = c_3^* - C_B^* x_3 = 5 + 5\lambda - (0,0,0)\begin{bmatrix}1\\2\\0\end{bmatrix} = 5 + 5\lambda$

$$\Rightarrow \quad c_3 - z_3 = 5, c_3' - z_3' = 5$$

Clearly, $c_3 - z_3 = 5 > 0 \Rightarrow$ corresponding vector x_3 can be introduced in the basis and by minimum ratio rule s_2 (x_5) is the outgoing vector.

Simplex table-2

B.V.	C_B^*	x_B	x_1	x_2	x_3	s_1 (x_4)	s_2 (x_5)	s_3 (x_6)	Min. Ratio
	$c_j \rightarrow$		3	2	5	0	0	0	
s_1	0	200	$-1/2$	②	0	1	$-1/2$	0	200/2 (min) →
x_3	$5+5\lambda$	230	3/2	0	1	0	1/2	0	—
s_3	0	420	1	4	0	0	0	1	420/4
$Z = C_B^* X_B$ $= 1150(1+\lambda)$	$c_j^* \rightarrow$		$3-6\lambda$	$2-2\lambda$	$5+5\lambda$	0	0	0	
$c_j^* - z_j^*$	$c_j - z_j$		$-9/2$	2	0	0	$-5/2$	0	
	$c_j' - z_j'$		$-27/2$	-2	0	0	$-5/2$	0	
			↑	↓					

In the above table, we have calculated $c_j^* - z_j^*$ for $j = 1, 2, 3, 5$ (not in the basis) such that

$$c_1^* - z_1^* = c_1^* - C_B^* x_1 = (3 - 6\lambda)(0, 5 + 5\lambda, 0)\begin{bmatrix} -1/2 \\ 3/2 \\ 1 \end{bmatrix} = (3 - 6\lambda) - \frac{3}{2}(5 + 5\lambda)$$

$$= \frac{-9}{2} - \frac{27}{2}\lambda = (c_1 - z_1) + \lambda(c_1' - z_1')$$

$$c_1 - z_1 = -\frac{9}{2}, c_1' - z_1' = \frac{-27}{2}$$

Also, $c_2^* - z_2^* = c_2^* - C_B^* x_2 = (2 - 2\lambda) - (0, 5 + 5\lambda, 0)\begin{bmatrix} 2 \\ 0 \\ 4 \end{bmatrix}$

$$= 2 - 2\lambda = (c_2 - z_2) + \lambda(c_2' - z_2')$$

$$\Rightarrow \quad c_2 - z_2 = 2, c_2' - z_2' = -2$$

and $c_5^* - z_5^* = c_5^* - C_B^* x_5 = 0 - (0, 5 + 5\lambda, 0)\begin{bmatrix} -1/2 \\ -1/2 \\ 0 \end{bmatrix}$

$$= -\frac{5}{2} - \frac{5}{2}\lambda = (c_5 - z_5) + \lambda(c_5' - z_5')$$

$$\Rightarrow \quad c_5 - z_5 = \frac{-5}{2} \text{ and } c_5' - z_5' = \frac{-5}{2}$$

We observe that $c_2 - z_2 = 2 > 0$ therefore, corresponding vector x_2 can be introduced in the basis and also by minimum ratio rule $s_1(x_4)$ is the outgoing vector.

Simplex table-3

B.V.	C_B^*	x_B	x_1	x_2	x_3	s_1 (x_4)	s_2 (x_5)	s_3 (x_6)	Min. Ratio x_B/x_4
	$c_j \rightarrow$		3	2	5	0	0	0	
x_2	$2 - 2\lambda$	100	$-1/4$	1	0	①/2	$-1/4$	0	$100/(1/2)$ (min)→
x_3	$5 + 5\lambda$	230	3/2	0	1	0	1/2	0	—
s_3	0	20	2	0	0	-2	1	1	—
$Z = C_B^* X_B$ $= 1350 + 950\lambda$	$c_j^* \rightarrow$		$3 - 6\lambda$	$2 - 2\lambda$	$5 + 5\lambda$	0	0	0	
$c_j^* - z_j^*$	$c_j - z_j$		-4	0	0	-1	-2	0	
	$c_j' - z_j'$		-14	0	0	1	-3	0	
			↓			↑			

In the above table, we have compute $c_j^* - z_j^*$ for $j = 1, 4, 5$ (not in the basis) as follows:

$$c_1^* - z_1^* = c_1^* - \mathbf{C_B^* x_1} = (3-6\lambda) - (2-2\lambda, 5+5\lambda, 0)\begin{bmatrix} -1/4 \\ 3/2 \\ 2 \end{bmatrix}$$

$$= (3-6\lambda) + \frac{1}{4}(2-2\lambda) - \frac{3}{2}(5+5\lambda) = -4 - 14\lambda = (c_1 - z_1) + \lambda(c_1' - z_1')$$

$\Rightarrow \qquad c_1 - z_1 = -4 \text{ and } c_1' - z_1' = -14$

Also, $\qquad c_4^* - z_4^* = c_4^* - \mathbf{C_B^* x_4} = 0 - (2-2\lambda, 5+5\lambda, 0)\begin{bmatrix} 1/2 \\ 0 \\ -1/2 \end{bmatrix}$

$$= -1 + \lambda$$

$\Rightarrow \qquad c_4 - z_4 = -1 \text{ and } c_4' - z_4' = 1$

and $\qquad c_5^* - z_5^* = c_5^* - \mathbf{C_B^* x_5} = 0 - (2-2\lambda, 5+5\lambda, 0)\begin{bmatrix} -1/4 \\ 1/2 \\ 1 \end{bmatrix}$

$$= \frac{1}{4}(2-2\lambda) - \frac{1}{2}(5+5\lambda) = -2 - 3\lambda$$

$\Rightarrow \qquad c_5 - z_5 = -2 \text{ and } c_5' - z_5' = -3$

Here, we observe that in simplex table-3 all $c_j - z_j < 0$ so the solution at this stage is optimal for $\lambda = 0$ and is given by

$\qquad x_1 = 0, x_2 = 100, x_3 = 230 \text{ and max. } Z = 1350$

This solution is optimal and feasible for those value of λ for which $\underline{\lambda} \le \lambda \le \overline{\lambda}$ where

$$\underline{\lambda} = \max_{(c_j' - z_j') < 0}\left[-\frac{(c_j - z_j)}{(c_j' - z_j')}\right] = \max\left\{-\frac{c_1 - z_1}{c_1' - z_1'}, -\frac{c_5 - z_5}{c_5' - z_5'}\right\}$$

$$= \max\left\{-\frac{-4}{-14}, -\frac{-2}{-3}\right\} = \max\left\{-\frac{2}{7}, -\frac{2}{3}\right\} = -\frac{2}{7}$$

and $\qquad \overline{\lambda} = \min_{(c_j' - z_j') > 0}\left\{-\frac{c_j - z_j}{c_j' - z_j'}\right\} = \min\left\{-\frac{c_4 - z_4}{c_4' - z_4'}\right\} = -\frac{(-1)}{1} = 1$

$\Rightarrow$ This solution is optimal and feasible for those value of λ for which $-\frac{2}{7} \le \lambda \le 1$.

Now, since $\overline{\lambda} = 1$, which is finite, so we can improve the range of λ beyond $\overline{\lambda} = 1$.

Now, for $\lambda > 1$, $c_4^* - z_4^* > 0$ and $a_{14} = \frac{1}{2} > 0$, so that $x_1 (s_1)$ can be introduced into the basis and by minimum ratio rule, s_1 is the outgoing vector.

Simplex table-4

B.V.	C_B^*	x_B	x_1	x_2	x_3	s_1 (x_4)	s_2 (x_5)	s_3 (x_6)
$c_j \rightarrow$			3	2	5	0	0	0
s_1	0	200	$-1/2$	2	0	1	$-1/2$	0
x_3	$5+5\lambda$	230	$3/2$	0	1	0	$1/2$	0
s_3	0	420	1	4	0	0	0	1
$Z = C_B^* X_B$ $= 1150(1+\lambda)$	$c_j^* \rightarrow$		$3-6\lambda$	$2-2\lambda$	$5+5\lambda$	0	0	0
$c_j^* - z_j^*$	$c_j - z_j$		$-9/2$	2	0	0	$-5/2$	0
	$c_j' - z_j'$		$-27/2$	-2	0	0	$-5/2$	0

Here, we observe that all $c_j^* - z_j^* \leq 0$ for $\lambda > 1$ and no $c_j' - z_j' > 0$

$\Rightarrow$ For $\lambda > 1$, optimal solution exists

$\therefore$ We conclude that the solution is optimal and feasible for those value of λ for which

$$-\frac{27}{7} \leq \lambda \leq \infty$$

Hence, $x_1 = 0, x_2 = 100, x_3 = 230$ and max. $Z = 1350 + 950\lambda$ is the optimal and

feasible solution for those value of λ for which $-\frac{2}{7} \leq \lambda \leq 1$ and $x_1 = 0, x_2 = 0,$

$x_3 = 230$ and max. $Z = 1150(1 + \lambda)$ is the optimal and feasible solution for those value of λ for which $1 < \lambda < \infty$.

EXAMPLE 3. *For the following LPP,*

$$min. \ Z = \lambda x_1 - \lambda x_2 - x_3 + x_4$$

subject to the constraints

$$3x_1 - 3x_2 - x_3 + x_4 \geq 5$$
$$2x_1 - 2x_2 + x_3 - x_4 \leq 3$$

and $\quad x_1, x_2, x_3, x_4 \geq 0$

Find the range of λ for which the solution remains basic feasible and optimal.

SOLUTION. Introducing surplus variable s_1, slack variable s_2 and artificial variable A, the given problem becomes

$$max. \ Z = -\lambda x_1 + \lambda x_2 + x_3 - x_4 + 0s_1 + 0s_2 - MA$$
s.t. $\quad 3x_1 - 3x_2 - x_3 + x_4 - s_1 + A = 5$
$$2x_1 - 2x_2 + x_3 - x_4 + s_2 = 3$$

and $\quad x_1, x_2, x_3, x_4, s_1, s_2, A \geq 0$

Now proceeding in a usual manner, we have the following simplex table.

Simplex table-1

B.V.	C_B^*	X_B	$c_j \rightarrow$ 0 x_1	0 x_2	1 x_3	-1 x_4	0 s_1 (x_5)	0 s_2 (x_6)	$-M$ A (x_7)	Min. Ratio X_B/x_4
A	$-M$	5	③	-3	-1	1	-1	0	1	$5/3$
s_2	0	3	2	-2	1	-1	0	1	0	$3/2$ (min) $\rightarrow$
$Z = C_B^* X_B$ $= -5M$	$c_j^* \rightarrow$		$-\lambda$	$-\lambda$	1	-1	0	0	$-M$	
$c_j^* - z_j^*$	$c_j - z_j$		$3M$	$-3M$	$1-M$	$M-1$	$-M$	0	0	
	$c_j' - z_j'$		-1	1	0	0	0	0	0	
			$\uparrow$			$\downarrow$				

In the above table, we have

We compute $c_j^* - z_j^*$ for $j = 1, 2, 3, 4, 5$ (not in the basis)

Here, $\qquad c_1^* - z_1^* = c_1^* - C_B^* x_1 = -\lambda - (-M, 0)\begin{bmatrix} 3 \\ 2 \end{bmatrix} = -\lambda + 3M$

$\Rightarrow \qquad c_1 - z_1 = 3M,\ c_1' - z_1' = -1$

Similarly, $\qquad c_2^* - z_2^* = c_2^* - C_B^* x_2 = \lambda - (-M, 0)\begin{bmatrix} -3 \\ 2 \end{bmatrix} = \lambda - 3M$

$\Rightarrow \qquad c_2 - z_2 = -3M,\ c_2' - z_2' = 1$

Also, $\qquad c_3^* - z_3^* = c_3^* - C_B^* x_3 = 1 - (-M, 0)\begin{bmatrix} -1 \\ 1 \end{bmatrix} = 1 - M$

$\Rightarrow \qquad c_3 - z_3 = 1 - M,\ c_3' - z_3' = 0$

Now, $\qquad c_4^* - z_4^* = c_4^* - C_B^* x_4 = -1 - (-M, 0)\begin{bmatrix} -1 \\ 1 \end{bmatrix} = -1 + M$

$\Rightarrow \qquad c_4 - z_4 = -1 + M,\ c_4' - z_4' = 0$

and $\qquad c_5^* - z_5^* = c_5^* - C_B^* x_5 = 0 - (-M, 0)\begin{bmatrix} -1 \\ 0 \end{bmatrix} = -M$

$\Rightarrow \qquad c_5 - z_5 = -M,\ c_5' - z_5' = 0$

We observe that not all $c_j - z_j \leq 0$ and not all $c_j^* - z_j^* \leq 0$. Thus the initial solution is not optimal.

Also, corresponding to $c_1 - z_1 = 3M$, the vector x_1 will enter in the basis and by minimum ratio rule we find that s_2 is the outgoing vector.

Simplex table-2

B.V.	C_B^*	x_B	$c_j \rightarrow$ 0 x_1	0 x_2	1 x_3	−1 x_4	0 s_1 (x_5)	0 s_2 (x_6)	−M A (x_7)	Min. Ratio X_B/x_4
A	−M	1/2	0	0	−5/2	⑤/2	−1	−3/2	1	(1/2)/(5/2)(min)→
x_1	−λ	3/2	1	−1	1/2	−1/2	0	1/2	0	—

$Z = C_B^* X_B$

$= \dfrac{-M}{2} - \dfrac{3\lambda}{2}$ $c_j^* \rightarrow$ −λ   λ   1   −1   0   0   −M

$c_j^* - z_j^*$	$c_j - z_j$	0	0	$1 - \dfrac{5}{2}M$	$\dfrac{5M}{2} - 1$	−M	−3M/2	0
	$c_j' - z_j'$	0	0	1/2	−1/2	0	1/2	0

↑

In the above table, we compute $c_j^* - z_j^*$ for $j = 2, 3, 4, 5, 6$ (not in the basis), we have

$$c_2^* - z_2^* = c_2^* - \mathbf{C_B^*}\mathbf{x_2} = \lambda - (-M, -\lambda)\begin{bmatrix} 0 \\ -1 \end{bmatrix}$$

$$= \lambda - \lambda = 0$$

$$\Rightarrow \qquad c_2 - z_2 = 0, c_2' - z_2' = 0$$

$$c_3^* - z_3^* = c_3^* - \mathbf{C_B^*}\mathbf{x_3} = 1 - (-M, -\lambda)\begin{bmatrix} -5/2 \\ 1/2 \end{bmatrix}$$

$$= 1 + \frac{5M}{2} + \frac{\lambda}{2}$$

$$\Rightarrow \qquad c_3 - z_3 = 1 - \frac{5M}{2}, c_3' - z_3' = \frac{1}{2}$$

Similarly we may get

$$c_4 - z_4 = -1 + \frac{5}{2}M, c_4' - z_4' = -\frac{1}{2}$$

$$c_5 - z_5 = -M, c_5' - z_5' = 0$$

$$c_6 - z_6 = -\frac{3M}{2}, c_6' - z_6' = \frac{1}{2}$$

Again not all $c_j - z_j \leq 0$ and not all $c_j^* - z_j^* \leq 0$ so the solution in this table is not optimal.

Now, corresponding to $c_4 - z_4 = \dfrac{5M}{2} - 1$, the vector x_4 is incoming vector and by minimum ratio rule, A is outgoing vector.

Now proceed in a usual manner, we have the third simplex table.

Simplex table-3

B.V.	C_B^*	x_B	x_1	x_2	x_3	x_4	s_1 (x_5)	s_2 (x_6)
$c_j \rightarrow$			0	0	1	-1	0	0
x_4	-1	1/5	0	0	-1	1	$-2/5$	$-3/5$
x_1	$-\lambda$	8/5	1	-1	0	0	$-1/5$	1/5
$c_j^* \rightarrow$			$-\lambda$	λ	1	-1	0	0
$c_j - z_j$			0	0	0	0	$-2/5$	$-3/5$
$c_j' - z_j'$			0	0	0	0	$-1/5$	1/5

$Z = C_B^* X_B$

$$= -\frac{1}{5} - \frac{8\lambda}{5}$$

In the above table we compute $c_j^* - z_j^*$ for $j = 2, 3, 4, 5, 6$ (not in the basis).

Here, $\quad c_2^* - z_2^* = c_2^* - \boldsymbol{C_B^* x_2} = \lambda - (-1 - \lambda)\begin{bmatrix} 0 \\ -1 \end{bmatrix} = \lambda - \lambda = 0$

$$= (c_2 - z_2) + \lambda(c_2' - z_2')$$

$\Rightarrow \qquad c_2 - z_2 = 0, c_2' - z_2' = 0$

Similarly, we obtain the other values of $c_j - z_j$ and $c_j' - z_j'$ as shown in the above table.

Further, the above table shows that all $c_j - z_j \leq 0$

$\Rightarrow$ solution is optimal for $\lambda = 0$

Also, this optimal solution is given by

$$x_1 = \frac{8}{5}, x_2 = x_3 = 0, x_4 = \frac{1}{5}$$

and $\qquad \min . Z = -\max . Z' = \frac{1}{5}$

Now, this solution will remain optimal for those values of λ for which

$$\underline{\lambda} \leq \lambda \leq \overline{\lambda}$$

where $\quad \underline{\lambda} = \max_{(c_j' - z_j') < 0} \left\{ -\frac{c_j - z_j}{c_j' - z_j'} \right\} = \max \left\{ -\frac{c_5 - z_5}{c_5' - z_5'} \right\}$ $[\because$ only $c_5' - z_5' < 0]$

$$= -\frac{-2/5}{-1/5} = -2$$

and $\qquad \overline{\lambda} = \min_{(c_j' - z_j') > 0} \left\{ -\frac{c_j - z_j}{c_j' - z_j'} \right\} = \min \left\{ -\frac{c_6 - z_6}{c_6' - z_6'} \right\}$ $[\because$ only $c_6' - z_6' > 0]$

$$= -\frac{-3/5}{1/5} = 3$$

$\Rightarrow$ The solution $x_1 = \dfrac{8}{5}, x_2 = x_3 = 0, x_4 = \dfrac{1}{5}$ will remain optimal for all values of λ for which $-2 \le \lambda \le 3$.

Further, $\bar{\lambda} = \min\left\{-\dfrac{c_6 - z_6}{c'_6 - z'_6}\right\} = -\dfrac{c_k - z_k}{c'_k - z'_k}$

$\Rightarrow \quad k = 6$

$\because \quad \lambda = \bar{\lambda} = 3$, finite and $a_{26} = \dfrac{1}{5} > 0$ therefore, range of λ can be improved beyond $\bar{\lambda} = 3$.

For $\lambda > 3, c^*_6 - z^*_6 > 0$, so that the vector $x_6 (= s_2)$ is the incoming vector and by minimum ratio rule, x_1 is the outgoing vector and key element $a_{26} = \dfrac{1}{5}$.

Now, we have the following simplex table

Simplex table-4

		$c_j \rightarrow$	0	0	1	-1	0	0
B.V.	C^*_B	X_B	x_1	x_2	x_3	x_4	s_1 (x_5)	s_2 (x_6)
x_4	-1	5	3	-3	-1	1	-1	0
x_2	0	8	5	-5	0	0	-1	1
$Z = C^*_B X_B$ $= -5$		$c^*_j \rightarrow$	$-\lambda$	λ	1	-1	0	0
$c^*_j - z^*_j$	$c_j - z_j$		3	-3	0	0	-1	0
	$c'_j - z'_j$		-1	1	0	0	0	0

In the above table, we have compute

$c^*_j - z^*_j$ for $j = 1, 2, 3, 5$ such that

$$c^*_1 - z^*_1 = c^*_1 - C^*_B x_1 = -\lambda - (-1, 0)\begin{bmatrix} 3 \\ 5 \end{bmatrix} = -\lambda + 3 = (c_1 - z_1) + \lambda(c'_1 - z'_1)$$

$\Rightarrow \qquad c_1 - z_1 = 3, c'_1 - z'_1 = -1$

also, $\quad c^*_2 - z^*_2 = c^*_2 - C^*_B x_2 = \lambda - (1, 0)\begin{bmatrix} -3 \\ -5 \end{bmatrix} = \lambda - 3$

$\Rightarrow \qquad c_2 - z_2 = -3, c'_2 - z'_2 = 1$

Also, $\quad c^*_3 - z^*_3 = c^*_3 - C^*_B x_3 = 1 - (-1, 0)\begin{bmatrix} -1 \\ 0 \end{bmatrix} = 1 - 1 = 0$

$\Rightarrow \qquad c_3 - z_3 = 0, c'_3 - z'_3 = 0$

and $\quad c^*_5 - z^*_5 = c^*_5 - C^*_B x_5 = 0 - (-1, 0)\begin{bmatrix} -1 \\ 1 \end{bmatrix} = -1$

$$\Rightarrow \qquad c_5 - z_5 = -1, c_5' - z_5' = 0$$

Clearly, for $\lambda > 3$, $c_2^* - z_2^* > 0$ so that we shall have to introduce x_2 in the basis. But corresponding to x_2 all $a_{i2} \leq 0$ so, x_2 can not be introduced in the basis and hence no optimal solution exists for $\lambda > 3$.

9.4 SYSTEMETIC LINEAR VARIATION IN b_i

Consider the parametric linear programming

$$\max . Z = \boldsymbol{Cx}$$

subject to the constraints

$$Ax \leq \boldsymbol{b} + \lambda \boldsymbol{b}' = \boldsymbol{b}(\lambda) \qquad \qquad \qquad ...(1)$$

and $$\boldsymbol{x} \geq 0$$

where λ is a parameter, $\boldsymbol{b}$, $\boldsymbol{b}'$, $\boldsymbol{C}$ and $\boldsymbol{A}$ are known vectors. We have to find the range of λ and the family of optimal solutions in their respective range of λ from $-\infty$ to ∞.

Firstly, find the optimal solution of (1) for $\lambda = 0$ by simplex method. Let $\boldsymbol{B}$ and $\boldsymbol{x_B}$ be the optimal basis and optimal basis feasible solution for $\lambda = 0$.

Now, since $\Delta_j = c_j - z_j = c_j - \boldsymbol{C_B}\boldsymbol{B}^{-1}\alpha_j$, therefore any change in λ does not affect the optimality of the solution.

At $\lambda = 0$ since $\boldsymbol{x_B} = \boldsymbol{B}^{-1}\boldsymbol{b}$ therefore, when λ changes $-\infty$ to ∞, then the new basic variable reduces to

$$\hat{\boldsymbol{x}}_{\boldsymbol{B}} = \boldsymbol{B}^{-1} \cdot \boldsymbol{b}(\lambda)$$

$$= \boldsymbol{B}^{-1}(\boldsymbol{b} + \lambda \boldsymbol{b}') = \boldsymbol{B}^{-1}\boldsymbol{b} + \lambda \boldsymbol{B}^{-1}\boldsymbol{b}'$$

$$= \boldsymbol{x_B} + \lambda \boldsymbol{x_B'}, \boldsymbol{x_B} = \boldsymbol{B}^{-1}\boldsymbol{b} \text{ and } \boldsymbol{x_B'} = \boldsymbol{B}^{-1} \cdot \boldsymbol{b}'$$

Clearly, the solution $\hat{\boldsymbol{x}}_{\boldsymbol{B}}$ will remain feasible if $\hat{\boldsymbol{x}}_{\boldsymbol{B}} \geq \boldsymbol{0}$ so for a given solution we find the range of λ within which the solution remains optimal and feasible.

Further, the new $\hat{\boldsymbol{x}}_{\boldsymbol{B}}$ will remain basic feasible if

$$\hat{\boldsymbol{x}}_{\boldsymbol{B}} \geq 0$$

$$\Rightarrow \quad \boldsymbol{x_B} + \lambda \boldsymbol{x_B'} \geq 0, i.e., x_{B_i} + \lambda x_{B_i}' \geq 0 \text{ for } i = 1, 2, ..., m.$$

Thus, $$\max_{x_{B_i} > 0}\left\{-\frac{x_{B_i}}{x_{B_i}'}\right\} \leq \lambda \leq \min_{x_{B_i} < 0}\left\{-\frac{x_{B_i}}{x_{B_i}'}\right\} \qquad \qquad ...(2)$$

which shows that the solution $\hat{\boldsymbol{x}}_{\boldsymbol{B}}$ will remain basic feasible and optimal when λ satisfy the above condition (2)

Now, there are following two possibilities:

(i) If there is no $x_{B_i}' < 0$, then λ has no upper bound.

(ii) If there is no $x_{B_i}' > 0$, then λ has no lower bound.

Further, if at least one of $x_{B_i}' < 0$ then from (2) we obtain the upper bound of λ (say $\bar{\lambda}$) and if at least one of $x_{B_i}' > 0$ then again from (2), we obtain the lower bound of λ (say $\underline{\lambda}$).

Hence, $\hat{\boldsymbol{x}}_{\boldsymbol{B}}$ will remain basic feasible and optimal for $\underline{\lambda} \leq \lambda \leq \bar{\lambda}$.

Also, for $\lambda > \bar{\lambda}$ (or $\lambda < \underline{\lambda}$) the basic feasible variable $\boldsymbol{x_B}$ is negative, where $x_{B_r}' < 0$. Then applying dual simplex method we can find the new optimal solution for $\lambda > \bar{\lambda}$.

Solved Examples

EXAMPLE 1. *For the following LPP,*

$$\text{max. } Z = 3x_1 + 2x_2 + 5x_3$$

subject to the constraints

$$x_1 + 2x_2 + x_3 \le 430 + 100\lambda$$
$$3x_1 + 2x_3 \le 460 - 200\lambda$$
$$x_1 + 4x_2 \le 420 + 400\lambda$$

and $x_1, x_2, x_3 \ge 0$

find the range of λ for which the solution remain optimal basic and feasible.

SOLUTION. Using slack variables s_1, s_2, s_3, we can write the given problem as

$$\text{max. } Z = 3x_1 + 2x_2 + 5x_3 + 0s_1 + 0s_2 + 0s_3$$

s.t.
$$x_1 + 2x_2 + x_3 + s_1 = 430 + 100\lambda$$
$$3x_1 + 2x_3 + s_2 = 460 - 200\lambda$$
$$x_1 + 4x_2 + s_3 = 420 + 400\lambda$$

and $x_1, x_2, x_3, s_1, s_2, s_3 \ge 0$

Here, we have

$$b(\lambda) = \begin{bmatrix} 430 + 100\lambda \\ 460 - 200\lambda \\ 420 + 400\lambda \end{bmatrix} = \begin{bmatrix} 430 \\ 460 \\ 420 \end{bmatrix} + \lambda \begin{bmatrix} 100 \\ -200 \\ 400 \end{bmatrix}$$

$$\Rightarrow \qquad b = \begin{bmatrix} 430 \\ 460 \\ 420 \end{bmatrix} \text{ and } b' = \begin{bmatrix} 100 \\ -200 \\ 400 \end{bmatrix}$$

In case of $\lambda = 0$, apply the simplex method, we obtain the final simplex table given as under

Simplex table-1

		$c_j \rightarrow$	3	2	5	0	0	0
B.V.	C_B	x_B	x_1	x_2	x_3	s_1	s_2	s_3
x_2	2	100	−1/4	1	0	1/2	−1/4	0
x_3	5	230	3/2	0	1	0	1/2	0
s_3	0	20	2	0	0	−2	1	1
$Z = C_B x_B = 1350$		$\Delta_j \rightarrow$	− 4	0	0	−1	−2	0

We observe that

$$x_B = \begin{bmatrix} 100 \\ 230 \\ 20 \end{bmatrix} = \begin{bmatrix} x_{B_1} \\ x_{B_2} \\ x_{B_3} \end{bmatrix} = \begin{bmatrix} x_2 \\ x_3 \\ s_3 \end{bmatrix}$$

$$B = [x_2, x_3, s_3] \quad \Rightarrow \quad B^{-1} = [s_1 \quad s_2 \quad s_3] = \begin{bmatrix} 1/2 & -1/4 & 0 \\ 0 & 1/2 & 0 \\ -2 & 1 & 1 \end{bmatrix}$$

$$\Rightarrow \qquad x'_B = B^{-1}b'$$

$$= \begin{bmatrix} 1/2 & -1/4 & 0 \\ 0 & 1/2 & 0 \\ -2 & 1 & 1 \end{bmatrix} \begin{bmatrix} 100 \\ -200 \\ 400 \end{bmatrix}$$

$$= \begin{bmatrix} 100 \\ -100 \\ 0 \end{bmatrix} = \begin{bmatrix} x'_{B_1} \\ x'_{B_2} \\ x'_{B_3} \end{bmatrix}$$

So, new solution is given by

$$\hat{x}_B = x_B + \lambda x'_B = \begin{bmatrix} 100 \\ 230 \\ 20 \end{bmatrix} + \lambda \begin{bmatrix} 100 \\ -100 \\ 0 \end{bmatrix} = \begin{bmatrix} 100 + 100\lambda \\ 230 - 100\lambda \\ 20 \end{bmatrix} = \begin{bmatrix} x_2 \\ x_3 \\ s_3 \end{bmatrix}$$

$$\Rightarrow \quad x_2 = 100 + 100\lambda, \; x_3 = 230 - 100\lambda, \; s_3 = 20$$

Now, $$\qquad \bar{\lambda} = \min_{x_{B_i} < 0} \left\{ -\frac{x_{B_i}}{x'_{B_i}} \right\} = \left\{ -\frac{x_{B_2}}{x'_{B_2}} \right\} = \frac{-230}{-100} = 23$$

and $$\qquad \underline{\lambda} = \max_{x_{B_i} > 0} \left\{ -\frac{x_{B_i}}{x'_{B_i}} \right\} = \max \left\{ -\frac{x_{B_1}}{x'_{B_1}} \right\} = \max \left\{ -\frac{100}{100} \right\} = -1$$

Thus, the solution $x_2 = -100 + 100\lambda$, $x_3 = 230 - 100\lambda$, $s_3 = 20$ and max. $Z = C_B \cdot \hat{x}_B = 1350 - 300\lambda$ is optimal basic feasible for
$$-1 \le \lambda \le 23$$
For $\lambda > 23$, x_3 is negative which correspond to x'_{B_2} so all $a_{2j} \ge 0$.
$\Rightarrow \quad$ No optimal solution exists for $\lambda > 23$
and for $\lambda < -1$, then x_2 is negative, so the column vector corresponding to x_2 will be outgoing vector.
Since, x_2 corresponds to $\hat{x}_{B_1}$, so $r = 1$

Now, $$\qquad \frac{\Delta_k}{a_{rk}} = \min_j \left\{ \frac{\Delta_j}{a_{rj}}, a_{rj} < 0 \right\}$$

$$\Rightarrow \qquad \frac{\Delta_k}{a_{1k}} = \min_j \left\{ \frac{\Delta_j}{a_{1j}}, a_{1j} < 0 \right\}$$

$$= \min \left\{ \frac{\Delta_1}{a_{11}}, \frac{\Delta_5}{a_{15}} \right\} = \min \left\{ \frac{-4}{-1/4}, \frac{-2}{-1/4} \right\}$$

$$= \min\{16, 8\} = 8 = \frac{\Delta_5}{a_{15}}$$

$$\Rightarrow \qquad k = 5$$

$\Rightarrow \quad s_2 \; (= x_5)$ is the incoming vector and key element $= a_{15} = -\dfrac{1}{4}$

Now, again apply simplex method in a usual manner, we have the next optimal table given as below.

Simplex table-2

B.V.	C_B	x_B	$c_j \rightarrow$ x_1 (3)	x_2 (2)	x_3 (5)	s_1 (0)	s_2 (0)	s_3 (0)
s_2	0	-400	1	-4	0	-2	1	0
x_3	5	430	1	2	1	1	0	0
s_3	0	420	1	4	0	0	0	1
$Z = C_B x_B = 2150$			$\Delta_j \rightarrow$ -2	-8	0	-5	0	0

From the above table, we observe that

$$x_B = \begin{bmatrix} -400 \\ 420 \\ 420 \end{bmatrix} = \begin{bmatrix} x_{B_1} \\ x_{B_2} \\ x_{B_3} \end{bmatrix} = \begin{bmatrix} s_2 \\ x_3 \\ s_3 \end{bmatrix}$$

$$B = [s_2, x_3, s_3] \quad \Rightarrow \quad B^{-1} = [s_1 \quad s_2 \quad s_3] = \begin{bmatrix} -2 & 1 & 0 \\ 1 & 0 & 0 \\ 0 & 0 & 1 \end{bmatrix}$$

So,

$$x_B' = B^{-1}b' = \begin{bmatrix} -2 & 1 & 0 \\ 1 & 0 & 0 \\ 0 & 0 & 1 \end{bmatrix} \begin{bmatrix} 100 \\ -200 \\ 400 \end{bmatrix} = \begin{bmatrix} -400 \\ 100 \\ 400 \end{bmatrix} = \begin{bmatrix} x_{B_1}' \\ x_{B_2}' \\ x_{B_3}' \end{bmatrix}$$

So, the new solution is given by

$$\hat{x}_B = x_B + \lambda x_B'$$

$$\Rightarrow \qquad \hat{x}_B = \begin{bmatrix} -400 \\ 430 \\ 420 \end{bmatrix} + \lambda \begin{bmatrix} -400 \\ 100 \\ 400 \end{bmatrix} = \begin{bmatrix} -400 - 400\lambda \\ 430 + 100\lambda \\ 420 + 400\lambda \end{bmatrix} = \begin{bmatrix} s_2 \\ x_3 \\ s_3 \end{bmatrix}$$

$$\Rightarrow \quad s_2 = -400 - 400\lambda, \ x_3 = 430 + 100\lambda, \ s_3 = 420 + 400\lambda$$

Now,

$$\overline{\lambda} = \min_{x_{B_i} < 0} \left\{ -\frac{x_{B_i}}{x_{B_i}'} \right\} = \min \left\{ -\frac{x_{B_1}}{x_{B_1}'} \right\} = -\frac{-400}{-400} = -1$$

and

$$\underline{\lambda} = \max_{x_{B_i} > 0} \left\{ -\frac{x_{B_i}}{x_{B_i}'} \right\} = \max \left\{ -\frac{x_{B_2}}{x_{B_2}'}, -\frac{x_{B_3}}{x_{B_3}'} \right\} = \max \left\{ -\frac{430}{100}, \frac{-420}{400} \right\}$$

$$= -\frac{420}{400} = -\frac{21}{20}$$

Therefore, the solution $s_2 = -400 - 400\lambda$, $x_3 = 430 + 100\lambda$, $s_3 = 420 + 400\lambda$ is

optimal feasible for $\dfrac{-21}{20} \leq \lambda \leq -1$.

If $\lambda < -\dfrac{21}{20}$, then s_3 is negative and which corresponds to x_{B_3}' and all $a_{3j} \geq 0$.

Hence, no optimal solution exists for $\lambda < -\dfrac{21}{20}$.

$\Rightarrow$ The solution $x_1 = 0$, $x_2 = 100 + 100\lambda$, $x_3 = 230 - 100\lambda$, max. $Z = 1350 - 300\lambda$ is feasible and optimal for $-1 \leq \lambda \leq 23$ and $x_1 = 0$, $x_3 = 430 + 100\lambda$, max $Z = 2150 + 500\lambda$ is feasible and optimal for $-\dfrac{21}{20} \leq \lambda \leq -1$.

EXAMPLE 2. *For the LPP given by*

$$\text{max. } Z = 4x_1 + 6x_2 + 2x_3$$

subject to the constraints

$$x_1 + x_2 + x_3 \leq 3 + 3\lambda$$
$$x_1 + 4x_2 + 7x_3 \leq 9 - 3\lambda$$

and $\quad x_1, x_2, x_3 \geq 0$

Find the range of λ for which the solution remains basic optimal and feasible.

SOLUTION. Using slack variables s_1 and s_2, the given LPP can be written as

$$\text{max. } Z = 4x_1 + 6x_2 + 2x_3 + 0s_1 + 0s_2$$
s.t. $\quad x_1 + x_2 + x_3 + s_1 = 3 + 3\lambda$
$$x_1 + 4x_2 + 7x_3 + s_2 = 9 - 3\lambda$$
and $\quad x_1, x_2, x_3, s_1, s_2 \geq 0$

Clearly, $b(\lambda) = \begin{bmatrix} 3 + 3\lambda \\ 9 - 3\lambda \end{bmatrix} = \begin{bmatrix} 3 \\ 9 \end{bmatrix} + \lambda \begin{bmatrix} 3 \\ -3 \end{bmatrix}$

$\Rightarrow \quad b = \begin{bmatrix} 3 \\ 9 \end{bmatrix}$ and $b' = \begin{bmatrix} 3 \\ -3 \end{bmatrix}$

Now, for $\lambda = 0$, the optimal solution of the given problem is given in the following table.

Simplex table-1

			$c_j \rightarrow$	4	6	2	0	0
B.V.	C_B	x_B		x_1	x_2	x_3	s_1 (x_4)	s_2 (x_5)
x_1	4	1		1	0	−1	4/3	−1/3
x_2	6	2		0	1	2	−1/3	1/3
$Z = C_B x_B = 16$		$\Delta_j \rightarrow$		0	0	−6	−10/3	−2/3

Here, clearly we have

$$x_B = \begin{bmatrix} 1 \\ 2 \end{bmatrix} = \begin{bmatrix} x_{B_1} \\ x_{B_2} \end{bmatrix} = \begin{bmatrix} x_1 \\ x_2 \end{bmatrix}$$

and $\quad B = [x_1, x_2] \quad \Rightarrow \quad B^{-1} = [s_1 \ \ s_2] = \begin{bmatrix} 4/3 & -1/3 \\ -1/3 & 1/3 \end{bmatrix}$

So, $\quad x_B' = B^{-1} \cdot b' = \begin{bmatrix} 4/3 & -1/3 \\ -1/3 & 1/3 \end{bmatrix} \begin{bmatrix} 3 \\ -3 \end{bmatrix} = \begin{bmatrix} 5 \\ 2 \end{bmatrix} \begin{bmatrix} x_{B_1}' \\ x_{B_2}' \end{bmatrix}$

$\therefore$ the new solution

$$\hat{x}_B = x_B + \lambda x_B'$$

$$= \begin{bmatrix} 1 \\ 2 \end{bmatrix} + \lambda \begin{bmatrix} 5 \\ -2 \end{bmatrix} = \begin{bmatrix} 1+5\lambda \\ 2-2\lambda \end{bmatrix} = \begin{bmatrix} x_1 \\ x_2 \end{bmatrix}$$

$\Rightarrow \quad x_1 = 1 + 5\lambda, x_2 = 2 - 2\lambda$

Now, since $x'_{B_1} = 5, x'_{B_2} = -2$, therefore,

$$\overline{\lambda} = \min_{x_{B_i} < 0} \left\{ -\frac{x_{B_i}}{x'_{B_i}} \right\} = \min \left\{ -\frac{x_{B_2}}{x'_{B_2}} \right\} = \frac{-2}{-2} = 1$$

and $\quad \underline{\lambda} = \max_{x_{B_i} > 0} \left\{ -\frac{x_{B_i}}{x'_{B_i}} \right\} = \max \left\{ -\frac{x_{B_2}}{x'_{B_2}} \right\} = -\frac{1}{5}$

So, solution $x_1 = 1 + 5\lambda, x_2 = 2 - 2\lambda, x_3 = 0$,

$\qquad$ max. $Z = C_B \hat{x}_B = 4(1+5\lambda) + 6(2-2\lambda) = 16 + 8\lambda$

is optimal basic and feasible for $-\dfrac{1}{5} \le \lambda \le 1$

Further, for $\lambda > 1$, x_2 is negative, so vector x_2 will be the outgoing vector.

$\because \qquad x'_{B_2} = -2 < 0$ then $r = 2$

Now, since

$$\frac{\Delta_k}{a_{rk}} = \min_j \left\{ \frac{\Delta_j}{a_{rj}}, a_{rj} < 0 \right\}$$

$\Rightarrow \qquad \dfrac{\Delta_k}{a_{2k}} = \min_j \left\{ \dfrac{\Delta_j}{a_{2j}}, a_{2j} < 0 \right\} = \min \left\{ \dfrac{\Delta_4}{a_{24}} \right\} = \dfrac{\Delta_4}{a_{24}}$

$\Rightarrow \qquad k = 4$

$\Rightarrow \quad x_4 (= s_1)$ is the incoming vector and key element $= a_{24} = -\dfrac{1}{3}$

Now, apply the simplex method in a usual manner, we have the final optimal simplex table.

Simplex table-2

		$c_j \rightarrow$	4	6	2	0	0
B.V.	C_B	x_B	x_1	x_2	x_3	s_1	s_2
x_1	4	9	1	4	7	0	1
s_1	0	-6	0	-3	-6	1	-1
$Z = C_B x_B = 36$		$\Delta_j \rightarrow$	0	-10	-26	0	

We have $\boldsymbol{x_B} = \begin{bmatrix} 9 \\ -6 \end{bmatrix} = \begin{bmatrix} x_{B_1} \\ x_{B_2} \end{bmatrix} = \begin{bmatrix} x_1 \\ s_1 \end{bmatrix}$

$$\boldsymbol{B} = [x_1, s_1] \qquad \Rightarrow \qquad B^{-1} = [s_1, s_2] = \begin{bmatrix} 0 & 1 \\ 1 & -1 \end{bmatrix}$$

Therefore, $\hat{\boldsymbol{x}}_{\boldsymbol{B}} = \boldsymbol{B}^{-1} \cdot \boldsymbol{b'} = \begin{bmatrix} 0 & 1 \\ 1 & -1 \end{bmatrix} \begin{bmatrix} 3 \\ -3 \end{bmatrix} = \begin{bmatrix} -3 \\ 6 \end{bmatrix} = \begin{bmatrix} x'_{B_1} \\ x'_{B_2} \end{bmatrix}$

The new solution $\hat{x}_B$ is given by

$$\hat{x}_B = x_B + \lambda x_B' = \begin{bmatrix} 9 \\ -6 \end{bmatrix} + \lambda \begin{bmatrix} -3 \\ 6 \end{bmatrix} = \begin{bmatrix} 9 - 3\lambda \\ -6 + 6\lambda \end{bmatrix} = \begin{bmatrix} x_1 \\ s_1 \end{bmatrix}$$

$\Rightarrow$ $\qquad x_1 = 9 - 3\lambda,\ s_1 = -6 + 6\lambda$

Since, $x_{B_1}' = -3$ and $x_{B_2}' = 6$

Now, $\qquad \bar{\lambda} = \min_{x_{B_i} < 0} \left\{ -\frac{x_{B_i}}{x_{B_i}'} \right\} = \min \left\{ -\frac{x_{B_1}}{x_{B_1}'} \right\} = -\frac{9}{-3} = 3$

and $\qquad \underline{\lambda} = \max_{x_{B_i} > 0} \left\{ -\frac{x_{B_i}}{x_{B_i}'} \right\} = \max \left\{ -\frac{x_{B_2}}{x_{B_2}'} \right\} = -\frac{6}{6} = 1$

Thus, the solution $x_1 = 9 - 3\lambda$, $x_2 = x_3 = 0$, $s_1 = -6 + 6\lambda$, $s_2 = 0$ and max. $Z = 36 - 12\lambda$ is optimal basic and feasible for $1 \le \lambda \le 3$.

Now, for $\lambda > 3$, x_1 is negative which is corresponding to $x_{B_1}' = -3 < 0$ and all $a_{1j} \ge 0 \Rightarrow$ optimal solution exists for $\lambda > 3$.

Again for $\lambda < -\dfrac{1}{5}$ then $x_1 < 0 \Rightarrow x_1$ is the outgoing vector.

Clearly, x_1 correspond to x_{B_1}' so $r = 1$.

Now, $\qquad \dfrac{\Delta_k}{a_{rk}} = \min_j \left\{ \dfrac{\Delta_j}{a_{rj}}, a_{rj} < 0 \right\} = \min_j \left\{ \dfrac{\Delta_j}{a_{1j}}, a_{1j} < 0 \right\}$

$$= \min \left\{ \frac{\Delta_3}{a_{13}}, \frac{\Delta_5}{a_{15}} \right\} = \min \left\{ \frac{-6}{-1}, \frac{-2/3}{-1/3} \right\}$$

$$= \min\{6, 2\} = 2 = \frac{\Delta_5}{a_{15}}$$

$\Rightarrow \qquad k = 5$

$\Rightarrow$ x_5 (s_2) is the incoming vector and key element $a_{15} = -\dfrac{1}{3}$. Now, by applying the simplex method in a usual manner we get the following simplex table.

Simplex table-3

		$c_j \rightarrow$	4	6	2	0	0
B.V.	C_B	x_B	x_1	x_2	x_3	s_1 (x_4)	s_2 (x_5)
s_2	0	-3	-3	0	3	-4	1
x_2	6	3	1	1	1	1	0
$Z = C_B x_B = 18$		$\Delta_j \rightarrow$	-2	0	-4	-6	0

In the above table, we have

$$x_B = \begin{bmatrix} -3 \\ 3 \end{bmatrix} = \begin{bmatrix} x_{B_1} \\ x_{B_2} \end{bmatrix} = \begin{bmatrix} s_2 \\ x_2 \end{bmatrix}$$

$$B = [s_2, x_2]$$

$$\Rightarrow \qquad B^{-1} = [s_1, s_2] = \begin{bmatrix} -4 & 1 \\ 1 & 0 \end{bmatrix}$$

Therefore, $\quad x_B' = B^{-1}b' = \begin{bmatrix} -4 & 1 \\ 1 & 0 \end{bmatrix}\begin{bmatrix} 3 \\ -3 \end{bmatrix} = \begin{bmatrix} -15 \\ 3 \end{bmatrix} = \begin{bmatrix} x_{B_1}' \\ x_{B_2}' \end{bmatrix}$

Hence, the new solution is given by

$$\hat{x}_B = x_B + \lambda x_B'$$

$$= \begin{bmatrix} -3 \\ 3 \end{bmatrix} + \lambda \begin{bmatrix} -15 \\ 3 \end{bmatrix} = \begin{bmatrix} -3 - 15\lambda \\ 3 + 3\lambda \end{bmatrix} = \begin{bmatrix} s_2 \\ x_2 \end{bmatrix}$$

$$\Rightarrow \qquad x_2 = 3 + 3\lambda, \; s_2 = -3 - 15\lambda$$

Now, since $\quad x_{B_1}' = -15, x_{B_2}' = 3$

So, $$\bar{\lambda} = \min_{x_{B_i}' < 0}\left\{-\frac{x_{B_i}}{x_{B_i}'}\right\} = \min\left\{-\frac{x_{B_1}}{x_{B_1}'}\right\} = \min\left\{-\frac{-3}{-15}\right\} = -\frac{1}{5}$$

and $$\underline{\lambda} = \max_{x_{B_k}' > 0}\left\{-\frac{x_{B_k}}{x_{B_k}'}\right\} = \max\left\{-\frac{x_{B_2}}{x_{B_2}'}\right\} = \frac{-3}{3} = -1$$

$\Rightarrow$ the solution $x_1 = 0$, $x_2 = 3 + 3\lambda$, $x_3 = 0$, $s_1 = 0$, $s_2 = -3 - 15\lambda$ and max. $Z = 18 + 18\lambda$ is optimal basis and feasible for the range $-1 \le \lambda \le -\dfrac{1}{5}$.

Further, for $\lambda < -1$, $x_2 < 0$ which is corresponding to x_{B_2}' also from above table-3 all $a_{2j} \ge 0$. Hence, no optimal solution exists for $\lambda < -1$.

Hence, we conclude that the solution $x_1 = 1 + 5\lambda$, $x_2 = 2 - 2\lambda$, $x_3 = 0$, max. $Z = 16 + 8\lambda$ is optimal feasible for $-\dfrac{1}{5} \le \lambda \le 1$. The solution $x_1 = 9 - 3\lambda$, $x_2 = 0$, $x_3 = 0$ and max. $Z = 36 - 12\lambda$ is optimal feasible for $1 \le \lambda \le 3$ and the solution is $x_1 = 0$, $x_2 = 3 + 3\lambda$, $x_3 = 0$, max. $Z = 18 + 18\lambda$ is optimal feasible for $-1 \le \lambda \le -\dfrac{1}{5}$.

 Exercise-9.1

1. Perform a complete parametric LPP
   max. $Z = \lambda x_1 - x_2$
   subject to the constraints
   $$3x_1 - 3x_2 \ge 5$$
   $$2x_1 + x_2 \le 3$$
   and $\quad x_1, x_2 \ge 0$
   for $\quad -\infty \le \lambda \le \infty$.

2. Find the critical value of λ for which the solution of the following LPP is optimal basic and feasible
   max. $Z = (3 + 3\lambda)x_1 + 2x_2 + (5 - 6\lambda)x_3$
   s.t. $\quad x_1 + 2x_2 + x_3 \le 430$
   $$3x_1 + 2x_3 \le 460$$
   $$x_1 + 4x_2 \le 420$$
   and $\qquad x_1, x_2 \ge 0$

3. Solve the following LPP for all values of λ
   min. $Z = 2\lambda x_1 + (1 - \lambda)x_2 - 3x_3 + \lambda x_4$
   $\qquad\qquad + 2x_5 - 3\lambda x_6$
   subject to the constraints
   $$x_1 + 3x_2 - x_3 + 2x_5 = 7$$
   $$- 2x_2 + 4x_3 + x_4 = 12$$
   $$- 4x_2 + 3x_3 + 8x_5 + x_6 = 10$$
   and $\qquad x_1, x_2, ..., x_6 \ge 0$

4. Find the variation in optimal solution of the following parametric LPP
   min. $Z = 4x_1 + x_2 + 2x_3$
   subject to the constraints
   $$3x_1 + x_2 + 2x_3 = 3 + 3\lambda$$
   $$4x_1 + 3x_2 + 2x_3 \ge 6 + 2\lambda$$
   $$x_1 + 2x_2 + 5x_3 \le 4 - \lambda$$
   and $x_1, x_2, x_3 \ge 0$ and $\lambda \ge 0$

5. If the LPP given by

max. $Z = x_1 + 6x_2$

subject to the constraints

$$x_1 \le 4 + 8\lambda$$
$$3x_1 + 2x_2 \le 18 - 24\lambda$$

and $x_1, x_2 \ge 0$

has the optimal solution for $l = 0$ as follows:

$c_j \rightarrow$			1	6	2	0
B.V.	C_B	x_B	x_1	x_2	s_1	s_2
s_1	0	4	1	0	1	1
x_2	6	9	3/2	1	0	1/2
$Z = C_B x_B$ = 36		$\Delta_j \rightarrow$	-8	0	0	-3

Find the range of λ for which the above solution remains basic feasible and optimal $(\lambda \ge 0)$.

6. Find the range of λ over which the solution remain basic and optimal for the following LPP

max. $Z = \lambda x_1 - \lambda x_2 - x_3 + x_4$

subject to the constraints

$$3x_1 - 3x_2 - x_3 + x_4 \ge 5$$
$$2x_1 - 2x_2 + x_3 - x_4 \le 3$$

and $x_1, x_2, x_3, x_4 \ge 0$

7. Find the range of λ (≥ 0) over which the following LPP

max. $Z = 3x_1 + 2x_2 + 5x_3$

subject to the constraints

$$x_1 + 2x_2 + x_3 \le 40 - \lambda$$
$$3x_1 + 2x_3 \le 60 + 2\lambda$$
$$x_1 + 4x_2 \le 30 - 7\lambda$$

and $x_1, x_2, x_3 \ge 0$

has optimal and feasible solution.

8. Find the variation in optimal solution of the following parametric LPP given by $(\lambda \ge 0)$

min. $Z = (4 - \lambda)x_1 + (1 - 3\lambda)x_2 + (2 - 2\lambda)x_3$

subject to the constraints

$$3x_1 + x_2 + 2x_3 = 3$$
$$4x_1 + 3x_2 + 2x_3 \ge 6$$
$$x_1 + 2x_2 + 5x_3 \le 4$$

and $x_1, x_2, x_3 \ge 0$

ANSWERS

1. $x_1 = \dfrac{8}{5}, x_2 = -\dfrac{1}{5}, \min Z = \dfrac{1}{5} + \dfrac{8}{5}\lambda$ for $-2 \le \lambda \le 3$ and for $\lambda = 3$, multiple solution exists

2. $x_1 = 0, x_2 = 100, x_3 = 230, \max Z = 1350 - 1380\lambda$ for $0 \le \lambda \le \dfrac{1}{3}$

$x_1 = 10, x_2 = 1025, x_3 = 215, \max Z = 1310 - 1260\lambda$ for $\dfrac{1}{3} \le \lambda \le \dfrac{5}{12}$

$x_1 = \dfrac{460}{3}, x_2 = \dfrac{200}{3}, x_3 = 0, \max Z = \dfrac{1780}{3} + 460\lambda$ for $\lambda \ge \dfrac{5}{12}$

3. $x_1 = 0, x_2 = 4, x_3 = 5, x_4 = x_5 = 0, x_6 = 11, \min Z = -11 - 37\lambda$ for $-\dfrac{1}{27} \le \lambda \le 2$

$x_1 = 0, x_2 = \dfrac{7}{3}, x_3 = 0, x_4 = \dfrac{50}{3}, x_5 = 0, x_6 = \dfrac{58}{3}, \min Z = \dfrac{7}{3} - 217\lambda$ for $2 \le \lambda \le \infty$

4. $x_1 = \dfrac{3}{2}, x_2 = \dfrac{9}{2}, \max Z = \dfrac{45}{2}$ and $-3 \le \theta \le \dfrac{3}{7}$

5. For $0 \le \lambda \le \dfrac{3}{8}, x_1 = \dfrac{3 + 7\lambda}{5}, x_2 = \dfrac{6 - 11\lambda}{5}, \min Z = \dfrac{21 - 11\lambda}{5}$

For $\dfrac{3}{8} \le \lambda \le \dfrac{2}{5}, x_1 = -3 + 11\lambda, x_2 = 6 - 15\lambda, \min Z = 6 + 3\lambda$

6. $x_1 = \dfrac{8}{5}, x_2 = x_3 = 0, x_4 = \dfrac{1}{5}, \min Z = \dfrac{1}{5} + \dfrac{8}{5}\lambda$ for $-2 \le \lambda \le 3$ and for $\lambda > 3$ no solution exists

7. $x_1 = 0, x_2 = 5 - \lambda, x_3 = 30 + \lambda, \max. Z = 160 + 3\lambda$ for $0 \le \lambda \le \dfrac{10}{3}$

$x_1 = 0, x_2 = \dfrac{30 - 7\lambda}{4}, x_3 = 30 + \lambda, \max. Z = 165 + \dfrac{3}{2}\lambda$ for $\dfrac{10}{3} \le \lambda \le \dfrac{30}{7}$

and for $\lambda > \dfrac{30}{7}$ no feasible solution exists

8. $x_1 = \dfrac{2}{5}, x_2 = \dfrac{9}{5}, x_3 = 0, \min Z = \dfrac{17}{5} - \dfrac{29}{5}\lambda$

Glossary

- **Parametric Linear Programming:** The investigation which deals with the effect of simultaneous changes of all components of **C** or **b** in the optimal solution of the problem is called parametric linear programming.

REVIEW QUESTIONS

1. What do you understand by parametric linear programming.
2. What is the parametric linear programming. How does it different from sensitivity analysis?
3. Write the types of parametric LPP.
4. Explain the different solution process of parametric linear programming.

Integer Programming

10.1 INTRODUCTION

In linear programming problem, we observe that each decision variables, slack and surplus variables, can take any real or fractional values. But there are some situations in which the fractional values of these variables has no significance. The integer programming is a linear programming in which some or all variables $x_1, x_2, ..., x_n$ are permitted to take only integral values.

Definition. *Integer linear programming are those in which some or all of the variables are restricted to integer or discrete values.*

10.2 NEED OF INTEGER LINEAR PROGRAMMING

The integer linear programming has important applications in business and industry. Actually in any situation involving the decision of the type "either-or", *i.e.*, either to do the activity or not to do can be viewed as integer linear programming. Besides these, in the manufacturing decisions of trucks, of cars etc. the quantity manufactured can be a whole number. Actually all the allocation problems requiring the allocations of men, machines or vehicles etc. to activities in a programming problem will be an integer linear programming as such things can be assigned in integer quantities not in fraction.

10.3 TYPES OF INTEGER LINEAR PROGRAMMING PROBLEMS

There are following three types of integer linear programming problem:

(1) **Pure Integer Linear Programming Problem:** An integer linear programming is said to be a pure integer linear programming when all its decision variables are restricted to be integer.

(2) **Mixed Integer Linear Programming Problem:** An integer linear programming problem is said to be mixed integer programming when some, but not all of its decision variables are restricted to be integer.

(3) **Zero-one Integer Linear Programming Problem:** An integer linear programming problem in which all the decision variables are restricted to integer of 0 or 1 is called zero-one integer linear programming problem.

10.4 METHODS TO SOLVE AN INTEGER LINEAR PROGRAMMING PROBLEM

There are following two methods:

10.4.1 GOMORY'S CUTTING PLANE METHOD

Consider a linear integer programming problem. A systematic procedure for solving all pure integer linear programming problem was first developed by R.E. Gomory in 1958. Later, he extended the procedure to solve the mixed integer programme.

First find the optimal solution to the usual linear programming problem by the simplex method ignoring the integer-valued restriction. If in the optimal solution all the variables have integer values, then it is also the optimal solution of the given LPP. But if not, then modify the LPP by systematically introducing a new constraints called secondary or Gomory's constraints which essentially represents necessary conditions for integrability and eliminates some non-integer solutions without loosing any integral solution. After adding the second constraints, the problem is solved by dual simplex algorithm to get an optimal integral solution. If optimal integer valued solution is obtained, then the problem is solved to get an integer valued optimum solution. Repeat this process iteratively until one obtains the required integer valued optimal solution. In this method, our main work is to enlarge the continuous ILPP obtained by ignoring the integral valued restriction of the ILPP by introducing the Gomory's constraints (Gomory's cut) and hence the construction of such constraints is very important job in this method, and need special attention.

10.4.2 CONSTRUCTION OF GOMORY'S CONSTRAINTS

To find the Gomory's constraints, use the fact that a solution satisfy the constraints of the given IPP also satisfies any other constraints obtained by adding or subtracting two more given constraints or obtained by multiplying a constraints by a non-zero real number.

Let $x_B = (x_{B_1}, x_{B_2}, \ldots, x_{B_m})$ be the optimal solution obtained by regular simplex method by ignoring integer condition on the variables of maximization LPP. Then we get the final simplex table as given below.

B.V.	C_B	x_B	x_1 (β_1)	x_2 (β_2)	...	x_i (β_i)	...	x_m (β_m)	x_{m+1}	...	x_n
x_1	C_{B_1}	x_{B_1}	1	0	...	0	...	0	$a_{1(m+1)}$	...	a_{1n}
x_2	C_{B_2}	x_{B_2}	0	1		:		:	$a_{2(m+1)}$	...	a_{2n}
:	:	:	:	:	:	:	:	:	:		:
x_i	C_{B_i}	x_{B_i}	0	0		1		:	$a_{i(m+1)}$	...	a_{in}
:	:	:	:	:	:	:	:	:	:		:
x_m	C_{B_m}	x_{B_m}	0	0		0		1	$a_{m(m+1)}$	...	a_{mn}
		x_j	x_{B_1}	x_{B_2}	...	x_{B_i}	...	x_{B_m}	0	...	0

Let us suppose x_i is not an integer. Then from the above table (i^{th} row) we have

$$x_{B_i} = 0 \cdot x_1 + 0 \cdot x_2 + \ldots + 1 \cdot x_i + \ldots + 0 \cdot x_m + a_{i(m+1)} x_{m+1} + \ldots + a_{in} x_n$$

$$\Rightarrow \qquad x_{B_i} = x_i + \sum_{j=m+1}^{n} a_{ij}x_j$$

$$\Rightarrow \qquad x_i = x_{B_i} - \sum_{j=m+1}^{n} a_{ij}x_j \qquad\qquad \text{...(1)}$$

Since, we have x_{B_i} is a non-integer, therefore, we have

$$x_{B_i} = [x_{B_i}] + f_{B_i} \quad \text{and} \quad a_{ij} = [a_{ij}] + f_{ij} \qquad\qquad \text{...(2)}$$

where $[\cdot]$ denote the integral part and f_{B_i} and f_{ij} are the positive fractional parts of x_{B_i}

and a_{ij} respectively.

Therefore,

$$\left.\begin{array}{l} [x_{B_i}] \le x_{B_i}, 0 \le f_{B_i} < 1 \\ [a_{ij}] \le a_{ij}, 0 \le f_{ij} < 1 \end{array}\right\} \qquad\qquad \text{...(3)}$$

Using (2) in (1), we get

$$x_i = [x_{B_i}] + f_{B_i} - \sum_{j=m+1}^{n} ([a_{ij}] + f_{ij})x_j$$

$$\Rightarrow \qquad x_i - [x_{B_i}] + \sum_{j=m+1}^{n} f_{ij}; x_j = f_{B_i} - \sum_{j=m+1}^{n} [a_{ij}]x_j$$

$$\text{or} \quad x_i - [x_{B_i}] + \sum_{j=m+1}^{n} [a_{ij}]x_j = f_{B_i} - \sum_{j=m+1}^{n} f_{ij}x_j \qquad\qquad \text{...(4)}$$

Since, for optimal integer solution x_i for $i = 1, 2, \ldots, m$ are all integers and also for x_j for $j = m +1, m + 2, \ldots, n$ are all integers, then from (4)

$$f_{B_i} - \sum_{j=m+1}^{n} f_{ij}x_j \quad \text{must be an integer}$$

Now, since, $0 < f_{ij} < 1$ and $x_{ij} \ge 0$ then

$$\sum_{j=m+1}^{n} f_{ij} \cdot x_j > 0$$

$$\Rightarrow \qquad f_{B_i} - \sum_{j=m+1}^{n} f_{ij} \cdot x_j \le f_{B_i}$$

But $0 \le f_{B_i} < 1$, therefore, $f_{B_i} - \sum_{j=m+1}^{n} f_{ij} \cdot x_j < 1$

Also, $f_{B_i} - \sum_{j=m+1}^{n} f_{ij}x_j$ is an integer.

$\Rightarrow$ It should be either 0 or negative integer.

So, $$f_{B_i} - \sum_{j=m+1}^{n} f_{ij}x_j \leq 0$$

$$\Rightarrow \qquad - \sum_{j=m+1}^{n} f_{ij}x_j \leq -f_{B_i} \qquad \qquad ...(5)$$

The inequality given above by (5) is called Gomory's constraints. Adding a non-negative slack variable s_i, Gomory's constraints become

$$- \sum_{j=m+1}^{n} f_{ij}x_j + s_i = -f_{B_i} \qquad \qquad ...(6)$$

It is called Gomory's cutting plane (fractional cut)

Now, since x_j $(j = m + 1, ..., n)$ are non-basic variables, so they are all zero which implies that

$$s_i = -f_{B_i}, \text{ which is infeasible.}$$

Now, adding the Gomory constraints equation (6) to the table (1) we get the new table 2 with one extra column of s_i $(\beta_{m + 1})$.

B.V.	C_B	x_B	x_1 (β_1)	x_2 (β_2)	...	x_i (β_i)	...	x_m (β_m)	$x_{m + 1}$	...	x_n	s_i $(\beta_{m + 1})$
x_1	C_{B_1}	x_{B_1}	1	0	...	0	...	0	$a_{1(m + 1)}$	...	a_{1n}	0
x_2	C_{B_2}	x_{B_2}	0	1	...	0	...	0	$a_{2(m + 1)}$	...	a_{2n}	0
$\vdots$	$\vdots$	$\vdots$	$\vdots$	$\vdots$	...	$\vdots$	...	$\vdots$	$\vdots$		$\vdots$	$\vdots$
x_i	C_{B_i}	x_{B_i}	0	0	...	i	...	0	$a_{i(m + 1)}$	...	a_{in}	0
$\vdots$	$\vdots$	$\vdots$	$\vdots$	$\vdots$	...	$\vdots$			$\vdots$		$\vdots$	$\vdots$
x_m	C_{B_m}	x_{B_m}	0	0	...	0		0	$a_{m(m + 1)}$	...	a_{mn}	0
s_i	0	$-f_{B_i}$	0	0	...	0		0	$f_{i(m + 1)}$	...	f_{in}	1
	x_j		x_{B_1}	x_{B_2}	...	x_{B_i}	...	x_{B_m}	0	...	0	$-f_{B_i}$

Now since s_i is negative, then the optimal solution given by above table is not feasible, so dual simplex method can be applied to clear this feasibility and get new optimal solution. If this solution has integer values of all decision variables, then this is the required solution and we end the process, otherwise we construct a new Gomory's constraints as obtained earlier to find further new optimal solution. Repeat the above process until an integer solution is obtained.

STEP 1. Write the given LPP into maximization LPP.

STEP 2. Make all $b_i's$ positive.

STEP 3. Apply the regular simplex method, by ignoring integer condition on variables and find optimal solution.

STEP 4. If the obtained optimal solution have all integer solution, then obtained solution will be the required solution. But if at least one of the variable is not an integer, go to step 5.

STEP 5. (i) If only one variable is not an integer, then from the simplex table of step 3, the row corresponding to non-integer, variable will generate Gomory's constraints.

(ii) If more than one variable are non-integer then select that variable which has largest fractional value. In case of tie, select the constraints having the lowest contribution for maximization or highest cost for minimization problem.

Alternatively, select the constraints with

$$\max \left\{ \frac{f_{B_i}}{\sum\limits_{j=m+1}^{n} f_{ij}} \right\}$$

STEP 6. Add the Gomory's constraints (obtained in the above step) with non-negative slack variable s_i into the final simplex table.

STEP 7. Obtain the optimal solution of the table in the above table by dual simplex method, so that s_i is the outgoing vector.

STEP 8. If the obtained solution (obtained in step 7) has all integral values then it is the required solution of the given integer programming. If it is not an integer solution, then repeat step 6 to 8 until we get an optimal feasible integer solution.

10.4.3 PROPERTIES OF GOMORY'S ALGORITHM

The Gomory's algorithm has the following properties:

(1) Additional linear constraints never cut off that portion of the original feasible solution space which contains a feasible integer solution to the original problem.

(2) Each new additional constraints cuts-off the current non-integer optimal solution to the linear programming problem.

☛ REMARK

• If the given integer programming has a constraints with fractional coefficients, then transform the constraints such that the coefficients are whole numbers.

Solved Examples

EXAMPLE 1. *Find the optimum integer solution to the following LPP by Gomory technique.*

$$\max. Z = x_1 + 2x_2$$
subject to the constraints
$$2x_2 \leq 7$$
$$x_1 + x_2 \leq 7$$
$$2x_1 \leq 11$$

and $x_1, x_2 \geq 0$ and x_1, x_2 are integers.

SOLUTION. Clearly, the given LPP of maximization is in standard form with all $b_i's \geq 0$

Proceeding as usual introducing slack variables s_1, s_2 and s_3, we can write

$$\text{max. } Z = x_1 + 2x_2 + 0s_1 + 0s_2 + 0s_3$$

subject to the constraints

$$0x_1 + 2x_2 + s_1 = 7$$
$$x_1 + x_2 + s_2 = 7$$
$$2x_1 + s_3 = 11$$

and $\quad x_1, x_2, s_1, s_2, s_3 \geq 0$

Now apply the simplex method in a usual manner, we get the following simplex table.

B.V.	C_B	x_B	$c_j \rightarrow$ 1 x_1	2 x_2	0 s_1 (x_3)	0 s_2 (x_4)	0 s_3 (x_5)	Min Ratio
s_1	0	7	0	2	1	0	0	7/2 (min)→
s_2	0	7	1	1	0	1	0	7/1
s_3	0	11	2	0	0	0	1	—
$Z = C_B x_B = 0,$		$\Delta_j \rightarrow$	1	2	0	0	0	
x_2	2	7/2	0	1	1/2	0	0	—
s_2	0	7/2	1	1	−1/2	1	0	(7/2)/1 (min)→
s_3	2	11	2	0	0	0	1	11/2
$Z = C_B x_B = 7,$		$\Delta_j \rightarrow$	1	0	−1	0	0	
x_2	2	7/2	0	1	1/2	0	0	
x_1	1	7/2	1	1	−1/2	1	0	
s_3	0	4	0	0	1	−2	1	
$Z = C_B x_B = 21/2,$		$\Delta_j \rightarrow$	0	0	−1/2	−1	0	

From above table, we observe that all $\Delta_j \leq 0$ and $x_1 = x_2 = \dfrac{7}{2}$. Therefore, the solution under the above table is optimal and feasible but not integer.

Now, apply Gomory technique to find the integer solution

$$\because \quad x_1 = \frac{7}{2} = 3 + \frac{1}{2} \quad \text{and} \quad x_2 = \frac{7}{2} = 3 + \frac{1}{2}$$

So, $f_{B_1} = \dfrac{1}{2}, f_{B_2} = \dfrac{1}{2} \Rightarrow$ fractional part for both x_1 and x_2 are same. Select x_1-row arbitrarily.

Now, we find first Gomory's constraints as given under

$$\because \quad -\sum_{j=m+1}^{n} f_{ij} \cdot x_j \leq -f_{B_i}$$

Here, we have $i = 1, m = 2, n = 5, x_3 = s_1, x_4 = s_2, x_5 = s_3$, so

$$-\sum_{j=3}^{5} f_{ij} x_j \leq -f_{B_1}$$

$$\Rightarrow f_{13}x_3 - f_{13}x_4 - f_{15}x_5 \leq -f_{B_1}$$

Now, from x_1-row, we have

$$a_{13} = -\frac{1}{2} = -1 + \frac{1}{2} \quad \Rightarrow \quad f_{13} = \frac{1}{2}$$

$$a_{14} = 1 \Rightarrow f_{14} = 0$$

$$a_{15} = 0 \Rightarrow f_{15} = 0$$

Thus $-\frac{1}{2}s_1 \leq -\frac{1}{2}$, which is the first Gomory's constraint.

Now, introducing a non-negative slack variable s', the first Gomory's constraint becomes

$$-\frac{1}{2}s_1 + s' = -\frac{1}{2}$$

Now, adding this Gomory's constraint equation into the above table, we get the table as given below.

B.V.	C_B	x_B	x_1 (β_2)	x_2 (β_1)	s_1 (x_3)	s_2 (x_4)	s_3 (β_3)	s' (β_4)
$c_j \rightarrow$			1	2	0	0	0	0
x_2	2	7/2	0	1	1/2	0	0	0
x_1	1	7/2	1	0	$-1/2$	1	0	0
s_1	0	4	0	0	1	-2	1	0
s'	0	$-1/2$	0	0	$-1/2$	0	0	1
$Z = C_B x_B = 21/2,$ $\Delta_j \rightarrow$			0	0	$-1/2$	-1	0	0
					$\uparrow$			

Clearly $s' = -\frac{1}{2} < 0$ then we apply dual simplex method. We can find the new optimal solution. Here, by dual simplex method, the vector s' (β_4) will be the outgoing vector.

$$\Rightarrow \qquad r = 4$$

Now, by minimum ratio rule

$$\frac{\Delta_k}{a_{rk}} = \min_j \left\{ \frac{\Delta_j}{a_{rj}}, a_{rj} < 0 \right\}$$

$$\Rightarrow \qquad \frac{\Delta_k}{a_{4k}} = \min_j \left\{ \frac{\Delta_j}{a_{4j}}, a_{4j} < 0 \right\} = \min \left\{ \frac{\Delta_3}{a_{43}} \right\} = \frac{\Delta_3}{a_{43}}$$

$$\Rightarrow \qquad k = 3$$

$$\Rightarrow \quad x_3 \ (s_1) \text{ is the incoming vector and hence the key element } = a_{43} = \frac{1}{2}$$

Now again apply the simplex method in a usual manner, we have the following simplex table.

$c_j \rightarrow$			1	2	0	0	0	0
B.V.	C_B	x_B	x_1 (β_2)	x_2 (β_1)	s_1 (β_4)	s_2 (x_4)	s_3 (x_5)	s' (x_6)
x_2	2	3	0	1	0	0	0	1
x_1	1	4	1	0	0	1	0	-1
s_3	0	3	0	0	0	-2	1	2
s_1	0	1	0	0	1	0	0	-2
$Z = C_B x_B = 10,$	$\Delta_j \rightarrow$		0	0	0	-1	0	-1

From the above table, we observe that all $\Delta_j \leq 0$ and $x_1 = 4$, $x_2 = 3$ and max. $Z = 10$ which shows that this solution is optimal, feasible and integer.

EXAMPLE 2. *Solve the following integer linear programming problem*

$$max. \ Z = 5x_1 + 7x_2$$
subject to the constraints
$$-2x_1 + 3x_2 \leq 6$$
$$6x_1 + x_2 \leq 30$$
and $x_1, x_2 \geq 0$ and integer.

Interpret graphically.

SOLUTION. Firstly, we solve the given problem by simplex method by ignoring integer condition as given under.

Introducing the slack variables s_1, s_2 the given maximization IPP with all $b_i's > 0$ becomes

$$max. \ Z = 5x_1 + 7x_2 + 0s_1 + 0s_2$$
s.t. $\qquad -2x_1 + 3x_2 + s_1 = 6$
$$6x_1 + x_2 + s_2 = 30$$
and $\qquad x_1, x_2, s_1, s_2 \geq 0$

Apply the simplex method in a usual manner, we have the following simplex table.

$c_j \rightarrow$			5	7	0	0	Min Ratio
B.V.	C_B	x_B	x_1	x_2	s_1	s_2	
s_1	0	6	-2	③	1	0	6/3 (min)→
s_2	0	30	6	1	0	1	30/1
$Z = C_B x_B = 0,$	$\Delta_j \rightarrow$		5	7 ↑	0	0	
x_2	7	2	-2/3	1	1/3	0	—
s_2	0	28	⟨20/3⟩	0	-1/3	1	28/(20/3) →
$Z = 14,$	$\Delta_j \rightarrow$		29/3 ↑	0	-7/3	0	
x_2	7	24/5	0	1	3/10	1/10	
x_1	5	21/5	1	0	-1/20	3/20	
$Z = C_B x_B = 273/5,$	$\Delta_j \rightarrow$		0	0	-37/20	-29/20	

We observe that all $\Delta_j \le 0 \Rightarrow$ solution is feasible and optimal and is given by

$$x_1 = \frac{21}{5}, x_2 = \frac{24}{5}, \max Z = \frac{273}{5}, \text{ which is not an integer solution.}$$

Clearly, we can write

$$x_1 = 4 + \frac{1}{5}, x_2 = 4 + \frac{4}{5}$$

Now, to construct Gomory's constraints, select x_2-row as it is and has the larger fractional part which is $\frac{4}{5}$. The Gomory's constraints is given by

$$- \sum_{j=m+1}^{n} f_{ij} x_j \le f_{B_i} \qquad \qquad \dots(1)$$

Clearly, we have $i = 2$ (x_2-row), $m = 2$, $n = 4$

Then from (1), we have

$$- \sum_{j=3}^{4} f_{2j} x_j \le -f_{B_2}$$

$$\Rightarrow \quad -f_{23} x_3 - f_{24} x_4 \le -f_{B_2}$$

$$\Rightarrow \quad -f_{23} s_1 - f_{24} s_2 \le -f_{B_2} \qquad \qquad \dots(2)$$

From the above table, we have

$$\left. \begin{aligned} x_{B_2} &= \frac{24}{5} \text{ therefore } f_{B_2} = \frac{4}{5} \\[2mm] a_{23} &= \frac{3}{10} \text{ therefore } f_{23} = \frac{3}{10} \\[2mm] a_{24} &= \frac{1}{10} \text{ therefore } f_{24} = \frac{1}{10} \end{aligned} \right] \qquad \dots(3)$$

Using (3) in (2) we get

$$-\frac{3}{10} s_1 - \frac{1}{10} s_2 \le -\frac{4}{5} \qquad \qquad \dots(4)$$

which is the first Gomory's constraint.

Now introducing a non-negative slack variable s', the Gomory's constraint becomes

$$-\frac{3}{10} s_1 - \frac{1}{10} s_2 + s' = -\frac{4}{5}$$

Adding this constraint, we get the following simplex table

		$c_j \rightarrow$	5	7	0	0	0
B.V.	C_B	x_B	x_1 (β_2)	x_2 (β_1)	s_1 (x_3)	s_2 (x_4)	s' (β_3)
x_2	7	24/5	0	1	3/10	1/10	0
x_1	5	21/5	1	0	–1/20	3/20	0
s'	0	– 4/5	0	0	⊖3/10	–3/10	1
		$\Delta_j \rightarrow$	0	0	–37/20	–29/20	0
					$\uparrow$		

Since, $s' = -\dfrac{4}{5} < 0$, thus the solution obtained in above table is not feasible so we apply dual simplex method to obtain the new optimal solution.

By dual simplex method, the outgoing vector is s', i.e., β_3

$\Rightarrow \qquad\qquad r = 3$

Then by minimum ratio rule

$$\frac{\Delta_k}{a_{rk}} = \min_j\left\{\frac{\Delta_j}{a_{rj}}, a_{rj} < 0\right\}$$

$\Rightarrow \qquad \dfrac{\Delta_k}{a_{3k}} = \min_j\left\{\dfrac{\Delta_j}{a_{3j}}, a_{3j} < 0\right\} = \left\{\dfrac{\Delta_3}{a_{33}}, \dfrac{\Delta_4}{a_{34}}\right\}$

$$= \left\{\frac{-37/20}{-3/10}, \frac{-29/20}{-1/10}\right\} = \left\{\frac{37}{6}, \frac{29}{2}\right\} = \frac{37}{6} = \frac{\Delta_3}{a_{33}}$$

$\Rightarrow \qquad\qquad k = 3$, i.e., $x_3 (= s_1)$ is the incoming vector.

Thus, the key element $= a_{33} = \dfrac{-3}{10}$

Now, proceed the simplex method in a usual manner, we have the following simplex table

	$c_j \rightarrow$		5	7	0	0	0
B.V.	C_B	x_B	x_1 (β_2)	x_2 (β_1)	s_1 (β_3)	s_2 (x_4)	s' (x_5)
x_2	7	4	0	1	0	0	1
x_1	5	13/3	1	0	0	1/6	–1/6
s_1	0	8/3	0	0	1	1/3	–10/3
$Z = C_B x_B = 149/3$,		$\Delta_j \rightarrow$	0	0	0	–5/6	–37/6

In the above table, we observe that all $\Delta_j \leq 0$, thus the solution is optimal and is given by

$$x_1 = \frac{13}{3}, x_2 = 4 \quad \text{and max. } Z = \frac{149}{3}$$

Here, it is also clear that this solution is not integer $\left(\because x_1 = \dfrac{13}{3}\right)$

So, we have to construct Gomory's constraints

We have $\qquad -\displaystyle\sum_{j=m+1}^{n} f_{ij}x_j \leq f_{B_i}$

Here, $i = 1, m = 3, n = 5$ $\qquad\qquad\qquad$ (m = No. of rows in the table)

Thus, we have

$$-\sum_{j=4}^{5} f_{ij}x_j \leq -f_{B_1}$$

$\Rightarrow \qquad -f_{14}x_4 - f_{15}x_5 \leq -f_{B_1}$ or $-f_{14}s_2 - f_{15}s' \leq -f_{B_1}$ $\qquad$ ...(1)

Using the table

$$\left.\begin{array}{l} x_{B_1} = \dfrac{13}{3} = 4 + \dfrac{1}{3} \Rightarrow f_{B_1} = \dfrac{1}{3} \\[2mm] a_{14} = \dfrac{1}{6} \Rightarrow f_{14} = \dfrac{1}{6} \\[2mm] a_{15} = -\dfrac{1}{6} \Rightarrow a_{15} = -1 + \dfrac{5}{6} \Rightarrow f_{15} = \dfrac{5}{6} \end{array}\right] \qquad \dots(2)$$

Putting all these values in (1) we get

$$-\frac{1}{6}s_2 - \frac{5}{6}s' \le -\frac{1}{3} \qquad \dots(3)$$

which is the required Gomory's second constraint.

Further, introducing a slack variable s'', (3) becomes $-\dfrac{1}{6}s_2 - \dfrac{5}{6}s' + s'' = -\dfrac{1}{3}$

Now, adding (3) to the above table we get the new table as given below.

B.V.	C_B	x_B	x_1 (β_2)	x_2 (β_1)	s_1 (β_3)	s_2 (x_4)	s' (x_5)	s'' (β_4)
	$c_j \rightarrow$		5	7	0	0	0	0
x_2	7	4	0	1	0	0	1	0
x_1	5	13/3	1	0	0	1/6	−1/6	0
s_1	0	8/5	0	0	1	1/3	−10/3	0
s''	0	−1/3	0	0	0	(−1/6)	−5/6	1
		$\Delta_j \rightarrow$	0	0	0	−5/6 ↑	−37/6	0

Here, $s'' = -\dfrac{1}{3} < 0 \Rightarrow$ obtained solution is optimal but not feasible. So again apply the dual simplex method.

Clearly, $\quad s''$ is the outgoing vector.
$\Rightarrow \qquad \beta_4$ is the outgoing vector.
$\Rightarrow \qquad r = 4$
Now, by minimum ratio rule

$$\frac{\Delta_k}{a_{rk}} = \min_j \left\{ \frac{\Delta_j}{a_{rj}}, a_{rj} < 0 \right\}$$

$$\Rightarrow \qquad \frac{\Delta_k}{a_{4k}} = \min_j \left\{ \frac{\Delta_j}{a_{4j}}, a_{4j} < 0 \right\} \; \forall j, \text{ not in the basis}$$

$$= \min \left\{ \frac{\Delta_4}{a_{44}}, \frac{\Delta_5}{a_{45}} \right\} = \min \left\{ \frac{-5/6}{-1/6}, \frac{-37/6}{-5/6} \right\}$$

$$= \min \left\{ 5, \frac{37}{6} \right\} = 5 = \frac{\Delta_4}{a_{44}}$$

$$\Rightarrow \qquad k = 4$$

$\Rightarrow \quad x_4 \; (= s_2)$ is the incoming vector and key element $= a_{44} = -\dfrac{1}{6}$

Now, apply the simplex method in a usual manner, we have the following table.

B.V.	C_B	x_B	x_1 (β_2)	x_2 (β_1)	s_1 (β_3)	s_2 (β_4)	s' (x_5)	s'' (x_6)
$c_j \rightarrow$			5	7	0	0	0	0
x_2	7	4	0	1	0	0	1	0
x_1	5	4	1	0	0	0	-1	1
s_1	0	2	0	0	1	0	-5	2
s_2	0	2	0	0	0	1	5	-6
$Z = C_B x_B = 48,$ $\Delta_j \rightarrow$			0	0	0	0	-2	-5

We observe that here all $\Delta_j \le 0$

$\Rightarrow$ solution is optimal and is given by

$x_1 = 4, x_2 = 4$ and max. $Z = 48$, integer solution

Graphical interpretation:

In the adjoining figure, the solution space is shown by the region $OABC$ and the non-integer solution is obtained at $B\left(\dfrac{21}{5}, \dfrac{24}{5}\right)$. To find integer solution, we add two Gomory's constraints one by one as follows:

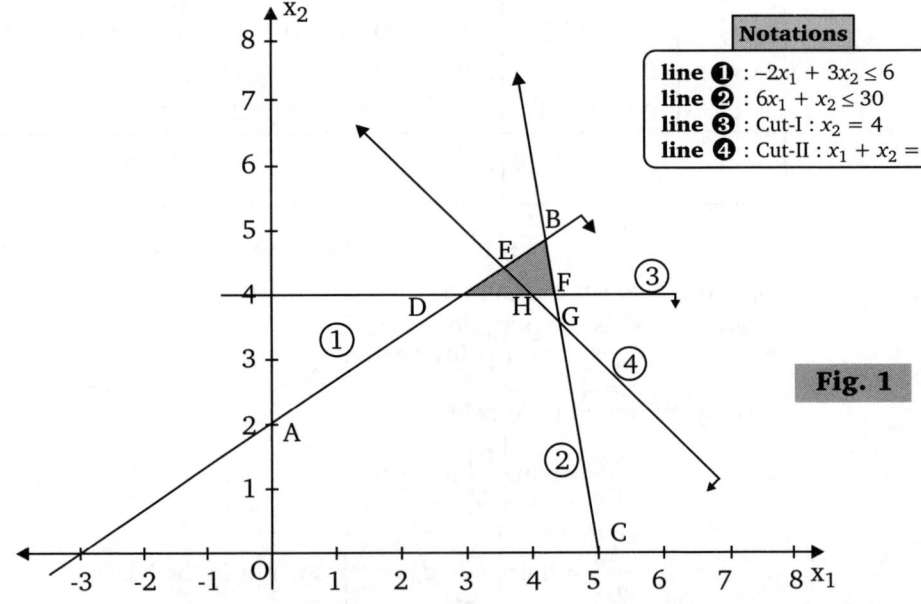

Notations

line ❶ : $-2x_1 + 3x_2 \le 6$
line ❷ : $6x_1 + x_2 \le 30$
line ❸ : Cut-I : $x_2 = 4$
line ❹ : Cut-II : $x_1 + x_2 = 8$

Fig. 1

(i) First constraint (Cut-I) reduces the given solution to the optimal solution at $F\left(\dfrac{13}{3}, 4\right)$ and $\max Z = \dfrac{149}{3}$

(ii) The second constraints (Cut-II) reduces to the solution thus obtained Cut-I to the optimal solution at $H(4, 4)$ and max. $Z = 48$.

In the above figure Cut-I and Cut-II reduces to the region $OABC$ to the region $EBFH$, shown by the shaded area.

Also, Cut-I: $-\dfrac{3}{10}s_1 - \dfrac{1}{10}s_2 \le -\dfrac{4}{5} \Rightarrow 3s_1 + s_2 \ge 8$

$\Rightarrow \quad 3(6 + 2x_1 - 3x_2) + 30 - 6x_1 - x_2 \ge 8$

$\Rightarrow \quad -10x_2 + 48 \ge 8, \text{ i.e., } x_2 \le 4$

Similarly, Cut-II is $x_1 + x_2 \le 8$.

EXAMPLE 3. *Solve the following LPP by Gomory's technique*

$$\text{max. } Z = 3x_2$$

subject to the constraints

$$3x_1 + 2x_2 \le 7$$
$$x_1 - x_2 \ge -2$$

and $x_1, x_2 \ge 0$ and are integers.

Interpret graphically. [MEERUT–2005, 11]

SOLUTION. Write the given LPP in standard form (with all $b_i's > 0$) as follows:

max. $Z = 0x_1 + 3x_2$

s.t.

$$3x_1 + 2x_2 \le 7$$
$$-x_1 + x_2 \le 2$$

and $\qquad x_1, x_2 \ge 0$

Now, introducing slack variables s_1 and s_2, we can write

max $Z = 0x_1 + 3x_2 + 0s_1 + 0s_2$

s.t.

$$\left. \begin{array}{l} 3x_1 + 2x_2 + s_1 = 7 \\ -x_1 + x_2 + s_2 = 2 \\ x_1, x_2, s_1, s_2 \ge 0 \end{array} \right] \qquad \text{...(1)}$$

and

Apply simplex method in a usual manner, we have the following table.

B.V.	C_B	x_B	$c_j \to$ 0 x_1	3 x_2	0 s_1	0 s_2	Min Ratio
s_1	0	7	3	2	1	0	7/2
s_2	0	2	–1	①	0	1	2/1 (min)→
Z = 0,		$\Delta_j \to$	0	3	0	0	
s_1	0	3	5	0	1	–2	3/5 (min)→
x_2	3	2	⊖	1	0	1	—
$Z = C_B x_B = 6,$		$\Delta_j \to$	3	0	0	–3	
x_1	0	3/5	1	0	1/5	–2/5	
x_2	3	13/5	0	1	1/5	3/5	
$Z = C_B x_B = 39/5,$		$\Delta_j \to$	0	0	–3/5	–9/5	

We observe that all $\Delta_j \le 0 \Rightarrow$ solution is optimal and is given by

$$x_1 = \frac{3}{5}, x_2 = \frac{13}{5}, \max Z = \frac{39}{5}$$

Here, x_1, x_2 are not integer, so we can write

$$x_1 = 0 + \frac{3}{5} \quad \text{and} \quad x_2 = 2 + \frac{3}{5} \Rightarrow f_{B_1} = f_{B_2} = \frac{3}{5}$$

Since, the fractional value of x_1 and x_2 are same, so to select x_1-row or x_2-row, we

choose the row with maximum $\dfrac{f_{B_i}}{\displaystyle\sum_{j=m+1}^{n} f_{ij}}$

Here, we have $m = 2$, $n = 4$, $f_{B_1} = f_{B_2} = \dfrac{3}{5}$

For x_1-row

$$\frac{f_{B_1}}{\displaystyle\sum_{j=3}^{4} f_{1j}} = \frac{f_{B_1}}{f_{13} + f_{14}}$$

Since $\qquad a_{13} = \dfrac{1}{5} \Rightarrow f_{13} = \dfrac{1}{5}$

$$a_{14} = -\frac{2}{5} = -1 + \frac{3}{5} \Rightarrow f_{14} = \frac{3}{5}$$

$$\therefore \qquad \frac{f_{B_1}}{\displaystyle\sum_{j=3}^{4} f_{1j}} = \frac{3/5}{\dfrac{1}{5} + \dfrac{3}{5}} = \frac{3/5}{4/5} = \frac{3}{4}$$

For x_2-row

$$\frac{f_{B_2}}{\displaystyle\sum_{j=3}^{4} f_{2j}} = \frac{f_{B_2}}{f_{23} + f_{24}}$$

$$\because \qquad a_{23} = \frac{1}{5} \Rightarrow f_{23} = \frac{1}{5}; a_{24} = \frac{3}{5} \Rightarrow f_{24} = \frac{3}{5}$$

$$\therefore \qquad \frac{f_{B_2}}{\displaystyle\sum_{j=3}^{4} f_{2j}} = \frac{3/5}{\dfrac{1}{5} + \dfrac{3}{5}} = \frac{3}{4} \qquad \Rightarrow \qquad \max\left\{\frac{f_{B_i}}{\displaystyle\sum_{j=3}^{4} f_{ij}}\right\} = \frac{3}{4}$$

So we may select at random any one of these.

Let us suppose we select x_1-row. We have to construct Gomory's constraints
The Gomory's constraints is given by

$$-\sum_{j=m+1}^{n} f_{ij} x_j \le -f_{B_i}$$

Here, we have $i = 1$, $m = 2$, $n = 4$

$$\therefore \qquad \sum_{j=3}^{4} f_{ij} x_j \le -f_{B_1}$$

$$\Rightarrow \qquad -f_{13} x_3 - f_{14} x_4 \le -f_{B_1} \qquad \Rightarrow \qquad -f_{13} s_1 - f_{14} s_2 \le -f_{B_1}$$

$$\Rightarrow \qquad -\frac{1}{5} s_1 - \frac{3}{4} s_2 \le -\frac{3}{5}$$

which is the first Gomory's constraint.

Further, introducing a non-negative slack variable s', then Gomory's constraints becomes

$$-\frac{1}{5}s_1 - \frac{3}{5}s_2 + s' = -\frac{3}{5}$$

Also, adding this new constraints in the last simplex table we get

B.V.	C_B	x_B	$c_j \rightarrow$	0	3	0	0	0
				x_1 (β_1)	x_2 (β_2)	s_1 (x_3)	s_2 (x_4)	s' (β_3)
x_1	0	3/5		1	0	1/5	–2/5	0
x_2	3	13/5		0	1	1/5	3/5	0
s'	0	–3/5		0	0	–1/5	–3/5	1
			$\Delta_j \rightarrow$	0	0	–3/5	–9/5	0
						$\uparrow$		$\downarrow$

Here, $s' = -\frac{3}{5} < 0$, which is not feasible. Thus we have to apply dual simplex method to find the new optimal solution.

By dual simplex method, the vector $s' (= \beta_3)$ is the outgoing vector.

$\Rightarrow$ $\qquad\qquad\qquad r = 3$

Then by minimum ratio rule

$$\frac{\Delta_k}{a_{rk}} = \min_j \left\{ \frac{\Delta_j}{a_{rj}}, a_{rj} < 0 \right\}$$

$\Rightarrow$

$$\frac{\Delta_k}{a_{3k}} = \min_j \left\{ \frac{\Delta_j}{a_{3j}}, a_{3j} < 0 \right\} = \min \left\{ \frac{\Delta_3}{a_{33}}, \frac{\Delta_4}{a_{34}} \right\}$$

$$= \min \left\{ \frac{-3/5}{-1/5}, \frac{-9/5}{-3/5} \right\} = \min\{3, 3\} = 3 = \frac{\Delta_3}{a_{33}} \text{ or } \frac{\Delta_4}{a_{34}}$$

$\Rightarrow$ $\qquad\qquad\qquad k = 3 \text{ or } 4$

Case 1   When $k = 3$

In this case $x_3 (= s_1)$ is the incoming vector and the key element $= a_{33} = -\frac{1}{5}$

Then apply the simplex method in a usual manner, we get the following simplex table.

B.V.	C_B	x_B	$c_j \rightarrow$	0	3	0	0	0
				x_1 (β_1)	x_2 (β_2)	s_1 (β_3)	s_2 (x_4)	s' (x_5)
x_1	0	0		1	0	0	–1	1
x_2	3	2		0	1	0	0	1
s_1	0	3		0	0	1	3	–5
$Z = C_B x_B = 6,$			$\Delta_j \rightarrow$	0	0	0	0	–3

We observe that all $\Delta_j \le 0$ and $x_1 = 0$, $x_2 = 2$ and max. $Z = 6$

$\Rightarrow$ $\qquad$ solution is optimal and integer.

Case 2 When $k = 4$

In this case $x_4 (= s_2)$ will be the incoming vector and key element $= a_{34} = -\dfrac{3}{5}$
Then again apply simplex method in a usual manner, we have the following simplex table.

B.V.	C_B	x_B	$c_j \rightarrow$ 0 x_1 (β_1)	3 x_2 (β_2)	0 s_1 (β_3)	0 s_2 (x_4)	0 s' (x_5)
x_1	0	1	1	0	1/3	0	−2/3
x_2	3	2	0	1	0	0	1
s_1	0	1	0	0	1/3	1	−5/3
$Z=C_Bx_B=6,$		$\Delta_j \rightarrow$	0	0	0	0	−3

We observe that all $\Delta_j \leq 0$ and $x_1 = 0$, $x_2 = 2$ and max. $Z = 6$
$\Rightarrow$ solution is optimal and integer.

Geometrical Interpretation:
In the adjoining figure, the solution space is given by $OABC$.

The non-integer optimal solution is obtained at the point $B\left(\dfrac{3}{5}, \dfrac{13}{5}\right)$.

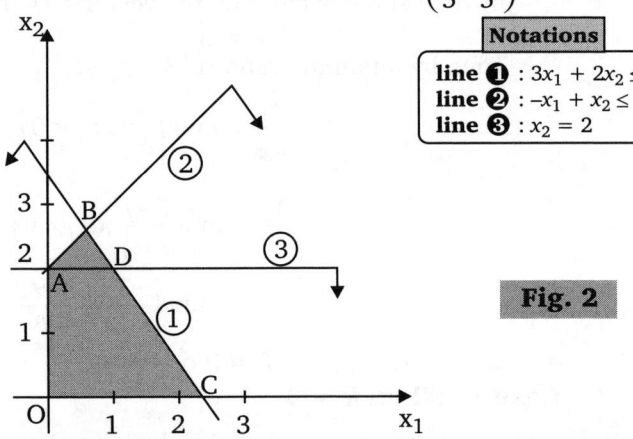

Notations
line ❶ : $3x_1 + 2x_2 \leq 7$
line ❷ : $-x_1 + x_2 \leq 2$
line ❸ : $x_2 = 2$

Fig. 2

The Gomory's constraints is given by
$$-\frac{1}{5}s_1 - \frac{3}{5}s_2 \leq -\frac{3}{5} \quad \Rightarrow \quad s_1 + 3s_2 \geq 3$$
Since, $s_1 = 7 - 3x_1 - 2x_2$ and $s_2 = 2 + x_1 - x_2$ (From (1))
Then Gomory's constraints reduces to
$$(7 - 3x_1 - 2x_2) + 3(2 + x_1 - x_2) \geq 3$$
$$\Rightarrow \quad -5x_2 + 13 \geq 3, i.e., x_2 \leq 2$$
On drawing $x_2 = 2$, the region $OABC$ reduces to the feasible region $OADC$ and the point $A(0, 2)$ and the point $D(1, 2)$ give the required integer solution given by
(i) $x_1 = 0$, $x_2 = 2$, max. $Z = 6$
(ii) $x_1 = 1$, $x_2 = 2$, max. $Z = 6$

10.5 PROBLEM ON MIXED INTEGER LINEAR PROGRAMMING

There are some situations in which only some variables are restricted to be an integer. To solve such type of problem, we proceed as follows:

STEP 1. Solve the given problem as regular LPP by simplex method by ignoring integer condition.

STEP 2. Use Gomory's technique corresponding to the integer variables one by one as given in step 3.

STEP 3. If only one of the integer restricted variables has the fractional value then corresponding to the row in which this fractional variables lies in the optimal simplex table from the Gomory's constraint as given below:

$$- \sum_{j \in \mathbf{R}^+} y_{ij} x_j - \left(\frac{f_{B_i}}{f_{B_i} - 1} \right) \sum_{j \in \mathbf{R}^-} y_{ij} x_j \leq -f_{B_i}$$

where $x_{B_i} = [x_{B_i}] + f_{B_i}$

$[x_{B_i}] = $ largest integral part of x_{B_i}

$f_{B_i} = $ largest fractional part of x_{B_i} such that $0 \leq f_{B_i} < 1$

$\mathbf{R}^+ = [j : y_{ij} \geq 0]$ $\mathbf{R}^- = [j : y_{ij} < 0]$

and $R = [R^-, R^+]$ set of indices corresponding to all non-basic variables. Here, the Gomory's cutting plane is given by

$$- \sum_{j \in R^+} y_{ij} x_j - \left(\frac{f_{B_i}}{f_{B_i} - 1} \right) \sum_{j \in R^-} y_{ij} x_j + s_i = -f_{B_i}$$

where s_i is the slack variable.

☛ **REMARK**
- Here the value of the objective function in the optimal solution of mixed integer programming is always superior to or at least equal to that of all integer LPP and is always inferior to or euqal to that of the original LPP.

Solved Examples

EXAMPLE 1. *Solve the following LPP by Gomory's technique*

$$\max. Z = 4x_1 + 6x_2 + 2x_3$$

subject to the constraints

$$4x_1 + 4x_2 \leq 5$$
$$-x_1 + 6x_2 \leq 5$$
$$-x_1 + x_2 + x_3 \leq 5$$

and $x_1, x_2, x_3 \geq 0$ and are integers.

SOLUTION. The given maximization problem is in standard form. Introducing slack variables s_1, s_2, s_3 in the given LPP, we get

$$\max Z = 4x_1 + 6x_2 + 2x_3 + 0s_1 + 0s_2 + 0s_3$$

s.t.

$$4x_1 - 4x_2 + s_1 = 5$$
$$-x_1 + 6x_2 + s_2 = 5$$
$$-x_1 + x_2 + x_3 + s_3 = 5$$

and $x_1, x_2, x_3, s_1, s_2, s_3 \geq 0$

...(1)

Apply the simplex method in a usual manner, we get the following simplex table.

B.V.	C_B	x_B	x_1	x_2	x_3	s_1 (x_4)	s_2 (x_5)	s_3 (x_6)	Min Ratio
		$c_j \rightarrow$	4	6	2	0	0	0	
s_1	0	5	4	4	0	1	0	0	—
s_2	0	5	−1	6	0	0	1	0	5/6 (min)→
s_3	0	5	−1	1	1	0	0	1	5/1
$Z = C_B x_B = 0$, $\Delta_j \rightarrow$			4	6	2	0	0	0	
s_1	0	25/3	10/3	0	0	1	2/3	0	5/2 (min)→
x_2	6	5/6	−1/6	1	0	0	1/6	0	
s_3	0	25/6	−5/6	0	1	0	−1/6	1	
$Z = C_B x_B = 5$, $\Delta_j \rightarrow$			5	0	2	0	−1	0	
x_1	4	5/2	1	0	0	3/10	1/5	0	
x_2	6	5/4	0	1	0	1/20	1/5	0	
s_3	0	25/4	0	0	1	1/4	0	1	
$Z = C_B x_B = 35/2$, $\Delta_j \rightarrow$			0	0	2	−3/2	−2	0	
x_1	4	5/2	1	0	0	3/10	1/5	0	
x_2	6	5/4	0	1	0	1/20	1/5	0	
x_3	2	25/4	0	0	1	1/4	0	1	
$Z = C_B x_B = 30$, $\Delta_j \rightarrow$			0	0	0	−2	−2	−2	

Here, we observe that all $\Delta_j \leq 0$ and $x_1 = \dfrac{5}{2}, x_2 = \dfrac{5}{4}, x_3 = \dfrac{25}{4}$ and max. $Z = 25$.

Therefore, solution is optimal feasible and non-integer.

Here, x_1, x_3 are constraint to be an integer.

Now, $x_1 = \dfrac{5}{2} = 2 + \dfrac{1}{2}, x_3 = \dfrac{25}{4} = 6 + \dfrac{1}{4}$

$\Rightarrow$ $f_{B_1} = \dfrac{1}{2}, f_{B_3} = \dfrac{1}{4}$

$\Rightarrow$ $f_{B_1} > f_{B_3}$

$\Rightarrow$ x_1-row is selected to construct first Gomory's constraints

First Gomory's constraints

$\because$ $-\sum\limits_{j=m+1}^{n} f_{ij} x_j \leq -f_{B_i}$

Here, $i = 1$, $m = 3$, $n = 6$

$-\sum\limits_{j=4}^{6} f_{ij} x_j \leq -f_{B_1}$ $\Rightarrow$ $-f_{14}x_4 - f_{15}x_5 - f_{16}x_6 \leq -f_{B_1}$

From x_1-row, we have

$$a_{14} = \frac{3}{10} \quad \Rightarrow \quad f_{14} = \frac{3}{10}$$

$$a_{15} = \frac{1}{5} \quad \Rightarrow \quad f_{15} = \frac{1}{5}$$

$$a_{16} = 0 \quad \Rightarrow \quad f_{16} = 0$$

and $x_4 = s_1, x_5 = s_2, x_6 = s_3$

$$\Rightarrow \quad -\frac{3}{10}s_1 - \frac{1}{5}s_2 \le -\frac{1}{2},$$ which is the first Gomory's constraints.

Now introducing a non-negative slack variable s' to the last table of the above simplex table we get the table as given below.

B.V.	C_B	x_B	$c_j \rightarrow$	4	6	2	0	0	0	0
				x_1 (β_1)	x_2 (β_2)	x_3 (β_3)	s_1 (x_4)	s_2 (x_5)	s_3 (x_6)	s' (β_4)
x_1	4	5/2		1	0	0	3/10	1/5	0	0
x_2	6	5/4		0	1	0	1/20	1/5	0	0
x_3	2	25/4		0	0	1	1/4	0	1	0
s'	0	−1/2		0	0	0	−3/10	−1/5	0	1
$Z = C_B x_B = 30, \Delta_j \rightarrow$				0	0	0	−2	−2	−2	0
							↑			↓

Clearly, $s' = -\frac{1}{2} < 0$. Thus, we apply dual simplex method.

Here, s' $(= \beta_4)$ will be the outgoing vector.

$$\Rightarrow \qquad r = 4$$

Then by minimum ratio rule, we have

$$\frac{\Delta_k}{a_{rk}} = \min_j \left\{ \frac{\Delta_j}{a_{rj}}, a_{rj} < 0 \right\}$$

$$\Rightarrow \qquad \frac{\Delta_k}{a_{4k}} = \min_j \left\{ \frac{\Delta_j}{a_{4j}}, a_{4j} < 0 \right\} = \min \left\{ \frac{\Delta_4}{a_{44}}, \frac{\Delta_5}{a_{45}} \right\}$$

$$= \min \left\{ \frac{-2}{-3/10}, \frac{-2}{-1/5} \right\} = \min \left\{ \frac{20}{3}, 10 \right\}$$

$$= \frac{20}{3} = \frac{\Delta_4}{a_{44}}$$

$\Rightarrow$ $k = 4$, i.e., x_4 $(= s_1)$ is the incoming vector.

and key element $= a_{44} = \frac{-3}{10}$

Now proceeding as usual, we get the following simplex table

B.V.	C_B	x_B	$c_j \to$ 4 x_1 (β_1)	6 x_2 (β_2)	2 x_3 (β_3)	0 s_1 (β_4)	0 s_2 (x_5)	0 s_3 (x_6)	0 s' (x_7)	Min Ratio
x_1	4	2	1	0	0	0	0	0	1	
x_2	6	7/6	0	1	0	0	1/6	0	1/6	
x_3	2	35/6	0	0	1	0	–1/6	1	5/6	
s_1	0	5/3	0	0	0	1	2/3	0	–10/3	
$Z=C_Bx_B=80/3,$		$\Delta_j \to$	0	0	0	0	–2/3	–2	–20/3	

Here, we observe that all $\Delta_j \le 0$ and $x_1 = 2, x_2 = \dfrac{7}{6}, x_3 = \dfrac{35}{6}$ and $\max.Z = \dfrac{80}{3}$.

Clearly this solution is optimal, feasible but not integer.

Now, select x_3-row to construct second Gomory's constraints

$$\because \quad x_3 = \frac{35}{6} = 5 + \frac{5}{6} \Rightarrow f_{B_3} = \frac{5}{6}$$

Now, $-\sum\limits_{j=m+1}^{n} f_{ij} \cdot x_j \le -f_{B_i}$

Here, $i = 3, m = 4, n = 7$

So, $-\sum\limits_{j=5}^{7} f_{3j}x_j \le -f_{B_3}$

$$\Rightarrow \quad -f_{35}x_5 - f_{36}x_6 - f_{37}x_7 \le -\frac{5}{6}$$

Clearly, from x_3-row above table, we have

$$a_{35} = -\frac{1}{6} = -1 + \frac{5}{6} \qquad \Rightarrow \qquad f_{35} = \frac{5}{6}$$

$$a_{36} = 1 \qquad \Rightarrow \qquad f_{36} = 0$$

$$a_{37} = \frac{5}{6}$$

$$\Rightarrow \qquad f_{37} = \frac{5}{6}$$

$$x_5 = s_2, x_6 = s_3, x_7 = s'$$

So, second Gomory's constraints is given by

$$-\frac{5}{6}s_2 - \frac{5}{6}s' \le -\frac{5}{6}$$

Now introducing a non-negative slack variable s'', the above constraints becomes

$$-\frac{5}{6}s_2 - \frac{5}{6}s' + s'' = -\frac{5}{6}$$

Also by adding the constraints to the above table, we get the following table.

$c_j \rightarrow$			4	6	2	0	0	0	0	0
B.V.	C_B	x_B	x_1	x_2	x_3	s_1	s_2	s_3	s'	s''
x_1	4	2	1	0	0	0	0	0	1	0
x_2	6	7/6	0	1	0	0	1/6	0	1/6	0
x_3	2	35/6	0	0	1	0	–1/6	1	5/6	0
s_1	0	5/3	0	0	0	1	2/3	0	–10/3	0
s''	0	–5/6	0	0	0	0	–5/6	0	–5/6	1
$Z = C_B x_B = 80/3$,	$\Delta_j \rightarrow$		0	0	0	0	–2/3	–2	–20/3	0
							$\uparrow$			

We observe that $s'' = -\dfrac{5}{6} < 0$, then dual simplex method can be applied to get the new optimal solution.

By dual simplex method, the vector s'' ($= \beta_5$) will leave the basis.

Clearly, β_5 is the outgoing vector then $r = 5$.

Then, by minimum ratio rule

$$\frac{\Delta_k}{a_{rk}} = \min_j \left\{ \frac{\Delta_j}{a_{rj}}, a_{rj} < 0 \right\}$$

$\Rightarrow$

$$\frac{\Delta_k}{a_{5k}} = \min_j \left\{ \frac{\Delta_j}{a_{5j}}, a_{5j} < 0 \right\} = \min \left\{ \frac{\Delta_5}{a_{55}}, \frac{\Delta_7}{a_{57}} \right\}$$

$$= \min \left\{ \frac{-2/3}{-5/6}, \frac{-20/3}{-5/6} \right\} = \min \left\{ \frac{4}{5}, 8 \right\} = \frac{4}{5} = \frac{\Delta_5}{a_{55}}$$

$\Rightarrow$ $\qquad k = 5$

$\Rightarrow$ $\qquad x_5$ ($= s_2$) is the incoming vector and key element $= a_{55} = \dfrac{5}{6}$

Again apply simplex method in a usual manner, we have the following simplex table.

$c_j \rightarrow$			4	6	2	0	0	0	0	0
B.V.	C_B	x_B	x_1 (β_1)	x_2 (β_2)	x_3 (β_3)	s_1 (β_4)	s_2 (β_5)	s_3 (x_6)	s' (x_7)	s'' (x_8)
x_1	4	2	1	0	0	0	0	0	1	0
x_2	6	1	0	1	0	0	0	0	0	1/5
x_3	2	6	0	0	1	0	0	1	1	–1/5
s_1	0	1	0	0	0	1	0	0	– 4	4/5
s_2	0	1	0	0	0	0	1	0	1	–6/5
$Z = C_B x_B = 26$,	$\Delta_j \rightarrow$		0	0	0	0	0	–2	–6	– 4/5

Here, we observe that all $\Delta_j \leq 0$ and $x_1 = 2, x_2 = 1, x_3 = 6$ and max. $Z = 26$ which shows that the solution is optimal feasible and integer.

10.6 THE BRANCH AND BOUND TECHNIQUE

This technique is applicable to both the IPP, pure as well as mixed and involve the continuous integer programming problem ignoring the integer valued condition. The branch and bound technique was originally developed by A.H. Land and A.G. Doig and later was modified by R.S. Dakin.

In this method, we first obtain optimum solution in a usual manner, then divide the feasible region into smaller regions by deleting parts that contain no feasible integer solution.

This is the most general technique for the solution of an IPP in which only a few or all the variables are constraints by their upper or lower bound or both.

In a maximization problem, the value of the objective function at the LPP optimum will always be an upper bound on the optimal integer programming objective. Also, any integer feasible point is always a lower bound on the optimal LP objective value. This is the basic idea behind the development of Branch and bound technique.

WORKING PROCEDURE

To solve the given integer programming problem by Branch and bound technique, we use the following steps:

STEP 1. Solve the given integer programming problem by ignoring integer condition on the variables by graphical or simplex method.

STEP 2. Check the following points of optimality

(i) If the variables in the optimum solution are all integers, then this is the required solution.

(ii) If the value of the some variables in the optimum solution is not an integer then go to the next step.

STEP 3. If some variables say x_j, is not an integer then

$$[x_j] < x_j < [x_j] + 1, \text{ where } [\cdot] \text{ is the integral value.}$$

But any feasible integer value of x_j must satisfy one of the two conditions given by

$$x_j \leq [x_j] \text{ or } x_j \geq [x_j] + 1$$

Now, since the variables has no integer value between $[x_j]$ and $[x_j] + 1$ then these two conditions are mutually exclusive and when they are applied separately to the given LPP, we get two different subproblems (called nodes)

(i) **Subproblem I:** Given LPP with $x_j \leq [x_j]$.

(ii) **Subproblem II:** Given LPP with $x_j \geq [x_j] + 1$.

STEP 4. Solve both the subproblems and test the integrability of the optimal solution and observe that

(i) if the optimal solution of both the subproblems are integral valued, then required solution is that which give larger value of the objective function.

(ii) if any one subproblem has integral valued optimal solution and the other subproblem has no feasible optimal solution, then the required solution is that of the subproblem having integral valued optimal solution.

(iii) if one subproblem has optimal integral valued and the other subproblem has non-integer valued solution then record the integral valued solution and repeat step 3 and step 4 to the other subproblem till all the integral valued solutions are recorded.

STEP 5. From all the recorded integral valued solutions, choose that integral valued solution which gives the largest value of the objective function. This is the required optimal solution of the given IPP.

☛ REMARKS
- The above technique is easily extended to solve mixed integer problem. Subdivision then are generated solely by the integral variables.
- If any subdivisions needs not to be subdivided then such subdivision is comprehend.
- If the given IPP is of minimization the procedure remains the same except that upper bound are used. Therefore, the value of the first integer valued solution becomes an upper bound for the given problem and the programs are eliminated when their objective function values are greater than the current upper bound.

 Solved Examples

EXAMPLE 1. *Use Branch and bound technique, solve the following LPP*

$$\text{max. } Z = 7x_1 + 9x_2$$

subject to the constraints

$$-x_1 + 3x_2 \le 6$$
$$7x_1 + x_2 \le 35$$
$$x_1 \le 7$$
$$x_2 \le 7$$

and $x_1, x_2 \ge 0$ and are integers.

SOLUTION. Ignoring integer condition, we may solve the given problem by graphical method. Here, we have the following graph.

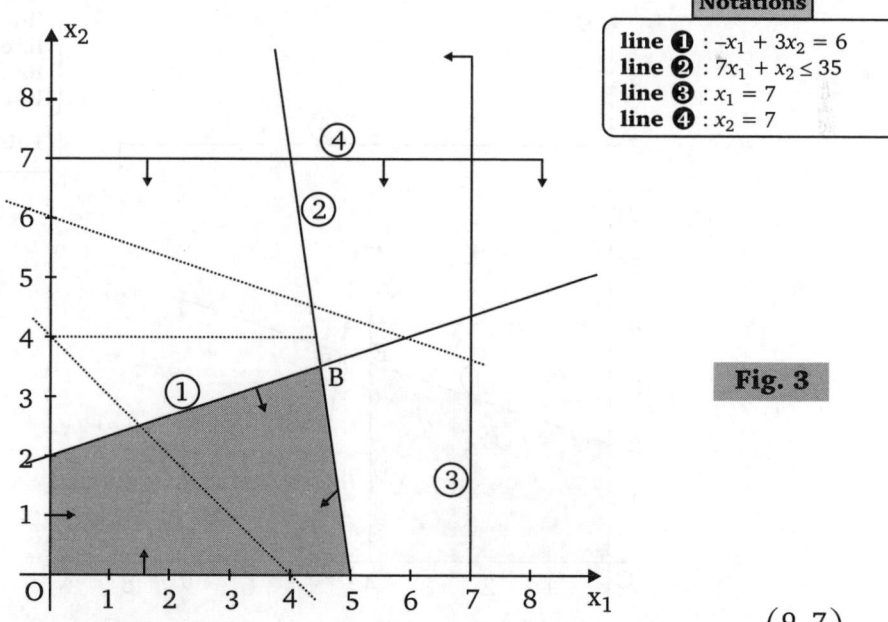

Notations
line ❶ : $-x_1 + 3x_2 = 6$
line ❷ : $7x_1 + x_2 \le 35$
line ❸ : $x_1 = 7$
line ❹ : $x_2 = 7$

Fig. 3

The optimal solution of the given problem is obtained at the point $B\left(\dfrac{9}{2}, \dfrac{7}{2}\right)$ at B

and max. $Z = 63$ which is taken as an upper bound of Z.

Clearly, the optimal solution is not integer valued

i.e., $\qquad x_1 = \dfrac{9}{2}, x_2 = \dfrac{7}{2}$

Let us suppose x_1 is selected for branching, then

$$x_1 = \frac{9}{2}$$

$\Rightarrow \qquad 4 < x_1 < 5$

Now, we have following two subproblems.

subproblem-1	subproblem-2
max. $Z = 7x_1 + 9x_2$ s.t. $\quad -x_1 + 3x_2 \le 6$ $\quad 7x_1 + x_2 \le 35$ $\quad\quad x_1 \le 4$ $\quad\quad x_2 \le 7$ and $x_1, x_2 \ge 0$ and are integers	max. $Z = 7x_1 + 9x_2$ s.t. $\quad -x_1 + 3x_2 \le 6$ $\quad 7x_1 + x_2 \le 35$ $\quad\quad 5 \le x_1 \le 7$ $\quad\quad x_2 \le 7$ and $x_1, x_2 \ge 0$ and are integers

Now, we solve these subproblem by graphical method separately.

The graph of subproblem 1 is given below.

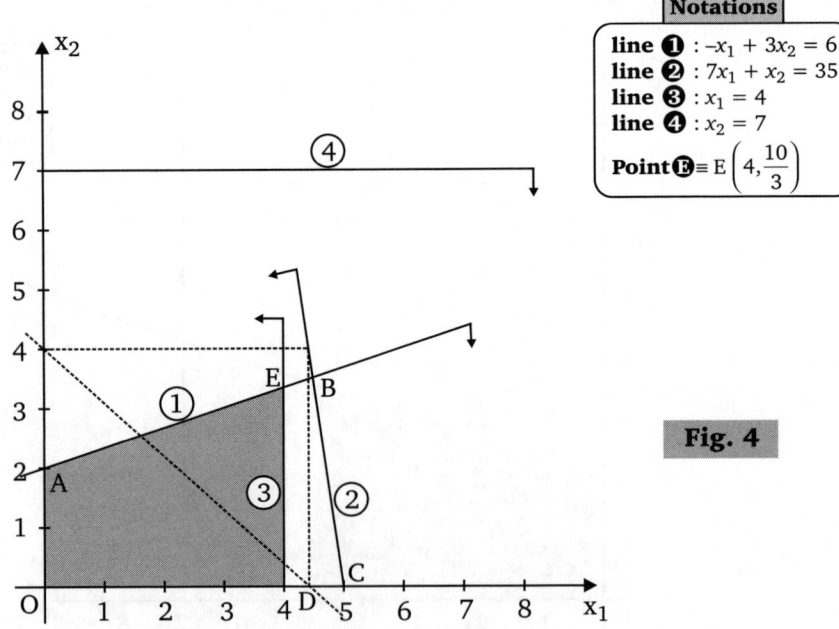

Notations

line ❶ : $-x_1 + 3x_2 = 6$
line ❷ : $7x_1 + x_2 = 35$
line ❸ : $x_1 = 4$
line ❹ : $x_2 = 7$

Point ❺ $\equiv$ E $\left(4, \dfrac{10}{3}\right)$

Fig. 4

From the above graph, the optimal solution of subproblem 1 is given by

$$x_1 = 4, x_2 = \frac{10}{3}, \text{max.} Z = 58$$

Similarly, the graph of subproblem 2 is given below

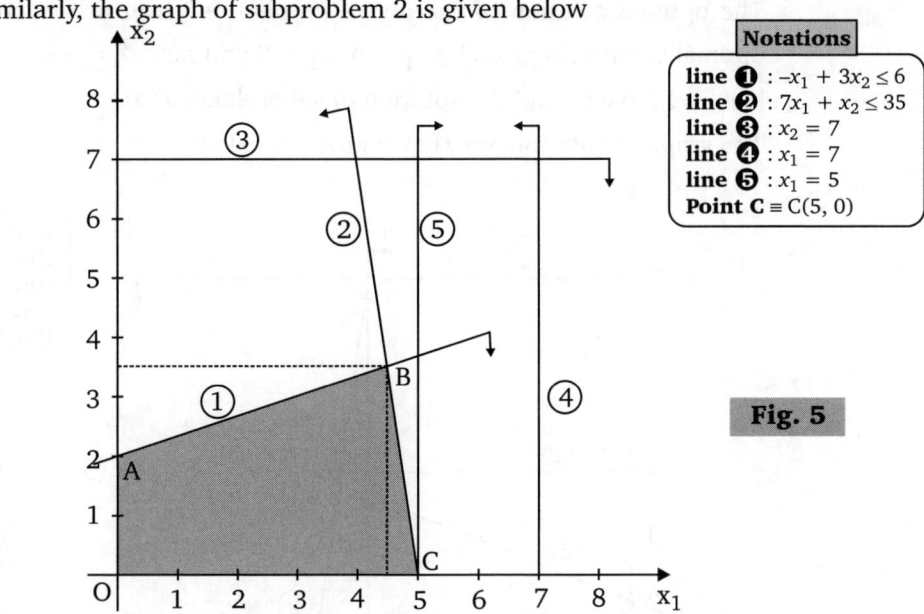

Notations
line ❶ : $-x_1 + 3x_2 \le 6$
line ❷ : $7x_1 + x_2 \le 35$
line ❸ : $x_2 = 7$
line ❹ : $x_1 = 7$
line ❺ : $x_1 = 5$
Point C $\equiv$ C(5, 0)

Fig. 5

From the above figure the optimal solution of subproblem 2 is obtained at $B(5, 0)$ and is given by $x_1 = 5$, $x_2 = 0$ and max. $Z = 35$

Here, we observe that subproblem 2 has integral valued solution and subproblem 1 has not integral valued solution. So, in subproblem 1, x_2 is the branching variable. Thus, subproblem 1 has two subproblems as given below.

subproblem 1(a)	subproblem 1(b)
max. $Z = 7x_1 + 9x_2$ s.t. $-x_1 + 3x_2 \le 6$ $7x_1 + x_2 \le 35$ $x_1 \le 4$ $x_2 \le 3$ and $x_1, x_2 \ge 0$ and are integers.	max. $Z = 7x_1 + 9x_2$ s.t. $-x_1 + 3x_2 \le 6$ $7x_1 + x_2 \le 35$ $x_1 \le 4$ $4 \le x_2 \le 7$ and $x_1, x_2 \ge 0$ and are integers.

Now, graph of subproblem 1(a) is given as follows.

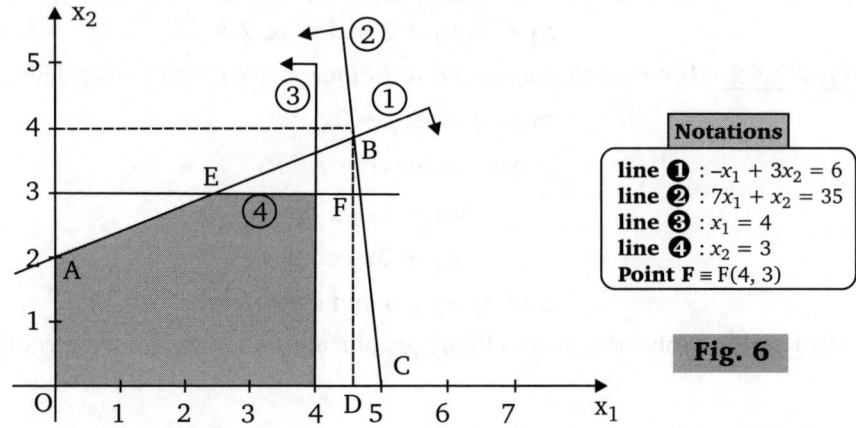

Notations
line ❶ : $-x_1 + 3x_2 = 6$
line ❷ : $7x_1 + x_2 = 35$
line ❸ : $x_1 = 4$
line ❹ : $x_2 = 3$
Point F $\equiv$ F(4, 3)

Fig. 6

The optimal solution of the above problem is given at the point $F(4, 3)$ and optimal solution is given by $x_1 = 4$, $x_2 = 3$ and max. $Z = 55$.

Now, we have to find the solution of subproblem 1(b).

The graph of subproblem 1(b) is given as below.

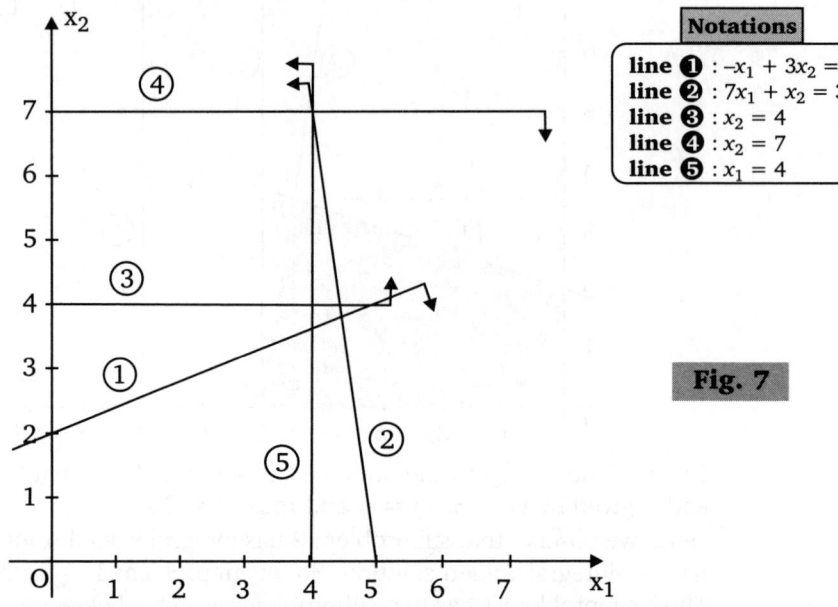

Notations
line ❶ : $-x_1 + 3x_2 = 6$
line ❷ : $7x_1 + x_2 = 35$
line ❸ : $x_2 = 4$
line ❹ : $x_2 = 7$
line ❺ : $x_1 = 4$

Fig. 7

In the above problem, no feasible region is obtained. Hence, problem 1(b) has no feasible solution.

Finally, we conclude that, in the subproblems we get the following integral valued solutions:

(i) $x_1 = 5$, $x_2 = 0$, max. $Z = 35$

(ii) $x_1 = 4$, $x_2 = 3$, max. $Z = 55$

In which the larger value of $Z = 55$

Hence, the required solution is given by
$$x_1 = 4, x_2 = 3 \text{ and max. } Z = 55$$

EXAMPLE 2. *Use Branch and bound technique, solve the following LPP.*

$$max. \ Z = 2x_1 + 3x_2$$

subject to the constraints

$$6x_1 + 5x_2 \leq 25$$

$$x_1 + 3x_2 \leq 10$$

and $x_1, x_2 \geq 0$ and are integers.

SOLUTION. Solve the given LPP by graphical method by ignoring conditions.

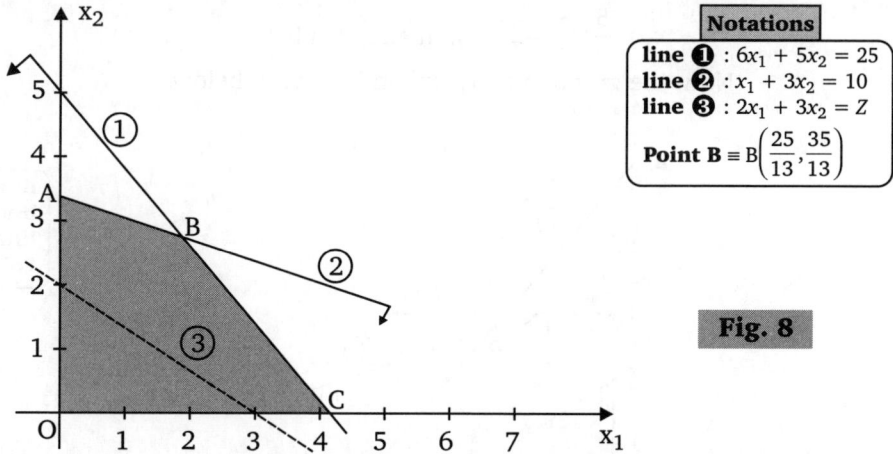

Notations

line ❶ : $6x_1 + 5x_2 = 25$
line ❷ : $x_1 + 3x_2 = 10$
line ❸ : $2x_1 + 3x_2 = Z$

Point $B \equiv B\left(\dfrac{25}{13}, \dfrac{35}{13}\right)$

Fig. 8

Clearly, the optimal solution can be obtained at the point $B\left(\dfrac{25}{13}, \dfrac{35}{13}\right)$ and is given by

$$x_1 = \frac{25}{13}, x_2 = \frac{35}{13} \text{ and max. } Z = 12$$

We observe that both x_1 and x_2 are non-integers. Thus, select one variable arbitrarily say x_2 for branching then

$$x_2 = \frac{35}{13} \implies 2 < x_2 < 3$$

Now, we form the following two subproblems by adding new constraints either $x_2 \le 2$ or $x_2 \ge 3$ to the original LPP.

subproblem-1	subproblem-2
max $Z = 2x_1 + 3x_2$ s.t. $\quad 6x_1 + 5x_2 \le 25$ $\quad x_1 + 3x_2 \le 10$ $\quad\quad x_2 \le 2$ and $x_1, x_2 \ge 0$ are integers	max $Z = 2x_1 + 3x_2$ s.t. $\quad 6x_1 + 5x_2 \le 25$ $\quad x_1 + 3x_2 \le 10$ $\quad\quad x_2 \ge 3$ and $x_1, x_2 \ge 0$ are integers

The graph of the subproblem 1 is given as below.

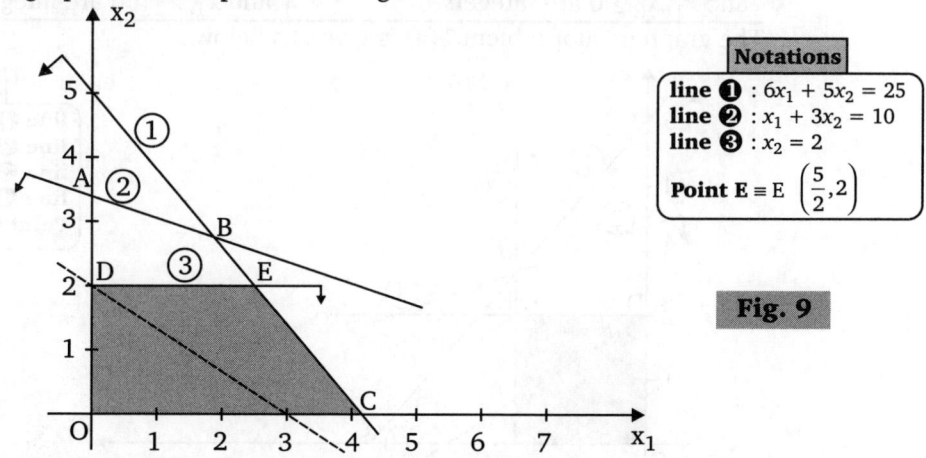

Notations

line ❶ : $6x_1 + 5x_2 = 25$
line ❷ : $x_1 + 3x_2 = 10$
line ❸ : $x_2 = 2$

Point $E \equiv E\left(\dfrac{5}{2}, 2\right)$

Fig. 9

Clearly, the optimal solution of the above problem is given below.

$$x_1 = \frac{5}{2}, x_2 = 2 \text{ and max } Z = 11$$

Now, the graph of subproblem 2 is given below.

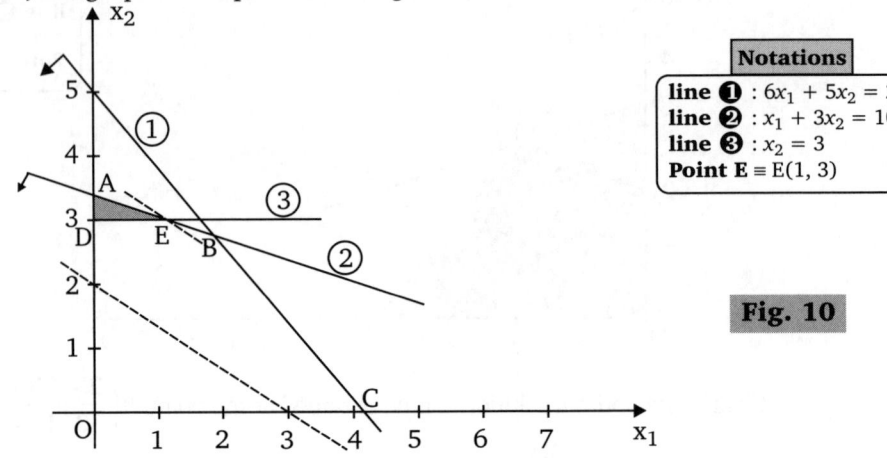

line ❶ : $6x_1 + 5x_2 = 25$
line ❷ : $x_1 + 3x_2 = 10$
line ❸ : $x_2 = 3$
Point E ≡ E(1, 3)

Fig. 10

From the above graph, the optimum solution of subproblem 2 is given by
$$x_1 = 1, x_2 = 3 \text{ and max. } Z = 11$$

Clearly, it is an integer valued solution and max. $Z = 11$ is taking as lower bound of Z. (Here, no need to branch subproblem 2)

Now subproblem 1 is further divided into two subproblems as given below.

In subproblem 1 $x_1 = \dfrac{5}{2} \implies x_1 \le \left[\dfrac{5}{2}\right]$ or $x_1 \ge \left[\dfrac{5}{2}\right] + 1$

$\implies \qquad x_1 \le 2 \text{ or } x_1 \ge 3$

subproblem 1(a)	subproblem 1(b)
max. $Z = 2x_1 + 3x_2$ subject to the constraints $6x_1 + 5x_2 \le 25$ $x_1 + 3x_2 \le 10$ $x_2 \le 2$ $x_1 \le 2$ and $x_1, x_2 \ge 0$ are integers	max. $Z = 2x_1 + 3x_2$ subject to the constraints $6x_1 + 5x_2 \le 25$ $x_1 + 3x_2 \le 10$ $x_2 \le 2$ $x_1 \ge 3$ and $x_1, x_2 \ge 0$ are integers

The graph of subproblem 1(a) is given as below.

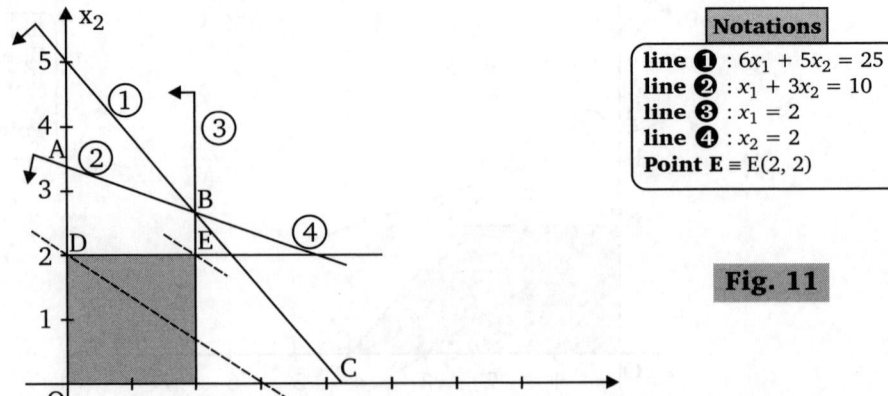

line ❶ : $6x_1 + 5x_2 = 25$
line ❷ : $x_1 + 3x_2 = 10$
line ❸ : $x_1 = 2$
line ❹ : $x_2 = 2$
Point E ≡ E(2, 2)

Fig. 11

From the above graph, we have
$$x_1 = 2, x_2 = 2 \text{ and max. } Z = 10$$
Similarly, the graph of subproblem 1(b) is given as below.

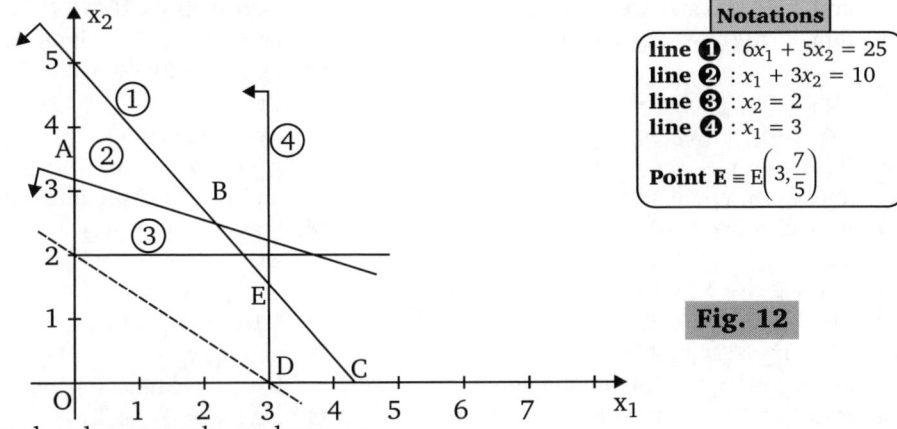

Notations

line ❶ : $6x_1 + 5x_2 = 25$
line ❷ : $x_1 + 3x_2 = 10$
line ❸ : $x_2 = 2$
line ❹ : $x_1 = 3$

Point E $\equiv E\left(3, \dfrac{7}{5}\right)$

Fig. 12

From the above graph, we have
$$x_1 = 3, x_2 = \frac{7}{5} \text{ and max.} Z = \frac{51}{5}$$

We observe that subproblem 1(a) has integer solution with max. $Z = 10$ which is less than lower bound so that it is comprehend.

Subproblem 1(b) has one non-integer solution $x_2 = \dfrac{7}{5}$.

Now, subproblem 1(b) can be further branched with x_2 as the branching variable but in this problem max. $Z = \dfrac{51}{5}$ which is less than lower bound 1, which does not assume a solution better than one already obtained. This subproblem 1(b) is comprehend.

Hence, we conclude that, the solution of subproblems are as follows.

　(i)　$x_1 = 1, x_2 = 3$ and max. $Z = 11$ for subproblem 2

　(ii)　$x_1 = 2, x_2 = 2$ and max. $Z = 10$ for subproblem 1(a)

Clearly, the larger $Z = 11$

Hence, the required solution is given by
$$x_1 = 1, x_2 = 3 \text{ and max. } Z = 11$$

Exercise-10.1

Using Gomory's cutting plane method solve the following integer programming problems.

1. max. $Z = x_1 + x_2$
　s.t.　$3x_1 + 2x_2 \le 5$
　　　　　$x_2 \le 5$
　and $x_1, x_2 \ge 0$ are integer.

2. max. $Z = 2x_1 + x_2$
　s.t. $2x_1 + 5x_2 \le 17$
　　　$3x_1 + 2x_2 \le 10$
　and $x_1, x_2 \ge 0$ are integers.

3. max. $Z = 3x_1 + 4x_2$
　s.t.
　$x_1 + x_2 \le 4$
　$\dfrac{3}{5}x_1 + x_2 \le 3$
　and $x_1, x_2 \ge 0$ are integers.

4. max. $Z = 4x_1 + 3x_2$
　s.t.
　$x_1 + 2x_2 \le 4$
　$2x_1 + x_2 \le 6$
　and $x_1, x_2 \ge 0$ are integers.

5. max. $Z = x_1 + 5x_2$
   s.t. $x_1 + 10x_2 \leq 20$
                     $x_1 \leq 2$
   and $x_1, x_2 \geq 0$ and are integers.

6. min. $Z = 20x_1 + 22x_2 + 18x_3$
   s.t.
      $4x_1 + 6x_2 + x_3 \geq 54$
      $4x_1 + 4x_2 + 6x_3 \geq 65$
      $0 \leq x_1, x_2, x_3 \leq 7$
   and x_1, x_2, x_3 are integers.

7. max. $Z = 2x_1 + 10x_2 + x_3$
   s.t.
      $5x_1 + 2x_2 + x_3 \leq 15$
      $2x_1 + x_2 + 7x_3 \leq 20$
      $x_1 + 3x_2 + 2x_3 \leq 25$
   and $x_1, x_2, x_3 \geq 0$ and are integers.

8. max. $Z = 3x_1 + 2x_2 + 5x_3$
   s.t.
      $5x_1 + 3x_2 + 7x_3 \leq 28$
      $4x_1 + 5x_2 + 5x_3 \leq 30$
   and $x_1, x_2, x_3 \geq 0$ are integers.

9. max. $Z = 3x_1 + 4x_2$
   s.t. $3x_1 + 2x_2 \leq 8$
            $x_1 + 4x_2 \geq 10$
   and $x_1, x_2 \geq 0$ are integers.

10. max. $Z = x_1 + x_2$
   s.t. $2x_1 + 5x_2 \leq 16$
          $6x_1 + 5x_2 \leq 30$
   and $x_1, x_2 \geq 0$ and x_1 is an integer.

Using branch and bound technique, solve the following LPP

11. max. $Z = x_1 + x_2$
   s.t.
      $3x_1 + 2x_2 \leq 12$
              $x_2 \leq 2$
   and $x_1, x_2 \geq 0$ are integers

12. max. $Z = 6x_1 + 8x_2$
   s.t.
      $4x_1 + 16x_2 \leq 32$
      $14x_1 + 4x_2 \leq 28$
   and $x_1, x_2 \geq 0$ are integers.

13. max. $Z = 3x_1 + 4x_2$
   s.t.
      $7x_1 + 16x_2 \leq 52$
       $3x_1 - 2x_2 \leq 18$
   and $x_1, x_2 \geq 0$ are integers.

14. max. $Z = x_1 + 4x_2$
   s.t.
        $x_1 + x_2 \leq 5$
      $10x_1 + 6x_2 \leq 45$
   and $x_1, x_2 \geq 0$
   Use x_1 as branching variable.

17. max. $Z = 3x_1 + 2x_2$
   s.t.
      $2x_1 + 2x_2 \leq 7$
              $x_1 \leq 2$
              $x_2 \leq 2$
   and $x_1, x_2 \geq 0$ are integers.

18. max. $Z = 3x_1 + 4x_2$
   s.t. $3x_1 - x_2 + x_3 = 12$
         $3x_1 + 11x_2 + x_4 = 66$
   and $x_1, x_2, x_3, x_4 \geq 0$ are integers.

19. max. $Z = x_1 + x_2$
   s.t.
      $4x_1 - x_2 \leq 10$
      $2x_1 + 5x_2 \leq 10$
      $4x_1 - 3x_2 \leq 6$
   and $x_1, x_2 \geq 0$ are integers.

20. max. $Z = 2x_1 + 20x_2 - 10x_3$
   s.t. $2x_1 + 20x_2 + 4x_3 \leq 15$
         $6x_1 + 20x_2 + 4x_3 = 20$
   and $x_1, x_2, x_3 \geq 0$ are integers.

📝 Glossary

- **Pure Integer Linear Programming:** An integer linear programming is said to be a pure integer linear programming when all its decision variables are restricted to be integer.
- **Mixed Integer Linear Programming Problem:** An integer linear programming problem is said to be mixed integer programming when some, but not all of its decision variables are restricted to be integer.
- **Zero-one Integer Linear Programming Problem:** An integer linear programming in which all the decision variables are restricted to integer of 0 or 1 is called zero-one integer linear programming problem.

🔍 REVIEW QUESTIONS

1. What do you mean by integer linear programming.
2. Explain all integer linear programming.
3. Discuss the need of integer linear programming.
4. Write the general form of integer linear programing.
5. Write a short note on Integer linear programming.
6. Write the short note on branch and bound technique.   [MEERUT–2008, 10, 12]
7. Write the short note on Gomory's cutting method.

📖 MULTIPLE CHOICE QUESTIONS (CHOOSE THE MOST APPROPRIATE ONE)

1. In mixed-integer programming problem: [MEERUT–2013]
   (a) all of decision variables require integer solution
   (b) few of the decision variable require integer solution
   (c) all decision variables are restricted to integer values of either 0 or 1
   (d) all of the above

2. In an I.P.P. rounding off solution values of decision variables in L.P.P. may not be acceptable, because: [MEERUT–2013]
   (a) it may violates non-negative condition
   (b) it does not satisfy the constraints
   (c) objective function value of the I.P.P. is more than the Objective function of the L.P.P.
   (d) none of the above

3. An Integer Programming Problem is a LPP with: [MEERUT–2013]
   (a) some variables take non-negative integer
   (b) all variables take non-negative integer
   (c) some/all variables take non-negative integer
   (d) none of the above

4. Which of the following is not correct? [MEERUT–2013]
   (a) an IPP that have only one constraint is called Knapsack problem
   (b) an IPP that has no constraint is known as a Knapsack problem

   (c) a travelling salesman problem can be solved using Branch and Bound method
   (d) variables in an IPP that are not integer constrained are called continuous variables

5. In the Branch and Bound approach to a maximization integer LP problem, a node is terminated if: [MEERUT–2013]
   (a) a node has an infeasible solution
   (b) a node yields a solution that is feasible but not an integer
   (c) upper bound is less than the current subproblem's lower bound
   (d) all of the above

6. Branch and Bound algorithm is applicable to: [MEERUT–2013]
   (a) mixed I.P.P.   (b) all I.P.P.
   (c) both (a) and (b)   (d) none of the above

7. In bounded variable algorithm: [MEERUT–2013]
   (a) lower bound of a decision variable can never be converted into non-negative decision variable
   (b) upper bound of a decision variable can always be converted into non-negative decision variable
   (c) lower and Upper bounds of a decision variable are 0 and ∞ respectively in the case of unbounded variables
   (d) the lower bound constraints $l_j \le x_j$ is converted as $x'_j (= x_j - l_j) \le 0$, where x_j are slack variables

8. Lower bound constraints are handled by substituting: [MEERUT–2013]
   (a) $x_j = u_j - x'_j$ (b) $x_j = u_j + x'_j$
   (c) $x_j = l_j + x'_j$ (d) $x_j = l_j - x'_j$

9. Branch and Bound method divides the feasible solution space into smaller parts by:
   (a) enumerating   (b) bounding
   (c) branching   (d) all of the above

10. The use of Cutting plane method: [MEERUT–2013]
    (a) require the use of standard linear programming approach between each cutting plane application
    (b) yields better value of the objective function
    (c) reduces the number of constraints in the given problem
    (d) both (b) and (c)

11. The 0-1 integer programming problem:
    (a) requires the decision variables to have values between Zero and One
    (b) requires that the constraints all have coefficients between Zero and One
    (c) requires that the decision variables have coefficients between Zero and One
    (d) all of the above

12. The part of the feasible solution space eliminated by putting a cut contains:
    (a) only non integer solution
    (b) only integer solution
    (c) both (a) and (b)
    (d) none of the above

13. While solving IP problem any non-integer variable in the solution is picked upto:
    (a) obtain the cut constraint
    (b) enter the solution
    (c) leave the solution
    (d) none of the above

14. Rounding off solution values of decision variables in an LP problem may not be acceptable because:
    (a) it does not satisfy constraints
    (b) it violates non-negativity conditions
    (c) objective function value is less than the objective function value of LP
    (d) none of the above

15. Which of the following is the consequence of adding a new cut constraint to an optimal simplex table:
    (a) addition of a new variable to the table
    (b) makes the previous optimal solution infeasible
    (c) eliminates non-integer solution from the solution space
    (d) all of the above

16. A non-integer variable is chosen in the optimal simplex table of integer LP problem to:
    (a) leaves the basis
    (b) enter the basis
    (c) to construct the Gomory cut
    (d) none of the above

17. The corners of the reduced feasible region of integer LP problem contains:
    (a) only integer solution
    (b) optimal integer solution
    (c) only non-integer solution
    (d) all of the above

18. In a branch and bound minimization tree, the lower bounds on Objective function value:
    (a) do not decrease in value
    (b) do not increase in value
    (c) remain constant
    (d) None of these

19. Modifications made for the mixed integer cutting plane method are:
    (a) top most rows of the simplex table contains integer variables
    (b) values of the Objective function is bounded
    (c) row corresponding to an integer variable serve as a source row
    (d) all of the above

20. The situation of multiple solutions arises with:
    (a) cutting plane method
    (b) branch and bound method
    (c) both (a) and (b)
    (d) none of these

21. While applying cutting plane method, dual simplex is used to maintain:
    (a) optimality   (b) feasibility
    (c) both (a) and (b)   (d) none of the above

ANSWERS

1. (c)	**2.** (d)	**3.** (b)	**4.** (c)	**5.** (c)	**6.** (c)	**7.** (c)	**8.** (c)	**9.** (d)
10. (a)	**11.** (a)	**12.** (a)	**13.** (a)	**14.** (d)	**15.** (d)	**16.** (c)	**17.** (a)	**18.** (b)
19. (c)	**20.** (d)	**21.** (b)						

The Transportation Problem

11.1 INTRODUCTION

The transportation problem are concerned with the distribution of certain product from several sources to numerous localities at a minimum cost. It is one of the types of linear programming problems. The origin of transportation problem is concerned with two contributions. First was the consequence of the study entitled "The distribution of a product from several sources to numerous locations" presented by F.L. Hitchock in 1941 and second was the presentation of "Optimum utilization" by T.C. Koopmans in 1947.

From the name itself, it is clear that the transportation problem means a problem where something is to be transferred.

Let us suppose we have a product which is to be transformed from a number of centres called 'origin' or 'sources' to a number of places called 'destinations'. The cost of transportation along different routes are different and known. The main objective is to minimize the cost of associated with such transportation from the place of supply to their destination. These special type of linear programming problems are called 'transportation problem'.

11.2 MATHEMATICAL FORMULATION OF TRANSPORTATION PROBLEM

To design the mathematical formulation of a transportation problem, we have the following assumptions.

   (i) Product can be transported easily from sources to the destination.
  (ii) Total quantity of the product available at different sources is equal to the requirements at different destinations.
 (iii) The unit transportation cost of the product from all sources to the destination is well known.
 (iv) The transportation cost on a given route is directly proportional to the number of units transported by that route.
  (v) The objective is to minimize the total cost of transportation from all sources to the destinations and not for individual supply and destination centres.

Let us have m origins and n destinations (n may or may not be equal to m) and

$$a_i : \text{quantity of product available at origin } i$$
$$b_j : \text{quantity of product required at destination } j$$
$$c_{ij} : \text{cost of transporting from origin } i \text{ to destination } j$$
$$x_{ij} : \text{quantity transported from origin } i \text{ to destination } j$$

Then by second assumption

$$\sum_{i=1}^{m} a_i = \sum_{j=1}^{n} b_j \qquad \qquad ...(1)$$

This is the case when demand is fully met from the supply.

The problem can be stated in the form of LPP as

$$\min . Z = \sum_{i=1}^{m} \sum_{j=1}^{n} c_{ij} x_{ij}$$

such that

$$\sum_{j=1}^{n} x_{ij} = a_i \text{ for } i = 1, 2, ..., m \qquad \qquad ...(2)$$

$$\sum_{i=1}^{m} x_{ij} = b_j \text{ for } j = 1, 2, ..., n \qquad \qquad ...(3)$$

and $\qquad \qquad x_{ij} \geq 0 \ \forall i = 1,2,...m; j = 1,2,...,n$

The above transportation problem has two set of constraints given by (2) and (3) which will be consistent if

$$\sum_{i=1}^{m} a_i = \sum_{j=1}^{n} b_j \qquad \qquad ...(4)$$

which shows that the necessary and sufficient condition for a transportation problem to have a feasible solution is given by (4).

☛ **REMARK**

- A transportation problem that satisfies the condition

$$\sum_{i=1}^{m} a_i = \sum_{j=1}^{n} b_j$$

is called balanced transportation problem.

11.2.1 TABULAR REPRESENTATION OF TRANSPORTATION PROBLEM

Let there be m sources (origin) and n destination. Then we have the following tabular representation.

Sources \ Destinations	D_1	D_2	D_3	...	D_n	Supply
O_1	c_{11}	c_{12}	c_{13}	...	c_{1n}	a_1
O_2	c_{21}	c_{22}	c_{23}	...	c_{2n}	a_2
O_3	c_{31}	c_{32}	c_{33}	...	c_{3n}	a_3
⋮	⋮					⋮
O_m	c_{m1}	c_{m2}	c_{m3}	...	c_{mn}	a_m
Demand	b_1	b_2	b_3	...	b_n	$\sum_{i=1}^{m} a_i = \sum_{j=1}^{n} b_j$

11.2.2 RELATED DEFINITIONS

(1) **Feasible solution.** A feasible solution to a transportation problem is a set of non-negative individual allocations ($x_{ij} \geq 0$) which satisfies the row and column sum restrictions (*i.e.*, equation (2) and (3)).

(2) **Basic Feasible solution.** A feasible solution of an $m \times n$ transportation problem is said to be basic if the total number of positive allocations is equal to $m + n - 1$, i.e., one less than the sum of no. of rows and columns.

(3) **Optimal solution.** A feasible solution, not necessarily basic is said to be optimum if it minimize the total transportation cost.

(4) **Non-degenerate Basic Feasible solution.** A basic feasible solution to a transportation problem of m sources and n destination is said to be non-degenerate basic feasible if
   (i)   the total no. of non-negative allocations x_{ij} is exactly equal to $m + n - 1$.
   (ii) these allocations should be in independent position (i.e., it is always impossible to form any closed circuit by joining these allocation by horizontal and vertical lines only).

(5) **Degenerate-basic Feasible solution.** A basic feasible solution to a transportation problem of m sources and n destinations is said to be degenerate if the total number of non-negative allocations is less than $(m + n - 1)$.

11.2.3 SOME THEOREMS

THEOREM 1. **(Existence of Feasible solution)** *The necessary and sufficient condition for the existence of feasible solution of a transportation problem is*

$$\Sigma a_i = \Sigma b_j : i = 1, 2, ..., m, j = 1, 2, ..., n$$

PROOF. Necessary Condition:

Let us suppose, a feasible solution of the transportation problem exists. Then

$$\sum_{j=1}^{n} a_{ij} = a_i : i = 1, 2, ..., m$$

and

$$\sum_{i=1}^{m} x_{ij} = b_j, j = 1, 2, ..., n$$

Now, summing over all i and j respectively, we get

$$\sum_{i=1}^{m} \sum_{j=1}^{n} x_{ij} = \sum_{i=1}^{m} a_i \quad \text{and} \quad \sum_{j=1}^{n} \sum_{i=1}^{m} x_{ij} = \sum_{j=1}^{n} b_j$$

which shows that

$$\sum_{i=1}^{m} a_i = \sum_{j=1}^{n} b_j$$

Sufficient Condition. Let $\sum\limits_{i=1}^{m} a_i = \sum\limits_{j=1}^{n} b_j = k$ (say)

If $x_{ij} = \lambda_i b_j \ \forall \ i, j, \lambda_i \neq 0$ is any real number. Then

$$\sum_{j=1}^{n} x_{ij} = \sum_{j=1}^{n} \lambda_i b_j = \lambda_i \sum_{j=1}^{n} b_j = k\lambda_i$$

$$\Rightarrow \qquad \lambda_i = \frac{1}{k} \sum_{j=1}^{n} x_{ij} = \frac{a_i}{k}$$

Therefore, $x_{ij} = \lambda_i b_j = \dfrac{a_i b_j}{k}$ for all i and j.

$$\Rightarrow \qquad x_{ij} \geq 0 \ \forall \ i \text{ and } j \qquad\qquad (\because \ a_i \geq 0, b_j > 0)$$

Hence, a feasible solution exists.

☛ REMARK

- A balanced transportation problem has a feasible solution.

THEOREM 2. **(Existence of BFS)** *Out of $(m + n)$ equations, there are only $m + n - 1$ independent equations in a transportation problem, m and n being the number of origins and destinations and any one equation can be dropped as the redundant equation.*

PROOF. Let us consider m row equations and $(n - 1)$ columns equations of the transportation problem as

$$\sum_{j=1}^{n} x_{ij} = a_i, i = 1, 2, \ldots, m \qquad\qquad \ldots(1)$$

and

$$\sum_{i=1}^{m} x_{ij} = b_j, j = 1, 2, \ldots, n - 1 \qquad\qquad \ldots(2)$$

Now, adding m origin constraints given in (1), we get

$$\sum_{i=1}^{n-1} \sum_{j=1}^{n} x_{ij} = \sum_{i=1}^{m} a_i \qquad\qquad \ldots(3)$$

Further, adding $(n - 1)$ destination constraints given in (2), we get

$$\sum_{j=1}^{n-1} \sum_{i=1}^{m} x_{ij} = \sum_{j=1}^{n-1} b_j \qquad\qquad \ldots(4)$$

Now, subtracting (4) from (3) we get

$$\sum_{i=1}^{m} \sum_{j=1}^{n} x_{ij} - \sum_{j=1}^{n-1} \sum_{i=1}^{m} x_{ij} = \sum_{i=1}^{m} a_i - \sum_{j=1}^{n-1} b_j$$

$$\Rightarrow \qquad \sum_{i=1}^{m} \left[\sum_{j=1}^{n} x_{ij} - \sum_{j=1}^{n-1} x_{ij} \right] = \sum_{j=1}^{n} b_j - \sum_{j=1}^{n-1} b_j \qquad (\because \ \Sigma a_i = \Sigma b_j)$$

$$\Rightarrow \qquad \sum_{i=1}^{m} x_{in} = b_n, \text{ which is the } n^{\text{th}} \text{ destination constraints.}$$

$\Rightarrow$ if $(m + n - 1)$ constraints are satisfied then the $(m + n)^{\text{th}}$ constraints will be satisfied because $\Sigma a_i = \Sigma b_j$.

Therefore, we have only $(m + n - 1)$ linearly independent equations out of $(m + n)$ equations, one is redundant.

☛ REMARK

- A basic feasible solution will contain $(m + n - 1)$ positive variables, others being zero.

THEOREM 3. **(Existence of optimal solution)** *There always exists an optimal solution to a balanced transportation problem.*

PROOF. We know that the necessary and sufficient condition for a feasible solution is

$$\sum_{i=1}^{m} a_i = \sum_{j=1}^{n} b_j \qquad\qquad \ldots(1)$$

We know that for a balanced transportation problem, we always have (1)

$\Rightarrow$ A feasible solution exists of the problem, i.e., $x_{ij} \geq 0 \ \forall \ i, j$

Now, for the constraints of the problem each $x_{ij} \leq \min\{a_i, b_j\}$

Therefore, $0 \leq x_{ij} \leq \min[a_i, b_j]$

$\Rightarrow$ The feasible region of the problem is non-empty, closed and bounded.

Hence, there exists an optimal solution.

11.3 SOLUTION OF THE TRANSPORTATION PROBLEM

To find the solution of a transportation problem, we have to find following two solutions.

(1) Initial basic feasible solution

(2) Optimal solution

(1) Methods of finding initial solution. The initial basic feasible solution can be obtained by using any one of the following three methods.

(i) North West Corner Rule

(ii) Least Cost Entry method or Matrix-minima method

(iii) Vogel's Approximation Method (VAM) or unit cost penalty method

(1) NORTH WEST CORNER METHOD

In this method we apply the following steps:

STEP 1. Start with the cell (1, 1) at the upper left (north-west) corner of the matrix and allocate it as much as possible amount equal to the minimum of the supply-demand values, i.e., we allocate x_{11} to the cell (1, 1) where

$$x_{11} = \min\{a_1, b_1\}$$

where a_1 is the supply amount for the first row and b_1 is the demand for the first column.

STEP 2. (i) If $a_1 > b_1$, then move to the cell (1, 2) and allocate x_{12} where

$$x_{12} = \min\{a_1 - x_{11}, b_2\}.$$

(ii) If $a_1 < b_1$, then move to the cell (2, 1) and allocate it as x_{21} where

$$x_{21} = \min(a_2 - b_1 - x_{11})$$

STEP 3. Continue this process step by step till an allocation is made in the south east corner of the cell, i.e., until all available amount is exhausted.

☛ REMARK

- Sometime, during the process of making allocation at a particular cell, supply equal demands, then next allocation of magnitude zero can be made in a cell either in the next row or column. This is called degeneracy.

 Solved Examples

EXAMPLE 1. *Find an IBFS of the following transportation problem by north-west corner rule.*

		D_1	D_2	D_3	D_4	Supply
	O_1	1	2	3	4	6
Origins	O_2	4	3	2	0	8
	O_3	0	2	2	1	10
	Demand	4	6	8	6	24

SOLUTION. We have $m = 3, n = 4, a_1 = 6, a_2 = 8, a_3 = 10$

$b_1 = 4, b_2 = 6, b_3 = 8, b_4 = 6$

Also, $\Sigma a_i = \Sigma b_j = 24 \Rightarrow$ Problem is balanced.

Now to find IBFS, we proceed as follows:

STEP 1. Start with the cell $(1, 1)$ at the north-west corner of the above table and allocate it as x_{11} such that

$$x_{11} = \min\{a_1, b_1\} = \min\{6, 4\} = 4$$

$\Rightarrow D_1$ column is exhausted.

STEP 2. Clearly, $a_1 > b_1$ so we move towards the cell $(1, 2)$ and allocate it as x_{12}, where

$$x_{12} = \min\{a_1 - x_{11}, b_2\} = \min\{6 - 4, 6\} = \min\{2, 6\} = 2$$

i.e., allocate 2 to the cell $(1, 2)$

$\Rightarrow O_1$ row is exhausted.

Then we move downward to the cell $(2, 3)$ to allocate it x_{23} such that

$$x_{23} = \min\{a_2 - x_{23}, b_3\} = \min\{8 - 4, 8\} = \min\{4, 8\} = 4$$

i.e., allocate 4 to the cell $(2, 3)$.

$\Rightarrow O_2$ row is exhausted.

Then we move downward to the cell $(3, 3)$ and allocate it as x_{33}, where

$$x_{33} = \min\{a_3, b_3 - x_{23}\} = \min\{10, 8 - 4\} = \min\{10, 4\} = 4$$

i.e., allocate 4 to the cell $(3, 3)$

$\Rightarrow D_3$ column is exhausted.

Then we move towards to the cell $(3, 4)$ and allocate it as x_{34} where

$$x_{34} = \min\{a_3 - x_{33}, b_4\} = \min\{10 - 4, 6\} = \min\{6, 6\} = 6$$

i.e., allocate 6 to the cell $(3, 4)$

$\Rightarrow O_3$ row and D_4 column are exhausted.

Thus all the row and columns are exhausted.

The above steps of allocations can be summarized in the table given below.

④	②		
1	2	3	4
	④	④	
3	3	2	8
		④	⑥
0	2	2	1

Hence, the initial basic feasible solution of the given transportation problem is given by

$$\text{Total cost} = 1 \times 4 + 2 \times 2 + 4 \times 3 + 4 \times 2 + 4 \times 2 + 6 \times 1 = ₹42$$

EXAMPLE 2. *Find an IBFS of the following transportation problem by north-west corner rule.*

			Supply
2	7	4	5
3	3	1	8
5	4	7	7
1	6	2	14
Demand 7	9	18	34

SOLUTION. Proceed same as in example 1, we have the following allocation table

⑤ 2	7	4
② 3	⑥ 3	1
5	③ 4	④ 7
1	6	⑭ 2

Hence, total cost $= 5 \times 2 + 2 \times 3 + 6 \times 3 + 3 \times 4 + 4 \times 7 + 14 \times 2 = ₹102$

(2) LOWEST COST ENTRY METHOD OR MATRIX MINIMA METHOD

In lowest cost entry method, we use the following steps:

STEP 1. Identify the cell with lowest cost. Let it be (i, j). Then allocate x_{ij} to the cell (i, j) such that

$$x_{ij} = \min\{a_i, b_j\}$$

If lowest cost cell is not unique, we may select any one of them.

STEP 2. If $x_{ij} = a_i$ then remove the ith row from the table and then demand b_j is reduced to $(b_j - a_i)$. Then go to step 3.

If $x_{ij} = b_j$, then remove the jth column from transportation table and the supply a_i is reduced to $a_i - b_j$, then go to step 3.

If $x_{ij} = b_j$, then remove either i^{th} row or j^{th} column but not both.

STEP 3. Repeat the above steps with reduced transportation table thus obtained in step 2 until all the available amount is exhausted.

Solved Examples

EXAMPLE 1. *Find an initial basic feasible solution of the following transportation problem by lowest cost entry method.*

	D_1	D_2	D_3	D_4	Supply
O_1	6	4	1	5	14
O_2	8	9	2	7	16
O_3	4	3	6	2	5
Demand	6	10	15	4	35

SOLUTION. Here, we have

$m = 3, n = 4$

$a_1 = 14, a_2 = 16, a_3 = 5$

$b_1 = 6, b_2 = 10, b_3 = 15, b_4 = 4$

Clearly, $\Sigma a_i = \Sigma b_j = 35 \Rightarrow$ Problem is balanced

Since, 1 is the lowest cost (at the cell (1, 3))

$\therefore$ We allocate x_{13} such that

$$x_{13} = \min\{a_1, b_3\} = \min\{14, 15\} = 14 = a_1$$

After allocating 14 to the cell $(1, 3)$ the O_1 row is deleted and then first reduced transportation table is given as under

	D_1	D_2	D_3	D_4	
O_2	8	9	2	7	16
O_3	4	3	6	2	5
	6	10	1	4	21

In the above table, the lowest cost is 2 which is at the $(2, 3)$ and the cell $(3, 4)$. Choose $(3, 4)$ cell arbitrarily and allocate x_{34} to this cell such that

$$x_{34} = \min\{a_3, b_4\} = \min\{5, 4\} = 4$$

After allocating 4 to the cell $(3, 4)$, the D_4 column is deleted and the second reduced table is given below.

	D_1	D_2	D_3	
O_2	8	9	2	16
O_3	4	3	6	1
	6	10	1	17

In this table, the lowest cost is 2 at the cell $(2, 3)$. Allocate x_{23} to this cell such that

$$x_{23} = \min\{a_2, b_3\} = \min\{16, 1\} = 1$$

After allocating 1 to the cell $(2, 3)$, the D_3 column is deleted and the third reduced table is given as below.

	D_1	D_2	
O_2	8	9	15
O_3	4	3	1
	6	10	16

In the above table, the lowest cost is 3 at the cell $(3, 2)$. Allocate x_{32} to this cell such that

$$x_{32} = \min\{a_3, b_2\} = \min\{1, 10\} = 1$$

After allocating 1 to the cell $(3, 2)$, the O_3 row is deleted and only O_2 row is remaining in which 8 is minimum at the cell $(2, 1)$. So, we allocate x_{21} to the cell such that

$$x_{21} = \min\{a_2, b_1\} = \min\{15, 6\} = 6$$

and finally allocate x_{22} to the remaining element 9 which is at the cell $(2, 2)$ such that

$$x_{22} = \min\{a_2, b_2\} = \min\{9, 9\} = 9$$

Therefore, all the rows and columns are exhausted.

Finally, the summary of all the allocations is given in the table given below.

6	4	⑭ 1	5
⑥ 8	⑨ 9	① 2	7
4	① 3	6	④ 2

Hence, total cost $= 1 \times 14 + 6 \times 8 + 9 \times 9 + 1 \times 2 + 1 \times 3 + 4 \times 2 = ₹156$

EXAMPLE 2. *Find the initial basic feasible solution of the following transportation problem by lowest cost entry method.*

			Supply
2	7	4	5
3	3	1	8
5	4	7	7
1	6	2	14
Demand 7	9	18	34

SOLUTION. Proceeding same as in example 1, we have the following allocation table:

	②		③
2		7	4
			⑧
3		3	1
	⑦		
5		4	7
⑦			⑦
1		6	2

Hence, the total cost $= 2 \times 7 + 3 \times 4 + 7 \times 4 + 8 \times 1 + 7 \times 1 + 7 \times 2 = ₹83$

(3) VOGEL'S APPROXIMATION METHOD (UNIT COST PENALTY METHOD)

To find the IBFS by Vogel's approximation method, we use the following steps.

STEP 1. Find the smallest and next to smallest costs for each row of the transportation table and then find the difference between them for each row. Write these difference (Penalties) alongside the transportation table against the respective rows. Similar exercise will be done on case of columns.

STEP 2. Select the maximum Penalty among the rows and columns penalties and if there is a tie, choose any one arbitrarily.

STEP 3. Allocate the maximum possible amount to the cell with lowest cost in that particular row or column.

Let the largest penalty correspond to i^{th} row and let c_{ij} be the smallest cost in the ith row. Allocate the amount.

$$x_{ij} = \min\{a_i, b_j\} \text{ in the cell } (i, j)$$

and then cross out i^{th} row and j^{th} column and obtain reduced matrix.

STEP 4. Now compute the row and column penalties for the reduced table and repeat step 2 and 3.

Continue this process until all the available quantity is exhausted or all the requirements are satisfied.

Solved Examples

EXAMPLE 1. *Find the initial basic feasible solution to the following transportation problem by Vogel's approximation method.*

	D_1	D_2	D_3	D_4	Supply
F_1	3	3	4	1	100
F_2	4	2	4	2	125
F_3	1	5	3	2	75
Demand	120	80	75	25	300

SOLUTION. We have $m = 3, n = 4$
$$a_1 = 100, a_2 = 125, a_3 = 75$$
$$b_1 = 120, b_2 = 80, b_3 = 75, b_4 = 25$$

$\Rightarrow \quad \sum\limits_{i=1}^{m} a_i = 300 = \sum\limits_{j=1}^{n} b_j$

$\Rightarrow$ given transportation problem is a balanced problem.

Now, we have to find the penalties as given below

Penalties

				Penalties
3	3	4	1	$3 - 1 = (2)$
4	2	4	2	$2 - 2 = (0)$
1	5	3	2	$2 - 1 = (1)$

Penalties $3 - 1 = (2)$ $3 - 2 = (1)$ $4 - 3 = (1)$ $2 - 1 = (1)$

Among all the above penalties, the maximum penalty $= 2$, which is corresponding to D_1 column and F_1 row (We can select any one of these)

Suppose we select F_1 row. In F_1 row, the minimum cost is 1 which is at the cell $(1, 4)$, so allocate x_{14} to the cell $(1, 4)$ such that
$$x_{14} = \min\{100, 25\} = 25 \text{ (demand for } D_4)$$
Delete D_4, the reduced matrix is given by

Penalties

			Penalties
3	3	4	$3 - 3 = (0)$
4	2	4	$4 - 2 = (2)$
1	5	3	$3 - 1 = (2)$

Penalties $3 - 1 = (2)$ $3 - 2 = (1)$ $4 - 3 = (1)$

Here, the maximum penalty $= 2$, which is corresponding to F_2, F_3 rows and D_1 column, so we can select anyone of these.

Now, select F_3 row in which the least cost is 1 at the cell $(3, 1)$. Therefore allocate x_{31} to the cell $(3, 1)$ such that
$$x_{31} = \min\{75, 120\} = 75, \text{ supply for } F_3$$
Now, delete F_3 after allocating 75 to the cell $(3, 1)$

We get the following reduced table

	D_1	D_2	D_3	Supply	Penalty
F_1	3	3	4	75	$4 - 3 = (1)$
F_2	4	2	4	125	$4 - 2 = (2)$
Demand	45	80	75	200	
Penalty	$4 - 3 = (1)$	$3 - 2 = (1)$	$4 - 4 = (0)$		

Clearly, the maximum penalty = 2, (corresponding to the cell (2,2))

$\therefore \qquad x_{22} = \min\{125, 80\} = 80$

So, allocate 80 to the cell (2, 2) and delete D_2 column.

The third reduced table is given as under

	D_1	D_3	Supply	Penalty
F_1	3	4	75	4 – 3 = (1)
F_2	4	4	45	4 – 4 = (0)
Demand	45	75	120	

Penalty   4 – 3 = (1)   4 – 4 = (0)

Here, the maximum penalty = 1, *i.e.*, corresponding to the F_1 row and D_1 column. Therefore, we can select any of these for allocation.

Let us suppose we select F_1 row for allocation.

Clearly, in F_1 row the lowest cost is 3 which is at the cell (1, 1). Thus,

$$x_{11} = \min\{75, 45\} = 45, \text{ demand for } D_1$$

So, allocate 45 to the cell (1, 1) and delete D_1.

Then we have the following table

	D_3	Supply	Penalty
F_1	4	30	(4)
F_2	4	45	(4)
Demand	75	75	

Penalty   4 – 4 = (0)

Here, the maximum penalty = 4 which is corresponding to F_1 and F_2 rows, so we may select any one of these.

Let us suppose F_1 is selected in which 4 is the only cost which is at the cell (1, 3). Therefore, allocate x_{13} to the cell (1, 3) where

$$x_{13} = \min\{30, 75\} = 30, \text{ supply for } F_1$$

Then delete F_1 row and get the following reduced table

		Supply
F_2	4	45
Demand	45	45

Since, there is only single element, *i.e.*, 4, so allocate 45 to this cell.

Now, all the above allocation can be shown in a single table as given bleow.

	D_1	D_2	D_3	D_4	Supply
	㊺ 3	3	㉚ 4	㉕ 1	100
	4	㉠⁸⁰ 2	㊺ 4	2	125
	㉥⁷⁵ 1	5	3	2	75
Demand	120	80	75	25	300

We observe that here total no. of allocation is 6 which is equal to $(m + n - 1)$

$\Rightarrow$ solution is non-degenerate.

Finally, the initial basic feasible solution is given by

$$= 1 \times 25 + 1 \times 75 + 2 \times 80 + 3 \times 45 + 4 \times 30 + 4 \times 45$$
$$= ₹ 695, \text{ which is the required total cost}$$

EXAMPLE 2. *Solve the following transportation problem to find IBFS.*

[MEERUT–2006, 07, 08; KANPUR–2010, 11]

O \ D	1	2	3	a_i
1	2	7	4	5
2	3	3	1	8
3	5	4	7	7
4	1	6	2	14
b_j	7	9	18	34

SOLUTION. We find the penalties as follows:

				Penalties
2	7	4		2
3	3	⑧ 1		2
5	4	7		1
1	6	2		1
Penalties 1	1	1		

			Penalties
⑤ 2	7	4	2
5	4	7	1
1	6	2	1
Penalties 1	2	2	

			Penalties
5	4	7	1
1	6	⑩ 2	1
Penalties 4	2	5	

5	⑦ 4
② 1	② 6

So, solution set is

5	0	0
0	0	8
0	7	0
2	2	10

Using all the above tables, the all allocations can be shown in the single table as given below

			a_i
⑤ 2	7	4	5
3	3	⑧ 1	8
⑦ 5	⑧ 4	7	7
② 1	② 6	⑩ 2	14
7	9	18	

$$\begin{aligned}
\text{Min. cost} &= 5\times 2 + 2\times 1 + 7\times 4 + 6\times 2 + 8\times 1 + 10\times 2\\
&= 10 + 2 + 28 + 12 + 8 + 20\\
&= 80
\end{aligned}$$

Example 3. *Solve the following transportation problem.*

	D_1	D_2	D_3	D_4	D_5	D_6	a_i
O_1	1	2	1	4	4	2	30
O_2	3	3	2	1	4	3	50
O_3	4	2	5	9	6	2	75
O_4	3	1	7	3	4	6	20
b_j	20	40	30	10	50	25	

SOLUTION. Apply the usual procedure as above, we can find the penalties as follows:

㉒ 1	2	1	4	4	2	1
3	3	2	1	4	3	1
4	2	5	9	6	2	2
3	1	7	3	4	6	2
2	1	1	2	2	1	

2	1	4	4	2	1
3	2	1	4	3	1
④⓪ 2	5	9	6	2	3
1	7	3	4	6	2
1	1	2	2	1	

1	4	4	2	1
2	1	4	3	1
5	9	6	㉕ 2	3
7	3	4	6	1
1	2	2	1	

1	4	4	3
2	1	4	1
5	9	6	1
7	3	4	1
1	2	2	

⑳ 2	1	4	1
5	9	6	1
7	3	4	1
3	2	2	

10 1	4	3
9	6	3
3	4	1
2	2	

⑳   4
⑩   6
⑳   4

So, solution set is
$$\begin{bmatrix} 20 & 0 & 10 & 0 & 0 & 0 \\ 0 & 0 & 20 & 10 & 20 & 0 \\ 0 & 40 & 0 & 0 & 10 & 25 \\ 0 & 0 & 0 & 0 & 20 & 0 \end{bmatrix}$$

Therefore, all allocations in a single table is given as follows:

	D_1	D_2	D_3	D_4	D_5	D_6	a_i
O_1	⑳ 1	2	⑩ 1	4	4	2	30
O_2	3	3	⑳ 2	⑩ 1	⑳ 4	3	50
O_3	4	㊵ 2	5	9	⑩ 6	㉕ 2	75
O_4	3	1	7	3	⑳ 4	6	20
b_j	20	40	30	10	50	25	

Min. cost

$$= 20 \times 1 + 10 \times 1 + 20 \times 2 + 10 \times 1 + 20 \times 4 + 40 \times 2 + 10 \times 6 + 25 \times 2 + 20 \times 4$$
$$= 20 + 10 + 40 + 10 + 80 + 80 + 60 + 50 + 80 \;=\; 430$$

EXAMPLE 4. *Solve the following transportation problem to find IBFS.*

	D_1	D_2	D_3	D_4	a_i
O_1	10	7	3	6	3
O_2	1	6	8	3	5
O_3	7	4	5	3	7
b_j	3	2	6	4	**15**

SOLUTION. First we find out penalties corresponding to each row and column by applying the usual procedure.

Penalties

				Penalties
10	7	3	6	3
③ 1	6	8	3	2
7	4	5	3	1

Penalties   6   2   2   3

			Penalties
③ 7	3	6	3
6	8	3	3
4	5	3	1

2   2   3

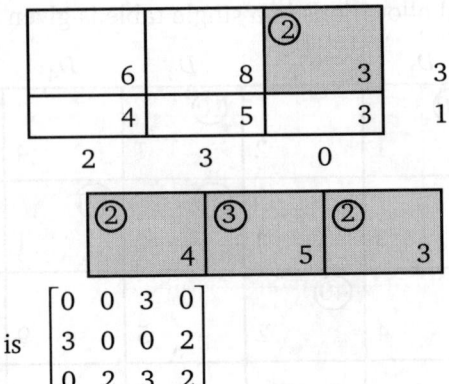

		②		
6	8	3	3	
4	5	3	1	
2	3	0		

②	③	②
4	5	3

So, solution set is
$$\begin{bmatrix} 0 & 0 & 3 & 0 \\ 3 & 0 & 0 & 2 \\ 0 & 2 & 3 & 2 \end{bmatrix}$$

Therefore, all allocations in a single table are given as below

	D_1	D_2	D_3	D_4
O_1	10	7	③ 3	6
O_2	③ 1	6	8	② 3
O_3	7	② 4	③ 5	② 3

Least cost $= 3 \times 1 + 2 \times 4 + 3 \times 3 + 3 \times 5 + 2 \times 3 + 2 \times 3 = 47$

EXAMPLE 5. *Solve the following transportation problem.*

	D_1	D_2	D_3	D_4	D_5	D_6	a_i
O_1	9	12	9	6	9	10	5
O_2	7	3	7	7	5	5	6
O_3	6	5	9	11	3	11	2
O_4	6	8	11	2	2	10	9
b_j	4	4	6	2	4	2	

SOLUTION. We use VAM here to solve the given transportation problem :

	D_1	D_2	D_3	D_4	D_5	D_6	Penalties
O_1	9	12	9	6	9	10	3
O_2	7	3	7	7	5	② 5	2
O_3	6	5	9	11	3	11	2
O_4	6	8	11	2	2	10	4
Penalties	1	2	2	4	1	5	

Penalties

9	12	9	6	9	3
7	3	7	7	5	2
6	5	9	11	3	2
6	8	11	② 2	2	4
1	2	2	4	1	

9	12	9	9	3
7	3	7	5	2
6	5	9	3	2
6	8	11	④ 2	4
1	2	2	1	

9	12	9	3
7	④ 3	7	4
6	5	9	1
6	8	11	2
1	2	2	

9	9	0
7	7	0
6	9	3
③ 6	11	5
1	2	

9	9	0
7	7	0
① 6	9	3
1	2	

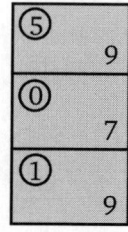

$$\text{Solution set is } \begin{bmatrix} 0 & 0 & 5 & 0 & 0 & 0 \\ 0 & 4 & 0 & 0 & 0 & 2 \\ 1 & 0 & 1 & 0 & 0 & 0 \\ 3 & 0 & 0 & 2 & 4 & 0 \end{bmatrix}$$

Min. cost $= 1 \times 6 + 3 \times 6 + 4 \times 3 + 5 \times 9 + 1 \times 9 + 2 \times 2 + 4 \times 2 + 2 \times 5 = 112$

EXAMPLE 6. *Solve*

	O_1	O_2	O_3	O_4	**Requirement**
D_1	19	14	23	11	11
D_2	15	16	12	21	13
D_3	30	25	16	39	19
Available	6	10	12	15	**43**

SOLUTION. **Calculation of penalty by VAM Method:**

Penalty

			⑪	Penalty
19	14	23	11	3
15	16	12	21	3
30	25	16	39	9

Penalty 4 2 4 10

			④	
15	16	12	21	3
30	25	16	39	9

15 9 4 18

⑥			
15	16	12	3
30	25	16	9

15 9 4

③		
16	12	4
⑦	⑫	
25	16	9

9 4

$$\text{So, solution set is } \begin{bmatrix} 0 & 0 & 0 & 11 \\ 6 & 3 & 0 & 4 \\ 0 & 7 & 12 & 0 \end{bmatrix}$$

Therefore, we have the following allocation table

	O_1	O_2	O_3	O_4	
D_1	19	14	23	⑪ 11	11
D_2	⑥ 15	③ 16	12	④ 21	13
D_3	30	⑦ 25	⑫ 16	39	19
	6	10	12	15	

Minimum cost $= 11 \times 11 + 6 \times 15 + 3 \times 16 + 4 \times 21 + 7 \times 25 + 12 \times 16 = 710$.

EXAMPLE 7. *Solve the following problem.*

	D_1	D_2	D_3	D_4	a_i
O_1	2	5	4	7	4
O_2	6	1	2	5	6
O_3	4	6	2	4	8
b_j	3	7	6	2	

SOLUTION. **Calculation of penalties:**

Penalty

				Penalty
2	5	4	7	2
6	⑥ 1	2	5	1
4	6	2	4	2

Penalty 2 4 2 1

2	5	4	7	2
4	6	2	② 4	2

2 1 2 3

③ 2	5	4	2
4	6	2	2

2 1 2

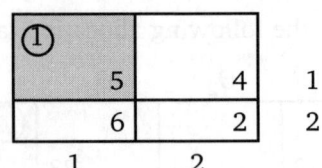

Solution set is $\begin{bmatrix} 3 & 1 & 0 & 0 \\ 0 & 6 & 0 & 0 \\ 0 & 0 & 6 & 2 \end{bmatrix}$

Therefore, we have the following allocation table

	D_1	D_2	D_3	D_4	a_i
O_1	③ 2	① 5	4	7	4
O_2	6	⑥ 1	2	5	6
O_3	4	6	⑥ 2	② 4	8
b_j	3	7	6	2	

Minimum cost $= 3\times2+1\times5+6\times1+6\times2+2\times4 = 37$.

EXAMPLE 8. *Solve the following transportation problem.*

	D_1	D_2	D_3	D_4	a_i
O_1	6	4	1	5	14
O_2	8	7	2	7	16
O_3	4	3	6	2	5
b_j	6	10	15	4	

SOLUTION. **Calculation of penalty:**

Penalty

				Penalty
6	4	1	5	3
8	7	⑮ 2	7	5
4	3	6	2	1
Penalty 2	1	1	3	

			Penalty
6	4	5	1
8	7	7	1
4	3	④ 2	1
2	1	2	

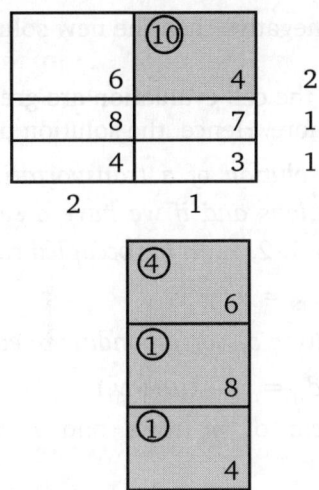

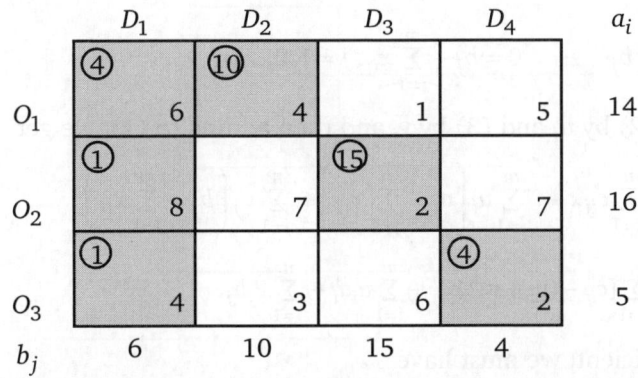

Solution set is $\begin{bmatrix} 4 & 10 & 0 & 0 \\ 1 & 0 & 15 & 0 \\ 1 & 0 & 0 & 4 \end{bmatrix}$

Therefore,

	D_1	D_2	D_3	D_4	a_i
O_1	④ 6	⑩ 4	1	5	14
O_2	① 8	7	⑮ 2	7	16
O_3	① 4	3	6	④ 2	5
b_j	6	10	15	4	

Minimum cost $= 4 \times 6 + 10 \times 4 + 1 \times 8 + 15 \times 2 + 1 \times 4 + 4 \times 2 = 114$.

11.4 TEST FOR OPTIMALITY

After obtaining the initial basic feasible solution of the given transportation problem by any method (discussed in previous section) we test the optimality of this solution. This test is applied to any IBFS with $(m + n - 1)$ allocations.

11.4.1 STEPPING STONE MEHTOD

In this method, we find the cell evaluation corresponding to each empty cell. For this, start with empty cell (with no allocation) and allocate 1 unit to this cell and then maintain the row and column sums by doing some necessary adjustments.

Due to such adjustments in the solution, the net change in the total cost of the transportation problem is known as cell evaluation of the cell. Then we have the following observation

(i) If the cell evaluation is positive, then the new solution increases the total transportation cost.

(ii) If the cell evaluation is negative then the new solution reduces to the transportation cost.

Thus, we conclude that if all the cell evaluation are greater than or equal to zero then we can not decrease the total cost more. Hence, the solution under test is optimal.

THEOREM I. *If a basic feasible solution of a transportation problem consist $(m + n - 1)$ independent allocations and if we have a set of arbitrary numbers u_i and v_j, $i = 1, 2, ..., m; j = 1, 2, ..., n$ for occupied cell (r, s) such that*

$$c_{rs} = u_r + v_s$$

Then the cell evaluation d_{ij} corresponding to each unoccupied cell (i, j) is given by

$$d_{ij} = c_{ij} - (u_i + v_j)$$

PROOF. In transportation problem of m rows and n columns we have to determine $x_{ij} \geq 0$, which

$$\text{minimize } Z = \sum_{i=1}^{m} \sum_{j=1}^{n} c_{ij} x_{ij} \qquad \text{...(1)}$$

subject to the restrictions

$$\sum_{j=1}^{n} x_{ij} = a_i \implies 0 = a_i - \sum_{j=1}^{n} x_{ij}, i = 1, 2, ..., m \qquad \text{...(2)}$$

and $$\sum_{i=1}^{m} x_{ij} = b_j \implies 0 = b_j - \sum_{i=1}^{m} x_{ij}, j = 1, 2, ..., n \qquad \text{...(3)}$$

Now, multiplying (2) by u_i and (3) by v_j and then adding to (1), we get

$$Z = \sum_{i=1}^{m} \sum_{j=1}^{n} c_{ij} x_{ij} + \sum_{i=1}^{m} u_i \left(a_i - \sum_{j=1}^{n} x_{ij} \right) + \sum_{j=1}^{n} v_j \left(b_j - \sum_{i=1}^{m} x_{ij} \right)$$

$$\implies Z = \sum_{i=1}^{m} \sum_{j=1}^{n} [c_{ij} - (u_i + v_j)] x_{ij} + \sum_{i=1}^{m} u_i a_i + \sum_{j=1}^{n} v_j b_j \qquad \text{...(4)}$$

Now, for zero coefficient, we must have

$$c_{rs} = u_r + v_s \qquad \text{...(5)}$$

for each occupied cell (r, s)

Since, there are $(m + n - 1)$ occupied cells, so there are $(m + n - 1)$ equation of form (5) in $(m + n)$ unknowns u_i and v_j.

If one of these unknowns is assigned a value arbitrarily then the cost of $(m + n - 1)$ unknowns can be solved algebraically. As any one of u_i or v_j can be assigned so we choose the u_i which has the largest number of allocations in its row and assign it the value of zero. Now, since $c_{ij} = u_i + v_j$, therefore v_j can be obtained for those columns having allocations. Further, suppose that (i, j) is the empty cell. Allocate $+1$ unit to this empty cell, then total number of allocations becomes $(m + n)$ which are independent positions, so a closed loop is formed by joining the empty cell (i, j) to the occupied cells as shown in the following figure.

	D_1	D_2	...	D_j	...	D_s	...	D_n
O_1			...		...		...	
O_2			...		...		...	
$\vdots$			...		...		...	
O_i			...	(+1)	...	(+1)	...	
$\vdots$			...		...		...	
O_r			...	(+1)	...	(+1)	...	
$\vdots$			...		...		...	
O_m			...		...		...	

The loop in the above table can be shown as follows

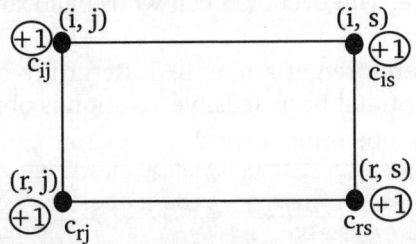

We observe that the cells (i, s), (r, s) and (r, j) are occupied cells. Then we have

$$c_{is} = u_i + v_s; \quad c_{rs} = u_r + v_s; \quad c_{rj} = u_r + v_j$$

The above loop can be formed as follows:

"Allocate $+1$ unit to the empty cell (i, j) then to maintain row and column sum unchanged we allocate -1 unit to the occupied cells (i, s) and (r, j) and $+1$ unit to the occupied cells (r, s).

Now, d_{ij} = cost difference between the new solution and the original solution

$$= c_{ij} - c_{is} + c_{rs} - c_{rj}$$
$$= c_{ij} - (u_i + v_s) + (u_r + v_s) - (u_r + v_j)$$
$$= c_{ij} - (u_i + v_j)$$

Here, d_{ij} gives the cell evaluation connecting empty cell (i, j) to the occupied cells by a square or rectangular shaped loop.

Similarly, we may generalise above process for an arbitrary shaped loop connecting empty cell (i, j) to the occupied cells.

Hence, the cell evaluation d_{ij} for each empty cell (i, j) is given by

$$d_{ij} = c_{ij} - (u_i + v_j)$$

11.4.2 Modified Distribution Method (Modi Method)

After determining the initial basic feasible solution of a transportation problem to check the optimality, we use the following steps:

STEP 1. Find a set of $(m + n)$ numbers u_i for $i = 1, 2, ..., m$ and v_j for $j = 1, 2, ..., n$ such that $c_{rs} = u_r + v_s$ for each occupied cell (r, s).

STEP 2. Calculate cell evaluation d_{ij} by using

$$d_{ij} = c_{ij} - (u_i + v_j)$$

for each empty cell (i, j) and enter at the upper right corner of that cell.

STEP 3.   If

         (i) $d_{ij} > 0$, then the solution under test is optimal and unique.

        (ii) $d_{ij} \geq 0$, then the solution under test is optimal but not unique (an alternate optimal solution exists)

       (iii) $d_{ij} < 0$, then solution is not optimal and further improvement is needed.

STEP 4.   In case at least one $d_{ij} < 0$, then select the empty cell for which the cell evaluation d_{ij} is most negative.

         Allocate an unknown quantity $+\theta$ to the empty cell and construct a closed loop connecting this cell to the other occupied cells. In order to maintain rows and column sums assign $-\theta$ and $+\theta$ altenatively to the occupied cells which are used in forming the loop (at the corner point only).

STEP 5.   Assign the largest possible value to θ in such a way that the value of at least one occupied cell turns to zero and in the other occupied cells, the allocations will remain non-negative. The occupied cell whose allocation becomes zero will leave the basis.

STEP 6.   From step 5, we therefore get a new BFS. Return to step 3 with new BFS, repeat the process till an optimal basic feasible solution is obtained.

    Here following points should be remembered:

(1) Every loop has an even number of cells and at least 4.

(2) Closed loops may or may not be square in shape

(3) The allocations are said to be independent position if it is not possible to increase or decrease any independent individual allocation without changing the positions of these allocations or violating the RIM conditions, a closed loop can not be formed through these allocations.

(4) Each row and column in the transportation table should have only one plus and minus sign. All cells that have a + or − sign, except the starting unoccupied cells, must be occupied cells.

Solved Examples

EXAMPLE 1.   *In the following transportation problem, find the basic feasible solution by North-West corner rule, Matrix Minima Method and by Vogel's Method and test the optimality for each solution giving the cost of transportation for the solution. Cost of transportation in per 1000 ton in units of ₹ 1000.*

Steel mills

		S_1	S_2	S_3	S_4	Availability of coal (in 1000 tons)
	I	14	56	48	27	13
Mines	II	82	35	21	81	19
	III	99	31	71	63	16
Requirement in 1000 tons		7	14	21	6	48

SOLUTION. (i) North-West Corner Rule

Steel mills

		S_1	S_2	S_3	S_4	Availability of coal (in 1000 tons)
	I	⑦ 14	⑥ 56	48	27	13 (– 7 = 6)
Mines	II	82	⑧ 35	⑪ 21	81	19 (– 8 = 11)
	III	99	31	⑩ 71	⑥ 63	16 (– 10 = 6)
Requirement in 1000 tons		7	14 – 6 = 8	21 – 11 = 10	6 – 6 = 0	48

Test for optimality. Find auxiliary number u_i and v_j such that $c_{ij} = u_i + v_j$ for each occupied cell

Cell (1, 1) $c_{11} = u_1 + v_1 = 14$ Let us choose $u_1 = 0$, $v_1 = 14$

Cell (1, 2) $u_1 + v_2 = 56$ $v_2 = 56$

Cell (2, 2) $u_2 + v_2 = 35$ $u_2 = -21$

Cell (2, 3) $u_2 + v_3 = 21$ $v_3 = 42$

Cell (3, 3) $u_3 + v_3 = 71$ $u_3 = 29$

Cell (3, 4) $u_3 + v_4 = 63$ $v_4 = 34$

Calculate Δ_{ij} for every non-occupied cell $\Delta_{ij} = c_{ij} - (u_i + v_j)$

Cell (1, 3) $\Delta_{13} = 48 - (u_1 + v_3) = 48 - (0 + 42) = 6$

Cell (1, 4) $\Delta_{14} = c_{14} - (u_1 + v_4) = 27 - (0 + 34) = -7$

Cell (2, 1) $\Delta_{21} = c_{21} - (u_2 + v_1) = 82 - (-21 + 14) = 89$

Cell (2, 4) $\Delta_{24} = 81 - (u_2 + v_4) = 81 - (-21 + 34) = 68$

Cell (3, 1) $\Delta_{31} = 99 - (29 + 14) = 99 - 43 = 56$

Cell (3, 2) $\Delta_{32} = 31 - (29 + 56) = -54$

Since all Δ_{ij} are not positive so the current solution is not optimal.

The cost of assignment

$$= 14 \times 7 + 56 \times 6 + 35 \times 8 + 21 \times 11 + 71 \times 10 + 63 \times 6$$
$$= 98 + 336 + 280 + 231 + 710 + 378 = 2033 \ \text{(IBFS)}$$

(ii) Matrix-Minima Method

Steel mills

		S_1	S_2	S_3	S_4	Availability
	I	⑦ 14	56	48	⑥ 27	13
Mines	II	82	35	⑲ 21	81	19
	III	99	⑭ 31	② 71	63	16
Demand		7	14	21 – 19 = 2	6 – 6 = 0	48

The initial solution is degenerate. Since only five cells are occupied which is less than $(3 + 4 - 1) = 6$. So to make the number of occupied cells as 6, assign a quantity $\varepsilon \to 0$ to the remaining cell of minimum cost so that loop is not formed. Assign a quantity ε in cell (1, 3) to make the number of occupied cell 6.

Test for optimality. Find auxiliary number u_i and v_j for occupied cells such that $c_{ij} = u_i + v_j$

Cell (1, 1) $\qquad u_1 + v_1 = 14$ Let us choose $v_1 = 0$, $u_1 = 14$

Cell (1, 3) $\qquad u_1 + v_3 = 48 \qquad\qquad v_3 = 34$

Cell (1, 4) $\qquad u_1 + v_4 = 27 \qquad\qquad v_4 = 13$

Cell (2, 3) $\qquad u_2 + v_3 = 21 \qquad\qquad u_2 = -13$

Cell (3, 2) $\qquad u_3 + v_2 = 31 \qquad\qquad v_2 = -6$

Cell (3, 3) $\qquad u_3 + v_3 = 71 \qquad\qquad u_3 = 37$

Now, for each non-occupied cell calculate $\qquad \Delta_{ij} = c_{ij} - (u_i + v_j)$

Cell (1, 2) $\quad \Delta_{12} = 56 - (14 - 6) = 48$

Cell (2, 1) $\quad \Delta_{21} = 82 - (-13 + 0) = 95$

Cell (2, 2) $\quad \Delta_{22} = 35 - (u_2 + v_2) = 35 - (-13 - 6) = 35 + 19 = 54$

Cell (2, 4) $\quad \Delta_{24} = 81 - (-13 + 13) = 81 - 0 = 81$

Cell (3, 1) $\quad \Delta_{31} = 99 - (37 + 14) = 99 - 51 = 48$

Cell (3, 4) $\quad \Delta_{34} = 63 - (37 + 13) = 63 - 50 = 13$

Since all $\Delta_{ij} \geq 0$, therefore current solution is optimal.

Minimum cost of assignment per 1000 tons (in units of ₹ 1000)

$$= 14 \times 7 + 27 \times 6 + 21 \times 19 + 31 \times 14 + 71 \times 2$$
$$= 98 + 162 + 399 + 434 + 142 = 1235$$

(iii) Vogel's method

Steel mills

Mines		S_1	S_2	S_3	S_4	Availability
	I	⑦ 14	56	48	⑥ 27	$13 - 7 = 6$
	II	82	35	⑲ 21	81	19
	III	99	⑭ 31	② 71	63	16
Demand		7	14	21	6	**48**
Difference		68	4	27	36	

In this case, test for optimality is same as in matrix minima method. So current solution is optimal.

Minimum cost of assignment (in units of ₹ 1000)

$$= 14 \times 7 + 27 \times 6 + 21 \times 19 + 31 \times 14 + 71 \times 2 = 1235$$

EXAMPLE 2. *Find the optimal basic feasible solution to the following transportation problem*

			Available
50	30	220	1
90	45	170	3
250	200	50	4

Requirement   4   2   2 $\boxed{8}$ ← Total

SOLUTION. Applying lowest cost entry method, we can obtain the initial basic feasible solution as follows

	1 (30)		1
2 (90)	1 (45)		3
2 (250)		2 (50)	4
4	2	2	

Here, the cost is given by

Cost $= 1 \times 3 + 2 \times 90 + 1 \times 45 + 2 \times 250 = 855$

To check the optimality of this solution, we proceed as example (1) and obtain the following table

			u_i
	(30)		−15
(90)	(45)		0
(250)		(50)	160

v_j 90   45   −110

(Matrix for set of u_i and v_j)

50	•	(220)
•	•	(170)
•	(200)	•

(Matrix c_{ij} for empty cells)

			u_i
75	•	−125	−15
•	•	−110	0
•	205	•	160

v_j 90   45   −110

(Matrix $(u_i + v_j)$ for empty cells)

−25	•	345
•	•	280
•	−5	•

(Matrix $c_{ij} - (u_i + u_j)$ for empty cells)

Here, all cell evaluation $\Delta_{ij} = c_{ij} - (u_i + v_j)$ are non-negative. Therefore, solution is not optimal.

Here, the most negative cell evaluation is $\Delta_{11} = -25$. Therefore, assume θ allocation in this cell and proceed as follows

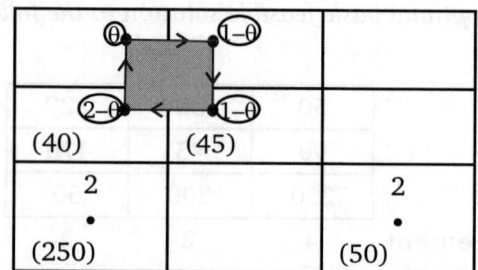

Here $\min\{1-\theta, 2-\theta\} = 0$

$\Rightarrow \quad \theta = 1$.

Therefore, the improved solution given by

1 (50)			1
1 (90)	2 (45)		3
2 (250)		2 (50)	4
4	2	2	

Now, we check the optimality of this improved solution as follows

			u_i
(50)			50
(90)	(45)		90
(250)		(50)	250
v_j $\quad$ 0	-45	-200	

(Matrix for set of u_i and v_j)

•	(30)	(220)
•	•	(170)
•	(200)	•

(Matrix c_{ij} for empty cells)

			u_i
•	5	-150	50
•	•	-110	90
•	205	•	250
v_j $\quad$ 0	-45	-200	

(Matrix for set of u_i and v_j)

•	25	370
•	•	280
•	-5	•

(Matrix $c_{ij} - (u_i + u_j)$ for empty cells)

Here, the cell evaluation $\Delta_{32} = -5$, therefore, this solution is not optimal.

Again, to find the optimality, proceeding same as above we get the following table.

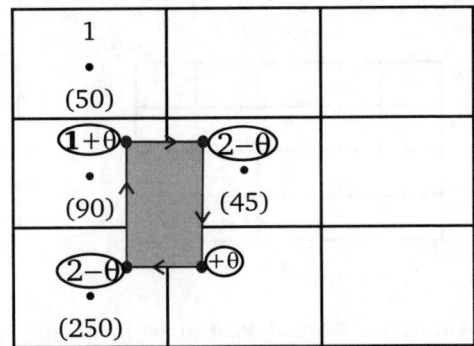

Here, $\theta = 2$. Therefore, the next improved solution is given by

1 (50)			1
3 (90)	0 (45)		3
	2 (200)	2 (50)	4
4	2	2	

By proceeding as above, we can easily see that all cell evaluation are non-negative. Hence, the solution is optimal and is given by
$$z = 1 \times 50 + 3 \times 90 + 0 \times 45 + 2 \times 200 + 2 \times 50 = 820$$

EXAMPLE 3. *The cost requirement table for the transportation problem is given as below :*

	ω_1	ω_2	ω_3	ω_4	ω_5	**Available**
F_1	4	3	1	2	6	40
F_2	5	2	3	4	5	30
F_3	3	5	6	3	2	20
F_4	2	4	4	5	3	10
Required	30	30	15	20	5	100

SOLUTION. By 'North-West Corner rule', the non-degenerate initial solution is given by following table

	ω_1	ω_2	ω_3	ω_4	ω_5	**Available**
F_1	(30) 4	10 3				40
F_2		(20) 2	(10) 3			30
F_3			(5) 6	(15) 3		20
F_4				(5) 5	(5) 3	10
Required	30	30	15	20	5	

Now, test this solution for optimality, we get the following table

					u_i
4	3				0
	2	3			-1
		6	3		2
			5	3	4

v_j 4   3   4   1   -1

(Matrix for set of u_i and u_j)

•	•	(1)	(2)	(6)
(5)	•	•	(4)	(5)
(3)	(5)	•	•	
(2)	(4)	(4)	•	•

(Matrix $[c_{ij}]$ for empty cells)

•	•	-3	1	7
2	•	•	4	7
-3	0	•	•	1
-6	-3	-4	•	•

Matrix $[c_{ij} - (u_i + u_j)]$ for empty cells

Since the largest negative cell evaluation is $\Delta_{41} = -6$, allocate as much as possible to this cell (4, 1). This necessitates shifting of 5 units to this cell (4, 1) as directed by the closed loop in next table.

					Available
(30 − θ) (4)	(10 + θ) (3)				40
	2 (20−θ)	(10+θ) (3)			30
		(5 − θ) (6)	(5 + θ) (3)		20
(θ)			(5 − θ) (5)	5 •	10

Required   30   30   15   20   5

Here maximum possible value of θ is obtained by

$$\min \{30 - \theta, 20 - \theta, 5 - \theta, 5 - \theta\} = 0,$$

i.e., $\theta = 5$ units.

Now, the revised solution is

					Availability
(25) 4	(15) 3				40
	(15) 2	(15) 3			30
		(0) 6	(20) 3		20
			(0) 5	(5) 3	10

Requirement   30   30   15   20   5

In this solution, the number of allocations becomes less than $m+n-1$ on account of simultaneous allocation of two cells $[(3, 3), (4, 4)]$. Hence, this is a degenerate solution. Now, this degeneracy may resolve by adding Δ to one of the recently vacated cells $[(3, 3)$ or $(4, 4)]$. But in minimization problem, add Δ to recently vacated $(4, 4)$ only, because it has the lowest shipping cost of ₹ 5 per unit.

The rest of procedure will be exactly the same as explained earlier. This way, the optimum solution can be obtained.

					Available
⑤ 4		⑮ 1	⑳ 2		40
	㉚ 2				30
⑮ 3				⑤ 2	20
	⑩ 2				10
Required 30	30	15	20	5	

EXAMPLE 4. *Solve the following transportation problem*

			Supply
2	7	4	5
3	3	1	8
5	4	7	7
1	6	2	14
Requirement 7	9	18	34

SOLUTION. By using Vogel's approximation method, we obtained the following initial basic feasible solution

⑤ (2)		
		⑧ (1)
	⑦ (4)	
② (1)	② (6)	⑩ (2)

Now, we check the optimality of this solution in the following manner

			u_i
(2)			1
		(1)	−1
	(4)		−2
(1)	(6)	(2)	0

v_j : 1, 6, 2

•	(7)	(4)
(3)	(3)	•
(5)	•	(7)
•	•	•

(Matrix for set of u_i and v_j)   (Matrix u_j for empty cells)

•	7	2
0	5	•
−1	•	0
•	•	•

(Matrix $(u_i + v_j)$ for empty cells)

•	0	1
3	−2	•
6	•	7
•	•	•

(Matrix $c_{ij} - (u_i + u_j)$ for empty cells)

Here $\Delta_{22} = -2$ is most negative value.

Hence, we can improve the solution as follows

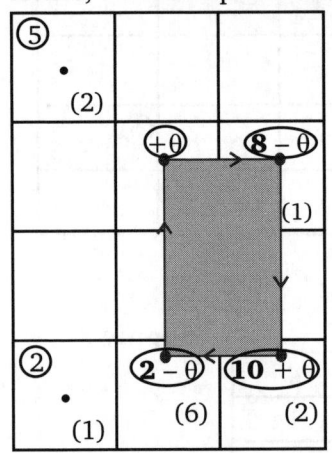 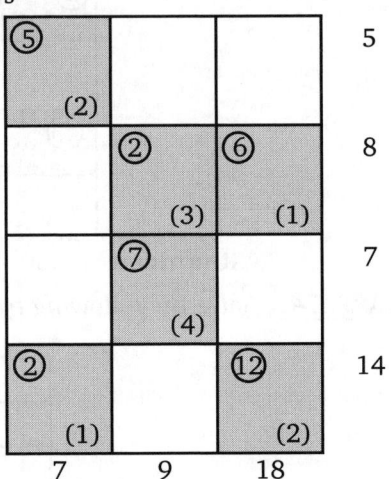

Here $\min\{8 - \theta, 2 - \theta\} = 0 \quad \Rightarrow \quad \theta = 2$

Proceeding as above, we get the following matrices :

			u_i
(2)			1
	(3)	(1)	−1
	(4)		0
(1)		(2)	0
v_j 1	4	2	

(Matrix for set of u_i and v_j)

•	(7)	(4)
(3)	•	•
(5)	•	(7)
•	(6)	•

(Matrix c_{ij} for empty cells)

•	5	3
0	•	•
1	•	2
•	4	•

(Matrix $(u_i + v_j)$ for empty cells)

•	2	1
3	•	•
4	•	5
•	2	•

(Matrix $c_{ij} - (u_i + u_j)$ for empty cells)

Since in the final table, all Δ_{ij}'s are positive, therefore, solution is optimal.

The optimal solution (minimum cost) is given by

Minimum cost = $5(2) + 2(3) + 6(1) + 7(4) + 2(1) + 12(2) = 76$.

EXAMPLE 5. *Solve the following transportation problem*

Villages

	T_1	T_2	T_3	Capacity
B_1	8	6	5	150
B_2	6	6	6	150
B_3	10	8	4	150
B_4	8	6	4	150
Demand	200	200	200	600

Air bases : B_1, B_2, B_3, B_4

SOLUTION. **Vogel's method:**

	T_1	T_2	T_3	Capacity	Diff.
B_1	(50) −8	(50) −6	(50) −5	150	1
B_2	−6	−6	(150) −6	150	0
B_3	(150) −10	−8	−4	150	2
B_4	−8	(150) −6	−4	150	2
Demand	200	200	200	600	
Diff.	2	2	1		

Test for optimality. There are $6 = (4 + 3 - 1)$ Calculate u_i and v_j for each occupied cell such that $c_{ij} = u_i + v_j$

Cell (1, 1) $u_1 + v_1 = -8$ Let us choose $u_1 = 0$, $v_1 = -8$
Cell (1, 2) $u_1 + v_2 = -6$ $v_2 = -6$
Cell (2, 3) $u_2 + v_3 = -6$ $u_2 = -1$
Cell (3, 1) $u_3 + v_1 = -10$ $u_3 = -2$
Cell (4, 2) $u_4 + v_2 = -6$ $u_4 = 0$

Now, compute $\Delta_{ij} = c_{ij} - (u_i + v_j)$ for each vacant cell

Cell (2, 1) $\Delta_{21} = -6 - (-1 - 8) = 3$
Cell (2, 2) $\Delta_{22} = -6 - (-1 - 6) = 1$
Cell (3, 2) $\Delta_{32} = -8 - (-2 - 6) = 0$
Cell (3, 3) $\Delta_{33} = -4 - (-2 - 5) = 3$
Cell (4, 1) $\Delta_{41} = -8 - (0 - 8) = 0$
Cell (4, 3) $\Delta_{43} = -4 - (0 - 5) = 1$

Since all $\Delta_{ij} \geq 0$, therefore current assignment is optimal. The assignments are

Cell (1, 1) = 50   Cell(2, 2) = 50   Cell (1, 3) = 50
Cell (2, 3) = 150   Cell(3, 1) = 150   Cell (4, 2) = 150

∴ Maximum cost of assignment

$= 8 \times 50 + 6 \times 50 + 5 \times 150 + 10 \times 150 + 6 \times 150 = 4250$

11.5 DEGENERACY IN TRANSPORTATION PROBLEMS

The solution procedure for non-degenerate basic feasible solution with exactly $m+n-1$ strictly positive allocations in independent positions has been discussed so far. However, sometimes it is not possible to get such initial feasible solution to start with. Thus degeneracy occurs in the transportation problem whenever a number of occupied cells is less than $m+n-1$.

Basic feasible solution to an m-origin and n-destination transportation problem can have at most $m+n-1$ number of positive (non-zero) basic variables. If this number is exactly $m+n-1$, the BFS is said to be non-degenerate; and if less than $m+n-1$ the basic solution degenerates. It follows that whenever the number of basic cells is less than $m+n-1$, the transportation problem is a degenerate one.

Degeneracy in transportation problems can occur in two ways:

1. Basic feasible solutions may be degenerate from the initial stage onward.
2. They may become degenerate at any intermediate stage

11.5.1 RESOLUTION OF DEGENERACY

(1) Among the unoccupied cells, select one occupied cell at independent position having the least cost. If such cells are more than one, then select anyone arbitrarily.

(2) The transportation problem may also become degenerate during the solution stages. This happens when most favourable quantity is allocated to the empty cell having the largest negative cell-evaluation resulting in simultaneous vacation of two or more of currently occupied cells. To resolve degeneracy, allocate Δ to one or more of recently vacated cells so that the number of occupied cells is $m+n-1$ in the new solution.

 Solved Examples

EXAMPLE 1. *Find the optimal solution of the following transportation problem.*

			Available
50	30	220	1
90	45	170	3
250	200	50	4
Required 4	2	2	**8**

SOLUTION. Applying the Vogel's approximation method in a usual manner, the IBFS is given by

①		
50	30	220
③		
90	45	170
	②	②
250	200	50

We observe that, the total no. of allocation = 4 which is less than $m + n - 1 = 5$.

⇒ solution is degenerate.

To resolve this degeneracy, we allocate a very small amount ε to some suitable cell. Here, we allocate ε to the cell (1, 2) getting 5 allocations at independent positions.

①	ε	
50	30	220
③		
90	45	170
	②	②
250	200	50

Now, we check the optimality as given below.

			u_i
①	ε	−120 +340	−170
50	30	220	
③	70 −25	−80 +250	−130
90	45	170	
220 **30**	②	②	0
250	200	50	

v_j 220   200   50

Clearly, $d_{22} = -25 < 0 \Rightarrow$ solution is not optimal.

Therefore, we take ε from cell (1, 2) to the cell (2, 2) and form the new table as given below

			u_i
①	25 **+5**	−145 +365	−195
50	30	220	
③	ε	−105 +275	−155
90	45	170	
245 **+5**	②	②	0
250	200	50	

v_j 245   200   50

We observe that for all empty cells, all $d_{ij} > 0$

$\Rightarrow$ solution is optimal.

and is given by

$x_{11} = 1, x_{21} = 3, x_{32} = 2, x_{33} = 2$

and minimum cost $= 50 \times 1 + 90 \times 3 + 200 \times 2 + 50 \times 2 = 820$

EXAMPLE 2. *Find the optimal basic feasible solution to the following transportation problem.*

				Capacity
1	2	3	4	6
4	3	2	0	8
0	2	2	1	10
Demand 4	6	8	6	24

SOLUTION. Apply Vogel's approximation method, an IBFS is given in the following table

	⑥		
1	2	3	4
		②	⑥
4	3	2	0
④		⑥	
0	2	2	1

Clearly, total no. of allocations = 5, which is less than $m + n - 1 = 6$.

$\Rightarrow$ solution is degenerate

To resolve this degeneracy, we allocate a very small quantity ε to the cell (1, 1) as follows

								u_i
⑤		⑥		3	**0**	1	**+3**	0
	1		2		3		4	
0	**+4**	1	**+2**	②		⑥		−1
	4		3		2		0	
④		1	**+1**	⑥		0	**+1**	−1
	0		2		2		1	
v_j	1		2		3		1	

We observe that all $d_{ij} \geq 0$ for each empty cell.

$\Rightarrow$ solution is optimal.

Hence, the optimal solution is given by

$$x_{12} = 6, x_{23} = 2, x_{24} = 6, x_{31} = 4, x_{33} = 6$$

and the minimum transportation cost is given by

$$= 6 \times 2 + 2 \times 2 + 6 \times 0 + 4 \times 0 + 6 \times 2 = ₹\ 28$$

EXAMPLE 3. *Solve the following transportation problem*

			Available
7	4	0	5
6	8	0	15
3	9	0	9
Requirement 15	6	8	29

SOLUTION. Applying north-west corner rule, the IBFS is given by

⑤		
7	4	0
⑩	⑤	
6	8	0
	①	⑧
3	9	0

Now, to check the optimality, we proceed as follows.

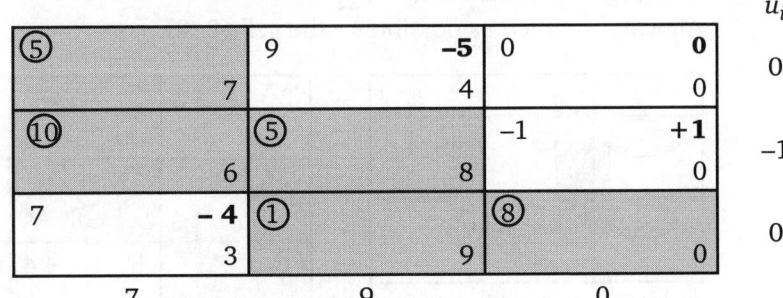

$\because$ all d_{ij} not $\geq 0 \Rightarrow$ solution is not optimal.

Now, the largest negative cell evaluation is $d_{12} = -5$

$\therefore$

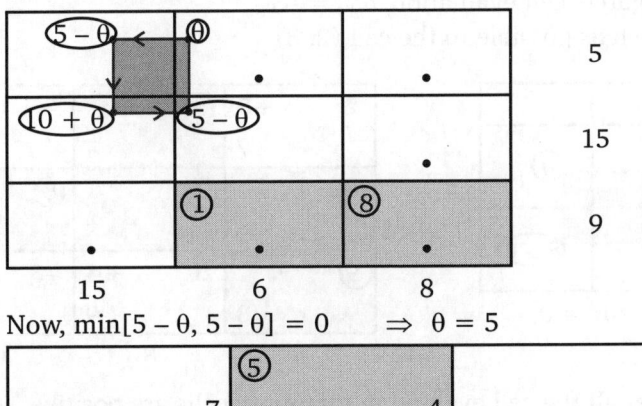

Now, min$[5 - \theta, 5 - \theta] = 0 \quad \Rightarrow \theta = 5$

Here, two cell vacate, therefore, no. of allocations becomes less than $m + n - 1$ (= 5)

$\Rightarrow$ There is a degenerate solution.

Introduce a very small quantity say ε to the cell (3, 1), although least cost independent cell is (2, 3).

Clearly, all cell evaluation are not $\geq 0 \Rightarrow$ solution is not optimal

The largest negative cell $= d_{22} = -4$

Allocate as much as possible to the cell (2, 2).

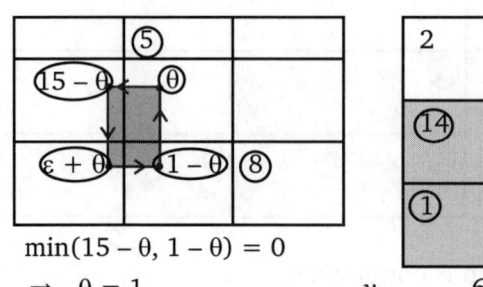 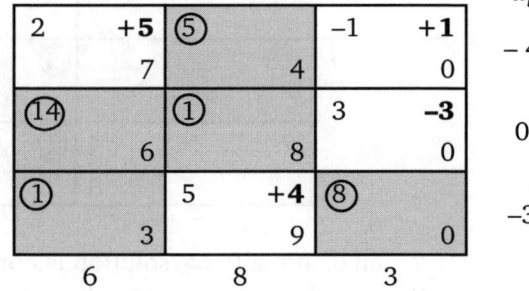

$$\min(15 - \theta, 1 - \theta) = 0$$
$$\Rightarrow \quad \theta = 1$$

Clearly all d_{ij} are not ≥ 0.

The largest negative cell evaluation, $d_{23} = -3$.

Allocate as much as possible to the cell (2, 3)

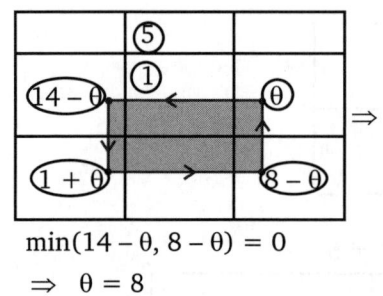 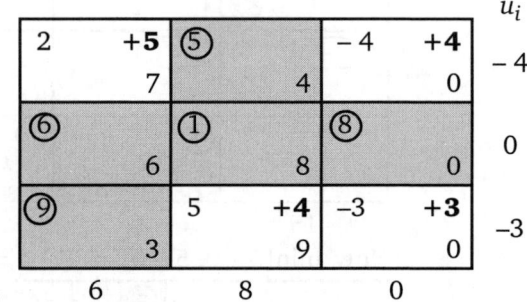

$$\min(14 - \theta, 8 - \theta) = 0$$
$$\Rightarrow \quad \theta = 8$$

We observe that all the cell evaluation for empty cells are positive.

$\Rightarrow$ solution is optimal

and is given by

$$x_{12} = 5, x_{21} = 6, x_{22} = 1, x_{23} = 8, x_{31} = 9$$

and minimum transportation cost

$$= 4 \times 5 + 6 \times 6 + 1 \times 8 + 8 \times 0 + 3 \times 9$$
$$= ₹91$$

EXAMPLE 4. *Find the optimal solution of the following problem.*

				Requirement
8	10	7	6	50
12	9	4	7	40
9	11	10	8	30
Availability 25	32	40	23	**120**

SOLUTION. Using Vogel's approximation method in a usual manner, we have the following IBFS

㉕ 8	② 10	7	㉓ 6
12	9	㊵ 4	7
9	㉚ 11	10	8

Now, we have to check the optimality of this solution.

Here, we observe that, there are total 5 allocations which is less than $m+n-1 = 6$

$\Rightarrow$ solution is degenerate

To resolve this degeneracy, allocate a small quantity ε to the cell $(2, 4)$.

$u_i\downarrow$

				$u_i\downarrow$
㉕ 8	② 10	3 **+4** 7	㉓ 6	0
9 **+3** 12	11 **−2** 9	㊵ 4	ⓔ 7	1
9 **0** 9	㉚ 11	4 **+6** 10	7 **+1** 8	1

$v_j\rightarrow$ 8   10   3   6

Clearly, $d_{22} = -2 < 0$

$\Rightarrow$ solution is not optimal

Therefore, taking ε from cell $(2, 4)$ to $(2, 2)$, we form the new table as given below.

				$u_i\downarrow$
㉕ 8	② 10	5 **+2** 7	㉓ 6	0
7 **+5** 12	ⓔ 9	㊵ 4	5 **+2** 7	−1
9 **0** 9	㉚ 11	6 **+4** 10	7 **+1** 8	1

$v_j\rightarrow$ 8   10   5   6

Here, all $d_{ij} \geq 0$ $\Rightarrow$ solution is optimal and is given by

$$x_{11} = 25, \ x_{12} = 2, \ x_{14} = 23, \ x_{23} = 40, \ x_{32} = 30$$

and transportation cost

$$= 25 \times 8 + 2 \times 10 + 23 \times 6 + 40 \times 4 + 30 \times 11$$
$$= ₹\,848$$

EXAMPLE 5. *Solve the following transportation problem*

				Available
5	3	6	5	15
10	7	12	4	11
7	5	8	4	13
Demand 8	12	13	6	**39**

SOLUTION. Apply Vogel's approximation method, the IBFS is given by

⑧	⑦		
5	3	6	5
	⑤		⑥
10	7	12	4
		⑬	
7	5	8	4

Clearly, there are 5 no. of allocations which is less than $m + n - 1 = 6$
Introduce a small allocation ε as follows

$u_i\downarrow$

⑧	⑦	4 **+2**	0 **+5**	0
5	3	6	5	
9 **+1**	⑤	8 **+4**	⑥	4
10	7	12	4	
9 **−2**	7 **−2**	⑬	⑤	4
7	5	8	4	

$v_j\rightarrow$ 5   3   4   0

Clearly, all d_{ij} not $\geq 0 \Rightarrow$ solution is a degenerate solution.
Here, $d_{ij} = -2 < 0$ in both the cells, so we take ε in one of these two cells say in (3, 1) and again test the optimality we have the following table.

$u_i\downarrow$

⑧	⑦	6 **0**	0 **+5**	0
5	3	6	5	
9 **+1**	⑤	10 **+2**	⑥	4
10	7	12	4	
⑤	5 **0**	⑬	2 **+2**	2
7	5	8	4	

$v_j\rightarrow$ 5   3   6   0

In the above table, we observe that all $d_{ij} \geq 0 \Rightarrow$ solution is optimal and is given
by $x_{11} = 8, x_{12} = 7, x_{22} = 5, x_{24} = 6, x_{33} = 13$
and the minimum transportation cost
$$= 8 \times 5 + 7 \times 3 + 5 \times 7 + 6 \times 4 + 13 \times 8 = ₹\ 224$$

11.6 UNBALANCED TRANSPORTATION PROBLEM

A transportation problem is said to be unbalanced if

$$\sum_{i=1}^{m} a_i \neq \sum_{j=1}^{n} b_j$$

An unbalanced problem having the following two forms.

(i) **When** $\displaystyle\sum_{i=1}^{m} a_i < \sum_{j=1}^{n} b_j$

In such type of problem, we introduce a dummy source (row) in the cost matrix with zero cost and the excess demand over the supply is entered for this dummy row.

(ii) **When** $\displaystyle\sum_{i=1}^{m} a_i > \sum_{j=1}^{n} b_j$

In such type of problem, we introduce a dummy demand (column) in the cost matrix with zero cost and the excess supply over the demand is entered for this dummy column.

 Solved Examples

EXAMPLE 1. *Solve the following transportation problem:*

				Supply
6	1	9	3	70
11	5	2	8	55
10	12	4	7	70
Demand 85	35	50	45	

SOLUTION. Total supply = 70 + 55 + 70 = 195

Total demand = 85 + 35 + 50 + 45 = 215

$\Rightarrow \quad \Sigma a_i < \Sigma b_j$

To convert this unbalanced problem into balanced, introduce a dummy row with zero cost and the excess demand 20 (215 – 195) is entered for this dummy row. Then we have

6	1	9	3	70
11	5	2	8	55
10	12	4	7	70
0	0	0	0	20
85	35	50	45	215

Now, apply, Vogel's approximation method, the IBFS is given in the following table

(65) 6	(5) 1	9	3
11	(30) 5	(25) 2	8
10	12	(25) 4	(45) 7
(20) 0	0	0	0

The IBFS is $= 65 \times 6 + 5 \times 1 + 30 \times 5 + 25 \times 2 + 25 \times 4 + 45 \times 7 + 20 \times 0 = 1010$

Now, we have to check the optimality of this solution

Clearly, the occupied cell $= (1, 1), (1, 2), (2, 2), (2, 3), (3, 3), (3, 4)$ and $(4, 1)$

Taking $u_1 = 0$ we get

$$c_{11} = u_1 + v_1 \Rightarrow 6 = 0 + v_1 \Rightarrow v_1 = 6$$

Similarly, $v_2 = 1, u_2 = 4, v_3 = -2, u_3 = 6, v_4 = 1, u_4 = -6$

Further, we have to find d_{ij} for each unoccupied cell (i, j) by using $d_{ij} = c_{ij} - (u_i + v_j)$.

Here, the unoccupied cells are $(1, 3), (1, 4), (2, 1), (2, 4), (3, 1), (3, 2), (4, 2),$ $(4, 3)$ and $(4, 4)$. Then

$$d_{13} = c_{13} - (u_1 + v_3) = 9 - (0 - 2) = 9 + 2 = 11$$
$$d_{14} = c_{14} - (u_1 + v_4) = 3 - (0 + 1) = 3 - 1 = 2$$

Similarly, $d_{21} = 1, d_{24} = 3, d_{31} = -2, d_{32} = 5, d_{42} = 5, d_{43} = 8, d_{44} = 5$

Clearly, all d_{ij} are not $\geq 0 \Rightarrow$ solution is not optimal.

Since, $d_{31} = -2$, then allocate an unknown positive quantity θ to the cell $(3, 1)$ and construct a closed loop connecting the cell $(3, 1)$ and occupied cells $(1, 1),$ $(1, 2), (2, 2), (2, 3)$ and $(3, 3)$. So to maintain row and column sum, we assign $-\theta$ to $(1, 1) + \theta$ to $(1, 2), -\theta$ to $(2, 2) + \theta$ to $(2, 3)$ and $-\theta$ to $(3, 3)$. This loop is shown as below.

65 − θ ←	5 + θ		
	30 − θ ←	25 + θ	
θ →		25 − θ	

Now, min $(65 - \theta, 25 - \theta) = 0 \Rightarrow \theta = 25$. Therefore, revised table is given by

40 6	30 1	9	3
11	5 5	50 2	8
25 10	12	4	45 7
20 0	0	0	0

Again we check the optimality of this solution

				u_i
40 6	30 1	−2 +11 9	3 0 3	6
10 +1 11	5 5	50 2	7 +1 8	10
25 10	5 +7 12	2 4	45 7	10
20 0	−5 +5 0	−8 +8 0	−3 +3 0	0
v_j 0	−5	−8	−3	

Clearly, all $d_{ij} \geq 0 \Rightarrow$ solution is optimal and is given by

$$x_{11} = 40, x_{12} = 30, x_{21} = 5, x_{22} = 50, x_{31} = 25, x_{34} = 45, x_{41} = 20$$

and the minimum transportation cost is given by

$$= 40 \times 6 + 30 \times 1 + 5 \times 5 + 50 \times 2 + 25 \times 10 + 45 \times 7 + 20 \times 0$$
$$= ₹\ 960$$

EXAMPLE 2. *Solve the following transportation problem*

	M_1	M_2	M_3	M_4	M_5	Available
F_1	4	2	3	2	6	8
F_2	5	4	5	2	1	12
F_3	6	5	4	7	3	14
Required	4	4	6	8	8	

SOLUTION. Clearly,

Requirement $= 4 + 4 + 6 + 8 + 8 = 30$

and Availability $= 8 + 12 + 14 = 34$

$\Rightarrow$ total requirement < total availability

To convert this unbalanced problem to balanced problem, we introduce a fictious M_6 with zero cost and essential requirement $= 34 - 30 = 4$

4	2	3	2	6	0	8
5	4	5	2	1	0	12
6	5	4	7	3	0	14
4	4	6	8	8	4	34

Now, apply Vogel's approximation method, we have the IBFS given in the following table.

$④$	$④$				
4	2	3	2	6	0
			$⑧$	$④$	
5	4	5	2	1	0
		$⑥$		$④$	$④$
6	5	4	7	3	0

Now, we have to check the optimality of the above solution. Since, there are total no. of allocations = 7 which is less than $m + n - 1$ (= 8). So, introduce a small allocation ε in the cell (1, 4).

						u_i
$④$ 4	$④$ 2	2 +1 / 3	$ⓔ$ 2	1 +5 / 6	-2 +2 / 0	-2
4 +1 / 5	2 +2 / 4	2 +3 / 5	$⑧$ 2	$④$ 1	-2 +2 / 0	-2
6 0 / 6	4 +1 / 5	$⑥$ 4	4 +3 / 7	$④$ 3	$④$ 0	0
v_j 6	4	4	4	3	0	

Here, we observe that all $d_{ij} \geq 0$

$\Rightarrow$ solution is optimal and is given by
$$x_{11} = 4, x_{12} = 4, x_{24} = 8, x_{25} = 4, x_{33} = 6, x_{35} = 4$$
and minimum cost is given by
$$= 4 \times 4 + 4 \times 2 + 8 \times 2 + 4 \times 1 + 6 \times 4 + 4 \times 3 = 80$$

11.7 SOME MISCELLANEOUS SOLVED PROBLEMS

EXAMPLE 1. *Solve the transportation problem as*

		Plant				
		1	**2**	**3**	**4**	**Demand**
Warehouse	**1**	6	−2	3	9	80
	2	6	−2	2	6	110
	3	1	−4	2	4	150
	4	−1	1	−3	8	100
	5	2	−1	−1	3	150
Capacity		150	200	175	100	

SOLUTION. We have

		Plant				
		1	**2**	**3**	**4**	**Demand**
Warehouse	**1**	6	−2	3	9	80
	2	6	−2	2	6	110
	3	1	−4	2	4	150
	4	−1	1	−3	8	100
	5	2	−1	−1	3	150
Capacity		150	200	175	100	

To make the problem balance, we take a fictitious warehouse 6 with demand 35 and zero transportation cost throughout changing this maximization problem into minimization problem by making each entry negative.

Iteration-1

To find auxiliary number u_i and v_j for occupied cells

Cell	Equation	Value
Cell (1, 4)	$u_1 + v_4 = -9$	Let $v_2 = 0$, $u_1 = -9$
Cell (2, 1)	$u_2 + v_1 = -6$	$u_2 = -4$
Cell (3, 3)	$u_3 + v_3 = -2$	$u_3 = -2$
Cell (4, 2)	$u_4 + v_2 = -1$	$u_4 = -1$
Cell (4, 4)	$u_4 + v_4 = -8$	$v_4 = -7$
Cell (5, 1)	$u_5 + v_1 = -2$	$v_1 = -2 - 1 = -3$
Cell (5, 2)	$u_5 + v_2 = 1$	$u_5 = 1$
Cell (6, 2)	$u_6 + v_2 = 0$	$v_6 = 0$
Cell (6, 3)	$u_6 + v_3 = 0$	$v_3 = 0$

Since Δ_{ij} is not all positive so present b.f.s. is not optimal. So assign θ to cell (1, 1). Now, cell (5, 1) leaves the basis. Now $\theta = 40$.

Iteration-2

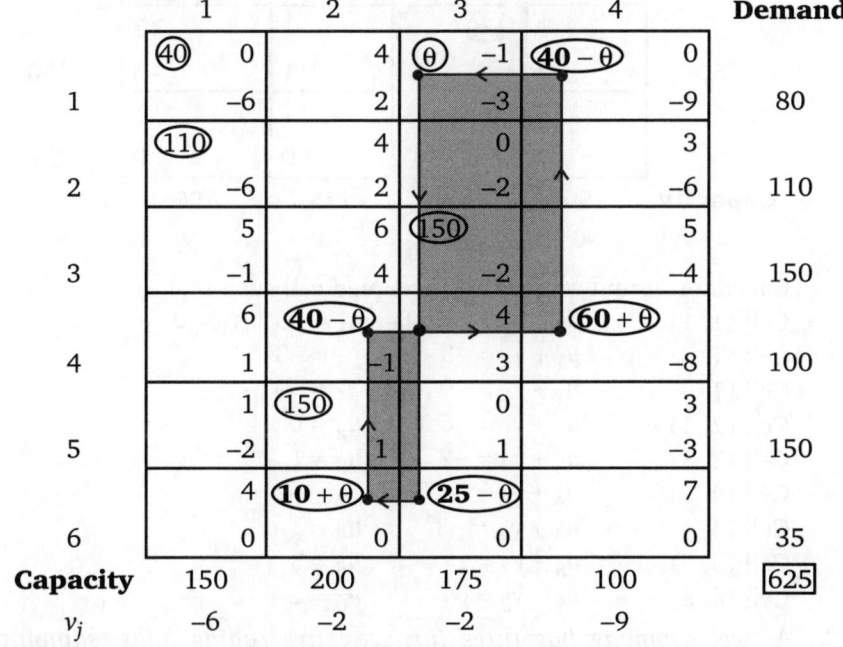

Calculate u_i and v_j for each occupied cell such that $c_{ij} = u_i + v_j$

Cell	Equation	Value
Cell (1, 1)	$u_1 + v_1 = -6$	Let $u_1 = -6$, $v_1 = 0$
Cell (1, 4)	$u_1 + v_4 = -9$	$v_4 = -9$
Cell (2, 1)	$u_2 + v_1 = -6$	$u_2 = 0$
Cell (3, 3)	$u_3 + v_3 = -2$	$v_3 = 0$
Cell (4, 2)	$u_4 + v_2 = -1$	$v_2 = -2$
Cell (4, 4)	$u_4 + v_4 = -8$	$u_4 = 1$
Cell (5, 2)	$u_5 + v_2 = 1$	$u_5 = 3$
Cell (6, 2)	$u_6 + v_2 = 0$	$u_6 = 2$
Cell (6, 3)	$u_6 + v_3 = 0$	$v_3 = -2$

Since all Δ_{ij} are not positive, assign a quantity θ in the cell of most negative Δ_{ij}, i.e., cell (1, 3). Now, $\theta = 25$. Cell (6, 3) leaves.

Iteration-3

Plant

	1	2	3	4	Demand	u_i
1	(40) −6	4 2	(25) −3	15 −9	80	0
2	(110) −6	4 2	1 −2	3 −6	110	0
3	4 −1	5 4	(150) −2	4 −4	150	1
4	4 1	(15) −1	5 3	(85) −8	100	1
5	1 −2	(150) 1	1 1	3 −3	150	3
6	4 0	(35) 0	1 0	1 0	35	2
Capacity	150	200	175	100		
v_j	−6	−2	−3	−9		

Calculate u_i and v_j for each occupied cell such that $c_{ij} = u_i + v_j$

Cell (1, 1) $u_1 + v_1 = -6$ Let $u_1 = -6$, $v_1 = 0$
Cell (1, 3) $u_1 + v_3 = -3$ $v_3 = -3$
Cell (1, 4) $u_1 + v_4 = -9$ $v_4 = -9$
Cell (2, 1) $u_2 + v_1 = -6$ $u_2 = 0$
Cell (3, 3) $u_3 + v_3 = -2$ $u_3 = 1$
Cell (4, 2) $u_4 + v_2 = -1$ $v_2 = -2$
Cell (4, 4) $u_4 + v_4 = -8$ $u_4 = 1$
Cell (5, 2) $u_5 + v_2 = 1$ $u_5 = 3$
Cell (6, 2) $u_6 + v_2 = 0$ $u_6 = 2$

EXAMPLE 2.   *A steel company has three furnaces, five rolling mills, shipping cost of steel from furnace to the mills is shown in the following table. Obtain the optimal transportation schedule.*

Plant

	M_1	M_2	M_3	M_4	M_5	Capacity in 100 quintal
F_1	4	2	3	2	6	8
F_2	5	4	5	2	1	12
F_3	6	5	4	7	3	14
Demand in units of 100 quintal	4	4	6	8	8	34

SOLUTION. **Iteration-1.** To make problem balance, make a fictitious column M_6 with demand 4 and cost 0 throughout. Make initial assignment by Matrix Minima Method

Plant

	M_1	M_2	M_3	M_4	M_5	M_6	Capacity
F_1	3 ④		4	0	5 ④		
	4	2	3	2	6	0	8
F_2	4	2	6 (4+θ)		(ε-θ)		
	5	4	5	2	1	0	12
F_3	④	-2 ⑥		(4-θ)	-3	⑨θ -5	
	6	5	4	7	3	0	14
Demand	4	4	6	8	8	4	34
v_j	6	7	4	7	6	5	

Number of occupied cell $= 7 < 9 - 1 = 8$.

So assign a quantity $\varepsilon \to 0$ in the cell of minimum cost, i.e., in cell (2, 6).

Calculate u_i and v_j for each occupied cell such that $c_{ij} = u_i + v_j$

Cell (1, 2)	$u_1 + v_2 = 2$	Let $v_2 = 7$, $u_3 = 0$
Cell (1, 4)	$u_1 + v_4 = 0$	$u_1 = -5$
Cell (2, 4)	$u_2 + v_4 = 2$	$u_2 = -5$
Cell (2, 5)	$u_2 + v_5 = 1$	$v_5 = 6$
Cell (2, 6)	$u_2 + v_6 = 0$	$v_6 = 5$
Cell (3, 1)	$u_3 + v_1 = 6$	$v_1 = 6$
Cell (3, 4)	$u_3 + v_4 = 4$	$v_3 = 4$
Cell (3, 4)	$u_3 + v_4 = 7$	$v_4 = 7$

Since all Δ_{ij} are not positive therefore assign a quantity θ in the cell of most negative Δ_{ij}, i.e., in cell (3, 6).

Here $\theta = \varepsilon \to 0$.

Iteration-2

Plant

	M_1	M_2	M_3	M_4	M_5	M_6	Capacity	u_i
F_1	-2	4	-1	⓪ -5	0	④-θ		
	4	2	3	2	6	0	8	0
F_2	4	7	6 4	8		5		
	5	4	5	2	1	0	12	-5
F_3	4	3 6		(4-θ)	-3	(ε+θ)		
	6	5	4	7	3	0	14	0
Demand	4	4	6	8	8	4		
v_j	6	2	4	7	6	0		

Calculate u_i and v_j for each occupied cell

Cell (1, 2)	$u_1 + v_2 = 2$	Let $u_3 = 0$, $v_2 = 2$
Cell (1, 6)	$u_1 + v_6 = 0$	$u_1 = 0$
Cell (2, 4)	$u_2 + v_4 = 2$	$u_2 = -5$
Cell (3, 5)	$u_2 + v_5 = 1$	$v_5 = 6$
Cell (3, 1)	$u_3 + v_1 = 6$	$v_1 = 6$
Cell (3, 3)	$u_3 + v_3 = 4$	$v_3 = 4$
Cell (3, 4)	$u_3 + v_4 = 7$	$v_4 = 7$
Cell (3, 6)	$u_3 + v_6 = 0$	$v_6 = 0$

Since all Δ_{ij} are not positive, assign a quantity θ in the cell of most negative Δ_{ij} such that loop is formed, i.e., in the cell (1, 4). Here $\theta = 4$.

Iteration-3

	M_1	M_2	M_3	M_4	M_5	M_6	Capacity	u_i
F_1	(θ) −2 | 4 4	2	−1 | 4 3	2	5 6	(ε−θ) 0	8	0
F_2	−1 5	2 4	6 5	4 | 2	8 | 1	0	12	0
F_3	(4−θ) 6	3 | 6 5	6 4	5 | 7	2 | 3	(4+θ) 0	14	0
Demand	4	4	6	8	8	4		
v_j	6	2	4	2	1	0		

Number of occupied cell $= 7 < 8$, therefore assign a quantity $\varepsilon \to 0$ in a cell of minimum cost such that loop is not formed, *i.e.*, in cell (1, 6).

Calculate u_i and v_j for each occupied cell

Cell (1, 2)	$u_1 + v_2 = 2$	Let $v_2 = 2$, $u_1 = 0$
Cell (1, 4)	$u_1 + v_4 = 2$	$v_4 = 2$
Cell (1, 6)	$u_1 + v_6 = 0$	$v_6 = 0$
Cell (2, 4)	$u_2 + v_4 = 2$	$u_2 = 0$
Cell (2, 5)	$u_2 + v_5 = 1$	$v_5 = 1$
Cell (3, 1)	$u_3 + v_1 = 6$	$v_1 = 6$
Cell (3, 3)	$u_3 + v_3 = 4$	$v_3 = 4$
Cell (3, 6)	$u_3 + v_6 = 0$	$u_3 = 0$

Since all Δ_{ij} are not positive, assign a quantity θ in cell (1, 1).

Here, $\theta = \varepsilon \to 0$.

Iteration-4

	M_1	M_2	M_3	M_4	M_5	M_6	Capacity	u_i
F_1	ⓔ 4	④ 2	1 3	④ 2	5 6	2 0	8	0
F_2	1 5	2 4	3 5	④ 2	⑧ 1	2 0	12	0
F_3	④ 6	1 5	⑥ 4	3 7	0 3	④ 0	14	2
Demand	4	4	6	8	8	4		
v_j	4	2	2	2	1	–2		

Calculate u_i and v_j for each occupied cell

Cell (1, 1) $\quad u_1 + v_1 = 4 \quad$ Let $v_1 = 4$, $u_1 = 0$
Cell (1, 2) $\quad u_1 + v_2 = 2 \qquad v_2 = 2$
Cell (1, 4) $\quad u_1 + v_4 = 2 \qquad v_4 = 2$
Cell (2, 4) $\quad u_2 + v_4 = 2 \qquad u_2 = 0$
Cell (2, 5) $\quad u_2 + v_5 = 1 \qquad v_5 = 1$
Cell (3, 1) $\quad u_3 + v_1 = 6 \qquad u_3 = 2$
Cell (3, 3) $\quad u_3 + v_3 = 4 \qquad v_3 = 2$
Cell (3, 6) $\quad u_3 + v_6 = 0 \qquad v_6 = -2$

Since all Δ_{ij} are positive, optimal reaches.

Minimum cost $= 4 \times 2 + 4 \times 2 + 4 \times 2 + 8 \times 1 + 4 \times 6 + 6 \times 4 + 4 \times 0 = 80$.

EXAMPLE 3. *Solve the following transportation problem of minimize cost starting with degenerate bases $x_{12} = 30$, $x_{21} = 40$, $x_{32} = 20$ and $x_{43} = 60$.*

	D_1	D_2	D_3	Availability
O_1	4	5	2	30
O_2	4	1	3	40
O_3	3	6	2	20
O_4	2	3	7	60
Demand	40	50	60	

SOLUTION. Iteration-1

	D_1	D_2	D_3	Availability	u_i
O_1	–4 4	㉚–θ 5	⊕θ 2	30	0
O_2	㊵–θ 4	ⓔ₁+θ 1	5 3	40	–4
O_3	–6 3	20 6	–1 2	20	1
O_4	⓸+11 2	–7 3	㊿60–θ 7	60	5
Demand	40	50	60	150	
v_j	8	5	2		

Number of occupied cell = 4 < 4 + 3 − 1 = 6. So degeneracy occurs.

To make the number of occupied cells and assign a quantity $\varepsilon \to 0$ to the two remaining cells of minimum cost, so that cell may not contain a loop.

Let ε_1 assign in cell (2, 2) and (1, 3).

Since all Δ_{ij} are not positive therefore assign a quantity θ in the cell of most negative Δ_{ij}, i.e., in cell (4, 1). Here $\theta = 30$ cell (1, 2) is leaving cell.

Iteration-2

	D_1	D_2	D_3	Availability	u_i
O_1	7 / 4	11 / 5	30 / 2	30	−5
O_2	(10 − θ) / 4	(30 + θ) / 1	−6 / 3	40	2
O_3	−6 / 3	(20 − θ) / 6	θ / −12 / 2	20	7
O_4	(30 + θ) / 2	4 / 3	(30 − θ) / 4 / 7	60	0
Demand	40	50	60	150	
v_j	2	−1	7		

Since all Δ_{ij} are not positive therefore assign a quantity θ in cell (3, 3).

Here $\theta = 10$.

Iteration-3

	D_1	D_2	D_3	Availability	u_i
O_1	7 / 4	−1 / 5	③⓪ / 2	30	−5
O_2	12 / 4	④⓪ / 1	6 / 3	40	−10
O_3	6 / 3	(10 − θ) / 6	(10 + θ) / 2	20	5
O_4	40 / 2	(θ) / −9 / 3	(20 − θ) / 7	60	0
Demand	40	50	60		
v_j	2	11	7		

Calculate u_i and v_j for each occupied cell

Cell (1, 3) $\qquad u_1 + v_3 = 2 \qquad$ Let $u_1 = -5$, $u_4 = 0$

Cell (2, 2) $u_2 + v_2 = 1$ $u_2 = -10$
Cell (3, 2) $u_3 + v_2 = 6$ $v_2 = 11$
Cell (3, 3) $u_3 + v_3 = 2$ $u_3 = 5$
Cell (4, 1) $u_4 + v_1 = 2$ $v_1 = 2$
Cell (4, 3) $u_4 + v_3 = 7$ $v_3 = 7$

Since all Δ_{ij} are not positive therefore assign a quantity θ in cell (4, 2).

Here $\theta = 10$. Cell (3, 2) is leaving cell.

Iteration-4

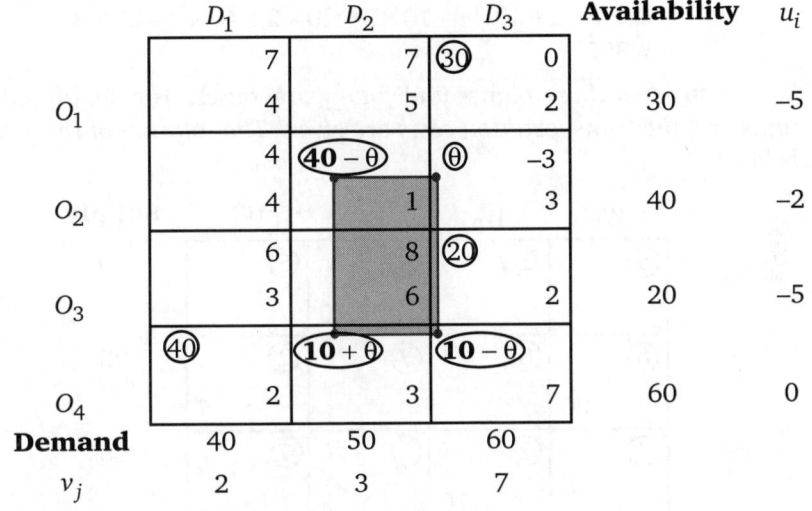

Calculate u_i and v_j for each occupied cell
Cell (1, 3) $u_1 + v_3 = 2$ Let $u_1 = -5$, $u_4 = 0$
Cell (2, 2) $u_2 + v_2 = 1$ $u_2 = -1$
Cell (3, 3) $u_3 + v_3 = 2$ $u_3 = -5$
Cell (4, 1) $u_4 + v_1 = 2$ $v_1 = 2$
Cell (4, 2) $u_4 + v_2 = 3$ $v_2 = 3$
Cell (4, 3) $u_4 + v_3 = 7$ $v_3 = 7$

Since all Δ_{ij} are not positive therefore assign a quantity θ in cell (2, 3).

Iteration-5

	D_1	D_2	D_3	Availability	u_i
O_1	5 / 4	5 / 5	㉚ / 2	30	−3
O_2	4 / 4	㉚ / 1	⑩ / 3	40	−2
O_3	4 / 3	6 / 6	⑳ / 2	20	−3
O_4	㊵ / 2	⑳ / 3	2 / 7	60	0
Demand	40	50	60		
v_j	2	3	5		

Calculate u_i and v_j for each occupied cell

Cell (1, 3) $u_1 + v_3 = 2$ Let $v_1 = -3$, $u_4 = 0$

Cell (2, 2) $u_2 + v_2 = 1$ $u_2 = -2$

Cell (2, 3) $u_2 + v_3 = 3$ $v_3 = 5$

Cell (3, 3) $u_3 + v_3 = 2$ $u_3 = -3$

Cell (4, 2) $u_4 + v_2 = 2$ $v_1 = 2$

Cell (4, 2) $u_4 + v_2 = 3$ $v_2 = 3$

Since all Δ_{ij} are positive, hence optimal is reached.

The minimum cost of assignment

$$= 30 \times 2 + 30 \times 1 + 10 \times 3 + 20 \times 2 + 40 \times 2 + 20 \times 3$$
$$= 300$$

EXAMPLE 4. *A company has three plants and four warehouses. The supply and demand in units and the transportation costs are given. The solution of the problem is given below :*

	W_1	W_2	W_3	W_4	**Supply**
O_1	⑤	⑩	④ 10	⑤	10
O_2	⑥ 20	⑧	⑦	② 5	25
O_3	④ 5	② 10	③ 5	⑦	20
Demand	25	10	15	5	

Answer the following questions :

  (i) *Is the solution feasible?*

  (ii) *Does the solution degenerate?*

 (iii) *Is this solution optimum.*

 (iv) *Does this problem have more than one optimum solution. If so, show all of them.*

  (v) *If the cost of route O_2W_3 is reduced from ₹ 7 to ₹ 6 per unit, what will be the optimum solution?*

SOLUTION.   In the above question, North-West rule applied, so we always move to the right or down, so no loop can be formulated by drawing horizontal and vertical lines to the allocations.

  (i) Solution is feasible because we can not get more than $(m+n-1)$ individual positive allocations, $(3+4-1) = 6$.

  (ii) We always get a non-degenerate basic feasible solution by the North-West Corner rule.

 (iii) The optimality test is applicable to a non-degenerate B.F.S. to a F.S. consisting of $(m+n-1)$ allocations in independent positions. Since there are $m.n - (m+n-1) - (m-1)(n-1)$ empty cell, so the solution is optimum.

(iv) No further possibility to obtain an optimal solution by making successive improvements to initial basic feasible solution until no further decrease in the transportation cost is possible.

(v) The solution is not feasible if replaced ₹ 7 to ₹ 6 per unit, so there is not possible to find optimum.

Exercise-11.1

1. Determine an initial basic feasible solution to the following transportation problem using VAM or Vogel's method

(i)

Destination

	D_1	D_2	D_3	D_4	D_5	Capacity
A	2	11	10	3	7	4
B	1	4	7	2	1	8
C	3	9	4	8	12	9
Demand	3	3	4	5	6	21 total

Origin A, B, C

(ii)

Destination

	A_1	B_1	C_1	D_1	Supply
A	1	2	1	4	30
B	3	3	2	1	50
C	4	2	5	9	20
Demand	20	40	30	10	100 total

Origin A, B, C

2. Determine the optimal solution to each of the following degenerate transportation problem.

(i)

Destination

	D_1	D_2	D_3	D_4	Capacity
O_1	2	3	11	7	6
O_2	1	0	6	1	1
O_3	5	5	15	10	10
Demand	7	5	3	2	17 total

Origin O_1, O_2, O_3

(ii)

To

	D_1	D_2	D_3	D_4	a_i
O_1	10	7	3	6	3
O_2	1	6	7	3	5
O_3	7	4	5	3	7
b_j	3	2	6	4	15 total

From O_1, O_2, O_3

3. Solve the following transportation problem whose cost matrix is given below

				Capacity
5	5	4	7	5
6	4	1	2	5
5	9	1	4	6
8	3	2	4	4
6	5	3	1	6
Demand 5	8	3	10	

4. Obtain the optimal solution by using the best starting solution

Destination

					a_i
10	20	5	7	10
13	9	12	8	20
4	15	7	9	30
14	7	1	0	40
3	12	6	19	50
b_j 60	60	20	10	

Origin

5. A company has 4 warehouses and 6 stores. The capacity in the warehouse, the demand of the stores and costs (in ₹) of transporting are unit of the commodity from warehouses i to the store j are given below. How should the commodity be transported so that the total transportation cost is a minimum? Obtain the initial program by applying the

Vogel's method.

Store

		1	2	3	4	5	6	Capacity
Warehouse	1	7	5	9	5	10	7	30
	2	7	8	24	7	9	13	40
	3	4	10	5	6	10	4	20
	4	11	8	12	7	12	11	80
Demand		30	30	60	20	10	20	170

6. An oil corporation has got three refineries A, B and C and it has to send petrol to four different depots P, Q, R and S, the cost of shipping 1 gal of petrol and the available petrol at the refineries are given in the table. The demand of the depots and the capacity petrol at the refineries are also given. Find the minimum cost of shipping after obtaining an initial solution by VAM.

Depot

		P	Q	R	S	Capacity
Refinery	A	10	12	15	8	130
	B	11	11	9	10	150
	C	20	9	7	18	170
Demand		90	100	140	120	

ANSWERS

1. (i) $x_{14} = 4, x_{22} = 2, x_{25} = 6, x_{31} = 3, x_{32} = 1, x_{33} = 4, x_{34} = 1$, cost = ₹ 68

   (ii) $x_{11} = 20, x_{13} = 10, x_{22} = 20, x_{23} = 20, x_{24} = 10, x_{32} = 20$, cost = ₹ 180

2. (i) $x_{12} = 5, x_{13} = 1, x_{24} = 1, x_{31} = 7, x_{33} = 2, x_{34} = 1$, Min. cost = ₹ 102

   (Alternative solutions also exist)

   (ii) $x_{13} = 3, x_{21} = 3, x_{24} = 2, x_{32} = 2, x_{33} = 3, x_{34} = 2$, Minimum cost = ₹ 47

3. $x_{11} = 2, x_{12} = 3, x_{22} = 1, x_{24} = 4, x_{31} = 3, x_{33} = 3, x_{42} = 4, x_{54} = 6$, Min. cost = ₹ 73

4. $x_{13} = 10, x_{22} = 20, x_{31} = 30, x_{42} = 20, x_{43} = 10, x_{44} = 10, x_{51} = 30, x_{52} = 20$,

   Minimum cost = ₹ 830

5. $x_{12} = 10, x_{16} = 20, x_{21} = 30, x_{25} = 10, x_{33} = 20, x_{42} = 20, x_{43} = 40, x_{44} = 20$,

   Minimum cost = ₹ 1,370

6. $x_{11} = 90, x_{14} = 40, x_{23} = 70, x_{24} = 80, x_{32} = 100, x_{33} = 70$, Minimum cost = ₹ 3960

Glossary

- **Feasible solution:** A feasible solution to a transportation problem is a set of non-negative individual allocations ($x_{ij} \geq 0$) which satisfies the row and column sum restrictions.

- **Basic Feasible solution.** A feasible solution of an m by n transportation problem is said to be basic if the total number of positive allocations is equal to $m + n - 1$, i.e., one less than the sum of no. of rows and columns.

- **Optimal solution.** A feasible solution, not necessarily basic is said to be optimum if it minimize the total transportation cost.

- **Non-degenerate Basic Feasible solution.** A basic feasible solution to a transportation problem of m sources and n destination is said to be non-degenerate basic feasible if

  (i) the total no. of non-negative allocations x_{ij} is exactly equal to $m + n - 1$.

  (ii) these allocations should be in independent position (i.e., it is always impossible to form any closed circuit by joining these allocation by horizontal and vertical lines only).

- **Degenerate-basic Feasible solution.** A basic feasible solution to a transportation problem of m sources and n destinations is said to be degenerate if the total number of non-degenerate allocations is less than $(m + n - 1)$.

REVIEW QUESTIONS

1. Give a mathematical formulation of transportation problem.

2. Write the characteristic of transportation problem of linear programming.

3. Explain various method to find the initial basic feasible solution of transportation problem.

4. Write a short note on transportation problem.

5. Write the computational procedure of optimality test in transportation problem.

MULTIPLE CHOICE QUESTIONS (CHOOSE THE MOST APPROPRIATE ONE)

1. In the optimal table of a transportation problem, a zero in the North-west corner rule shows that: [MEERUT–2013]

   (a) An alternative optimal solution exists

   (b) The optimal solution is degenerated

   (c) Both (a) and (b) are true

   (d) None of the above

2. The dummy source or destination in a transportation problem is introduced to:
   [MEERUT–2013]

   (a) prevents solution to become degenerate

   (b) satisfy rim conditions

   (c) ensure that total cost does not exceed a limit

   (d) solve the balanced transportation problem

3. In a T.P., the solution under test will be optimal if all the cell evaluation are: [MEERUT–2013]

   (a) > 0 (b) < 0

   (c) ≥ 0 (d) ≤ 0

4. Transportation problem

	W_1	W_2	W_3	W_4	**Available**
F_1	19	30	50	10	7
F_2	70	30	40	60	9
F_3	40	8	70	20	18
Requirement	5	8	7	14	

has optimal solution: [MEERUT–2013]

(a) 1015   (b) 814

(c) 779   (d) 743

5. The total number of allocations in a basic feasible solution of transportation problem of $m \times n$ size is equal to: [MEERUT–2013]

   (a) $m \times n$ (b) $m + n$

   (c) $m + n + 1$ (d) $m + n - 1$

6. Which of the following is not correct?

   (a) The transportation problem is a distribution problem

   (b) A closed loop would always involve an even number of cells, subject to a minimum of 4

   (c) The u_i and v_j values may be determined by initially inserting any finite number which may be positive, negative or zero to a row/column

(d) Units sents from a dummy source to various markets represent the shortfall in supply to those markets

7. The transportation problem is balanced if:

(a) total demands equals total supply irrespective of the number of sources and destinations.

(b) number of sources matches with the number of destinations.

(c) total demand and total supply are equal and the number of sources equals the number of destinations.

(d) None of the route is prohibited.

8. Which of the following is not correct?

(a) A degenerate solution may or may not be optimum.

(b) A transportation problem solution is said to be degenerate if the number of occupied cells is smaller than the number of rows plus the number of columns minus one.

(c) To remove degeneracy, an infinitesimally small quantity is placed in each of the required number of independent cells.

(d) Once non-optimum degenerate solution is obtained, the next solution is bounded to be degenerate.

9. The transportation problem deals with the transportation of:

(a) single product from a source to several destinations.

(b) a single product from several source to the several destination.

(c) a multi-product from several sources to several destinations.

(d) a single product from several sources to a destination.

10. For a transhipment problem, choose the statements which is not correct:

(a) a transhipment problem is not likely to involve a lower cost than a transportation problem, in a given situation.

(b) an 'm' source, 'n' destination transportation problem when written as a transhipment problem would have $m + n$ sources and n destinations.

(c) there is no real distinction between sources and destinations.

(d) the problem allows for the shipment of goods from one sources to another and from one destination to another.

11. Which of the following is not correct?

(a) If some $u_i + v_j - c_{ij}$ is equal to zero in the optimum solution, then the problem has multiple optimum solutions

(b) If all the cost elements c_{ij} are multiplied by a cosntant, the total cost of transportation in optimum solution shall be multiplied by the same constant

(c) The number of occupied cells involved in a closed path is always even

(d) In the transportation problem, if certain routes are prohibited then their cost elements are replaced by M (an extremely large value)

12. Which of the following is not correct?

(a) It is possible that in some cases both, the dummy source and dummy destination, may be required to convert an unbalanced transportation problem into a balanced one

(b) The cost elements in a dummy row/column shall always be taken equal to zero

(c) An unbalanced transportation problem must be converted into a balanced problem before solving it

(d) It is not necessary for the aggregate demand to be equal to the aggregate supply in a transportation problem

13. Which of the following is not correct?

(a) A maximization transportation problem is first converted into a minimization one by subtracting each value of the given matrix from the largest value

(b) If each cost element in a transportation problem is increased or decreased by a constant amount, it will not effect the optimum solution of the problem

(c) Multiple optimum solutions are indicated if there are multiple zeroes for u_i and v_j values

(d) For an optimum solutions to a transportation problem, the u_i and v_j values represent the optimum values of the dual problem

14. Which of the following is correct?

(a) The least cost method does not provide the least cost solution to a transportation problem

(b) The initial solution obtained by the NW corner rule would invariably be optimum

(c) If some routes are prohibited, VAM cannot be used to find an initial solution to a transportation problem

(d) The cost difference in the Vogel's approximation method indicate the penalties for not using the respective least cost routes

15. Which of the following is not correct?

(a) The $u_i + v_j - c_{ij}$ value of an unoccupied cells indicates the net change in cost of re-allocating one unit through the routh involved

(b) In time minimization problem, the cost c_{ij} is replaced by the unit time t_{ij}

(c) Any of the $m + n - 1$ number of occupied cells would allow determining whether a given solution is optimum or not

(d) The maximum quantity that can be re-allocated in a closed path is equal to the minimum quantity in the cells bearing negative sign

16. During an iteration while moving from one solution to the next, degeneracy may occurs when:

(a) two or more occupied cells on the closed path with minus sign are tied for the lowest circled value

(b) two or more occupied cells are on the closed path but neither of them represents a corner of the path

(c) the closed path indicates a diagonal move

(d) either of the above

17. To find a B.F.S. of a transportation problem by matrix minima method which cell we choose first:

(a) highest cell   (b) lowest cell

(c) zero cost   (d) none of these

18. To improve the current basic F.S. of a transportation problem if it is not optimal we allocate to the cell for which d_{ij} is:

(a) 0

(b) maximum and negative

(c) minimum and negative

(d) positive

19. A feasible solution of a T.P. is said to be optimal if it:

(a) maximizes the total trasportation cost

(b) minimizes the total transportation cost

(c) balance the total transportation cost

(d) none of these

20. An assignment problem is a special case of a $l \times n$ transportation problem in where:

(a) $l - 2n$ (b) $l = n$

(c) $l = 3n$ (d) None of these

21. In a Transportation Problem a loop may be defined as an ordered set of at least:

(a) 3 cells   (b) 4 cells

(c) 5 cells   (d) 6 cells

22. If a Transportation Problem $\Sigma a_i < \Sigma b_j$ then we introduce a dummy origin to make it a:

(a) Unbalanced T.P.   (b) Balanced T.P.

(c) L.P.P.   (d) All of these

23. For a Transportation Problem for testing the optimality we find the u_i and v_j that for all the occupied cells $c_{ij} =$

(a) $u_i - v_j$ (b) $u_i + v_j$

(c) u_i / v_j (d) $u_i \cdot v_j$

24. In transportation problem, to find BFS we start with the cell (1, 1) following some rule that rule is:

(a) N-W corner rule

(b) Lowest cost entry method

(c) Vogel's approximation method

(d) None of the above

25. A feasible solution is said to be optimal if it _____ the total transportation cost the blank space can be filled by:
    (a) maximizes
    (b) minimizes
    (c) either maximizes or minimizes
    (d) none of the above

26. To find initial BFS of transportation problem by matrix revenue method, we first choose the cell with:
    (a) zero cost   (b) lowest cost
    (c) highest cost   (d) none of these

27. If a B.F.S. of a $m \times n$ Transportation Problem, the number of positive allocations is at most:
    (a) $m + n$ (b) $m + n - 1$
    (c) $m - n$ (d) None of these

28. A solution is not a basic feasible solution in a transportation problem if after allocations:
    (a) there is degeneracy
    (b) there is closed loop
    (c) there is no closed loop
    (d) total number of allocations is one less than sum of the number of sources and destinations

29. One disadvantage of using North-west corner rule to find initial solution to the transportation problem is that:
    (a) it leads to adequate initial solution
    (b) it is complicated to use
    (c) it does not take into account the cost of transportation
    (d) all of the above

30. The dummy source or destination in a transportation problem is added to:
    (a) ensure that total cost does not exceed a limit
    (b) prevent solution from becoming degenerate
    (c) satisfy rim conditions
    (d) none of the above

31. In transportation problem, the materials are transported from 3 plants to 5 warehouses. The basic feasible solution must contain exactly which one of the following allocated cells?
    (a) 8   (b) 7
    (c) 5   (d) 3

32. The calculation of opportunity cost in the MODI method is analogous to a:
    (a) variable in the B-columns in the Simplex method
    (b) value of a variable in x_B-column of the simplex method
    (c) $z_j - c_j$ value for non-basic variable columns in the simplex method
    (d) none of the above

33. In a 6×6 transportation problem, degeneracy would arise, if the number of filled sports were:
    (a) less than eleven   (b) equal to twelve
    (c) more than twelve   (d) equal to thirty six

34. The occurance of degeneracy while solving a transportation problem means that:
    (a) the few allocation become negative
    (b) the solution so obtained is not feasible
    (c) total supply equals total demand
    (d) none of the above

35. The solution of a transportation model (of dimension $m \times n$) is said to be degenerate if it has:
    (a) $(m + n)$ allocation
    (b) more than $(m + n - 1)$ allocation
    (c) fewer than $(m + n - 1)$ allocation
    (d) exactly $(m + n - 1)$ allocation

36. An unbounded cell in the transportation method is analogous to a:
    (a) value in the x_B-column in the Simplex method
    (b) variable not in the B-column in the Simplex method
    (c) variable in the B-column in the Simplex method
    (d) $z_j - c_j$ value in the Simplex method

37. The initial solution of a transportation problem can be obtained by applying any known method. However, the only condition is that:
    (a) the solution may not be degenerate
    (b) the RIM condition are satisfied
    (c) the solution be optimal
    (d) all of the above

38. An alternative optimal solution to a minimization transportation problem exists whenever opportunity cost corresponding to unused route of transportation is:

    (a) negative with at least one is equal to zero

    (b) positive and greater than zero

    (c) positive with at least one is equal to zero

    (d) none of the above

39. The solution to a transportation problem with m rows (supplies) and n columns (destinations) is feasible if number of positive allocations are:

    (a) $m + n$ (b) $m \times n$

    (c) $m + n - 1$ (d) $m + n + 1$

40. If we were to use opportunity cost value for an unused cell to test optimality, it should be:

    (a) most positive number

    (b) equal to zero

    (c) most negative number

    (d) any value

41. When there are 'm' rows and 'n' columns in a transportation problem, degeneracy is said to occur when the number of allocation is:

    (a) less than $m - n - 1$

    (b) equal to $m + n - 1$

    (c) greater than $m + n - 1$

    (d) less than $m + n - 1$

42. In a transportation problem, obtaining the starting BFS by VAM or any other method, a column and a row are satisfied together. This shows that:

    (a) there is no feasible solution

    (b) at least one basic variable is at zero level

    (c) at least two basic variable is at zero level

    (d) none of these

43. In the optimal table of a transportation problem, a zero in the S.W. corner shows that:

    (a) the optimal solution is degenerate

    (b) an alternate optimal solution exists

    (c) an optimal solution does not exist

    (d) None of the above

44. In a balanced transportation problem with three sources and four destinations and with availabilities 40 at each source and demand 30 at each destination, the dual variables in the optimal table corresponding to sources and destinations are respectively $-1, 2, 3$ and $0, 2, -1, 4$. Then the optimal value is:

    (a) 310   (b) 320

    (c) 300   (d) None of these

45. In transportation problem, one of the dual variables is assigned an arbitrary value, because:

    (a) then a solution is obtained immediately

    (b) one of the constraints is redundant in a T.P.

    (c) this facilitates construction of the loop

    (d) None of these

46. In a transportation problem while obtaining the starting BFS by VAM or any other method, a column and row are satisfied together. Then treating both as satisfied (that is not writting 0 either at row or column) will give a solution:

    (a) which is not basic

    (b) which will be degenerate

    (c) which is basic

    (d) None of these

47. In a transportation problem (T.P.) the dual variables u_i and v_j are unrestricted in sign because:

    (a) the TP is a minimization problem

    (b) the TP is with all equality constraints

    (c) in TP all decision variables are ≥ 0

    (d) none of these

48. In a balanced transportation problem with m sources and n destinations, the number of linearly independent constraints is:

    (a) $m + n$ (b) $m + n + 1$

    (c) $m + n - 1$ (d) $m + n$

49. In a balanced transportation problem with m sources and n destinations the coefficient matrix has rank:

    (a) $m + n + 1$ (b) $m + n$

    (c) $m + n - 1$ (d) $m - n$

50. In an unbalanced transportation problem with m sources and n destinations the number of basic variables is:

    (a) $m + n + 1$ (b) $m + n$

    (c) $m + n - 1$ (d) $m + n + 2$

51. If in using VAM or any other method to obtain a starting solution the allocation made in a cell is not maximum possible, then the resulting solution:

    (a) will not be BFS

    (b) will be a BFS

    (c) will be a basic solution but not BFS

    (d) None of these

52. In Transportation Problem, if less allocation than given by loop criteria is made in the cell corresponding to entering vairable, then the resulting solution will:

    (a) not be a BFS

    (b) be a BFS

    (c) be a basic solution but not BFS

    (d) None of these

53. In Transportation Problem, if two dual variables are assigned arbitrary values, then:

    (a) the method will yield correct solution

    (b) the method will not yield correct solution

    (c) the method may or may not yield correct solution

    (d) None of these

54. In T.P., the values of dual variables are not unique in the sense that the values depend on the variable assigned arbitrarily and the values assigned:

    (a) $z_{ij} - c_{ij}$ will also change

    (b) $z_{ij} - c_{ij}$ are unique

    (c) $z_{ij} - c_{ij}$ will not change

    (d) None of these

ANSWERS

1. (a)	2. (d)	3. (b)	4. (a)	5. (d)	6. (c)	7. (a)	8. (d)	9. (b)
10. (b)	11. (c)	12. (a)	13. (c)	14. (d)	15. (c)	16. (a)	17. (b)	18. (c)
19. (b)	20. (b)	21. (b)	22. (b)	23. (b)	24. (a)	25. (b)	26. (b)	27. (b)
28. (c)	29. (c)	30. (c)	31. (a)	32. (c)	33. (a)	34. (b)	35. (c)	36. (b)
37. (c)	38. (c)	39. (c)	40. (c)	41. (d)	42. (b)	43. (b)	44. (a)	45. (b)
46. (a)	47. (b)	48. (a)	49. (c)	50. (b)	51. (a)	52. (a)	53. (b)	54. (b)

Assignment Problems

12.1 INTRODUCTION

An assignment problem is a special case of transportation problem. It arises because available resources have varying degrees of efficiency for performing different activities, *i.e.*, cost, profit or time of performing the different activities is different. This type of problem where the main motive is to allot a number of origins to the equal number of destinations at a least cost are called assignment problems.

Let us suppose there are n people and they are to be assigned to n jobs such that each person can do each job at a time with varying degree of efficiency. Let c_{ij} be the cost of assigning i^{th} person to the jth job, then the objective of assignment problem is to find an assignment (*i.e.*, which job should be assigned to which particular person) that minimize the total cost of doing all the jobs.

12.2 MATHEMATICAL REPRESENTATION OF ASSIGNMENT PROBLEM

[GORAKHPUR–2010; KANPUR–2012]

The assignment problem can be stated in the form of $n \times n$ matrix. This, in general can be defined as the square transportation problem in which the no. of origins equal the number of destinations.

It can be stated mathematically as follows:

"To determine x_{ij} for i = 1, 2, ..., n and j = 1, 2, ..., n which minimize the total cost

$$Z = \sum_{i=1}^{n} \sum_{j=1}^{n} c_{ij} x_{ij}$$

subject to the constraints

$$\sum_{j=1}^{n} x_{ij} = 1 \quad \text{(only one job is done by } i^{th} \text{ person)}$$

and

$$\sum_{i=1}^{n} x_{ij} = 1 \quad \text{(only one person should be assigned to } j^{th} \text{ job)}$$

where

$$x_{ij} = \begin{cases} 1, & \text{if } i^{th} \text{ person is assigned to } j^{th} \text{ job} \\ 0, & \text{otherwise} \end{cases}$$

In matrix form, an assignment problem can be written as follows.

	1	2	...	j	...	n
1	c_{11}	c_{12}	...	c_{1j}	...	c_{1n}
2	c_{21}	c_{22}	...	c_{2j}	...	c_{2n}
$\vdots$	$\vdots$	$\vdots$				
i	c_{i1}	c_{i2}	...	c_{ij}	...	c_{in}
$\vdots$	$\vdots$	$\vdots$				
n	c_{n1}	c_{n2}	...	c_{nj}	...	c_{nn}

12.3 DIFFERENCE BETWEEN TRANSPORTATION AND ASSIGNMENT PROBLEM

[MEERUT–2006, 07; KANPUR–2011, 12]

The assignment problem is a variation of the transportation problem with two characteristics
  (i) The cost matrix is a square matrix
  (ii) The optimal solution for the problem would always be such that there would be only one assignment in a given row or column of the cost matrix.

In tabular form, these differences can be written as follows:

	Transportation Problem		Assignment Problem
1.	There are m sources and n destinations.	1.	There should be m persons (sources) and n jobs (destinations) (*i.e.*, m = n).
2.	For i^{th} source, the supply is a_i and for j^{th} destination the demand is b_j.	2.	Here, $a_i = 1$ for each person i and $b_j = 1$ for each job j.
3.	The quantity $x_{ij} \geq 0$	3.	Here, $x_{ij} = 1$ when i^{th} person is assigned to j^{th} job and $x_{ij} = 0$ when i^{th} person is not assigned to j^{th} job.
4.	It is unbalanced if $\sum\limits_{i=1}^{m} a_i \neq \sum\limits_{j=1}^{n} b_j$	4.	It is unbalanced if the number of persons is not equal to the number of jobs.

12.4 THEOREMS ON ASSIGNMENT PROBLEM

THEOREM 1. **(Reduction Theorem)** *In an assignment problem, if a constant is added to or subtracted from every element of any row or column in the given cost matrix, an assignment that minimize the total cost in one matrix also minimize the total cost in the other matrix.*

(OR)

If $x_{ij} = X_{ij}$ minimize $Z = \sum\limits_{i=1}^{n} \sum\limits_{j=1}^{n} c_{ij}x_{ij}$ over all $x_{ij} = 0$ or 1 such that

$$\sum_{i=1}^{n} x_{ij} = 1, \sum_{j=1}^{n} x_{ij} = 1 \ then \ x_{ij} = X_{ij}$$

also minimize $Z' = \sum\limits_{i=1}^{n} \sum\limits_{j=1}^{n} c'_{ij}x_{ij}$ when $c'_{ij} = c_{ij} \pm a_i \pm b_j$ where $a_i, b_j \in \mathbf{R}$.

PROOF. Consider
$$Z' = \sum_{i=1}^{n} \sum_{j=1}^{n} c'_{ij}x_{ij} = \sum_{i=1}^{n} \sum_{j=1}^{n} (c_{ij} \pm a_i \pm b_j)x_{ij}$$

$$= \sum_{i=1}^{n} \sum_{j=1}^{n} c_{ij}x_{ij} \pm \sum_{i=1}^{n} \sum_{j=1}^{n} a_i x_{ij} \pm \sum_{i=1}^{n} \sum_{j=1}^{n} b_j x_{ij}$$

$$= Z \pm \sum_{i=1}^{n} a_i \sum_{j=1}^{n} x_{ij} \pm \sum_{j=1}^{n} b_j \sum_{i=1}^{n} x_{ij}$$

$$= Z \pm \sum_{i=1}^{n} a_i \cdot 1 \pm \sum_{j=1}^{n} b_j \cdot 1 = Z \pm \sum_{i=1}^{n} a_i \pm \sum_{j=1}^{n} b_j$$

Since $\sum_{i=1}^{n} a_i$ and $\sum_{j=1}^{n} b_j$ are independent on x_{ij}.

$\Rightarrow Z'$ is minimized whenever Z is minimized and conversely.

Hence, $x_{ij} = X_{ij}$ which minimize Z will also minimize Z'.

THEOREM 2. *In an assignment problem with cost matrix c_{ij} if all $c_{ij} \geq 0$ then a feasible solution x_{ij} which satisfies $\sum_{i=1}^{n} \sum_{j=1}^{n} c_{ij}x_{ij} = 0$ is an optimal solution for the problem.*

PROOF. Since, we have all $c_{ij} \geq 0$ and all $x_{ij} \geq 0$ therefore

$$Z = \sum_{i=1}^{n} \sum_{j=1}^{n} c_{ij}x_{ij} \geq 0$$

$\Rightarrow$ The minimum value of $Z = 0$

Hence, any feasible solution x_{ij} satisfying $\sum_{i=1}^{n} \sum_{j=1}^{n} c_{ij} \cdot x_{ij} = 0$ will be optimal.

12.5 SOLUTION OF ASSIGNMENT PROBLEM : HUNGARIAN METHOD

This method was developed by Hungarian mathematician D. Konig. It works on the principle of reducing the given cost matrix to a matrix of opportunity cost, which shows the relative penalties associated with assigning a resource to an activity as opposed to making the best or least cost assignment.

In this method, we use the following procedure:

WORKING PROCEDURE

STEP 1. **(Find the opportunity cost table)**

Subtract the smallest element of each row of the cost matrix $[c_{ij}]$ from all the elements of the respective rows and then modify the reduced cost matrix by subtracting the smallest element of each column from all the elements of the respective columns. These two operations on rows and columns create zeroes.

STEP 2. **(Make the assignments)**

➡ Identify rows successively from top to bottom until a row with exactly one zero element is found. Make an assignment to this single zero by making a square (□) around it and crossed off (×) all other zeroes in the corresponding column.

➡ Identify column successively from left to right with exactly one zero element that has not been assigned. Make assignment to the single zero by making a square (□) around it and crossed-off (×) all other zero elements in the corresponding rows.

➡ If a row (or column) has two or more unmarked zeroes and one cannot be chosen by inspection, then select the zero cell arbitrarily for assignment.

➡ Repeat the above steps successively until all the zeroes have been examined.

STEP 3. **(Optimality check)**

(i) If all zero elements in the matrix are either marked with square (□) or cross-off

(×) and there is exactly one assignment in each row and column then it is an optimal solution. Here the total cost associated with this solution is obtained by adding original cost figures in the occupied cells.

(ii) If a zero element in a row or column was chosen arbitrarily for assignment in the above step (i), there exist an alternate optimal solution.

(iii) If there is no assignment in a row (or column) then it implies that the total number of assignments are less than the no. of rows (or columns) in the square matrix. In such cases go to the next step.

STEP 4. **(Revision of opportunity cost matrix)**

We draw a set of horizontal and vertical lines to cover all the zeroes in the revised cost matrix obtained from step 3 by using the following procedure.

(i) For each row with no assignments mark a tick (✓).

(ii) Examine the marked rows. If any zero element occurs in those rows, mark a tick (✓) to the respective columns containing those zeroes.

(iii) Examine marked columns. If any assigned zero elements occur in those columns tick (✓) the respective rows containing those assigned zeroes.

(iv) Repeat this process until no more rows or columns can be marked.

(v) Draw a straight line through each marked column and each unmarked row.

If the no. of lines drawn (or total assignments) is equal to the number of rows (or columns), the current solution is the optimal solution otherwise go to the next step.

STEP 5. **(Find the new Revised Opportunity Cost matrix)**

(i) Choose the smallest element (say m) among the cells not covered by any line.

(ii) Subtract m from every element in the cell not covered by a line.

(iii) Add m to every element in the cell covered by two lines, *i.e.*, intersection of two lines.

(iv) Elements in cells covered by one line remains unchanged.

STEP 6. Repeat steps 3 to 5 until an optimal solution is obtained.

 Solved Examples

EXAMPLE 1. *A department head has four subordinates and four tasks have to be performed. Subordinates differ in efficiency and tasks differ in their intrinsic difficulties. Time each man would take to perform each task is given in the effectiveness matrix. How the tasks should be allocated to each person so as to minimize the total man-hour?* [KANPUR–2007]

	I	II	III	IV
A	8	26	17	11
B	13	28	4	26
C	38	19	18	15
D	19	26	24	10

SOLUTION. (i) **Row Reduced Matrix:** Subtract the smallest element of each row from all the elements of the respective rows, we get the following row-reduced matrix.

$$\begin{vmatrix} 0 & 18 & 9 & 3 \\ 9 & 24 & 0 & 22 \\ 23 & 4 & 3 & 0 \\ 9 & 16 & 14 & 0 \end{vmatrix}$$

(ii) **Column reduced matrix:** Subtract the smallest element of each column from all the elements of respective column, we get

$$\begin{vmatrix} 0 & 14 & 9 & 3 \\ 9 & 20 & 0 & 22 \\ 23 & 0 & 3 & 0 \\ 9 & 12 & 14 & 0 \end{vmatrix}$$

(iii) Now we have to allot zero assignments, starting from the first row. Now, we allot the [0] and cancel the corresponding zeroes in row and column

	I	II	III	IV
A	[0]	14	9	3
B	9	20	[0]	22
C	23	[0]	3	X
D	9	12	14	[0]

(a) In the first row a [0] is allotted and corresponding row has been cancelled.

(b) In the second row a [0] is allotted and in the same manner corresponding row has been cancelled.

(c) But in the third row, two [0]'s are present. One is in the III column and other in the IV column.

We have chosen the zero in the III row as our [0] assignment. This is because if we have taken zero in the IV row as our [0] assignment, then zero in the IV row must have also cancelled.

(iv) As we have allotted the [0] assignments, we examine that [0] equals number of row. The solution becomes $A \rightarrow I$, $B \rightarrow III$, $C \rightarrow II$, $D \rightarrow IV$.

EXAMPLE 2. *Solve the assignment problem represented by the following matrix* [KANPUR–2009]

	I	II	III	IV	V	VI
A	9	22	58	11	19	27
B	43	78	72	50	63	48
C	41	28	91	37	45	33
D	74	42	27	49	39	32
E	36	11	57	22	25	18
F	3	56	53	31	17	28

SOLUTION. STEP-1. **Row Reduced Matrix**

	I	II	III	IV	V	VI
A	0	13	49	2	10	18
B	0	35	29	7	20	5
C	13	0	63	9	17	5
D	4	15	0	22	12	5
E	25	0	46	11	14	7
F	0	53	50	28	14	25

STEP-2. Column Reduced Matrix

	I	II	III	IV	V	VI
A	0	13	49	0	0	13
B	0	35	29	5	10	0
C	13	0	63	7	7	0
D	4	15	0	20	2	0
E	25	0	46	9	4	2
F	0	53	50	26	4	20

STEP-3. Assignment Matrix

	I	II	III	IV	V	VI
A	⊠	13	49	⓪	⊠	13
B	⊠	35	29	5	10	⓪
C	13	0	63	7	7	⊠
D	4	15	⓪	20	2	⊠
E	25	⓪	46	9	4	2
F	⓪	53	50	26	4	20

Here, row 3 and column 5 have no assignments.

STEP-4. To draw the minimum number of lines, we proceed as follows:

(i) We tick (✓) row 3 in which there is no assignment.

(ii) We tick (✓) column 2 and 5 which have zeroes in ticked row 3.

(iii) We tick (✓) rows 5 and 2 which have assignments in the ticked columns 2 and 6.

(iv) Then we tick column 1 (not already ticked) which has zero in the ticked row 2.

(v) Then, we tick row 6 which has assignment in the ticked column 1.

(vi) Draw lines through all ticked columns 1, 2, 6. Then draw lines through unticked row 1 and 4 having zeros through which there is no line. Therefore, we get 5 lines to cover all the zeros.

	L_1	L_2			L_3		
L_4	⊠	13	49	⓪	⊠	13	
	⊠	35	29	5	10	⓪	✓(5)
	13	⊠	63	7	7	⊠	✓(1)
L_5	4	15	⓪	20	2	⊠	
	25	⓪	46	9	4	2	✓(4)
	⓪	53	50	26	4	20	✓(7)
	✓(6)	✓(2)			✓(3)		

STEP-5. Select smallest element, *i.e.*, 4 that do not have a line through them. Subtracting this element from all the elements that do not have a line through them and adding to every element that lies at the intersection of two lines and leaving the remaining elements unchanged. Further applying step 1 to 3, we get the following assignment matrix

	I	II	III	IV	V	VI
A	4	17	49	⓪	⊠	17
B	⓪	35	25	1	6	⊠
C	13	⊠	59	3	3	0
D	51	19	⓪	20	2	4
E	25	⓪	42	5	⊠	2
F	⊠	53	46	22	⓪	20

Hence, the optimal assignment is given by
$$A \to IV, \ B \to I, \ C \to VI, \ D \to III, \ E \to II, \ F \to V$$
with Minimum cost $= 11 + 43 + 33 + 27 + 11 + 17 = 142$.

EXAMPLE 3. *Solve the following assignment problems*

Man

		(1)	(2)	(3)	(4)	(5)
	A	11	17	8	16	20
	B	9	7	12	6	15
Job	C	13	16	15	12	16
	D	21	24	17	28	26
	E	24	10	12	11	15

SOLUTION. (i) **Row reduced matrix:** First we find out the row reduced matrix by subtracting from each element the minimum element of its row.

	(1)	(2)	(3)	(4)	(5)
A	3	9	0	8	12
B	3	1	6	0	9
C	1	4	3	0	4
D	4	7	0	11	9
E	14	0	2	1	5

(ii) **Column reduced matrix.** This can be found out by subtracting the minimum element from other elements of every column by taking row reduced matrix as the primary matrix

	(1)	(2)	(3)	(4)	(5)
A	2	9	0	8	8
B	2	1	6	0	5
C	0	4	3	0	0
D	3	7	0	11	5
E	3	0	2	1	1

(iii) **Assignment matrix**

	(1)	(2)	(3)	(4)	(5)	
A	2	9	[0]	8	8	✓
B	2	1	6	[0]	5	✓
C	[0]	4	3	0	⊠	
D	3	7	0	11	5	✓
E	3	[0]	2	1	1	

Since the number of [0] assignments is less than 5, *i.e.*, the number of rows. Hence, the above is not the required solution.

So, now we find out the minimum element from the untouched elements of the assignment matrix or model. Subtract this element from all the untouched elements. Add this element to those elements which are at the intersection of two lines. Rest of the elements remain the same.

So, minimum element = 1.

The new assignment model becomes

	(1)	(2)	(3)	(4)	(5)	
A	[1]	8	[0]	8	7	✓
B	1	⊠	6	[0]	4	
C	[0]	4	4	1	⊠	
D	2	6	⊠	11	4	✓
E	3	[0]	3	2	⊠	

Number of 0 assignments ≠ 5.

So, again proceeding in the same way, we have Minimum element = 1

	(1)	(2)	(3)	(4)	(5)	
A	[0]	7	⊠	7	6	
B	1	⊠	7	[0]	4	
C	⊠	4	5	1	[0]	✓
D	1	5	[0]	10	3	
E	3	[0]	4	2	⊠	

Number of $\boxed{0}$ assignments = 5.

The required solution becomes

$$A \to I, \ B \to IV, \ C \to V, \ D \to III, \ E \to II$$

EXAMPLE 4. *Five wagons are available at five stations 1, 2, 3, 4, 5. These are required at five stations I, II, III, IV and V. The mileages between various stations are given by the following matrix:*

To station

	I	II	III	IV	V
1	10	5	9	18	11
2	13	9	6	12	14
3	3	2	4	4	5
4	18	9	12	17	15
5	11	6	14	19	10

From station

How should the wagons be transported so as to minimize the total mileage covered?

SOLUTION. **Row reduced matrix**

	I	II	III	IV	V
1	5	0	4	13	6
2	7	3	0	6	8
3	1	0	2	2	3
4	9	0	3	8	6
5	5	0	8	13	4

Column reduced matrix

	I	II	III	IV	V
1	4	0	4	11	3
2	6	3	0	4	5
3	0	0	2	0	0
4	8	0	3	6	3
5	4	0	8	11	1

Assignment model

	I	II	III	IV	V
1	4	$\boxed{0}$	4	11	3
2	6	3	$\boxed{0}$	4	5
3	$\boxed{0}$	0	2	0	0
4	8	0	3	6	3
5	4	0	8	11	1

Number of $\boxed{0}$ assignments ≠ 5.

Again, proceeding for $\boxed{0}$ assignments, we have minimum element = 1.

New Assignment Model

The row reduced and column reduced matrix are same as the assignment model.

So, we are not writing them here again.

	I	II	III	IV	V
1	3	0	3	10	2
2	6	4	0	4	5
3	0	1	2	0	0
4	7	0	2	5	2
5	3	0	7	10	0

Again, number of 0 assignments $\neq$ 5.

So, the new minimum element = 2. Subtract 2 from all the untouched elements. Add this element which are at the intersection of two lines. Rest of the elements remains the same.

New Assignment Model

	I	II	III	IV	V
1	1	0	1	8	2
2	6	6	0	4	5
3	0	3	2	0	0
4	6	0	0	3	0
5	3	2	7	10	0

Again, number of 0 assignments $\neq$ 5.

$\therefore$ Minimum element = 3.

New Assignment Model becomes

	I	II	III	IV	V
1	1	3	4	8	3
2	3	6	0	1	5
3	0	3	3	0	0
4	3	0	0	0	0
5	0	2	4	7	0

But the above matrix is not in the reduced form. So, we proceed as follows

Row reduced Matrix

	I	II	III	IV	V
1	0	2	3	7	2
2	3	6	0	1	5
3	0	3	3	0	0
4	3	0	0	0	0
5	0	2	4	7	0

Since the column reduced matrix and the assignment model for the row reduced matrix were same, so we have given 0 assignments in the row reduced matrix itself.

$\therefore$ $1 \to I$, $2 \to III$, $3 \to IV$, $4 \to II$, $5 \to V$

Minimum distance = 10 + 6 + 4 + 9 + 10

                   = 39 miles.

EXAMPLE 5. *An airline that operates seven days a week, has the time table shown below. Crews must have a minimum layover time of 1 hour between flights. Obtain the pairing of flights that minimizes layover time away from home. For any given pairing, the crew will be based at the city that results in the smaller layover. For each pair, also mention the town where crew should be based.*

(i)	Delhi-Srinagar		Srinagar-Delhi		
Flight No.	Depart	Arrive	Flight No.	Depart	Arrive
1	7:30 A.M.	9:00 A.M.	2	7:00 A.M.	10:00 A.M.
3	8:45 A.M.	9:45 A.M.	4	7:45 A.M.	10:45 A.M.
5	2:00 P.M.	3:30 P.M.	6	11:00 A.M.	2:00 P.M.
7	5:45 P.M.	7:15 P.M.	8	6:00 P.M.	9:00 P.M.
9	7:00 P.M.	8:30 P.M.	10	7:30 P.M.	10:30 P.M.

(ii)	Delhi-Calcutta		Calcutta-Delhi		
Flight No.	Depart	Arrive	Flight No.	Depart	Arrive
1	7:00 A.M.	9:00 A.M.	101	9:00 A.M.	11:00 A.M.
2	9:00 A.M.	11:00 A.M.	102	10:00 A.M.	12:00 Noon
3	1:30 P.M.	3:30 P.M.	103	3:30 P.M.	5:30 P.M.
4	7:30 P.M.	9:30 P.M.	104	8:00 P.M.	10:00 P.M.

The minimum layover time is 6 hours for part (ii).

SOLUTION. (i) First, we must construct the assignment model for both the flights when their layover time is 1 hour.

Crew based at Delhi

	2	4	6	8	10
1	22	22.75	2	9	10.5
3	21.25	22	1.25	8.25	9.75
5	15.5	16.25	19.25	2.5	4
7	11.75	12.5	15.75	22.75	23.75
9	10.5	11.25	14.5	21.5	23

Crew based at Srinagar

	2	4	6	8	10
1	20.5	20.75	17.5	10.5	9
3	22.75	21	18.75	11.75	10.25
5	4	3.25	24	17	15.5
7	7.75	7	3.75	20.75	19.25
9	9	8.25	5	22	20.5

Now, we combine the above two matrices, choosing the base which gives minimum layover time. In the composite matrix, the cell marked (*) means crew based at Srinagar, otherwise it is Delhi.

	2	4	6	8	10
1	20.5 *	20.75 *	2	9	9 *
3	21.25	21 *	1.25	8.25	9.75
5	4 *	3.25 *	19.25	2.5	4
7	7.75 *	7 *	3.75 *	20.75 *	19.25 *
9	9 *	8.25 *	5 *	21.50	20.5 *

Now, we proceed as usual to solve the obtained assignment matrix

Row reduced matrix

	2	4	6	8	10
1	18.5*	18.75*	0	7	7*
3	20	19.75*	0	12.75	11.25
5	1.5*	0.75*	16.75	0	1.5
7	4*	3.25*	0*	17*	15.5*
9	4*	3.25*	0*	16.5	15.5*

Column reduced matrix

	2	4	6	8	10
1	17.0*	18	0	7	5.5*
3	18.5	19*	0	12.75	9.75
5	0*	0	16.75	0	0
7	2.5*	2.5*	0*	17.0*	14.0*
9	2.5*	2.5*	0*	16.5	14.0*

Assignment matrix

	2	4	6	8	10
1	17.0*	18	[0]	7	5.5*
3	18.5	19*	0	12.75	9.75
5	[0]*	0	16.75	0	0
7	2.5*	2.5*	0*	17.0*	14.0*
9	2.5*	2.5*	0*	16.5	14.0*

Since number of [0] assignment ≠ 5. Hence, minimum element = 2.5.
New Assignment Model can be obtained

	2	4	6	8	10
1	14.5*	15.5*	[0]	4.5	3*
3	16.0	16.5*	0	10.25	7.25
5	0*	0*	19.25	[0]	0
7	[0]*	0*	0*	14.5*	11.5*
9	0*	[0]*	0*	14.0	11.5*

Again, number of [0] assignments ≠ 5.
So, minimum element = 3.
We again form new assignment model by applying usual procedure.

	2	4	6	8	10
1	14.5*	15.5*	0	1.5	0*
3	16	16.5*	0	7.25	4.25
5	3*	3*	22.25	0	0
7	3*	3*	3*	14.5*	11.5*
9	3*	3*	3*	14	11.5*

But, it is not in the reduced form. So, we first reduce it to Row reduced Matrix and Column Reduced Matrix and then allot the assignments :

Assignment matrix

	2	4	6	8	10
1	14.5*	15.5*	0	1.5	[0]*
3	16	16.5*	[0]	7.25	4.25
5	3*	3*	22.25	[0]	0
7	[0]*	0*	0*	11.5*	8.5*
9	0*	[0]*	0*	11	8.5*

Since Row Reduced matrix, Column Reduced matrix and Assignment model are same, so we have not mentioned them here.

So, $1 \to 10$, $3 \to 6$, $5 \to 8$, $7 \to 2$, $9 \to 4$.

Minimum layover time = 28.75 hours.

(ii) So, proceeding as usually, we first construct the tables for the layover time between two flights

Layover time in hours

When crew based at Delhi

	101	102	103	104
1	24	25	6.5	11
2	22	23	28.5	9
3	17.5	18.5	24	28.5
4	11.5	12.5	18	22.5

When crew based at Calcutta

	101	102	103	104
1	20	22	26.5	8.5
2	19	21	25.5	7.5
3	13.5	15.5	20	26
4	9	11	15.5	21.5

Now, we combine the above two tables, choosing that base which gives a lesser layover time for each pair. But before that, we multiply the above matrices by 4, to reduce the fractional complications and considering the layover time for 4 weeks.

When crew based at Delhi

	101	102	103	104
1	96	100	26	44
2	88	92	114	36
3	70	74	96	114
4	46	50	72	90

When crew based at Calcutta

	101	102	103	104
1	80	76	54	36
2	88	84	62	44
3	106	102	80	62
4	34	30	104	86

In the composite the cell marked (*) means that the crew is based at Calcutta.

	101	102	103	104
1	80*	76	26	36*
2	88*	84*	62*	36
3	70	74	80*	110*
4	34*	30*	72	86*

Now, we follow our usual method of solving the assignment problem.

Using general method of solving the assignment problem, we get the final assignment model as

54*	50*	0	10*
52*	48*	26*	0
0	4	10*	40*
4*	0*	42	56*

Allotting the 0 assignments, we have

54 *	50 *	[0]	10 *
52 *	48 *	26 *	[0]
[0]	4	10 *	40 *
4 *	[0] *	42	56 *

Hence, the optimal assignment with the base of the crew is as follows:

$$1 \rightarrow 103, \quad 2 \rightarrow 104, \quad 3 \rightarrow 101, \quad 4 \rightarrow 103$$

| Crew at | (Delhi) | (Delhi) | (Delhi) | (Calcutta) |

12.6 THE MAXIMAL ASSIGNMENT PROBLEM

So far, we have been dealing with the assignment problems which involve minimization. But, sometimes, it may happen that instead of being minimized, the problem utters to be maximized. In this case, we may follow the slightly different but simple method for the solution, we can proceed as given below:

(a) We must first select the greatest element of our given assignment problem and must subtract each element from this element to obtain a new matrix. The new matrix obtained can now be solved by the usual method of assignment technique used for minimal problems.

(b) One another method is also used for the conversion of maximization to minimization. Place minus sign before each element of the given assignment matrix to obtain a new matrix. The new matrix thus can be treated as the usual assignment minimal problem.

Solved Examples

EXAMPLE 1. *Alpha Corporation has four plants each of which can manufacture any of the four products. Production costs differ from plant to plant as do sales revenue. From the following data, obtain which product each plant should produce to maximize profit?*

Sales revenue (₹'000)				Production cost (₹'000)			
Product				Product			
Plant ↓	1	2	3	4	Plant ↓	1	2	3	4
A	50	68	49	62	A	49	60	45	61
B	60	70	51	74	B	55	63	45	69
C	55	67	53	70	C	52	62	49	68
D	58	65	54	69	D	55	64	48	66

SOLUTION.   In this case, we must first find out the Profit Matrix

$$\text{Profit Matrix} = \text{Sales revenue} - \text{Production cost}$$

	1	2	3	4
A	1	8	4	1
B	5	7	6	5
C	3	5	3	2
D	3	1	6	3

Since profit matrix is a maximization problem. Hence, first we convert it into minimization.

Minimization Problem

	1	2	3	4
A	7	0	4	7
B	3	1	2	3
C	5	3	5	6
D	5	7	2	5

Row Reduced Matrix

	1	2	3	4
A	7	0	4	7
B	2	0	1	2
C	2	0	2	3
D	3	5	0	3

Column Reduced Matrix

	1	2	3	4
A	5	0	4	5
B	0	0	1	0
C	0	0	2	1
D	1	5	0	1

Assignment Model

	1	2	3	4
A	5	[0]	4	5
B	0	0	1	[0]
C	[0]	0	2	1
D	1	5	[0]	1

Since number of assignments = 4. Hence, the required solution is

$$A \to 2, \quad B \to 4, \quad C \to 1, \quad D \to 3$$

$$\text{Profit} = (8 + 5 + 3 + 6) \times 1000 = ₹\ 22000.$$

12.7 UNBALANCED ASSIGNMENT PROBLEM

Sometimes, during a problem, the number of jobs is not equal to number of persons, that means our assignment matrix is not square. Such kind of problems are known as unbalanced assignment problem.

These type of problems are solved by the introduction of a supposed or dummy rows or columns in which each element has zero cost to form a square matrix. The new square matrix can be solved by the usual method of balanced minimal assignment problem.

[MEERUT–2004, 07, 09, 12]

 Solved Examples

EXAMPLE 1. *A company has 4 machines on which to do 3 jobs. Each job can be assigned to one and only one member. The cost of each job on each machine is given in the following table*

Job↓	Machine			
	W	X	Y	Z
A	18	24	28	32
B	8	13	17	19
C	10	15	19	22

What are the job assignments that will minimize the cost?

SOLUTION. The given problem is an unbalanced problem, since the number of jobs does not equals the number of machines. So, we add a dummy row (Job) with elements of zero cost. So, we have

$$
\begin{array}{c|cccc}
 & W & X & Y & Z \\
\hline
A & 18 & 24 & 28 & 32 \\
B & 8 & 13 & 17 & 19 \\
C & 10 & 15 & 19 & 22 \\
D & 0 & 0 & 0 & 0 \\
\end{array}
$$

Row Reduced Matrix

$$
\begin{array}{c|cccc}
 & W & X & Y & Z \\
\hline
A & 0 & 6 & 10 & 14 \\
B & 0 & 5 & 9 & 11 \\
C & 0 & 5 & 9 & 12 \\
D & 0 & 0 & 0 & 0 \\
\end{array}
$$

Column Reduced Matrix

$$
\begin{array}{c|cccc}
 & W & X & Y & Z \\
\hline
A & 0 & 1 & 1 & 3 \\
B & 0 & 0 & 0 & 0 \\
C & 0 & 0 & 0 & 1 \\
D & 0 & 0 & 0 & 0 \\
\end{array}
$$

Assignment Model

$$
\begin{array}{c|cccc}
 & W & X & Y & Z \\
\hline
A & [0] & 1 & 1 & 3 \\
B & 0 & [0] & 0 & 0 \\
C & 0 & 0 & [0] & 1 \\
D & 0 & 0 & 0 & [0] \\
\end{array}
$$

So, solution becomes $A \rightarrow W,\ B \rightarrow X,\ C \rightarrow Y,\ D \rightarrow Z$
But, D is only assumed job. Hence, we have
$$A \rightarrow W,\ B \rightarrow X,\ C \rightarrow Y$$

Exercise-12.1

1. Solve the following assignment problems :

(a)

Men

Tasks	I	II	III	IV	V
A	1	3	2	8	8
B	2	4	3	1	5
C	5	6	3	4	6
D	3	1	4	2	2
E	1	5	6	5	4

(b)

Persons

Tasks	1	2	3	4
A	10	12	19	11
B	5	10	7	8
C	12	14	13	11
D	8	15	11	9

(c)

Men

Tasks	1	2	3	4	5
I	12	8	7	15	4
II	7	9	17	14	10
III	9	6	12	6	7
IV	7	6	14	6	10
V	9	6	12	10	6

2. Find the minimum cost solution for the 5×5 assignment problem whose cost coefficients are as given below :

	1	2	3	4	5
1	−2	−4	−8	−6	−1
2	0	−9	−5	−5	−4
3	−3	−8	0	−2	−6
4	−4	−3	−1	0	−3
5	−9	−5	−8	−9	−5

3. The owner of a small machine shop has four machinists available to assign to jobs for the day. Five jobs are offered with the expected profit (in ₹) for each machinist on each job being as follows :

Mechinist	A	B	C	D	E
1	6.20	7.80	5.00	10.10	8.20
2	7.10	8.40	6.10	7.30	5.90
3	8.70	9.20	11.10	7.10	8.10
4	4.80	6.40	8.70	7.70	8.00

Find out assignment of mechinists to jobs that will result in a maximum profit. Which job should be declined?

4. Use Hungarian method to solve the following cost minimizing assignment problem.

	1	2	3	4
I	20	22	28	15
II	16	20	12	13
III	19	23	14	25
IV	10	16	12	10

5. Solve

Jobs

Persons	J_1	J_2	J_3	J_4	J_5
P_1	7	8	6	5	9
P_2	9	6	7	6	10
P_3	8	7	9	5	6

6. Find the optimal assignment

Programs

Programmers	A	B	C
1	120	100	80
2	70	90	110
3	110	140	120

Assign the programs to the programmers so that the total time taken is least. Here, the elements of the matrix are the time in minutes.

7. Solve the assignment problem

Location

Project	I	II	III	IV	V
A	15	21	6	4	9
B	3	40	21	10	7
C	9	6	5	8	10
D	14	8	6	9	3
E	21	16	18	7	4

8. Solve the assignment problem

Machine

		M_1	M_2	M_3
	J_1	8	7	6
Job	J_2	5	7	8
	J_3	6	8	7

ANSWERS

1. (a) $A \to I, B \to IV, C \to III, D \to II, E \to V$, Min. cost = 10.

  (b) $A \to 2, B \to 3, C \to 4, D \to I$, Minimum cost = 38.

  (c) (i) $I \to 3, II \to 1, III \to 2, IV \to 4, V \to 5$

     (ii) $I \to 3, II \to 1, III \to 4, IV \to 2, V \to 5$

2. $1 \to 3, 2 \to 2, 3 \to 5, 4 \to 4, 5 \to 1$ or $1 \to 4, 2 \to 2, 3 \to 3, 4 \to 5, 5 \to 1$, Min. cost = 36

3. $1 \to D, 2 \to B, 3 \to C, 4 \to E, 5 \to A$, Minimum cost = ₹ 37.60, Job A should be declined.

4. $1 \to 2, II \to 4, III \to 3, IV \to 1$ or $I \to 4, II \to 2, III \to 3, IV \to 1$

5. $P_1 \to J_4, P_2 \to J_2, P_3 \to J_5$; Minimum cost = 17, Jobs J_1 and J_2 left undone.

6. $1 \to C, 2 \to B, 3 \to A$, Minimum cost = 280.

7. $A \to IV, B \to I, C \to II, D \to III, E \to V$.

8. $J_1 \to M_3, J_2 \to M_2, J_3 \to M_1$ or $J_1 \to M_3, J_2 \to M_1, J_3 \to M_2$, Min. cost = 19

12.8 SOME MISCELLANEOUS SOLVED EXAMPLES

EXAMPLE 1. *Solve the following minimal assignment problem.*

Men

		1	2	3	4
	I	12	30	21	15
Job	II	18	33	9	31
	III	44	25	24	21
	IV	23	30	28	14

SOLUTION. **Row Reduced Matrix.** Subtracting the smallest element of each row from every element of the corresponding row, we get the following row reduced matrix

	1	2	3	4
I	0	18	9	3
II	9	24	0	22
III	23	4	3	0
IV	9	16	14	0

Column Reduced Matrix. Subtracting smallest element of each column from every element of the corresponding column, we get the following column reduced matrix

	1	2	3	4
I	0	14	9	3
II	9	20	0	22
III	23	0	3	0
IV	9	12	14	0

Assignment Matrix. Starting with row I, we mark □ in the row containing only one zero and cross (×) the zeroes in the corresponding column in which □ lies.

	1	2	3	4
I	0̄	9	14	3
II	9	20	0̄	22
III	23	0	3	✗
IV	9	12	14	0̄

Further, starting with column 1, we mark □ in the column containing only one unmarked or uncrossed zero in the above table and cross out the zeroes in the corresponding row in which the assignment □ is marked. Then, we get the following table.

	1	2	3	4
I	0̄	14	9	3
II	9	20	0	22
III	23	0̄	3	✗
IV	9	12	14	0̄

In this table, every row and column have one assignment, therefore we have the following optimal assignment.

$$I \to 1, \ II \to 3, \ III \to 2, \ IV \to 4$$

EXAMPLE 2. *Solve the minimal assignment problem whose effectiveness matrix is given below*

	1	2	3	4
I	2	3	4	5
II	4	5	6	7
III	7	8	9	8
IV	3	5	8	4

SOLUTION. **Row Reduced Matrix**

	1	2	3	4
I	0	1	2	3
II	0	1	2	3
III	0	1	2	1
IV	0	2	5	1

Column Reduced Matrix

	1	2	3	4
I	0	0	0	2
II	0	0	0	2
III	0	0	0	0
IV	0	1	3	0

Assignment Matrix

```
       1  2  3  4              1  2  3  4              1  2  3  4
   I  | ⊠  0  0  2 |       I  | ⊠  0  0  2 |       I  | ⊠  0  0  2 |
  II  | ⊠  0  0  2 |  ⇒   II  | ⊠  0  0  2 |  ⇒   II  | ⊠  ⊠  0  2 |
 III  | ⊠  0  0  0 |      III | ⊠  ⊠  ⊠  0 |      III | ⊠  ⊠  ⊠  0 |
  IV  | 0  1  3  ⊠ |       IV | 0  1  3  ⊠ |       IV | 0  1  3  ⊠ |
```

Hence, optimal assignment is given by

$I \to 2,\ II \to 3,\ III \to 4,\ IV \to 1$ and minimum cost $= 3 + 6 + 8 + 3 = 20$.

EXAMPLE 3. *Solve the following assignment problem* [MEERUT–2009, 12]

```
     I  II  III  IV  V
 A | 1   3   2   3   6 |
 B | 2   4   3   1   5 |
 C | 5   6   3   4   6 |
 D | 3   1   4   2   2 |
 E | 1   5   6   5   4 |
```

SOLUTION. The row and column reduced matrix is given by

```
     I  II  III  IV  V
 A | 0   2   1   2   4 |
 B | 1   3   2   0   3 |
 C | 2   3   0   1   2 |
 D | 2   0   3   1   0 |
 E | 0   4   5   4   2 |
```

The assignment matrix is given by

```
     I  II  III  IV  V
 A | [0]  2   1   2   4 |
 B | 1   3   2  [0]  3 |
 C | 2   3  [0]  1   2 |
 D | 2  [0]  3   1   ⊠ |
 E | ⊠   4   5   4   2 |
```

Since row 4 and column 5 have no assignments, so we proceed as follows

```
  L₁
 ┌──────────────────┐
 │[0]  2   1   2   4 │  ✓(3)
 │ 1┼─ 3   2  [0] ─3 │  L₂
 │ 2┼─ 3  [0]  1 ─2  │  L₃
 │ 2┼─[0]  3   1 ─⊠  │  L₄
 │ ⊠   4   5   4   2 │  ✓(3)
 └──────────────────┘
    ✓(3)
```

The smallest element that do not contain line through them is 1. Subtracting 1 from the elements that do not have a line through them, adding to every element that lies at the intersection of two lines and leaving the remaining elements unchanged and then repeating the above steps, we get the following matrix

	I	II	III	IV	V
A	⊠	1	⊠	1	3
B	2	3	2	0̄	3
C	3	3	0̄	1	2
D	3	0̄	3	1	⊠
E	0̄	3	4	3	1

Here row 1 and column 5 do not contain any assignments. Therefore, we again repeat the above steps. We draw the lines as follows

The minimum number of lines can be drawn as follows

$$
\begin{array}{c}
L_1 \qquad\quad L_2 \\
\begin{array}{|ccccc|c}
\hline
\cancel{⊠} & 1 & \cancel{⊠} & 1 & 3 & ✓(1)\\
2 & 3 & 2 & \boxed{0} & 3 & L_3\\
3 & 3 & \boxed{0} & 1 & 2 & ✓(4)\\
3 & \boxed{0} & 3 & 1 & 0 & L_4\\
\boxed{0} & 3 & 4 & 3 & 1 & ✓(5)\\
\hline
\end{array}\\
✓(3) \qquad ✓(2)
\end{array}
$$

Here, we see that even now row 1 and column 5 do not contain any assignment. So, we proceed as follows

Take the minimum element, *i.e.*, 1 and repeat the same process as above, we get

	I	II	III	IV	V
A	0̄	⊠	⊠	⊠	2
B	3	3	2	0̄	3
C	3	2	0̄	⊠	1
D	4	0̄	4	1	⊠
E	⊠	2	4	2	0̄

Hence, the optimal assignment is given by

$$A \to I,\ B \to IV,\ C \to III,\ D \to II,\ E \to V$$

with Minimum cost = $1 + 1 + 3 + 1 + 4 = 10$.

EXAMPLE 4. *Find the optimal assignment for the following problem*   [MEERUT–2007]

	I	II	III	IV
A	5	3	1	8
B	7	9	2	6
C	6	4	5	7
D	5	7	7	6

SOLUTION.   **Row reduced matrix.** Find the smallest element of each row and subtract it from each element of the row, we get

$$
\begin{array}{c c c c c}
 & I & II & III & IV \\
A & 4 & 2 & 0 & 7 \\
B & 5 & 7 & 0 & 4 \\
C & 2 & 0 & 1 & 3 \\
D & 0 & 2 & 2 & 1 \\
\end{array}
$$

Column reduced matrix. Find the smallest element of each column and subtract it from each element of that column, we get

$$
\begin{array}{c c c c c}
 & I & II & III & IV \\
A & 4 & 2 & 0 & 6 \\
B & 5 & 7 & 0 & 3 \\
C & 2 & 0 & 1 & 2 \\
D & 0 & 2 & 2 & 0 \\
\end{array}
$$

Assignment matrix. Making an assignment to single zero, we get

$$
\begin{array}{c c c c c}
 & I & II & III & IV \\
A & 4 & 2 & \boxed{0} & 6 \\
B & 5 & 7 & \cancel{0} & 3 \\
C & 2 & \boxed{0} & 1 & 2 \\
D & \boxed{0} & 2 & 2 & \cancel{0} \\
\end{array}
$$

Clearly row B has no assignment. So, tick that row which has no assignment corresponding to that row, tick that column which has cross zero and corresponding to that column, tick that row which has assigned zero

$$
\begin{array}{c c c c c c}
 & I & II & III & IV \\
A & 4 & 2 & 0 & 6 & \checkmark \\
B & 5 & 7 & \cancel{0} & 3 & \checkmark \\
C & 2 & \boxed{0} & 1 & 2 & \\
D & \boxed{0} & 2 & 2 & \cancel{0} & \\
 & & & \checkmark & &
\end{array}
$$

Cross unticked rows and ticked column. Find smallest element from the remaining matrix. Subtract this smallest element from the unticked elements. Then proceeding same as usual, we get the next reduced matrix as follows :

$$
\begin{array}{c c c c c}
 & I & II & III & IV \\
A & 2 & 0 & 0 & 4 \\
B & 3 & 5 & 0 & 1 \\
C & 2 & 0 & 3 & 2 \\
D & 0 & 2 & 4 & 0 \\
\end{array}
$$

Then, we have the following single zero assignment matrix.

$$
\begin{array}{c c c c c c}
 & I & II & III & IV \\
A & 2 & \cancel{0} & \cancel{0} & 4 & \checkmark \\
B & 3 & 5 & \boxed{0} & 1 & \checkmark \\
C & 2 & \boxed{0} & 3 & 2 & \checkmark \\
D & \boxed{0} & 2 & 4 & \cancel{0} & \\
\end{array}
$$

$$
\begin{array}{c c c c c}
 & I & II & III & IV \\
A & 1 & 0 & \boxed{0} & 3 \\
\Rightarrow \quad B & 2 & 5 & 0 & \boxed{0} \\
C & 1 & \boxed{0} & 3 & 1 \\
D & \boxed{0} & 3 & 5 & 0
\end{array}
$$

Hence, final assignment is given by
$$A \to III, \ B \to IV, \ C \to II, \ D \to I.$$

EXAMPLE 5. *Solve the following assignment problem*

[GORAKHPUR–2007; MEERUT–2008; KANPUR–2012]

$$
\begin{array}{c c c c c}
 & 1 & 2 & 3 & 4 \\
A & 10 & 12 & 19 & 11 \\
B & 5 & 10 & 7 & 8 \\
C & 12 & 14 & 13 & 11 \\
D & 8 & 15 & 19 & 9
\end{array}
$$

SOLUTION. **Row reduced matrix.**

$$
\begin{array}{c c c c c}
 & 1 & 2 & 3 & 4 \\
A & 0 & 2 & 9 & 1 \\
B & 0 & 5 & 2 & 3 \\
C & 1 & 3 & 2 & 0 \\
D & 0 & 7 & 3 & 1
\end{array}
$$

Column reduced matrix.

$$
\begin{array}{c c c c c}
 & 1 & 2 & 3 & 4 \\
A & 0 & 0 & 7 & 1 \\
B & 0 & 3 & 0 & 3 \\
C & 1 & 1 & 0 & 0 \\
D & 0 & 5 & 1 & 1
\end{array}
$$

Assignment matrix.

$$
\begin{array}{c c c c c}
 & 1 & 2 & 3 & 4 \\
A & \cancel{0} & \boxed{0} & 7 & 1 \\
B & \cancel{0} & 3 & \boxed{0} & 3 \\
C & 1 & 1 & \cancel{0} & \boxed{0} \\
D & \boxed{0} & 5 & 1 & 1
\end{array}
$$

Hence, the required assignment is given by $A \to 2, \ B \to 3, \ C \to 4, \ D \to 1$.

EXAMPLE 6. *A company is faced with the problem of assigning six different machines to five different jobs. The cost are estimated as follows*

Jobs

	1	2	3	4	5
1	2.5	5	1	6	1
2	2	5	1.5	7	3
Machine 3	3	6.5	2	8	3
4	3.5	7	2	9	4.5
5	4	7	3	9	6
6	6	9	5	10	6

Solve the problem assuring that the objective is to minimize total cost.

[MEERUT–2006]

SOLUTION. It is observed that the given matrix is not square, therefore we must add one fictitious job 6 (sixth column) to make it a square matrix. Therefore, we have the following matrix

	1	2	3	4	5	6
1	2.5	5	1	6	1	0
2	2	5	1.5	7	3	0
3	3	6.5	2	8	3	0
4	3.5	7	2	9	4.5	0
5	4	7	3	9	6	0
6	6	9	5	10	6	0

Rows and Column Reduced Matrix

Subtracting the smallest element of each row from every element of the corresponding row and then subtracting smallest element of each column from every element of the corresponding column, we get the following row and column reduced matrix

	1	2	3	4	5	6
1	0.5	0	0	0	0	0
2	0	0	0.5	1	2	0
3	1	1.5	1	2	2	0
4	1.5	2	1	3	3.5	0
5	2	2	2	3	5	0
6	4	4	4	4	5	0

Assignment Matrix

	1	2	3	4	5	6
1	0.5	[0]	⊠	⊠	⊠	⊠
2	[0]	0	0.5	1	2	⊠
3	1	1.5	1	2	2	[0]
4	1.5	2	1	3	3.5	⊠
5	2	2	2	3	5	⊠
6	4	4	4	4	5	⊠

Observe that the rows 3, 4, 5 and column 4, 5, 6 have no zero assignments. Now we draw the minimum number of lines

$$
\begin{array}{c}
\quad\quad\quad\quad\quad\quad\quad\quad L_3 \\
\begin{array}{l|cccccc}
L_1 & 0.5 & \boxed{0} & \cancel{0} & \cancel{0} & \cancel{0} & \cancel{0} \\
L_2 & \boxed{0} & 0 & 0.5 & 1 & 2 & \cancel{0} \\
 & 1 & 1.5 & 1 & 2 & 2 & \boxed{0} \\
 & 1.5 & 2 & 1 & 3 & 3.5 & \cancel{0} \\
 & 2 & 2 & 2 & 3 & 5 & \cancel{0} \\
 & 4 & 4 & 4 & 4 & 5 & \cancel{0}
\end{array}
\end{array}
$$

✓(5)
✓(1)
✓(2)
✓(3)
✓(4)

The smallest element among all uncovered elements in the above table is 1. Subtracting this element 1 from all uncovered elements, adding to every element that lies at the intersection of two lines. Remaining elements remain unchanged. Then reduced matrix is given below

	1	2	3	4	5	6
1	0.5	0	0	0	0	1
2	0	0	0.5	1	2	1
3	0	0.5	0	1	1	0
4	0.5	1	0	2	2.5	0
5	1	1	1	2	4	0
6	3	3	3	3	4	0

Proceeding same as above

$$
\begin{array}{c}
\quad\quad\quad\quad\quad\quad\quad\quad L_5 \\
\begin{array}{l|cccccc}
L_1 & 0.5 & \cancel{0} & \cancel{0} & 0 & \cancel{0} & 1 \\
L_2 & 0 & 0 & 0.5 & 1 & 2 & 1 \\
L_3 & 0 & 0.5 & \cancel{0} & 1 & 1 & \cancel{0} \\
L_4 & 0.5 & 1 & 0 & 2 & 2.5 & \cancel{0} \\
 & 1 & 1 & 1 & 2 & 4 & 0 \\
 & 3 & 3 & 3 & 3 & 4 & \cancel{0}
\end{array}
\end{array}
$$

✓(3)
✓(1)
✓(2)

Here, smallest element among uncovered element is 1.

$$
\begin{array}{c}
\quad L_2 \quad L_3 \quad L_4 \quad\quad\quad L_5 \\
\begin{array}{l|cccccc}
L_1 & 0.5 & 0 & 0 & 0 & 0 & 2 \\
 & 0 & 0 & 0.5 & 1 & 2 & 2 \\
 & 0 & 0.5 & 0 & 1 & 1 & 1 \\
 & 0.5 & 1 & 0 & 2 & 2.5 & 1 \\
 & 0 & 0 & 0 & 1 & 3 & 0 \\
 & 2 & 2 & 2 & 2 & 3 & 0
\end{array}
\end{array}
$$

⇒

✓ ✓ ✓ ✓ ✓

✓ ✓ ✓ ✓

Here, smallest element among uncovered element is 1.

	1	2	3	4	5	6
1	1.5	1	1	[0]	⊠	3
2	[0]	⊠	0.5	⊠	1	2
3	⊠	0.5	⊠	⊠	[0]	1
4	0.5	1	[0]	1	1.5	1
5	⊠	[0]	⊠	⊠	2	⊠
6	2	2	2	1	2	[0]

Hence, the optimal solution is given by

$$1 \to 4, \; 2 \to 1, \; 3 \to 5, \; 4 \to 3, \; 5 \to 2$$

with Minimum total cost = 20, i.e., ₹2000.

☛ REMARK

- We can also obtain other optimal solutions of the above problem as given below

$$1 \to 4, \; 2 \to 2, \; 3 \to 5, \; 4 \to 3, \; 5 \to 1$$
$$1 \to 5, \; 2 \to 1, \; 3 \to 4, \; 4 \to 3, \; 5 \to 2$$
$$1 \to 5, \; 2 \to 2, \; 3 \to 1, \; 4 \to 3, \; 5 \to 4$$
$$1 \to 5, \; 2 \to 2, \; 3 \to 4, \; 4 \to 3, \; 5 \to 1$$
$$1 \to 5, \; 2 \to 4, \; 3 \to 1, \; 4 \to 3, \; 5 \to 2$$

EXAMPLE 7. *A company has four territories open and four salesmen available for assignment. The territories are not equally rich in their sales potential. It is estimated that a typical salesman on operating in each territory would bring in the following annual sales*

Territory	I	II	III	IV
Annual sales CPs	60000	50000	40000	30000

The four salesmen are also considered to differ in ability, it is estimated that working under the same conditions, their yearly sales would be proportionally as follows

Salesman	A	B	C	D
Proportion	7	5	5	4

If the criterion is maximum expected total sales, the intuitive answer is to assign the best salesman to the richest territory, the next best salesman to the second richest and so on. Verify the answer by assignment technique.　　　[KANPUR–2010]

SOLUTION.　Firstly, we shall construct the effectiveness matrix.

Sum of the proportion of sales of four salesmen = 7 + 5 + 5 + 4 = 21.

Now, taking the salesmen in the four territories sales are as follows :

For A: $\dfrac{7}{21} \times 6, \; \dfrac{7}{21} \times 5, \; \dfrac{7}{21} \times 4, \; \dfrac{7}{21} \times 3$, i.e., 42, 35, 28, 21 (By avoiding fractions)

For B: $\dfrac{5}{21} \times 6, \; \dfrac{5}{21} \times 5, \; \dfrac{5}{21} \times 4, \; \dfrac{5}{21} \times 3$, i.e., 30, 25, 20, 15

For C: $\dfrac{5}{21} \times 6, \; \dfrac{5}{21} \times 5, \; \dfrac{5}{21} \times 4, \; \dfrac{5}{21} \times 3$, i.e., 30, 25, 20, 15

For D: $\dfrac{4}{21} \times 6,\ \dfrac{4}{21} \times 5,\ \dfrac{4}{21} \times 4,\ \dfrac{4}{21} \times 3$, *i.e.*, 24, 20, 16, 12

Therefore, the effectiveness matrix which make the total sales maximum is given by

	I	II	III	IV
A	42	35	28	21
B	30	25	20	15
C	30	25	20	15
D	24	20	16	12

Now, we convert this maximization problem into minimization problem

	I	II	III	IV
A	−42	−35	−28	−21
B	−30	−25	−20	−15
C	−30	−25	−20	−15
D	−24	−20	−16	−12

Proceeding in the usual manner, we get

L_1

0	3	6	9
0	1	2	3
0	1	2	3
0	0	0	0	L_2

Minimum element = 1.

L_1 L_2

⇒

0	2	5	8	✓(4)
⊠	0	1	2	✓(5)
⊠	⊠	1	2	✓(1)
1	⊠	0	⊠	L_3

✓(2) ✓(3)

Minimum element = 1.

⇒

	I	II	III	IV
A	⓪	2	4	7
B	⊠	⓪	⊠	1
C	⊠	⊠	⓪	1
D	2	1	⊠	⓪

or

	I	II	III	IV
A	⓪	2	4	7
B	⊠	⊠	⓪	1
C	⊠	⓪	⊠	1
D	2	1	⊠	⓪

Hence, we get the following two optimal solutions

(i) $A \rightarrow I,\ B \rightarrow II,\ C \rightarrow III,\ D \rightarrow IV$ (ii) $A \rightarrow I,\ B \rightarrow III,\ C \rightarrow II,\ D \rightarrow IV$

Exercise-12.2

1. Solve the following assignment problems :

	I	II	III	IV
A	1	4	6	3
B	9	7	10	9
C	4	5	11	7
D	8	7	8	5

2. There are five jobs to be assigned one each to five machines and the associated cost matrix is as follows :

Machine

		1	2	3	4	5
	A	11	17	8	16	20
	B	9	7	12	6	15
Job	C	13	16	15	12	16
	D	21	24	17	28	26
	E	14	10	12	11	15

Find the optimal solution of this minimal assignment problem.

3. An automobile dealer wishes to put four repairmen to four different jobs. The repairmen have somewhat different kinds of skills and they exhibit different levels of efficiency from one job to the another. The dealer has estimated the number of man hour that would be required for each job-man combinations. This is given in the following matrix.

Job

		A	B	C	D
	1	5	3	2	8
Man	2	7	9	2	6
	3	6	4	5	7
	4	5	7	7	8

Find the optimal solution of this minimum assignment problem.

4. A computer centre has got three expert programmers. The centre needs three application programs to be developed. The head of the computer centre, after studying carefully the programs to be developed, estimates the computer time in minutes required by the experts to the application programs as follows

Program

		A	B	C
	1	120	100	80
Programmer	2	70	90	110
	3	110	140	120

Assign the programmers to the programs in such a way that the total computer time is least.

5. Solve the following assignment problem :

	I	II	III	IV	V
A	6	5	8	11	16
B	1	13	16	1	10
C	16	11	8	8	8
D	9	14	12	10	16
E	10	13	11	8	16

6. Find the optimal solution of the following minimal assignment problem

	I	II	III	IV
A	30	25	26	28
B	26	32	24	20
C	20	22	18	27
D	23	20	21	11

7. Find the minimum cost solution of the following assignment problem

	I	II	III	IV	V
1	−2	−4	−8	−6	−1
2	0	−9	−5	−5	−4
3	−3	−8	−9	−2	−6
4	−4	−3	−1	0	−3
5	−9	−5	−8	−9	−5

8. Solve the following minimal assignment problem

	1	2	3	4	5
A	8	4	2	6	1
B	0	9	5	5	4
C	3	8	9	2	6
D	4	3	1	0	3
E	9	5	8	9	5

9. Solve the following minimal assignment problem

	1	2	3	4
A	10	12	19	11
B	5	10	7	8
C	12	14	13	11
D	8	15	11	9

10. Solve the following minimal assignment problem

	1	2	3	4	5
I	12	8	7	15	14
II	7	9	17	14	10
III	9	6	12	6	7
IV	7	6	14	6	10
V	9	6	12	10	6

11. Solve the following assignment problem having the following cost matrix

	1	2	3	4	5	6	7
A	35	20	60	41	27	52	44
B	51	39	42	33	65	47	58
C	25	32	53	41	50	36	43
D	32	28	40	46	33	55	49
E	43	36	45	63	57	49	42
F	27	18	31	46	35	42	34
G	48	50	72	59	43	64	58

12. A car hire company has one car at each of five depots a, b, c, d and e. A customer requires a car in each town namely A, B, C, D and E. Distances (in Kms) between depots (origins) and towns (destinations) are given in the following distance matrix.

	a	b	c	d	e
A	160	130	175	190	200
B	135	120	130	160	175
C	140	110	155	170	185
D	50	50	80	80	110
E	54	34	70	80	105

How should car be assigned to customers so as to minimize the distance travelled.

13. Solve the following cost minimizing jobs problem

	1	2	3	4	5
A	11	10	18	5	9
B	14	13	12	19	6
C	5	3	4	2	4
D	15	18	17	9	12
E	10	11	19	6	14

14. Find the optimal solution of the following problem

Machine

		1	2	3
Job	1	5	7	9
	2	14	10	12
	3	15	13	16

15. Use the Hungarian method to find which of the two jobs should be left undone when each of the four persons will do only one job in the following cost minimizing assignment problem.

Job

		J_1	J_2	J_3	J_4	J_5	J_6
Person	P_1	10	9	11	12	8	5
	P_2	12	10	9	11	9	4
	P_3	8	11	10	7	12	6
	P_4	10	7	8	10	10	5

16. There are three persons P_1, P_2 and P_3 and five jobs $J_1, J_2, ..., J_5$. Each person can do only one job and a job is to be done by one person only. Using Hungarian method, find which two jobs should be left undone in the following cost minimizing problem.

	J_1	J_2	J_3	J_4	J_5
P_1	7	8	6	5	9
P_2	9	6	7	6	10
P_3	8	7	9	5	6

17. In a machine shop a supervisor wishes to assign five jobs among six machines. Any one of the jobs can be processed completely by any one of the machines as given below

Machines

		A	B	C	D	E	F
Job	1	13	13	16	23	19	9
	2	11	19	26	16	17	18
	3	12	11	4	9	6	10
	4	7	15	9	14	14	13
	5	9	13	12	8	14	11

The assignment of jobs to machines be on a one-to-one basis. Assign the jobs to machines so that the total cost is minimum. Find the minimum total cost.

18. The owner of a small machine shop has four persons available to assign to jobs for the day. Five jobs are offered with the expected profit in rupees for each person on each job being as follows

Jobs

		A	B	C	D	E
	1	6.20	7.80	5.00	10.10	8.20
Person	2	7.10	8.40	6.10	7.30	5.90
	3	8.70	9.20	11.10	7.10	8.10
	4	4.80	6.40	8.70	7.70	8.00

Find the assignments of persons to jobs that will result in a maximum profit. Which job should be declined?

19. Solve the following maximum assignment problem

	A	B	C	D	E
1	32	38	40	28	40
2	40	24	28	21	36
3	41	27	33	30	37
4	22	38	41	36	36
5	29	33	40	65	39

20. A company has five jobs to be done. The following matrix shows the return in Rupees assigning i^{th} ($i = 1, 2, 3, ..., 5$) machine to

the jth job ($j = 1, 2, ..., 5$). Assign the five jobs to the five machines so as to maximize the total return.   [MEERUT–2008, 10, 11, 12]

Machine

		1	2	3	4	5
	1	5	11	10	12	4
	2	2	4	6	3	5
Job	3	3	12	5	14	6
	4	6	14	4	11	7
	5	7	9	8	12	5

21. Five engineers are available to design four projects. Engineer 2 is not competent to design the project B. Given the following time estimates needed to each engineer to design a given project, find how should the engineers be assigned to projects so as to minimize the total design time of four projects.

Project

		A	B	C	D
	1	16	14	14	12
Engineer	2	16	–	17	13
	3	11	15	21	9
	4	8	10	9	7

ANSWERS

1. $A \to I, B \to III, C \to II, D \to IV$, Minimum cost = 10.

2. $A \to 1, B \to 4, C \to 5, D \to 3, E \to 2$, Minimum cost = 60.

3. $1 \to B, 2 \to C, 3 \to D, 4 \to A$; $1 \to C, 2 \to D, 3 \to B, 4 \to A$, Min. time = 17 hrs.

   $1 \to C, 2 \to B, 3 \to A$, Minimum time = 280 min.

5. $A \to II, B \to I, C \to V, D \to III, E \to IV$

   or $A \to II, B \to IV, C \to V, D \to I, E \to III$, Minimum cost = 34.

6. $A \to II, B \to IV, C \to I, D \to III$ or $A \to III, B \to IV, C \to I, D \to II$,

   Minimum cost = 80.

7. $1 \to III, 2 \to II, 3 \to V, 4 \to I, 5 \to IV$

   or $1 \to IV, 2 \to II, 3 \to III, 4 \to V, 5 \to I$, Minimum cost = 36.

8. $A \to 5, B \to 1, C \to 4, D \to 3, E \to 2$, Minimum cost = 9.

9. $A \to 2, B \to 3, C \to 4, D \to 1$, Minimum cost = 38.

10. $I \to 3, II \to 1, III \to 2, IV \to 4, V \to 5$ or $I \to 3, II \to 1, III \to 4, IV \to 2, V \to 5$.

11. $A \to 2, B \to 4, C \to 6, D \to 1, E \to 7, F \to 3, G \to 5$, Minimum cost = 237 units.

12. $a \to D, b \to C, c \to B, d \to E, e \to A$, Minimum distance = 570 Kms.

13. $A \to II, B \to V, C \to III, D \to IV, E \to I$, Minimum cost = 39.

14. $1 \to 1, 2 \to 3, 3 \to 2$.

15. $P_1 \to J_5, P_2 \to J_6, P_3 \to J_4, P_4 \to J_2$, Jobs J_1 and J_3 are left undone.

16. $P_1 \to J_3, P_2 \to J_2, P_3 \to J_4$, J_1 and J_2 are left undone.

17. $1 \to F, 2 \to A, 3 \to E, 4 \to C, 5 \to D$, Minimum cost = 43.

18. $1 \to D$, $2 \to B$, $3 \to C$, $4 \to E$, Job A should be declined.

19. $1 \to B$, $2 \to A$, $3 \to E$, $4 \to C$, $5 \to D$, Maximum sales = ₹ 221.

20. $1 \to 5$, $2 \to 4$, $3 \to 1$, $4 \to 3$, $5 \to 2$, Maximum return = 50 units.

21. $1 \to B$, $2 \to D$, $3 \to A$, $4 \to C$, Minimum total time = 47 hours.

12.9 TRAVELLING SALESMAN PROBLEM

In the travelling salesman problem, a salesman visits a certain number of cities and between every pair of cities the distance is known to him. He start from his home city, passes through each city once and returns to his home city in such a way that the route chosen by him must be shortest in distance.

A travelling slaesman wants to minimize the total distance travelled during his visit of n cities. A similar problem arises when n items say A_i, $i = 1, 2, ..., n$, are to be produced on a machine in continuation, given that c_{ij} $(i, j = 1, 2, ..., n)$ is the set up cost of the machine when item A_i is followed by A_j. Note that $c_{ij} = \infty$, when $i = j$, i.e., we don't produce the item A_i again after A_j. The individual set up costs can be arranged in the form of the adjacent square matrix.

$$
\begin{array}{c c}
& \begin{matrix} A_1 & \quad A_1 \,...\, A_i \,...\, A_n \end{matrix} \\
\begin{matrix} A_1 \\ A_1 \\ \vdots \\ A_1 \\ \vdots \\ A_i \\ \vdots \\ A_n \end{matrix} &
\begin{vmatrix}
\infty & c_{12} \cdots & c_{1i} \cdots c_{1n} \\
c_{21} & \infty \cdots & c_{2i} \cdots c_{2n} \\
\vdots & \vdots & \vdots \quad \vdots \\
c_{21} & \infty \cdots & c_{2i} \cdots c_{2n} \\
\vdots & \vdots & \vdots \quad \vdots \\
c_{i2} & c_{i2} \cdots & \infty \cdots c_{in} \\
\vdots & \vdots & \vdots \quad \vdots \\
c_{n1} & c_{n2} \cdots c_{ni} & \cdots \infty
\end{vmatrix}
\end{array}
$$

Our problem is to determine a set of n elements of this matrix, one of each row and one in each column, so as to minimize the sum of the elements determined above. Here two extra restrictions are imposed. One restriction is that we can not select the element in the leading diagonal as we have already assumed the elements of leading diagonal to be infinity. The other restriction is that we don't produce an item again until all the items are produced once.

Definition 1: *A travelling salesman problem is said to be symmetric or asymmetric according as cost matrix is symmetric or not. Thus it is symmetric if the cost from A_i to A_j is the same as that from A_j to A_i.*

Definition 2: A travelling salesman problem may be stated as follows:

Like assignment problem, if we represent" going" and "not going" of the salesman from A_i to A_j station by saying $x_{ij} = 1$ and $x_{ij} = 0$ respectively, then we wish to determine x_{ij}, $i, j = 1, ..., n$, which minimizes

$$z = \sum_{i=1}^{n} \sum_{j=1}^{n} c_{ij} x_{ij}$$

s.t.
$$\sum_{j=1}^{n} x_{ij} = 1, \quad i = 1, 2, ..., n$$

$$\sum_{j=1}^{n} x_{ij} = 1, \quad j = 1, 2, ..., n$$

$$x_{ij} = 0 \quad \text{or} \quad 1,$$

and one extra restriction that the x_{ij} must be so chosen that no city is visited twice until the tour of all the cities is completed. Note that as we have written ∞ in the leading diagonal, x_{ij} cannot be 1 when $i = j$, because then z will not be minimum. Thus the second restriction of not going to A_i again just after A_i is automatically satisfied.

12.9.1 SOLUTION OF TRAVELLING SALESMAN PROBLEM

The problem can be solved by assignment technique.

The solution thus obtained may not be feasible for this problem. For example, if we choose the elements $c_{15}, c_{23}, c_{34}, c_{42}, c_{51}$, then the corresponding solution gives $A_1 \rightarrow A_5, A_5 \rightarrow A_1$, i.e., A_1 is followed by A_5 and A_5 by A_1. This violates our restriction. In such cases after solving the given problem by assignment technique, we use the method of enumeration by assigning the next minimum element of the matrix in place of zero. This is best explained by the following example.

Such problems occur in the field of postal deliveries, school bus routing, television relays assembly lines, production of several items by one machine.

Solved Examples

EXAMPLE 1. *Given the matrix of set up costs below, show how to sequence production so as to minimize the total set up cost per cycle.*

	A_1	A_2	A_3	A_4	A_5
A_1	∞	2	5	7	1
A_2	6	∞	3	8	2
A_3	8	7	∞	4	7
A_4	12	4	6	∞	5
A_5	1	3	2	8	∞

SOLUTION. Solving the problem by assignment technique we get the following matrix, showing a solution in terms of marked zeros.

	A_1	A_2	A_3	A_4	A_5
A_1	∞	1	3	6	[0]
A_2	4	∞	[0]	6	0
A_3	4	3	∞	[0]	3
A_4	8	[0]	1	∞	1
A_5	[0]	2	0	7	∞

The matrix does not provide the solution of the original problem as it gives $A_1 \rightarrow A_5, A_5 \rightarrow A_1$, while A_5 cannot be followed by A_1 until A_2, A_3, A_4 are produced.

Now we try to find the next best solution which also satisfies this extra restriction. The next minimum (non zero) element in the matrix is 1. We try to bring 1 into the solution. The cost 1 also occurs at three places. We shall consider all the cases one by one.

In case we bring $c_{12} = 1$ in the solution then as no other assignment can be made in the first row and second column, the resulting feasible solution will be $A_1 \rightarrow A_2, A_2 \rightarrow A_3, A_3 \rightarrow A_4, A_4 \rightarrow A_5, A_5 \rightarrow A_1$.

The selected elements for this solution are placed in the squares of the following matrix :

	A_1	A_2	A_3	A_4	A_5
A_1	∞	①	3	6	0
A_2	4	∞	⓪	6	0
A_3	4	3	∞	⓪	3
A_4	8	0	1	∞	①
A_5	⓪	2	0	7	∞

The cost in this reduced matrix corresponding to this feasible point is 2.

On the other hand, if we select the element $c_{43} = 1$, in the solution, then no feasible solution is available in terms of zero's or which gives the cost less than 2.

Hence the best programme is

$$A_1 \rightarrow A_2 \rightarrow A_3 \rightarrow A_4 \rightarrow A_5 \rightarrow A_1.$$

The total setup cost will be $2 + 3 + 4 + 5 + 1 = 15$.

EXAMPLE 2. *Solve the travelling salesman problem given by the following data:*

$c_{12} = 20, c_{13} = 4, c_{14} = 10, c_{23} = 5, c_{34} = 6, c_{25} = 10, c_{35} = 6, c_{45} = 20$, *where* $c_{ij} = c_{ji}$ *and there is no route between cities i and j if a value for* c_{ij} *is not given above.* [MEERUT–2005, 06]

SOLUTION. The cost matrix is given as follows :

	1	2	3	4	5
1	∞	20	4	10	∞
2	20	∞	5	∞	10
3	4	5	∞	6	6
4	10	∞	6	∞	20
5	∞	10	6	20	∞

Note that in the above matrix, we have taken $c_{15} = \infty$, etc. It is taken to avoid the possibility of going from first station to 5th station as there is no such route.

Note that c_{15} is not given in the data. Similarly, we have taken

$$c_{24} = \infty, c_{51} = \infty, c_{42} = \infty.$$

Now, solving the problem by assignment algorithm, we reach at the following assignment plan in terms of zero's :

	1	2	3	4	5
1	∞	12	[0]	⊠	∞
2	11	∞	⊠	∞	[0]
3	⊠	1	∞	[0]	1
4	[0]	∞	⊠	∞	9
5	∞	[0]	⊠	8	∞

In this assignment plan, the route is as follows :
$$1 \to 3 \to 4 \to 1 \to 2 \to 5 \to 2$$
This is not feasible route of the salesman problem. We bring $c_{32} = 1$ in the solution. A new assignment plan is as follows :

	1	2	3	4	5
1	∞	12	[0]	⊠	∞
2	11	∞	⊠	∞	[0]
3	⊠	[1]	∞	⊠	1
4	[0]	∞	⊠	∞	9
5	∞	⊠	⊠	[8]	∞

The optimal route is $1 \to 3 \to 2 \to 5 \to 4 \to 1$.

The corresponding total cost is $= 4 + 5 + 10 + 20 + 10 = 49$.

EXAMPLE 3. *Solve the following travelling salesman problem :*

	A	B	C	D	E
A	∞	4	7	3	4
B	4	∞	6	3	4
C	7	6	∞	7	5
D	3	3	7	∞	7
E	4	4	5	7	∞

SOLUTION.   Following the general procedure of solving the assignment matrix, we have :

Row reduced matrix

	A	B	C	D	E
A	∞	1	4	0	1
B	1	∞	3	0	1
C	2	1	∞	2	0
D	0	0	4	∞	4
E	0	0	1	3	∞

Column reduced matrix

	A	B	C	D	E
A	∞	1	3	0	1
B	1	∞	2	0	1
C	2	1	∞	2	0
D	0	0	3	∞	4
E	0	0	0	3	∞

Assignment model

	A	B	C	D	E	
A	∞	1	3	[0]	1	✓
B	1	∞	2	✗	1	✓
C	2	1	∞	2	[0]	
D	[0]	✗	3	∞	4	
E	✗	[0]	✗	3	∞	
	✓	✓		✓		

Since the number of assignments = 4
Minimum element = 1.

	A	B	C	D	E
A	∞	0	2	0	0
B	1	∞	1	0	0
C	3	1	∞	3	0
D	1	0	3	∞	4
E	1	0	0	4	∞

Row reduced matrix

	A	B	C	D	E
A	∞	0	2	0	0
B	1	∞	1	0	0
C	3	1	∞	3	0
D	1	0	3	∞	4
E	1	0	0	4	∞

Column reduced matrix

	A	B	C	D	E
A	∞	0	2	0	0
B	0	∞	1	0	0
C	2	1	∞	3	0
D	0	0	3	∞	4
E	0	0	0	4	∞

Assignment model

	A	B	C	D	E
A	∞	0	2	⊠	[0]
B	⊠	∞	1	[0]	⊠
C	2	1	∞	3	⊠
D	[0]	⊠	3	∞	4
E	0	⊠	[0]	4	∞

Since there is no assignment in third row, so we can allot the next minimum (*i.e.,* 1) in row C.

	A	B	C	D	E
A	∞	0	2	⊠	[0]
B	⊠	∞	1	[0]	⊠
C	2	[1]	∞	3	⊠
D	[0]	⊠	3	∞	4
E	⊠	⊠	[0]	4	∞

Required route is
$$A \to E, \quad B \to D, C \to B, D \to A, E \to C, \quad i.e., \quad A \to E \to C \to B \to D \to A$$
Total cost $= 4 + 5 + 6 + 3 + 3 = 21$.

EXAMPLE 4. *Solve the following travelling sales problem :*

	A_1	A_2	A_3	A_4	A_5
A_1	∞	2	5	7	1
A_2	6	∞	3	8	2
A_3	8	7	∞	4	7
A_4	12	4	6	∞	5
A_5	1	3	2	8	∞

SOLUTION. **Row reduced matrix**

	A_1	A_2	A_3	A_4	A_5
A_1	∞	1	4	6	0
A_2	4	∞	1	7	0
A_3	4	3	∞	0	3
A_4	8	0	2	∞	1
A_5	0	2	1	7	∞

Column reduced matrix

	A_1	A_2	A_3	A_4	A_5
A_1	∞	1	3	6	0
A_2	4	∞	0	7	0
A_3	4	3	∞	0	3
A_4	8	0	1	∞	1
A_5	0	2	0	7	∞

Assignment model

	A_1	A_2	A_3	A_4	A_5
A_1	∞	1	3	6	$\boxed{0}$
A_2	4	∞	$\boxed{0}$	7	$\cancel{0}$
A_3	4	3	∞	$\boxed{0}$	3
A_4	8	$\boxed{0}$	1	∞	1
A_5	$\boxed{0}$	2	$\cancel{0}$	7	∞

$$A \rightarrow E, B \rightarrow C, C \rightarrow D, D \rightarrow B, E \rightarrow A.$$

So the route obtained

$$A \rightarrow E \rightarrow A, B \rightarrow C \rightarrow D \rightarrow B.$$

But this is not a feasible route.

So, giving assignment to the next minimum to make the feasible route, we have

	A_1	A_2	A_3	A_4	A_5
A_1	∞	$\boxed{1}$	3	6	$\cancel{0}$
A_2	4	∞	$\boxed{0}$	7	$\cancel{0}$
A_3	4	3	∞	$\boxed{0}$	3
A_4	8	$\cancel{0}$	1	∞	$\boxed{0}$
A_5	$\boxed{0}$	2	$\cancel{0}$	7	∞

$$A \rightarrow B, \; B \rightarrow C, \; C \rightarrow D, \; D \rightarrow E, \; E \rightarrow A.$$

So, $$A \rightarrow B \rightarrow C \rightarrow D \rightarrow E \rightarrow A$$

EXAMPLE 5. *Solve the following travelling sales problem :*

		To item				
		A	B	C	D	
From item	A	∞	4	7	3	
	B	4	∞	6	3	
	C	7	6	∞	7	
	D	3	3	7	∞	

SOLUTION. **Row reduced matrix**

	A	B	C	D
A	∞	1	4	0
B	1	∞	3	0
C	1	0	∞	1
D	0	0	4	∞

Column reduced matrix

	A	B	C	D
A	∞	1	1	0
B	1	∞	0	0
C	1	0	∞	1
D	0	0	1	∞

Assignment model

	A	B	C	D
A	∞	1	1	[0]
B	1	∞	[0]	⦻
C	1	[0]	∞	1
D	[0]	⦻	1	∞

$A \to D, B \to C, C \to B, D \to A$.

So, (1) $A \to D \to A$ (2) $B \to C \to B$.

But these are not the feasible path.

Hence, we give assignments to the next minimum element. We have,

	A	B	C	D
A	∞	1	[1]	⦻
B	1	∞	⦻	[0]
C	1	[0]	∞	1
D	[0]	⦻	1	∞

$A \to C, B \to D, C \to B, D \to A$.

Hence, $A \to C \to B \to D \to A$. Cost = 19

EXAMPLE 6. *Solve the following travelling salesman problem :*

	A	B	C	D	E		
A	∞	6	12	6	4	8	1
B	6	∞	10	5	4	3	3
C	8	7	∞	11	3	11	8
D	5	4	11	∞	5	8	6
E	5	2	7	8	∞	4	7
F	6	3	11	5	4	∞	2
G	2	3	9	7	4	3	∞

SOLUTION. **Row reduced matrix**

	A	B	C	D	E		
A	∞	5	11	5	3	7	0
B	3	∞	7	2	1	0	0
C	5	4	∞	8	0	8	5
D	1	0	7	∞	1	4	2
E	3	0	5	6	∞	2	5
F	4	1	9	3	2	∞	0
G	0	1	7	5	2	1	∞

Column reduced matrix

	A	B	C	D	E		
A	∞	5	6	3	3	7	0
B	3	∞	2	0	1	0	0
C	5	4	∞	6	0	8	5
D	1	0	2	∞	1	4	2
E	3	0	0	4	∞	2	5
F	4	1	4	1	2	∞	0
G	0	1	2	3	2	1	∞

Assignment model

	A	B	C	D	E		
A	∞	5	6	3	3	7	[0]
B	3	∞	2	[0]	1	⊠	⊠
C	5	4	∞	6	[0]	8	5
D	1	[0]	2	∞	1	4	2
E	3	0	[0]	4	∞	2	5
F	4	1	4	1	2	∞	⊠
G	[0]	1	2	3	2	1	∞

Since the number of ⎕0 assignments ≠ 7. So we take minimum element = 1. Then we have the following new assignment model

New assignment model

	A	B	C	D	E		
A	∞	5	5	2	2	6	[0]
B	3	∞	2	⊠	1	[0]	1
C	5	5	∞	6	[0]	8	6
D	⊠	[0]	1	∞	⊠	3	2
E	3	1	[0]	4	∞	2	6
F	3	1	3	[0]	1	∞	⊠
G	[0]	2	2	3	2	1	∞

The number of $\boxed{0}$ assignments = 7 (required) the path becomes
$$A \to G \to A, B \to F \to D \to B, \ C \to E \to C.$$
But it is not a feasible path. So, we must make it feasible as follows :

	A	B	C	D	E		
A	∞	5	5	$\boxed{2}$	2	6	0
B	3	∞	2	0	1	0	$\boxed{1}$
C	$\boxed{5}$	5	∞	6	0	8	6
D	0	$\boxed{0}$	1	∞	0	3	2
E	3	1	$\boxed{0}$	4	∞	2	6
F	3	1	3	0	$\boxed{1}$	∞	0
G	0	2	2	3	2	$\boxed{1}$	∞

The above path has been made feasible by allotting the next minimum instead of 0 in A, B, C rows.
Thus, we get
$$A \to D \to B \to G \to F \to E \to C \to A$$
Minimum cost = 6 + 4 + 3 + 3 + 4 + 7 + 8 = 35.

Exercise-12.3

1. Solve the travelling salesman problem.

	1	2	3	4	5
1	∞	6	12	6	4
2	6	∞	10	5	4
3	8	7	∞	11	3
4	5	4	11	∞	5
5	5	2	7	8	∞

2. Solve the travelling salesman problem.

	1	2	3	4	5	6
1	∞	20	23	27	29	34
2	21	∞	19	26	31	24
3	26	28	∞	15	36	26
4	25	16	25	∞	23	18
5	23	40	23	31	∞	10
6	27	18	12	35	16	∞

3. Solve the travelling salesman problem given in the following data.

$c_{12} = 4, c_{13} = 7, c_{14} = 3, c_{23} = 6, c_{24} = 3, c_{34} = 7$ where $c_{ij} = c_{ji}$.

4. Solve the travelling salesman problem.

	A	B	C	D	E
A	—	7	6	8	4
B	7	—	8	5	6
C	6	8	—	9	7
D	8	5	9	—	8
E	4	6	7	8	—

ANSWERS

1. $1 \to 3 \to 5 \to 2 \to 4 \to 1$, min. cost = 27
2. $1 \to 5 \to 6 \to 3 \to 4 \to 2 \to 1$, min. cost = 103
3. $1 \to 3 \to 2 \to 4 \to 1$, min. cost = 19
4. $A \to C \to D \to B \to E \to A$ and $A \to E \to B \to D \to C \to A$, total cost = 30

REVIEW QUESTIONS

1. Define assignment problem.
2. Explain the Hungarian method to solve an assignment problem.
3. Explain the difference between a transportation problem and an assignment problem.
4. Write a short note on unbalanced assignment problem.
5. What is travelling salesman problem?
6. Formulate the travelling salesman problem as an assignment problem.
7. How can the travelling salesman problem be solved using assignment algorithm.

MULTIPLE CHOICE QUESTIONS (CHOOSE THE MOST APPROPRIATE ONE)

1. For solving an assignment problem, we modify cost matrix: [MEERUT–2013]
   (a) by creating zero   (b) by creating infinite
   (c) by creating one   (d) none of these
2. An assignment problem is a special case of $m \times n$ transportation problem in which:
   [MEERUT–2013]
   (a) $m = n$ (b) $2m = n$
   (c) $m = 2n$ (d) None of these
3. An assignment problem is solved by:
   [MEERUT–2013]
   (a) Simplex method   (b) Graphical method
   (c) Vector method   (d) Hungarian method
4. Due to some restriction the assignment of a particular facility to a particular job is not given, then we take that cost as:
   [MEERUT–2013]
   (a) 0   (b) –1
   (c) ∞ (d) None of these
5. In an unbalanced assignment problem to form a square matrix, fictitious rows or columns are added in the matrix with costs: [MEERUT–2013]
   (a) 0   (b) 1
   (c) ∞ (d) None of these
6. If the cost matrix of an assignment problem is not a square matrix, the assignment problem is called: [MEERUT–2013]
   (a) balanced   (b) unbalanced
   (c) maximization   (d) none of these
7. In an assignment problem with m jobs and m persons, the number of basic variables at zero level in a BFS is:
   (a) m (b) $m - 1$
   (c) $m + 1$ (d) None of these
8. The cost matrix in assignment problem is a:
   [MEERUT–2013]
   (a) square matrix   (b) rectangle matrix
   (c) diagonal matrix   (d) none of these
9. Maximization assignment problem is transformed in a minimization problem by:
   (a) deduct smallest element from all other elements of matrix
   (b) deduct all element of the row from highest element of the row
   (c) all elements of the matrix are deducted from the highest elements in the matrix
   (d) none of the above
10. Solve the following assignment problem:

		Person			
		1	2	3	4
Task	A	10	12	19	11
	B	5	10	7	8
	C	12	14	13	11
	D	8	15	11	9

   (a) $A \to 1, B \to 2, C \to 3, D \to 4$
   (b) $A \to 2, B \to 3, C \to 4, D \to 1$
   (c) $A \to 2, B \to 3, C \to 1, D \to 4$
   (d) $A \to 3, B \to 2, C \to 1, D \to 4$
11. The assignment problem is a:
   (a) non-linear programming problem
   (b) dynamic programming problem
   (c) integer linear programming problem
   (d) integer non-linear programming problem
12. For an $n \times n$ assignment problem the number of possible solution will be:
   (a) $2n$ (b) $(n-1)!$
   (c) $n!$ (d) n^2
13. For solving an assignment problem we modify the cost matrix:
   (a) by creating zero   (b) one
   (c) (a) and (b) both   (d) none of these
14. The complete optimal assignment is obtained if in the reduced cost matrix of order n the number of marked '□' zero is:
   (a) less than n (b) greater than n
   (c) exactly n (d) none of these
15. In travelling salesman problem, elements of leading diagonal of the cost:
   (a) finite   (b) constant
   (c) variable   (d) infinite
16. If in a given problem a constant is added or subtracted to every element of a row of the cost matrix c_{ij} then an assignment which minimize the total cost for one matrix also minimizes the total cost for other matrix then the problem is called:

(a) an assignment bvp transpite
(b) an assignment problem transpite
(c) (a) and (b) both
(d) none of these

17. In an unbalanced assignment problem to form a square matrix fictitious row or columns and are added in the matrix with costs:
(a) 0   (b) 1
(c) 2   (d) ∞

18. A salesman wants to visit n cities then the number of possible routes is:
(a) $n!$ (b) $(n-1)!$
(c) $(n+1)!$ (d) n

19. The complete optimal assignment is obtained if in the reduced cost matrix of order n the no. of assignments is:
(a) n (b) $> n$
(c) $< n$ (d) None of these

20. In travelling salesman problem, the elements of the leading diagonal of the cost are taken to be:
(a) finite   (b) infinite
(c) constant   (d) variable

21. An optimal assignment exists if the total reduced cost of the assignment is:
(a) 0   (b) 1
(c) 2   (d) 3

22. How many types travelling salesman problem?
(a) one   (b) two
(c) three   (d) four

23. If there were n workers and n jobs, there would be:
(a) $n!$ solutions   (b) $(n-1)!$ solutions
(c) $(n!)^n$ solutions   (d) n solutions

24. An assignment problem is considered as a particular case of a transportation problem, because:
(a) the number of rows equals the number of columns
(b) all $x_{ij} = 0$ or 1
(c) all rim conditions are 1
(d) all of the above

25. Maximization assignment problem is transformed into a minimization problem by:
(a) adding each entry in a column from the maximum value in that column
(b) subtracting each entry in a column from the maximum value in that column

(c) subtracting each entry in the table from the maximum value in that table
(d) any one of the above

26. While solving assignment problem an activity is assigned to a resource through a square with zero opportunity cost because the objective is to:
(a) reduce the cost of assignment to zero
(b) reduce the cost of that particular assignmen is zero
(c) minimize the cost of assignment
(d) all of the above

27. An assignment problem can be solved by:
(a) Simplex method
(b) Transportation method
(c) Both (a) and (b)
(d) None of these

28. In an assignment problem to obtain optimal assignment we draw minimum no. of lines to cover all the zero of reduced matrix then these line must through:
(a) unmarked rows
(b) unmarked columns
(c) unmarked rows and marked columns
(d) None of these

29. If an optimal assignment of an assignment problem exists then total reduced cost of the assignment will be:
(a) > 1 (b) < 1
(c) 0   (d) None of these

30. If distance between any pair of cities is independent of the direction of journey the problem is called:
(a) symmetrical   (b) asymmetrical
(c) degeneracy   (d) None of these

31. The assignment problem:
(a) requires that only one activity be assigned to each resource
(b) is a special case of transportation problem
(c) can be used to maximize resources
(d) all of the above

32. The purpose of a dummy row or column in an assignment problem is to:
(a) obtain balance between total activities and total resources
(b) prevent a solution from becoming degenerate
(c) provide the means of representing a dummy problem
(d) None of the above

ANSWERS

1. (a)	2. (a)	3. (d)	4. (c)	5. (a)	6. (b)	7. (b)	8. (a)	9. (c)
10. (b)	11. (a)	12. (c)	13. (a)	14. (c)	15. (d)	16. (a)	17. (a)	18. (b)
19. (a)	20. (b)	21. (a)	22. (b)	23. (a)	24. (d)	25. (c)	26. (c)	27. (c)
28. (c)	29. (c)	30. (a)	31. (d)	32. (a)				

Goal Programming

13.1 INTRODUCTION

Goal programming is an approach used for solving a multi objective optimization problem that balances trade off in conflicting objectives. It is an approach of deriving a best possible 'satisfactory' level of goal attainment. The business management has to achieve multiple goals or objectives, *i.e.*, the decision criteria of the management involves multiple goals. Hence, the goal programming assumes greater importance as a powerful tool to handle multiple decision criteria.

13.2 CONCEPTS OF GOAL PROGRAMMING

The concept of goal programming was introduced by **Charnes** and **Coopor** in 1961. They suggested a method for solving an infeasible linear programming arising from various resource constraints or goals. **Ijiri** developed the concept of different priority levels to the goals and different weights for goals at same priority level in 1965. **Lee (1972)** and **Ignizis (1976)** have written text books on the subject of goal programming.

In goal programming model, the decision variables of the model are to be different first. The goal related to the problem are to be listed down and ranked in order of priority. Since it is not possible to achieve every goal to the extent desired by the decision maker attempts are made to achieve each goal sequentially rather than simultaneously upto a satisfactory level rather than optimal level.

Thus a goal programming is often referred to as a lexico graphic method which consists of formulating an objective function in which the various goals are satisfied in order of their relative importance.

The goal programming technique is used in optimization of multiple objective goals by minimizing the derivation for each of the objective (goals) from the desired target that are set according to the priorities.

Goal programming tries to minimize the deviations from the targets that are set according to their priorities. It begins with the most important goal and continues until the achievement of a less important goal.

13.3 GOAL PROGRAMMING MODEL FORMULATION

The fomulation of goal programming is similar to that of LP model. The main difference between L.P. and G.P. is that L.P. optimize (maximize or minimize) a single objective function whereas G.P. minimize the deviations between the target values of the objectives and the realized results.

The general form of goal programming model is as follows:

$$\text{minimize } Z = \sum_{i=1}^{m} w_i(d_i^- + d_i^+)$$

subject to the constraints

$$\sum_{j=1}^{n} a_{ij}x_j + d_i^- - d_i^+ = b_i, \quad i = 1, 2, ..., m$$

and $x_j, d_i^-, d_i^+ \geq 0 \forall i, j$

where

d_i^- = negative deviation from i^{th} goal (under-achievement of the profit goal)

d_i^+ = positive deviation from the ith goal (over-achievement of the profit goal)

Since, both under and over-achievement of a goal cannot be achieved simultaneously (either one or both of these deviational variables will be equal to zero. i.e., $d_i^- \times d_i^+ = 0$. Hence, it either assumes a positive value or the other must be zero and vice-versa.

☛ REMARK
- The lower order goals are considered only after the higher goals are achieved.

13.3.1 SINGLE GOAL MODELS

Let us suppose one unit of effort applied to activity x_j might contribute an amount a_{ij} towards the ith goal.

If the target level for the ith level is fully achieved, the ith constraints is written as

$$\sum_{j=1}^{n} a_{ij}.x_j = b_i$$

where d_i^- = negative deviation from i^{th} goal (amount below the target value)

d_i^+ = positive deviation from the ith goal (amount above the target value)

In this case, the above stated i^{th} goal can be written as

$$\sum_{j=1}^{n} a_{ij}.x_j + d_i^- - d_i^+ = b_i, \quad i = 1, 2, ... n$$

(value of the objective) + (amount below the goal) – (amount above the goal) = Goal

☛ REMARKS
- The goal deviational variables must be non-negative.
- The deviational variables in goal programming model are equivalent to slack and surplus variables in linear programming model.
- The deviational variable d_i^+ is removed from the objective function of goal programming which over achievement is acceptable.
- The deviational variable d_i^- is removed from the objective function of the goal programming when under achivement is acceptable.
- If exact attainment of the goal is desired, then both d_i^- and d_i^+ are included in the objective function and ranked according to their pre-emptive priority factor from most important to the least, i.e., in goal programming, the lower order goals are considered only after the higher goals are achieved.

Solved Examples

EXAMPLE 1. *A manufacturing firm produces two type of products say A and B. The unit profit from product A is ₹ 100 and that from product B is ₹ 50. The goal of the firm is to earn a total profit of exactly ₹ 1700 in the next week. Formulate the problem.*

SOLUTION. Let x_1, x_2 be the no. of units of products A and B produced respectively. Then the LP formulation of the given problem is

$$\text{max. } Z = 100x_1 + 50x_2$$

subject to the constraints

$$100x_1 + 5x_2 = 1700$$

and $x_1, x_2 \geq 0$

Now, we have to formulate the goal programming

Clearly, the goal of the firm is to earn a profit of ₹ 1700 per week then

$d_i^- = $ under achievement of the profit of goal of ₹ 1700

$d_i^+ = $ over achievement of the profit of goal of ₹ 1700

So, the only constraints of the problem is

$$100x_1 + 50x_2 + d_i^- - d_i^+ = 1700$$

Hence, the required goal programming is given by

$$\text{min. } Z = d_i^- + d_i^+$$

subject to the constraints

$$100x_1 + 50x_2 + d_i^- - d_i^+ = 1700$$

and $x_1, x_2, d_i^-, d_i^+ \geq 0$

13.3.2 Multiple Goal Models

Multiple goal models are of the following three types.
 (i) Multiple goal models with equal priorities (or non-priorities)
 (ii) Multiple goal models with unequal priorities.
 (iii) Multiple goal models with priorities and weights.

(i) Multiple Goals with Equal Priorities

The multiple goal model with equal priorities may exist with less probability of all the three type but it is easy to deal mathematically.

To make it clear, consider the following examples.

 Solved Examples

EXAMPLE 1. *An office equipment manufacturer produces two kinds of products chairs and lamps. Production of either a chair or a lamp requires one hour of production capacity in the plant. The plant has a maximum production capacity of 50 hours per week because of the limited sales capacity, the maximum number of chairs and lamps that can be sold are 6 and 8 respectively. The gross margin from the sale of a chair is ₹ 90 and ₹ 60 for a lamp. The plant manager desires to determine the no. of units of each product that should be produced per week in consideration of the following equally ranked goals.*

Goal 1 *Available production capacity should be utilized as much as possible but not exceeded.*

Goal-2 *Sales of two products should be as much as possible.*

Goal-3 *Overtime should not exceed 20 percent of available production time.*

Formulate this problem as a G.P. model so that the plant manager may achieve his goals as clearly as possible.

SOLUTION. Let us suppose the no. of units of the product chairs and lamps per week be x_1 and x_2 respectively.

Then as per given,

$$x_1 + x_2 + d_1^- - d_1^+ = 50 \quad \text{(according to the first goal)}$$

The constraints according to the second goal are

$$x_1 + d_2^- = 6$$
$$x_2 + d_3^- = 8$$

Similarly, constraints according to the third goal is given by

$$d_1^+ + d_4^- - d_4^+ = 10$$

Hence, the given problem can be stated as a goal programming model as

$$\text{min. } Z = d_1^+ + d_2^- + d_3^- + d_4^-$$

subject to the constraints

$$x_1 + x_2 + d_1^- - d_1^+ = 50$$
$$x_1 + d_2^- = 6$$
$$x_2 + d_3^- = 8$$
$$d_1^+ + d_4^- - d_4^+ = 10$$

and $\quad x_1, x_2, d_1^-, d_1^+, d_2^-, d_3^-, d_4^+, d_4^- \geq 0$

Here, we have used the following symbols

d_1^- = under utilization of product capacity of 50 hours.

d_1^+ = over utilization of product capacity of 50 hours.

d_2^- = under achievement of sales goals of chairs per week.

d_3^- = under achievement of sales goals of lamps per week.

d_4^- = over achievement of overtime hours per week.

d_4^+ = under achievement of overtime hours per week

EXAMPLE 2. *An office equipment manufacturer produces two kinds of products, computer covers and flopply boxes production of either a computer cover or a floppy box requires 1 hour of production capacity in the plant. The plant has a maximum production capacity of 10 hours per day. The gross margin from the sale of a computer cover is ₹180 and ₹140 for a floppy box.*

Formulate as a G.P. with the following equally ranked goals

(i) to earn a profit of ₹ 800 per day.

(ii) because of the limted sales capacity the maximum number of computer covers and flopy boxes that can be sold are 6 and 8 per day respectively.

SOLUTION. Let x_1 and x_2 be the number of units of the product computer covers and floppy boxes produced per day respectively.

Then linear programing problem can be stated as

$$\text{max. } Z = 180x_1 + 140x_2$$

subject to be constraints

$$x_1 + x_2 \leq 10$$
$$x_1 \leq 6$$

$$x_2 \leq 8$$

and $x_1, x_2 \geq 0$

Now, the goals for the GP are

(i) To earn the profit of ₹ 800

(ii) The maximum no. of computer covers and floppy boxes that can be sold 6 and 8 per day.

Therefore, G.P. constraints for Goal (i) is

$$80x_1 + 40x_2 + d_1^- - d_1^+ = 800$$

and G.P. constraints for goal (ii) are

$$x_1 + d_2^- = 6$$

$$x_2 + d_3^- = 8$$

where,

d_1^- = under achievement of the profit of ₹ 800

d_1^+ = over achievement of the profit of ₹ 800

d_2^- = under achievement of sales target 6 of the computer covers

d_3^- = over achievement of sales target 8 of the floppy box

Hence, the required goal programming (G.P) is given as below:

min. $Z = d_1^- + d_2^- + d_3^-$

subject to the constraints

$$80x_1 + 40x_2 + d_1^- - d_1^+ = 800$$

$$x_1 + x_2 \leq 10$$

$$x_1 + d_2^- = 6$$

$$x_2 + d_3^- = 8$$

and $x_1, x_2, d_1^-, d_1^+, d_2^-, d_3^- \geq 0$

(II) MULTIPLE GOALS WITH PRIORITIES

In any firm, when management has multiple goals, then they put their goals in order of their priorities. i.e. they want the most important goal to be achieved fully or very near to the full satisfaction in comparison to the other.

To form the goal programming in such situation, we assign the priority coefficients $P_1, P_2, \dots$ with highest priority, 2nd priority 3rd priority ...etc., respectively. These priority coefficients have no numerical value, they simply represent the level of priority.

Solved Examples

EXAMPLE 1. *A firm manufactures two products. Each product requires time in two production departments. Product 1 requires 20 hours in department-1 and 10 hour in department-2. Product 2 requires 10 hours in department-1 and 10 hours in department-2. Production time is limited in department-1 to 160 hours and 140 hours in department-2. Contribution to profits by two products is ₹40 and ₹80 respectively. Management has established the following goal priorities.*

Priority-1 (P_1) *To meet production goals of 2 units of each product*

Priority-2 (P_2) *To earn profit of ₹4000.*

SOLUTION. Let x_1, x_2 be the no. of units of product-1 and product-2 produced respectively. Then we can formulate the LPP as given below.

max. $Z = 40x_1 + 80x_2$

subject to the constraints

$$20x_1 + 10x_2 \leq 160 \qquad \text{(department-1)}$$
$$10x_1 + 10x_2 \leq 140 \qquad \text{(department-2)}$$

and $x_1, x_2 \geq 0$

Now, to formulate a G.P., the constraints for the goal priority P_1 are

$$x_1 + d_1^- - d_1^+ = 2$$

and $x_2 + d_2^- - d_2^+ = 2$

and the constraints for the goal priority P_2 is given by

$$40x_1 + 80x_2 + d_3^- - d_3^+ = 4000$$

clearly, to achieve first goal d_1^- and d_2^- should be minimized and to achieve second goal d_3^- should be minimized.

Hence, the formulation of G.P. is given by

$$\text{min. } Z = P_1 d_1^- + P_1 d_2^- + P_2 d_3^-$$

subject to the constraints

$$20x_1 + 10x_2 \leq 160$$
$$10x_1 + 10x_2 \leq 140$$
$$x_1 + d_1^- - d_1^+ = 2$$
$$x_2 + d_2^- - d_2^+ = 2$$
$$40x_1 + 80x_2 + d_3^- - d_3^+ = 4000$$

and $x_1, x_2, d_1^-, d_1^+, d_2^-, d_2^+, d_3^-, d_3^+ \geq 0$

(III) MULTIPLE GOALS WITH PRIORITIES AND WEIGHTS

Sometimes, there are some problem in which two or more goals have same level of priorities but the different improtance. In such type of cases, different weights are used to reflect the difference of their weights within the same level of priorities.

 Solved Examples

EXAMPLE 1. *A production manager is faced with the problem of job allocation to his two production teams. The production rate of team-1 is 8 units per hour while the production rate of team-2 is 5 units per hour. The normal working hours for each of the teams is 40 hours per week. The production manager has prioritized the following goals for the coming week.*

P_1: *Avoid under achievement of the desired production level of 650 units.*

P_2: *Overtime operation of team-1 is limited to 5 hours.*

P_3: *The total overtime for both teams should be minimized.*

P_4: *Any under utilization of regular working hours of the teams should be avoided assign different weights according to the relative productivity of the two teams.*

Formulate the problem as G.P. model.

SOLUTION: Let x_1 and x_2 be the number of hours of working of team-1 and team-2 per week respectively. Since the maximum production level is 650 units, then production volume constraint is given by

$$8x_1 + 5x_2 + d_1^- - d_1^+ = 650$$

Where $\quad d_1^- = $ under-achievement of the production target

$\qquad d_1^+ = $ over-achievement of the production target

Further, since the normal working hours of each team are 40 hours per week. So the overtime constraints for the two team is given by

$$x_1 + d_2^- - d_2^+ = 40 \qquad \text{(for team-1)}$$

and $\qquad x_2 + d_3^- - d_3^+ = 40 \qquad \text{(for team-2)}$

where, $\quad d_2^- = $ under-achievement of the normal working hours by team-1

$\qquad d_2^+ = $ over-achievement of the normal working hours by team-1

$\qquad d_3^- = $ under-achievement of the usual working hours by team-2

$\qquad d_3^+ = $ over-achievement of the usual working hours by team-2

Now, the overtime constraints for the team-1 is given by

$$d_2^+ + d_4^- - d_4^+ = 5$$

where d_4^- and d_4^+ are the under-achievement and over-achievement of overtime per week by team-1 respectively.

Since the production rate of team-1 per hour is 8 units and that of team-2 per hour is 5 units. Also the utilization of regular working hours of each team should be minimized in the ratio 8:5 of the relative productivity of team-1 and team-2. Now, we have to form the objective function.

We observe that the first goal priority P_1 of the manager is to avoid under achievement of the desired production level of 650 units then $P_1 d_1^-$ is the team of goal-1 in the objective function. Now in second priority P_2, the overtime operation of team-1 is restricted to 5 hours, then $P_2 d_4^+$ is the term of goal-2 in the objective function.

Also, in third priority P_3, the total overtime for both teams should be minimized, therefore $P_3(d_2^+ + d_3^+)$ is the objective function term for goal-3.

Similarly, in P_4, we have $8P_4 d_2^- + 5P_4 d_3^-$ is the objective function term. Hence the required goal programming is stated as

$$\text{min. } Z = P_1 d_1^- + P_2 d_4^+ + P_3(d_2^+ + d_3^+) + P_4(8d_2^- + 5d_3^-)$$

subject to the constraints

$$8x_1 + 5x_2 + d_1^- - d_1^+ = 650$$

$$x_1 + d_2^- - d_2^+ = 40$$

$$x_2 + d_3^- - d_3^+ = 40$$

$$d_2^+ + d_4^- - d_4^+ = 5$$

and $\qquad x_1, x_2, d_1^-, d_1^+, d_2^-, d_2^+, d_3^-, d_3^+, d_4^-, d_4^+ \geq 0$

13.4 GENERAL FORM OF GOAL PROGRAMMING PROBLEM

Let us suppose a problem has m goals, p structural constraints, n decision variables and k-level of priorities.

Then a general GP can be written as

$$\text{min. } Z = \sum_{i=1}^{m} \sum_{r=1}^{k} P_r(w_{i,r}^+ \, d_i^+ + w_{i,r}^- \, d_i^+)$$

subject to the constraints

$$\sum_{j=1}^{n} a_{ij} \cdot x_j + d_i^- - d_i^+ = b_i \qquad \text{for } i = 1, 2, \ldots, m$$

$$\sum_{j=1}^{n} a_{ij} x_j (\le, =, \ge) b_i \qquad \text{for } i = m + 1, m + 2, \ldots m + p$$

and $\qquad x_j, d_i^-, d_i^+ \ge 0$ for $i = 1, 2, 3, \ldots m, j = 1, 2, 3, \ldots n$

where, $\qquad P_r$ = The priority coefficient for r^{th} priority

$\qquad w_{i,r}^-$ = The relative weight of d_i^- variable in the rth priority level.

$\qquad w_{i,r}^+$ = The relative weight of the d_i^+ variable in the rth priority level.

☞ **REMARKS**

- Two type of variables are taken in the formulation namely decision variables (x_i) and the deviational variables d^- and d^+.
- Two classes of constraints can exist in a given GP model, structural constraints, which are not directly related to the goals and goal constraints which are directly related to the goals.
- In most cases, a goal constraints will have both the deviational variables even when both deviational variables do not appear simultaneously in the objective function.

13.5 METHOD OF SOLUTION OF A GOAL PROGRAMMING (GP) PROBLEM

To solve a GP problem, we have following two methods

(i) Graphical method

(ii) GP algorithm or modified simplex method

13.5.1 GRAPHICAL METHOD

We used graphical method when the GP model involve two decision variables. The method is quite similar to the graphical method of an LPP. In LPP the objective function of one goal only is to be optimized whereas in the GP models, the objective function of the deviational variables from multiple goals in respective order of priorities are minimized so that the achievement of the goals of higher order are not affected.

To solve a GP model by graphical method, we use the following procedure.

WORKING PROCEDURE

STEP 1. Graph (Plot) all the structural constraints and identify the feasible region. If there is no structural constraints then the feasible region is the first quadrant (*i.e.* $x_1 \ge 0, x_2 \ge 0$). If no feasible region exists then there is no solution to the problem.

STEP 2. Draw the lines corresponding to the goal constraints by setting the deviational variables in the goal constraints equal to zero.

STEP 3. Identify the top-priority solution which is obtained by determining the points within the feasible region so obtained in step 1 that satisfy the highest priority goal.

STEP 4. Sequentially consider the remaining goals and the point that satisfy them to the greatest extent possible. Make sure that a lower priority goal is not achieved by reducing the degree of achievement of higher priority goals.

STEP 5. Repeat the step 4 until all levels of priority have been investigated.

Solved Examples

EXAMPLE 1. *An office equipment manufacturer produces two kinds of products: chairs and lamps. Production of either a chair or a lamp requires one hour of production* *capacity in the plant. The plant has a maximum capacity of 10 hours per week. The gross margin from the sale of a chair is ₹80 and ₹40 for that of a lamp. Formulate and solve the problem as a GP model with the following equally ranked goals.*

(i) to earn a profit of ₹800 per week

(ii) the maximum no. of chairs and lamps that can be sold are 6 and 8 per week respectively. [MEERUT–2004, 06]

SOLUTION: Clearly, the G.P. model of the given problem is

$$\min. \ Z = d_1^- + d_2^- + d_3^-$$

subject to the constraints

$$80x_1 + 40x_2 + d_1^- - d_1^+ = 800$$
$$x_1 + d_2^- = 6$$
$$x_1 + d_3^- = 8$$

and $x_1, x_2, d_1^-, d_1^+, d_2^-, d_3^- \geq 0$

Now, we have to solve the above GP by graphical method.

STEP 1. Let us take $d_1^- = d_1^+ = d_2^- = d_3^- = 0$ and then plot all the constraints. The arrows (← or →) are then associated with each line to represent under achievement or over achievement.

STEP 2. Because, all the goals are equally ranked and the objective of the problem is to minimize d_1^-, d_2^- and d_3^-. So, we set $d_1^- = 0, d_2^- = 0$ and $d_3^- = 0$. Then we get an intersection point $P(6, 8)$. At $P(6, 8)$ the profit of the firm is ₹800 and min. $Z = 0$

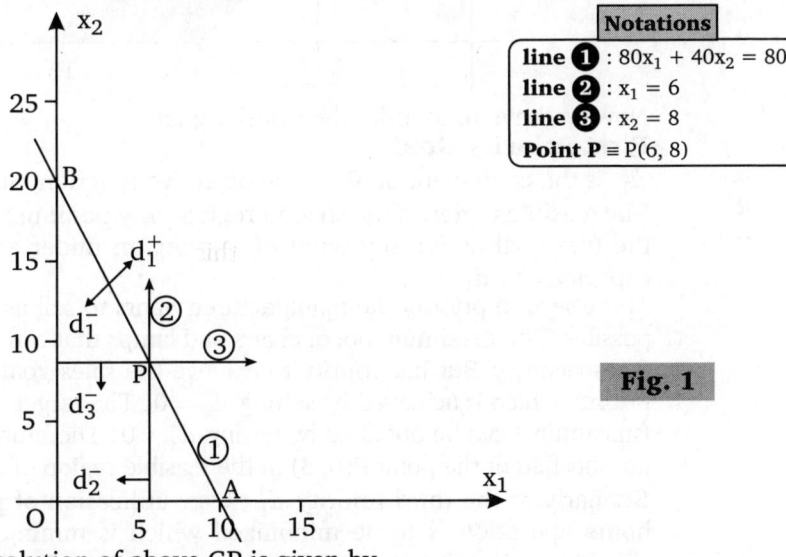

Notations

line **①** : $80x_1 + 40x_2 = 800$
line **②** : $x_1 = 6$
line **③** : $x_2 = 8$
Point P ≡ P(6, 8)

Fig. 1

Hence, the solution of above GP is given by

$$x_1 = 6, x_2 = 8 \text{ and min.} Z = 0$$

EXAMPLE 2. *Solve the following G.P. by graphical method*

$$\min. \ Z = P_1 d_1^- + P_2(2d_2^- + d_3^-) + P_3 d_1^+$$

subject to the constraints

$$x_1 + x_2 + d_1^- - d_1^+ = 10 \hspace{3cm} ...(1)$$

$$x_1 + d_2^- = 6 \qquad\qquad \ldots(2)$$

$$x_2 + d_3^- = 8 \qquad\qquad \ldots(3)$$

and $\qquad x_1, x_2, d_1^-, d_1^+, d_2^-, d_3^- \geq 0$

SOLUTION. To solve the above problem, we use the following steps

STEP 1 Taking $d_1^- = d_1^+ = d_2^- = d_3^- = 0$ and then plot the goal constraints as shown in the the following figure

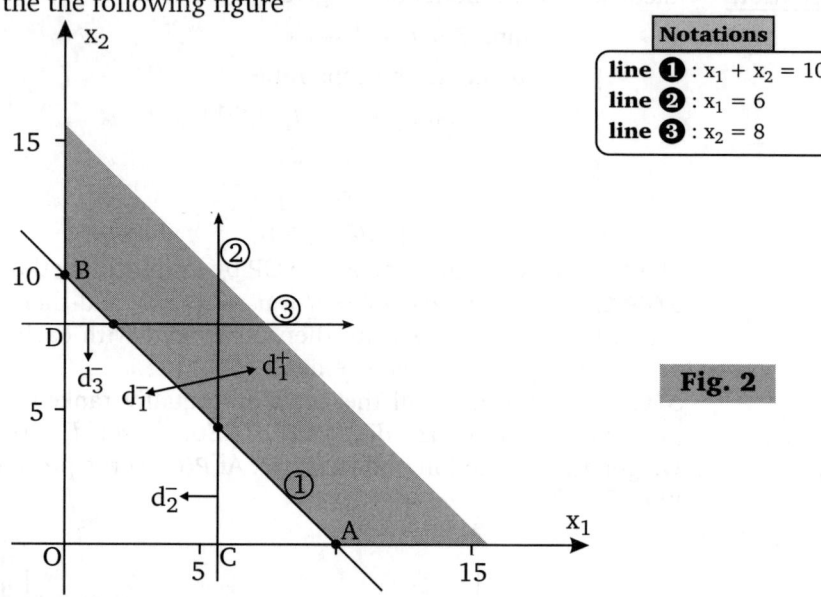

Notations
line ❶ : $x_1 + x_2 = 10$
line ❷ : $x_1 = 6$
line ❸ : $x_2 = 8$

Fig. 2

Now we have to identify the priority goal

First Priority Goal

d_1^- is the coefficient of P_1 in the objective function. Setting $d_1^- = 0$

The feasible region is the shaded region. Any point in this shaded region satisfies the first goal as for any point of this region under achievement of production capacity, *i.e.*, $d_1^- = 0$.

In the second priority the manufacturer wants to sell as many chairs and lamps as possible. The maximum no. of chairs and lamps that can be sold per week are 6 and 8 respectively. But his priority to achieve the sales goals of chairs to 6 is the first priority which is achieved by setting $d_2^- = 0$. The target of sales goal of lamps to 10 (maximum) can be obtained by setting $d_3^- = 0$. Therefore, the second priority goals are satisfied at the point $P(6, 8)$ in the feasible region of the top priority.

Similarly in the third priority d_1^+, over utilization of production capacity of 10 hours operation is to be minimized which is minimum when $d_1^+ = 0$. Clearly $d_1^+ = 0$ at all points on the line segment AB at which the first priority goal is achieved. But at any point of the line segment AB, the second priority goal can not be achieved. Hence we can not achieve the third goal without sacrificing the second goal. So, keeping d_1^+ minimum possible, the solution of the goal programming problem is achieved at $P(6, 8)$ at which the first two top priorties goals are achieved fully, but the third priority goal is achieved as much as possible.

For $\quad d_2^- = 0, d_3^- = 0 \quad$ from (2) and (3) $\quad x_1 = 6, x_2 = 8$

$\therefore \quad$ using $d_1^- = 0, x_1 = 6, x_2 = 8$ in (1) we get $d_1^+ = 4$

Finally, we conclude that the manufacturer should produce 6 chairs and 8 lamps per week, so that his first two goals are fully achieved and the third goal is the overtime operation of the plant is minimized to 4 hours per week.

13.5.2 MODIFIED SIMPLEX METHOD TO SOLVE A GOAL PROGRAMMING MODEL

The steps of modified simplex method for a GP model are as follows:

STEP 1. **(Formulation of Initial Table)**

Construct the initial table in the same way as that for LPP. The top most row of the table will have the coefficients of decision variables and deviational weights c_j, in the objective function. Below the entries of c_j, the coefficient of x_j's and d_i^- and d_i^+ are placed in the appropriate column. Now draw a horizontal line below these entries and write the pre-empty priorities factor $P_1, P_2, ...$ in X_B column, starting from bottom to top i.e., first priority P_1 written at the bottom and the least priority is written at the top. In GP model, we minimize the unattained portion of goals as much as possible by minimizing the deviational variables.

STEP 2. **(Test of Optimality)**

Compute the values z_j and $c_j - z_j$ seperately by each of the ranked goals, $P_1, P_2, ...$ etc., in the same way as in usual simplex method. The optimal criterion z_j or $c_j - z_j$ becomes a matrix of order k×n, where k represents the no. of pre-empty priority levels and n denotes the number of total variables i.e., decision variable and deviational variables.

Further, check $c_j - z_j$ for top priority goal P_1. If all $c_j - z_j \geq 0$ in P_1 row or there is a zero in P_1 row in X_B-column then goal P_1 is achieved and then we go to step 5. But if at least one of the entries is negative and there is no zero in X_B column in P_1-row, then goal P_1 is not achieved and we go to the next step.

STEP 3. **(To Determine Incoming and Outgoing Vector)**

Select the most negative entry in P_1 row of matrix $c_j - z_j$. The variable in the column corresponding to the most negative entry is the incoming variable. If there is a tie, then check the next lower priority level. The column corresponding to the most negative entry in the lower priority level, out of the columns in which there is a tie in P_1-row is selected as incoming variable.

When we have obtained incoming vector, then we find the outgoing vector as in usual simplex method in LPP by minimum ratio rule and the element at the intersection of the incoming vector column and the minimum ratio row is called key element.

STEP 4. Reduce the key element to 1 as in usual simplex method and with its help using row operation, all other elements in the corresponding column are reduced to zero. So we get a new reduced matrix. From this reduced matrix, find the value of z_j or $c_j - z_j$ for each of the ranked goals $P_1, P_2, ...$. Now again examine $c_j - z_j$ for top priority goal P_1. If all the entries in P_1-row are positive, then P_1 is achieved. If at least one of the entries in P_1 is negative then P_1 still not achieved. Then to achieve P_1, we repeat step 3 and 4.

STEP 5. If the first priority goal P_1 is achieved, then we proceed as above to achieve the next priority goal P_2. The goal P_2 can not be further improved from its present level if there is positive entry in P_1 row below the most negative entry in P_2 row. Continue the above process until the lowest priority goal say P_k is also fully achieved or is achieved nearest to the satisfaction. Again the goal P_k can not be further improved from its present level if there is a positive entry in the higher priority goals $P_1, P_2, \ldots P_{k-1}$ rows below the most negative entry in P_k row.

Solved Examples

EXAMPLE 1. *An office equipment manufacturer produces two kinds of products: chairs and lamps. Production of either a chair or a lamp requires one hour of production capacity in a plant. The plant has a maximum production capacity of 10 hours per week. Because of the limited sales capacity, the maximum number of chairs and lamps that can be sold in 6 and 8 per week respectively. The gross margin from the sale of a chair is ₹ 80 and ₹ 40 for that of a lamp.*

The plant manager has set the following goals arranged in order of importance.

(i) *He wants to avoid any under utilization of production capacity.*

(ii) *He wants to sell as many chairs and lamps as possible since the gross margin from the sale of a chair is set at double the amount of profit from a lamp, he has double as much as desire to achieve the sales goal, for chairs as for lamps.*

(iii) *He wants to minimize overtime operation of the plant as much as possible.*

Formulate and solve this problem as a GP problem so that the plant manager makes a decision that will achieve his goals as closely as possible.

SOLUTION. Following the usual procedure, the G.P. model of the above problem is as follows.

$$\text{min. } Z = P_1 d_1^- + P_2 (2d_2^- + d_3^-) + P_3 d_1^+$$

subject to the constraints

$$x_1 + x_2 + d_1^- - d_1^+ = 10$$
$$x_1 + d_2^- = 6$$
$$x_2 + d_3^- = 8$$

and $\qquad x_1, x_2, d_1^-, d_1^+, d_2^-, d_3^- \geq 0$

Now to solve it, we proceed as follows:

Taking $x_1 = 0$, $x_2 = 0$ and $d_1^+ = 0$ we get $d_1^- = 10$, $d_2^- = 6$, $d_3^- = 8$ which is the initial basic feasible solution

STEP 1 (Construction of initial table)

	$c_j \rightarrow$		O	O	P_1	P_3	$2P_2$	P_2	Min. ratio X_B / x_1
B.V.	C_B	X_B	x_1	x_2	d_1^- (x_3)	d_1^+ (x_4)	d_2^- (x_5)	d_3^- (x_6)	
d_1^-	P_1	10	1	1	1	−1	0	0	$\dfrac{10}{1}$
d_2^-	$2P_2$	6	1	0	0	0	1	0	
d_3^-	P_2	8	0	1	0	0	0	1	$\dfrac{6}{1}$ (min) →
$c_j - z_j$	P_3	0	0	0	0	1	0	0	
	P_2	20	−2	−1	0	0	0	0	
	P_1	10	−1	−1	0	1	0	0	

↑

In the above table

$$c_1 - z_1 = c_1 - C_B x_1 = 0 - (P_1, 2P_2, P_2)\begin{pmatrix} 1 \\ 1 \\ 0 \end{pmatrix} = -P_1 - 2P_2$$

$$c_2 - z_2 = c_2 - C_B x_2 = 0 - (P_1, 2P_2, P_2)\begin{pmatrix} 1 \\ 0 \\ 1 \end{pmatrix} = -P_1 - P_2$$

$$c_4 - z_4 = c_4 - C_B x_4 = P_3 - (P_1, 2P_2, P_2)\begin{pmatrix} -1 \\ 0 \\ 0 \end{pmatrix} = P_3 + P_1$$

and $Z = C_B x_B = (P_1, 2P_2, P_2)\begin{pmatrix} 10 \\ 6 \\ 8 \end{pmatrix} = 10P_1 + 20P_2$

The $c_j - z_j$ matrix is obtained by the coefficients of P_1, P_2 and P_3 in $c_j - z_j$ values.

STEP 2.

The most negative entry in P_1 - row is –1 which occurs for column x_1 as well as for column x_2. So there is a tie, so we check the lower priority P_2. In P_2 - row, the most negative entry is –2 which is in x_1 - column thus x_1 - column is selected as key element i.e., x_1 is the incoming vector and by minimum ratio rule d_2^- is the outgoing vector. Hence, the key element is 1 i.e. a_{21}.

STEP 3.

Since the key element is already 1. So using row transformation, the elements other than the key element are reduced to zero. Then we have the following reduced table.

Reduced Table-2

			0	0	P_1	P_3	$2P_2$	P_2	Min. ratio
		$c_j \rightarrow$							X_B / x_2
B.V.	C_B	x_B	x_1 (d_2^-)	x_2	(x_3) d_1^-	d_1^+ (x_4)	x_5	d_3^- (x_6)	
d_1^-	P_1	4	0	①	1	–1	–1	0	$\frac{4}{1}$(min.) →
x_1	0	6	1	0	0	0	1	0	
d_3^-	P_2	8	0	1	0	0	0	1	$\frac{8}{1}$
$c_j - z_j$	P_3	0	0	0	0	1	0	0	
	P_2	8	0	–1	0	0	2	0	
	P_1	4	0	–1	0	1	1	0	
				↑					

In the above table

$$c_2 - z_2 = c_2 - C_B x_2 = 0 - (P_1, 0, P_2)\begin{bmatrix} 1 \\ 0 \\ 1 \end{bmatrix} = -P_1 - P_2$$

$$c_4 - z_4 = c_4 - C_B x_4 = P_3 - (P_1, 0, P_2) \begin{bmatrix} -1 \\ 0 \\ 0 \end{bmatrix} = P_3 + P_1$$

$$c_5 - z_5 = c_5 - C_B x_5 = 2P_2 - (P_1, 0, P_2) \begin{bmatrix} -1 \\ 1 \\ 0 \end{bmatrix} = 2P_2 + P_1$$

and $Z = C_B x_B = (P_1, 0, P_2) \begin{bmatrix} 4 \\ 6 \\ 8 \end{bmatrix} = 4P_1 + 8P_2$

Now from the above table we observe that in P_1-row, the most negative element is -1 which occurs in column x_2 i.e., x_2 is the incoming vector and by minimum ratio rule, d_1^- is the outgoing vector and the key element is 1, i.e., a_{12}.

Further, using row transformation as usual, the next reduced table is as given below.

Reduced Table-3

B.V.	C_B	x_B	x_1 (d_2^-)	x_2 (d_1^-)	(x_3)	d_1^+ (x_4)	(x_5)	d_3^- (x_6)	Min. ratio X_B / d_1^+
$c_j \rightarrow$			0	0	P_1	P_3	$2P_2$	P_2	
x_2	0	4	0	1	1	-1	-1	0	$-$
x_1	0	6	1	0	0	0	1	0	$-$
d_3^-	P_2	4	0	0	-1	①	1	1	$\dfrac{4}{1} \rightarrow$
$c_j - z_j$ $\;\; P_3$	0	0	0	0	1	0	0	
P_2	4	0	0	1	-1	1	0	
P_1	0	0	0	1	0	0	0	
						$\uparrow$			

In the above table, we compute $c_j - z_j$ for $j = 3, 4, 5$ (not in the basis)

$$c_3 - z_3 = c_3 - C_B x_3 = P_1 - (0, 0, P_2) \begin{bmatrix} 1 \\ 0 \\ -1 \end{bmatrix} = P_1 + P_2$$

$$c_4 - z_4 = c_4 - C_B x_4 = P_3 - (0, 0, P_2) \begin{bmatrix} -1 \\ 0 \\ 1 \end{bmatrix} = P_3 - P_2$$

$$c_5 - z_5 = c_5 - C_B x_5 = 2P_2 - (0, 0, P_2) \begin{bmatrix} -1 \\ 1 \\ 1 \end{bmatrix} = P_2$$

and $Z = C_B x_B = (0, 0, P_2) \begin{bmatrix} 4 \\ 6 \\ 4 \end{bmatrix} = 4P_2 = 0 \cdot P_3 + 4P_2 + 0 \cdot P_1$

From the above table we observe that there is zero in x_B column in P_1 row so the goal P_1 is fully achieved.

STEP 4

As goal P_1 is achieved, we move to goal P_2. In P_2 row of table 3, the most negative element is -1 which occurs in d_1^+ column, so d_1^+ is the incoming vector and by minimum ratio rule, d_3^- is the outgoing vector and the key element is 1, *i.e.*, a_{34}. Now repeat step 3, we get the next reduced table 4.

Reduced Table-4

	$c_j \rightarrow$		0	0	P_1	P_3	$2P_2$	P_2	Min. ratio
B.V.	C_B	x_B	x_1 (d_2^-)	x_2 (d_1^-)	(x_3)	(x_4) (d_3^-)	(x_5)	(x_6)	
x_2	0	8	0	1	0	0	0	1	
x_1	0	6	1	0	0	0	1	0	
d_1^+	P_3	4	0	1	-1	1	1	1	
$c_j - z_j$	P_3	4	0	0	1	0	-1	-1	
	P_2	0	0	0	0	0	2	1	
	P_1	0	0	0	1	0	0	0	

In the above table, we have computed $c_j - z_j$ for $j = 3, 5, 6$ (non-basic variable)

$$c_3 - z_3 = c_3 - C_B x_3 = P_1 - (0,0,P_3)\begin{bmatrix} 0 \\ 0 \\ -1 \end{bmatrix} = P_1 + P_3$$

$$c_5 - z_5 = c_5 - C_B x_5 = 2P_2 - (0,0,P_3)\begin{bmatrix} 0 \\ 1 \\ 1 \end{bmatrix} = 2P_2 - P_3$$

$$c_6 - z_6 = c_6 - C_B x_6 = P_2 - (0,0,P_3)\begin{bmatrix} 1 \\ 0 \\ 1 \end{bmatrix} = P_2 - P_3$$

and $Z = C_B x_B = (0,0,P_3)\begin{bmatrix} 8 \\ 6 \\ 4 \end{bmatrix} = 4P_3$

In the above table, we observe that in P_1 and P_2 rows all the elements are non-negative and these are zero in x_B column in P_1 and P_2 rows. Therefore, the goals P_1 and P_2 are fully achieved.

Now, we move to goal P_3 and observe that in P_3 -row the most negative element is -1 which occurs in column x_5 and x_6. But in higher priority goal P_2, in P_2 row there are positive entries in column x_5 and x_6 below the most negative element.

Hence goal P_3 can not be achieved.

Again from table 4 we have
$$x_1 = 6, x_2 = 8, d_1^+ = 4, d_1^- = 0, d_2^- = 0, d_3^- = 0$$
Hence, the optimal solution of present goal programming is
$$x_1 = 6, x_2 = 8, d_1^+ = 4$$

EXAMPLE 2. *A company produces two kind of products: A and B. Production of either A or B requires 3 hours of production capacity in the plant. The plant has a maximum production capacity of 30 hours per week to manufacture these two products, has set the following goals arranged in the order of importance.*

  (i) *To avoid any underutilization of production capacity.*
  (ii) *To limit the overtime to 5 hours.*
  (iii) *To minimize the overtime operations of the plant as much as possible.*
Formulate this problem as a goal programming and then solve it by modified simplex method.

SOLUTION. Let x_1 and x_2 be the number of units of product A and B produced respectively. As per given
  no. of hours per unit to produce product $A = 3$ hours
  no. of hours per unit to produce product $B = 3$ hours
maximum normal production capacity per week = 30 hours
upper limit for overtime hours per week = 5 hours.
Then clearly, production capacity constraints is given by
$$3x_1 + 3x_2 + d_1^- - d_1^+ = 30$$
where d_1^- and d_1^+ are the under achievement and over achievement of the production target respectively.
Also, the overtiem constraints is given by
$$d_1^+ + d_2^- - d_2^+ = 5$$
where d_2^- and d_2^+ are respectively the under achievement and over achievement of overtime target.
Now, we have to define the objective function.
Let us suppose that P_1, P_2 and P_3 be the goal priorities.
Clearly, the first priority P_1 of the manager is to avoid underutilization of production capacity so that d_1^- is to be minimized, then $P_1 d_1^-$ is the objective function term for goal 1.
Now, the second goal priority P_2 of the manager is to limit the overtime hours to 5 hours so that d_2^+ is to be minimized, then $P_2 d_2^+$ is the objective function term for goal 2. The third goal priority P_3 is to be minimized overtime operation of the plant as much as possible so that d_1^+ is to be minimized. Therefore $P_3 d_1^+$ is the objective function term for goal 3.
  $\therefore$ the objective function is
$$\text{min. } Z = P_1 d_1^- + P_2 d_2^+ + P_3 d_1^+$$
Thus, the required GPP is given by
$$\text{min. } Z = P_1 d_1^- + P_2 d_2^+ + P_3 d_1^+$$
subject to the constraints
$$3x_1 + 3x_2 + d_1^- - d_1^+ = 30$$
$$d_1^+ + d_2^- - d_2^+ = 5$$
and
$$x_1, x_2, d_1^-, d_1^+, d_2^-, d_2^+ \geq 0$$

Now, we apply the modified simplex method and get the initial table as given below.

Initial Table-1

B.V.	C_B	x_B	c_j → x_1 0	x_2 0	d_1^- (x_3) P_1	d_2^+ (x_4) P_3	d_2^- (x_5) 0	d_2^+ (x_6) P_2	Min. Ratio
d_1^-	P_1	30	③	3	1	−1	0	0	$\dfrac{30}{3}$ (min) →
d_2^-	0	5	0	0	0	1	1	−1	—
$c_j - z_j$	P_3	0	0	0	0	1	0	0	
	P_2	0	0	0	0	0	0	1	
	P_1	30	−3	−3	0	1	0	0	
			↑			↓			

In the above table, we have to compute $c_j - z_j$ for $j = 1, 2, ..., 6$

$$c_1 - z_1 = c_1 - C_B x_1 = 0 - (P_1, 0)\begin{bmatrix}3\\0\end{bmatrix} = -3P_1$$

$$c_2 - z_2 = c_2 - C_B x_2 = 0 - (P_1, 0)\begin{pmatrix}3\\0\end{pmatrix} = -3P_1$$

$$c_3 - z_3 = c_3 - C_B x_3 = P_1 - (P_1, 0)\begin{bmatrix}1\\0\end{bmatrix} = P_1 - P_1 = 0$$

$$c_4 - z_4 = c_4 - C_B x_4 = P_3 - (P_1, 0)\begin{bmatrix}-1\\1\end{bmatrix} = P_3 + P_1$$

$$c_5 - z_5 = c_5 - C_B x_5 = 0 - (P_1, 0)\begin{bmatrix}0\\1\end{bmatrix} = 0$$

$$c_6 - z_6 = c_6 - C_B x_6 = P_2 - (P_1, 0)\begin{bmatrix}0\\-1\end{bmatrix} = P_2$$

∴ The criterion matrix is obtained by entering the coefficient of P_1, P_2 and P_3 in each value of $c_j - z_j$.

Now, $Z = C_B x_B = (P_1, 0)\begin{pmatrix}30\\5\end{pmatrix} = 30P_1 = 30P_1 + 0P_2 + 0P_3$

The coefficient of P_1, P_2 and P_3 in Z are written in column x_B in the criterion matrix $c_j - z_j$.

Further, since the most negative entry in P_1 row is −3 which occurs for column x_1 as well as the column x_2. Also in P_2 and P_3 rows, there is zero in x_B column in $c_j - z_j$ matrix so that P_1 goal is the only unattained goal, so there is no other row above P_1-row with attainment so that tie is broken randomly and the column x_1 is selected as incoming vector only by minimum ratio rule d_1^- in the outgoing vector and the key element is 3, i.e., a_{11}.

Now reduce the key element to 1 by dividing the d_1^- row by 3 and using the row operation, reduce other elements in c_1-column equal to zero so that the next reduced table is as given below.

B.V.	c_j		0	0	P_1	P_3	0	P_2
	C_B	x_B	x_1	x_2	d_1^- (x_3)	d_1^+ (x_4)	d_2^- (x_5)	d_2^+ (x_6)
x_1	0	10	1	1	1/3	−1/3	0	0
d_2^-	0	5	0	0	0	1	1	−1
$c_j - z_j$	P_3	0	0	0	0	1	0	0
	P_2	0	0	0	0	0	0	1
	P_1	0	0	0	0	1	0	0

In the above table we have computed $c_j - z_j$ for (non-basic) $j = 2, 3, 4, 6$ as follows:

$$c_2 - z_2 = c_2 - C_B x_2 = 0 - (0,0)\begin{bmatrix}1\\0\end{bmatrix} = 0 = 0P_1 + 0P_2 + 0P_3$$

$$c_3 - z_3 = c_3 - C_B x_3 = P_1 - (0,0)\begin{bmatrix}1/3\\0\end{bmatrix} = P_1 = P_1 + 0P_2 + 0P_3$$

$$c_4 - z_4 = c_4 - C_B x_4 = P_3 - (0,0)\begin{bmatrix}-1/3\\0\end{bmatrix} = P_3 = 0P_1 + 0P_2 + P_3$$

$$c_6 - z_6 = c_6 - C_B x_6 = P_2 - (0,0)\begin{bmatrix}0\\-1\end{bmatrix} = P_2 = 0P_1 + P_2 + 0P_3$$

and $Z = C_B x_B = (0,0)\begin{bmatrix}10\\5\end{bmatrix} = 0 = 0P_1 + 0P_2 + 0P_3$

Also, from the above table, we observe that in criterion matrix $c_j - z_j$, there is zero in column x_B in P_3-row, in P_2-row, in P_1-row.

Hence all the priorities of the goals are fully achieved

Hence, the optimal solution of GPP is given by

$$x_1 = 10, x_2 = 0, d_1^- = 0, d_1^+ = 0, d_2^- = 5, d_2^+ = 0 \text{ and min. } Z = 0$$

EXAMPLE 3. *An office equipment produces two kinds of products chairs and lamps. Production of either a chair or a lamp requires one hour of production capacity in the plant. The plant has a maximum production capacity in the plant. The plant has a maximum production capacity of 50 hours per week.*

Because of the limited sales capacity, the maximum no. of chairs and lamps that can be sold are 6 and 8 per week respectively. The gross margin from the sale of a chair is ₹90 and ₹60 for a lamp. The plant manager desires to determine the no. of units of each product that should be produced per week in consideration of the following set of goals.

(i) *Available production capacity should be utilized as much as possible but not exceed.*

(ii) *Sales of the two products should be as much as possible.*

(iii) *Overtime should not exceed* 20 *percent of available production time.*

Formulate and solve this problem as a G.P. model so that the plant manager may achieve his goals as closely as possible.

SOLUTION. Let x_1, x_2 be the no. of chairs and lamps produced per week and

$d_1^- =$ Time in hours by which the production capacity is under utilized.

$d_1^+ =$ Time in hours by which the production capacity is over utilized.

$d_2^- =$ Number by which the sales of six chairs is under achieved.

$d_3^- =$ Number by which the sales of eight lamps is under achieved.

$d_{12}^- =$ Time in hours by which the overtime of 10 hours (20% of 50) is underachieved.

$d_{12}^+ =$ Time in hours by which the overtime of 10 hours is over achieved.

Also, there is no priority of 3 goals to be achieved.

Then the formulation of the given problem as a G.P. model is given as follows:

$$\text{min. } Z = d_1^+ + d_2^- + d_3^- + d_{12}^+$$

subject to the constraints

$$x_1 + x_2 + d_1^- - d_1^+ = 50$$
$$x_1 + d_2^- = 6$$
$$x_2 + d_3^- = 8$$
$$d_1 + d_{12}^- - d_{12}^+ = 10$$

and $x_1, x_2, d_1^-, d_1^+, d_2^-, d_3^-, d_{12}^-, d_{12}^+ \geq 0$

Further, since there is no priority in the 3 goals to be achieved, therefore it can be solved by usual simplex method. Now applying the simplex method in a usual manner, we have the following simplex table.

		$c_j \rightarrow$	0	0	0	1	1	1	0	1	Min. Ratio
B.V.	C_B	x_B	x_1	x_2	d_1^-	d_1^+	d_2^-	d_3^-	d_{12}^-	d_{12}^+	x_B / x_1
d_1^-	0	50	1	1	1	−1	0	0	0	0	50
d_2^-	1	6	①	0	0	0	1	0	0	0	6(min.) →
d_3^-	1	8	0	1	0	0	0	1	0	0	—
d_{12}^-	0	10	0	0	0	1	0	0	1	−1	—
$Z = 14$	$c_j - z_j$		−1 ↑	−1	0	1	0	0	0	1	x_B / x_2
d_1^-	0	44	0	1	1	−1	−1	0	0	0	44
x_1	0	6	1	0	0	0	1	0	0	0	—
d_3^-	1	8	0	①	0	0	0	1	0	0	8(min.) →
d_{12}^-	0	10	0	0	0	1	0	0	1	−1	—
$Z = 8$	$c_j - z_j$		0	−1 ↑	0	1	1	0	0	0	
d_1^-	0	36	0	0	1	−1	−1	−1	0	0	
x_1	0	6	1	0	0	0	1	0	0	0	
x_2	0	8	0	1	0	0	0	1	0	0	
d_{12}^-	0	10	0	0	0	1	0	0	1	−1	
$Z = 0$	$c_j - z_j$		0	0	0	1	1	1	0	1	

Clearly, in the last row of the above table all $c_j - z_j \geq 0$

$\Rightarrow$ solution is optimal and is given by

$$x_1 = 6, \, x_2 = 8, \, d_1^- = 36, \, d_{12}^- = 10$$

Here, $d_1^- = 36$. Therefore, production capacity is under utilized by 36 hours. Hence there is no question of overtime.

EXAMPLE 4. *A textile company produces two types of material A and B. The average production rates for the material A and B are identical at 1000 m/hrs. By running two shifts the operational capacity of the plant is 80 hours per week. The marketing department report that maximum estimated sales for the following week is 70,000 meters of material A and 45,000 meters of material B. According to the account department the profit from one meter of material A is 2.50 and from one meter of material B is 1.50. The management of the company decide that a stable employment level is the primary goal for the firm. Thus, whenever there is a demand exceeding normal production capacity. The management simply expands production capacity by providing overtime. However management feels that overtime operation of the plant of more than 10 hours per week should be avoided because of the accelerating costs. The management has the following goals.*

Goal (1) *The first goal is to avoid any under utilization of production capacity. i.e., to maintain stable employment at normal capacity.*

Goal (2) *To limit the overtime operation of the plant to 10 hours.*

Goal (3) *To achieve the sale 70,000 meters of material A and 45000 meters of material B.*

Goal (4) *To minimize the overtime operation of the plant as much as possible.*

Formulate and solve the problem as a GPP to help the management for the best decision.

SOLUTION. Let x_1, x_2 be the no. of hours per week spent for producing the material A and B respectively.

Also,

$d_1^- =$ under utilization of the production capacity of 80 hours

$d_1^+ =$ over utilization of the production capacity of 80 hours

$d_2^- =$ under achievement of the sales of 70000 of material A in meter

$d_3^- =$ under achievement of the sales of 45000 of material B in meter

$d_4^- =$ under achievement of overtime of 10 hours

$d_4^+ =$ over achievement of overtime of 10 hours

Thus, the production capacity constraints is given by

$$x_1 + x_2 + d_1^- - d_1^+ = 80$$

Also, sales constraints are given by

$$1000x_1 + 1000d_2^- = 70000 \Rightarrow x_1 + d_2^- = 70$$

$$1000x_2 + 1000d_3^- = 45000 \Rightarrow x_2 + d_3^- = 45$$

and, overtime operation constraints is given by

$$d_1^+ + d_4^- - d_4^+ = 10$$

Now, we have to find the objective function in the following manner

(i) The first goal priority P_1 of the plant is to avoid under utilization of production capacity i.e. d_1^- is to be minimized and $P_1 d_1^-$ is the objective function term for goal 1.

(ii) The second goal priority P_2 of the plant is to limit the overtime operation to 10 hours, *i.e.*, d_4^+ is to minimized and $P_2 d_4^+$ is the objective function term for goal 2.

(iii) The third goal priority P_3 of the plant is to achieve sales goals of material A and material B i.e., d_2^- and d_3^- are to be minimized. But the profit from one meter of material A is ₹ 2.50 and ₹ 1.50 from one meter of material B. Also the production rate for both the material A and B is the same as 1000 meter per hour, so the hourly profit of A and B is in the ratio of 2.50 : 1.50 or 5 : 3 then $P_3(5d_2^- + 3d_3^-)$ is the objective function term for goal 3.

(iv) The fourth goal priority P_4 is to minimize the overtime operation of the plant as much as possible i.e., d_1^+ is to minimized then $P_4 d_1^+$ is the objective function term for goal 4.

Keeping in mind the above fact, the required GPP is given by

$$\min.\ Z = P_1 d_1^- + P_2 d_4^+ + P_3(5d_2^- + 3d_3^-) + P_4 d_1^+$$

subject to the constraints

$$x_1 + x_2 + d_1^- - d_1^+ = 80$$
$$x_1 + d_2^- = 70$$
$$x_2 + d_3^- = 45$$
$$d_1^+ + d_4^- - d_4^+ = 10$$

and $\quad x_1, x_2, d_1^-, d_1^+, d_2^-, d_3^-, d_4^+, d_4^- \geq 0$

To solve the above GPP, we proceed as follows:

Firstly we construct the initial table as given below.

Initial table-1

B.V.	C_B	x_B	0 x_1	0 x_2	P_1 d_1^- (x_3)	P_4 d_1^+ (x_4)	$5P_3$ d_2^- (x_5)	$3P_3$ d_3^- (x_6)	0 d_4^- (x_7)	P_2 d_4^+ (x_8)	Min. Ratio x_B / x_1
d_1^-	P_1	80	1	1	1	–1	0	0	0	0	80/1
d_2^-	$5P_3$	70	①	0	0	0	1	0	0	0	75/1(min.)→
d_3^-	$3P_3$	45	0	1	0	0	0	1	0	0	—
d_4^-	0	10	0	0	0	1	0	0	1	–1	—
$c_j - z_j$ P_4	P_4	0	0	0	0	1	0	0	0	0	
	P_3	485	–5	–3	0	0	0	0	0	0	
	P_2	0	0	0	0	0	0	0	0	1	
	P_1	80	–1	–1	0	1	0	0	0	0	
			↑								

In the above table, we have computed $c_j - z_j$ for $j = 1, 2, 4, 8$ (not in the basis)

$$c_1 - z_1 = c_1 - C_B x_1 = 0 - (P_1, 5P_3, 3P_3, 0) \begin{bmatrix} 1 \\ 1 \\ 0 \\ 0 \end{bmatrix} = -P_1 - 5P_3$$

$$c_2 - z_2 = c_2 - C_B x_2 = 0 - (P_1, 5P_3, 3P_3, 0) \begin{bmatrix} 1 \\ 0 \\ 1 \\ 0 \end{bmatrix} = -P_1 - 3P_3$$

$$c_4 - z_4 = c_4 - C_B x_4 = P_4 - (P_1, 5P_3, 3P_3, 0) \begin{bmatrix} -1 \\ 0 \\ 0 \\ 1 \end{bmatrix} = P_4 + P_1$$

and $Z = C_B x_B = (P_1, 5P_3, 3P_3, 0) \begin{bmatrix} 80 \\ 70 \\ 45 \\ 10 \end{bmatrix} = 80P_1 + 485P_3$

The above criterian matrix $c_j - z_j$ is obtained by the coefficients of P_1, P_2, P_3 and P_4 and x_B column is obtained by the coefficients of P_1, P_2, P_3, P_4 in Z-value.

Now, in P_1-row, the most negative element is –1. Thus solution is not optimal.

Then following the usual procedure, we have the following reduced table.

Reduced Table-1

B.V.	c_j		0	0	P_1	P_4	$5P_3$	$3P_3$	0	P_2	Min. Ratio
	C_B	x_B	x_1	x_2	d_1^-	d_1^+	d_2^-	d_3^-	d_4^-	d_4^+	x_B / x_2
d_1^-	P_1	10	0	①	1	–1	–1	0	0	0	10/1(min.)→
x_1	0	70	1	0	0	0	1	0	0	0	—
d_3^-	$3P_3$	45	0	1	0	0	0	1	0	0	45/1
d_4^-	0	10	0	0	0	1	0	0	1	–1	—
$c_j - z_j$	P_4	0	0	0	0	1	0	0	0	0	
	P_3	135	0	–3	0	0	5	0	0	0	
	P_2	0	0	0	0	0	0	0	0	1	
	P_1	10	0	–1	0	1	1	0	0	0	
				↑							
x_2	0	10	0	1	1	–1	–1	0	0	0	
x_1	0	70	1	0	0	0	1	0	0	0	
d_3^-	$3P_3$	35	0	0	–1	1	1	1	0	0	
d_4^-	0	10	0	0	0	①	0	0	1	–1	

$c_j - z_j$										
$c_j - z_j$	P_4	0	0	0	1	1	0	0	0	0
	P_3	105	0	0	3	-3	2	0	0	0
	P_2	0	0	0	0	0	0	0	0	1
	P_1	0	0	0	1	0	0	0	0	0
x_2	0	20	0	1	1	0	-1	0	1	-1
x_1	0	70	1	0	0	0	1	0	0	0
d_3^-	$3P_3$	25	0	0	-1	0	1	1	-1	1
d_1^+	P_4	10	0	0	0	1	0	0	1	-1
$c_j - z_j$	P_4	10	0	0	0	0	0	0	-1	1
	P_3	75	0	0	3	0	2	0	3	-3
	P_2	0	0	0	0	0	0	0	0	1
	P_1	0	0	0	1	0	0	0	0	0

From the above table we observe that the first two goals are fully achieved. In third priority goal P_3, the most negative entry is –3 which occurs in column d_4^+. But in P_2 (higher priority) row, the element below –3 is positive so that we can not improve P_3 and similarly P_4.

∴ solution is given by

$$x_1 = 70, x_2 = 20, d_1^+ = 10, d_3^- = 25, d_1^- = 0, d_2^- = 0, d_4^- = 0, d_4^+ = 0$$

and in P_1 and P_2 rows there is zero in column x_B so that the goals P_1 and P_2 are fully achieved and there is 10 hours $(d_1^+ = 10)$ over achievement of the plants and $d_3^- = 25$. So that 25000 meters under achievement in the sales goal of material B is obtained.

Hence, the required solution is given by

$$x_1 = 70, x_2 = 20$$

⇒ company should produce 70000 meters of material A and 20,000 meters of material B.

EXAMPLE 5. *A company manufacturers two products radios and transistors which must be processed through assembly and finishing department. Assembly has 90 hours available, finishing can handle upto 72 hours of work. Manufacturing one radio requires 6 hours in assembly and 3 hours in finishing. Each transistor requires 3 hours in assembly and 6 hours in finishing. The profit is ₹120 per radio and ₹90 per transistor. The company has established the following goals and has assigned them priorities P_1, P_2, P_3 (P_1 is most important) as follows:*

P_1 : Produce to meet a radio goal of 13

P_2 : Reach a profit goal of ₹1950

P_3 : Produce to meet a transistor goal of 5

Formulate the problem as a GPP and find the optimum solution.

SOLUTION: Let x_1, x_2 be the number of radios and transistors manufactured respectively. Also let,

d_1^- = amount by which the profit goal is under achieved

d_1^+ = amount by which the profit goal is over achieved

d_2^- = amount by which the radio goal is under achieved

d_2^+ = amount by which the radio goal is over achieved

d_3^- = amount by which the transistor goal is under achieved

d_3^+ = amount by which the transistor goal is over achieved

Then proceeding as usual, the given problem can be formulated as a GPP as follows:

$$\text{min. } Z = P_1 d_2^- + P_2 d_1^- + P_3 d_3^-$$

subject to the constraints

$$120x_1 + 90x_2 + d_1^- - d_1^+ = 1950$$
$$x_1 + d_2^- - d_2^+ = 13$$
$$x_2 + d_3^- - d_3^+ = 5$$
$$6x_1 + 3x_2 \le 90$$
$$3x_1 + 6x_2 \le 72$$

and $\quad x_1, x_2, d_1^-, d_1^+, d_2^-, d_2^+, d_3^-, d_3^+ \ge 0$

Now to solve the above GPP, we use the slack variables s_1 and s_2 such that

$$\text{min. } Z = P_1 d_2^- + P_2 d_1^- + P_3 d_3^-$$

subject to the constraints

$$120x_1 + 90x_2 + d_1^- - d_1^+ = 1950$$
$$x_1 + d_2^- - d_2^+ = 13$$
$$x_2 + d_3^- - d_3^+ = 5$$
$$6x_1 + 3x_2 + s_1 = 90$$
$$3x_1 + 6x_2 + s_2 = 72$$

and $\quad x_1, x_2, d_1^-, d_1^+, d_2^-, d_2^+, d_3^-, d_3^+, s_1, s_2 \ge 0$

Now we formulate the starting table as follows:

Initial table-1

B.V.	C_B	X_B	x_1	x_2	s_1	s_2	d_1^-	d_1^+	d_2^-	d_2^+	d_3^-	d_3^+	Min. Ratio X_B/x_1
	$c_j \rightarrow$		0	0	0	0	P_2	0	P_1	0	P_3	0	
d_1^-	P_2	1950	120	90	0	0	1	−1	0	0	0	0	1950/120
d_2^-	P_1	13	①	0	0	0	0	0	1	−1	0	0	13/1 (min.)→
d_3^-	P_3	5	0	1	0	0	0	0	0	0	1	−1	—
s_1	0	90	6	3	1	0	0	0	0	0	0	0	90/6
s_2	0	72	3	6	0	1	0	0	0	0	0	0	72/3
$c_j - z_j$	P_3	5	0	1	0	0	0	0	0	0	0	1	
	P_2	1950	−120	−90	0	0	0	−1	0	0	0	0	
	P_1	13	−1	0	0	0	0	0	0	1	0	0	
			↑										

In the above table

$$c_1 - z_1 = 0 - (P_2, P_1, P_3, 0, 0) \begin{bmatrix} 120 \\ 1 \\ 0 \\ 6 \\ 3 \end{bmatrix} = -120P_2 - P_1 + 0P_3$$

$$c_2 - z_2 = 0 - (P_2, P_1, P_3, 0, 0) \begin{bmatrix} 90 \\ 0 \\ 1 \\ 3 \\ 6 \end{bmatrix} = -90P_2 + 0P_1 + 1P_3$$

$$c_3 - z_3 = 0 - (P_2, P_1, P_3, 0, 0) \begin{bmatrix} 0 \\ 0 \\ 0 \\ 1 \\ 0 \end{bmatrix} = 0P_2 + 0P_1 + 0P_3$$

$$c_4 - z_4 = 0 - (P_2, P_1, P_3, 0, 0) \begin{bmatrix} 0 \\ 0 \\ 0 \\ 0 \\ 1 \end{bmatrix} = 0P_2 + 0P_1 + 0P_3$$

$$c_5 - z_5 = P_2 - (P_2, P_1, P_3, 0, 0) \begin{bmatrix} 1 \\ 0 \\ 0 \\ 0 \\ 0 \end{bmatrix} = 0P_2 + 0P_1 + 0P_3$$

$$c_6 - z_6 = 0 - (P_2, P_1, P_3, 0, 0) \begin{bmatrix} -1 \\ 0 \\ 0 \\ 0 \\ 0 \end{bmatrix} = P_2 + 0P_1 + 0P_3$$

$$c_7 - z_7 = P_1 - (P_2, P_1, P_3, 0, 0) \begin{bmatrix} 0 \\ 1 \\ 0 \\ 0 \\ 0 \end{bmatrix} = 0P_2 + 0P_1 + 0P_3$$

$$c_8 - z_8 = 0 - (P_2, P_1, P_3, 0, 0) \begin{bmatrix} 0 \\ -1 \\ 0 \\ 0 \\ 0 \end{bmatrix} = 0P_2 + 1 \cdot P_1 + 0P_3$$

$$c_9 - z_9 = P_3 - (P_2, P_1, P_3, 0, 0) \begin{bmatrix} 0 \\ 0 \\ 1 \\ 0 \\ 0 \end{bmatrix} = 0P_2 + 0P_1 + 0P_3$$

and $$c_{10} - z_{10} = 0 - (P_2, P_1, P_3, 0, 0) \begin{bmatrix} 0 \\ 0 \\ -1 \\ 0 \\ 0 \end{bmatrix} = 0P_2 + 0P_1 + 1 \cdot P_3$$

and $$Z = C_B x_B = (P_2, P_1, P_3, 0, 0) \begin{bmatrix} 1950 \\ 13 \\ 5 \\ 90 \\ 72 \end{bmatrix} = 1950P_2 + 13P_1 + 5P_3$$

Now the most negative entry in P_1 is –1 in the first column

$\Rightarrow x_1$ is the incoming variable and by minimum ratio rule d_2^- is the outgoing variable

So, the key element = 1 i.e., a_{21}

Now proceeding as usual, we have the following revised table

Revised table

B.V.	$c_j \rightarrow$		0	0	0	0	P_2	0	P_1	0	P_3	0	Min. Ratio
	C_B	x_B	x_1	x_2	s_1	s_2	d_1^-	d_1^+	d_2^-	d_2^+	d_3^-	d_3^+	x_B / d_2^+
d_1^-	P_2	390	0	90	0	0	1	–1	–120	120	0	0	390/120
x_1	0	13	1	0	0	0	0	0	1	–1	0	0	—
d_3^-	P_3	5	0	1	0	0	0	0	0	0	1	–1	—
s_1	0	12	0	3	1	0	0	0	–6	⑥	0	0	12/6(min.)→
s_2	0	33	0	0	0	1	0	0	–3	3	0	0	33/3
$c_j - z_j$	P_3	5	0	–1	0	0	0	0	0	0	0	1	
	P_2	390	0	–90	0	0	0	1	120	–120	0	0	
	P_1	0	0	0	0	0	0	0	1	0	0	0	
										↑			

In the above table

$$c_2 - z_2 = 0 - (P_2, 0, P_3, 0, 0) \cdot \begin{bmatrix} 90 \\ 0 \\ 1 \\ 3 \\ 6 \end{bmatrix} = -90P_2 - P_3$$

$$c_6 - z_6 = 0 - (P_2, 0, P_3, 0, 0) \begin{bmatrix} -1 \\ 0 \\ 0 \\ 0 \\ 0 \end{bmatrix} = P_2$$

$$c_7 - z_7 = P_1 - (P_2, 0, P_3, 0, 0) \begin{bmatrix} -120 \\ -1 \\ 0 \\ -6 \\ -3 \end{bmatrix} = P_1 + 120P_2$$

$$c_8 - z_8 = 0 - (P_2, 0, P_3, 0, 0) \begin{bmatrix} 120 \\ -1 \\ 0 \\ 6 \\ 3 \end{bmatrix} = -120P_2$$

$$\text{and } c_{10} - z_{10} = 0 - (P_2, 0, P_3, 0, 0) \begin{bmatrix} 0 \\ 0 \\ -1 \\ 0 \\ 0 \end{bmatrix} = P_3$$

The value of $c_j - z_j$ in P_1, P_2, P_3 rows may also be found easily by making 1 at the place of key element use it to reduce all entries in P_1, P_2, P_3 rows corresponding to the column of key element to zero.
Also,

$$Z = C_B x_B = (P_2, 0, P_3, 0, 0) \begin{bmatrix} 390 \\ 13 \\ 5 \\ 12 \\ 33 \end{bmatrix} = 390P_2 + 5P_3$$

Clearly all entries in P_1 row are ≥ 0, so the priority goal P_1 is achieved. Now we have to proceed to achieve the goal P_2 without affecting the achievement of top priority goal P_1.
In the P_2-row of the above table most negative value is -120 in column

corresponding to variable d_2^+, which is taken as entering variable. Now by minimum ratio rule, x_3 in 4th row is the outgoing vector

$\Rightarrow$ $6(=a_{48})$ is the key element

Now following the usual procedure, we have the following reduced table.

Reduced table-1

c_j			0	0	0	0	P_2	0	P_1	0	P_3	0	Min. Ratio
B.V.	C_B	x_B	x_1	x_2	s_1	s_2	d_1^-	d_1^+	d_2^-	d_2^+	d_3^-	d_3^+	x_B / x_2
d_1^-	P_2	150	0	30	−20	0	1	−1	0	0	0	0	150/30
x_1	0	15	1	1/2	1/6	0	0	0	0	0	0	0	15/(1/2)
d_3^-	P_3	5	0	1	0	0	0	0	0	0	1	−1	5/1
d_2^+	0	2	0	①/2	1/6	0	0	0	−1	1	0	0	2/(1/2)(min)→
s_2	0	27	0	9/2	−1/2	1	0	0	0	0	0	0	27/(9/2)=6
$c_j - z_j$	P_3	5	0	−1	0	0	0	0	0	0	0	1	
	P_2	150	0	−30	20	0	0	1	0	0	0	0	
	P_1	0	0	0	0	0	0	0	1	0	0	0	
			↑										

In the above table, for non-basic variables

$$c_2 - z_2 = 0 - (P_2,0,P_3,0,0)\begin{bmatrix} 30 \\ 1/2 \\ 1 \\ 1/2 \\ 9/2 \end{bmatrix} = -30P_2 - P_3$$

$$c_3 - z_3 = 0 - (P_2,0,P_3,0,0)\begin{bmatrix} -20 \\ 1/6 \\ 0 \\ 1/6 \\ -1/2 \end{bmatrix} = 20P_2$$

$$c_6 - z_6 = 0 - (P_2,0,P_3,0,0)\begin{bmatrix} -1 \\ 0 \\ 0 \\ 0 \\ 0 \end{bmatrix} = P_2$$

$$c_7 - z_7 = P_1 - (P_2,0,P_3,0,0)\begin{bmatrix} 0 \\ 0 \\ 0 \\ -1 \\ 0 \end{bmatrix} = P_1$$

$$c_{10} - z_{10} = 0 - (P_2, 0, P_3, 0, 0) \begin{bmatrix} 0 \\ 0 \\ -1 \\ 0 \\ 0 \end{bmatrix} = P_3$$

$$\text{and} \quad Z = C_B x_B = (P_2, 0, P_3, 0, 0) \begin{bmatrix} 150 \\ 15 \\ 5 \\ 2 \\ 27 \end{bmatrix} = 150 P_2 + 5 P_3$$

Again in P_2- row, $c_2 - z_2$ is negative $\Rightarrow$ solution is not optimal from P_2 point of view. So, we take x_2 in second column corresponding to most negative entry in P_2 row as entering vector, by minimum ratio rule d_2^+ in 4th row is the outgoing vector, therefore the key element is $\frac{1}{2} (= a_{42})$.

Now following the usual procedure, we have the following reduced table.

Reduced table-2

B.V.	C_B	x_B	$c_j \rightarrow$ 0 x_1	0 x_2	0 s_1	0 s_2	P_2 d_1^-	0 d_1^+	P_1 d_2^-	0 d_2^+	P_3 d_3^-	0 d_3^+	Min. Ratio
d_1^-	P_2	30	0	0	-30	0	1	-1	60	-60	0	0	
x_1	0	13	1	0	0	0	0	0	1	-1	0	0	
d_3^-	P_3	1	0	0	-1/3	0	0	0	2	-2	1	-1	
x_2	0	4	0	1	1/3	0	0	0	-2	2	0	0	
s_2	0	9	0	0	-2	1	0	0	9	-9	0	0	
$c_j - z_j$	P_3	5	0	0	1/3	0	0	0	-2	2	0	1	
	P_2	150	0	0	30	0	0	1	-60	60	0	0	
	P_1	0	0	0	0	0	0	0	1	0	0	0	

Now,

$$c_3 - z_3 = 0 - (P_2, 0, P_3, 0, 0) \begin{bmatrix} -30 \\ 0 \\ -1/3 \\ 1/3 \\ -2 \end{bmatrix} = 30 P_2 + \frac{1}{3} P_3$$

$$c_6 - z_6 = 0 - (P_2, 0, P_3, 0, 0) \begin{bmatrix} -1 \\ 0 \\ 0 \\ 0 \\ 0 \end{bmatrix} = P_2$$

$$c_7 - z_7 = P_1 - (P_2, 0, P_3, 0, 0) \begin{bmatrix} 60 \\ 1 \\ 2 \\ -2 \\ 9 \end{bmatrix} = P_1 - 60P_2 - 2P_3$$

$$c_8 - z_8 = 0 - (P_2, 0, P_3, 0, 0) \begin{bmatrix} -60 \\ -1 \\ -2 \\ 2 \\ -9 \end{bmatrix} = 60P_2 + 2P_3$$

$$c_{10} - z_{10} = 0 - (P_2, 0, P_3, 0, 0) \begin{bmatrix} 0 \\ 0 \\ -1 \\ 0 \\ 0 \end{bmatrix} = P_3$$

$$\text{and} \quad Z = C_B x_B = (P_2, 0, P_3, 0, 0) \begin{bmatrix} 150 \\ 15 \\ 5 \\ 2 \\ 27 \end{bmatrix} = 150P_2 + 5P_3$$

We observe that in the above table -60 is the negative entry in P_2. But P_2 can not be improved further as there is a positive entry below this element in P_1 - row (top priority). Similarly if we move to improve P_3 then it is also not possible as there is positive entry in row P_1 below the negative entry in row P_3.

$\Rightarrow$ P_2 and P_3 can not be improved further.

Therefore, the solution of the above GPP is given as below:

     $x_1 = 13, x_2 = 4$ $d_1^- = 30, d_3^- = 1$ $d_1^+ = 0 = d_2^- = d_2^+ = d_3^-$

$\Rightarrow$ Total 13 radios and 4 transistors should be manufactured.

Also, we observe that the first priority goal P_1 is fully achieved. The second priority goal P_2 is missed by ₹30 ($\because$ Profit $= 120 \times 13 + 90 \times 4 = 1920$ and 1950–1920=30) and the last priority goal P_3 is also missed by 1 transistor ($\because$ 5–4=1).

EXAMPLE 6.   *A manufacturer produces two products A and B. Each product requires time in two production departments. Product A requires 20 hours in department 1 and 10 hours in department 2. Product B requires 10 hrs in department 1 and 10 hours in department 2. Production time is limited in department 1 to 60 hrs and in department 2 to 40 hours. Contribution to profit for the two products is ₹40 and ₹80 respectively. Managements objectives are*

   (i) to maximize profit

(ii) at least two units of each product are desired

Management considers the deviation of ₹1 from the profit goal equal to one unit deviation from the product goal.

Formulate the above problem as GPP. Also, solve it by graphical as well as modified simplex method.

SOLUTION. Let x_1 and x_2 be the no. of units of product A and B produced respectively. Also, let

d_1^- = amount by which the maximum profit of ₹1000 is under achieved

d_1^+ = amount by which the maximum profit of ₹1000 is over achieved

d_2^- = no. of units by which the product of 2 units of product A is under achieved

d_2^+ = no. of units by which the product of 2 units of product A is over achieved

d_3^- = no. of units by which the product of 2 units of product B is under achieved

d_3^+ = no. of units by which the product of 2 units of product B is over achieved

Then following the usual procedure, we have the following GPP

$$\text{min. } Z = d_1^- + d_2^- + d_3^-$$

subject to the constraints

$$20x_1 + 10x_2 \le 60 \qquad \text{[time in dep.1 constraints]}$$

$$10x_1 + 10x_2 \le 40 \qquad \text{[time in dep.2 constraints]}$$

$$40x_1 + 80x_2 + d_1^- - d_1^+ = 1000 \quad \text{[max. profit goal]}$$

$$x_1 + d_2^- - d_2^+ = 2 \qquad \text{[Production goal of unit A]}$$

$$x_2 + d_3^- - d_3^+ = 2 \qquad \text{[Production goal of unit B]}$$

and $x_1, x_2, d_1^-, d_1^+, d_2^-, d_2^+, d_3^-, d_3^+ \ge 0$

Further since, priorities of the goals is not given and the deviation of ₹1 from the profit goal equals to one unit deviation from the product goal, therefore this problem can be solved as usual by graphical or simplex method.

(i) Solution by Graphical Method

Taking the deviational variables equal to zero, plot all the lines on the graph as shown in the figure and attach the arrows associated with the line to represent under and over achievement of the goals.

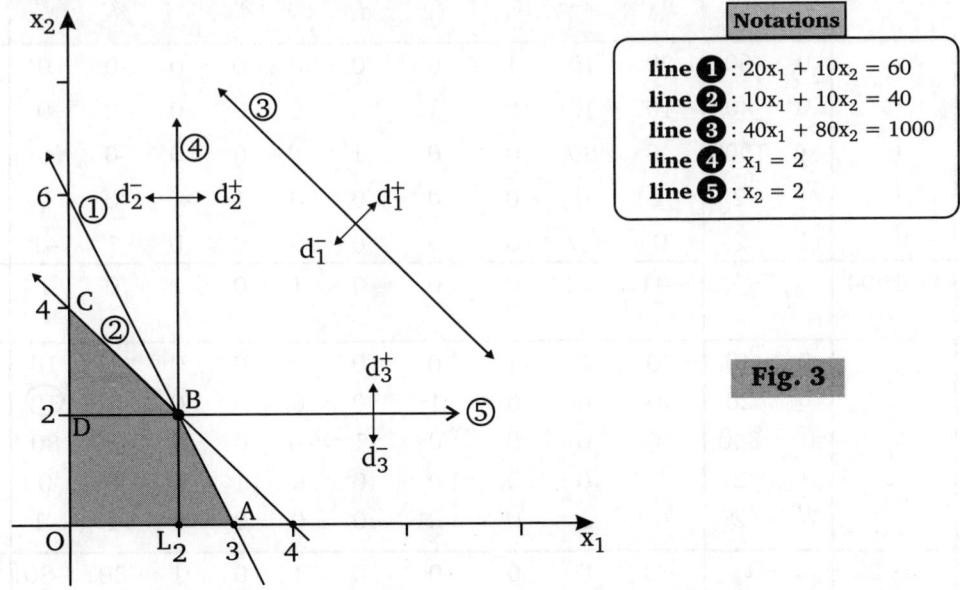

Notations
line ① : $20x_1 + 10x_2 = 60$
line ② : $10x_1 + 10x_2 = 40$
line ③ : $40x_1 + 80x_2 = 1000$
line ④ : $x_1 = 2$
line ⑤ : $x_2 = 2$

Fig. 3

Clearly $OABCO$ is the permissible region satisfying (1) and (2). The goals of the problem is to minimize d_1^-, d_2^-, d_3^-.

From the figure, it is clear that d_2^- is 0 (minimum) on the right of the line LB and d_3^- is 0 (minimum) above the line BD while d_1^- is 0 in the region not containing origin divided the line 3, which is far away from the permissible region $OABCO$.

So, $d_1^- \neq 0$, and $d_1^+ = 0$

Moving towards the permissible region, we observe that point C of the permissible region is nearest to the line 3.

So, at C, $x_1 = 0, x_2 = 4$ and at C the maximum profit is missed by $d_1^- = 1000 - 0 - 80 \times 4 = $ ₹680, which can be taken as the satisfactory optimal solution.

(ii) Solution by Simplex Method

Introducing slack variables s_1 and s_2, the given problem becomes

$$\text{min. } Z = d_1^- + d_2^- + d_3^-$$

subject to the constraints

$$20x_1 + 10x_2 + s_1 = 60$$
$$10x_1 + 10x_2 + s_2 = 40$$
$$40x_1 + 80x_2 + d_1^- - d_1^+ = 1000$$
$$x_1 + d_2^- - d_2^+ = 2$$
$$x_2 + d_3^- - d_3^+ = 2$$

and $x_1, x_2, s_1, s_2, d_1^-, d_1^+, d_2^-, d_2^+, d_3^-, d_3^+ \geq 0$

Now apply the simplex method in a usual manner, we have the following computational table.

Simplex table

B.V.	C_B	x_B	x_1	x_2	s_1	s_2	d_1^-	d_1^+	d_2^-	d_2^+	d_3^-	d_3^+	Min. Ratio
	c_j		0	0	0	0	1	0	1	0	1	0	x_B / x_2
s_1	0	60	20	10	1	0	0	0	0	0	0	0	60/10
s_2	0	40	10	10	0	1	0	0	0	0	0	0	40/10
d_1^-	1	1000	40	80	0	0	1	-1	0	0	0	0	1000/80
d_2^-	1	2	1	0	0	0	0	0	1	-1	0	0	—
d_3^-	1	2	0	①	0	0	0	0	0	0	1	-1	2/1(min.)→
$Z=1004$	$c_j - z_j$		-41	-81 ↑	0	0	0	1	0	1	0	1	x_B / d_3^+
s_1	0	40	20	0	1	0	0	0	0	0	-10	10	40/10
s_2	0	20	10	0	0	1	0	0	0	0	-10	⑩	20/10(min)→
d_1^-	1	840	40	0	0	0	1	-1	0	0	-80	80	840/80
d_2^-	1	2	1	0	0	0	0	0	1	-1	0	0	—
x_2	0	2	0	1	0	0	0	0	0	0	1	-1	—
$Z=922$	$c_j - z_j$		-41	0	0	0	0	1	0	1	80	-80 ↑	

s_1	0	20	10	0	1	−1	0	0	0	0	0	0
d_3^+	0	2	1	0	0	1/10	0	0	0	0	−1	1
d_1^-	1	680	−40	0	0	−8	1	−1	0	0	0	0
d_2^-	1	2	1	0	0	0	0	0	1	−1	0	0
x_2	0	4	1	1	0	1/10	0	0	0	0	0	0
$Z=762$	$c_j - z_j$		39	0	0	8	0	1	0	1	0	0

We observe that in the last row of the above table all $c_j - z_j \geq 0$

$\Rightarrow$ solution is optimal and is given by

$$x_1 = 0, x_2 = 4, \quad d_1^- = 680, d_2^- = 2, d_3^+ = 2, \quad d_1^+ = d_2^+ = d_3^- = 0$$

Hence, we conclude that the solution does not achieve the production target of at least two units of each product A and B and also the profit target is missed by ₹680.

EXAMPLE 7. *A manufacturing firm produces two types of product A and B. According to past experience, production of either product A or B requires an average of one hour in the plant. The plant has a normal production capacity of 400 hours a month. The marketing department of the firm report that because of the limited market, the maximum number of product A and B that can be sold in a month is 240 and 300 respectively. The net profit from the sale of products A and B are ₹800 and ₹400 respectively. The manager of the firm has set the following goals arranged in the order of importance.*

P_1 : *He wants to avoid any under-utilization of normal production capacity*

P_2 : *He wants to sale possible units of products A and B. Since the net profit from the sale of product A is twice the amount from product B, therefore the manager has twice as much desire to achieve sales for product A as for product B.*

P_3 : *He wants to minimize the overtime operation of the plant as much as possible.*

Formulate the problem as the GPP. Also solve it by graphical as well as modified simplex method.

SOLUTION. Let x_1 and x_2 be the number of products A and B respectively

Also, let

d_1^- = hours by which the production capacity is under utilized

d_1^+ = hours by which the production capacity is over utilized

d_2^- = number by which sale of product A is under achieved

d_3^- = number by which sale of product B is under achieved

Then, following the usual procedure, the required GPP is given by

$$\text{min. } Z = P_1 d_1^- + P_2(2d_2^- + d_3^-) + P_3 d_1^+$$

subject to the constraints

$$x_1 + x_2 + d_1^- - d_1^+ = 400 \qquad \text{(production goal)}$$
$$x_1 + d_2^- = 240 \qquad \text{(sale of unit 1 goal)}$$
$$x_2 + d_3^- = 300 \qquad \text{(sale of unit 2 goal)}$$

Since, maximum number of sale of product A and B are given, so there is no need of over achievements d_2^+ and d_3^+.

Solution by Graphical Method

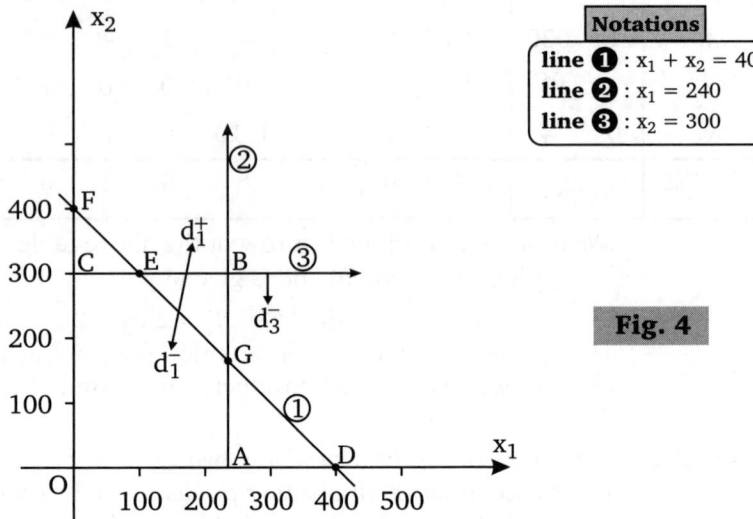

Notations

line ❶ : $x_1 + x_2 = 400$
line ❷ : $x_1 = 240$
line ❸ : $x_2 = 300$

Fig. 4

Clearly, the first priority goal $d_1^- = 0$ is fully achieved on the line DF and in the above it away from the origin. In the second priority goal, the sales goal of product A is completely achieved on the line AB and on the right of it while the sale goal of product is completely achieved on the line BC and above it. So first two goals are completely achieved at the point $B(240, 300)$. At this point the third priority is not achieved and it cannot be achieved at the expense of the first two goals.

Hence, the solution of the problem is given by

$$x_1 = 240, x_2 = 300, \ d_1^- = 0 = d_2^- = d_3^-$$
$$d_1^+ = 240 + 300 + 0 - 400 = 140$$

Solution by Modified Simplex Method

Following the usual procedure we have the following initial table.

Initial table

	$c_j \rightarrow$		0	0	P_1	P_3	$2P_2$	P_2	Min. Ratio
B.V.	C_B	x_B	x_1	x_2	d_1^-	d_1^+	d_2^-	d_3^-	x_B / x_1
d_1^-	P_1	400	1	1	1	−1	0	0	400/1
d_2^-	$2P_2$	240	①	0	0	0	1	0	240/1(min.)→
d_3^-	P_2	300	0	1	0	0	0	1	—
$c_j - z_j$	P_3	0	0	0	0	1	0	0	
	P_2	780	−2	−1	0	0	0	0	
	P_1	400	−1	−1	0	−1	0	0	
			↑						

In the above table (for non-basic variables)

$$c_1 - z_1 = 0 - (P_1, 2P_2, P_2) \begin{bmatrix} 1 \\ 1 \\ 0 \end{bmatrix} = -P_1 - 2P_2$$

$$c_2 - z_2 = 0 - (P_1, 2P_2, P_2) \begin{bmatrix} 1 \\ 0 \\ 1 \end{bmatrix} = -P_1 - P_2$$

$$c_4 - z_4 = P_3 - (P_1, 2P_2, P_2) \begin{bmatrix} 1 \\ 0 \\ 0 \end{bmatrix} = -P_1 + P_3$$

$$\text{and} \quad Z = C_B x_B = (P_1, 2P_2, P_2) \begin{bmatrix} 400 \\ 240 \\ 300 \end{bmatrix} = 400P_1 + 780P_2$$

Key element is $= 1(a_{21})$

Now we have the following reduced table

Reduced table-1

B.V.	$c_j \rightarrow$		0	0	P_1	P_3	$2P_2$	P_2	Min. Ratio
	C_B	x_B	x_1	x_2	d_1^-	d_1^+	d_2^-	d_3^-	x_B / x_2
d_1^-	P_1	160	0	①	1	−1	−1	0	160/1(min.)→
x_1	0	240	1	0	0	0	1	0	—
d_3^-	P_2	300	0	1	0	0	0	1	300/1
$c_j - z_j$	P_3	0	0	0	0	1	0	0	
	P_2	300	0	−1	0	0	2	0	
	P_1	160	0	−1	0	1	1	0	
				↑					

In the above table (for non-basic variables)

$$c_2 - z_2 = 0 - (P_1, 0, P_2) \begin{bmatrix} 1 \\ 0 \\ 1 \end{bmatrix} = -P_1 - P_2;$$

$$c_4 - z_4 = P_3 - (P_1, 0, P_2) \begin{bmatrix} -1 \\ 0 \\ 0 \end{bmatrix} = P_1 + P_3$$

$$c_5 - z_5 = 2P_2 - (P_1, 0, P_2) \begin{bmatrix} -1 \\ 1 \\ 0 \end{bmatrix} = P_1 + 2P_2$$

and $Z = C_B x_B = (P_1, 0, P_2) \begin{bmatrix} 160 \\ 240 \\ 300 \end{bmatrix} = 160 P_1 + 300 P_2$

The key element $= 1(a_{12})$

Further, the next reduced table is given as under.

Reduced table-2

B.V.			$c_j \rightarrow$	0	0	P_1	P_3	$2P_2$	P_2	Min. Ratio
	C_B	x_B		x_1	x_2	d_1^-	d_1^+	d_2^-	d_3^-	x_B / d_1^+
d_1^-	P_1	400		1	1	1	−1	0	0	−ve
d_2^-	$2P_2$	240		①	0	0	0	1	0	—
d_3^-	P_2	300		0	1	0	0	0	1	140/1(min.)→
$c_j - z_j$	P_3	0		0	0	0	1	0	0	
	P_2	140		0	0	1	−1	1	0	
	P_1	0		0	0	1	0	0	0	
							↑			

In the above table (for non-basic variables)

$$c_3 - z_3 = P_1 - (0,0,P_2) \begin{bmatrix} 1 \\ 0 \\ -1 \end{bmatrix} = P_1 + P_2, c_4 - z_4 = P_3 - (0,0,P_2) \begin{bmatrix} -1 \\ 0 \\ 1 \end{bmatrix} = -P_2 + P_3$$

$$c_5 - z_5 = 2P_2 - (0,0,P_2) \begin{bmatrix} -1 \\ 1 \\ 1 \end{bmatrix} = P_2 \text{ and } Z = C_B x_B = (0,0,P_2) \begin{bmatrix} 160 \\ 240 \\ 140 \end{bmatrix} = 140 P_2$$

Then key element $= 1(a_{21})$

Then next reduced table is given as below

Reduced table-3

B.V.			$c_j \rightarrow$	0	0	P_1	P_3	$2P_2$	P_2
	C_B	x_B		x_1	x_2	d_1^-	d_1^+	d_2^-	d_3^-
x_2	0	300		0	1	0	0	0	1
x_1	0	240		1	0	0	0	1	0
d_1^+	P_3	140		0	0	−1	1	1	1
$c_j - z_j$	P_3	140		0	0	1	0	−1	−1
	P_2	0		0	0	0	0	2	1
	P_1	0		0	0	1	0	0	0
							↑		

Here, $c_3 - z_3 = P_1 - (0,0,P_3)\begin{bmatrix} 0 \\ 0 \\ -1 \end{bmatrix} = P_1 + P_3$

$$c_5 - z_5 = 2P_2 - (0,0,P_3)\begin{bmatrix} 0 \\ 1 \\ 1 \end{bmatrix} = 2P_2 - P_3$$

$$c_6 - z_6 = P_2 - (0,0,P_3)\begin{bmatrix} 1 \\ 0 \\ 1 \end{bmatrix} = P_2 - P_3$$

and $Z = C_B x_B = (0,0,P_3)\begin{bmatrix} 400 \\ 240 \\ 140 \end{bmatrix} = 140 P_3$

In the last table we observe that all $c_j - z_j \geq 0$ in P_1 and P_2 rows and 0,0 in x_B column. Therefore the priority goals P_1 and P_2 are fully achieved. In P_3 - row all $c_j - z_j$ are not non-negative and the most negative entries are –1, –1 in this row. But we cannot improve P_3 further because below –1, –1 in the last two columns, there are positive entries in higher priority P_2.

Also, $x_1 = 300, x_2 = 240, d_1^+ = 140, d_1^- = 0, d_2^- = d_3^- = 0$

Hence, the optimal solution is given by

$$x_1 = 300, x_2 = 240$$

i.e., first two goals are fully achieved while third priority goal is missed by 140 hours (140 hours overtime is required)

Exercise-13.1

1. Find x_1, x_2 to minimize $Z = (d_1^-, d_2^-)$ subject to the constraints

$$x_1 + x_2 + d_1^- - d_1^+ = 20$$
$$4x_1 + 5x_2 + d_2^- - d_2^+ = 150$$

and $x_1, x_2, d_1^-, d_1^+, d_2^-, d_2^+ \geq 0$

2. Find x_1, x_2 to minimize $Z = (3d_1^+ + 2d_2^+, d_3^- d_4^-)$ subject to the constraints

$$x_1 + x_2 + d_1^- - d_1^+ = 8$$
$$x_1 + d_2^- - d_2^+ = 3$$
$$3x_1 + 5x_2 + d_3^- - d_3^+ = 65$$
$$x_1 + x_2 + d_4^- - d_4^+ = 65$$

and $x_1, x_2, d_1^-, d_1^+, d_2^-, d_2^+, d_3^-, d_3^+, d_4^-, d_4^+ \geq 0$

3. Solve the following GPP by modified simplex method

min. $Z = P_1 d_1^- + P_2 d_4^+ + P_3(2d_2^-$

$\qquad + d_3^-) + P_4 d_1^+$

subject to the constraints

$$x_1 + x_2 + d_1^- - d_1^+ = 10$$
$$x_1 + d_2^- = 6$$
$$x_2 + d_3^- = 8$$
$$d_1^+ + d_4^- - d_4^+ = 2$$

and $x_1, x_2, d_1^-, d_1^+, d_2^-, d_3^-, d_4^-, d_4^+ \geq 0$

4. A company manufactures two products radio and transistor which must be proceed through assembly and finishing department. Assembly has 90 hours available. Finishing can handle 72 hours of work. Manufacturing one radio requires 6 hours is assembly and 6 hours in finishing. If the profit is ₹12 per radio and ₹90 per transistor. Determine the best combination of radio and transistors to realize a maximum profit of ₹2000. Formulate the problem as a GPP. Also solve it by graphical as well as simplex method.

5. A camera company manufactures two types of cameras. The production process for manufacturing the camera is such that two departmental operations are required. To produces their standard camera requires 2 hours of production time in department 1 and 3 hours in department 2. To produce their deluxe model requires 4 hours of production time in department 1 and 3 hours in department 2. This labour time is a same what restrictive factor since the company has a general policy of avoiding overtime, if possible. The manufacturer's profit on each standard camera is ₹30 while the profit on the deluxe model is ₹40. The management has set the following goals arranged in the order of importance (pre-emptive factors).

P_1 : Avoid overtime operation in each department.

P_2 : Prior-sales records indicate that on the average, a minimum of 10 standard and 10 deluxe cameras can be sold weekly. Management would like to meet these sales goal. Since production time may limit producing the number of each camera and since the deluxe camera has a higher profit margin, the sales goals should be weighted by the profit contribution for the respective camera's i.e., ₹30 for the standard camera and ₹40 for the deluxe camera.

P_3 : To maximize profit.

Formulate the problem as a GPP. Also solve it by graphical and modified simplex methods.

6. Suppose two products are to be produced in a given department of a manufacturer, quantities of two products are denoted by x_1 and x_2 respectively. A product mix is to be obtained by utilizing two limited resources: labour and raw material. Each unit of the first product requires two hours of labour and three units of raw material. Each unit of the second product requires 4 hours of labour and 4 units of raw material. Every day 20 hours of labour and 24 units of raw material are available. The goals before the management according to priorities are as follows:

(i) The profit per day should be at least ₹36 assuming that the profit per unit of the products are ₹8 and ₹6 respectively.

(ii) Because of marketing condition as well as production substitutability, the number of units of product 1 should be double the number of units of product 2.

(iii) The labour should be fully utilized.

Formulate this problem as GPP and solve it.

ANSWER

1. $x_1 = \dfrac{75}{2}, x_2 = 0$, min. $Z = (0,0), d_1^+ = \dfrac{35}{2}$ (or) $x_1 = 0, x_2 = 30, \text{min}.Z = (0,0), d_1^+ = 10$

2. $x_1 = 0, x_2 = 8, \text{min}.Z = (0,25,27)$ **3.** $x_1 = 6, x_2 = 6, d_3^- = 2, d_1^+ = 2$

4. GPP: min. $Z = d_1^-$ subject to the constraints $120x_1 + 90x_2 + d_1^- - d_1^+ = 2000$, $6x_1 + 3x_2 \le 90$,

$3x_1 + 6x_2 \le 72$ and $x_1, x_2, d_1^-, d_1^+ \ge 0$, optimal solution: $x_1 = 12, x_2 = 6, \text{min}.Z = 20$, i.e., profit target

is missed by ₹20

5. min. $Z = P_1(d_1^+ + d_2^+) + P_2(3d_3^- + 4d_4^-) + P_3 d_5^-$ subject to $2x_1 + 4x_2 + d_1^- - d_1^+ = 80$,

$3x_1 + 3x_2 + d_2^- - d_2^+ = 80$, $x_1 + d_3^- - d_3^+ = 10$, $x_2 + d_4^- - d_4^+ = 10$, $30x_1 + 40x_2 + d_5^- - d_5^+ = 1500$

and $x_1, x_2, d_1^-, d_1^+, d_2^-, d_2^+, d_3^-, d_3^+, d_4^-, d_4^+, d_5^-, d_5^+ \ge 0$, solution is $x_1 = \dfrac{40}{3}$, $x_2 = \dfrac{40}{3}$. First two priorities

are achieved fully second priority goal is over achieved by $\dfrac{10}{3}$ unit each, third is missed by $\dfrac{1700}{3}$

rupees **6.** min. $Z = P_1 d_1^- + P_2(d_2^- + d_2^+) + P_3 d_3^-$ subject to $8x_1 + 6x_2 + d_1^- - d_1^+ = 36$

$x_1 - 2x_2 + d_2^- - d_2^+ = 0$, $2x_1 + 4x_2 + d_3^- = 20$, $3x_1 + 4x_2 + d_4^- = 24$ and $x_1, x_2, d_i^-, d_i^+ \ge 0$. The optimal

solution is $x_1 = \dfrac{24}{5}, x_2 = \dfrac{12}{5}$, the total profit is $\dfrac{264}{5}$. Goal 1 is over achieved with the quantum of

over achievement $d_1^+ = \dfrac{84}{5}$. Goal 2 has been achieved since $d_1^+ = d_1^- = 0$ and $x_1 = 2x_2$. Goal 3 has

not been achieved. The degree of under achievement is $d_3^- = \dfrac{4}{5}$. Hence, there is an under utilization

of available labour to the extent of $\dfrac{4}{5}$ hours.

⚙ REVIEW QUESTIONS

1. Write the application of goal programming.
2. What is goal programming? Distinguish it from linear programming.
3. Define the following:
   (i) Differential weights
   (ii) Deviational variables
   (iii) Priority factors
4. Identify the importance areas where GP can be used effectively.

📖 MULTIPLE CHOICE QUESTIONS (CHOOSE THE MOST APPROPRIATE ONE)

1. Deviation Variables in G.P. model must satisfy the following conditions: [MEERUT 2013]
   (a) $d_i^+ \times d_i^- = 0$ (b) $d_i^+ - d_i^- = 0$
   (c) $d_i^+ + d_i^- = 0$ (d) None of the above

2. If the targeted value of each goal in the solution value, X_B Column is zero, then it indicates: [MEERUT 2013]
   (a) multiple solution
   (b) optimum solution
   (c) infeasible solution
   (d) none of the above

3. The concept of goal programming was introduced by: [MEERUT 2013]
   (a) Chames
   (b) Cooper
   (c) Chames and Cooper
   (d) Lee and Ignizio

4. Goal Programming:
   (a) requires only that decision maker knows whether the goal is direct profit maximization or cost minimization.
   (b) allows you to have multiple goals, with or without priorities.
   (c) is an approach to achieve goal of a solution to all integer LP Problems.
   (d) None of the above

5. The GP approach attempts to achieve each objective:
   (a) sequentially   (b) simultaneously
   (c) both (a) and (b)   (d) none of these

6. In GP problem, a constraint having unachieved variable is expressed as:
   (a) an equality constraint
   (b) a less than or equal to type constraints
   (c) a greater than or equal to type constraint
   (d) all of the above

7. The use of GP model preferred when:
   (a) goals are satisfied in an ordinal sequence
   (b) goals are multiple incommensurable
   (c) more than one objective is set to achieve
   (d) all of the above

8. In optimal simplex table of GP problem, two or more $c_j - z_j$ rows indicate:
   (a) unequal priority goals
   (b) equal priority goals
   (c) priority goals
   (d) unattainable goals

9. Consider a goal with constraints: $g_1(x_1, x_2, ..., x_n) + d_1^- \geq b_1(d_1^- \geq 0)$ with d_1^- in the objective function then:
   (a) the goal is to minimize under achievement
   (b) the constraint is achieve provided $d_1^- > 0$
   (c) both (a) and (b)
   (d) none of the above

10. In G.P. problem, a goal constraint having over achievement variable is expressed as a:
    (a) $\geq$ constraint   (b) $\leq$ constraint
    (c) = constraint   (d) all of the above

11. In GP, at optimality which of the following conditions indicated that a goal has been exactly satisfied:
    (a) positive deviational variable is in the solution mix with a negative value.
    (b) both positive and negative deviational variables are in the solution mix.
    (c) both positive and negative deviational variables are not in the solution mix.
    (d) none of the above

12. In simplex method of goal programming, the variable to enter the solution mix is selected with:
    (a) lowest priority row and most negative $c_j - z_j$ value in it
    (b) lowest priority row and largest positive $c_j - z_j$ value in it
    (c) higest priority row and most negative $c_j - z_j$ value in it
    (d) higest priority row and most positive $c_j - z_j$ value in it

13. The deviational variable in the basis of the initial simplex table of GP problem is:
    (a) positive deviational variable
    (b) negative deviational variable
    (c) both (a) and (b)
    (d) artificial variable

14. Consider a goal with constraint $g_1(x_1, x_2, ..., x_n) + d_1^- - d_1^+ = b_1$ and the term $3d_1^- + 2d_1^+$ in the objective function the decision maker:

(a) prefers $g_1(x_1, x_2, ..., x_n) \geq b_1$, rather than $\leq b_1$

(b) prefers $g_1(x_1, x_2, ..., x_n) \leq b_1$, rather than $\geq b_1$

(c) not concerned with either $\leq$ or $\geq$

(d) none of the above

15. In GP problem goals are assigned priorities such that:

(a) higher priority goals must be achieved before lower priority goals

(b) goals may not have equal priority

(c) goals of greatest importance are given lowest priority

(d) all of the above

16. For applying a GP approach decision maker must

(a) set targets for each of the goals

(b) assign pre-empvitive priority to each goal

(c) assume that linearity exists in the use of resources to achieve goals

(d) all of the above

17. Which of the following is a step of algorithm to formulate GP model:

(a) Identify the goals and constraints on availability of resources (or constraints) which may restrict achievement of the goals (targets)

(b) Determine priority to be associated with

each goal in such a way that goals with priority level P_1 are most important, those with priority level P_2 are next most important, and so on.

(c) Define the decision variables

(d) all are true

18. Which of the following is a step of algorithm to formula GP model:

(a) Formulate the constraints in same manner as in LP model.

(b) For each constraint, develop an equation by adding deviational variable d_i^- and d_i^+. These variables indicate the possible deviations below or above the target value (right hand side of each constraint).

(c) Write the objective function in terms of minimizing a prioritized function of the deviational variables

(d) All are true

19. Which of the following is a step of obtain graphical solution of goal prgramming:

(a) graph all system constraints and identify the feasible solution space.

(b) graph the straight lines corresponding to the goal constraints, labelling the deviational variables.

(c) written the feasible solutions space identified, in determine the point or points that best satisfy the highest priority goal.

(d) All are true

ANSWERS

1. (a)	**2.** (b)	**3.** (c)	**4.** (a)	**5.** (a)	**6.** (c)	**7.** (d)	**8.** (c)	**9.** (c)
10. (c)	**11.** (c)	**12.** (c)	**13.** (a)	**14.** (a)	**15.** (a)	**16.** (d)	**17.** (d)	**18.** (d)
19. (d)								

❑❑❑❑

Game Theory 14

14.1 INTRODUCTION

Game Theory is a special type of decision theory. It is also known as the competition strategy. A great variety of competitive situations is commonly seen in everyday life, for example, in an election, all the candidates fighting are interested to secure more votes than all the others and in military battle everyone wants to win.

Now at first, we want to define a game. "A game is an activity under a set of rules between two or more players in which each player get some gain or loss."

Definition. *Game theory is a type of decision theory in which one's choice of action is determined after assuming all possible outcomes.*

Some Basic Definitions

1. **n-person game.** When number of players are n where $n \geq 2$, then for $n = 2$, the game is known as 2 person game and for $n > 2$ the game is known as n person game.

2. **Zero sum and non-zero sum game.** A game is known as zero sum game in which the net gain or not profit after the game is zero means nothing comes from outside, payment is always between the players, the loss of one player is the gain of others, and a game which is not zero sum game is known as non-zero sum game.

3. **Competitive game.** A game is said to be competitive if it has the following four properties :
   (i) There should be finite number of players means $n \geq 2$.
   (ii) Each player has a finite list of his possible course of action.
   (iii) A play is said to be played when each player choose one of his course of action and no player knows the choice of action of the other player until he has decided his own.
   (iv) When each player chooses his activity, then this combination of activities gives a result according which each player gains a payment which may be –ve, +ve or zero.

4. **Strategy.** For a given player, the strategy is given by the set of rules which specify that which of the available course of action he should make at each play. The strategy may be of two types :
   (a) pure strategy   (b) mix strategy
   (a) **Pure strategy.** A pure strategy is that in which one player knows what the other player is going to do. In this case, the player always choose a particular course of action.

(b) Mixed strategy. When a player does not know exactly what the other player is going to do, a probabilistic situation is obtained and such type of strategy is known as mixed strategy. Mathematically, let x_i be the probability to choose the ith activity, then we define the set

$$X = (x_1, x_2, x_3, ..., x_n)$$
$$\text{s.t.} \quad x_1 + x_2 + x_3 + ... + x_n = 1$$
$$\text{and } x_1, x_2, x_3, ..., x_n \geq 0.$$

5. Payoff Matrix

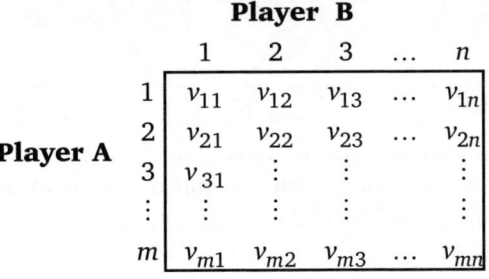

A's payoff matrix

Player B

Player A

	1	2	3	...	n
1	v_{11}	v_{12}	v_{13}	...	v_{1n}
2	v_{21}	v_{22}	v_{23}	...	v_{2n}
3	v_{31}	:	:		:
:	:	:	:		:
m	v_{m1}	v_{m2}	v_{m3}	...	v_{mn}

B's payoff matrix

	1	2	3	...	n
1	$-v_{11}$	$-v_{12}$	$-v_{13}$	...	$-v_{1n}$
2	$-v_{21}$	$-v_{22}$	$-v_{23}$	...	$-v_{2n}$
3	$-v_{31}$	:	:		:
:	:	:	:		:
m	$-v_{m1}$	$-v_{m2}$	$-v_{m3}$	...	$-v_{mn}$

In A's payoff matrix, the cell entry v_{ij} is the payment to player A when A choose the ith activity and B chooses the jth.

In a rectangular game or two person zero sum game, the player B's payoff matrix will be the negative of A's payoff matrix. Thus the net gain will be zero.

6. Minimax and Maxmin Principles. If a player lists his worst possible outcomes of all his potential strategy, then he will choose the best strategy among all these outcomes. Such a principle is known as maxmin principle or optimal strategy.

7. Saddle Point. A point which is minimum in its row and maximum in its column is known as the saddle point.

8. Optimal Strategy and Value of the Game. If in a given payoff matrix (i, j) is the saddle point, then the player A and B are said to have the optimal strategy i and j. The value of the $(i, j)^{th}$ cell is known as the value of the game and is denoted by v.

WORKING PROCEDURE

We can determine a saddle point by using the following steps :

STEP 1. In the given pay-off matrix, find the minimum element in each row and circle them.

STEP 2. Find the maximum element in each column and mark them by □.

STEP 3. Now find the element which have both the sign □ and ○.
This point is the required saddle point.

Solved Examples

EXAMPLE 1. *The player A's payoff matrix is given. Find the optimal strategy by using maxmin or minimax criterion.*

$$
\begin{array}{c}
& & \text{B} \\
& & \begin{array}{ccc} I & II & III \end{array} \\
A & \begin{array}{c} I \\ II \\ III \end{array} & \left[\begin{array}{ccc} -3 & -2 & -3 \\ 2 & 0 & 2 \\ 5 & -2 & -4 \end{array}\right]
\end{array}
$$

SOLUTION. According to maxmin criterion, first of all we find the minimum elements in each row and then find its maximum element. Now, we find the maximum element in column and then find its minimum.

In a game player A wants to maximum the value of the game v_{ij} by choosing one of his activity, while the player B wants to minimize the A's gain as mush as possible.

Now, in the given problem min $(-3, -2, -3) = -3$
$$\text{min } (2, 0, 2) = 0$$
$$\text{min } (2, 0, 2) = 0$$
$$\text{min } (5, -2, -4) = -4$$

and max $(-3, 0, -4) = 0$

Now, max $(-3, 2, 5) = 5$
$$\text{max } (-2, 0, -2) = 0$$
$$\text{max } (-3, 2, -4) = 2$$
$$\text{min } (5, 0, 2) = 0$$

$\Rightarrow$ min max $v_{ij}(\overline{v}) = 0$

Here, $\underline{v} = \overline{v} = 0$

The max min strategy used by player A = II
The min max strategy used by player B = II
The corresponding gain $(v) = 0$.

Thus the strategy used is known as optimal strategy and the gain is known as the value of the game and the point 0 is also known as saddle point.

EXAMPLE 2. *The pay off matrix for a two person zero sum game is given by*

Player B

$$
\begin{array}{c}
& & \begin{array}{ccc} I & II & III \end{array} \\
\textbf{Player A} & \begin{array}{c} I \\ II \\ III \end{array} & \left[\begin{array}{ccc} 15 & 2 & 3 \\ 6 & 5 & 7 \\ -7 & 4 & 0 \end{array}\right]
\end{array}
$$

Find the optimal strategy and value of the game.

SOLUTION. First of all, find the minimum element in each row and max element in each column.

	I	II	III	Row min
I	15	②	3	2
II	6	⑤	7	5
III	⊖7	4	0	−7
Column max	15	5	7	

Circle the element which is minimum in each row and mark the square to the element which is maximum in each column.

The entry which have both the sign is known as saddle point.

Now the optimal strategy used by player A = II.

The optimal strategy used by player B = II.

and value of the game $v = 5$ for player A

$$v = -5 \text{ for player B.}$$

EXAMPLE 3. *Find the saddle point and hence find the value of the game.*

Player B

	I	II	III	IV	V
I	9	3	1	8	0
Player A *II*	6	5	4	6	7
III	2	4	4	3	8
IV	5	6	2	2	1

SOLUTION. For finding the saddle point, find the minimum element in each row and maximum element in each column.

Table for saddle point

Player B

Player A	I	II	III	IV	V	Row min
I	9	3	1	8	⓪	0
II	6	5	④	6	7	4
III	②	4	4	3	8	2
IV	5	6	2	2	①	1
Column max	9	6	4	8	8	

The element having both the marks (O and □) = v_{23}.

So the optimal strategy used by player A = II.

The optimal strategy used by player B = III.

and value of the game is given by = 4.

So, optimal strategy = (II, III) and value of the game $v = 4$.

EXAMPLE 4. *Find the saddle point and hence find the value of the game for the given pay off matrix.*

$$(a) \begin{bmatrix} -5 & 5 & 0 & 7 \\ 2 & 6 & 1 & 8 \\ -4 & 0 & 1 & -3 \end{bmatrix} \qquad (b) \begin{bmatrix} 1 & 7 & 3 & 4 \\ 5 & 6 & 4 & 5 \\ 7 & 2 & 0 & 3 \end{bmatrix}$$

<u>SOLUTION.</u>   (a) Find the minimum element in each row and maximum element in each column and then find the maximum element in the row minimum elements and find the minimum in column maximum elements.

Table for saddle point

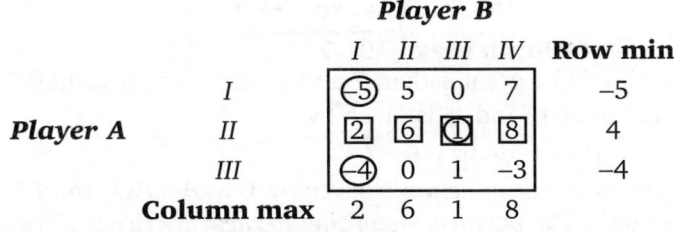

So, the required saddle point = 1.

The optimal strategy used by player A = II.

The optimal strategy used by player B = III.

The value of the game = 1.

So, optimal strategy = (II, III)

and value v_{ij} = 1 for player A

$\qquad\qquad$ = $-$ 1 for player B.

(b) First find the min element in each row and max element in each column and mark them.

Table for saddle point

$$\text{Player B}$$

		I	II	III	IV	**Row min**
	I	①	⑦	3	4	1
Player A	II	5	6	④	⑤	4
	III	⑦	2	⓪	3	0
Column max		7	7	4	5	

The point 4 having both marks (○ and ☐). Therefore, it is the required saddle point.

The optimal strategy used by player A = II.

The optimal strategy used by player B = III.

So, optimal strategy = (II, III)

Value of the game (v) = 4.

<u>EXAMPLE 5.</u>   *Find the range of p and q under which the given pay off matrix gives the entry (II, II) as a saddle point.*

$$\text{Player B}$$

		I	II	III
	I	2	4	5
Player A	II	10	7	q
	III	4	p	6

<u>SOLUTION.</u>   Since 7 is given to be a saddle point, so 7 will be min in II row and max in II column and after it, it will be max in all three min elements of row and min in

the max elements of column. So

	I	II	III	**Row min**
I	②	4	5	2
II	⏹10	⑦	*q*	7
III	4	*p*	6	

Column max   10   7

Here, q should be greater than 7 and p should be less than 7.

So, the range of p and q is given by

$$p \leq 7 \quad \text{and} \quad q \geq 7$$

EXAMPLE 6. *The player A's choice of action is given by (A_1, A_2, A_3) and B's choice of action is (B_1, B_2) only. The payment according to their activity is given as follow. Find the optimal strategy and value of the game*

Activity selected	Payment made
(A_1, B_1)	Player A pays ₹ 1 to player B
(A_2, B_1)	Player B pays ₹ 2 to player A
(A_3, B_1)	Player A pays ₹ 2 to player B
(A_1, B_2)	Player B pays ₹ 6 to player A
(A_2, B_2)	Player B pays ₹ 4 to player A
(A_3, B_2)	Player A pays ₹ 6 to player B

SOLUTION.   Firstly, make the pay off matrix with the help of given data.

Player B

	B_1	B_2
A_1	−1	6
A_2	2	4
A_3	−2	−6

Player A

Table for the saddle point

	B_1	B_2	**Row min**
A_1	⊖1	⏹6	−1
A_2	②	4	2
A_3	−2	⊖6	−6

Column max   2   6

So, 2 is the required saddle point.

Optimal strategy is given by (A_2, B_1).

Value of the game $v = 2$.

EXAMPLE 7. *Find the optimal strategy and value of the game whose pay-off matrix is given by*

Player B

	I	II	III	IV	V
I	−2	0	0	5	3
Player A II	3	2	1	2	2
III	−4	−3	0	−2	6
IV	5	3	−4	2	−6

SOLUTION.   **Table for saddle point**

Player B

	I	II	III	IV	V	Row min
I	⊝2	0	0	⊡5	3	−2
II	3	2	⊙1	2	2	1
Player A III	⊝4	−3	0	−2	⊡6	−4
IV	⊡5	⊡3	−4	2	⊝6	−6
Column max	5	3	1	5	6	

So, the required saddle point is given by = 1.

The optimal strategy is given by (II, III).

The value of the game = 1.

EXAMPLE 8. *Find out the saddle point and the value of the game*

Player B

	I	II	III	IV
Player A I	−5	2	1	20
II	5	5	4	6
III	4	−2	0	−5

SOLUTION.   **Table for the saddle point**

Player B

	I	II	III	IV	Row min
I	⊝5	2	1	⊡20	−5
Player A II	⊡5	⊡5	⊙4	6	4
III	⊝4	−2	0	⊝5	−5
Column max	5	5	4	20	

So, saddle point is given by = 4.

The optimal strategy is given by (II, III).

The value of the game = 4.

EXAMPLE 9. *Determine the optimal strategy and value of the game*

(a)

	I	II	III
I	6	8	6
II	4	12	2

(b)

	I	II	III
I	3	0	−3
II	2	3	1
III	−4	2	−1

SOLUTION. (a) **Table for saddle point**

	I	II	III	Row min.
I	⑥	8	⑥	6
II	4	12	②	2
Column max	6	12	6	

So, saddle point is given by 6.

Optimal strategy is given by (I, I) or (I, III).

Value of the game = 6.

(b) **Table for saddle point**

	I	II	III	Row min.
I	3	0	-3	-3
II	2	3	①	1
III	-4	2	-1	-4
Colum max	3	3	1	

So, the required saddle point is given by 1.

The optimal strategy = (II, III). Value of the game = 1.

EXAMPLE 10. *Find the saddle point in the following case and also find the game value*

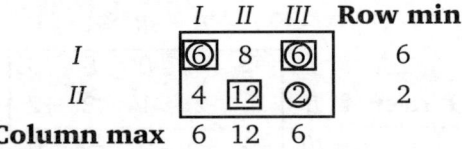

		I	II	III
A	I	1	14	11
	II	-9	5	-11
	III	0	-3	14

SOLUTION. Given pay-off matrix is

		I	II	III	Row min.
	I	1	14	11	1
A	II	-9	5	-11	-11
	III	0	-3	14	-3
Column max.		1	14	14	

To find the saddle point, find the smallest element in each row and mark it ○. Find maximum element in each column and mark it as □. Saddle point is the intersection of row minima and column maxima.

		I	II	III
	I	①	14	11
A	II	-9	5	-11
	III	0	-3	14

Hence, we conclude that

Strategy of A is I.

Strategy of B is I.

Value of game is 1.

14.2 SOLUTION OF A RECTANGULAR GAME IN TERMS OF MIXED STRATEGIES

In any game, if there is no saddle point, the two players can not use maxmin, minimax strategies (pure) as their optimal strategies, then best strategies are mixed strategies. Now, two players, instead of selecting pure strategies only, may play their plays according to the predetermined set consisting probabilities corresponding to each of their pure strategies.

Thus, consider a rectangular game played by two players A (maximizing player) and B with pay off matrix $[a_{ij}]_{m \times n}$. The players A and B have m and n pure strategies respectively.

Player B

Probabilities		y_1	y_2	$\cdots$	y_j	$\cdots$	y_x
	Pure strategies	1	2	$\cdots$	j	$\cdots$	n
x_1	1	a_{11}	a_{12}	$\cdots$	a_{1j}	$\cdots$	a_{1n}
x_2	2	a_{21}	a_{22}	$\cdots$	a_{2j}	$\cdots$	a_{2n}
$\vdots$	$\vdots$	$\vdots$	$\vdots$	$\vdots$	$\vdots$	$\vdots$	$\vdots$
x_i	i	a_{i1}	a_{i2}	$\cdots$	a_{ij}	$\cdots$	a_{in}
$\vdots$	$\vdots$	$\vdots$	$\vdots$	$\cdots$	$\vdots$	$\cdots$	$\vdots$
x_m	m	a_{m1}	a_{m2}	$\cdots$	a_{mj}	$\cdots$	a_{mn}

Player A (row label)

Let the mixed strategies of players A and B be respectively given by $X = (x_1, x_2, ..., x_m)$ and $Y = (y_1, y_2, ..., y_n)$.

Here, $x_1, x_2, ..., x_m$ and $y_1, y_2, ..., y_n$ are the probabilities of selecting pure strategies by A and B respectively.

Also, $\quad \sum_{i=1}^{m} x_i = 1$ and $\sum_{j=1}^{n} y_j = 1$, $x_i \geq 0$, $y_j \geq 0$, $\forall\, i = 1, 2, ..., m$, $j = 1, 2, ..., n$

Further expected gain to A is

$$a_{11}x_1 + a_{21}x_2 + ... + a_{i1}x_i + ... + a_{m1}x_m = \sum_{i=1}^{m} a_{i1}x_i$$

(If B uses strategy 1 with probability y_1)

$$a_{12}x_1 + a_{22}x_2 + ... + a_{i2}x_i + ... + a_{m2}x_m = \sum_{i=1}^{m} a_{i2}x_i$$

(If B uses strategy 2 with probability y_2)

$$\cdots \qquad \cdots \qquad \cdots \qquad \cdots \qquad \cdots \qquad \cdots$$
$$\cdots \qquad \cdots \qquad \cdots \qquad \cdots \qquad \cdots \qquad \cdots$$

$$a_{1j}x_1 + a_{2j}x_2 + ... + a_{ij}x_i + ... + a_{mj}x_m = \sum_{i=1}^{m} a_{ij}x_i$$

(If B uses strategy j with probability y_j)

$$\cdots \qquad \cdots \qquad \cdots \qquad \cdots \qquad \cdots \qquad \cdots$$
$$\cdots \qquad \cdots \qquad \cdots \qquad \cdots \qquad \cdots \qquad \cdots$$

$$a_{1n}x_1 + a_{2n}x_2 + ... + a_{in}x_i + ... + a_{mn}x_m = \sum_{i=1}^{m} a_{in}x_i$$

(If B uses strategy n with probability y_n)

Then, expected gain to A is

$$E(X, Y) = \sum_{i=1}^{m} \sum_{j=1}^{n} a_{ij} x_i y_j \qquad \qquad ....(1)$$

Using maxmin-minimax criterion, A selected A_i $\left(x_i \geq 0, \sum_{i=1}^{m} x_i = 1 \right)$ which will maximize his minimum expected gain, i.e., A selects x_i which will

$$Max \left[Min \left\{ \sum_{i=1}^{m} a_{i1} x_i, \sum_{i=1}^{m} a_{i2} x_i, ..., \sum_{i=1}^{m} a_{in} x_i \right\} \right]$$

This value is known as maxmin $(\underline{v})$ expected value for player A. In a similar manner, B selects y_j $\left(y_j \geq 0, \sum_{i=1}^{n} y_j = 1 \right)$ which will minimize his maximum expected loss.

Therefore, B selects y_j which will

$$Min \left[Max \left\{ \sum_{j=1}^{n} a_{1j} y_j, \sum_{j=1}^{n} a_{2j} y_j, ..., \sum_{j=1}^{n} a_{nj} y_j \right\} \right]$$

This value is known as minimax $(\overline{v})$ expected value for B. Hence, we conclude that for player A, the best strategy is that which maximizes the $Min \sum_{j}^{n} \sum_{i=1}^{n} a_{ij} x_i$ and for player B best strategy is that which minimizes the $Max \sum_{j=1}^{n} a_{ij} x_j$.

☛ Remarks

- In case of pure strategies, $\underline{v} \leq \overline{v}$.
- Fundamental theorem of rectangular games assumes that there always exists optimum strategies such that $\underline{v} = \overline{v}$.

14.3 PROPERTIES OF OPTIMAL MIXED STRATEGIES

Some important properties of optimal mixed strategies are as follows :

1. If one of the players adheres to his optimal mixed strategy and other deviates from his optimal strategy, then deviating players can only increase his yield and can not increase in any case. At most it may be equal.
2. If one of the players adheres to his optimal strategy, then the value of the game does not change if other uses his supporting strategies only either singly or mixture.
3. If a fixed number a is added to each element of the pay off matrix, then the optimal strategies remain unchanged while the value of the game increases by a .
4. If every element of the pay off matrix is multiplied by a constant a , then the optimal strategies do not change while the value of the game becomes a times the value of the original game.

14.4 SOLUTION OF 2 x 2 GAMES WITHOUT SADDLE POINTS

THEOREM 1. *For any zero sum two persons game where the optimal strategies are not pure strategies and for which A's pay off matrix is*

$$B$$

$$\begin{array}{c}\quad\quad I\,(y_1)\quad II\,(y_2)\\ A\quad\begin{array}{c}I\,(x_1)\\ II\,(x_2)\end{array}\begin{array}{|cc|}\hline a_{11} & a_{12}\\ a_{21} & a_{22}\\ \hline\end{array}\end{array}$$

The optimal strategies (x_1, x_2) and (y_1, y_2) are given by

$$\frac{x_1}{x_2} = \frac{a_{22} - a_{21}}{a_{11} - a_{12}} \quad and \quad \frac{y_1}{y_2} = \frac{a_{22} - a_{12}}{a_{11} - a_{21}}$$

Also, the value of the game to A is given by

$$v = \frac{a_{11}a_{22} - a_{12}a_{21}}{(a_{11} + a_{22}) - (a_{12} + a_{21})}$$

PROOF. Let (x_1, x_2) and (y_1, y_2) be the mixed strategies for players A and B respectively. Then, clearly, we have

$$x_1 + x_2 = 1 \qquad \qquad \qquad \text{...(1)}$$
$$y_1 + y_2 = 1 \qquad \qquad \qquad \text{...(2)}$$

where, $x_1 \geq 0,\ x_2 \geq 0,\ y_1 \geq 0,\ y_2 \geq 0$.

Then, expected gain to A is given by $a_{11}x_1 + a_{21}x_2$, when B uses strategy I and expected gain to A is $a_{12}x_1 + a_{22}x_2$ when B uses strategy II.

Also, expected loss to B is $a_{11}y_1 + a_{12}y_2$ when A uses strategy I.

and expected loss to B is $a_{21}y_1 + a_{22}y_2$ when A uses strategy II.

Further, let v be the value of the game, then since A expects to get at least v, so

$$\left.\begin{array}{c}a_{11}x_1 + a_{21}x_2 \geq v\\ a_{12}x_1 + a_{22}x_2 \geq v\end{array}\right\} \qquad \qquad \text{...(3)}$$

Further, B expects to loose at most v, so

$$\left.\begin{array}{c}a_{11}y_1 + a_{12}y_2 \leq v\\ a_{21}y_1 + a_{22}y_2 \leq v\end{array}\right\} \qquad \qquad \text{...(4)}$$

Converting (3) and (4) into equation form, we get

$$a_{11}x_1 + a_{21}x_2 = v \qquad \qquad \text{...(5)}$$
$$a_{12}x_1 + a_{22}x_2 = v \qquad \qquad \text{...(6)}$$
$$a_{11}y_1 + a_{21}y_2 = v \qquad \qquad \text{...(7)}$$
$$a_{21}y_1 + a_{12}y_2 = v \qquad \qquad \text{...(8)}$$

From (5) and (6), we get

$$(a_{11} - a_{12})\,x_1 = (a_{22} - a_{21})\,x_2$$

$$\Rightarrow \qquad \frac{x_1}{x_2} = \frac{a_{22} - a_{21}}{a_{11} - a_{12}} \qquad \qquad \text{...(9)}$$

Also, from (8) and (7), we get

$$\frac{y_1}{y_2} = \frac{a_{22} - a_{12}}{a_{11} - a_{21}} \qquad \qquad \text{...(10)}$$

Equation (9) can be written as

$$x_2 = \frac{a_{11} - a_{12}}{a_{22} - a_{21}}\,x_1$$

Putting this value in (1), we get

$$x_1 \left[1 + \frac{a_{11} - a_{12}}{a_{22} - a_{21}} \right] = 1$$

$$\Rightarrow \qquad x_1 = \frac{a_{22} - a_{21}}{(a_{11} + a_{22}) - (a_{12} + a_{21})} \qquad \qquad ...(11)$$

and $\qquad x_2 = \frac{a_{11} - a_{12}}{(a_{11} + a_{22}) - (a_{12} + a_{21})} \qquad \qquad ...(12)$

Similarly, from (2) and (10), we get

$$y_1 = \frac{a_{22} - a_{12}}{(a_{11} + a_{22}) - (a_{12} + a_{21})} \qquad \qquad ...(13)$$

and $\qquad y_2 = \frac{a_{11} - a_{21}}{(a_{11} + a_{22}) - (a_{12} + a_{21})} \qquad \qquad ...(14)$

Putting all these values in (5), we get

$$v = \frac{a_{11} a_{22} - a_{12} a_{21}}{(a_{11} + a_{22}) - (a_{12} + a_{21})} \qquad \qquad ...(15)$$

Further, we have to prove that x_1, x_2, y_1, y_2 all are non-negative. Since game has no saddle point, thus the largest and second largest elements must lie on one of the diagonals. Thus, there are only the following possible ordering of the elements of the matrix.

$$a_{11} \geq a_{22} \geq a_{12} \geq a_{21}$$
$$a_{11} \geq a_{22} \geq a_{21} \geq a_{12}$$
$$a_{22} \geq a_{11} \geq a_{12} \geq a_{21}$$
$$a_{22} \geq a_{11} \geq a_{21} \geq a_{12}$$
$$a_{12} \geq a_{21} \geq a_{11} \geq a_{22}$$
$$a_{12} \geq a_{21} \geq a_{22} \geq a_{11}$$
$$a_{21} \geq a_{12} \geq a_{11} \geq a_{22}$$
$$a_{21} \geq a_{12} \geq a_{22} \geq a_{11}$$

Clearly, all ordering of the elements of the pay off matrix x_1, x_2, y_1, y_2 given by (11), (12), (13), (14) are all non-negative.

☛ **Remark**

- The above theorem is not always true for a 2×2 game with a saddle point. Thus, this formula should be applied in case of 2×2 game without saddle point.

Solved Examples

EXAMPLE I. *In a game of matching coins with two players A wins ₹ 2 when two heads occur, wins nothing when two tails occur, and looses ₹ 1 when there are one head and one tail. Determine the pay off matrix and best strategies for player A and B and also find the value of the game.*

SOLUTION. The pay off matrix of the given problem will be

$$
\begin{array}{cc}
 & B \\
 & \begin{array}{cc} H & T \end{array} \\
A \begin{array}{c} H \\ T \end{array} & \begin{array}{|cc|} \hline 2 & -1 \\ -1 & 0 \\ \hline \end{array}
\end{array}
$$

Now, the probability for choosing his strategy by player A

$$x_1 = \frac{a_{22} - a_{21}}{(a_{11} + a_{22}) - (a_{12} + a_{21})} = \frac{0+1}{(2+0) - (-1-1)} = \frac{1}{4}$$

$$x_2 = \frac{a_{11} - a_{12}}{(a_{11} + a_{22}) - (a_{12} + a_{21})} = \frac{2+1}{(2+0) - (-1-1)} = \frac{3}{4}$$

Probability for choosing his strategy by player B

$$y_1 = \frac{a_{22} - a_{12}}{(a_{11} + a_{22}) - (a_{12} + a_{21})} = \frac{0-(-1)}{(2+0) - (-1-1)} = \frac{1}{4}$$

$$y_2 = \frac{a_{11} - a_{21}}{(a_{11} + a_{22}) - (a_{12} + a_{21})} = \frac{2-(-1)}{(2+0) - (-1-1)} = \frac{3}{4}$$

Now, value of the game is given by

$$v = \frac{a_{11}a_{22} - a_{12}a_{21}}{(a_{11} + a_{22}) - (a_{12} + a_{21})}$$

$$= \frac{2 \times 0 - (-1)(-1)}{(2+0) - (-1-1)} = \frac{1}{4}$$

$$v = \frac{1}{4}$$

Hence, Player A's optimal strategy $= \left(\frac{1}{4}, \frac{3}{4}\right)$

Player B's optimal strategy $= \left(\frac{1}{4}, \frac{3}{4}\right)$

and value of the game, $v = \frac{1}{4}$.

EXAMPLE 2. *Find the optimal strategy for each of the player and the value of the game*

(a)

	B I	II
A I	6	−3
II	−3	0

(b)

	B I	II
A I	1	3
II	4	2

(c)

	B I	II
A I	−4	6
II	2	−3

(d)

	B I	II
A I	1	7
II	6	2

(e)

	B I	II
A I	6	−3
II	−3	0

SOLUTION. (a) Probability for choosing his strategy by player A

$$x_1 = \frac{a_{22} - a_{21}}{(a_{11} + a_{22}) - (a_{12} + a_{21})} = \frac{0+3}{(6+0) - (-3-3)} = \frac{1}{4}$$

$$x_2 = \frac{a_{11} - a_{12}}{(a_{11} + a_{22}) - (a_{12} + a_{21})} = \frac{6-(-3)}{(6+0) - (-3-3)} = \frac{3}{4}$$

Probability of choosing his strategy by player B

$$y_1 = \frac{a_{22} - a_{12}}{(a_{11} + a_{22}) - (a_{12} + a_{21})} = \frac{0 - (-3)}{(6 + 0) - (-3 - 3)} = \frac{1}{4}$$

$$y_2 = \frac{a_{11} - a_{21}}{(a_{11} + a_{22}) - (a_{12} + a_{21})} = \frac{6 - (-3)}{(6 + 0) - (-3 - 3)} = \frac{3}{4}$$

So, the optimal strategy for player A $= \left(\frac{1}{4}, \frac{3}{4} \right)$

the optimal strategy for player B $= \left(\frac{1}{4}, \frac{3}{4} \right)$.

Now, value of the game is given by

$$v = \frac{a_{11}a_{22} - a_{12}a_{21}}{(a_{11} + a_{22}) - (a_{12} + a_{21})}$$

$$= \frac{6 \times 0 - (-3)(-3)}{(6 + 0) - (-3 - 3)} = -\frac{3}{4}$$

$$\Rightarrow \quad v = -\frac{3}{4}$$

(b)

$$\begin{array}{c} & & B \\ & & \begin{array}{cc} I & II \end{array} \\ A & \begin{array}{c} I \\ II \end{array} & \begin{array}{|cc|} \hline 1 & 3 \\ 4 & 2 \\ \hline \end{array} \end{array}$$

Optimal strategy used by player A $= (x_1, x_2)$

$$x_1 = \frac{2 - 4}{(1 + 2) - (3 + 4)} = \frac{-2}{3 - 7} = \frac{1}{2}; x_2 = \frac{1 - 3}{(1 + 2) - (3 + 4)} = \frac{-2}{-4} = \frac{1}{2}$$

Optimal strategy used by player B $= (y_1, y_2)$

$$y_1 = \frac{2 - 3}{(1 + 2) - (3 + 4)} = \frac{1}{4};$$

$$y_2 = \frac{1 - 4}{(1 + 2) - (3 + 4)} = \frac{-3}{-4} = \frac{3}{4}$$

So, the optimal strategy for player A $= \left(\frac{1}{2}, \frac{1}{2} \right)$

the optimal strategy for player B $= \left(\frac{1}{4}, \frac{3}{4} \right)$.

And, value of the game is given by

$$v = \frac{2 \times 1 - 3 \times 4}{(2 + 1) - (3 + 4)} = \frac{2 - 12}{3 - 7} = \frac{5}{2} \quad \Rightarrow \quad v = \frac{5}{2}$$

(c)

$$\begin{array}{c} & & B \\ & & \begin{array}{cc} I & II \end{array} \\ A & \begin{array}{c} I \\ II \end{array} & \begin{array}{|cc|} \hline -4 & 6 \\ 2 & -3 \\ \hline \end{array} \end{array}$$

Let the optimal strategy used by player A $= (x_1, x_2)$

and the optimal strategy used by player B = (y_1, y_2)

Now, $x_1 = \dfrac{-3-2}{(-4+(-3))-(6+2)} = \dfrac{-5}{-7-8} = \dfrac{1}{3}$

$x_2 = \dfrac{-4-6}{(-4-3)-(6+2)} = \dfrac{-10}{-7-8} = \dfrac{2}{3}$

$y_1 = \dfrac{-3-6}{(-4-3)-(6+2)} = \dfrac{-9}{-7-8} = \dfrac{3}{5}$

$y_2 = \dfrac{-4-2}{(-4-3)-(6+2)} = \dfrac{-6}{-15} = \dfrac{2}{5}$

Value of the game is given by, $v = \dfrac{(-4)(-3)-6\times2}{(-4-3)-(6+2)} = 0$

So, Optimal strategy used by player A = $\left(\dfrac{1}{3}, \dfrac{2}{3}\right)$

Optimal strategy used by player B = $\left(\dfrac{3}{5}, \dfrac{2}{5}\right)$

and value of the game, $v = 0$.

(d)

$$\begin{array}{cc} & \textbf{\textit{B}} \\ & \begin{array}{cc} I & II \end{array} \\ \textbf{\textit{A}} \ \begin{array}{c} I \\ II \end{array} & \begin{array}{|cc|} \hline 1 & 7 \\ 6 & 2 \\ \hline \end{array} \end{array}$$

Let the optimal strategy used by player A = (x_1, x_2)

and the optimal strategy used by player B = (y_1, y_2)

Now, $x_1 = \dfrac{2-6}{(1+2)-(7+6)} = \dfrac{-4}{3-13} = \dfrac{2}{5}$;

$x_2 = \dfrac{1-7}{(1+2)-(7+6)} = \dfrac{-6}{3-13} = \dfrac{3}{5}$

$y_1 = \dfrac{2-7}{(1+2)-(7+6)} = \dfrac{-5}{-10} = \dfrac{1}{2}$;

$y_2 = \dfrac{1-6}{(1+2)-(7+6)} = \dfrac{-5}{-10} = \dfrac{1}{2}$

and value of the game is given by

$v = \dfrac{1\times2-7\times6}{(1+2)-(7+6)} = \dfrac{2-42}{3-13} = 4$

So, Optimal strategy used by player A = $\left(\dfrac{2}{5}, \dfrac{3}{5}\right)$

Optimal strategy used by player B = $\left(\dfrac{1}{2}, \dfrac{1}{2}\right)$

and value of the game, $v = 4$.

14.5 SOLUTION BY LINEAR PROGRAMMING

Let a rectangular game be played by two players A (max player) and B (minimum player) with pay off matrix $[a_{ij}]_{m \times n}$. The mixed strategies for A and B be respectively given by $X = [x_1, x_2, ..., x_m]$ and $Y = [y_1, y_2, ..., y_n]$. We know that A selects his optimal mixed strategies which is given by

$$\underset{x_i}{Max}\left[Min\left\{\sum_{i=1}^{m} a_{i1}x_i, \sum_{i=1}^{m} a_{i2}x_i, ..., \sum_{i=1}^{m} a_{in}x_i\right\}\right]$$

such that $x_1 + x_2 + ... + x_m = 1$, $x_i \geq 0$, $\forall i = 1, 2, ..., m$.

Further, B selects his optimal strategies which will be

$$\underset{y_j}{Min}\left[Max\left\{\sum_{j=1}^{n} a_{1j}y_j, \sum_{j=1}^{n} a_{2j}y_j, ..., \sum_{j=1}^{n} a_{mj}y_j\right\}\right]$$

such that $y_1 + y_2 + ... + y_n = 1$, $y_j \geq 0$, $\forall j = 1, 2, ..., n$

Now, if $$Min\left\{\sum_{i=1}^{n} a_{i1}x_i, \sum_{i=1}^{m} a_{i2}x_i, ..., \sum_{i=1}^{m} a_{in}x_i\right\} = v$$

then, A expects to gain at least v. Now, since v is minimum of all expected gains, therefore, we have

$$\sum_{i=1}^{m} a_{i1}x_i \geq v, \sum_{i=1}^{m} a_{i2}x_i \geq v, ..., \sum_{i=1}^{m} a_{in}x_i \geq v$$

Therefore, A's problem to determine $x_1, x_2, ..., x_n$ to minimize $z = v$, subject to the constraints

$$\left.\begin{array}{l} a_{11}x_1 + a_{21}x_2 + ... + a_{m1}x_m \geq v \\ a_{12}x_1 + a_{22}x_2 + ... + a_{m2}x_m \geq v \\ \quad ... \qquad ... \qquad ... \qquad ... \\ \quad ... \qquad ... \qquad ... \qquad ... \\ a_{1n}x_1 + a_{2n}x_2 + ... + a_{mn}x_m \geq v \end{array}\right\} \quad ....(1)$$

$$x_1 + x_2 + ... + x_m = 1, \quad x_1, x_2, x_m \geq 0$$

For $v > 0$, it is enough to prove that all the elements of the pay-off matrix are positive. If all are not positive, then we can add a sufficient large quantity to every element of the pay-off matrix so that they all become positive. Then, we can take v as positive.

Putting $\dfrac{x_1}{v} = X_1, \dfrac{x_2}{v} = X_2, ..., \dfrac{x_m}{v} = X_m$ in (1), we get

$$a_{11}X_1 + a_{21}X_2 + ... + a_{m1}X_m \geq 1$$
$$a_{12}X_1 + a_{22}X_2 + ... + a_{m2}X_m \geq 1$$
$$\quad ... \qquad ... \qquad ... \qquad ...$$
$$\quad ... \qquad ... \qquad ... \qquad ...$$
$$a_{1n}X_1 + a_{2n}X_2 + ... + a_{mn}X_m \geq 1$$

and $$X_1 + X_2 + ... + X_m = \frac{1}{v}, \quad X_1, X_2, ..., X_m \geq 0$$

Now, $$Max. \ v = Min\left(\frac{1}{v}\right)$$

$$= Min\left\{\frac{x_1 + x_2 + ... + x_m}{v}\right\}$$

$$= Min \ (X_1 + X_2 + ... + X_m)$$

Hence, the given rectangular game reduces to the following linear programming.

$$Min \ x^* = \frac{1}{v} = X_1 + X_2 + ... + X_m$$

subject to the constraints

$$\left.\begin{array}{l} a_{11}X_1 + a_{21}X_1 + ... + a_{m1}X_m \geq 1 \\ a_{12}X_1 + a_{22}X_2 + ... + a_{m2}X_m \geq 1 \\ ... \quad\quad ... \quad\quad ... \quad\quad ... \\ ... \quad\quad ... \quad\quad ... \quad\quad ... \\ a_{1n}X_1 + a_{2n}X_2 + ... + a_{mn}X_m \geq 1 \\ X_1 \geq 0, \ X_2 \geq 0, ..., X_m \geq 0 \end{array}\right] \qquad ...(2)$$

and

For B's point of view, we want to minimize v, by solving the following linear programming

$$Maximize \ y^* = \frac{1}{v} = Y_1 + Y_2 + ... + Y_n$$

subject to the constraints

$$\left.\begin{array}{l} a_{11}Y_1 + a_{12}Y_1 + ... + a_{1n}Y_n \geq 1 \\ a_{21}Y_1 + a_{22}Y_1 + ... + a_{2n}Y_n \geq 1 \\ ... \quad\quad ... \quad\quad ... \quad\quad ... \\ ... \quad\quad ... \quad\quad ... \quad\quad ... \\ a_{m1}Y_1 + a_{m2}Y_2 + ... + a_{mn}Y_n \geq 1 \\ Y_1, Y_2, ..., Y_m \geq 0. \end{array}\right] \qquad ...(3)$$

and

where, $y^* = \dfrac{1}{v}$, $Y_1 = \dfrac{y_1}{v}$, $Y_2 = \dfrac{y_2}{v}$, ..., $Y_n = \dfrac{y_n}{v}$

☞ **Remark**
- The linear programming problems given by (2) and (3) are the duals of each other. Thus, if one problem is solved, then other is solved automatically.

14.6 MINIMAX THEOREM : FUNDAMENTAL THEOREM OF GAME THEORY

THEOREM 1. *Every game can be solved in terms of mixed strategies, i.e., if mixed strategies are adopted, there always exists a value of the game, i.e., $\underline{v} = v = \overline{v}$, where $\underline{v}$ and $\overline{v}$ are maxmin and minimax values of v.*

PROOF. If $X = (x_1, x_2, ..., x_n)$, $Y = (y_1, y_2, ..., y_n)$ are the mixed strategies of two players where $x_1, x_2, ..., x_m$, $y_1, y_2, ..., y_n$ are the probabilities with which they choose their pure strategies then A's problem is

To minimize $x^* = X_1 + X_2 + ... + X_m$

subject to the constraints

$$a_{11}X_1 + a_{21}X_2 + ... + a_{m1}X_m \geq 1$$

$$a_{12}X_1 + a_{22}X_2 + ... + a_{m2}X_m \geq 1$$

$$... \quad ... \quad ... \quad ...$$
$$a_{1n}X_1 + a_{2n}X_2 + ... + a_{mn}X_m \geq 1$$
$$X_1, X_2, ..., X_m \geq 0 \quad \text{and} \quad X_i = \frac{x_i}{v}$$

Similarly, the problem for player B is given by

$$\text{Maximize } y^* = Y_1 + Y_2 + ... + Y_n$$

subject to the constraints

$$a_{11}Y_1 + a_{12}Y_2 + ... + a_{1n}Y_n \leq 1$$
$$a_{21}Y_1 + a_{22}Y_2 + ... + a_{2n}Y_n \leq 1$$
$$...\quad ... \quad ... \quad ...$$
$$...\quad ... \quad ... \quad ...$$
$$a_{m1}Y_1 + a_{m2}Y_2 + ... + a_{mn}Y_n \leq 1$$

$$Y_1, Y_2, ..., Y_n \geq 0 \quad \text{and} \quad Y_j = \frac{y_j}{v}$$

Clearly, B's problem is the dual of A's problem. By duality theorem, it is known that If either the primal or the dual has a finite optimal solutions, then the other has a finite optimal solution and the optimal values of two objective functions are equal. Hence,

$$\text{Max } y^* = \text{Min } x^*$$

or

$$\underline{v} = v = \overline{v}.$$

WORKING PROCEDURE

If a problem has no saddle point, then convert it to a linear programming problem for player B. Then solve it by simplex method. From the final simplex table of the solution, we can read the solution for player A by the dual method.

Solved Examples

EXAMPLE 1. *Solve the following game by Simplex method. Find the optimal strategy and value of the game*

$$
\begin{array}{c}
 & B \\
A & \begin{bmatrix} 1 & -1 & 3 \\ 3 & 5 & -3 \\ 6 & 2 & -2 \end{bmatrix}
\end{array}
$$

SOLUTION. First of all we find out maxmin and minimax element from row and column. From this, we can find the range in which the value of the game will lie

$$
\begin{array}{ccc}
 & & \text{Row min} \\
1 & \boxed{-1} \quad \boxed{3} & \\
3 & \boxed{5} \quad \ominus{3} & -1 \\
\boxed{6} & 2 \quad \ominus{2} & -3 \\
\end{array}
\left.\begin{array}{c} \\ \end{array}\right\} \max = -1
$$

Column max $\underbrace{6 \quad 5 \quad 3}$ $\quad -2$

$$\text{min} = 3$$

$$\Rightarrow \quad -1 \leq v \leq 3$$

This implies that value of the game may be negative or zero. So, we add a constant such that all entry in the given payoff matrix become +ve.
So, we add $k = 4$. Then the new payoff matrix will be

$$\begin{vmatrix} 5 & 3 & 7 \\ 7 & 9 & 1 \\ 10 & 6 & 2 \end{vmatrix}$$

Let the strategy used by player A $= (x_1, x_2, x_3)$
The strategy used by player B $= (y_1, y_2, y_3)$
Now, B's linear programming problem will be
Min v'
s.t. $5y_1 + 3y_2 + 7y_3 \leq v'$
$7y_1 + 9y_2 + y_3 \leq v'$
$10y_1 + 6y_2 + 2y_3 \leq v'$
and $y_1 + y_2 + y_3 = 1$, where $y_1, y_2, y_3 \geq 0$

Dividing all the equations by v' and putting $\dfrac{y_1}{v'} = Y_1, \quad \dfrac{y_2}{v'} = Y_2, \quad \dfrac{y_3}{v'} = Y_3$
Now, the problem reduces to

$$Max\ z = \frac{1}{v'} = \frac{y_1 + y_2 + y_3}{v'} = \frac{y_1}{v'} + \frac{y_2}{v'} + \frac{y_3}{v'}$$

$Max\ z = Y_1 + Y_2 + Y_3$
s.t. $5Y_1 + 3Y_2 + 7Y_3 \leq 1$
$7Y_1 + 9Y_2 + Y_3 \leq 1$
$10Y_1 + 6Y_2 + 2Y_3 \leq 1$, where, $Y_1, Y_2, Y_3 \geq 0$
$Max\ z = Y_1 + Y_2 + Y_3 + 0s_1 + 0s_2 + 0s_3$
s.t. $5Y_1 + 3Y_2 + 7Y_3 + s_1 = 1$
$7Y_1 + 9Y_2 + Y_3 + s_2 = 1$
$10Y_1 + 6Y_2 + 2Y_3 + s_3 = 1$

B.V.	C_B	x_B	$c_j \rightarrow$ Y_1	1 Y_2	1 Y_3	0 s_1	0 s_2	0 s_3	Min Ratio X_B / Y_j
s_1	0	1	5	3	7	1	0	0	1/7 ←
s_2	0	1	7	9	1	0	1	0	1
s_3	0	1	10	6	2	0	0	1	1/2
			1	1	1				
Y_3	1	1/7	5/7	3/7	1	1/7	0	0	1/3
s_2	0	6/7	44/7	60/7	0	–1/7	1	0	1/10 ←
s_3	0	5/7	60/7	36/7	0	–2/7	0	1	5/36
			2/7	4/7 ↑	0	–1/7	0	0	

Y_3	1	1/10	2/5	0	1	3/20	–1/20	0
Y_2	1	1/10	11/15	1	0	–1/60	7/60	0
s_3	0	1/5	24/5	0	0	–1/5	–3/5	1
Z		1/5	–2/15	0	0	–2/15	–1/15	0

Since all terms are ≤ 0. So the solution obtained will be optimal.

$$Y_1 = 0, \ Y_2 = 1/10, \ Y_3 = 1/10.$$

$$\text{Max } z = c_B X_B = 1/5$$

Now, $\qquad \text{Max } z = \dfrac{1}{v'}$

So, $\qquad \dfrac{1}{v'} = \dfrac{1}{5}$

$\Rightarrow \qquad \text{Min } v' = 5$.

and $\quad y_1 / v' = Y_1, \quad \dfrac{y_2}{v'} = Y_2, \quad \dfrac{y_3}{v'} = Y_3$.

$$y_1 = Y_1 v' = 0$$

$$y_2 = Y_2 v' = \dfrac{1}{10} \times 5 = \dfrac{1}{2}$$

$$y_3 = Y_3 v' = \dfrac{1}{10} \times 5 = \dfrac{1}{2}.$$

Now, to find the value of the game, we subtract the previous added constant $k = 4$ from the v'.

Then, $\qquad v = v' - 4 = 5 - 4 = 1$

$$v = 1.$$

Now, since player A's strategy will be the dual of the strategy of player B.

So, X_1, X_2 and X_3 will be the value of Δ_4, Δ_5 and Δ_6.

So, $\quad X_1 = 2/15 \quad X_2 = 1/15 \quad X_3 = 0$

$$X_1 = \dfrac{x_1}{v'} \qquad X_2 = \dfrac{x_2}{v'} \quad X_3 = \dfrac{x_3}{v'}$$

$$x_1 = X_1 v' \qquad x_2 = X_2 v' \qquad X_3 = X_3 v'$$

$$x_1 = \dfrac{2}{15} \times 5 \quad x_2 = \dfrac{1}{15} \times 5 \quad x_3 = 0 \times 5$$

$$x_1 = \dfrac{2}{3} \qquad\qquad x_2 = \dfrac{1}{3} \qquad\qquad x_3 = 0.$$

Hence, The optimal strategy for player A $= \left(\dfrac{2}{3}, \dfrac{1}{3}, 0 \right)$

The optimal strategy for player B $= \left(0, \dfrac{1}{2}, \dfrac{1}{2} \right)$

and value of the game $v = 1$.

EXAMPLE 2. *Solve the following game by linear programming problem whose pay-off matrix is given by*

Player B

I   II   III

$$
\textbf{Player A} \quad
\begin{array}{c}
I \\
II \\
III
\end{array}
\begin{array}{|ccc|}
\hline
3 & -1 & -3 \\
-3 & 3 & -1 \\
-4 & -3 & 3 \\
\hline
\end{array}
$$

SOLUTION. First of all we find out maxmin and minimax element from row and column. From this, we can find the range in which the value of the game will lie

Row min

$$
\begin{array}{|ccc|}
\hline
3 & -1 & -3 \\
-3 & 3 & -1 \\
-4 & -3 & 3 \\
\hline
\end{array}
\left.\begin{array}{c}
-3 \\
-3 \\
-4
\end{array}\right\} \quad \max = -3
$$

Column max   3   3   3

$$\underbrace{\qquad\qquad\qquad}_{\min = 3}$$

$\Rightarrow \qquad -3 \le v \le 3$

This implies that value of the game may be negative or zero. So, we add a constant such that all entry in the given payoff matrix become +ve.

So, we add $k = 5$. Then the new payoff matrix will be

$$
\begin{array}{|ccc|}
\hline
8 & 4 & 2 \\
2 & 8 & 4 \\
1 & 2 & 8 \\
\hline
\end{array}
$$

Let the strategy used by player A $= (x_1, x_2, x_3)$

The strategy used by player B $= (y_1, y_2, y_3)$

Now, B's problem is

Min v'

s.t. $\qquad 8y_1 + 4y_2 + 2y_3 \le v'$

$\qquad\qquad 2y_1 + 8y_2 + 4y_3 \le v'$

$\qquad\qquad y_1 + 2y_2 + 8y_3 \le v'$

and $\qquad y_1 + y_2 + y_3 = 1$

where $\qquad y_1, y_2, y_3 \ge 0$

Dividing all the equations by v' and putting

$$\frac{y_1}{v'} = Y_1, \quad \frac{y_2}{v'} = Y_2, \quad \frac{y_3}{v'} = Y_3$$

Now, the problem reduces to

$\qquad Max\ z = Y_1 + Y_2 + Y_3$

s.t. $\quad 8Y_1 + 4Y_2 + 2Y_3 \le 1$

$\qquad\quad 2Y_1 + 8Y_2 + 4Y_3 \le 1$

$\qquad\quad Y_1 + 2Y_2 + 8Y_3 \le 1$

where, $\qquad Y_1, Y_2, Y_3 \ge 0$

$\qquad\qquad Max\ z = Y_1 + Y_2 + Y_3 + 0s_1 + 0s_2 + 0s_3$

s.t. $\qquad 8Y_1 + 4Y_2 + 2Y_3 + s_1 = 1$

$\qquad\qquad 2Y_1 + 8Y_2 + 4Y_3 + s_2 = 1$

$\qquad\qquad Y_1 + 2Y_2 + 8Y_3 + s_3 = 1$

where, $Y_1, Y_2, Y_3, s_1, s_2, s_3 \geq 0$

B.V.	C_B	x_B	$c_j \rightarrow$ 1 Y_1	1 Y_2	1 Y_3	0 s_1	0 s_2	0 s_3	Min Ratio X_B / Y_j
s_1	0	1	$\boxed{8}$	4	2	1	0	0	$1/8 \leftarrow$
s_2	0	1	2	8	4	0	1	0	1/2
s_3	0	1	1	2	8	0	0	1	1
			$1\uparrow$	1	1	0	0	0	
Y_1	1	1/8	1	1/2	1/4	1/8	0	0	1/2
s_2	0	3/4	0	7	7/2	−1/4	1	0	3/14
s_3	0	7/8	0	3/2	$\boxed{31/4}$	−1/8	0	1	$7/62 \leftarrow$
			0	1/2	$3/4\uparrow$	−1/8	0	−1	
Y_1	1	3/31	1	14/31	0	4/31	0	−1/31	3/14
s_2	0	11/31	0	$\boxed{196/31}$	0	−6/31	1	−14/31	$11/196 \leftarrow$
Y_3	1	7/62	0	6/31	1	−1/62	0	4/31	7/12
			0	$11/31\uparrow$	0	−7/62	0	−3/31	
Y_1	1	1/14	1	0	0	1/7	1/14	0	
Y_2	1	11/196	0	1	0	−3/98	31/196	−1/14	
Y_3	1	5/49	0	0	1	−1/98	−3/98	1/7	
		$Z = 45/196$	0	0	0	−5/49	−11/196	−1/14	

Since all terms are ≤ 0. So the solution obtained will be optimal.
$$Y_1 = 1/14, \quad Y_2 = 11/196, \quad Y_3 = 5/49.$$

$$\text{Max } z = \frac{45}{196}$$

Now, $\qquad \text{Max } z = \frac{1}{v'} = \frac{45}{196}$

$\Rightarrow \qquad \text{Min } v' = \frac{196}{45}.$

and $\dfrac{y_1}{v'} = Y_1, \quad \dfrac{y_2}{v'} = Y_2, \quad \dfrac{y_3}{v'} = Y_3.$

$$y_1 = Y_1 v', \qquad y_2 = Y_2 v', \qquad y_3 = Y_3 v'$$

$\Rightarrow \qquad y_1 = \dfrac{1}{14} \times \dfrac{196}{45} = \dfrac{14}{45}$

$\qquad\qquad y_2 = \dfrac{11}{196} \times \dfrac{196}{45} = \dfrac{11}{45}$

$$y_3 = \frac{5}{49} \times \frac{196}{45} = \frac{20}{45}.$$

and value of the game $v = v' - 5 = \frac{196}{45} - 5 = -\frac{29}{45}$

$$v = -\frac{29}{45}$$

Now, by the dual of this LPP, we can find out the optimal strategy of player A.

Hence, A's optimal strategy $= \left(\frac{20}{45}, \frac{11}{45}, \frac{14}{45} \right)$

B's optimal strategy $= \left(\frac{14}{45}, \frac{11}{45}, \frac{20}{45} \right)$

and value of the game, $v = -\frac{29}{45}.$

 **Exercise-14.1**

1. Convert the given game into linear programming problem and solve them

(a) $\begin{bmatrix} 5 & 3 & 7 \\ 7 & 9 & 1 \\ 10 & 6 & 2 \end{bmatrix}$ (b) $\begin{bmatrix} 3 & -2 & 4 \\ -1 & 4 & 2 \end{bmatrix}$ (c) $\begin{bmatrix} 0 & -1 & 1 \\ 1 & 1 & -1 \\ 1 & -1 & 0 \end{bmatrix}$ (d) $\begin{bmatrix} 3 & -2 & 4 \\ -1 & 4 & 2 \\ 2 & 2 & 6 \end{bmatrix}$

Answers

1. (a) $\left(\frac{2}{3}, \frac{1}{2}, 0 \right), \left(0, \frac{1}{2}, \frac{1}{2} \right)$ and $v = 5$. (b) $\left(\frac{1}{2}, \frac{1}{2} \right), \left(\frac{3}{5}, \frac{2}{5}, 0 \right)$ and $v = 1$.

(c) $\left(\frac{1}{2}, \frac{1}{2}, 0 \right), \left(0, \frac{1}{2}, \frac{1}{2} \right)$ and $v = 0$. (d) $(0, 0, 1), \left(\frac{4}{5}, \frac{1}{5}, 0 \right)$ and $v = 2$.

14.7 DOMINANCE PROPERTY

14.7.1 Principle of Dominance

If one pure strategy of a player is better or superior than another one, then the inferior strategy may be simply ignored by assigning a zero probability while searching for optimal strategies.

14.7.2 Rule of Dominance

RULE (1). If each element in one row, say i^{th} of the pay off matrix are less than or equal to the corresponding elements of the other row (say j^{th} row), then player A will never choose the ith strategy, i.e., i^{th} strategy is dominated by the jth strategy.

RULE (2). If each element in one column, say r^{th} is greater than or equal to the corresponding elements of the other column, say j^{th}, then the player B never use the r th strategy, i.e., j^{th} strategy dominates the rth strategy.

RULE (3). A pure strategy may be dominated if it is inferior to an average of two or more other pure strategy.

☛ **Remarks**

- The rules of dominance are especially used for evaluation of 2 person zero sum games without saddle points. Using these dominance properties, we try to reduce the size of pay off matrix.

- In case the pay off matrix reduces to the size $2 \times n$ or $m \times 2$, then we use the graphical method (discussed in next article) to find the solution of this game.

- If the dominance holds strictly, then value of the optimal strategies do coincide and when the dominance does not hold strictly, then optimal strategy may not coincide.

- These rules of dominance are used when the pay off matrix is a profit matrix for the player A and a loss matrix for player B, otherwise the rule gets reversed.

 Solved Examples

EXAMPLE 1. *Solve the game whose pay off matrix is given below*

$$\begin{bmatrix} 2 & 3 & 1/2 \\ 3/2 & 2 & 0 \\ 1/2 & 1 & 1 \end{bmatrix}$$

SOLUTION.

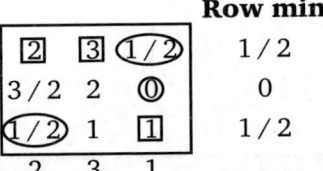

			Row min
2	3	1/2	1/2
3/2	2	0	0
1/2	1	1	1/2

Column max   2   3   1

There is no saddle point. To find the optimal strategy, first we will convert it into 2×2 matrix with the help of dominance property.

	I	II	III
I	2	3	1/2
II	3/2	2	0
III	1/2	1	1

Here, column III is inferior to II. So, column II will dominate the column III. So, we can delete II.

Then

	I	III
I	2	1/2
II	3/2	0
III	1/2	1

Here, I row is superior to II row. So, I row dominates the II. So, we can delete row II.

Now, our game reduces to 2×2 game which can be solved by arithmetic method without saddle point.

	I	III
I	2	1/2
III	1/2	1

If (x_1, x_2, x_3) and (y_1, y_2, y_3) be the optimal strategy used by player A and B respectively, then

$$x_1 = \frac{a_{22} - a_{21}}{(a_{22} - a_{21}) + (a_{11} - a_{12})} = \frac{1 - 1/2}{(1 - 1/2) + 1 - 1/2} = \frac{1/2}{1/2 + 3/2}$$

$$x_1 = 1/4$$

Thus $x_3 = 3/4$

So, the strategy used by player A $= \left(\frac{1}{4}, 0, \frac{3}{4}\right)$

Thus, $$y_1 = \frac{a_{22} - a_{12}}{(a_{22} - a_{12}) + (a_{11} - a_{21})} = \frac{1 - 1/2}{(1 - 1/2) + (2 - 1/2)} = \frac{1/2}{4/2}$$

$$y_1 = 1/4$$
$$y_3 = 3/4$$

So, the strategy used by player B $= \left(\frac{1}{4}, 0, \frac{3}{4}\right)$

Now, value of the game $= \dfrac{a_{22}a_{11} - a_{12}a_{21}}{(a_{22} + a_{11}) - (a_{12} + a_{21})}$

$$= \frac{1 \times 2 - \dfrac{1}{2} \times \dfrac{1}{2}}{(1 + 2) - \left(\dfrac{1}{2} + \dfrac{1}{2}\right)} = \frac{7}{8}$$

$$v = 7/8.$$

EXAMPLE 2. *Solve the following game*

B

		I	II	III
	I	1	7	2
A	II	6	2	7
	III	5	2	6

SOLUTION.

B

		I	II	III	Row min
	I	1	⑦	2	1
A	II	⑥	②	⑦	2
	III	5	②	⑥	2
Column max		6	7	7	

There is no saddle point. To find the optimal strategy, first we will convert it into 2×2 matrix with the help of dominance property.

		I	II	III
	I	1	7	2
	II	6	2	7
	III	5	2	6

Column III is inferior to III. So, we can delete the III column.

$$
\begin{array}{c}
& \begin{array}{cc} I & II \end{array} \\
\begin{array}{c} I \\ II \\ III \end{array} &
\left.\begin{array}{cc}
1 & 7 \\
6 & 2 \\
\cancel{5} & \cancel{2}
\end{array}\right.
\end{array}
$$

Row III is inferior to row II. So, we can delete the III row.

Now, our problem reduces to 2×2 matrix without saddle point.

$$
\begin{array}{c}
& \begin{array}{cc} I & II \end{array} \\
\begin{array}{c} I \\ II \end{array} &
\left.\begin{array}{cc}
1 & 7 \\
6 & 2
\end{array}\right.
\end{array}
$$

Let (x_1, x_2, x_3) be the optimal strategy used by A.

      (y_1, y_2, y_3) be the optimal strategy used by B.

$$
\frac{x_1}{x_2} = \frac{a_{22} - a_{21}}{a_{11} - a_{12}} = \frac{2-6}{1-7} = \frac{2}{3}
$$

$$
x_1 = \frac{2}{5}, \quad x_2 = \frac{3}{5}
$$

$$
\frac{y_1}{y_2} = \frac{a_{22} - a_{12}}{a_{11} - a_{21}} = \frac{2-7}{1-6} = \frac{5}{5} = 1
$$

$$
y_1 = \frac{1}{2}, \quad y_2 = \frac{1}{2}
$$

Value of the game, $v = \dfrac{v_{11}v_{22} - v_{12}v_{21}}{(v_{11} + v_{22}) - (v_{12} + v_{21})} = \dfrac{1 \times 2 - 7 \times 6}{(1+2) - (7+6)} = \dfrac{2-42}{3-13}$

$$
v = 4 .
$$

Optimal strategy used by A $= \left(\dfrac{2}{5}, \dfrac{3}{5}, 0\right)$

Optimal strategy used by B $= \left(\dfrac{1}{2}, \dfrac{1}{2}, 0\right)$

Value of the game = 4.

EXAMPLE 3.   *Solve the following game to find the optimal strategy*

$$
\begin{array}{c}
& \begin{array}{cccc} I & II & III & IV \end{array} \\
\begin{array}{c} I \\ II \\ III \\ IV \end{array} &
\left.\begin{array}{cccc}
3 & 2 & 4 & 0 \\
3 & 4 & 2 & 4 \\
4 & 2 & 4 & 0 \\
0 & 4 & 0 & 8
\end{array}\right.
\end{array}
$$

SOLUTION.   We use the dominance property to solve this game.

Row I is inferior to III. So, we can delete it.

$$
\begin{array}{c}
& \begin{array}{cccc} I & II & III & IV \end{array} \\
\begin{array}{c} I \\ II \\ III \\ IV \end{array} &
\left.\begin{array}{cccc}
\cancel{3} & \cancel{2} & \cancel{4} & \cancel{0} \\
3 & 4 & 2 & 4 \\
4 & 2 & 4 & 0 \\
0 & 4 & 0 & 8
\end{array}\right.
\end{array}
$$

Column I is superior to III.

	I	II	III	IV
II	3	4	2	4
III	4	2	4	0
IV	0	4	0	8

Since there is no row and column which are comparable. So, we use the average. Here, the average of II and III column is inferior to I. So, we can delete column-I.

	II	III	IV
II	4	2	4
III	2	4	0
IV	4	0	8

	III	IV
II	2	4
III	4	0
IV	0	8

Here, row I is equal to the average of II and III. So, we can delete row-I.
Now, the reduced pay off matrix is

	III	IV
III	4	0
IV	0	8

Since no saddle point is obtained. So, the optimal strategy will be the mixed strategy.

$$\frac{x_3}{x_4} = \frac{a_{22} - a_{21}}{a_{11} - a_{12}} = \frac{8-0}{4-0} = \frac{2}{1}$$

$$x_3 = \frac{2}{3}, \quad x_4 = \frac{1}{3}$$

$$\frac{y_3}{y_4} = \frac{a_{22} - a_{12}}{a_{11} - a_{21}} = \frac{8-0}{4-0} = \frac{2}{1}$$

$$y_3 = \frac{2}{3}, \quad y_4 = \frac{1}{3}$$

Optimal strategy used by A = $\left(0, 0, \dfrac{2}{3}, \dfrac{1}{3}\right)$

Optimal strategy used by B = $\left(0, 0, \dfrac{2}{3}, \dfrac{1}{3}\right)$

EXAMPLE 4. *Find the optimal strategy and value of the game*

	I	II	III
I	2	0	3
II	3	-1	1
III	5	2	-1

SOLUTION. Since there is no saddle point, so we use the principle of dominance.

$$
\begin{array}{c c c c}
 & I & II & III \\
I & 2 & 0 & 3 \\
II & 3 & -1 & 1 \\
III & 5 & 2 & -1
\end{array}
$$

$$
\begin{array}{c c c}
 & I & II \\
I & 0 & 3 \\
II & 1 & 1 \\
III & 2 & -1
\end{array}
$$

$$
\begin{array}{c c c}
 & I & II \\
I & 0 & 3 \\
III & 2 & -1
\end{array}
$$

Now, it is a 2×2 game without saddle point. So, the strategy will be mixed strategy.

$$\frac{x_1}{x_3} = \frac{a_{22} - a_{21}}{a_{11} - a_{12}} = \frac{-1 - 2}{0 - 3} = \frac{1}{1}$$

$$x_1 = \frac{1}{2}, \quad x_3 = \frac{1}{2}$$

So, the strategy used by A $= \left(\frac{1}{2}, 0, \frac{1}{2} \right)$

$$\frac{y_1}{y_2} = \frac{-1 - 3}{0 - 2} = \frac{-4}{-2} = \frac{2}{1}$$

$$y_1 = \frac{2}{3}, \quad y_2 = \frac{1}{3}$$

So, the strategy used by B $= \left(0, \frac{2}{3}, \frac{1}{3} \right)$

Value of the game, $v = \dfrac{a_{22}a_{11} - a_{21}a_{12}}{(a_{11} + a_{22}) - (a_{21} + a_{12})}$

$\Rightarrow \qquad v = 1$

EXAMPLE 5. *Given the following pay-off matrix of player A. Obtain the optimum strategies for both the players.*

$$
\begin{array}{c c c c}
 & & B & \\
 & 2 & 8 & 2 \\
A & 8 & 3 & 8 \\
 & 5 & 3 & 7
\end{array}
$$

SOLUTION. We try to find a saddle point of the game. The game does not have a saddle point. So apply the rule of dominance.

Every element of 3rd column is greater than or equal to the first column, therefore, by dominance rule, from player B's point of view, the pure strategy (3rd column) is dominated by the first. Thus, we get the reduced matrix

$$A \begin{array}{c} B \\ \begin{bmatrix} 2 & 8 \\ 8 & 3 \\ 5 & 3 \end{bmatrix} \end{array}$$

Again, every element of second row is greater than the third row, therefore by dominance rule, from player A's point of view, the pure 3rd row is dominated by two, so we get 2×2 matrix

$$A \begin{array}{c} B \\ \begin{bmatrix} 2 & 8 \\ 8 & 3 \end{bmatrix} \end{array}$$

Then by formulae for 2×2 game

$$x_1 = \frac{a_{22} - a_{21}}{(a_{11} + a_{22}) - (a_{12} + a_{21})} = \frac{3 - 8}{(2 + 3) - (8 + 8)} = \frac{5}{11}$$

$$x_2 = \frac{a_{11} - a_{12}}{(a_{11} + a_{22}) - (a_{12} + a_{21})} = \frac{2 - 8}{-11} = \frac{6}{11}$$

$$y_1 = \frac{a_{22} - a_{12}}{(a_{11} + a_{22}) - (a_{12} + a_{21})} = \frac{3 - 8}{-11} = \frac{5}{11}$$

$$y_2 = \frac{a_{11} - a_{21}}{(a_{11} + a_{22}) - (a_{12} + a_{21})} = \frac{2 - 8}{-11} = \frac{6}{11}$$

Hence, we conclude that

Optimal strategy for player A is $\left(\dfrac{5}{11}, \dfrac{6}{11}, 0 \right)$

Optimal strategy for player B is $\left(\dfrac{5}{11}, \dfrac{6}{11}, 0 \right)$.

 Exercise-14.2

1. Solve the following games

$$\begin{array}{c} & & B \\ & & \begin{array}{ccc} I & II & III \end{array} \\ & I & \begin{array}{|ccc|} \hline 8 & 5 & 8 \\ \end{array} \\ A & II & \begin{array}{ccc} 8 & 6 & 5 \\ \end{array} \\ & III & \begin{array}{ccc} 7 & 4 & 5 \\ \end{array} \\ & IV & \begin{array}{|ccc|} 6 & 5 & 6 \\ \hline \end{array} \end{array}$$

2. Solve the following game whose pay off matrix is given

		Player B				
		I	II	III	IV	V
Player A	I	4	4	2	−4	−6
	II	8	6	8	−4	0
	III	10	2	4	10	12

3. Solve the following game with the help of dominance property

(a) $\begin{bmatrix} 1 & 7 & 2 \\ 0 & 2 & 7 \\ 5 & 2 & 6 \end{bmatrix}$ (b) $\begin{bmatrix} 30 & 40 & -80 \\ 0 & 15 & -20 \\ 90 & 20 & 50 \end{bmatrix}$

4. Solve the following games whose pay off matrices are as follows

(a)
		Firm II		
		y_1	y_2	y_3
Firm I	x_1	60	50	40
	x_2	70	70	50
	x_3	80	60	75

(b)
$$\begin{bmatrix} -5 & 10 & 20 \\ 5 & -10 & -10 \\ 5 & -20 & -20 \end{bmatrix}$$

Answers

1. $\left(0, \dfrac{1}{4}, 0, \dfrac{3}{4}\right)$, $\left(0, \dfrac{3}{4}, \dfrac{1}{4}\right)$ and $v = 23/4$.

2. $\left(0, \dfrac{4}{9}, \dfrac{5}{9}\right)$, $\left(0, \dfrac{7}{9}, 0, \dfrac{2}{9}, 0\right)$ and $v = 34/9$.

3. (a) $\left(\dfrac{1}{3}, 0, \dfrac{2}{3}\right)$, $\left(\dfrac{5}{9}, \dfrac{4}{9}, 0\right)$ and $v = 11/3$ (b) $\left(\dfrac{1}{5}, 0, \dfrac{4}{5}\right)$, $\left(0, \dfrac{13}{15}, \dfrac{2}{15}\right)$ and $v = 24$.

4. (a) $\left(0, \dfrac{3}{7}, \dfrac{4}{7}\right)$, $\left(0, \dfrac{5}{7}, \dfrac{2}{7}\right)$ and $v = 450/7$ (b) $\left(\dfrac{1}{2}, \dfrac{1}{2}, 0\right)$, $\left(\dfrac{2}{3}, \dfrac{1}{3}, 0\right)$ and $v = 0$.

14.8 GRAPHICAL METHOD FOR THE SOLUTION OF 2 x n AND m x 2 GAMES

Graphical method is used to solve $2 \times n$ or $m \times 2$ games, *i.e.*, a game with mixed strategy which has only two pure strategies for one of the players. Optimal strategies for both the players assign non-zero probabilities to the same number of pure strategies. Clearly, if one player has only two strategies, the other will also use two strategies. Graphical method helps us to find which two strategies should be used. Hence, the game reduces to 2×2 which can be solved by any method discussed earlier.

(1) Graphical method for 2 x n Games

Consider a $2 \times n$ game. Assume that game has no saddle point. The pay off matrix of this game is as follows

$$
\begin{array}{c}
\\
\\
A
\end{array}
\begin{array}{cc}
 & \\
x_1 & 1 \\
x_2 & 2
\end{array}
\overset{\displaystyle B}{
\begin{array}{cccc}
1 & 2 & \dots & n \\
\hline
a_{11} & a_{12} & \cdots & a_{1n} \\
a_{21} & a_{22} & \cdots & a_{2n}
\end{array}}
$$

Since the player A has two strategies, then $x_1 + x_2 = 1 \Rightarrow x_2 = 1 - x_1$, $x_1 \geq 0$, $x_2 \geq 0$. Thus, for each of the pure strategies available to the player B, the expected pay-off for the player A are tabulated as follows :

Pure strategies used for player B	$E(v)$, expected pay off to player A
1	$a_{11}x_1 + a_{21}x_2 = a_{11}x_1 + a_{21}(1 - x_1) = (a_{11} - a_{21})x_1 + a_{21}$
2	$a_{12}x_1 + a_{22}x_2 = a_{12}x_1 + a_{22}(1 - x_1) = (a_{12} - a_{22})x_1 + a_{22}$
$\vdots$	$\vdots$
n	$a_{1n}x_1 + a_{2n}x_2 = a_{1n}x_1 + a_{2n}(1 - x_1) = (a_{1n} - a_{2n})x_1 + a_{2n}$

Clearly, A's expected pay off varies linearly with x_1. According to maximum criterion for mixed strategies game, the player A will select that value of x_1 which will maximize his minimum expected pay offs. To find this value, we plot the following straight lines.

$$E(v) = (a_{11} - a_{21})x_1 + a_{21}$$
$$E(v) = (a_{12} - a_{22})x_1 + a_{22}$$
$$\dots \quad\quad \dots \quad\quad \dots$$
$$\dots \quad\quad \dots \quad\quad \dots$$
$$E(v) = (a_{1n} - a_{2n})x_1 + a_{2n}$$

The lowest boundary of these lines will give the maximum expected pay off as function of

x_1. The highest point on this lowest boundary would give the maximum expected pay off and the optimum value of x_1.

WORKING PROCEDURE

To Plot the Above Lines

STEP 1. Draw two parallel lines one unit apart and make a scale on each. These two lines represent the two strategies available to A.

STEP 2. We draw lines to represent each to B's strategies. To represent B's 1st strategy we join a_{11} on scale I to a_{21} on scale II. This line will represent the expected pay off line $E(v) = (a_{11} - a_{21})x_1 + a_{21}$.

STEP 3. Similarly draw other pay off lines.

STEP 4. The lowest boundary to these lines will give the minimum expected pay off as function of x_1. The highest point on this lower boundary will give the maximum expected pay off to A and hence optimum value of x_1.

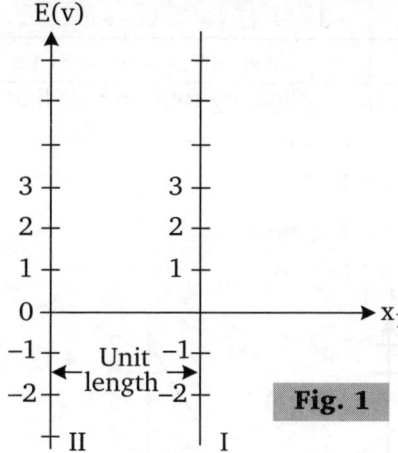

Fig. 1

In this case we determine only two strategies for player B corresponding to those two lines which pass through this maximum point P. In this way, the game is reduced to 2×2 game which can be solved by any method discussed earlier.

☛ Remarks

- If more than two lines pass through this point, then any two of them having opposite signs for their slopes will be alternative optimum solutions.
- We can solve $m \times 2$ games in the same manner, except the minimax point P is the lowest point on the boundary.

Solved Examples

EXAMPLE 1. *Solve the following game*

			I	II	III	IV
			B			
A	x_1	I	1	3	−3	7
	x_2	II	2	5	4	−6

SOLUTION. Clearly this game has no saddle point. Thus, we reduce this game into 2×2 using graphical method.

If x_1 and x_2 are the probabilities with which the player A uses his pure strategies, then

$$x_1 + x_2 = 1, \ x_1 \geq 0, \ x_2 \geq 0$$
$$x_2 = 1 - x_1$$

Then, expected pay off to player A for different pure strategies used by player B may be given in the following table.

Pure strategies used for player B	$E(v)$, A's expected pay off
I	$1x_1 + 2x_2 = x_1 + 2(1 - x_1) = -x_1 + 2$
II	$3x_1 + 5x_2 = 3x_1 + 5(1 - x_1) = -2x_1 + 5$
III	$-3x_1 + 4x_2 = -3x_1 + 4(1 - x_1) = -7x_1 + 4$
IV	$7x_1 - 6x_2 = 7x_1 - 6(1 - x_1) = 13x_1 - 6$

Thus, we have to draw following four pay-off lines

$$E(v) = -x_1 + 2 \qquad \qquad \dots(1)$$
$$E(v) = -2x_1 + 5 \qquad \qquad \dots(2)$$
$$E(v) = -7x_1 + 4 \qquad \qquad \dots(3)$$
$$E(v) = 13x_1 - 6 \qquad \qquad \dots(4)$$

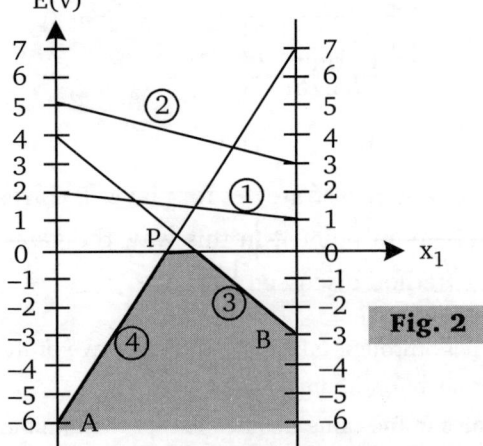

Fig. 2

From graph, lowest boundary APB to these lines give the minimum expected pay off to A. The highest point P on the lowest boundary will give the maximum expected pay-off and hence the expected value of x_1. Therefore, the best strategies for player B are III and IV pure strategies passing through point P. Thus, the game is reduced to 2×2 game given by the following pay off matrix.

$$\textbf{\textit{B}}$$

			y_3 III	y_4 IV
A	x_1	I	-3	7
	x_2	II	4	-6

Using the formula of 2×2 game without saddle point, we get

$$x_1 = \frac{1}{2}, \ x_2 = \frac{1}{2}, \ y_3 = \frac{13}{20}, \ y_4 = \frac{7}{20} \ \text{ and } \ v = \frac{1}{2}$$

Hence, the solution is given by

(i) For player A, optimal mixed strategies are $\left(\dfrac{1}{2}, \dfrac{1}{2} \right)$

(ii) For player B, optimal mixed strategies are $\left(0, 0, \dfrac{13}{20}, \dfrac{7}{20} \right)$

and (iii) Value of the game is $\dfrac{1}{2}$ for A and $-\dfrac{1}{2}$ for B.

EXAMPLE 2. *Solve the following game graphically*

$$\begin{array}{c} & \textbf{\textit{B}} \\ & \begin{array}{cc} y_1 & y_2 \\ I & II \end{array} \\ \textbf{\textit{A}} \ \begin{array}{c} I \\ II \\ III \end{array} & \begin{array}{|cc|} \hline 2 & 7 \\ 3 & 5 \\ 11 & 2 \\ \hline \end{array} \end{array}$$

SOLUTION. Clearly, there is no saddle point. If y_1, y_2 are the probabilities with which the player B uses his pure strategies, then

$y_1 + y_2 = 1, \quad y_1 \geq 0, \ y_2 \geq 0$
$y_2 = 1 - y_1$

Then, B's expected pay off for different pure strategies used by A may be given in the following table.

Pure strategies used for A	$E(v)$, B's expected pay off
I	$2y_1 + 7y_2 = 2y_1 + 7(1 - y_1) = -5y_1 + 7 \quad (1)$
II	$3y_1 + 5y_2 = 3y_1 + 5(1 - y_1) = -2y_1 + 5 \quad (2)$
III	$11y_1 + 2y_2 = 11y_1 + 2(1 - y_1) = 9y_1 + 2 \quad (3)$

Now, we draw these lines on the graph.

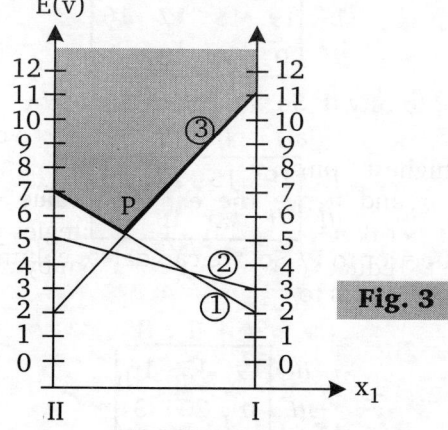

Fig. 3

Since this is a problem of $n \times 2$ type, thus minimax is the lowest point P on the upper boundary, which is the intersection of pay off lines.
$$E(v) = -5y_1 + 7 \quad \text{and} \quad E(v) = -9y_1 + 2$$
Thus, best strategies for player A are I and III. So, the given game reduces to the following 2×2 game

$$\mathbf{B}$$

$$
\begin{array}{cccc}
 & & y_1 & y_2 \\
 & & I & II \\
\mathbf{A} & x_1 \quad I & 2 & 7 \\
 & x_2 \quad II & 11 & 2
\end{array}
$$

If x_1 and x_2 are the probabilities with which player A chooses strategy I and III, then by using the method of 2×2 game without saddle point, we get
$$x_1 = \frac{9}{14}, \ x_2 = \frac{5}{14}, \ y_1 = \frac{5}{14}, \ y_2 = \frac{9}{14} \quad \text{and} \quad v = \frac{73}{14}$$
Hence, optimal solution is given by

(i) Optimal strategy for $A = \left(\dfrac{9}{14}, 0, \dfrac{5}{14}\right)$

(ii) Optimal strategy for $B = \left(\dfrac{5}{14}, 0, \dfrac{9}{14}\right)$

and (iii) Value of the game for $A = \dfrac{73}{14}$.

EXAMPLE 3. *Solve the game whose pay off matrix is given by*

$$
\begin{array}{cccc|c}
 & & & & \text{Row min} \\
8 & 15 & -4 & -2 & -4 \\
19 & 15 & 17 & 16 & 15 \\
0 & 20 & 15 & 5 & 0
\end{array}
$$

Column max $\quad 19 \quad 20 \quad 17 \quad 16$

SOLUTION. Since there is no saddle point. So, we will solve it with the help of dominance property.

$$
\begin{array}{c|cccc}
 & I & II & III & IV \\
\hline
I & 8 & 15 & -4 & -2 \\
II & 19 & 15 & 17 & 16 \\
III & 0 & 20 & 15 & 5
\end{array}
$$

Row I is inferior to row II. So, we can delete row I.

$$
\begin{array}{c|cccc}
 & I & II & III & IV \\
\hline
II & 19 & 15 & 17 & 16 \\
III & 0 & 20 & 15 & 5
\end{array}
$$

Column III is superior to IV. So, we can delete column III.
Now, our matrix reduces to

$$
\begin{array}{c|ccc}
 & I & II & IV \\
\hline
II & 19 & 15 & 16 \\
III & 0 & 20 & 5
\end{array}
$$

Now, the game can not be reduced further with the help of dominance property. Then we use the graphical method

$$
\begin{array}{c|ccc}
 & y_1 & y_2 & y_3 \\
\hline
x_1 & 19 & 15 & 16 \\
1-x_1 & 0 & 20 & 5 \\
\hline
 & ① & ② & ③
\end{array}
$$

Here A is the maximum point of lowest boundary and is the intersection point of y_1 and y_3. So our matrix reduces to

$$
\begin{array}{cc}
15 & 16 \\
20 & 5
\end{array}
$$

Now,
$$\frac{x_1}{x_2} = \frac{a_{22} - a_{21}}{a_{11} - a_{12}} = \frac{5 - 20}{15 - 16} = \frac{15}{1}$$

$$x_1 = \frac{15}{16}, \quad x_2 = \frac{1}{16}$$

$$\frac{y_1}{y_2} = \frac{a_{22} - a_{12}}{a_{11} - a_{21}} = \frac{5 - 16}{15 - 20} = \frac{11}{5}$$

$$y_1 = \frac{11}{16}, \quad y_2 = \frac{5}{16}$$

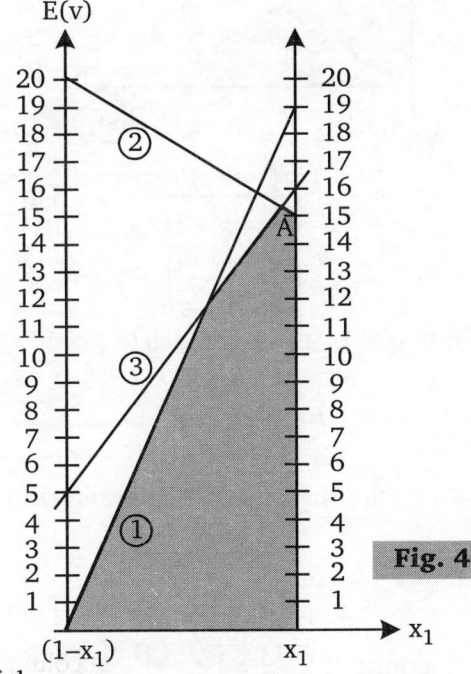

and value of the game,

$$v = \frac{v_{22}v_{11} - v_{21}v_{12}}{(v_{22} + v_{11}) - (v_{21} + v_{12})}$$

$$= \frac{5 \times 15 - 20 \times 16}{(5 + 15) - (20 + 16)} = \frac{245}{16}$$

Hence,   Optimal strategy used by A = $\left(\dfrac{15}{16}, 0, \dfrac{1}{16}\right)$

Optimal strategy used by B = $\left(\dfrac{11}{16}, 0, \dfrac{5}{16}, 0\right)$.

EXAMPLE 4. *Solve the following game*

$$B$$

			Row min
①	⑧	4	1
⑥	④	⑤	4
⑩	1	2	0

A

Column max   6   8   5

SOLUTION.   Since there is no saddle point. So, the optimal strategy will be mixed strategy. Now, we use the principle of dominance to solve the game.

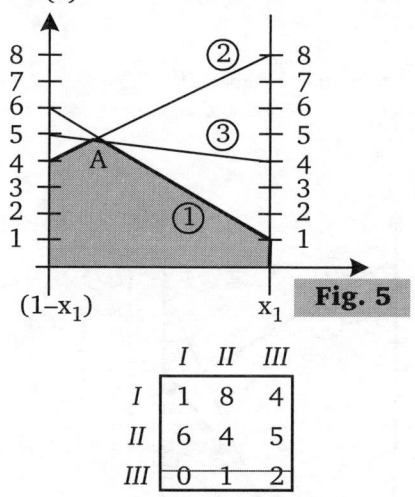

	I	II	III
I	1	8	4
II	6	4	5
III	0	1	2

Row III is inferior to row I. So, we can delete row III.

	I	II	III
I	1	8	4
II	6	4	5

Now, we will solve it with the help of graphical method.

	y_1	y_2	y_3
x_1	1	8	4
$1 - x_1$	6	4	5
	①	②	③

A is the maximum point of lowest boundary. It is the intersection of y_1 and y_2. So, our matrix reduces to

	y_1	y_2
x_1	8	4
x_2	4	5

$$\frac{x_1}{x_2} = \frac{a_{22} - a_{21}}{a_{11} - a_{12}} = \frac{5-1}{8-4} = \frac{1}{4}$$

$$x_1 = \frac{1}{5}, \quad x_2 = \frac{4}{5}$$

$$\frac{y_1}{y_2} = \frac{a_{22} - a_{12}}{a_{11} - a_{21}} = \frac{1}{4}$$

$$y_1 = \frac{1}{5}, \quad y_2 = \frac{4}{5}$$

So,   The strategy used by A $= \left(\frac{1}{5}, \frac{4}{5}, 0\right)$

The strategy used by B $= \left(\frac{1}{5}, \frac{4}{5}, 0\right)$

Now, the value of the game

$$v = \frac{a_{22}a_{11} - a_{12}a_{21}}{(a_{22} + a_{11}) - (a_{12} + a_{21})}$$

$$= \frac{5 \times 8 - 4 \times 4}{(5+8) - (4+4)} = \frac{26}{5}.$$

EXAMPLE 5.   *Solve the game whose pay-off matrix to the player A is given by the table*

B

		I	II	III
	I	1	7	2
A	II	6	2	7
	III	5	2	0

SOLUTION.

B

		I	II	III
	I	1	7	2
A	II	6	2	7
	III	5	2	0

Using dominance rule, all elements of II row all greater than or equal to all elements of III so dominating III row by II

B

		I	II	III
	I	1	7	2
A	II	6	2	7
	III	5	2	0

No row and column can be dominated now, so using graphical method

B's pure strategy	A's expected pay off $[E(x)]$
I	$E(x_1) = 1x_1 + 6(1 - x_1) = -5x_1 + 6$
II	$E(x_2) = 7x_1 + 2(1 - x_1) = 5x_1 + 2$
III	$E(x_3) = 2x_1 + 7(1 - x_1) = -5x_1 + 7$

Now, plotting these three lines

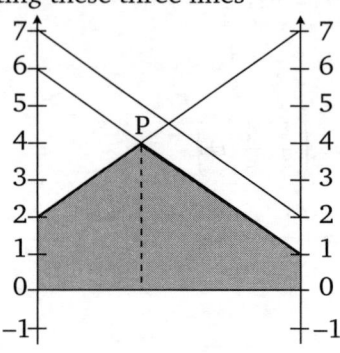

Fig. 6

Maxmin point occurs at point P. The highest point P on the lowest boundary will give the largest expected gain PN to A. So the best strategies for the player B are those which pass through the point P. The reduced game to 2×2 table is

$$\begin{array}{cc} & B \\ & \begin{array}{cc} I & II \end{array} \\ A \begin{array}{c} I \\ II \end{array} & \begin{array}{|cc|} \hline 1 & 7 \\ 6 & 2 \\ \hline \end{array} \end{array}$$

Now, let strategy of A is (x_1, x_2) and B is (y_1, y_2).

$$x_1 = \frac{2 - 6}{(1 - 7) + (2 - 6)} = \frac{-4}{-6 + (-4)} = \frac{2}{5}$$

$$x_2 = 1 - x_1 = 1 - \frac{2}{5} = \frac{3}{5}$$

$$y_1 = \frac{2 - 7}{(1 - 6) + (2 - 7)} = \frac{-5}{-10} = \frac{1}{2}$$

$$y_2 = 1 - y_1 = 1 - \frac{1}{2} = \frac{1}{2}$$

Thus,   Strategy of A is $\left(\frac{2}{5}, \frac{3}{5}, 0\right)$

Strategy of B is $\left(\frac{1}{2}, \frac{1}{2}, 0\right)$.

Value of game $V = \dfrac{(1 \times 2) - (6 \times 7)}{-10}$

$V = 4$ (for again).

14.9 ALGEBRAIC METHOD FOR THE SOLUTION OF A GENERAL GAME

In this method, first we convert the game into inequalities and then solve them.

Let $[a_{ij}]_{m \times n}$ be the pay off matrix of a rectangular game between two persons. Let us suppose $X = (x_1, x_2, ..., x_m)$, $Y = (y_1, y_2, ..., y_n)$ be the mixed optimal strategies of player A and B respectively. Then expected gains when B used his pure strategies 1, 2, ..., n respectively are

$$\sum_{i=1}^{m} a_{i1} x_i, \ \sum_{i=1}^{m} a_{i2} x_i, \ ......, \ \sum_{i=1}^{m} a_{in} x_i$$

Let v be the value of the game, then minimum expected gain of A is v.

Thus,
$$\sum_{i=1}^{m} a_{i1} x_i \geq v, \ \sum_{i=1}^{m} a_{i2} x_i \geq v, \ ......, \ \sum_{i=1}^{m} a_{in} x_i \geq v \qquad ...(1)$$

In a similar way, considering B's expected losses and considering the fact that the maximum loss of B is v, we get the following system of inequalities.

$$\sum_{j=1}^{n} a_{1j} y_j \leq v, \ \sum_{j=1}^{n} a_{2j} y_j \leq v, \ ......, \ \sum_{j=1}^{n} a_{mj} y_j \leq v \qquad ...(2)$$

and
$$\left. \begin{array}{l} x_1 + x_2 + ... + x_m = 1 \\ y_1 + y_2 + ... + y_n = 1 \end{array} \right] \qquad ...(3)$$

$$x_i \geq 0, \ x_j \geq 0, \ \forall \ i = 1, 2, ..., m \ \text{ and } \ j = 1, 2, ..., n.$$

So, we have to find the values of $x_1, x_2, ..., x_m, \ y_1, y_2, ..., y_n$ such that (1), (2) and (3) are satisfied. For this, we use the following steps :

WORKING PROCEDURE

STEP 1. Convert inequalities (1) and (2) as equalities. Try to solve them. If we get a solution satisfying (1), (2) and (3), then the given problem is solved completely.

STEP 2. If the system of equations obtained above is inconsistent, then we conclude that at least one of the inequalities is strict. Then, we try to solve by taking one or more inequalities as strict inequality and other as equalities until we get a solution.

(1) If for some i $(i = 1, 2, ..., m)$, $v_{i1} y_1 + v_{i2} y_2 + ... + v_{in} y_n < v$
Then, corresponding $x_i = 0$ and similarly, if for some j $(j = 1, 2, ..., n)$
$v_{1j} x_1 + v_{2j} x_2 + ... + v_{mj} x_m > v$, then corresponding $y_j = 0$.

(2) If A's optimal strategy is a mixed strategy in which exactly r pure strategies have non-zero probabilities, then the optimal strategies of other player also involves exactly r pure strategies.

Solved Examples

EXAMPLE 1. *Find the value of the game and the optimal strategies for both players*

$$
\begin{array}{c c}
 & B \\
 & \begin{array}{ccc} I & II & III \end{array} \\
A \begin{array}{c} I \\ II \\ III \end{array} & \begin{array}{|ccc|} \hline -1 & 2 & 1 \\ 1 & -2 & 2 \\ 3 & 4 & -3 \\ \hline \end{array}
\end{array}
$$

SOLUTION. Clearly this game has no saddle point and can not be reduced to 2×2 game by the dominance rule. So, we solve it by algebraic method.

Let (x_1, x_2, x_3) and (y_1, y_2, y_3) be the optimal mixed strategies of the two players A and B respectively and v the value of the game.

For player A

$$-1.x_1 + 1.x_2 + 3.x_3 \geq v \qquad \ldots(1)$$
$$2.x_1 - 2.x_2 + 4.x_3 \geq v \qquad \ldots(2)$$
$$1.x_1 + 2.x_2 - 3.x_3 \geq v \qquad \ldots(3)$$

For player B

$$-1.y_1 + 2.y_2 + y_3 \leq v \qquad \ldots(4)$$
$$1.y_1 - 2.y_2 + 2.y_3 \leq v \qquad \ldots(5)$$
$$3.y_1 + 4.y_2 - 3y_3 \leq v \qquad \ldots(6)$$

Also,
$$x_1 + x_2 + x_3 = 1 \qquad \ldots(7)$$
$$y_1 + y_2 + y_3 = 1 \qquad \ldots(8)$$

$$\left. \begin{array}{c} x_1, x_2, x_3 \geq 0 \\ y_1, y_2, y_3 \geq 0 \end{array} \right\} \qquad \ldots(9)$$

Now, we have to find the values $x_1, x_2, x_3, y_1, y_2, y_3$ such that all the above relations are satisfied.

Firstly, we consider the inequalities (1) to (6) as equation :

$$-x_1 + x_2 + 3x_3 = v, \quad 2x_1 - 2x_2 + 4x_3 = v, \quad x_1 + 2x_2 - 3x_3 = v$$
$$-y_1 + 2y_2 + y_3 = v, \quad y_1 - 2y_2 + 2y_3 = v, \quad 3y_1 + 4y_2 - 3y_3 = v$$

On solving these equations, we get

$$x_2 = \left(\frac{2}{3}\right)v; \qquad x_3 = \left(\frac{3}{10}\right)v; \qquad x_1 = \left(\frac{17}{30}\right)v$$

Putting all these values in (7), we get

$$v = \frac{15}{23} \quad \text{and} \quad x_1 = \frac{17}{46}, \; x_2 = \frac{10}{23}, \; x_3 = \frac{9}{46}.$$

 Exercise-14.3

1. Solve the following game graphically:

(a)
$$\begin{array}{c|ccc} & y_1 & y_2 & y_3 \\ \hline x_1 & 1 & 3 & 11 \\ 1-x_1 & 8 & 5 & 2 \end{array}$$

(b)
$$\begin{array}{c|cccc} & y_1 & y_2 & y_3 & y_4 \\ \hline x_1 & 19 & 6 & 7 & 5 \\ x_2 & 7 & 3 & 14 & 6 \\ x_3 & 12 & 8 & 18 & 4 \\ x_4 & 8 & 7 & 13 & -1 \end{array}$$

(c)
$$A\begin{array}{c|cccc} & & & B & \\ \hline & 2 & 2 & 3 & -1 \\ & 4 & 3 & 2 & 6 \end{array}$$

(d)
$$A\begin{array}{c|cc} & & B \\ \hline & 2 & 4 \\ & 2 & 3 \\ & 3 & 2 \\ & -2 & 6 \end{array}$$

Answers

1. (a) $\left(\frac{3}{11}, \frac{8}{11}\right), \left(0, \frac{2}{11}, \frac{9}{11}\right)$ and $v = \frac{49}{11}$

   (b) $\left(\frac{3}{4}, \frac{1}{4}, 0, 0\right), \left(0, \frac{1}{4}, 0, \frac{3}{4}\right)$ and $v = \frac{21}{4}$.

   (c) $\left(\frac{1}{2}, \frac{1}{2}\right), \left(0, 0, \frac{7}{8}, \frac{1}{8}\right)$ and $v = \frac{5}{2}$

   (d) $\left(\frac{1}{3}, 0, \frac{2}{3}, 0\right), \left(\frac{2}{3}, \frac{1}{3}\right)$ and $v = \frac{8}{3}$.

Glossary

- **n-person game:** When number of players are n where $n \geq 2$, then for $n = 2$, the game is known as 2 person game and for $n > 2$ the game is known as n person game.

- **Zero sum and non-zero sum game:** A game is known as zero sum game in which the net gain or not profit after the game is zero means nothing comes from outside, payment is always between the players, the loss of one player is the gain of others, and a game which is not zero sum game is known as non-zero sum game.

- **Competitive game.** A game is said to be competitive game if it has the following four properties :
  (i) There should be finite number of players means $n \geq 2$.
  (ii) Each player has a finite list of his possible course of action.
  (iii) A play is said to be played when each player choose one of his course of action

and no player knows the choice of action of the other player until he has decided his own.

  (iv) When each player chooses his activity, then this combination of activities gives a result according which each player gains a payment which may be –ve, +ve or zero.

- **Strategy:** For a given player the strategy is given by the set of rules which specify that which of the available course of action he should make at each play.

- **Minimax and Maxmin Principles:** If a player lists his worst possible outcomes of all his potential strategy, then he will choose the best strategy among all these outcomes. Such a principle is known as maxmin principle or optimal strategy.

- **Saddle Point:** A point which is minimum in its row and maximum in its column is known as the saddle point.

REVIEW QUESTIONS

1. Explain min-max and max-min principle in game theory.
2. Explain two-person zero sum game.
3. State major limitations of Game theory.
4. Define the following:
   (i) Competitive game
   (ii) pay off matrix
   (iii) pure and mixed strategies
   (iv) saddle point
   (v) Two person zero sum game
5. Write the assumptions made in the theory of game.
6. What is a game in game theory?

MULTIPLE CHOICE QUESTIONS (CHOOSE THE MOST APPROPRIATE ONE)

1. If the value of a game is zero, then the game is called:
   (a) pure game   (b) pure strategy
   (c) fair strategy   (d) none of these
2. Two persons zero sum game means that the:
   (a) sum of losses to one player equals the sum of gain to the other
   (b) sum of losses to one player is not equal to the sum of gains to the other
   (c) both (a) and (b)
   (d) None of these
3. The game with saddle points are:
   (a) probabilistic in nature
   (b) deterministic in nature
   (c) stochastic in nature
   (d) none of these
4. Game theory models are classified by the:
   (a) no. of players   (b) sum of all payoff
   (c) no. of strategies   (d) all are true
5. When minimax and maximin criterian meets then:
   (a) mixed strategies exist
   (b) fair game exist
   (c) saddle point exits
   (d) None of these
6. In case there is no saddle point in a game then the game is:
   (a) fair game
   (b) mixed strategic game
   (c) determinstic game
   (d) none of these
7. If there are more than two persons in a game

then it is called:

(a) open game   (b) big game

(c) multiplayer game (d) none of these

8. A competitive situation is known as:

(a) competition   (b) game

(c) marketing   (d) None of these

9. A game involving n persons is known as:

(a) n-person game   (b) multiplayer game

(c) not a game   (d) None of these

10. A saddle point exits when:

(a) maximum value =maximax value

(b) minimax value = minimum value

(c) minimax value = maximum value

(d) None of these

11. Linear programming method should be used to determine value of the game when size of pay off matrix is:

(a) 2×2 (b) 3×4

(c) $2 \times n$ (d) None of these

12. In a mixed strategy game:

(a) no saddle point exits

(b) each player always selects same strategy

(c) both (a) and (b) are true

(d) None of these

13. In a pure strategy game:

(a) any strategy may be selected arbitrarily

(b) a particular strategy is selected by each player

(c) both players select their optimal strategy

(d) None of these

14. A two person game is said to be zero-sum if:

(a) gain of one player is exactly matched by a loss to the other so that their sum is equal to zero

(b) gain of one player does not match the loss to he other

(c) both the players must have an equal no. of strategies

(d) None of these

Answers

1. (a)	**2.** (a)	**3.** (b)	**4.** (d)	**5.** (c)	**6.** (b)	**7.** (c)	**8.** (b)	**9.** (a)
10. (c)	**11.** (b)	**12.** (a)	**13.** (c)	**14.** (a)				

□□□□

BIBLIOGRAPHY

1.	Ackoff R.L. and M.W. Sasini	Fundamental of Operation Research, John Wilay and sons
2.	Dantzig, G.B.	Linear Programming and extensions, Princeton University Press, New Jersey
3.	Hadley, G.	Linear Programming, Addison-Wesley, Reading masses.
4.	Hadley, G.	Non-linear and Dynamic Programming, Reading mass.
5.	Lee, S.M.	Goal Programming for decision analysis, Auerbach Publishers, Philadelphia.
6.	Sharma, J.K.	Operation Research: Theory and Application, Macmilan, India Ltd., New Delhi.
7.	Sharma, J.K.	Quantitative techniques for Managerial Decision, Macmilan, India Ltd., New Delhi.
8.	Taha, H.A.	Operation Research – An Introduction, Prantice-Hall Inc., New Jersey.
9.	Vajda, S.	Theory of linear and non-linear programming, LongMan, London.
10.	Zoints, S.	Linear and Integer Programming, Practice-Hall England cliffs, New Jersey.

□□□□

Index